Lecture Notes in Computer Science 2611

Edited by G. Goos, J. Hartmanis, and J. van Leeuwen

Springer
Berlin
Heidelberg
New York
Barcelona
Hong Kong
London
Milan
Paris
Tokyo

Günther Raidl et al. (Eds.)

Applications of Evolutionary Computing

EvoWorkshops 2003: EvoBIO, EvoCOP, EvoIASP,
EvoMUSART, EvoROB, and EvoSTIM
Essex, UK, April 14-16, 2003
Proceedings

 Springer

Series Editors

Gerhard Goos, Karlsruhe University, Germany
Juris Hartmanis, Cornell University, NY, USA
Jan van Leeuwen, Utrecht University, The Netherlands

Cataloging-in-Publication Data applied for

A catalog record for this book is available from the Library of Congress

Bibliographic information published by Die Deutsche Bibliothek
Die Deutsche Bibliothek lists this publication in the Deutsche Nationalbibliografie;
detailed bibliographic data is available in the Internet at <http://dnb.ddb.de>.

CR Subject Classification (1998): D.1, F.2, I.2, G.2.1, I.4, I.5, G.1.6, J.5, J.3

ISSN 0302-9743
ISBN 3-540-00976-0 Springer-Verlag Berlin Heidelberg New York

Springer-Verlag Berlin Heidelberg New York
a member of BertelsmannSpringer Science+Business Media GmbH

http://www.springer.de

© Springer-Verlag Berlin Heidelberg 2003
Printed in Germany

Typesetting: Camera-ready by author, data conversion by PTP-Berlin GmbH
Printed on acid-free paper   SPIN 10872857   06/3142   5 4 3 2 1 0

Preface

Evolutionary Computation (EC) involves the study of problem solving, optimization, and machine learning techniques inspired by principles of natural evolution and genetics. EC has been able to draw the attention of an increasing number of researchers and practitioners in several fields. The number of applications and different disciplines that benefit from these techniques is probably the most immediate proof of EC's high flexibility and potential. In recent years, many studies and results have been reported in the literature documenting the capabilities of EC techniques in solving difficult problems in several domains.

EvoNet, the European Network of Excellence in Evolutionary Computing, organized its first events in 1998 as a collection of workshops that dealt with both theoretical and application-oriented aspects of EC. EuroGP soon became the main European event dedicated to Genetic Programming (GP). In 2000 this led to a reorganization of EvoNet events into two co-located independent parts: EuroGP became a single track conference, while the more application-oriented workshops were merged in a multitrack event: EvoWorkshops 2000.

This volume contains the proceedings of EvoWorkshops 2003 held in Essex, UK, on 14–16 April 2003, jointly with EuroGP 2003, the 6th European Conference on Genetic Programming. EvoWorkshops 2003 consisted of the following individual workshops:

- *EvoBIO*, the 1st European Workshop on Evolutionary Bioinformatics,
- *EvoCOP*, the 3rd European Workshop on Evolutionary Computation in Combinatorial Optimization,
- *EvoIASP*, the 5th European Workshop on Evolutionary Computation in Image Analysis and Signal Processing,
- *EvoMUSART*, the 1st European Workshop on Evolutionary Music and Art,
- *EvoROB*, the 4th European Workshop on Evolutionary Robotics,
- *EvoSTIM*, the 4th European Workshop on Scheduling and Timetabling.

EvoBIO was concerned with the exploitation of evolutionary computation, and advanced hybrids of evolutionary computation with other techniques, in addressing the very wide range of problems which occur in the understanding and analysis of biological data. In this area, evolutionary computation is playing an increasingly important role in the pharmaceutical, biotechnology and associated industries, as well as in scientific discovery.

Combinatorial optimization problems of academic and industrial interest were addressed in the EvoCOP workshop. In particular, problem analyses, studies of algorithmic techniques applied in evolutionary algorithms and related metaheuristics, the hybridization of different approaches, and performance evaluations were addressed.

EvoIASP celebrated its fifth edition in 2003. It has become a traditional appointment for European and non-European researchers in EC applications to image analysis and signal processing, after being the pioneering event dedicated to those fields in 1999.

EvoMUSART was dedicated to the application of evolutionary computation to the fields of music and art. The goals of this workshop were to present recent research results, to describe the development of systems in this area, to identify and explore directions for future research, to stimulate closer interaction between members of this scientific (and artistic) community, to explore the historical context for these systems and to lay the foundations for a more unified theory and body of work in this creative and complex application area.

EvoROB was concerned with the use of evolutionary computing techniques for the automatic design of adaptive robots. The aims of this workshop, which brought together active ER researchers and people from industry, were to assess the current state-of-the-art and to provide opportunities for fostering future developments and applications.

Scheduling and timetabling are amongst the most successful applications of evolutionary techniques. A related and growing field to which evolutionary methods are being applied is that of AI planning. EvoSTIM covered all aspects of these inter-related methods, including case studies, theoretical developments, hybrid techniques, and performance evaluations and comparisons.

EvoWorkshops 2003 has confirmed its tradition by providing researchers in these fields, as well as people from industry, with an opportunity to present their latest research and discuss current developments and applications, besides fostering closer future interaction between members of all scientific communities that may benefit from EC techniques.

The following numbers of papers submitted to EvoWorkshops 2003 show the liveliness of the scientific movement in the corresponding fields and made Evo-Workshops 2003 the largest event of its series. The acceptance rates indicate the high quality of the papers presented at the workshops and included in these proceedings. We would like to give credit to all members of the program committees, to whom we are very grateful for their quick and thorough work.

Workshop	submitted	accepted	acceptance ratio
EvoBIO	20	12	60.0%
EvoCOP	39	19	48.7%
EvoIASP	17	11	64.7%
EvoMUSART	13	10	76.9%
EvoROB	15	8	53.3%
EvoSTIM	5	3	60.0%
Total	109	63	57.8%

Volume Editors

Stefano Cagnoni
Dept. of Computer Engineering
University of Parma
Parco Area delle Scienze 181/a
43100 Parma, Italy
cagnoni@ce.unipr.it

Juan J. Romero Cardalda
Dept. of Information and
Communications Technologies
Faculty of Computer Science
University of A Coruña
A Coruña CP 15071, Spain
jj@udc.es

David W. Corne
School of Systems Engineering
University of Reading
PO Box 225, Whiteknights,
Reading RG6 6AY, UK
d.w.corne@reading.ac.uk

Jens Gottlieb
SAP AG
Neurottstrasse 16
69190 Walldorf, Germany
jens.gottlieb@sap.com

Agnès Guillot
AnimatLab, LIP6
8 rue du capitaine Scott
75015 Paris, France
agnes.guillot@lip6.fr

Emma Hart
Napier University
School of Computing
219 Colinton Road
Edinburgh EH14 1DJ, UK
e.hart@napier.ac.uk

Colin G. Johnson
Computing Laboratory
University of Kent
Canterbury, Kent, CT2 7NF, UK
c.g.johnson@ukc.ac.uk

Elena Marchiori
Dept. of Mathematics and
Computer Science
Free University of Amsterdam
de Boelelaan 1081a
1081 HV, Amsterdam
The Netherlands
elena@cs.vu.nl

Jean-Arcady Meyer
AnimatLab, LIP6
8 rue du capitaine Scott
75015 Paris, France
jean-arcady.meyer@lip6.fr

Martin Middendorf
Parallel Computing and Complex
Systems Group
University of Leipzig
Augustusplatz 10/11
04109 Leipzig, Germany
middendorf@informatik.uni-leipzig.de

Günther R. Raidl
Algorithms and Data Structures Group
Institute of Computer Graphics
Vienna University of Technology
Favoritenstrasse 9-11/186
1040 Vienna, Austria
raidl@ads.tuwien.ac.at

EvoWorkshops 2003 was sponsored by EvoNet. The organization of the event was made possible thanks to the active participation of many members of EvoNet. In particular, we want to thank Jennifer Willies, EvoNet's administrator, and Chris Osborne, EvoNet's technical manager, for their tremendous efforts. Evo-BIO, EvoIASP, EvoMUSART, EvoROB, and EvoSTIM are activities of the EvoNet working groups with the same names.

April 2003

Stefano Cagnoni
Juan J.R. Cardalda
David W. Corne
Jens Gottlieb
Agnès Guillot
Emma Hart
Colin G. Johnson
Elena Marchiori
Jean-Arcady Meyer
Martin Middendorf
Günther R. Raidl

Organization

EvoWorkshops 2003 were organized by EvoNet jointly with EuroGP 2003.

Organizing Committee

EvoWorkshops chair: Günther R. Raidl, Vienna University of Technology, Austria

Local co-chairs: Edward Tsang, University of Essex, UK

Riccardo Poli, University of Essex, UK

EvoBIO co-chairs: David Corne, University of Reading, UK

Elena Marchiori, Free University Amsterdam, The Netherlands

EvoCOP co-chairs: Jens Gottlieb, SAP AG, Germany

Günther R. Raidl, Vienna University of Technology, Austria

EvoIASP chair: Stefano Cagnoni, University of Parma, Italy

EvoMUSART co-chairs: Colin G. Johnson, University of Kent, UK

Juan Jesús Romero Cardalda, Universidade da Coruña, Spain

EvoROB co-chairs: Agnès Guillot, Université Paris 6, France

Jean-Arcady Meyer, Université Paris 6, France

EvoSTIM co-chairs: Emma Hart, Napier University, Edinburgh, UK

Martin Middendorf, University of Leipzig, Germany

Program Committees

EvoBIO Program Committee:

Jesus S. Aguilar-Ruiz, University of Seville, Spain
Wolfgang Banzhaf, University of Dortmund, Germany
Jacek Blazewicz, Institute of Computing Science, Poznan, Poland
Carlos Cotta, University of Malaga, Spain
Bogdan Filipic, Jozef Stefan Institute, Ljubljana, Slovenia
Gary B. Fogel, Natural Selection, Inc., USA
James Foster, University of Idaho, USA
Steven A. Frank, University of California, Irvine, USA

Jin-Kao Hao, LERIA, Université d'Angers, France
William Hart, Sandia National Labs, USA
Jaap Heringa, Free University Amsterdam, The Netherlands
Francisco Herrera, University of Granada, Spain
Daniel Howard, QinetiQ, UK
Antoine van Kampen, AMC University of Amsterdam, The Netherlands
Maarten Keijzer, Free University Amsterdam, The Netherlands
Douglas B. Kell, University of Wales, UK
William B. Langdon, UCL, UK
Bob MacCallum, Stockholm University, Sweden
Brian Mayoh, Aarhus University, Denmark
Andrew C.R. Martin, University of Reading, UK
Peter Merz, University of Tübingen, Germany
Martin Middendorf, University of Leipzig, Germany
Jason H. Moore, Vanderbilt University Medical Center, USA
Pablo Moscato, University of Newcastle, Australia
Martin Oates, British Telecom, Plc, UK
Jon Rowe, University of Birmingham, UK
Jem Rowland, University of Wales, UK
Vic J. Rayward-Smith, University of East Anglia, UK
El-ghazali Talbi, Laboratoire d'Informatique Fondamentale de Lille, France
Eckart Zitzler, Swiss Federal Institute of Technology, Switzerland

EvoCOP Program Committee:

Jürgen Branke, University of Karlsruhe, Germany
Edmund Burke, University of Nottingham, UK
David W. Corne, University of Reading, UK
Carlos Cotta, University of Malaga, Spain
Peter Cowling, University of Bradford, UK
David Davis, NuTech Solutions, Inc., MA, USA
Karl Doerner, University of Vienna, Austria
Marco Dorigo, Université Libre de Bruxelles, Belgium
Anton V. Eremeev, Omsk Branch of the Sobolev Institute of Mathematics, Russia
David Fogel, Natural Selection, Inc., CA, USA
Jens Gottlieb, SAP AG, Germany
Jin-Kao Hao, LERIA, Université d'Angers, France
Jano van Hemert, CWI, The Netherlands
Bryant A. Julstrom, St. Cloud State University, MN, USA
Dimitri Knjazew, SAP AG, Germany
Joshua D. Knowles, Université Libre de Bruxelles, Belgium
Gabriele Kodydek, Vienna University of Technology, Austria
Mario Köppen, FhG IPK, Germany
Jozef Kratica, Serbian Academy of Sciences and Arts, Yugoslavia
Ivana Ljubic, Vienna University of Technology, Austria

Elena Marchiori, Free University Amsterdam, The Netherlands
Dirk Mattfeld, University of Bremen, Germany
Helmut Mayer, University of Salzburg, Austria
Peter Merz, University of Tübingen, Germany
Zbigniew Michalewicz, NuTech Solutions, Inc., NC, USA
Martin Middendorf, University of Leipzig, Germany
Nenad Mladenovic, Serbian Academy of Sciences and Arts, Yugoslavia
Christine L. Mumford, Cardiff University, UK
Francisco J.B. Pereira, Universidade de Coimbra, Portugal
Günther R. Raidl, Vienna University of Technology, Austria
Marcus Randall, Bond University, Australia
Colin Reeves, Coventry University, UK
Marc Reimann, University of Vienna, Austria
Claudio Rossi, Polytechnic University of Madrid, Spain
Franz Rothlauf, University of Mannheim, Germany
Andreas Sandner, SAP AG, Germany
Marc Schoenauer, INRIA, France
Christine Solnon, University of Lyon I, France
Thomas Stützle, Darmstadt University of Technology, Germany
El-ghazali Talbi, Laboratoire d'Informatique Fondamentale de Lille, France
Edward P. Tsang, University of Essex, UK
Xin Yao, University of Birmingham, UK

EvoIASP Program Committee:

Giovanni Adorni, University of Genoa, Italy
Lucia Ballerini, Orebro University, Sweden
Wolfgang Banzhaf, University of Dortmund, Germany
Dario Bianchi, University of Parma, Italy
Alberto Broggi, University of Parma, Italy
Stefano Cagnoni, University of Parma, Italy
Ela Claridge, University of Birmingham, UK
Marc Ebner, University of Würzburg, Germany
Terry Fogarty, Napier University, UK
Daniel Howard, QinetiQ, UK
Mario Köppen, Fraunhofer IPK, Berlin, Germany
Evelyne Lutton, INRIA, France
Peter Nordin, Chalmers University of Technology, Sweden
Gustavo Olague, CICESE, Mexico
Riccardo Poli, University of Essex, UK
Conor Ryan, University of Limerick, Ireland
Jim Smith, University of Western England, UK
Giovanni Squillero, Turin Polytechnic, Italy
Andy Tyrrell, University of York, UK
Hans-Michael Voigt, GFaI – Center for Applied Computer Science, Germany

EvoMUSART Program Committee:

Peter Bentley, University College London, UK
Eleonora Bilotta, Università della Calabria, Italy
Amílcar Cardoso, Universidade de Coimbra, Portugal
Matthew Lewis, Ohio State University, USA
Penousal Machado, Universidade de Coimbra, Portugal
Eduardo R. Miranda, University of Glasgow, UK
Luigi Pagliarini, University of Southern Denmark, Denmark
Antonino Santos, University of A Coruña, Spain
Stephen Todd, IBM Research Laboratories, UK
Tatsuo Unemi, Soka University, Japan

EvoROB Program Committee:

Wolfgang Banzhaf, University of Dortmund, Germany
Marco Dorigo, Université Libre de Bruxelles, Belgium
Dario Floreano, EPFL, Switzerland
Takashi Gomi, AAI, Canada
John Hallam, University of Edinburgh, UK
Inman Harvey, University of Sussex, UK
Patrick Hénaff, Université de Versailles, France
Phil Husbands, University of Sussex, UK
Auke Jan Ijspeert, EPFL, Switzerland
Pier Luca Lanzi, Politecnico di Milano, Italy
Enrik Hautop Lund, University of Aarhus, Denmark
Stefano Nolfi, National Research Council, Italy
Peter Nordin, Chalmers University, Sweden
Rolf Pfeifer, University of Zürich, Switzerland
Olivier Sigaud, Université Paris 6, France
Tom Ziemke, University of Skövde, Sweden

EvoSTIM Program Committee:

Daniel Borrajo, Universidad Carlos III de Madrid, Spain
Emma Hart, Napier University, UK
Daniel Merkle, University of Leipzig, Germany
Martin Middendorf, University of Leipzig, Germany
Ben Paechter, Napier University, UK
Peter Ross, Napier University, UK
Peter Swann, Rolls-Royce, UK
Andrea Tettamanzi, Genetica-Soft, Italy

Sponsoring Institutions

- EvoNet, the Network of Excellence in Evolutionary Computing, funded by the European Commission's IST Programme.

- University of Essex, UK.

Table of Contents

EvoBIO Contributions

EvoCOP Contributions

EvoIASP Contributions

EvoMUSART Contributions

Artificial Immune System for Classification of Cancer

Shin Ando[1] and Hitoshi Iba[2]

[1] Dept. of Electronics Engineering, School of Engineering University of Tokyo
[2] Dept. of Frontier Informatics, School of Frontier Science, University of Tokyo
{ando, iba}@miv.t.u-tokyo.ac.jp

Abstract. This paper presents a method for cancer type classification based on microarray-monitored data. The method is based on artificial immune system(AIS), which utilizes immunological recognition for classification. The system evolutionarily selects important genes; optimize their weights to derive classification rules. This system was applied to gene expression data of acute leukemia patients to classify their cancer class. The primary result found few classification rules which correctly classified all the test samples and gave some interesting implications for feature selection principles.

1 Introduction

The analysis of human gene expression is an important topic in bioinformatics. The microarrays and DNA chips can measure the expression profile of thousands of genes simultaneously. Genes are expressed differently depending on its environment, such as their affiliate organs and external stimulation. Many ongoing researches try to extract information from the difference in expression profile given the stimulation or environmental change.

This paper focuses on gene expression data provided by Golub et al. [14]. They monitored the expression of more than 7,000 genes in cancerous cells collected from 72 patients of acute leukemia. Attempts were made to discriminate different cancer class (ALL/AML) from these profiles. Two independent data sets were provided: one being the training data set to learn the cancer classes, and the other being the test data set to evaluate its prediction. The reliable diagnoses were made by combination of clinical tests.

This paper describes the implementation of Artificial Immune System (AIS) [2] for ALL/AML classification. The AIS simulates the human immune system, which is a complex network structure, which responds to an almost unlimited multitude of foreign pathogens. It is considered to be potent in intelligent computing applications such as detection, pattern recognition, and classification.

In our AIS, classification rules are derived as hyperplanes, which divide the domain of sample vectors. The primary result show reduced complexity of the rules, while improving on the accuracy of prediction. The best result correctly classified training and test data set based on four automatically chosen genes.

S. Cagnoni et al. (Eds.): EvoWorkshops 2003, LNCS 2611, pp. 1–10, 2003.

2 ALL/AML Classification

Discovery and prediction of cancer class is important, as tumor's clinical transition and response to therapy vary depending on its class. In the case of acute leukemia, clinical outcome vary depending on whether the leukemia arise from lymphoid precursor(ALL) or from myeloid precursors(AML). These classes are solidly recognized by specific antibodies binding with surface of leukemia cells in each class. In clinical practice, distinction is made by a number of tests and experienced interpretation.

[14] provides a data set of 38 cell samples (27 ALL, 11 AML) obtained from different patients. Expression levels of 7109 genes were monitored by DNA microarray. Another set of data was collected independently, which contained 34 samples (20 ALL, 14 AML). The former samples were used as training set to discover classification rules, and the latter were test set to evaluate the obtained rules.

Important aspects in this problem are selection of informative genes(feature selection) and optimization of strength(weight) of each gene. Many ranking methods are used to rank genes and select features. [5] compares several ranking methods and results when combined with several machine learning methods.

Many studies [7, 13, 14], uses correlation metrics G_i (i) to rank the feature genes. Subset of genes with highest correlations is chosen as classifier genes. μ and σ are the mean and standard deviation for the expression levels of gene i, in ALL or AML class samples.

$$G_i=(\mu_{ALL}-\mu_{AML})/(\sigma_{ALL}+\sigma_{AML}) \qquad (i)$$

Few of the machine learning methods applied to this problem are: weighted vote cast [14], Bayesian Network, Neural Network, RBF Network [7], Support Vector Machine [6, 13], etc.. Table 1 shows comparison of performance by various machine learning techniques in ALL/AML classification. The results are cited from [6, 7, 13, 14]. Following paragraph describes the conditions of each result.

[14] uses correlation (i) to select and weight the genes. Since weights are not learned, training samples are misclassified. [13] uses 50, 100, and 200 genes of high correlation. Bayesian Network[8] uses 4 genes with higher correlations, though how the features were chosen is not clearly stated. Neural Network[8] creates different classifier in every run. The success rate shown in Table 1 is the best in 10 runs, while the average and worst success rate of test data was 4.9 and 9 respectively. [6] uses Recursive Feature Elimination to select informative genes. Result in Table 1 is obtained when 8, 16, 32 genes were chosen. AIS select combination of genes and weights in evolutionary recombination process. The result in Table 1 is the best of 20 runs, while average failure rate is 0.85. More detail will be given in later section.

Table 1. Comparison of performance by various machine learning techniques

Failure Rate	WVM[14]	SVM[13]	BN[7]	NN[7]	SVM[6]	AIS
Training Data	2/38	0/38	0/38	0/38	0/38	0/38
Test Data	5/34	2-4/34	2/34	1/34	0/34	0/34

3 Features of Immune System

The capabilities of natural immune system, which are to recognize, destroy, and remember almost unlimited multitude of foreign pathogens, have drawn increasing interest of researchers over the past few years. Application of AIS includes fields of computer security, pattern recognition, and classification.

The natural immune system responds to and removes intruders such as bacteria, viruses, fungi, and parasites. Substances that are capable of invoking specific immune responses are referred to as antigens(Ag).

Immune system learns the features of antigens and remembers successful responses to use against invasions by similar pathogens in the future. These characteristics are achieved by a class of white blood cells called lymphocytes, whose function is to detect antigens and assist in their elimination. The receptors on the surface of a lymphocyte bind with specific epitopes on the surfaces of antigens. These proteins related to immune system are called antibodies(Ab).

Immune system can maintain a diverse repertoire of receptors to capture various antigens, because the DNA strings which codes the receptors are subject to high probability crossover and mutation, and new receptors are constantly created.

Lymphocytes are subject to two types of selection process. Negative selection, which takes place in thymus, operates by killing all antibodies that binds to any self-protein in its maturing process. The clonal selection takes place in the bone marrow. Lymphocyte which binds to a pathogen is stimulated to copy themselves. The copy process is subject to a high probability of errors, i.e., hypermutation. The combination of mutation and selection amounts to an evolutionary algorithm that produces lymphocytes that become increasingly specific to invading pathogens.

During the primary response to a new pathogen, the organism will experience an infection while the immune system learns to recognize the epitope by evolutionary process. The memory of successful receptors is maintained to allow much quicker secondary response when same or similar pathogens invade thereafter.

There are several theories of how immune memory is maintained. The AIS in this paper stores successful antibodies in permanent memory cells to store adapted results.

4 Implementation of Artificial Immune System

In the application of artificial immune system to ALL/AML classification problem, the following analogy applies. The ALL training data sets correspond to Ag, and AML training data sets to the self-proteins. Classification rules represent Ab, which captures training samples(Ag or self-proteins) when its profile satisfies the conditions of the rule. A population of Ab goes through a cycle of invasion by Ag and selective reproduction. As successful Abs are converted into memory cells, ALL/AML class is learned.

4.1 Rule Encoding

Classification rules are linear separators, or hyper-planes, as shown in (ii)0. Vector x =$(x_1, x_2, ..., x_i, ... x_n)$ represents gene expression levels of a sample, and vector w=$(w_1,$

$w_2, \ldots, w_i, \ldots, w_n$) represents the weight of each gene. A hyperplane $W(x)=0$ can separate the domain and the samples into two classes. W determines the class of sample x; if $W(x)$ is larger than or equal to 0, sample x is classifies as ALL. If it is smaller than 0, the sample is classified as AML.

$$W(x) \geq 0 \quad \{ W(x) = w^T \cdot x \} \tag{ii}$$

Encoded rules are shown in (iii). It represents a vector w, where each locci consists of a pointer to a gene and weight value of that gene. It corresponds to a vector where unspecified gene weights are supplemented with 0. It is similar to messyGA[3] encoding of vector w.

$$(X_{123}, 0.5) \, (X_i, w_i) \, (\ldots) \, (\ldots) \, (\ldots) \, \text{(gene index, weight)} \tag{iii}$$

4.2 Initialization

Initially, rules are created by sequential creation of locus. For each locus, a gene and a weight value is chosen randomly. With probability P_i, next locus is created. Thus, average lengths of the initial rules are $\Sigma_n nP_i n$. Empirically, the number of initial rules should be in the same order as the number of genes to ensure sufficient building blocks for classification rules.

4.3 Negative Selection

All newly created rules will first go through negative selection. Each rule is met with set of AML training samples, $x_i (i=1,2,\ldots,N_{AML})$, as self-proteins. If a rule binds with any of the samples(satisfy (iv)0, it is terminated. The new rules are created until N_{Ab} antibodies pass the negative selection. These rules constitute population of antibodies $Ab_i (i=1,2,\ldots,N_{Ab})$.

$$\prod \delta(W(x)) \neq 1 \begin{cases} \delta(t) = 0 \mid t > 0 \\ \quad\;\; = 1 \mid t < 0 \end{cases} \tag{iv}$$

4.4 Memory Cell Conversion

The antibodies who endures the negative selection are met with invading antigens, $Ag_i (i=1,2,\ldots,N_{ALL})$, or ALL training samples. Antibodies which can capture many antigens are converted into memory cells $M_i (i=1,2,\ldots,N_{Mem})$.

Ab$_i$ are converted to memory cell in following conditions. A set of antigens captured by Ab$_i$ or a memory cell M_i is represented by $C(Ab_i)$, $C(M_i)$.

- M_i is removed if $C(M_i) \subset C(Ab_i)$.
- Ab$_i$ is converted to M_{N+1} if $C(Abi) \not\subset C(M_1) \cup C(M_2) \cup \ldots \cup C(M_N)$.
- Ab$_i$ is converted to M_{N+1}, if $C(Mi) = C(Abi)$

4.5 Clonal Selection

The memory cells and Abs which bind with Ags go through clonal selection for reproduction. This process is a cycle described as follows:

- First, an Ag is selected from the list of captured Ags. The probability of selection is proportional to $S(Ag_i)$, the concentration of Ag_i, which is initially 1.
- Randomly select two antibodies Ab_{p1} and Ab_{p2} from all the antibodies bound with the antigen.
- Ab_{p1} and Ab_{p2} are crossed over with probability P_c to produce offspring Ab_{c1} and Ab_{c2}.
  The crossover operation is defined as cut and splice operation.
  Crossover is followed by hypermutation, which is a series of copy mutation applied to w_{c1}, w_{c2}, and their copied offspring. There are several types of mutation. Locus deletion, deletes randomly selected locus. Locus addition, adds newly created locus to the antibody. Weight mutation changes the weight value of randomly chosen locus.
- With probability P_m, newly created antibody creates another mutated copy. Copy operation is repeated for $\Sigma nP_m{}^n$ times on average.
- Parents are selected from memory cells as well. Same crossover and hypermutation process is applied.
- The copied antigens go through negative selection as previously described. The reproduction processes are repeated until N_{Ab} antigens pass the negative selection.
- Finally, the score of each Ag is updated by (v)0. T is the score of Ag, s is the number of Ab bound to an Ag, and N is the total number of Ab. β is an empirically determined constant, 1.44 in this study. The concentration of Ag converges to 1 with appropriate β.

$$T'=\beta^{T-s/N} \tag{v}$$

The process goes back to Negative Selection to start a new cycle.

4.6 Summary of Experiment

AIS repeats the cycle as previously described. Flow of the system is described in Fig. 1. A single run is terminated after $N_c(=50)$ cycles. The AIS runs on parameters shown in Table 2. The results were robust to minor tuning of these parameters.

4.7 Generalization

Many rules with same C(Mem), set of captured antigens, are stored as memory cells. After a run is terminated, one memory cell is chosen to classify the test samples. A memory cell with smallest generality measure M (vi)0, is chosen.

$$M = \frac{\dfrac{1}{N_{ALL}}\sum_i^{N_{ALL}}\left(x_i-\tilde{x}_i\right)}{\dfrac{1}{N_{AML}}\sum_i^{N_{AML}}\left(y_i-\tilde{y}_i\right)} - \frac{W(\tilde{x}_i)}{W(\tilde{y}_i)} \tag{vi}$$

x_i and y_i are the ALL/AML sample vectors. $\tilde{x}_i$ and $\tilde{y}_i$ are the median of the ALL/AML samples. The first term in (vi)0 approximates the radius of distribution by average distance from the median. Second term is the average distance from the hyperplane to the sample vectors of each class. We assume proportion of two terms is close with enough training samples.

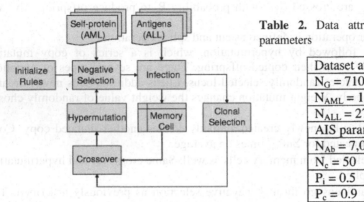

Fig. 1. Cycle of Artificial Immune System

Table 2. Data attributes and AIS parameters

Dataset attributes	
$N_G = 7109$	
$N_{AML} = 11$	
$N_{ALL} = 27$	
AIS parameters	
$N_{Ab} = 7,000$	
$N_c = 50$	
$P_i = 0.5$	
$P_c = 0.9$	
$P_m = 0.6$	

4.8 Results

In 20 runs, training data set was correctly classified. Table 3 shows the number of false positives(misclassified AML test data) and false negative(misclassified ALL test data) prediction on test samples in each run. Average and standard deviation is also shown.

Table 3. The number of misclassified samples in test data set

Run	1	2	3	4	5	6	7	8	9	10	11	12	13	14	15	16	17	18	19	20	Avg	Stdev
#FN	0	0	0	1	0	0	0	0	0	0	0	0	1	0	1	0	1	0	0	1	0	0.3	0.47
#FP	1	0	0	0	1	0	0	1	1	1	0	1	1	0	1	1	1	1	0	1	0	0.55	0.51

Fig. 2 and Fig. 3 show the learning process of AIS in a typical run. Fig. 2 shows the number of Ag caught by best and worst Abs. The average number of caught Ag is also shown. It shows training samples are learned by 20th cycle. It can be read from the graph that cycles afterward are spent to derive more general rules.

Fig. 3 shows the concentration of Ags at each cycle. Most Ags converge to 1, while few are slow to converge. These samples imply the border between the classes. Samples near the classification border are harder to capture, thus slower to converge. Empirically, the run should continue till the grasp of border is clear.

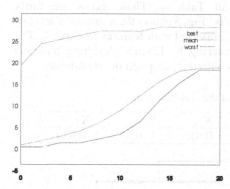

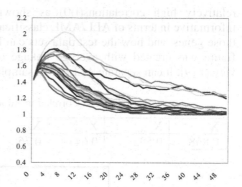

Fig. 2. # Of antigens caught by the best and worst antibodies

Fig. 3. Transition of antigen scores

5 Analyses of Selected Features

Fig. 4 shows classification rules which correctly classified the test data set. Each rule was different for each run. In this section, we further analyze the selected genes. Some of the genes appear repeatedly in the classification rules. Such genes have

A) $1.21896X_{3675} + -1.5858X_{4474} + 1.46134X_{1540} + -1.19885X_{2105} + 1.84803X_{757} + 1.82983X_{4034}$

B) $1.50797X_{6254} + 1.15777X_{757} + 3.95642X_{4810} + -1.60981X_{5667} + -1.57276X_{1826} + -1.39817X_{380} \quad 1.23648X_{1210} + 1.15777X_{757} + 1.32333X_{1506} + -1.28283X_{1966} + 1.84323X_{6030}$

C) $1.61061X_{1238} + -1.0343X_{5646} + -1.8181X_{6022} + 1.44975X_{6161}$

D) $1.25686X_{2328} + 1.49214X_{4038} + 1.48367X_{2190} + -1.61836X_{6022} + 1.66891X_{5265}$

E) $-1.4577X_{1385} + -1.57815X_{4363} + 1.31819X_{2317} + 1.75329X_{2328}$

F) $1.84455X_{4665} + -1.64613X_{759} + 1.23154X_{6216} + 1.8024X_{2855} + 1.92594X_{3086}$

G) $1.00809X_{1904} + 1.7706X_{6244} + -1.41034X_{4526} + -1.14542X_{759} + 1.94696X_{2723} + -1.34382X_{4875}$

H) $1.26745X_{3110} + 1.43941X_{1190} + 1.97632 + 1.74422X_{5519} + 1.79449X_{6874} + -1.44577X_{6022}$

Fig. 4. Examples of classification rules

relatively high correlations(i)0) as shown in Table 4. These genes are fairly informative in terms of ALL/AML classification. Fig. 5 shows the expression level of those genes, and how the test data sets can be clustered with features in Table 4. The figure was created with average linkage clustering by Eisen's clustering tool and viewer [9]. It can separate ALL/AML samples with the exception of one sample.

Table 4. Correlation value of the classifier genes

X_{757}	X_{1238}	X_{4038}	X_{2328}	X_{1683}	X_{6022}	X_{4363}
0.838	0.592	0.647	0.766	0.816	-0.837	-0.834

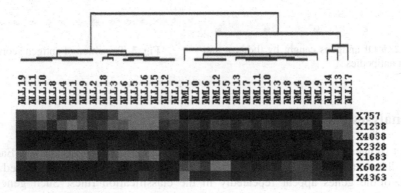

Fig. 5. Expression levels of informative genes and clustering based on those genes

The following section analyzes featured genes in rule A (Fig. 4). Each gene does not always have high correlation value as can be seen in Table 5.

```
Rule A) 1.22X_3675 + -1.59X_4474 + 1.46X_1540 + -1.20X_2105 +
1.85X_757 + 1.83X_4038
```

Table 5. Correlation of classifier genes in rule A

X_{3675}	X_{4474}	X_{1540}	X_{2105}	X_{757}	X_{4038}
0.151	0.371	0.418	-1.02	0.838	0.647

Fig. 6 shows the expression levels of feature genes in rule A. It implies that majority of the samples can be classified by few ALL(X_{2105}, X_{757}) and AML(X_{4038}) classifier genes. These classifier genes have relatively high correlation value.

Some of the samples(AML1, 6, 10, ALL11, 17, 18, 19) in Fig. 6 seem indistinguishable by those classifier genes. Functions of supplementary genes(X_{3675}, X_{4474}, X_{1540}) become evident when these samples are looked at especially.

Fig. 7 shows normalized expression levels of selected samples. In this selected group, X_{3675} and X_{4474} are highly correlated to ALL and AML respectively, while the classifier genes(X_{2105}, X_{757}, X_{4038}) became irrelevant to sample class.

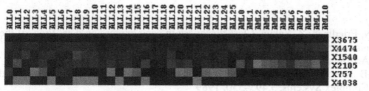

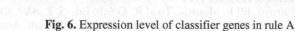

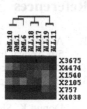

Fig. 6. Expression level of classifier genes in rule A

Fig. 7. Selected samples

6 Conclusion

Despite sparseness in training data, the accuracy of prediction was satisfactory, as test data were correctly classified 8 out of 20 times.

Regarding the feature selection, AIS chooses informative genes evolutionarily whilst optimizing its weight as well. Though gene subset chosen by AIS for classification differs in each run, the genes with strong correlation are chosen frequently as Table 4 shows. Similar results can be obtained by application Genetic Algorithms[8].

These genes with strong correlation, either selected by frequency or correlation, may not contain enough information to be a sufficient subset for classification, when many co-regulated genes are selected in the subset. Many genes are predicted to be co-expressed and those genes are expected to have similar rankings.

On the other hand, the result in Fig. 7 implies that selection of complementary genes, which are not necessary highly correlated, can be useful in classification. It might be suggested that the choice of feature gene subsets should be based not only on single ranking method, but also on redundancy and mutual information between the genes. Changing of ranking objective, when one feature is removed as a ranking criterion, has been suggested in [12]. It is interesting that AIS can choose primary and complementary feature genes by evolutionary process.

As future work to improve classification capability, use of effective kernel functions, and expressing relations between the genes, such as combining antibodies with AND/OR functions should be addressed.

Acknowledgement. This work was partially supported by the Grants-in-Aid for Scientific Research on Priority Areas (C), "Genome Information Sciences" (No.12208004) from the Ministry of Education, Culture, Sports, Science and Technology in Japan.

References

1. A. Ben-Dor, N. Friedman, Z. Yakini, Class discovery in gene expression data, Proc. of the 5th Annual International Conference on Computational Molecular Biology, 31–38, 2001.
2. D. Dasgupta. Artificial Immune Systems and Their Applications. Springer, 1999.
3. D. Goldberg, B. Korb and K. Deb, Messy Genetic Algorithms: Motivation, Analysis and First Results, Complex Systems, 3:493–530, 1989
4. Donna K. Slonim, Pablo Tamayo, Jill P. Mesirov, Todd R. Golub, Eric S. Lander, Class Prediction and Discovery Using Gene Expression Data, Proc. of the 4th Annual International Conference on Computational Molecular Biology(RECOMB), 263–272, 2000.
5. H. Liu , J. Li, L. Wong, A Comparative Study on Feature Selection and Classification Methods Using Gene Expression Profiles and Proteomic Patterns, in Proceeding of Genome Informatics Workshop, 2002
6. I. Guyon, J. Weston, S. Barnhill, V. Vapnik, Gene Selection for Cancer Classification using Support Vector Machines, Machine Learning Vol. 46 Issue 1–3, pp. 389–422, 2002
7. K.B. Hwang, D.Y. Cho, S.W. Wook Park, S.D. Kim, and B.Y. Zhang, Applying Machine Learning Techniques to Analysis of Gene Expression Data: Cancer Diagnosis, in Proceedings of the First Conference on Critical Assessment of Microarray Data Analysis, CAMDA2000.
8. L. Li, C. R. Weinberg, T. A. Darden, L. G. Pedersen, Gene selection for sample classification based on gene expression data: study of sensitivity to choice of parameters of the GA/KNN method, Bioinformatics, Vol. 17, No. 12, pp. 1131–1142, 2001
9. M. B. Eisen, P. T. Spellman, P. O. Brown, and D. Botstein. Cluster analysis and display of genome-wide expression patterns. Proceedings of the National Academy of Science, 85:14863–14868, 1998.
10. P. Baldi and A. Long, A Bayesian framework for the analysis of microarray expression data: Regularized t-test and statistical inferences of gene changes, Bioinformatics, 17:509–519, 2001.
11. P.J. Park, M. Pagano, and M. Bonetti, A nonparametric scoring algorithm for identifying informative genes from microarry data, PSB2001, 6:52–63, 2001.
12. R. Kohavi and G. H. John, Wrappers for Feature Subset Selection, Artificial Intelligence, vol.97, 1–2, pp273–324, 1997
13. T. S. Furey, N. Cristianini, N. Duffy, D. W. Bednarski, M. Schummer, and D. Haussler. Support vector machine classification and validation of cancer tissue samples using microarray expression data. Bioinformatics, 2001
14. T.R. Golub, D.K. Slonim, P. Tamayo, Molecular classification of cancer: class discovery and class prediction by gene expression monitoring. Science, 286:531–537, 1999.
15. U. Alon, N. Barkai, D. Notterman, K. Gish, S. Ybarra, D. Mack, and A. Levine. Broad patterns of gene expression revealed by clustering analysis of tumor and normal colon cancer tissues probed by oligonucleotide arrays. Cell Biology, 96:6745–6750, 1999.

Pattern Search in Molecules with FANS: Preliminary Results

Armando Blanco, David A. Pelta, and Jose-L. Verdegay

Depto. de Ciencias de la Computación e I.A.
Calle Periodista Daniel Saucedo Aranda s/n
Universidad de Granada, 18071 Granada, Spain
{armando,dpelta,verdegay}@ugr.es

Abstract. We show here how *FANS*, a fuzzy sets-based heuristic, is applied to a particular case of the Molecular Structure Matching problem: given two molecules A (the pattern) and B (the target) we want to find a subset of points of B whose set of intra-atomic distances is the most similar to that of A. This is a hard combinatorial problem because, first we have to determine a subset of atoms of B and then some order for them has to be established.

We analyze how the size of the pattern affects the performance of the heuristic, thus obtaining guidelines to approach the solution of real problems in the near future.

1 Introduction

FANS, an acronym for Fuzzy Adaptive Neighborhood Search [3], is a fuzzy sets-based heuristic that among other characteristics, can be considered as a framework of local search techniques in the sense that it allows to achieve the same qualitative behavior of other techniques. This characteristic is achieved through considering the neighborhood of a solution as a fuzzy set, and using the membership values to guide the search. *FANS* was successfully applied in knapsack problems [11] and the protein folding problem [12].

As an increasing number of protein structures become known, the need for algorithms to analyze 3D conformational structures increases as well. For example, the search for common substructures of proteins is of value in uncovering relationships among them for inferring similarities in function and discovering common evolutionary origins. There is now widespread agreement that similarities among distantly related proteins are often preserved at the level of three-dimensional form, even after very little similarity remains at the sequence level.

The comparison of the 3D structures of protein molecules poses a very complex algorithmic problem. The search for effective solution techniques for it, is justified because such tools can aid scientists in the development of procedures for drug design, in the identification of new types of protein architecture, in the organization of the known set of protein structures and can help to discover unexpected evolutionary relations between proteins [6,8].

S. Cagnoni et al. (Eds.): EvoWorkshops 2003, LNCS 2611, pp. 11–21, 2003.

Techniques for protein comparison include distance matrix alignment [5], dynamic programming [15], graph theory [4], genetic algorithm [10,14] and computer vision techniques [16,9], just to cite a few references.

In this article we will address a particular case of the so called molecular matching problem where the objective is to minimize the dissimilarity in corresponding atoms positions between two molecules A and B with sizes N_a, N_b respectively. Four cases are recognized of practical importance:

- $N_a = N_b$, the whole A is compared against the whole B
- $N_a \leq N_b$, the structure B is searched for a region that resembles only a defined region of A
- $N_a < N_b$, the structure B is searched for a region that resembles the whole of A
- $N_a \leq N_b$, the structure B is searched for a region that resembles any unspecified region of A.

We concentrate here on case 3 where we can consider molecule A as a pattern and we want to find a subset of atoms of B where A occurs with more "strength".

The main objective of this work is to analyze one factor that may affect the performance of algorithms. This factor is the size of the pattern A and we will analyze its influence in the performance of *FANS*. The experiments will be done on synthetic data with known optima.

The article is organized as follows: first, we define the problem of interest; then we describe the main aspects of *FANS*'s application to *MSM*. After that, we show how the test set is constructed and next, the experiments and results are presented. Finally, the conclusions appear.

2   Definition of the Problem

The problem faced here is defined as follows: given two proteins A, B and their associated intra-atomic distance matrices D_A, D_B, where $D_A(i,j)$ is the Euclidean distance between atoms i and j in the molecule A, the objective is to find a permutation matrix P of rows/cols of D_B which minimizes

$$min||D_A - P^T D_B P||. \tag{1}$$

This very general formulation was also used in [1,7]. The comparison of proteins by means of their distance matrices is also performed in [5]. A distance matrix is a 2D representation of a 3D structure. It is independent of the coordinate frame and contains enough information to reconstruct the 3D structure by distance geometry techniques.

In short, we are seeking for a subset of points of B whose set of intra distances is the most similar to that of A. The norm is computed only for the matched atoms. This is a hard combinatorial problem because, first we have to determine a subset of atoms of B and then some order for them has to be established.

3   The Fuzzy Adaptive Neighborhood Search

The Fuzzy Adaptive Neighborhood Search Method (*FANS*) [3,13] is a local search procedure which differs from other local search methods in two aspects. The first one is how solutions are evaluated; within *FANS* a fuzzy valuation representing some (maybe fuzzy) property P is used together with the objective function to obtain a "semantic evaluation" of the solution. In this way, we may talk about solutions satisfying P in certain degree. Under this view, the neighborhood of a solution effectively becomes a fuzzy set [17] with the neighbor solutions as elements and the fuzzy valuation as the membership function.

The fuzzy valuation enables the algorithm to achieve the qualitative behavior of other classical local search schemes [3]. *FANS* moves between solutions satisfying P with at least certain degree, until it became trapped in a local optimum. In this situation, the second novel aspect arise: the operator used to construct solutions is changed, so solutions coming from different neighborhoods are explored next. This process is repeated once for each of a set of available operators until some finalization criterion for the local search is met.

The scheme of *FANS* is shown in Fig. 1. The execution of the algorithm finishes when some external condition holds, here when the number of cost function evaluations reached a pre-specified limit. Each iteration begins with a call to the so called *neighborhood scheduler* NS, which is responsible for the generation and selection of the next solution in the optimization path. The call is done with parameters S_{cur} (the current solution), $\mu()$ (the fuzzy valuation), and $\mathcal{O}$ (a parameterized operator which is used to construct solutions). The neighborhood scheduler can return two alternative results; either a good enough (in terms of $\mu()$) solution (S_{new}) was found or not.

In the first case S_{new} is taken as the current solution and $\mu()$ parameters are adapted. In this way, the fuzzy valuation is changed as a function of the context or, in other terms, as a function of the state of the search. This mechanism allows the local search stages to adapt during the search. If NS failed to return an acceptable solution (no solution was good enough in the neighborhood induced by the operator), the parameters of the operator are changed. The strategy for this adaptation is encapsulated in the so called *operator scheduler OS*. The next time NS is executed, it will have a modified operator to search for solutions.

When the whole set of operators available was used and the search was still stagnated (*TrappedSituation = True*), a classical random restart procedure is applied, and *FANS* continues the search from the new solution.

The reader should note that what varies at each iteration are the parameters used in the NS call. The algorithm starts with NS $(s_0, \mathcal{O}^{t_0}, \mu_0)$. If NS could retrieve an acceptable neighborhood solution, the next iteration the call will be NS $(s_1, \mathcal{O}^{t_0}, \mu_1)$, the current solution is changed and the fuzzy valuation is adapted. If NS failed to retrieve an acceptable neighborhood solution (at certain iteration l), the operator scheduler will be executed returning a modified version of the operator, so the call will be NS $(s_l, \mathcal{O}^{t_1}, \mu_l)$.

```
Procedure FANS:
Begin
  While (not-end) Do
    /* The neighborhood scheduler NS is called */
    S_new = NS(O,μ(),S_cur);
    If (S_new is good enough in terms of μ()) Then
      S_cur := S_new;
      adaptFuzzyValuation(μ(),S_cur);
    Else
      /* NS failed to return a good solution with O */
      /* The operator scheduler will modify the operator */
      O := OpSchedul(O);
    Fi
    If (TrappedSituation()) Then
      doEscape();
    Fi
  Od
End.
```

Fig. 1. Scheme of *FANS*

3.1   Adapting *FANS* to *MSM*

In order to apply *FANS* to a specific problem, the user must provide definitions for the components. Next, we describe how solutions are represented, the characteristics of the modification operator, the operator scheduler and the neighborhood scheduler. Emphasis will be put in the description of the fuzzy valuation.

Representation of Solutions. A solution represents a permutation matrix, i.e. a square matrix verifying $\sum_{i=1}^{n} p_{ij} = 1$ for all $j = 1, 2, .., n$; $\sum_{j=1}^{n} p_{ij} = 1$ for all $i = 1, 2, ..., n$; and $p_{ij} = \{0, 1\}$ for all $i, j = 1, 2, ..., n$.

A permutation matrix can be represented using $\mathcal{O}(n)$ space. For example, given the permutation

$$P = \begin{pmatrix} 0 & 0 & 1 & 0 & 0 & 0 \\ 0 & 0 & 0 & 0 & 0 & 1 \\ 0 & 0 & 0 & 1 & 0 & 0 \\ 1 & 0 & 0 & 0 & 0 & 0 \\ 0 & 1 & 0 & 0 & 0 & 0 \\ 0 & 0 & 0 & 0 & 1 & 0 \end{pmatrix}$$

we represent it as $P = \{3, 6, 4, 1, 2, 5\}$ where $P[i]$ indicates the position of the 1 in row i.

Now suppose you are given a pattern $A = \{a_1, a_2, a_3\}$ of size $p = 3$ and a protein $B = \{b_1, b_2, b_3, b_4, b_5, b_6\}$ of size $n = 6$. The previous permutation leads to the following order for $B = \{b_3, b_6, b_4, b_1, b_2, b_5\}$. The cost function is only applied for the first p atoms. This subset is the *inner set* while the rest $n - p$ atoms made the *outer set*. Then, the matched atoms are $M = \{(a_1, b_3), (a_2, b_6), (a_3, b_4)\}$.

Modification Operator. The operator $\mathcal{O}$ constructs new solutions (permutations) through changes over a given solution (permutation) s. Using a parameter k and a solution s, new solutions $\hat{s}$ are obtained through the application of any of the following operations:

- 2-*Swap*: the values in two positions (randomly selected) (i, j) of the permutation are exchanged. This procedure is repeated k times. It is not allowed the exchange between two positions belonging to the outer set.
- MoveBlock: two randomly selected blocks of length k are exchanged. For example, if $k = 2$ and $s - \{1, 2, 3, 4, 5\}$ then, a possible outcome is $\hat{s} = \{4, 5, 3, 1, 2\}$. The first block always belongs to the inner set.
- InvertBlock: a randomly selected block of length k is inverted. For example, if $k = 2$ and $s = \{1, 2, 3, 4, 5\}$ then a possible outcome is $\hat{s} = \{3, 2, 1, 4, 5\}$. The initial position of the block must belong to the inner set.

The operators are selected according to certain predefined probabilities. Operator 2-*Swap* is selected with probability 0.7. If it is not selected, then either *MoveBlock* or *InvertBlock* could be applied with equal probability.

The value of the parameter k is modified as a function of the state of the search. The strategy used for its modification is defined within the operator scheduler component defined ahead.

Operator Scheduler. This component, called OS, is responsible for the variation of the parameter k of the operator. For these experiments, we use the simplest approach: each time OS is called, the value of k is decremented by 1. In this way, the algorithm will start with coarse modifications (which are associated with an exploration stage) making them finer as the execution progresses (corresponding to an exploitation stage).

Fuzzy Valuation. The fuzzy valuation is represented in *FANS* by a fuzzy set $\mu() : \mathcal{R} \rightarrow [0, 1]$. For example, having the fuzzy set of "good" solutions, we will consider the goodness of the solution of interest. In the same spirit, given two solutions $a, b \in \mathcal{S}$ we could think about how *Similar* a and b are, or how *Close* they are, or also how *Different* b is from a. *Similar, Close, Different* will be fuzzy sets represented by appropriated membership functions $\mu()$.

In this work, we will measure solutions in terms of their "Acceptability". The corresponding fuzzy valuation *"Acceptable"* reflects the following idea: with a solution at hand, those generated solutions improving the current cost, will have a higher degree of acceptability than those with worse cost. Solutions increasing the cost a little will be considered as "acceptable" but with a lower degree. Finally, those solutions increasing too much the current cost would not be considered as acceptable. These ideas are captured in the following definition:

$$\mu(s, \hat{s}) = \begin{cases} 1.0 & \text{if } f(\hat{s}) < f(s) \\ (f(\hat{s}) - \beta)/(f(s) - \beta) & \text{if } f(s) \le f(\hat{s}) \le \beta \\ 0.0 & \text{if } f(\hat{s}) > \beta \end{cases} \tag{2}$$

where f is the objective function, s the current solution, $\hat{s} \in N(s)$ a neighborhood solution, and β the limit for what is considered as acceptable. As a first approximation, $\beta = f(s)*(1+\alpha)$, where $\alpha \in [0..1]$. In this work, we use $\alpha = 0.01$.

Neighborhood Scheduler. This component is responsible for the generation and selection of a new solution from the neighborhood.

FANS uses two types of neighborhood: the *operational* and the *semantic* neighborhood of s. Given the current operator $\mathcal{O}$, the fuzzy valuation $\mu()$ and the current solution s, the operational neighborhood is:

$$N(s) = \{\hat{s}|\ \hat{s} = \mathcal{O}_i(s)\} \tag{3}$$

where $\mathcal{O}_i$ stands for the i-th application of $\mathcal{O}$ over s. The *semantic neighborhood* of s is defined as follows:

$$\hat{N}(s) = \{\hat{s} \in N(s)|\ \mu(s,\hat{s}) \geq \lambda\}. \tag{4}$$

Here, the neighborhood $\hat{N}(s)$ represents the λ-cut of the fuzzy set of solutions.

The value of λ is a key aspect in *FANS* because each value induce a particular behavior of the heuristic. If we have in mind the previous definition of "Acceptability", then a value of $\lambda = 1$ makes *FANS* to behave as a hillclimber: transitions only to improving solutions are allowed. When $\lambda = 0$, *FANS* behaves almost like a random search, although a deterioration limit is posed by the definition of the fuzzy valuation. Other values for λ lead to behaviors where non-improving moves have the opportunity of being selected.

Because several ways exist to define subjective concepts like "Acceptability" or "Similarity", then it is clear that a very wide range of behaviors could be obtained through appropriated combinations of fuzzy valuation definitions and λ values. In this sense, we consider *FANS* as a local search framework.

Focusing again in the neighborhood scheduler, we use here a simple definition called *First*: it returns the first solution $\hat{s}$ found such that $\hat{s} \in \hat{N}(s)$ using at most $maxTrials = 600$ attempts.

When the operator reaches $k = 1$ then a procedure is executed in order to look for an acceptable solution in a systematic way: a position i is randomly selected and then every exchange $(i,j), j > i$ is done. The process is repeated, if needed, for all i leading to a worst case complexity of $\mathcal{O}(n^2)$.

4   Description of the Data and Experiments

Because the problem of structure comparison does not have a unique answer and the definition of a similarity score between proteins is far from being solved (see [8] and references therein), we decided to test *FANS* over artificial test problems with known optima.

Table 1. Average values of $pctOK$ for each protein and λ (a), and pattern size (b).

Prot	Value of λ				Prot	Pattern Size			
	0.7	0.8	0.9	1.0		20%	40%	60%	80%
101M	42.79	48.58	53.43	54.74	101M	49.18	51.08	48.55	50.31
1ALB	36.51	48.16	40.46	46.66	1ALB	37.74	40.93	44.43	47.96
1FMB	80.02	71.01	72.62	75.39	1FMB	69.03	69.26	78.11	82.96
2LZM	64.86	65.96	62.36	68.08	2LZM	65.34	60.40	64.61	71.29
3DFR	80.82	93.19	89.47	92.32	3DFR	79.61	90.22	93.47	92.21
3HHB	52.09	60.15	51.67	56.97	3HHB	52.37	54.73	53.75	59.87

(a)   (b)

To construct the test set we proceed as follows: given a protein B with n atoms, we randomly select a subset of them of size p. This subset will be the pattern A. Then, the objective is to find the pattern A within B. This implies the selection of the correct subset in B and the determination of the right order: being n the size of the protein and p that of the pattern, we have $\binom{n}{p} * p!$ ways for choosing p atoms from n, and each one with the corresponding permutations. Test instances constructed in this way, have an optimum value of zero.

For our experiments, we took the C_α carbons of each residue of 6 proteins from the Protein Data Bank (*PDB*, http://www.rcsb.org); their description (PDB code, size) are the following ones: (101M,154), (1ALB,131), (1FMB,104), (2LZM,164), (3DFR,162), (3HHB,141). These proteins will represent B.

For each protein, we define 4 pattern sizes corresponding with $pSize = \{20, 40, 60, 80\}\%$ of the size of B. Then, for each pattern size, we generate 15 patterns. Finally, for each value of $\lambda = \{0.7, 0.8, 0.9, 1.0\}$, protein and pattern, we perform at most 3 trials of *FANS* to try to find the pattern (to obtain the optimum). Each run ended when the limit for the number of cost functions evaluations (or configurations tested) reached certain limit. This value is calculated as a function of the protein and patterns size as follows: $maxEvals = (n+p/2)*1200$. For each run we record two values: *Best*: the cost of the best solution ever found; and $pctOK$: the percentage of correctly matched atoms.

4.1 Results by Protein

First, we analyze the average value of $pctOK$ per protein and value of λ. These results are shown in Table 1(a). They confirm something expected: there is no single value of λ (particular algorithmic behavior) being the best one for every test case. This is because each test case has its own characteristics, i.e. the corresponding distribution of points which can produce internal symmetries or particular arrangements of points that can fool or help *FANS* to reach high quality solutions.

Table 2. Number of non solved patterns (out of 15) by value of λ and pattern size. A blank appears if all the patterns were solved. **T**= Total.

	101M							1ALB							1FMB				
%/λ	0.7	0.8	0.9	1	**T**		%/λ	0.7	0.8	0.9	1	**T**		%/λ	0.7	0.8	0.9	1	**T**
20	1	5	3	2	11		20	7	3	8	4	22		20	2	1	1		4
40	5	2	1	2	10		40	6	5	6	5	22		40	1		2	2	5
60	4	4	2	2	12		60	8	4	4	5	21		60					
80	7	2	4	2	15		80	7	3	6	2	18		80		1	1		2
T	17	13	10	8			**T**	28	15	24	16			**T**	3	2	4	2	

	2LZM							3DFR							3HHB				
%/λ	0.7	0.8	0.9	1	**T**		%/λ	0.7	0.8	0.9	1	**T**		%/λ	0.7	0.8	0.9	1	**T**
20		2	3	4	9		20	1				1		20	4	2	4	3	13
40	3	2	3	2	10		40							40	1	1	5	4	11
60	4	2	1	3	10		60							60	3	4	4	2	13
80	2	1	1	1	5		80							80	4	1	1	1	11
T	9	7	8	10			**T**					1		**T**	12	8	14	10	

The results in Table 1(b) show the average of $pctOK$ for each protein and every pattern size. In general, it is clear that as the pattern size increases, the average number of correctly matched atoms also increases.

In order to assess the performance of *FANS*, we need to check the number of patterns that were NOT solved. Remember that there are 15 patterns of each size. We consider a pattern as solved if at least one of the three trials performed, ended with $pctOK \geq 95$.

In Table 2 we show the number of patterns (out of 15) that were not solved ($pctOK < 95$) in at most three trials. A blank appears if all the patterns were solved. The results are separated by value of pattern size (first column) and by value of λ (first row). The total value for the rows and columns is 60. From the Tables, it can be deduced that *FANS* is quite effective. Setting apart the results for 1ALB, almost any pattern can be solved with a reduced effort. For example, for 101M, pattern size of 40%, and $\lambda = 0.7$, *FANS* not solved 5 patterns. However, when $\lambda = 0.9$ just 1 pattern remained unsolved. This situation also holds for the rest of proteins, where for every pattern size, a value for λ appears which enables *FANS* to solve more than 13 patterns (three or less than three patterns remain unsolved).

About the "hardness" of each protein in the test set, it can be seen that protein 3DFR is the easiest one. Whatever value of λ is used, *FANS* solved all the patterns. In the other extreme, 1ALB is the hardest case. Different values of λ lead to quite different results. For example, for a pattern size of 20%, the non solved patterns varied between 3 ($\lambda = 0.8$) and 8 ($\lambda = 0.9$).

4.2   Global Analysis of the Results

Now, in order to obtain general conclusions we will analyze the results in a global manner for the whole test set using the histograms in Fig. 2 where the x axis represents $pctOK$ while the y axis stands for the number of executions which ended in such percentage. Every column has a wide of 2.5%.

Figure 2 shows the distribution of cases grouped by pattern size. There is a direct relation between the size of the pattern and the value of $pctOK$, at least in terms of the number of executions that ended with more than 95% of correctly matched atoms. Bigger patterns are more easily solved than smaller ones. This seems to be in contradiction with the number of combinations available for each pattern size. Consider a protein B of size 100 and two patterns P_1, P_2 of sizes 20 and 80. Then, the number of combinations for each pair are $(B, P_1) = \binom{100}{20} * 20!$ and $(B, P_2) = \binom{100}{80} * 80!$; so $(B, P_1) < (B, P_2)$ and P_2 is easier to find than P_1.

However, when dealing with big patterns, the chance to select an incorrect atom decrease. In the extremal case where the pattern size equals that of the protein, the problem "reduces" to find the correct permutation for the atoms. So, as the pattern size increases, the problem of determining the correct subset is reduced but the problem of finding the correct permutation becomes bigger. Given that *FANS* is able to find bigger patterns than shorter ones, we can establish two facts: first, that algorithm is well suited to deal with huge search spaces, and second, that the problem of finding the correct subset of atoms is harder than that of finding the correct permutation.

Another thing to remark is the "gap" or "hole" that exists for values between $0.35 \leq pctOK \leq 0.75$. Almost none execution ended with such values of atoms correctly assigned. This implies that the algorithm gets it right with a high value of $pctOK$, or goes wrong, with a high number of bad atom assignments. In principle, this situation may be attributed to internal symmetries of the structures which guide the algorithm to local optimal solutions of good quality. A similar situation was also reported in [2] where the authors used simulated annealing to solve another type of *MSM*.

The easiest way to avoid this dichotomy is to perform several runs of *FANS*, each one starting from different initial solutions. This approach was used here and the results previously presented confirm its usefulness: almost none pattern remain unsolved after at most three runs of the algorithm.

For a total number of runs of 3022, the number of successful ones (those with $pctOK > 95\%$) was 1514 (50%), while the number of optimal ones (those with $pctOK = 100\%$) was 998 (33%).

5   Conclusions

In this work we presented the application of a fuzzy sets-based heuristics, called *FANS*, to a particular case of the molecular structure matching problem.

We analyzed how the influence of the pattern size affected the performance of *FANS* over the *MSM*. The results revealed that, on average, the algorithm was

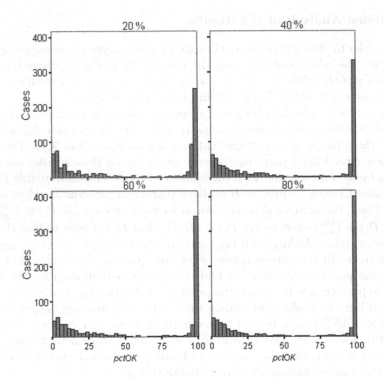

Fig. 2. Distribution of *pctOK* for each value of pattern size.

able to correctly match more atoms for bigger patterns than for smaller ones. Having in mind that the number of configurations for bigger patterns is much higher than that for smaller ones, this result should be considered as an evidence of the ability of *FANS* to deal with huge search spaces. As it was expected, as the size of the patterns increased, the effort needed to solve them also increased.

If we analyze the results from an optimization point of view, they have to be considered as very promising. The set of behaviors induced by the combination of the fuzzy valuation, together with different values of λ, were enough to solve almost every pattern in the test set.

The whole set of results encourage us to face, in the near future, real problems and comparisons against other algorithms which will confirm the potential of our fuzzy sets-based heuristic.

Acknowledgments. Research supported in part by Projects TIC2002-04242-CO3-02 and TIC 99-0563. David Pelta is a grant holder from Consejo Nacional de Investigaciones Científicas y Técnicas (CONICET), Argentina. The authors wish to thanks Dr. Hilario Ramirez for his help with this problem.

References

1. M. Barakat and P. Dean. Molecular structure matching by simulated annealing. I. A comparison between different cooling schedules. *Journal of Computer-Aided Molecular Design*, 4:295–316, 1990.

2. M. Barakat and P. Dean. Molecular structure matching by simulated annealing. II. An exploration of the evolution of configuration landscape problems. *Journal of Computer-Aided Molecular Design*, 4:317–330, 1990.

3. A. Blanco, D. Pelta, and J. Verdegay. A fuzzy valuation-based local search framework for combinatorial problems. *Journal of Fuzzy Optimization and Decision Making*, 1(2):177–193, 2002.

4. E. J. Gardiner, P. J. Artymiuk, and P. Willet. Clique-detection algorithms for matching three-dimensional molecular structures. *Journal of Molecular Graphics and Modelling*, 15:245–253, 1997.

5. L. Holm and C. Sander. Protein structure comparison by alignment of distance matrices. *Journal of Molecular Biology*, (233):123–138, 1993.

6. L. Holm and C. Sander. Mapping the protein universe. *Science*, 273:595–602, 1996.

7. R. Kincaid. A molecular structure matching problem. *Computers Ops Res.*, (1):25–35, 1997.

8. P. Koehl. Protein structure similarities. *Current Opinion in Structural Biology*, 11:348–353, 2001.

9. N. Leibowitz, Z. Fligerman, R. Nussinov, and H. Wolfson. Multiple structural alignment and core detection by geometric hashing. In T. L. Et.Al., editor, *Procs of 7th Intern. Conference on Intelligent Systems for Molecular Biology ISMB 99.*, pages 167–177. AAAI Press, 1999.

10. A. May and M. Johnson. Improved genetic algorithm-based protein structure comparisons: Pairwise and multiple superpositions. *Protein Engineering*, 8:873–882, 1995.

11. D. Pelta, A. Blanco, and J. L. Verdegay. A fuzzy adaptive neighborhood search for function optimization. In *Fourth International Conference on Knowledge-Based Intelligent Engineering Systems & Allied Technologies, KES 2000*, volume 2, pages 594–597, 2000.

12. D. Pelta, A. Blanco, and J. L. Verdegay. Applying a fuzzy sets-based heuristic for the protein structure prediction problem. *International Journal of Intelligent Systems*, 17(7):629–643, 2002.

13. D. Pelta, A. Blanco, and J. L. Verdegay. Fuzzy adaptive neighborhood search: Examples of application. In J. L. Verdegay, editor, *Fuzzy Sets based Heuristics for Optimization*, Studies in Fuzziness and Soft Computing. Physica-Verlag, 2003. to appear.

14. J. D. Szustakowsky and Z. Weng. Protein structure alignment using a genetic algorithm. *PROTEINS: Structure, Function, and Genetics*, 38:428–440, 2000.

15. W. Taylor. Protein structure comparison using iterated double dynamic programming. *Protein Science*, 8:654–665, 1999.

16. R. N. H. Wolfson. Efficient detection of three dimensional structural motifs in biological macromolecules by computer vision techniques. *Procs. of the National Academy of Sciences U.S.A.*, 88:10495–10499, 1991.

17. H. J. Zimmermann. *Fuzzy Sets Theory and Its Applications*. Kluwer Academic, 1996.

Applying Memetic Algorithms to the Analysis of Microarray Data

Carlos Cotta[1], Alexandre Mendes[2], Vinícius Garcia[2], Paulo França[2], and
Pablo Moscato[3]

[1] Dept. Lenguajes y Ciencias de la Computación, University of Málaga,
ETSI Informática, Campus de Teatinos, 29071 – Málaga, Spain

[2] Faculdade de Engenharia Elétrica e de Computação
Universidade Estadual de Campinas C.P. 6101, 13083-970, Campinas, Brazil

[3] School of Electrical Engineering and Computer Science
University of Newcastle, Callaghan, NSW, 2308, Australia

ccottap@lcc.uma.es

Abstract. This work deals with the application of Memetic Algorithms
to the Microarray Gene Ordering problem, a NP-hard problem with
strong implications in Medicine and Biology. It consists in ordering a
set of genes, grouping together the ones with similar behavior. We propose a MA, and evaluate the influence of several features, such as the
intensity of local searches and the utilization of multiple populations,
in the performance of the MA. We also analyze the impact of different
objective functions on the general aspect of the solutions. The instances
used for experimentation are extracted from the literature and represent
real biological systems.

1   Introduction

Due to the huge amount of data generated by the Human Genome Project,
as well as what is being compiled from other genomic initiatives, the task of
interpreting the functional relationships between genes appears as one of the
greatest challenges to be addressed by scientists [13]. The traditional approach
of Molecular Biology was based on the "one gene in one experiment" scenario.
This approach severely limits understanding "the whole picture", making the
ramified paths of gene function interactions hard to track. For this reason, new
technologies have been developed in the last years. In this sense, the so-called
DNA microarray technique [5,7] has attracted tremendous interests since it allows monitoring the activity of a whole genome in a single experiment.

In order to analyze the enormous amount of data that is becoming available,
reduction techniques are clearly necessary. As a matter of fact, it is believed
that genes are influenced on average by no more than eight to ten other genes
[3]. To achieve such a reduction, and allow molecular biologists concentrate on a
sensible subset of genes, clustering techniques such as k-means, or agglomerative

S. Cagnoni et al. (Eds.): EvoWorkshops 2003, LNCS 2611, pp. 22–32, 2003.

methods can be used (see [8,9] for example). However, there is still much room for improvement in the solutions they provide.

Memetic algorihms [16] (MAs) have been recently proposed as a tool for aiding in this process [15]. In this work we consider the application of MAs to a related variant of this clustering problem: the Gene Ordering problem. It consists of finding a high-quality rearrangement of gene-expression data, such that related (from the point of view of their expression level) genes be placed in nearby locations within a gene sequence. This problem is motivated by the usual bidimensional representation of microarray data, and hence it combines aspects of clustering and visualization. Here, we concentrate our efforts in analyzing the influence of several features of the MA –such as the intensity of local searches and the configuration of the population– in the performance of the algorithm. The impact of different objective functions on the quality of the solutions is also thoroughly tested. This is a critical issue, especially because the ordering quality is a relative attribute, strongly connected to the visual aspect.

The remainder of the article is organized in three sections: first, the definition of the problem, and the application of MAs to this domain are detailed in Section 2; next, the results of an extensive computational experimentation involving the parameters of the algorithm are reported in Section 3; finally, a brief summary of the main contributions of this work, and some prospects for future work are provided in Section 4.

2   Memetic Algorithms

In this section we will discuss the implementation of the Memetic Algorithm. The MA is a population-based algorithm that uses analogies to natural, biological and genetic concepts, very similarly to Genetic Algorithms (GAs) [4]. As in the latter, it consists of making a population of solutions evolve by mutation and recombination processes. Additionally, a distinctive feature of MAs is the use of local search operators in order to optimize individual solutions. The best fitted solutions of the population will survive and perpetuate their information, while the worse fitted will be replaced by the offspring. After a large number of generations, it is expected that the final population would be composed of highly adapted individuals, or in an optimization application, high-quality solutions of the problem at hand. Next, we briefly explain some specific MA aspects.

2.1   The Fitness Function

Before getting into the details of the fitness function, some comments must be made about the nature of the data defining the problem. For our purposes, the output of a microarray experimentation can be characterized by a matrix $G = \{g_{ij}\}$, $1 \leq i \leq n$, $1 \leq j \leq m$, where n is the number of genes under scrutiny, and m is the number of experiments per gene (corresponding to measurements under different conditions or at different time points). Roughly speaking, entry g_{ij} (typically a real number) provides an indication on the activity of gene i under condition j.

Now, as mentioned in Section 1, a good solution in the Gene Ordering problem (i.e., a good permutation of the genes) will have similar genes grouped together, in clusters. A notion of distance must thus be defined in order to measure similarity among genes. We have considered a simple measure, the Euclidean distance (other options are possible; see e.g [18]). We can thus construct a matrix of inter-gene distances. This matrix will be used for measuring the goodness of a particular gene ordering.

The most simple way of doing this is calculating the total distance between adjacent genes, similarly to what is done in the Traveling Salesman Problem. A drawback of such an objective function is that, since it only uses information of adjacent genes, it has a very narrow vision of the solution. For a better grouping of the genes, the use of 'moving windows' is a better choice. The total distance thus becomes a two-term sum, instead of a single-term one. The first term sums up the distances between the window's central gene and all others within the window's length. The second one sums up those partial distances as the window moves along the entire sequence. Moreover, in order to give more weight to the genes that are closer a multiplying term was added to the function. Let $\pi = \langle \pi_1, \pi_2, \cdots, \pi_n \rangle$ be the order of the n genes in a given solution. Then, the fitness function is:

$$fitness(\pi) = \sum_{l=1}^{n} \sum_{i=\min(l-s_w,1)}^{\max(l+s_w,n)} w(i,l) D[\pi_l, \pi_i] \qquad (1)$$

where $2s_w + 1$ is the window size (the number of genes involved in each partial distance calculation), and $w(i,l)$ is a scalar weighting the influence of the gene located at position l on the gene located at position i. We have considered weights proportional to $s_w - |l - i| + 1$ (i.e., linear in the distance between the genes), and normalized so as to have the sum of all weights involved in each application of the inner sum in Eq. (1) be 1. The use of this function gives higher ratings to solutions where genes are grouped in larger sets, with few discontinuities between them. As a matter of fact, the function makes in some sense the assumption that every gene could be the center of a cluster, subsequently measuring the similarity of nearby genes. The global optimization of this function is thus expected to produce robust solutions. Section 3 will provide results regarding the use of this function, and the effect of varying the window size.

2.2 Representation and Operators

The representation chosen for the Gene Ordering problem takes some ideas from hierarchical clustering. To be precise, solutions are represented as a binary tree whose leaves are the genes. By doing so, solutions are endowed with extra information regarding the level of relationship among genes, information that would be missing in other representations that only concentrated on the actual leaf order (e.g., permutations). This way, reproductive operators such as recombination and mutation can be less disruptive.

We have used a pre-order traversal of the tree to store it into individuals. More precisely, the chromosome is a string of integers in the $[-1, n-1]$ interval, where n is the number of genes to be ordered. Each of the genes (leaves of the tree) is identified with a number in $[0, n-1]$, and internal nodes of the tree with the value -1 (all of them are indistinguishable). It is easy to see that the length of the chromosome will thus be $2n - 1$ genes.

Having defined this representation, adequate operators must be defined to manipulate it. Considering firstly the recombination operator, we have used an approach similar to that used in the context of phylogeny inference [6]. To be precise, we perform recombination by selecting a subtree T from one of the parents, removing any leaf present in T from the second parent (to avoid duplications), and inserting T at a random point of the modified parent. As to mutation, it is based on gene-sequence flip. A subtree is selected uniformly at random, and its structure is flipped. This way, the entire sequence belonging to it is flipped too. This mutation preserves the gene grouping inside the subtree, introducing some disruption in its boundaries.

2.3   Population Structure

The use of hierarchically structured populations boosts the performance of the GA/MA (see [10]). There are two aspects that should be noticed. First, the placement of the individuals in the population structure according to their fitness. And second, a well-tailored selection mechanism that selects pairs of parents for recombination according to their position in the population. In our approach, the population is organized following a complete ternary tree structure. In contrast with a non-structured population, the complete ternary tree can also be understood as a set of overlapping 4-individual sub-populations (that we will refer to as *clusters*).

Each cluster consists of one *leader* and three *supporter* individuals. The leader individual always contains the best solution of all individuals in the cluster. This relation defines the population hierarchy. The number of individuals in the population is equal to the number of nodes in the complete ternary tree, i.e. we need 13 individuals to make a ternary tree with 3 levels, 40 individuals to have 4 levels, and so on. The general equation is $(3^n - 1)/2$, where n is the number of levels. Previous tests comparing the tree-structured population with non-structured approaches are present in [10]. They show that the ternary-tree approach leads to considerably better results, and with the use of much less individuals.

2.4   Reproduction

The selection of parents for reproduction is an important part of the algorithm, since the quality of the offspring will depend basically on this choice. Based on the assumption that a ternary tree is composed of several clusters, we adopted the restriction that recombination can only be made between a leader and one of its supporters within the same cluster. The recombination procedure thus selects any leader uniformly at random and then it chooses –also uniformly at random– one of the three supporters. Indirectly, this recombination is fitness-biased, since individuals situated at the upper nodes are better than the ones at the lower nodes. Therefore, it is unlikely that high-quality individuals recombine with low-quality one, although it might happen a few times.

The number of new individuals created every generation is equal to the number of individuals present in the population. This recombination rate, apparently high, is due to the offspring acceptance policy. The acceptance rule makes several new individuals be discarded. In our implementation, a new individual is inserted into the population only if it is better than one of its parents. This is a very restrictive rule and causes a quick loss of diversity among the population. Nevertheless, the impact on the MA of this scheme was noticeable, making the algorithm reach much better solutions using less CPU time. After each generation the population is restructured to maintain the hierarchy relation among the individuals. The adjustment is done comparing the supporters of each cluster with their leader. If any supporter is found to be better than its respective leader, they swap their places.

It must be emphasized that this scheme must be used together with a check procedure for premature population convergence, in order not to waste CPU time. The check procedure implemented verifies the number of generations without improvement of the incumbent solution. If more than 200 generations have passed and no improvement was obtained, we conclude that the population has converged and apply local search on all individuals. Moreover, the local search is also carried out every time a new incumbent solution is found through recombination/mutation. Such an event signalizes a new starting point, where the application of a local search could improve that incumbent solution even more. In this second case, it is also very likely that the rest of the population is well-fitted too, since it also usually requires several generations to find, just by recombination and mutation, a better solution than the incumbent, especially if the incumbent has already gone through at least one local search process.

2.5   Local Search

Generally, local searches utilize neighborhood definitions to determine which movements will be tested. A somewhat common neighborhood for sequences is the *all-pairs* [10]. In our case, it should be equivalent to test all possible position swaps for every gene, keeping the movements that improve the fitness. This local search turns out to be very computationally expensive since the evaluation of the fitness is costly. We thus tried the lightest swap local search type, the pairwise interchange, which only tests pairs of adjacent genes for swap. This neighborhood resulted in a reasonable computational cost, with an acceptable improvement in terms of solution quality.

The other local search implemented acts at the clustering tree structure level, by inverting the branches of every subtree present in the solution, much like it is done during mutation. Such an inversion is a good choice to make radical changes in the chromosome without loosing information about the gene grouping. As a matter of fact, this local search is very successful in joining together separated groups that contain similar genes. Both local searches are applied sequentially: the branch-inversion local search is applied on the initial solution, and the resulting individual goes through the gene-swap local search. Notice that each local search can be iterated a number of times until no change is detected in the individual (i.e., it is at a local optimum). However, notice that the fitness improvement achieved by performing repeated passes might not be worth the associated computational cost. For this reason, the number of passes each local search performs on an individual is an important parameter of the MA that has been studied in Section 3.

Another important matter is to decide which individuals should go through local search. In the Gene Ordering problem, for instance, the application of local search on every new individual is simply too costly and the algorithm wastes a lot of time optimizing individuals that are not worth it. In our implementation, the application of the local search takes place only after convergence of the population. This reduces the number of local searches and guarantees that most individuals are promising, since the population must have evolved for many generations before converging. Again, the influence of the number of individuals that are subject to local optimization will be empirically studied in Section 3.

2.6 Island Model

The use of the so-called island model [17] is known to be advantageous for several reasons: firstly, the use of semi-isolated populations helps preserving diversity, and increases the chances of escaping from local optima; secondly, it leads to a natural parallelization scheme, in which different populations are assigned to different processors. For these reasons, we have studied the influence of using multiple populations in the MA performance for this application. To be precise, we consider the effect of having an increasing number of populations cooperating with each other.

For the study of multi-population MAs, we had to define how individuals migrate from one population to another. The migration policy adopted states that populations are arranged in a ring structure. Moreover, migration occurs in all populations and the best individual of each one migrates to the population right next to it, replacing a randomly chosen individual (except the best one). This way, every population receives only one new individual, after all populations have converged and the local search phase has ended.

3 Computational Results

First of all, Table 1 describe the instances used in this work. They were extracted from real gene sequences available in the literature. Although the high number of configurations analyzed has precluded including larger instances in the test-suite, the different sizes of these instances allowed testing the algorithm in diverse optimizations scenarios.

The times allotted to these instances were 30s, 300s, and 500s respectively. These times are roughly proportional to n^2, where n is the number of genes, due to the growth trend of the local search cost. In each case, the initial population of the MA has been fed with the solutions provided by three classical clustering algorithms: complete-linkage, average-linkage, and single-linkage clustering (see

Table 1. Instances considered in this work.

instance	genes	experiments	description
HERPES	106	21	Kaposi's sarcoma-associated herpesvirus gene expression [12]
LYMPHOMA	380	19	Selectively expressed genes in diffuse large B-cell lymphoma [2]
FIBROBLAST	517	18	Response of human fibroblasts to serum [11]

[9] for example). The experiments have been done using the NP-Opt Framework [14], running on a PC (Pentium IV - 1.7 Ghz, 256 MB RAM) under Windows XP and Java build 1.4.1_01-b01.

The first study addresses the different window sizes (s_w), which were fixed at 1, 1%, 5%, 10% and 20% of the instance size. These tests tried to adjust the window size in order to find solutions with a good general aspect. Figure 1 shows some results for the FIBROBLAST data set. The results for the remaining data sets reflect the same behavior.

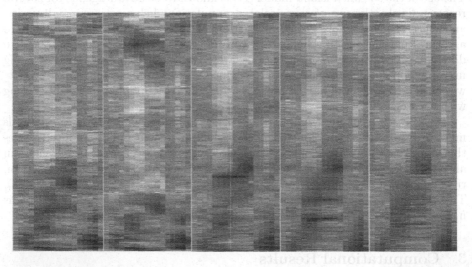

Fig. 1. Median results (over ten runs of the MA) for different sizes of the window (FIBROBLAST instance). From left to right, window size = 1, 1%, 5%, 10%, and 20%. Light (respectively dark) dots represent data points in which a particular gene is expressed (respectively non-expressed).

The tests indicated that the TSP-like fitness function, with a window size of 1, provide poor results. The problem is that, as said before, this function has a narrow view of the instance, what makes similar genes be grouped apart. This behavior can also be seen in the 1% window size. The best results came when the window is at least 5% of the instance. Up from this value, groups begin to be well-formed, with soft transitions between them. The larger window sizes provide good results, but we observed a light increase in the overall noise for very large window sizes, partly because the higher computational cost of the fitness function precludes a finer optimization within the allotted time. Given these results, we decided to use a window fixed at 5% of the instance size in the remaining experiments, as a good tradeoff between quality and computational cost.

The next tests were oriented to set the best policy for applying the local search methods. Given that the population is structured as a tree, we firstly considered to which levels local search was going to be applied. Secondly, we had to decide the number of passes n_p for each local searcher. Subsequently, the tests included applying local search according to the following configurations:

– Only on the root (i.e., just on the best individual), n_p =1, 4, and 13.
– On the two first levels (i.e., on the best four individuals), n_p =1, and 3.
– On all three levels (i.e., on the entire population), n_p = 1.

The local search was thus tested on a total of six configurations, being applied only when the population converges. The number of passes indicates the number of times that the gene-swap and the tree-swap local searches would be sequentially applied on the individual, at maximum. Logically, if a pass does not improve the present individual, the local search ends promptly. Since we used the same CPU time for all configurations, the maximum number of local search passes was fixed. Following this, every time a population converges, 13 passes were carried out, at maximum. The results are shown in Table 2.

Table 2. Results (averaged for ten runs) of the MA for different intensities of local search.

Instance	1 level			2 levels		3 levels
	$n_p = 1$	$n_p = 4$	$n_p = 13$	$n_p = 1$	$n_p = 3$	$n_p = 1$
HERPES	600.102	603.930	604.335	600.011	599.893	598.769
LYMPHOMA	2609.274	2607.666	2609.619	2610.593	2620.339	2622.640
FIBROBLAST	1376.804	1386.560	1390.353	1382.191	1398.143	1407.402

The configuration offering the best overall tradeoff was to apply the local search only on the best individual, making a single pass. This configuration is the one with less local search effort, leaving more CPU time for the recombination and mutation phases. This indicates that genetic operators are very important for the problem and contribute in a strong manner to find good solutions.

The last parameter to be tested was the number of populations. The tests were carried out for 2, 4, 6, 8 and 10 populations. These tests were performed in a sequential environment, and their focus was on the study of the algorithmic speedup of the MA. By algorithmic speedup we refer to an upper bound of the actual computational speedup that could be achieved were the MA deployed on a physically distributed environment. This bound is computed as follows: firstly,

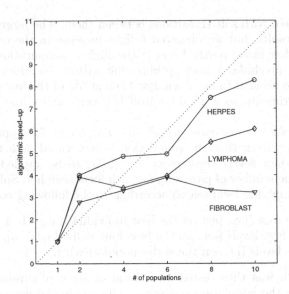

Fig. 2. Algorithmic speedup of the MA for different number of populations.

a desired fitness level is selected. subsequently, all configurations of the MA are run until a solution at least as good as desired is found. The corresponding times (in sequential model, i.e., by sequentially running each population for one step until a synchronization point is reached) are measured. By normalizing these times dividing by the number of populations we obtain the ideal time the MA would require to produce such a solution in a distributed environment, i.e., assuming negligible communication times. As mentioned before, this provides an upper bound of the achievable speedup with respect to the single-population MA, since in practice communications will have non-zero cost[1]. The results are shown in Fig. 2

As it can be seen, excellent algorithmic speedups can be obtained for up to four populations. Beyond this point, the performance starts to degrade for the larger instances, specially for FIBROBLAST. Nevertheless, very good speedups are possible for HERPES, and at least moderately good for LYMPHOMA. This suggests that migration plays an important role in the search process. In fact, migration helps in the way it reduces the effect of premature convergence. Moreover, given the 'genetic-drift' effect, larger portions of the search space can be scanned in a clever way. Notice also that for some configurations there is superlinear speedup. This is a known effect due to the different behavior of a multipopulation evolutionary algorithm (EA) with respect to a single-population EA [1]. As a matter of fact, such as behavior is precisely one of the reasons supporting the use of these non-panmictic MAs.

---

[1] We do not consider here other factors that may affect this speedup in a positive way. For example, the distribution of populations can result in a better use of local caches, thus making distributed populations run faster that sequential ones.

4   Conclusions

This work presented a memetic algorithm for finding the best gene ordering in microarray data. The main features of this MA are the use of a fitness function capturing global properties of the gene arrangement, the utilization of a tree representation of solutions, the definition of *ad-hoc* operators for manipulating this representation and the use of multiple populations in a island model migration scheme. Two different local search procedures have also been used.

Several configurations for the MA parameters were tested. The window size, used by the fitness function to measure the distance between a specific gene and its neighbors; the local search policy, which defined how local search should be applied on the population; and finally the number of populations utilized by the MA.

Besides showing how the MA can improve over classical agglomerative clustering algorithms, the results indicate that the window size should be proportional to the instance size. Moreover, the local search should be applied in a light way, giving time for the algorithm to complete several generations. This result reinforces the quality of the genetic operators, which apparently play important roles in the evolutionary process. Finally, the use of multiple populations results in better overall performance over the single population approach.

Future work will be directed to test the MA in a physically distributed environment, in order to confirm the results presented. This is important with respect to the deployment of the algorithms on larger problem instances too. We are actively working in this topic. Preliminary results support the scalability of the approach. We are also interested in using some smart weighting scheme within the fitness function, so as to giving lower relevance to outliers. Statistical tools should be used to this end. In this line, the possibility of utilizing some sort of *annealing* scheme to change the window size dynamically could be considered as well.

Acknowledgements. Carlos Cotta is partially supported by Spanish MCyT, and FEDER under contract TIC2002-04498-C05-02. Alexandre Mendes is supported by *Fundação de Amparo à Pesquisa do Estado de São Paulo* (FAPESP - Brazil). This work was also partially supported by *Conselho Nacional de Desenvolvimento Científico e Tecnológico* (CNPq - Brazil).

References

1. E. Alba. Parallel evolutionary algorithms can achieve super-linear performance. *Information Processing Letters*, 82(1):7–13, 2002.
2. A.A. Alizadeh et al. Distinct types of diffuse large b-cell lymphoma identified by gene expression profiling. *Nature*, 403:503–511, 2001.
3. A. Arnone and B. Davidson. The hardwiring of development: Organization and function of genomic regulatory systems. *Development*, 124:1851–1864, 1997.
4. T. Bäck, D.B. Fogel, and Z. Michalewicz. *Handbook of Evolutionary Computation*. Oxford University Press, New York NY, 1997.

5. P.O. Brown and D. Botstein. Exploring the new world of the genome with DNA microarrays. *Nature Genetics*, 21:33–37, 1999.

6. C. Cotta and P. Moscato. Inferring phylogenetic trees using evolutionary algorithms. In J.J. Merelo et al., editors, *Parallel Problem Solving From Nature VII*, volume 2439 of *Lecture Notes in Computer Science*, pages 720–729. Springer-Verlag, Berlin, 2002.

7. J.L. DeRisi, V.R. Lyer, and P.O Brown. Exploring the metabolic and genetic control of gene expression on a genomic scale. *Science*, 278:680–686, 1997.

8. M.B. Eisen, P.T. Spellman, P.O. Brown, and D. Botstein. Cluster analysis and display of genome-wide expression patterns. *Proceedings of the National Academy of Sciences of the USA*, 95:14863–14868, 1998.

9. D. Fasulo. An analysis of recent work on clustering algorithms. Technical Report UW-CSEO1-03-02, University of Washington, 1999.

10. P.M. França, A.S. Mendes, and P. Moscato. A memetic algorithm for the total tardiness single machine scheduling problem. *European Journal of Operational Research*, 132(1):224–242, 2001.

11. V.R. Iyer et al. The transcriptional program in the response of human fibroblasts to serum. *Science*, 283:83–87, 1999.

12. R.G. Jenner, M.M. Alba, C. Boshoff, and P. Kellam. Kaposi's sarcoma-associated herpesvirus latent and lytic gene expression as revealed by DNA arrays. *Journal of Virology*, 75:891–902, 2001.

13. E.V. Koonin. The emerging paradigm and open problems in comparative genomics. *Bioinformatics*, 15:265–266, 1999.

14. A.S. Mendes, P.M. França, and P. Moscato. NP-Opt: An optimization framework for NP problems. In *Proceedings of POM2001 - International Conference of the Production and Operations Management Society*, pages 82–89, 2001.

15. P. Merz. Clustering gene expression profiles with memetic algorithms. In J.J. Merelo et al., editors, *Parallel Problem Solving From Nature VII*, volume 2439 of *Lecture Notes in Computer Science*, pages 811–820. Springer-Verlag, Berlin, 2002.

16. P. Moscato and C. Cotta. A gentle introduction to memetic algorithms. In F. Glover and G. Kochenberger, editors, *Handbook of Metaheuristics*. Kluwer Academic Publishers, Boston, 2002.

17. R. Tanese. Distributed genetic algorithms. In J.D. Schaffer, editor, *Proceedings of the Third International Conference on Genetic Algorithms*, pages 434–439, San Mateo, CA, 1989. Morgan Kaufmann.

18. H.-K. Tsai, J.-M. Yang, and C.-Y. Kao. Applying genetic algorithms to finding the optimal gene order in displaying the microarray data. In W.B. Langdon et al., editors, *Proceedings og the 2002 Genetic and Evolutionary Computation Conference*. Morgan Kaufmann, 2002.

Gene Network Reconstruction Using a Distributed Genetic Algorithm with a Backprop Local Search

Mark Cumiskey, John Levine, and Douglas Armstrong

School of Informatics, University of Edinburgh,
5 Forest Hill, Edinburgh EH1 2QL, Scotland
{mcumiske,johnl,jda}@inf.ed.ac.uk

Abstract. With the first draft completion of multiple organism genome sequencing programmes the emphasis is now moving toward a functional understanding of these genes and their network interactions. Microarray technology allows for large-scale gene experimentation. Using this technology it is possible to find the expression levels of genes across different conditions. The use of a genetic algorithm with a backpropagation local searching mechanism to reconstruct gene networks was investigated. This study demonstrates that the distributed genetic algorithm approach shows promise in that the method can infer gene networks that fit test data closely. Evaluating the biological accuracy of predicted networks from currently available test data is not possible. The best that can be achieved is to produce a set of possible networks to pass to a biologist for experimental verification.

1 Introduction

The accurate reconstruction of gene networks has many possible benefits and currently is the focus of much active research. This interest was sparked off by the development of a new technology that allows the massive scale measuring of mRNA. The major areas of research that will benefit from gene network inference include: cancer research [12] (compared the profiles of healthy and cancerous tissue and identified differences in expression profiles); drug discovery [7]; toxicology [6]; disease prevention [3]. This paper investigates the use of genetic algorithms (GAs) in analysing the expression levels of genes toward constructing gene networks.

2 Background and Microarray Technology

It is possible to measure on a large scale the levels of mRNA within a biological sample at different time points, or across experimental events, producing an expression profile of gene activity across these stages. Such profiles can be used to infer gene networks. A gene network can be described as a group of interdependent genes, when expressed contribute to the occurrence of a specific

S. Cagnoni et al. (Eds.): EvoWorkshops 2003, LNCS 2611, pp. 33–43, 2003.

biological function. Also, 'gene network' includes any unknown variables that effect the operation of the network. The goal is ultimately to decipher the precise connections of the genetic network, identifying for each gene which other genes it influences.

Microarray technology provides the researcher with a snapshot of the levels of the protein precursor; the mRNA, within a biological sample. While it always has to be remembered that mRNA is only a precursor. The mRNA profile does give us an indirect measurement of gene activity in the sample. The data sets retrieved from microarray experiments are too large to infer anything meaningful by hand. The yeast cell for example has around 6000 genes. For this data to be useful a large number of readings must be processed.

3   Approach

A genetic algorithm with backpropagation local search was applied to the problem of inducing valid unidirectional networks from given microarray data. In the GA population, each individual is a possible gene network, consisting of a collection binary links between genes and weights on each link. A crossover operator has been developed which combines randomly chosen sub-networks from two parents in order to create a new network. In the initial networks found by the GA, the weights are set to coarsely grained values, to find a broadly acceptable configuration of gene interactions; these are then fine tuned by a backpropagation local search algorithm to find the weights which give the best match to the microarray data.

String Representation

An important part of the workings of a GA is the definition of the individuals' (chromosome) representation. Figure 1 shows a sample network. An arrow entering a gene signifies that the gene is regulated either positively of negatively by the genes originating the arrow. The weight value is a measure of the level of influence, with a minus value denoting a repressing effect. In this model a string is used to define the gene network topology; this configuration is a short hand version of the weight matrix method described below. It avoids having to store large sparse matrices. A topology is made up of a series of nodes with weights; representing a gene and the effect of other genes on it. Each node has an id, and a series of node id's with associated weights, identifying influencing nodes and the strength of their influence as a weight. Nodes that are effected by other nodes are separated using a comma; each of these nodes has its influencing connection information enclosed by angle brackets; within these brackets is the from node id with the weight value, which indicates the level of influence. This representation is shown in Figure 3. Using a weight matrix as shown in Figure 2, first used by Weaver [2] is a well known method of representing the gene interactions of a regulatory network. The weight matrix consists of $n \times n$ weight values, each of which indicates the influence of one specific gene on another. Each column

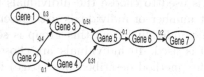

Fig. 1. A sample gene network with weights

$$
\begin{pmatrix}
0 & 0 & 0 & 0 & 0 & 0 & 0 & 0 \\
0 & 0 & 0 & 0 & 0 & 0 & 0 & 0 \\
0.3 & -0.4 & 0 & 0 & 0 & 0 & 0 & 0 \\
0 & 0.1 & 0 & 0 & 0 & 0 & 0 & 0 \\
0 & 0 & 0.51 & 0.31 & 0 & 0 & 0 & 0 \\
0 & 0 & 0 & 0 & 0.1 & 0 & 0 & 0 \\
0 & 0 & 0 & 0 & 0 & 0.2 & 0 & 0 \\
0 & 0 & 0 & 0 & 0 & 0 & 0 & 0
\end{pmatrix}
$$

Fig. 2. A weight matrix representation of the gene network shown in Fig 1.

details the influencing genes for each gene identified by the row index. Figure 2 is a matrix representation of the network in Figure 1. With the row and column indices starting at 1 it can be seen that gene 3 (row 3) is influenced by gene 1 (column 1) with a positive regulation of 0.3 and is repressed by gene 2 (column 2) by -0.4.

```
              From              To
   Weight     Node              Node
<1~0.3><2~-0.4>^3,<2~0.1>^4,<3~0.51><4~0.31>^5,<5~0.1>^6,<6~0.2>^7
              ↑
           Possible
           splice
           site
```

Fig. 3. A string representation of the gene network shown in Fig 1 and Fig 2.

The weight matrix can be further translated into a string as shown in Figure 3 representation. This is the representation used by the GA.

Selection

Tournament selection is used to choose the individuals for crossover. In tournament selection a set number of individuals are chosen at random from the population and the best individual from this group is selected as a parent. A tournament size of 10 is used. 10 individuals are chosen randomly from the population; the most fit (method described next page) individual is chosen to reproduce.

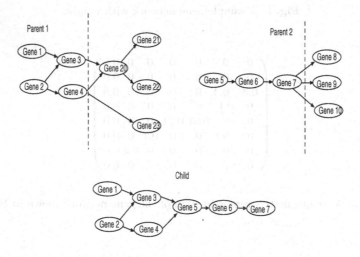

Fig. 4. Parent 1 and Parent 2 and a child network

Crossover

One point crossover is used to inter-combine the subnetworks within individuals. Crossover can occur at any gene that is affected by other genes. Genes preceding and following the crossover point are considered separate groups; these groups are combined with the groups from another individual. Validity checks are made to ensure that the configuration is a valid network, not just a gathering of random non-connected nodes. Figure 4 shows two networks of genes and their dependencies. After crossover the two parent networks may combine in a way that could possibly increases the individuals' fitness value. The resultant child network derived is shown below the parents. If a particular pattern has a higher fitness, it has a greater probability of surviving through future generations.

Mutation

A mutation setting of 0.001 is used to ensure that a misplaced discarded node connection can make a reappearance later on. This mutation is present on the connections only. Weight mutation does not occur. After each generation an individual has a 0.001 probability of one of its nodes changing connection to another randomly chosen node.

Evaluating Network Fitness

Each network topology is checked to see if it fits the training data. To evaluate a candidate network it must be translated into a form that can be measured against the test data. The idea is to find a network that explains the data. If a candidate network is correct than there is a relationship between the genes effecting a given gene, across all time steps. The procedure is; generate a matrix of data in the same form as the test data and get the error distance. The steps are: for each gene in the candidate network get the estimated expression level based on the weights and connections; pass this value through a sigmoid function; repeat for all genes in the candidate network, across all test data steps.

$$s_i(t+1) = \sum_{j=0}^{n} w_{ji}x_j(t) \qquad (1)$$

w - weights.
x - gene expression value.
s_i - gene expression estimate.

The above equation gets the estimated gene expression level for gene s at time $t+1$ by multiplying the expression level of each gene at time t by the proportional influence level of each dependent gene (from the gene dependency matrix). So, for each node in the network being evaluated that has dependency arrows arriving, the corresponding weight of these arrows is multiplied by the actual expression level (from test data) of the nodes (genes) originating the dependency arrows, this gets an expression level estimate for the gene receiving the dependency arrows. If the dependency weight values and the network connections are correct then the estimated expression value will be the same (or close) as the actual test data expression value.

$$x_i(t+1) = \frac{n}{1 + e^{-n(s_i(t+1))}} \qquad (2)$$

In weight matrix models the direct result of the step-wise calculation is often of an arbitrary magnitude, this being the result of the additive nature by which the input signals from all nodes are combined into the node of interest. To overcome this problem, it is necessary to scale this value to reflect what occurs in biological systems. For this reason the estimate gene expression value for each gene is passed through a sigmoid function [2].

$$Error = \sum_{i=0}^{n} \sum_{t=0}^{T} |y_i(t) - x_i(t)| + p/b \qquad (3)$$

$i = 0..n$ - number nodes in the possible network.
$t = 0..T$ - across time steps/events.
y - actual expression levels.
x - estimated expression levels.
p - number of nodes (genes in the topology).
b - network size bias.

Each sample network topology is evaluated against the target expression level, using equation 3. The estimated value x is found for each node in the network being evaluated using equation 1 and equation 2. Equation 3 gives a distance measure of how closely the estimated expression levels of the network being evaluated match the actual expression levels from the test data. The smaller the error distance the better the fitness. There is an imposed bias toward smaller networks as known biological networks are relatively small.

Local Search

When the GA has completed it passes the discovered gene interconnections with the coarse weight settings (-1,0,5, 1 etc) to the backpropagation component. Its task is to finely tune the weight settings. In standard BP each weight is adjusted in proportion to its contribution to the overall error. Here, since the expression levels are unrolled across time steps (or events) it is necessary to get each weights contribution across all steps. During the update stage, the weights across all layers need to be updated by the same amount. The sum of the weight error is used, and the weights are updated equally, that is, the sum of adjustments over all layers for the weight is used for each weight update. With this restriction in place the neural network will minimise the error by adjusting the same weight for each gene across all steps. The end result being the weight settings that most closely explain the data, based on the candidate network presented. To compensate for this larger than usual update a low learning rate is set. This is a common approach in bptt [8] (backpropagation through time). For a good description of bptt see [9] and [8].

Training

Training is the process of getting the weights on the connections between nodes to adapt to a training data set by applying a set of training data sequentially to the input layer of the net and comparing the output to the (known) correct result. Following this, the weights are updated to reduce the error. In backpropagation [11] the error value at each output node is passed back to the preceding layer. The backpropagation algorithm causes each preceding layer to adjust its weights to neutralise the error; this procedure is repeated causing each preceding layer to adjust its weights in a similar manner. The bptt restriction on this is that the same node connection weight must be updated by the same amount across all the steps.

Island Model

Studies have shown that parallelisation can significantly improve the performance of genetic algorithms [4]. The use of the island model also reduces the likelihood of premature convergence, as each island has its own separately sampled initial population.

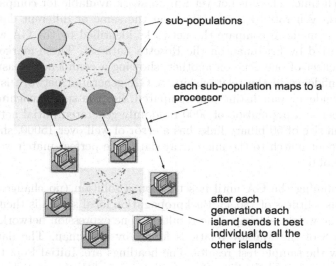

sub-populations

each sub-population maps to a processor

after each generation each island sends it best individual to all the other islands

Fig. 5. In the island model several islands of sub-populations evolve simultaneously, with each island mapped to one processor. Each processor simultaneously evolves its own private population; from which it then migrates fit individuals.

GA's are highly suitable for a parallel computation approach. Figure 5 shows the island model. Each island is run on a separate processor dealing with its own section of randomly generated individuals. At the end of each generation migrations occurs. Each island sends its best individual to all the other islands. In this model, the population is divided into multiple sub-populations that evolve isolated from each other and exchange individuals occasionally. This method keeps the network load light as bandwidth is only used to exchange fit individuals at the end of each generation. Also, less memory is used than a shared global population approach.

4 Results

For the experiments reported here, an island GA [13] running on 4 and 8 machines is compared with a GA running on a single machine. The microarray data used is the Rosetta Compendium [5] which is a data set consisting of 300 expression profiles of yeast (saccharomyces cerevisiae) mutants and chemical treatments. The results generated by the GA were evaluated in two ways: firstly, by how well they matched the microarray data, as judged by the fitness function, and secondly, by comparison with the Markov relations derived by Friedman [10] using his Bayesian network approach.

Testing whether a constructed network generated does in fact reflect what is happening biologically is difficult without submitting potential networks to a biologist for knock out experiments. This involves a biologist repeatedly deactivating specific genes within the genome and redoing the microarray expression process each time. There is not yet a framework available for comparing different networks generated by researchers on the same or different data sets. An effort here is made to compare the networks identified by the GA with the relations inferred by Friedman on the Rosetta data set. These markov relations detail the effect of one gene on another, showing a confidence measure signifying the confidence in the relationship. A GA generated network is translated into a dependency pair listing and compared. For the single machine case, the best member of a population of 5000 randomly generated initial networks with a maximum size of 50 binary links has a error of well over 15000, showing it to be a very poor match to the microarray data (the perfect match would have a error score of 0).

After running the GA until it settles on a solution (no changes), the best network has a fitness of 180.43. Backpropagation local search is then applied to fine tune the weight settings. The resulting gene expression network was found to contain 8 of the Markov relations found by Friedman. The data tables 1 and 2 show the sample test results. The headings are: Initial Pop, the starting population size; Initial Net Size, the maximum size of the generated network; GA Error, the lowest (best) values; BP Error, the lowest (best) error value; Markov matches, the number of corresponding pairwise connections found by Friedman; Time, execution time.

Table 1. GA plus backprop local search results on a single machine

Initial Pop	Initial Net Size	GA Error	BP Error	Markov matches	Time
2000	50	186.60	2.032	6	30 min
2000	20	257.16	1.200	0	2 mins
5000	50	180.43	1.200	8	35 min
5000	20	221.40	0.8323	6	5 min

For the island model, the results are better: using 8 machines with a population of 2000 on each machine, the best coarse-valued network found has a fitness of 122.60. This network was found to contain 9 of the Friedman's Markov relations. Altering the number of islands, the number candidate networks in each population and the maximum number of binary links allowed in the gene expression network. The results broadly show that the technique works best with more islands, large populations and larger network sizes. The method also is computationally feasible: it takes between 6 and 42 minutes to find a solution. A problem became apparent in testing the validity of any sub-networks

Table 2. GA plus backprop local search results on a single machine

Num Machines	Initial Pop	Initial Net Size	GA Error	BP Error	Markov matches	Time
4	2000	50	225.60	2.214	6	30 min
4	2000	20	257.16	1.334	2	7 mins
4	5000	50	223.43	2.200	4	42 min
4	5000	20	227.30	1.542	7	9 min
8	2000	50	122.60	2.635	9	30 min
8	2000	20	117.16	1.986	5	6 mins
8	5000	50	116.22	1.256	7	40 min
8	5000	20	132.30	0.623	5	11 min

inferred was difficult. Given that biological testing of these predicted networks is a long-term problem we simulated a network and ran the same algorithm on the corresponding expression data. Removing all noise sources demonstrated that the GA approach was predicting valid relationships between the simulated gene nodes. See Shin[1] for a description of a GA evaluating simulated networks.

5   Model Improvements

Multi-network Representation

The approach described previously was good at finding small fit networks, containing two or three genes. The model currently under development uses this to good advantage by allowing these fragment networks to combine and grow. In the new representation an individual is a network consisting of a series of interconnected sub-networks. For computational operations each candidate is represented as a list of matrices, with each matrix representing a sub-network. These matrices are encoded as strings as described previously.

Figure 6 shows a network consisting of interconnected sub-networks, this representation is used to allow a local search crossover operator to find the sub-network interconnection configuration that achieve the highest fitness.

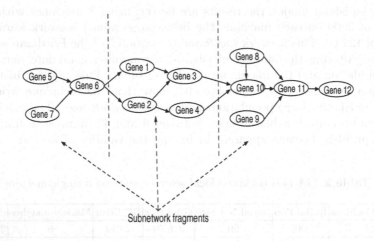

Subnetwork fragments

Fig. 6. A gene network consisting of three sub-networks (fragments)

Multi-level Crossover

Crossover occurs at two levels, the sub-network level and the individual level. The sub-network crossover involves a one point crossover between two randomly chosen sub-networks from each of the individuals involved in the crossover. All of the sub-networks of the two individuals are split and then inter-combined. The higher level crossover method uses a form of local search; this operates by breaking up and recombining the individuals. An individual is a combination of interconnected sub-networks (fragments). Following the sub-network (level 1) crossover the newly formed fragments are separated and then recombined. This occurs for each possible placement of a sub-network in the individual. Picture a shuffling of sub-components until the overall system is as fit as it can be. These configurations are then measured for fitness; the two configurations with the highest fitness values are chosen to proceed. The advantage of this approach is that it stops the GA from converging on a small very fit sub-network.

6   Conclusion

The techniques for the reconstruction of gene networks are still in their infancy and are lacking a framework for the comparison and reuse of research results and data. What is currently achievable is to produce a set of possible networks to pass to a biologist for verification. The immediate goal is to reduce this set. Highly fit networks generated by the genetic algorithm closely explain the test data; this reduces the set of networks that need further investigation. The use of separate populations (island model) markedly improves the fitness level achieved.

References

1. Shin Ando and Hitoshi Iba. Inference of gene regulatory model by genetic algorithms. *Proceedings of the 2001 IEEE Congress on Evolutionary Computation. Seoul Korea*, 2001.

2. C. Weaver et. al. Modeling regulatory networks with weight matrices, pacific symposium on biocomputing. *Journal of Computational Biology*, 4:112–123, 1999.

3. Carol A. et. al. Extraordinary bones functional and genetic analysis of the ext gene family. *Current Genomics*, 2:91–124, 2001.

4. Goldberg et. al. The design of innovation, lessons from genetic algorithms. *Technological Forecasting and Social Change.In press*, 2000.

5. Hughes et. al. Functional discovery via a compendium of expression profiles. *Cell*, 102:109–126, 2000.

6. Kim K et. al. Identification of gene sets from cdna microarrays using classification trees. Technical report, Biostatistics and Medical Informatics, University of Wisconsin Madison, Madison, USA, 2001.

7. Lawrence J. Lesko et. al. Pharmacogenomic-guided drug development regulatory perspective drug evaluation and research. Technical report, Center for Drug Evaluation and Research, 5600 Fishers Lane Rockville, Maryland 20852 USA, 2001.

8. Werbos P. J. Backpropagation through time, what it does and how to do it. *Proceedings of the 1990 IEEE*, 78:1550–1560, 1990.

9. J. Hertz A. Krogh and R.G. Palmer. *Introduction to the Theory of Neural Computation*. Addison-Wesley, 1991.

10. I. Nachman. Nir Firedman., with M. Linial. and D. Pe'er. Using bayesian networks to analyze expression data. *Journal of Computational Biology*, 7:601–620, 2000.

11. D. E. Rumelhart and J. L. McClelland. *Explorations in Parallel Distributed Processing*. MIT Press, 1988.

12. Richardson S. Mixture model for the identifying gene expression factors in survival analysis using microarray experiments. Technical report, Department of Epidemiology and Public Health, Imperial College, London, UK, 2001.

13. Darrell Whitley. A genetic algorithm tutorial. Technical report, Computer Science Department, Colorado State University, Colorado State University Fort Collins, CO 80523.

Promoter Prediction with a GP-Automaton

Daniel Howard and Karl Benson

Software Evolution Centre,
Knowledge and Information Systems Division,
QinetiQ Ltd, Malvern Technology Centre,
St Andrews Road, Malvern,
WORCS WR14 3PS, United Kingdom.
{dhoward,kabenson}@qinetiq.com

Abstract. A GP-automaton evolves motif sequences for its states; it moves the point of motif application at transition time using an integer that is stored and evolved in the transition; and it combines motif matches via logical functions that it also stores and evolves in each transition. This scheme learns to predict promoters in human genome. The experiments reported use 5-fold cross validation.

1 Promoter Prediction

A discussion on promoters is covered in more detail in [Lewin, 2000] on which the material of this section is based. The promoter is the sequence of DNA needed for RNA polymerase to bind to DNA strands and accomplish the initiation reaction. It is also a sequence of DNA whose function is to be recognized by proteins and is a classic example of a *cis*-acting site. Consensus sequences are defined as essential nucleotide sequences which are present in all promoters where they are conserved with some variation, and identified by maximizing homology. But it turns out that conservation of only very short consensus sequences is a typical feature of promoters in both prokaryotic and eukaryotic genomes.

1.1 Prokaryotic Promoters

A minimum length for the essential signal in bacterial promoters must be 12 bp because a shorter sequence could arise often by chance producing too many false signals, and this minimum length requirement increases for longer genomes. The pattern need not be unique at all 12 bp, and can contain "don't care" symbols.

Bacterial promoters have four conserved features: the startpoint is 90% of the time a purine; the -10 sequence $T_{80}A_{95}T_{45}A_{60}A_{50}T_{96}$ (subscripts indicate probability of occurrence); the -35 sequence $T_{82}T_{84}G_{78}A_{65}C_{54}A_{45}$; and the separation between them (15 to 20 bp) critical for the geometry of RNA polymerase (16 to 18 bp in 90% of promoters).

RNA polymerase is difficult to define precisely. It not only must incorporate nucleotides into RNA under the direction of a DNA template but it must also possess (or be complemented by) ancillary factors, which are required to initiate

S. Cagnoni et al. (Eds.): EvoWorkshops 2003, LNCS 2611, pp. 44–53, 2003.

the synthesis of RNA when the enzyme must associate with a specific site on DNA. Some of these ancillary factors or polypeptides are required to recognize promoters for all genes, but others specifically for initiation at particular genes. The former are regarded to be subunits forming part of RNA polymerase, and the latter as separate ancillary control factors.

In order to recognize the promoter a very important polymerase subunit is σ^{70}. This independent polypeptide (70 denotes its mass) has domains that recognize promoter DNA but its N-terminal region occludes the DNA binding domains when σ^{70} is free. The RNA polymerase core enzyme, however, has affinity to all DNA and is stored in DNA. When σ^{70} finds it, it binds behind it to form a holoenzyme. In this state, σ^{70} decreases the affinity of the core enzyme for all DNA, allowing the holoenzyme to move swiftly down the DNA. But free from its N-terminal region, σ^{70} has strong affinity to promoter sequences and when these are found the holoenzyme stops and binds tightly. RNA synthesis begins and the initiation phase ends as σ^{70} is released and the remaining core enzyme regains its mobility and affinity to all DNA, which helps it with elongation until a terminator sequence in the template strand causes transcription to end.

The startpoint is the first base pair that is transcribed to RNA. It follows from this account that the more upstream parts of the promoter are involved in initial recognition of RNA polymerase but that they are not required for the later stages of initiation. This is why in bacterial promoters the -35 sequence is known as the 'recognition domain' and the -10 sequence as the 'unwinding domain' (consist with the observation that a stretch of AT pairs is much easier to melt than of GC pairs). However, experiments show that σ^{70} contacts both the -10 and -35 sequences simultaneously. In particular, amino acids in the α-helix of σ^{70} contact specific bases in the non-template strand of the -10 sequence.

The strength of different promoters varies 100 fold. It not only depends on both consensus sequences (and associated up or down mutations) but also on the initially transcribed region from +1 to +30, which influences the rate at which RNA polymerase clears the promoter.

When either the -10 sequence or the -35 sequence is missing in rare promoters, RNA polymerase is usually too weak and recognition reaction requires intercession by ancillary proteins. Certain other σ polypeptides are activated in response to environmental or lifestyle change, e.g. 'heat shock genes' are turned on by σ^{32} in E. coli. Promoters for these other σ factors have the same size and location relative to startpoint and conserved -10 and -35 sequences. They differ from each other at either or both -10 and -35 sequences, enabling mutually exclusive transcription. The σ^{54} has a consensus sequence at -20 instead of at -35, behaving more like a eukaryotic regulator by binding to DNA independently of core polymerase, and by directing the pattern of regulation so that sites distant from the promoter region influence its activity.

1.2 Eukaryotic Promoters

Although remarkable similarities exist between bacteria and eukaryotic cells, e.g. the mitochondrial DNA is similar to prokaryotic DNA, eukaryotic genes in nuclear DNA are structurally very different from genes in bacteria. Among the many differences, mRNA molecules are capped and stable in eukaryotes. Moreover, the initiation of transcription and recognition of promoters follows different rules. Promoters utilized by RNA polymerase II (pol-II) are recognized principally by the separate accessory factors rather than principally by the enzyme. With eukaryotes the default state of a gene is inactive and pol-II can only initiate transcription when some or all *trans*-acting factors bind to *cis*-acting sites, which are usually, but not always, located upstream of the gene.

These transcription factors are recognized by short *cis*-acting sites spread out upstream of the startpoint usually over a region > 200 bp. And of these only those < 50 bp upstream of the startpoint are fixed in location. pol-II binds near the startpoint but does not directly contact the extended upstream region of the promoter. A great many factors can act in conjunction with pol-II and can be divided into three kinds:

- few called General Transcription Factors (GTF) that combine with RNA polymerase to form a complex near the startpoint that is required by all promoters to initiate RNA synthesis, which for this reason is called the *basal transcriptional apparatus*.
- many known as Upstream Factors (UF) required for a promoter to function at an adequate level, and which increase the efficiency of initiation. Their activity is not regulated and they act on any promoter which contains specific short consensus elements upstream of the startpoint. The precise set of these factors required for full expression is characteristic of any particular promoter.
- those responsible for control of transcription patterns in time and space, synthesized or activated at specific times, and which bind to short sequences known as response elements. They function similarly to UF and are known as Inducible Factors (IF).

The definition of a pol-II promoter is more elaborate than for bacteria. The major feature is the location of binding sites for transcription factors. A promoter is the region containing all of the short *cis*-acting sites to which a large number of transcription factors will bind, which can support transcription at the normal efficiency and with the proper control. This is in contrast with bacteria where the additional short, <10 bp, *cis*-acting regulatory sites[1] that are juxtaposed to, or interspersed with the promoter, and that regulate the promoter, are considered distinct from it.

Also by definition, sequence components of the promoter are required to be located in the general vicinity of the startpoint. There is also a distinct region known as the enhancer which is several kb upstream of the startpoint, and 100 bp in length, which contains several closely arranged sequences that stimulate initiation, some of which are quite similar to those of the promoter. These are bound to by proteins which interact with proteins bound at the promoter to form

---

[1] A regulator gene codes for a protein that controls transcription by binding to such sites. They are of two types: enhancer or repressor (a protein binding to the operator).

a large protein complex (DNA must be coiled or rearranged to allow this interaction). Enhancers assist initiation from great distances, functioning in either
orientation, and from either side of the gene.

A transcription factor is defined to be any protein needed for the initiation of
transcription regardless of the mechanics, e.g. it may bind to DNA at promoters
or enhancers, recognize pol-II, get incorporated into an initiation complex only
in the presence of other proteins, etc.

The eukaryotic pol-II promoter prediction problem which is addressed in this
paper is challenging because while some of these short *cis*-acting sequences and
the factors that recognize them are commonly found in a variety of promoters,
and are used constitutively (in housekeeping genes), there exist others which
only identify a specific class of gene and their use is regulated. These elements
occur in different combinations in different promoters.

1.3   Objectives of Promoter Prediction

The desired output from this research is an algorithm that can process the eukaryotic DNA sequence to identify regions that are likely to constitute promoter
regions. As discussed in the previous section, the promoter region is defined to be
a region proximate to the startpoint, and transcription factors may bind outside
of this region.

Hence, design of a promoter predictor developed from inductive learning of
bp information in such a short region, e.g. between 200 to 500 bp upstream of
startpoint is consensus for promoter length, is likely to contain a few promoters
which have no special information. Thus, it would be desirable not to learn such
"targets" or to miss them.

This is an argument for working at lower sensitivity and higher specificity.
The objective of Promoter Prediction software of this kind is to propose putative
cis-acting sequences, promoter regions, and genes to guide experimental work.

1.4   Solution Strategy

The solution strategy is to use evolutionary computation (EC) to discover the
predictor by learning this classification rule from a number of known promoter
regions and non-promoter regions.

It seems sensible to use the knowledge that this region consists of a number
of important and short *cis*-acting sequences, in order to design the architecture
of the solution which is evolved. That is, analyze the problem in advance using
human insight to determine that a certain decomposition is likely to be helpful
in solving the problem; establishing an advantageous architecture for the yet to
be evolved computer program [Koza, 1999].

The representation for the EC used in this paper is a Finite State Automaton.
And short sequence pattern gets evolved inside each of its states. The automaton combines these patterns. The approach presented in this paper should be
competitive with other popular methods [Hannenhalli and Levy, 2001], and including other evolutionary approaches, e.g. evolution of weight matrices.

1.5 Medical Importance of Understanding Promoters

The research of the type described in this paper has as objective not only the prediction of promoter sequences, but an understandable algorithm, i.e. one which can be subjected to white box analysis, to obtain an understanding increasingly required for drug discovery.

Promoters can be blocked by mutant proteins which can then damage the body and cause disease, as is suspected in a monogenetic disease called Huntington's disease [Cattaneo et. al., 2002].

For common complex disorders, a gene variant type may carry a different promoter, affecting the level of expression of a receptor molecule. Increased expression of the gene causes overabundant receptor molecules, i.e. one person's variant results in "normal" receptor expression but another person's variant results in over-expression of this receptor.

At equal Pharamacokinetic response, drugs designed to combat the effects of a hormone which binds to a receptor are more effective in individuals carrying this variant of the promoter. This is because overabundance of receptors in these individuals makes them sensitive to the actions of neuron hormones.

Similarly, over-expression of an enzyme, i.e. proteins, owing to differences in its promoter also plays a crucial role in Pharmacokinetics or how a patient's body handles to a particular drug in terms of absorption, inactivation and excretion. Genetic variants of the enzymes responsible for these functions may significantly affect the amount of drug that reaches the biological target. For example, over-expression causes the drug to be absorbed too rapidly when the body eliminates the drug before it can take effect [RGEP, 2002].

2 Algorithm Description

The combination of Genetic Programming within Finite State Automata (FSA) or GP-automaton was introduced by Ashlock [Ashlock, 1997]. This architecture was enhanced to include the explicit gathering of state information via the use of state logical functions and successfully applied to Machine Vision (Automatic Target Detection) problems [Benson, 2000a,Benson, 2000b]. Research reported in this paper uses this extended version of the algorithm.

The objective of the GP tree structure is to find motifs within the promoter and non-promoter regions. The function set consists of a single two valued function called GLUE, which concatenates the terminals to produce a motif, and behaves as a logical AND returning true if both conditions (i.e. inputs) are met.

The terminal set is composed of nucleotides A C G and T. For example the tree (GLUE G (GLUE A T)) is the motif GAT. This tree will return true if in the piece of chromosome evaluated there is a G in the first position, an A in the second, and a T in the third; else it returns false. To increase flexibility the terminal set is enriched to include terminals which return true for more than one given nucleotide, i.e. the wildcards listed in Table 1. For example (GLUE Y R) will match the patterns CA CG TA TG.

Table 1. Set of standard symbols for nucleotide wildcards.

M	R	W	S	Y	K	V	H	D	B	N
A	A	A	C	C	G	A	A	A	C	A
C	G	T	G	T	T	C	C	G	G	C
						G	T	T	T	G
										T

The FSA acts as the main program which has the ability to call functions. The functions it calls are the GP tree structures in each state. Each returns a boolean that indicates which of two transitions to make next, and consequently what function to call next. The FSA possesses a start state and an end state and can be halted after a pre-determined number of moves when it cycles.

Both a logical function, e.g. NAND, and an integer value that indicates how to move a DNA position pointer, e.g. +10, are associated with each transition of the FSA. Both of these undergo genetic modification. As transitions take place, the main program builds a logical relationship between the output of functions. For example, it uses the logical function of the forthcoming transition to logically combine the boolean returned by the last function with the current cumulative boolean result. Furthermore, when a transition occurs, the stored integer is used to move a pointer along the DNA string, moving the inspection site upstream (positive integer) or downstream (negative integer) to position where the next motif (the next GP tree at the destination state) is to be applied/evaluated. An example clarifies algorithmic execution with the aid of Figure 1.

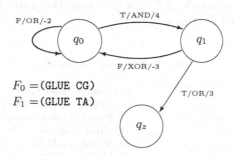

$$F_0 = (\text{GLUE CG})$$
$$F_1 = (\text{GLUE TA})$$

Fig. 1. Example algorithm. The notation T/AND/3 means that if the function returns a logical true, then AND this with the previous running answer and move the pointer 3 downstream.

The input data for the example is ACGTATAGCA; the start state is q_0, and the motif is F_0 and placed such that its first nucleotide (the C) points to the fifth position of the input data (an A) of the data. F_0 will return false since the motif does not match. The pointer is now moved two positions upstream, and since this is the first transition the current state output, which is false, is also the current answer because there is no previous state of answer to combine it with. The current state is still q_0. F_0 is now applied at the current site and it returns true. Now, the reading position is advanced 4 places and the current answer is combined logically with the last answer to give F AND T $\Rightarrow$ F, and the current state is now q_1. F_1 is now applied to the current site and returns true. The pointer is advanced 3 places and the current answer is combined logically with the last answer to give F OR T $\Rightarrow$ T. The current state is now q_z which is the halt state, thus the algorithm is terminated and the result is T.

The algorithm enjoys interesting properties:

1. Since the motifs are constructed via the GP tree structures they can be of various lengths, and since wildcards are used they can match more than one pattern in promoters and non-promoters.
2. The algorithm also incorporates a decision type process. In the example F_0 did not match the current location being inspected so it was looked for upstream. Only after this motif had been found in the DNA string was the second motif (F_1) searched for. Moreover, a given motif may be looked for in many places accounting for the possibility that a given motif may occur in more than one place.
3. The decision as to whether a promoter or non-promoter is being inspected is not based on just one criteria. The decision is based on a logical combination of the responses from the applied motifs.

Modification of the GP-automaton is achieved using various mutation operators (some of which are analogous to crossover of an equivalent GP tree) are given in Table 2 and since these are many, selecting the frequency at which they should be applied is difficult. Self-adaptive mutation rates is used to overcome this as fully explained in [Benson, 2000a].

Table 2. Mutation operators.

1. Add a State.	8. Exchange state GPs.
2. Delete a state.	9. Replace a GP with a randomly created one.
3. Mutate start of reading site.	10. Headless chicken crossover on a GP.
4. Mutate a transition.	11. Grow a subtree.
5. Cycle two states.	12. Shrink a subtree.
6. Mutate a state logical function.	13. Mutate a GP terminal
7. Headless chicken crossover on a state.	14. Mutate GP function.

Selection is achieved using a $(\mu+\lambda)$-EP scheme and the fitness function $f =$ Sensitivity+Specificity (see Equations 1 and 2). In this paper $\mu = 125$ and $\lambda = 125$. At each generation a tournament of size 4 takes place and the top μ individuals produce one offspring each via mutation giving a total of λ children. These $\mu+\lambda$ individuals are then carried over into the next generation.

3   The Experiment

To evaluate the effectiveness of the algorithm at identifying promoters in human DNA, we took a total of 565 promoters and 890 non-promoters from the database held at fruitfly.org[2]. The non-promoters used were taken from the coding sequence (CDS) data. Each sequence is 300 bp long extracted 250 bp upstream and 50 bp downstream of the TSS.

The data is organized to facilitate 5-fold cross validation thus allowing algorithms to be trained and tested on five sets of data. Each training set comprised of 452 promoters and 712 non-promoters, and each test set comprised of 113 promoters and 178 non-promoters.

Table 3. Parameters used in the experimentation.

Parameters	
Number of runs	50
Number of Generations	1000
Population size	250
Max states	5 + halt state
Min states	3 + halt state
Max motif length	16
Min motif length	1
Tree terminals	A, C, G, T,
	M, R, W, S, Y, K,
	V, H, D, B, N
Tree functions	GLUE

For each of the five data sets 50 runs of 1000 generations were performed with a population of 250 individuals. The best individual found on each of the 5 training sets was then applied to its test set and is the standard way to evaluate the likely performance of an algorithm. However, the evolutionary process produces many such algorithms at different generations. In this instance, no attempt was made to optimize the generalization of the evolutionary process, i.e. using a different generalization test set to judge how to stop it prematurely to avoid over-training, i.e. undertaking a search for the simplest model that fits the data. Thus, and assuming our data set is representative, our reported performance may even be conservative, i.e. had we carried the test of over-fit, it may have even been possible to select a more generalist individual.

Table 3 lists parameters used in the experiments and Table 4 lists the results of the 5-fold cross validation on the five test sets.

Statistical calculations on the results in Table 4 provide a measure of algorithm performance. In the context of this work, Sensitivity is the likelihood that

---

[2] http://www.fruitfly.org/seq_tools/datasets/Human/promoter/

Table 4. Results of five fold cross validation tests.

	True Positives TP	True Negatives TN	False Positives FP	False Negatives FN
Test 1	92	159	19	21
Test 2	87	165	13	26
Test 3	86	168	10	27
Test 4	91	167	11	22
Test 5	90	160	18	23
$\sum$	445	819	71	119

a promoter is detected given that it is present, and Specificity is the likelihood that a non-promoter is detected given that it is present. Equations 1 and 2 are mathematical definitions of Sensitivity and Specificity. Here true positives (TP) are correctly classified promoters; false negatives (FN) are promoters classified as non-promoters; true negatives (TN) are correctly classified non-promoters; and false positives (FP) are non-promoters classified as promoters.

$$\text{Sensitivity} = \frac{TP}{TP + FN} = 0.79 \tag{1}$$

$$\text{Specificity} = \frac{TN}{TN + FP} = 0.92 \tag{2}$$

One of the resulting algorithms from the 5-fold cross validation is shown in Figure 2. The algorithm appears to perform reasonable well on the problem of promoter prediction. However, at the time of writing no direct comparison of algorithm performance can be drawn between the one presented in this paper and others due to difference in data sets used.

4   Conclusions

GP-automata can address the problem of promoter prediction by evolving motifs of various lengths in automata states, and using values stored at transitions to apply motifs at relative positions in the test DNA string, and to then combine the evidence using logical functions. The division of control structure and code is a feature of the GP-automata, which should make for easy analysis of evolved programs. Addition of memory in various forms, global or local, and possibly of ADFs inside the GP trees may enhance the potential of this technique to help to model the more complicated *in vivo* problem.

Acknowledgements. We wish to thank Thomas Werner at Genomatix Software GmbH, Munich, Stephen Firth from QinetiQ, and David Corne from Reading University for interesting discussions about promoter prediction.

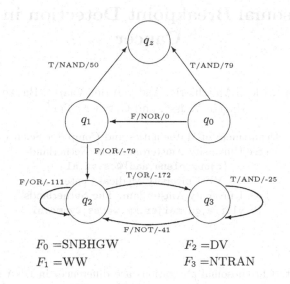

$$F_0 = \text{SNBHGW} \qquad F_2 = \text{DV}$$
$$F_1 = \text{WW} \qquad F_3 = \text{NTRAN}$$

Fig. 2. One of the resulting algorithms from the 5-fold cross validation.

References

[Ashlock, 1997] Ashlock, D. (1997). GP Automata for dividing the dollar. In Koza et. al. (Eds) *Genetic Programming: Proceedings of the Second Annual Conference*, 18–26, Stanford University.

[Benson, 2000a] Benson, K. A. (2000). Evolving Finite State Machines with Embedded Genetic Programming for Automatic Target Detection within SAR Imagery In *Proceedings of the Congress on Evolutionary Computation*, 1543–1549, La Jolla, San Diego, USA.

[Benson, 2000b] Benson, K. A. (2000). Performing Automatic Target Detection with evolvable finite state automata. In *Journal of Image and Vision Computing*, Volume 20, Issue 9-10, Elsevier.

[Cattaneo et. al., 2002] Cattaneo E., Rigamonti D.,Zuccato C. The Enigma of Huntington's Disease. Scientific American, December 2002.

[Hannenhalli and Levy, 2001] Hannenhalli S. and Levy S. (2001). Promoter prediction in the human genome. In *Proceedings of the 9th International Conference on Intelligent Systems for Molecular Biology*. Copenhagen, Denmark, July 21-25, 2001, Bioinformatics, Vol 17, Supplement 1, pp. S90–S96. ISSN: 1367-4803.

[Koza, 1999] John R. Koza (1999). *Genetic Programming III: Darwinian Invention and Problem Solving*, MIT Press.

[Lewin, 2000] Benjamin Lewin (2000). Genes VII, Oxford University Press.

[Orphanides et al., 1996] Orphanides, G., Lagrange, T., Reinberg, D.. The general transcription factors of RNA polymerase II. *Genes. Dev.*, vol. 10, 2657–2683, 1996.

[Pedersen et al., 1999] Pedersen, A. G., Baldi, P., Chauvin, Y., Brunak, S. The biology of eukaryotic promoter prediction - a review. *Computers and Chemistry*, vol. 23, 191–207, 1999.

[RGEP, 2002] Roche Genetics Education Program CD-ROM Scientific American, December 2002.

Chromosomal Breakpoint Detection in Human Cancer

Kees Jong[1], Elena Marchiori[1], Aad van der Vaart[1], Bauke Ylstra[2],
Marjan Weiss[2], and Gerrit Meijer[2]

[1] Department of Mathematics and Computer Science
Free University Amsterdam, The Netherlands
{cjong,elena,aad}@cs.vu.nl
[2] VU University Medical Center
Free University Amsterdam, The Netherlands
{b.ylstra,ga.meijer,mm.weiss}@vumc.nl

Abstract. Chromosomal aberrations are differences in DNA sequence copy number of chromosome regions [1]. These differences may be crucial genetic events in the development and progression of human cancers. Array Comparative Genomic Hybridization is a laboratory method used in cancer research for the measurement of chromosomal aberrations in tumor genomes. A recurrent aberration at a particular genome location may indicate the presence of a tumor suppressor gene or an oncogene. The goal of the analysis of this type of data includes detection of locations of copy number changes, called breakpoints, and estimate of the values of the copy number value before and after a change. Knowing the exact locations of a breakpoint is important to identify possibly damaged genes. This paper introduces genetic local search algorithms to perform this task.

1 Introduction

Array Comparative Genomic Hybridization (array-CGH) is an approach for genome-wide scanning of differences in chromosomal copy numbers. Normal human cells contain two copies of each of the 22 non-sex chromosomes. In tumor cells one or both copies of parts of chromosomes may be deleted or duplicated. Chromosomal copy numbers are defined to be 2 for normal cells, 1 or 0 for single and double deletions and 3 and higher for single copy gains and higher level amplifications. Ideally the purpose of array-CGH is to construct a graph of the copy numbers for a selection of clones (normal mapped chromosomal sequences, i.e. small pieces of DNA) as a function of position of the clone on the genome. DNA copy-number aberrations are used in cancer research, for instance, by searching for novel genes implicated in cancer by analyzing those genes located in regions

---

[1] Aberrations can occur without change of copy number, but these aberrations are not the subject of this paper.

S. Cagnoni et al. (Eds.): EvoWorkshops 2003, LNCS 2611, pp. 54–65, 2003.

with abnormal copy numbers. It is therefore of fundamental relevance to identify as precise as possible chromosomal regions with abnormal copy numbers.

Because copy numbers cannot be measured directly, tumor cells are compared to normal cells. A large number (up to 2500) of clones are printed on a glass slide (micro array), which is next treated with a mixture of DNA originating from tumor and normal cells, both cut into fragments. Before applying the DNA mixture to the micro array the two types of DNA are labelled red (Cy5) and green (Cy3), respectively. The labelled fragments hybridize ("stick") to a spot on the array with a matching DNA sequence. The measured red/green ratio for each of the spots on the array is roughly proportional to the quotient of copy numbers for tumor and normal tissue. This experiment is repeated for a number of tumors. A more elaborate introduction to array-CGH can be found in [5].

Unfortunately, a sample of cells taken from a tumor will generally consist of multiple cell types, which may differ in their chromosomal copy numbers. In particular, the sample usually consists of tumor cells and admixed normal cells. In some cases the sample may also contain tumor cells that are intermediate in the development of the tumor and have fewer copy number changes. It is assumed that the actual tumor cells occur by far the most in the sample. In our case the experimental samples were selected by the pathologist to have more than 70% tumor cells. The values found in the experiment then represent the copy numbers of the actual tumor cells plus some "noise" generated by the normal cells and some experimental noise.

When the observed relative copy numbers (or their logarithms) for the clones are ordered by location on the genome, the values form "clouds" with different means, supposedly reflecting different levels of copy numbers. We introduce a "smoothing" algorithm that tries to adjust the observed array-CGH values such that they represent the copy number of the most common tumor cells. That is, the algorithm tries to set the values to the means of the "clouds". Since the copy number of a clone is always quite small (normally 2, varying from 0 to about 10), we would like to set means of "clouds" that are close to the same value, because they represent the same copy number. Next we also want the number of value changes ("breakpoints") to be small.

The problem can be formalized as model fitting to search for most-likely-fit model given the data. A model describes a number of breakpoints, a position for each, and parameters of the distribution of copy number for each. Then one has to estimate the real parameters of the model from the observed array-CGH values.

We assume that the data are generated by a Gaussian process and use the maximum likelihood criterion for measuring the goodness of a partition, adjusted with a penalization term for taking into account model dimension. We introduce a local search procedure that searches for a most probable partition of the data using N breakpoints, for a given N. The procedure is incorporated into a genetic algorithm that evolves a population of partitions with possibly different number of breakpoints that may vary during execution. We design two algorithms based on this approach. The first one is a genetic local search algorithm that iteratively selects two 'good' chromosomes, generates two offspring using uniform crossover, applies mutation and the local search procedure to the offsprings and inserts them in the (worst chromosomes of the) population. The second algorithm generates only one offspring, applies local search and tries to further optimize the offspring with an ad-hoc procedure.

We analyze the performance of these algorithms on array-CGH measurements for 9 gastric cancer tumors. For each chromosome of a tumor, we compare the smoothing of the two GAs, the multi-start LS and the SA algorithm. The best algorithm we compare with the expert.

2 Breakpoint Detection

In our CGH experiments copy numbers are measured for approximately 2200 clones spread along the genome. We apply our algorithm to each of the 23 chromosomes separately. Denote by $x_1, \ldots, x_n$ the measured CGH values for a given chromosome. The main goal is to cluster these values in a small number of clusters $(x_1, \ldots, x_{y_1}), (x_{y_1+1}, \ldots, x_{y_2}), \ldots, (x_{y_N+1}, \ldots, x_n)$ such that the copy numbers of the clones in each cluster are identical. We refer to the indices $y_0 = 0 < y_1 < \cdots < y_N < n = y_{N+1}$ as breakpoints.

Our algorithm is motivated by the working hypothesis that the measured value x_j is equal to the relative copy number of clone j plus random noise that is independent across clones. Thus our model stipulates that for $y_{i-1} < j \leq y_i$ the observed CGH value x_j can be considered as drawn from a normal distribution with mean μ_i and variance σ_i^2 particular to the ith cluster. This leads to the likelihood function

$$\prod_{i=1}^{y_1} \frac{1}{\sigma_1 \sqrt{2\pi}} e^{-\frac{1}{2}\left(\frac{x_i - \mu_1}{\sigma_1}\right)^2} \cdots \prod_{i=y_N+1}^{n} \frac{1}{\sigma_{N+1} \sqrt{2\pi}} e^{-\frac{1}{2}\left(\frac{x_i - \mu_{N+1}}{\sigma_{N+1}}\right)^2}$$

The maximum likelihood estimators are the parameter values for which this expression is maximal. Given breakpoints $y_0 = 0 < y_1 < \cdots < y_N < n = y_{N+1}$ the maximization relative to the μ_i and σ_i^2 is equivalent to performing maximum likelihood estimation on each of the samples $x_{y_{i-1}+1}, \ldots, x_{y_i}$ separately, which leads to the usual estimates

$$\hat{\mu}_i = \frac{1}{y_i - y_{i-1}} \sum_{j=y_{i-1}+1}^{y_i} x_j, \qquad \hat{\sigma}_i^2 = \frac{1}{y_i - y_{i-1}} \sum_{j=y_{i-1}+1}^{y_i} (x_j - \hat{\mu}_i)^2.$$

Reinserting these values into the likelihood we are, after some simplification, left with

$$\frac{1}{\hat{\sigma}_1^{y_1} \sqrt{2\pi}^{y_1}} e^{-1/2} \cdots \frac{1}{\hat{\sigma}_{N+1}^{n-y_N} \sqrt{2\pi}^{n-y_N}} e^{-1/2}.$$

The next step is to find suitable breakpoints by maximizing this relative to $y_1, \ldots, y_N$. Equivalently, we minimize minus the logarithm, which up to an additive constant is equal to

$$\sum_{i=1}^{N+1} (y_{i+1} - y_i) \log \hat{\sigma}_i$$

Note here that the $\hat{\sigma}_i$ in this expression also depends on the choice of $y_1, \ldots, y_N$. However, it is obvious that the highest value of the likelihood is obtained by choosing the highest possible number of breakpoints, as this gives more flexibility in choosing the parameters μ_i and σ_i. The last minimization step is therefore not well defined. We remedy this by adding a penalty to the criterion, in order to discourage a large number of breakpoints. A simple penalty of the form λN, for λ a suitable constant, performed well in our experiments. This leads to the following function to be minimized.

$$f(y_1, \ldots, y_N) = \sum_{i=1}^{N+1} (y_{i+1} - y_i) \log \hat{\sigma}_i + \lambda N \tag{1}$$

If we consider there to be $3N$ parameters ($2N$ continuous parameters and N breakpoints), then the choices $\lambda - (3/2) \log n$ and $\lambda - 3$ correspond to Bayesian information criterion [7] and Akaike information criterion [1], respectively. In our experiments the choice $\lambda = 10$ was appropriate.

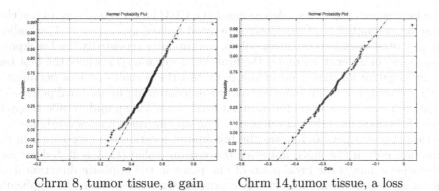

Chrm 8, tumor tissue, a gain   Chrm 14,tumor tissue, a loss

The assumptions of normality and independence may be slightly violated, as illustrated by the normal probability plots [6] (data are normally distributed when they lay near the dotted line). This holds also for normal tissues, as shown by the corresponding plot.

Nevertheless, in our experiments the resulting criterion gives adequate results.

The most obvious violations of the normality assumption are caused by amplifications. Results can be further improved if the clones in amplification areas are removed from the data. However, it is not easy to define an amplification unambiguously. An amplification area starts with a "big" increase of CGH value, last for only a "few" clones, after which the CGH values decrease "steeply". The number of clones for which an amplification lasts at most is a parameter. The increase and decrease of the value that are at least necessary to form an amplification depend on all value changes between consecutive clones. Say the average value change is $\overline{d}$ and the standard deviation of the value changes is s_d. Then the criterion is $\frac{d_i - \overline{d}}{s_d} \geq T$, where T is a parameter.

In the sequel, we do not apply pre-processing for dealing with amplifications.

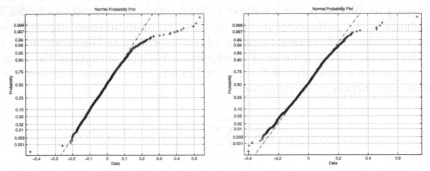

Chrms 1 to 22, tumor tissue, all normal   Chrms 1 to 22, normal tissue

3   The Genetic / Local Search Algorithms

The local search algorithm takes as input the CGH data $l = x_1 \ldots x_n$ of one chromosome, a number N of randomly generated indices $y_1, \ldots, y_N \in [1, n]$ indicating potential (locations of) breakpoints, and updates repeatedly the breakpoints (locations) in order to minimize the function f given in (1), where the first term is the negative log-likelihood of the data and the second one is a penalization with parameter λ which punishes partitions containing many breakpoints. The algorithm uses f as scoring function. At every iteration an update rule is applied to each breakpoint, selected randomly. The update rule chooses randomly a direction (left or right) and moves the breakpoint location of one position in that direction only if the move improves the scoring value (that is if f decreases), otherwise it moves the breakpoint of one place in the opposite direction if this yields an improvement. The iterative process terminates when the application of the update rule to each breakpoint does not improve the scoring. We call this algorithm LS. We use LS in a multi-start local search algorithm, and as local optimizer in the two heuristic algorithms described in the sequel.

Genetic local search algorithms, also called memetic algorithms [4], use local search for optimizing the population after the application of the genetic operators. So at each iteration of the evolutionary process the population consists of a set of local optima. We introduce the two memetic algorithms illustrated below for identifying breakpoints in array-CGH data of a chromosome, called GLS and GLSo, respectively.

In order to avoid confusion, in the sequel we say 'individual' instead of the standard genetic algorithms term 'chromosome' for indicating an element of the population.

```
GLS
{
  generate initial population
  while (termination criterion not satisfied)
  {
    select two parents from population using roulette wheel
    generate offsprings using uniform crossover
    apply mutation to each offspring
    apply LS to each offspring
    replace two worst individuals of population with offsprings
  }
}
```

```
GLSo
{
  generate initial population
  while (termination criterion not satisfied)
  {
    select two parents from population using roulette wheel
    generate offspring using OR crossover
    apply LS to offspring
    apply JOIN to offspring
    replace worst individual of population with offspring
  }
}
```

Our genetic algorithms use a representation where an individual is a bit string denoting chromosome locations with a 1 in each location containing a breakpoint and a 0 elsewhere. The fitness function to be minimized is the score function (1). The initial population is constructed as follows. For each N in a fixed range, a number k of elements is generated, where an element is a bit string with N 1's randomly placed. The local search LS is applied to each individual.

GLS uses (blind) uniform crossover, while mutation randomly decides whether to add or remove a breakpoint and then applies the chosen operation (that is flipping the value of the selected individual location). The 'remove' operation consists of removing the breakpoint that yields the best fitness function score. Note that this operation is applied even if it does not decrease the fitness of the individual. The 'add' operation selects the segment (a region between two consecutive ones) with relative chromosomal array-CGH region (set of clones values) having the highest standard deviation, and places a breakpoint in the middle of that region.

The termination criterion is satisfied when either a maximum number of iterations is reached or when the fitness of the best individual does not decrease and there is no pair of corresponding clones in the population having a difference in smoothed value of more than 0.01. The smoothed value of a clone is the mean value of the (chromosomal array-CGH region corresponding to the) segment containing that clone.

GLSo generates one offspring per iteration by selecting two individuals and constructing one offspring by taking the union of their breakpoints (by performing a bitwise OR of the two individuals). Then the offspring is optimized using LS and further optimized by removing breakpoints using the JOIN procedure. The JOIN procedure repeatedly selects the breakpoint whose removal yields the biggest improvement (decrease) of the fitness function, and continues until the fitness does not decrease anymore.

4   Experimental Results

Genomic DNA was isolated from snap-frozen tumor samples taken from gastrectomy specimens. The samples were obtained from the archives of the department of Pathology of the VU University Medical Center. Array-CGH experiments were performed according to [8] and ratio measurements according to [3]. The scanning array comprised DNA from 2275 BAC and P1 clones spotted in triplicate, evenly spread across the whole genome at an average resolution of 1.4 Mb. Chromosome X-clones were discarded from further analysis since all tumor samples were hybridized to male reference DNA, leaving 2214 clones per array to be evaluated. Each clone contains at least one STS for linkage to the sequence of the human genome. These data is analyzed in [9].

The 9 tumors used to test our method are all gastric tumors. A manual smoothing for these tumors, carried out by the expert B. Ylstra, is used to assess the performance of the algorithms and the maximum likelihood function as approximation for the expert. We run our algorithms on each chromosome of these 9 tumors, for a total of 207 chromosomes containing an average of about 100 clones.

The following GA parameter setting is chosen. The initialization generates 40 individuals containing N breakpoints, with N that varies from 1 to 10. An individual with 0 breakpoints is also added, thus giving a total of 401 individuals. The maximum number of iterations allowed is 100000. Crossover and mutation rates are equal to 1.

The multi-start LS performs 100000 plus 1 runs, where the number N of breakpoints varies from 1 to 20, with an equal number of runs assigned to each value of N. Also a run with 0 breakpoints is done. The final result is the solution with best score over the runs.

We compare the performance of GLSo, GLS, multi-start LS, and a multi-start variant of LS based on simulated annealing (SA). The annealing schedule of SA is as follows. The starting temperature is 100000. After 10000 changes of breakpoint location it cools down to 0.00001. After each change the actual temperature is divided by $10^{10^{-3}}$. After 10000 changes we make the algorithm behave exactly like LS. The other settings are similar to the multi-start LS, except that it only performs 2001 runs to make the comparison more fair in terms of computation time.

We compare the performance of the four algorithms in minimizing the function (1) by the median and mean values obtained for the $9 \times 23 = 207$ chromosomes in our gastric tumors. In the tables below mLS and mSA denote multi-start LS and SA, respectively. As shown in the following table method GLS performs best according to this criterion followed by the second genetic algorithm GLSo.

Algorithm	Median	Mean
mLS	-192.99	-218.77
mSA	-193.29	-220.07
GLSo	-194.47	-220.83
GLS	-196.00	-223.08

At closer inspection the nature of the differences in performance of the four algorithms vary considerably over the 207 chromosomes. For 22 chromosomes all four algoritms yield an identical smoothing, and for as many as 76 chromosomes the smoothings produced by at least three of the four algorithms are identical. For this reason we also investigated the differences in fitness for the 207 chromosomes for each pair of algorithms. Medians, means, 20% trimmed means, and the number of chromosomes with identical smoothing are given in the following table, together with the p-values of the sign test. The latter test (cf. [6]) indicates if the observed differences (in obtained fitness values) between the pairs of algorithms are statistically significant, under the assumption that the 207 chromosomes can be considered a random sample of chromosomes.

Algorithm	Median	Mean	Trimmed Mean	# Zeros	P-value
mLS-mSA	0.00	1.30	0.52	67	0
mLS-GLSo	0.00	2.06	0.22	50	0.63
mLS-GLS	0.70	4.31	1.84	81	0
mSA-GLSo	0.00	0.76	-0.08	31	0.47
mSA-GLS	0.79	3.01	1.66	62	0
GLSo-GLS	0.24	2.25	1.50	48	0

From these numbers we conclude again that GLS is best, followed by GLSo, mSA and mLS. Furthermore, the superior performance of GLS is statistically significant, whereas the observed differences between the other methods may not be replicable on data of additional tumors.

To illustrate the results of the four algorithms we show below pictures of the smoothing/breakpoints found by each of algorithms on one of the tumors in which various types of chromosomal aberrations (gain, loss and amplification) occur.

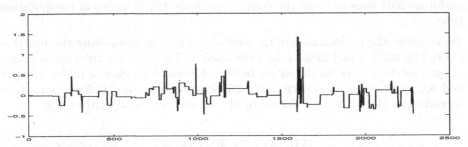

Tumor 2008c, dots are raw data, line is result of multi-start LS.

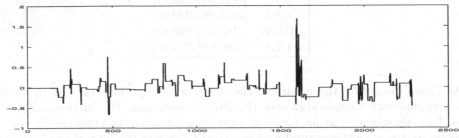

Tumor 2008c, dots are raw data, line is result of multi-start SA.

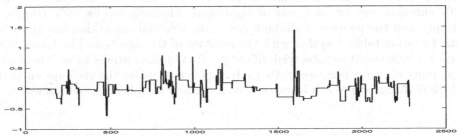

Tumor 2008c, dots are raw data, line is result of GLSo.

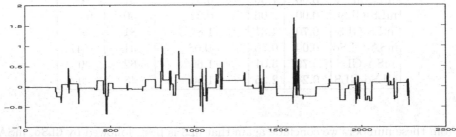

Tumor 2008c, dots are raw data, line is result of GLS.

In order to assess the convergence behaviour of the genetic algorithms we give below plots of a typical histogram of the distribution of breakpoints within the solutions of the pool after the stopping criterion is satisfied. The plots indicate that the evolutionary process ends with individuals with breakpoints in nearby locations. Observe that the stopping criterion is such that shifting a breakpoint within an area that has no clear breakpoint stops the iterations. In such a case there is "no clear breakpoint", meaning that the means of the two corresponding segments in all individuals are close. This may cause the algorithm to stop after a few iterations even if the individuals have breakpoints in different locations.

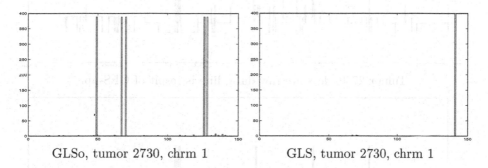

GLSo, tumor 2730, chrm 1   GLS, tumor 2730, chrm 1

Next, we compare the robustness of the genetic algorithms, that is the sensitivity of the outcome to the initialization and other random operators used. Below we plot a typical histogram of the location of breakpoints of the best individual of the final population over 100 runs of the genetic algorithms on chromosome 1.

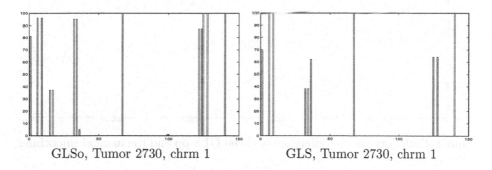

GLSo, Tumor 2730, chrm 1   GLS, Tumor 2730, chrm 1

Finally, we compare the smoothings and breakpoints obtained by GLS with those manually produced by the expert. The manual smoothings have been built under the assumption that there is a small number of different smoothing levels, reflecting the observation that few copy number values are present in chromosomes. In order to incorporate this constraint in our method, we perform a post processing step that joins close smoothing levels. To this aim the k-means algorithm is applied to the set of CGH values generated by running GLS over all the chromosomes of a tumor, and then the resulting smoothing levels that are closer than a fixed threshold are joined. Over all tumors the average difference between the values of the clones is 0.0513, indicating that GLS followed by post processing (denoted below by GLS-pp) is a satisfactory approximation of the manual smoothing. GLS seems more sensitive to outliers. This can be explained by the fact that the expert sometimes knows an outlier is meaningless and so ignores it.

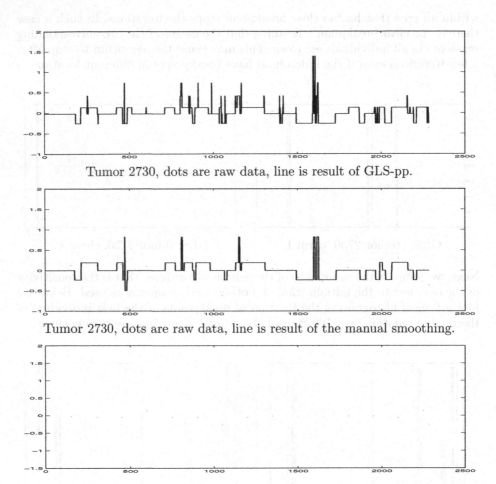

Tumor 2730, dots are raw data, line is result of GLS-pp.

Tumor 2730, dots are raw data, line is result of the manual smoothing.

Tumor 2730, dots are the difference between GLS-pp and the manual smoothing.

5 Discussion

The results of the experiments indicate that GLS performs better than the other algorithms in minimising function (1).

Both GAs converge within the maximum number of iterations in case the data contains clear breakpoints. The stopping criterion prevents the algorithm from searching for optimal locations of breakpoints that are not clear.

GLS-pp finds smoothings that are very similar to the manual smoothings. It should be noted that an expert produces smoothings based on more information than just the CGH values. An expert also keeps in mind information like misplacement of clones on the genome and recurring aberrations of clones due to known experimental artefacts. From the normality plots shown it seems that the final expert smoothings are reasonably well normally distributed. Lacking data combining CGH values with known copy numbers in cell types and frequencies of cell types in samples, we were not able to test the suitability of our model to remove noise from the experiment and some cells of types that occur in small numbers. This remains an open problem for future research.

We conclude with some words on related work. To the best of our knowledge, nothing has yet being published on automatic breakpoint detection and estimation of copy number values. We are aware of work in progress carried out at UCSF Cancer Center by Jane Fridlyand, who is trying to use Hidden Markov Models to tackle this problem, and at the Memorial Sloan-Kettering Cancer Center by Adam Olshen who is using change-points and Markovian methods [2].

References

1. H. Akaike. Information theory and an extension of the maximum likelihood principle. In B. Petrox and F. Caski, editors, *Second International Symposium on Information Theory*, page 267, 1973.
2. J. Fridlyand. Personal communication, 2002.
3. A. N. Jain, T. A. Tokuyasu, A. M. Snijders, R. Segraves, D. G. Albertson, and D. Pinkel. Fully automatic quantification of microarray image data. *Genome research*, 12:325–332, 2002.
4. P. Moscato. On evolution, search, optimization, genetic algorithms and martial arts: Towards memetic algorithms. Technical report, Caltech Concurrent Computation Program, Californian Institute of Technology, U.S.A., TR No790 1989.
5. D. Pinkel, R. Segraves, D. Sudar, S. Clark, I. Poole, D. Kowbel, C. Collins, W.L. Kuo, C. Chen, Y. Zhai, S.H. Dairkee, B.M. Ljung, J.W. Gray, and D.G. Albertson. High resolution analysis of dna copy number variation using comparative genomic hybridization to microarrays. *Nature Genetics*, 20, 1998.
6. J.A. Rice. *Mathematical Statistics and Data Analysis Second Edition*. Duxbury Press, 1995.
7. G. Schwarz. Estimating the dimension of a model. *The Annals of Statistics*, 6(2):461–464, 1978.
8. A.M. Snijders, N. Nowak, R. Segraves, S. Blackwood, N. Brown, J. Conroy, G. Hamilton, A.K. Hindle, B. Huey, K. Kimura, S. Law, K. Myambo, J. Palmer, B. Ylstra, J.P. Yue, J.W. Gray, A.N. Jain, D. Pinkel, and D.G. Albertson. Assembly of microarrays for genome-wide measurement of dna copy number by cgh. *Nature Genetics*, 29:263–264, 2001.
9. M.M. Weiss, E.J. Kuipers, C. Postma, A. M. Snijders, I. Siccama, D. Pinkel, J. Westerga, S.G.M. Meuwissen, D. G. Albertson, and G.A. Meijer. Genomic profiling of gastric cancer predicts lymph node status survival. *Oncogene*, 2003. In press.

Discovering Haplotypes in Linkage Disequilibrium Mapping with an Adaptive Genetic Algorithm

Laetitia Jourdan**, Clarisse Dhaenens, and El-Ghazali Talbi

LIFL, Université de Lille1
Bât M3 Cité Scientifique
59655 Villeneuve d'Ascq Cedex
FRANCE
jourdan@lifl.fr,
http://www.lifl.fr/~jourdan

Abstract. In this paper, we present an evolutionary approach to discover candidate haplotypes in a linkage disequilibrium study. This work takes place into the study of factors involved in multi-factorial diseases such as diabetes and obesity. A first study on the linkage disequilibrium problem structure led us to use a genetic algorithm to solve it. Due to the particular, but classical, evaluation function given by the biologists, we design our genetic algorithm with several populations. This model lead us to implement different cooperative operators such as mutation and crossover. Probabilities of application of those mechanisms are set adaptively. In order to introduce some diversity, we also implement a random immigrant strategy and to cover up the cost of the evaluation computation we parallelize it in a master / slave model. Different combinations of the presented mechanisms are tested on real data and compared in term of robustness and computation cost. We show that the most complete strategy is able to find the best solutions and is the most robust.

1 Introduction

Using single-nucleotide polymorphisms (SNPs) is currently a major tool in the search for genes involved in complex diseases. In order to find candidate haplotypes (combination of markers) of SNPs for multi-factorial diseases such as diabetes and obesity, we develop a study with the multi-factorial Disease Laboratory of Lille (France). In this study, we have to look for disease-associated haplotypes in a very large number of loci on different chromosomes. This generates a lot of data.

On a previous study of the problem structure [5], we have shown that classical algorithms are not adapted to deal with this search and that it is paramount to develop a method which is able to deal with a very large search space and which has a good exploration potential.

** This work is supported by the Nord-Pas de Calais region and the Genopole of Lille.

S. Cagnoni et al. (Eds.): EvoWorkshops 2003, LNCS 2611, pp. 66–75, 2003.

In this work, we propose to solve this problem with a dedicated genetic algorithm which is based on several subpopulations in order to cope with non commensurable solutions. This particular model lead us to use cooperative operators which are set adaptively. As the evaluation process imposed by biologists is time consuming, we also proposed a parallel implementation.

This paper is organized as follows. The second section of this paper gives some biological definitions, presents the data and the biological problem with its particular evaluation function. In the third section, the dedicated genetic algorithm and its specificities are presented. Then, the fourth section presents results obtained thanks to this algorithm. Finally the conclusion gives indications about exploitation of the results.

2 The Biological Problem

This work deals with the linkage disequilibrium for multi-factorial diseases and in particular with the studies of factors that are implied in diseases such as diabetes and obesity. We will firstly introduce some notions of biology [1,9] that are necessary to understand the problem and show what kind of data we have to exploit. Then we will formulate the biological problem.

2.1 Definitions and Data

Genetic markers are alleles of genes, or DNA polymorphisms, that are used as experimental probes to keep track of an individual, a tissue, a cell, a nucleus, a chromosome, or a gene. Stated another way, any character that acts as a signpost or signal of the presence or location of a gene or heredity characteristic for an individual in a population. There are 4 chromosome changes that do occur from generation to generation, and these are known as markers: indel, snips (SNP's), micro-satellites and mini-satellite. In our case we are interested in single nucleotide polymorphisms (SNPs).

SNPs are DNA sequence variations that occur when a single nucleotide (A,T,C, or G) in the genome sequence is changed. Most SNPs, actually about two over three SNPs, involve the replacement of cytosine (C) with thymine (T). SNPs occur every 100 to 300 bases along the human genome. SNPs are stable from an evolutionarily standpoint– not changing much from generation to generation –making them easier to follow in population studies. SNPs can have two forms a and b on each allele which leads to three different combinations: aa, ab and bb (alleles are not differentiated so $ab = ba$).

Data available for our study are, for all the individuals that have been examined, the form of all their SNPs and their status (affected / not affected) (see table 1). Then in order to discover associations between SNPs and the disease, we know for all the SNPs the mean frequency of each alternative (a and b) (see table 2).

An haplotype is a set of closely linked alleles (genes or DNA polymorphisms) inherited as a unit and is a contraction of the phrase "haploid genotype". Different combinations of polymorphisms are known as haplotypes. Two reasons lead

Table 1. Data on individuals.

Individual SNP	SNP_1	...	SNP_n	STATUS
IND_1	aa	...	bb	Not affected
IND_2	aa	...	ab	Affected
...	...	...	...	...
IND_{k-1}	ab	...	aa	Not affected
IND_k	aa	...	bb	Unknown

Table 2. Frequencies of SNPs.

Table 3. Disequilibrium between SNPs.

SNP	Freq. of a	Freq. of b
1	0.997	0.003
2	0.856	0.144
...	...	...
n-1	0.576	0.424
n	0.389	0.611

SNP_1	SNP_2	Disequilibrium
1	2	-1.00
1	3	0.67
...	...	...
n-1	n	0.42

us to use multi-locus analysis. First, haplotype are more specific of the ancestral chromosome where the mutation occurred. Moreover, it is possible that several loci (SNPs) have an effect on the disease risk.

Two markers are said to be in linkage equilibrium if for all SNP X and SNP Y which are between these two markers the frequency of the haplotype XY is equal to the product of the frequencies of the SNPs X and Y. In the other case, we will talk about linkage disequilibrium.

For example, let us compute the linkage disequilibrium for HLA (Human Leuko-cyte Antigen). The HLA has four SNPs (A, B, C and D). Let the frequencies $F(HLA-A)$ and $F(HLA-C)$ be the frequencies for markers A and C of HLA. $F(HLA-A) = 0.161$ and $F(HLA-C) = 0.153$. The frequency of HLA-A / HLA-C with no linkage disequilibrium is : $0.161 \times 0.153 = 0.0246$ but the observed frequency is 0.089. So the linkage disequilibrium is $D = 0.089 - 0.0246$ and equals to 0.064. The linkage disequilibrium is then tested with a χ^2 test for evaluating the relevance of the frequencies difference. In our study, we know for all pairs of SNPs their linkage disequilibrium (see table 3). This disequilibrium belongs to the interval in [-1,1] where -1 and 1 signify a huge disequilibrium whereas a value near 0 signifies no linkage disequilibrium.

2.2   Description of the Problem

The objective of our application is to find haplotypes that are able to explain the disease under study. Two SNPs of an haplotype must verify two properties (to be in disequilibrium). Firstly, their two by two disequilibrium must be less than a threshold S_1. Secondly, the difference between the smaller frequencies of the two alternatives must be greater than a threshold S_2. There will be several haplotypes that verify these constraints. In order to evaluate the quality of an haplotype biologists often use the two widely used statistical procedures EH-DIALL [10] and CLUMP [3].

EH-DIALL (EH is for Estimated Haplotype) is a procedure that determines the most probable distribution of alleles in an haplotype according to values of the SNPs. Given a sample coomposed of a large number of individuals collected at random from the population (see table 1), EH-DIALL program estimates allele frequencies for each marker. Haplotype frequencies are estimated with allelic association (Hypothesis H_1) and without allelic association (Hypothesis H_0).

CLUMP is a program designed to assess the significance of the departure of observed values in a contingency table from the expected values conditional on the marginal totals [3]. CLUMP produces several statistics. The one that corresponds to our problem is referred to as T_1. A good haplotype is an haplotype that is highly correlated with the disease, which corresponds to a high value of T_1.

The whole evaluation process for an haplotype is:

- Starting from a set of candidate SNPs forming an haplotype, estimate independently, for affected and unaffected people, the distribution of alleles in the haplotype thanks to EH-DIALL.
- Use CLUMP to evaluate the association haplotype-disease.

This process allows us to evaluate the quality of an haplotype in biological terms and our objective will be to find haplotypes that maximize this criterion of quality. We will consider this problem as a combinatorial optimization problem where the search space is composed of all the combinations of SNPs and the objective function is the maximization of the quality describes above.

3   The Specific Genetic Algorithm

We will present our steady state genetic algorithm to discover designed candidate haplotypes that are able to explain the disease. The general scheme of the algorithm is given in figure 1. To present the genetic algorithm adapted to this problem, we have to define the encoding, operators and advanced search mechanisms that are used.

3.1   Encoding

The encoding is very intuitive: an individual represents an haplotype (a set of SNPs). It is encoded with a structure that stores the size of the haplotype (number of SNP), a table of the SNPs that compose it and its fitness.

3.2   Multi-population

A characteristic of this problem is that haplotypes of different sizes are not directly comparable. For example, for one of the dataset, the best haplotypes of size 3 has a fitness of 58.81 whereas the best haplotype of size 4 has a fitness of 84.85.

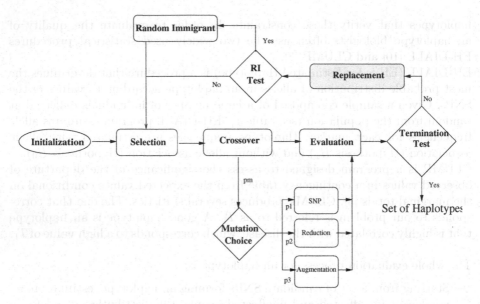

Fig. 1. The general scheme of our genetic algorithm.

These values have been found thanks to a complete enumeration of haplotypes of size three and four. This phenomena is general: the range of the fitness of an haplotype grows with the size of the haplotype. So, considering haplotypes of different sizes would always be in favor of the larger. Hence the global population of the genetic algorithm is divided into several cooperating subpopulations, where each subpopulation corresponds to a given size of haplotype. The number of individuals in each subpopulation are not

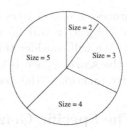

Fig. 2. An example of subpopulations repartition.

equal and increase with the size of the haplotypes in order to follow the growth of the size of the search space related to each size (see figure 2). Some cooperations between subpopulations will occur during mutation and crossover.

3.3 Operators

Mutation: We implement three kinds of mutation operators that slightly modify an individual:

- Mutation of a SNP: we randomly choose a SNP of the individual and re-place it by another randomly chosen SNP. This process is similar to a local search which allows to explore the neighborhood of the solution. We use this mutation several times in parallel and keep the best individual found.

- Reduction Mutation: we randomly choose a SNP of the individual and remove it. The individual has now a lower size. This operator allows to move individuals from a subpopulation to another. This operator constructs smaller
  haplotypes and tries to generalize the association.
- Augmentation Mutation: we add a randomly chosen SNP. This mutation
  constructs increasingly large size haplotypes and specialize the association
  of SNPs.

Probabilities of mutation are hard to set when we have several mutation
operators and are often set experimentally. To overcome this problem, we implement an adaptive strategy for calculating the rate of each mutation operator.
Many authors have worked on setting automatically probabilities of applying
operator [2,8,6]. In [7], authors proposed to compute the new rate of mutation
by calculating the progress of the jth application of mutation operator M_i, for
an individual ind mutated into an individual mut as follows:

$$progress_j(M_i) = Max(fitness(ind), fitness(mut)) - fitness(ind)$$

But here we have mutation operators that increase or decrease the number of
SNPs and we saw that the fitness function gave by the biologists is correlated to
the number of SNPs. Hence if individuals ind and mut are not of the same size,
their fitness must not be compared. In order to adapt the notion of progress to
our problem, we normalize the progress with the best individual and the worst
of the subpopulation corresponding to the individual (the best and the worst
individuals of the same size). The progress is:

$$progress_j(M_i) = Max(norm(ind), norm(mut)) - norm(ind)$$

Then for each mutation operator M_i, assume $Nb\_mut(M_i)$ applications of the
mutation are done during a given generation ($j = 1, .., Nb_{mut}(M_i)$). Then we
can compute the profit of a mutation M_k :

$$Profit(M_k) = \frac{\sum_j progress_j(M_k)/Nb\_mut(M_k)}{\sum_i \left(\sum_j progress_j(M_i)/Nb\_mut(M_i) \right)}$$

We set a minimum rate δ and a global mutation rate $p_{mutation}$ for N mutation
operators to apply. The new mutation ratio $p(M_i)$ for each M_i is calculated using
the following formula [7]:

$$p(M_i) = Profit(M_i) \times (p_{mutation} - N \times \delta) + \delta$$

The sum of all the mutation rates is equal to the global rate of mutation
$p_{mutation}$. The initial rate of each mutation operator is set to $p_{mutation}/N$.

Crossover: We use an uniform quadratic crossover: take the two strings of
SNPs of the parents and create two children by randomly shuffling the variables

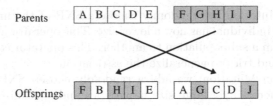

Fig. 3. Uniform crossover of individuals of size 5.

corresponding to the SNP at each site (see figure 3). Then we calculate the number of SNPs and the score.

We use two kinds of crossovers:

- Intra-population: only crossovers between individuals of a same subpopulation are allowed
- Inter-population: crossovers between individuals of different subpopulations are allowed.

The probabilities of application of each kind of crossover are also set in an adaptive manor. We use the same strategy used for mutation (see 3.3) to the case of the quadratic crossover. We define the improvement of a child e with regards to its parents p_1 and p_2 for intra-population crossover as:

$$2 \times Improve_{Intra}(e, p_1, p_2) = \begin{array}{c} \frac{Max(fitness(p_1), fitness(e)) - fitness(p_1)}{best\_of\_size(e.size)} \\ + \\ \frac{Max(fitness(p_2), fitness(e)) - fitness(p_2)}{best\_of\_size(e.size)} \end{array}$$

In this case, the three individuals e, p_1 and p_2 are of the same size. They can be compared.

We define the improvement for the inter-population crossover by only comparing the improvement between a child e and its parent of the same size:

$$Improve_{Inter}(e, p_1, p_2) = \begin{cases} \frac{Max(fitness(p_1), fitness(c)) - fitness(p_1)}{best\_of\_size(e.size)} & \text{If } size.p_1 = size.e \\ \frac{Max(fitness(p_2), fitness(c)) - fitness(p_2)}{best\_of\_size(e.size)} & \text{If } size.p_2 = size.e \end{cases}$$

Hence, the intra-improvement considers two comparisons and the inter-improvement only one. This is the reason why we must divide by two the intra-improvement.

So the global function for the improvement is:

$$Improve(e, p_1, p_2) = Improve_{Intra}(e, p_1, p_2) \ OR \ Improve_{Inter}(e, p_1, p_2)$$

The progress of the jth application of each crossover C_i, which mates two individuals p_1 and p_2 to obtain two children e_1 and e_2 is:

$$progress_j(C_i) = Improve_j(e_1, p_1, p_2) + Improve_j(e2, p_1, p_2)$$

Then for all the crossover operators C_i, assume $Nb\_cross(C_i)$ applications of the crossover are made during a given generation. Then the profit of a crossover C_k is:

$$Profit(C_k) = \frac{\sum_j progress_j(C_k)/Nb\_cross(C_k)}{\sum_i \left(\sum_j progress_j(C_i)/Nb\_cross(C_i)\right)}$$

We set a minimum rate δ and a global crossover rate $p_{crossover}$ for N crossover operators to apply. The new crossover ratio for each C_i is calculated using the same formula than in the paragraph 3.3 for the mutation by replacing $p_{mutation}$ by $p_{crossover}$.

3.4   Random Immigrant

We use random immigrant to introduce some diversity in the search and to avoid premature convergence. This mechanism replaces all the individuals that have a fitness under the mean fitness of the population by new generated individuals. It is setting on when the best individual has not been improved for a given number of generations.

3.5   Parallel Implementation

The evaluation function of the problem is time consuming. In order to run the algorithm in a reasonable time, we have made a synchronous parallel implementation of the evaluation phase.
The implementation is based on a master / slaves model. The slaves are initiated at the beginning and access only once to the data. During the evaluation phase, the master gives each slave an individual to evaluate. Then the slave computes the fitness of this individual and send it back to the master.
The programming environment used is C/PVM (Parallel Virtual Machine) on a network of PC under Linux [4].

4   Results

We test the proposed method on an extract of real data. The extract contains 51 SNPs for 133 individuals whereas the real data contains 248 SNPs for 133 individuals.
Table 4 presents the results obtained by different combinations of the mechanisms explain above. For each combination we made five runs and we indicate the best haplotype found over the five runs for each size and its fitness. We also report the mean fitness and its standard deviation (Dev.) for the five runs.

Table 4. Comparison of results obtained by the GA.

Scheme	Best Haplotype Found	Fitness	Mean	Dev	Min. Nb of Eval.	Mean
1. Simple GA without cooperative operators	8 12 15	58.814	58.8	0.0	175	766
	6 8 16 31	80.631	78.9	5.9	1938	2714.6
	8 12 16 33 43	123.108	123.1	0.0	2581	4271
	8 12 15 21 32 43	161.252	158.8	2.5	3762	11307.8
2. Simple GA + Random Immigrant	8 12 15	58.814	58.8	0.0	167	354.6
	8 18 26 50	84.856	82.5	2.4	1287	2254.8
	8 12 16 33 43	123.108	123.1	0.0	1375	4410.6
	8 12 15 21 32 43	161.252	160.2	1.0	6051	9074
3. Simple GA with cooperative non adaptive operators	8 12 15	58.814	58.8	0.0	187	425.8
	6 8 16 31	80.631	79.3	5.6	644	3584
	8 12 16 33 43	123.108	123.1	0.0	1778	6749
	8 12 15 21 32 43	161.252	161.2	0.0	6676	11620
4. Adaptive Mutation + Crossover intra population	8 12 15	58.814	58.8	0.0	302	603.6
	8 18 26 50	84.856	82.5	2.4	375	3164.8
	8 12 16 33 43	123.108	123.1	0.0	2397	4737.8
	8 12 15 21 32 43	161.252	161.2	0.0	3364	10074
5. Adaptive Mutation + Adaptive crossover	8 12 15	58.814	58.8	0.0	207	1110.2
	8 18 26 50	84.856	83.1	1.7	426	2809.6
	8 12 16 33 43	123.108	123.1	0.0	2227	3474.2
	8 12 15 21 32 43	161.252	161.2	0.0	4455	9851.2
6. Adaptive Mutation + Adaptive crossover + Random Immigrant	8 12 15	58.814	58.8	0.0	317	587.4
	8 18 26 50	84.856	84.8	0.0	1111	3238.2
	8 12 16 33 43	123.108	123.1	0.0	2994	5615.2
	8 12 15 21 32 43	161.252	161.2	0.0	11573	15464.6

Finally, we give the minimum number of required evaluations to obtain the solution and the mean over the five runs. As we said before the optimal solutions for size 3 and 4 are known (58.8514 and 84.856) thanks to a complete enumeration. For size 5 and 6, the best known are respectively 123.108 and 161.252.

We highlight in grey, the best haplotype found over the five runs when it does not correspond to the optimal one. So, for combinations 1. and 3., the best haplotype of size four has not been found in any of the five runs.

The column Dev. gives the standard deviation of the fitness found for the five runs. A zero indicates that each run has found the same solution. We can see that for combination 1. to 5., this deviation is not always equal to zero. Only combination 6. allows to find at each run and for the different size the best solution.

Moreover, the last column shows that the number of evaluations required for each combination is of the same range. This indicates that introducing complex mechanisms does not penalize the search time.

All these experiments show that the complete version (combination 6.) is the most interesting (best result in reasonable time).

5   Conclusion

In this paper, we have presented a parallel cooperating multi-population genetic algorithm for the search of candidate haplotypes in linkage disequilibrium study of multi-factorial diseases. Our aim was to respect the constraints given by the biologists and in particular to use a dedicated evaluation function. This function and its particularity of no possible comparison between haplotypes of different sizes lead us to design a multi-population genetic algorithm. We have implemented cooperative operators (mutations and crossovers) that allow individual of different sizes to cooperate whose rates are set adaptively and used also a random immigrant strategy. We have shown their performances on real datasets by comparing different combinations of operators and mechanisms. The complexity of the evaluation function has lead us to a parallel implementation of the algorithm which is based on a master / slave model.

Thanks to this method, biologists are able to test different data sets and to formulate hypotheses on genetic factors involved in diseases under study.

References

1. Bruce Alberts, Alexander Johnson, Julian Lewis, Martin Raff, Keith Roberts, and Peter Walter. *Molecular Biology of the Cell*. Garland Pub, 4 edition, March 2002.
2. L. Davis. Adapting operator probabilities in genetic algorithms. In J. D. Schaffer, editor, *Third International Conference on Genetic Algorithms*, pages 61–69. Morgan Kaufmann, 1989. San Mateo, CA.
3. P.C. Sham et D. Curtis. Monte carlo tests for associations between disease and alleles at hightly polymorphic loci. *Annal Human Genetic*, pages 97–105, 1995.
4. A. Geist, A. Beguelin, J. Dongarra, W. Jiang, R. Mancbek, and V. Sunderam. *PVM: Parallel Virtual Machine – A User's Guide and Tutorial for Networked Parallel Computing*,. MIT PRess, 1994.
5. Vermeersch Grégory. Algorithmes génétiques pour la bio-informatique. Master's thesis, University of Lille 1, LIFL, 2001.
6. Francisco Herrera and Manuel Lozano. Adaptation of genetic algorithm parameters based on fuzzy logic controllers. In F. Herrera and J. L. Verdegay, editors, *Genetic Algorithms and Soft Computing*, pages 95–125. Physica-Verlag, Heidelberg, 1996.
7. T. P. Hong, H.S. Wang, and W.C. Chen. Simultaneosly applying multiple mutation operators in genetic algorithms. *Journal of Heuristics*, 6:439–455, 2000.
8. Bryant A. Julstrom. What have you done for me lately? adapting operator probabilities in a steady-state genetic algorithm. In L. J. Eshelman, editor, *Proceedings of the sixth International Conference on Genetic Algorithms*, pages 81–87. Morgan Kaufmann, 1995. San Francisco, CA.
9. Department of Energy. The human genome program of the U.S. URL : http://www.ornl.gov/hgmis/, 2002.
10. J.D. Terwilliger and J. Ott. *Handbook of human genetic linkage*. Johns Hopkins University Press, Baltimore, June 1994. ISBN: 0801848032.

Genetic Algorithms for Gene Expression Analysis

Ed Keedwell and Ajit Narayanan

School of Engineering and Computer Science
University of Exeter
Exeter
EX4 4QF
{E.C.Keedwell, A.Narayanan}@ex.ac.uk

Abstract. The major problem for current gene expression analysis techniques is how to identify the handful of genes which contribute to a disease from the thousands of genes measured on gene chips (microarrays). The use of a novel neural-genetic hybrid algorithm for gene expression analysis is described here. The genetic algorithm identifies possible gene combinations for classification and then uses the output from a neural network to determine the fitness of these combinations. Normal mutation and crossover operations are used to find increasingly fit combinations. Experiments on artificial and real-world gene expression databases are reported. The results from the algorithm are also explored for biological plausibility and confirm that the algorithm is a powerful alternative to standard data mining techniques in this domain.

1 Introduction

Knowledge discovery from gene expression data is a highly topical area of research in Bioinformatics and can potentially yield some of the most important discoveries in the field. Gene expression data is created by a process known as microarraying which yields a set of floating point and absolute values. These values represent the activation level of every gene within an organism at a particular point in time and a typical dataset can often consist of tens of thousands of genes. If these microarrays are taken from several individuals with a disease and also from those who are normal, a database of gene expression records which fall into separate classes can be created.

The task for knowledge discovery algorithms is to then find the handful of genes out of the thousands measured which contribute, either by themselves or in combination, to a particular classification. If this knowledge can be discovered from the data, then the possibilities for either enhancing or inhibiting the expression of these genes could lead to more effective therapies for combating disease. There are several attributes of gene expression data though which make it less amenable to normal knowledge discovery than other types of data used for data mining. Values are largely continuous rather than discrete, and most datasets consist of a huge number of genes (or variables) and a small number of records. This is because there will typically be tens of thousands of genes sampled and only a small number of

S. Cagnoni et al. (Eds.): EvoWorkshops 2003, LNCS 2611, pp. 76–86, 2003.
© Springer-Verlag Berlin Heidelberg 2003

subjects used. This is a reversal of the traditional data mining dataset where there are typically more records than variables.

Research in this area has grown in recent years as the data has become more readily available. Several approaches are documented (see Narayanan *et al.* (2002) for an overview of current techniques) which describe systems of statistics (Golub *et al.*, 1999), a number of neural network approaches (Su *et al.*,2002; Xu *et al.*, 2002) and the use of a variety of algorithms including Support Vector Machines and Decision Trees (Ryu and Cho, 2002). However, there is no reference to genetic algorithms having been used for this purpose, although they have previously been used to identify gene regulatory networks (i.e. one gene in one timestep affects the activation of another gene in a subsequent timestep) from such data (Ando and Iba, 2001a, 2001b). Therefore the techniques described in this paper may well constitute a novel application of genetic algorithms.

The new neural-genetic hybrid described here attains very good results on extracting knowledge from gene expression data, comparable with existing techniques. The knowledge that is discovered contains more information and is more amenable to analysis by biologists. This new technique therefore represents a step towards a complete tool for knowledge discovery from gene expression data.

This paper firstly describes how the neural genetic model functions and how it has been implemented. Experiments on two readily available gene expression classification datasets (in papers by Golub *et al.* (1999) and Page *et al.* (2002)) are conducted and compared with the well-known data mining package See5. In addition to this, the results from the algorithm are briefly investigated to determine their biological plausibility.

2 Method

The neural-genetic system uses a genetic algorithm and backpropagation in conjunction to discover small combinations of genes which lead to the correct classification of the dataset. Each chromosome is evaluated by using a single layer step-function backpropagation component which "learns" an optimal set of weights between the selected genes and a single class value (classification). Each classification requires a separate run of the genetic algorithm. The following describes a run through the algorithm on an artificial dataset (the Sunburn dataset taken from Winston (1992)).

Before the algorithm can be applied, the categorical data often contained in these databases must be converted to a "field" representation. This process is shown in Figure 1. Whilst all gene expression data is continuous in nature, most of the data can be used in a 3-valued format with an absence (A), presence (P) and marginal (M) call given to each of the gene expression values[1].

---

[1] For more details on the Affymetrix microarray process please see http://www.affymetrix.com

Name	Hair	Height	Weight	Lotion	Result
Sarah	Blonde	Average	Light	No	Sunburned
Dana	Blonde	Tall	Average	Yes	Not sunburned
Alex	Brown	Short	Average	Yes	Not sunburned
Annie	Blonde	Short	Average	No	Sunburned
Emily	Red	Average	Heavy	No	Sunburned
Pete	Brown	Tall	Heavy	No	Not sunburned
John	Brown	Average	Average	No	Not sunburned
Katie	Blonde	Short	Light	Yes	Not sunburned

Name	Hair	Height	Weight	Lotion	Result
Sarah	1,0,0	1,0,0	1,0,0	1,0	1,0
Dana	1,0,0	0,1,0	0,1,0	0,1	0,1
Alex	0,1,0	0,0,1	0,1,0	0,1	0,1
Annie	1,0,0	0,0,1	0,1,0	1,0	1,0
Emily	0,0,1	1,0,0	0,0,1	1,0	1,0
Pete	0,1,0	0,1,0	0,0,1	1,0	0,1
John	0,1,0	1,0,0	0,1,0	1,0	0,1
Katie	1,0,0	0,0,1	1,0,0	0,1	0,1

Fig. 1. Field representation as pre-processed before input to the neural-genetic algorithm. This allows the patterns of attributes and classes to be matched by the neural component of the algorithm. Note that each attribute value in the original dataset is converted into a binary field, where a '1' in a specific bit position signifies a specific attribute value. Hence, the 11-bit field representation '10010010010' represents Sarah in field form.

Once this conversion has taken place, the algorithm proper can be applied to the data. The genetic algorithm chromosome representation takes the form of a user-limited number of alleles (or K-Value) in the chromosome, each of which represents a single variable in the dataset. The value 0 represents a missing variable for the GA and is ignored by backpropagation. This missing variable to a certain extent allows the GA to find solutions which have fewer variables than the total size of the chromosome. This is a crude, but moderately effective method of achieving multiple K-values within a fixed length chromosome. The general representation can be seen in Figure 2.

Once a set of attributes has been selected by the genetic algorithm, an objective function must be formulated which relates the chosen genes to the dataset. This is achieved through the use of backpropagation which acts directly on the data to develop a set of weights which minimises the difference between the current and observed output. For these experiments with categorical data, the continuous behaviour of the sigmoid function is not required, so a simple step or threshold function is used as the activation function. Backpropagation then minimises the difference between input and output and in doing so, determines the logical relationships between genes and the classification. It then returns the error to the genetic algorithm which the GA then uses to compute fitness. Fitness therefore in the following experiments is determined as 1/(errors) where "errors" is the number of errors incurred by the current gene and weight set over the dataset.

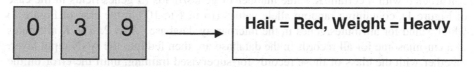

Fig. 2. Genetic algorithm chromosome representation of the dataset. '0' is ignored, '3' is the third attribute (Red Hair) and '9' is the ninth (Heavy Weight) in the total of 11 attributes seen in the dataset in Figure 1 .

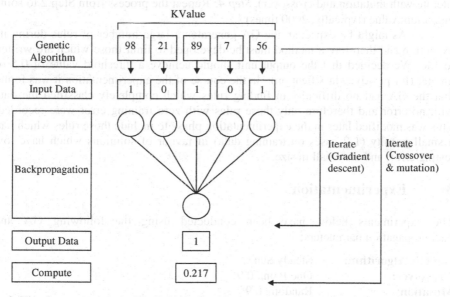

Fig. 3. Visual representation of the hybrid algorithm execution for one output. This is repeated for a population of chromosomes and for all classes in the dataset. Step 1 consists of generating a random population of chromomes with K different fields/'genes'. The chromosome (98 21 89 7 56) is depicted in the Figure (Kvalue=5), where the numbers refer to one of the hundred genes in the microarray data (i.e. genes 7, 21, 56, 89 and 98). Step 2 consists of feeding into the ANN just these five gene values for all records in the database one at a time (values 1, 0, 1, 0, 1 for genes 98, 21, 89, 7 and 56, respectively, are depicted in the Figure) and training the ANN to produce the output class of each record ('1' for 1, 0, 1, 0, 1) until the error at the output node is minimised or until some maximum number of presentations (typically 20 in our experiments) is exceeded. Step 3 is to use the error for each chromosome at the output layer as a fitness function to determine mutation, crossover and selection for the next generation of chromosomes. This process is repeated until a maximum number of generations is exceeded (typically 2000 in our experiments). See the text for further details.

In this neural-genetic hybrid, the genetic algorithm provides the main search engine for combinations of genes, whilst the backpropagation provides a set of weights which allow the attributes to be weighted according to their contribution to the class. Further details of interpreting rules are discussed later. Figure 3 provides an overview of the neural genetic architecture.[2] *Step 1:* A population of chromosome is constructed with a certain K value number of genes/fields (5 genes/fields in the case of Figure 1), containing random allele values (from 1 to 100 in the case of Figure 1) which stand for attributes/genes in the microarray database. *Step 2:* The K attributes of a chromosome for all records in the database are then fed into the ANN input layer, together with the class of those records for supervised training, until the error on the output nodes is as low as possible over the N records (iterate with gradient descent to some maximum value, typically, 20 per evaluation) for each chromosome. *Step 3:* The ANN's output value is used to calculate the fitness of each chromosome and for determining mutation and crossover to generate a new population of chromosomes (iterate with mutation and crossover). *Step 4:* Repeat the process from Step 2 to some maximum value (typically, 2000 times) [3].

As might be expected, the GA generates a large number of rules during its execution and therefore a method must be developed to limit those which are written to file. We decided that the output unit should achieve a threshold value of 0.5 or greater (for positive classification). In the course of the experimentation it was found that the GA had no difficulty in finding rules which completely classified the data with no error and therefore only those rules with zero training error were recorded. This was modified later in the experimentation phase to include those rules which had a small penalty (1-2 cases on training data) in favour of solutions which have low testing error, and are small in size.

3   Experimentation

The experiments below have been conducted using the following GA and backpropagation parameters:

Genetic Algorithm:	Steady State
Crossover:	One Point, 0.9
Mutation:	Random, 0.9
Selector:	Roulette Wheel
Replacer:	Weakest
Population Size:	50
Iterations:	Max 2000 per class

---

[2] In what follows, there may be some confusion between the 'genes' of a chromosome (using GA terminology), and the real genes in the microarray database. We will use 'field' or enclose 'gene' within quotation marks where possible when describing the 'genes' of a chromosome to distinguish them from the genes in the microarray database. Similarly, we will use 'alleles' in GA terminology to refer to the real genes in the microarray database. This possible confusion is an unfortunate side-effect of using genetic algorithms for analysing gene expression data.

[3] Due to the sparse nature of the networks, the entire process takes approximately 10-15 seconds per class on modern PCs.

Activation Function: Step Function – threshold 0.5
Iterations: 20
Beta: 1
K-Value: 4

3.1 Experiment 1 – Sunburn Dataset

This experiment on the toy sunburn dataset (as described in the Method section) is designed to show the effective operation of the algorithm on a traditional data mining dataset. The algorithm was run on the Sunburn dataset with the parameters as described above. The rules are described below:

```
Rule: 1
Lotion = Yes -1                   Rule: 350
Hair = Blonde 1                   Lotion = Yes 1
Height = Tall -1                  Hair = Brown 1
Hair = Red 1                      Height = Tall 1
       ->    Sunburned                   ->    NotSunburned
  0/3                               0/5
```

The interpretation of these rules is as follows. The figure which follows each attribute (e.g. –1 for 'Lotion=Yes') is the discrete weight (ranging from –10 to +10[4]) connecting that attribute to the output class and signifies the classificatory influence between that attribute and the class. Note that in Rule 1, two attributes (Hair=Blonde and Hair=Red) singly (by themselves) achieve the threshold figure of 0.5. Therefore, Rule 1 states that If (Hair = Blonde OR Hair = Red) AND (IF Lotion Used = NOT Yes AND Height = NOT Tall) THEN Sunburned. Similarly, Rule 350 states that if a person uses lotion OR has brown hair OR is tall, then that person is not sunburned. The figures below indicate that the rule achieved 0 misclassifications from 3 in the sunburned class and zero from 5 in the NotSunburned class.

Discussion

The algorithm has discovered the known interactions of Lotion and Hair Colour and describes them by giving a negative weight to the Lotion attribute for Sunburned (Rule 1), but a positive one for NotSunburned (Rule 350). These interactions indicate that the neural-genetic system is discovering the required information from the dataset. A surprising inclusion in the dataset is the Height = Tall element which appears to identify being tall with being not sunburned. This interaction is understandable when the dataset is viewed, as each instance of being tall is accompanied by the classification of NotSunburned. This shows the need to check for the plausibility of candidate rules by experts.

---

[4] Weights in ANNs trained through backpropagation tend to hover around the 0 value (positive and negative). Weights greater than or less than 2 or –2, respectively, therefore signify strong influence.

3.2   Experiment 2 – Leukaemia Dataset

The standard dataset used in many of the experiments regarding classification of gene expression data is that from Golub *et al.* (1999). The dataset details 7129 gene expression values (i.e. 7129 genes or attributes) for a total of 72 (38 training and 34 test) individuals who are suffering either from acute lymphoblastic leukaemia (ALL) or acute myeloid leukaemia (AML). The ratio of individuals to attributes is about 1%, which is hard for traditional data mining techniques to cope with. The data mining algorithm must, by using the gene measurements, differentiate between the two types of leukaemia.

Obviously this represents a considerable step up in size from the toy sunburn problem. As described in Ryu et al (2002) the use of standard feature reduction techniques can have a large effect on the final quality of the rules produced by algorithms and should therefore be avoided.

The neural-genetic algorithm when run on the Leukaemia dataset discovered these rules:

```
Rule: 1101                      Rule: 4688
S81737_s_at = A 1               U88629_at = P 2
U79297_at = P 3                 M23197_at = P 1
U46499_at = P -4                M63138_at = A -1
AF015913_at = P 1               D49728_at = M -3
D10495_at = P -1                     ->    AML
     ->    ALL                   0/11
 0/27                           TestError: 2:14
TestError: 1:20
```

Note that the genes U88629 and U79297 contribute strongly to classification with causal weights of 2 and 3 respectively on their connections with the output nodes. M23197 also contributes to classification in that it separately and singly achieve the threshold of 0.5 on the output node. Again, the training error is shown as the figure directly beneath the rule, but in addition to this, an error is given over a testset of examples. The test set shown here is that which was stipulated by Golub *et al* in the original paper.

As can be seen the test error is small, but not perfect. However this compares favourably with See5, a well-known and used data mining package written by Quinlan (1993) which develops this ruleset which has the same training and testing error:

```
Rule 1: (28/1, lift 1.3)
       L05424_cds2_at = A
       M84526_at = A
    ->   class ALL  [0.933]
```

```
Rule 2:  (7, lift 3.3)
      M84526_at = P
      -> class AML  [0.889]
```

```
Rule 3:  (4, lift 3.1)
      L05424_cds2_at = P
      -> class AML  [0.833]
```

```
Default class: ALL
```

Biological Plausibility

The genes found in AML, U88629 and M23197 have previously been identified as genes associated with myeloid leukaemia. U88629 has been identified in rats as coding for a "human ELL gene, a frequent target for translocations in acute myeloid leukaemia" Shilatifard *et al* (1997). M23197 has been identified as "coding for CD33, a differentiation antigen of myeloid progenitor cells" Simmons *et al* (1988),
The gene U79297 strongly associated with the ALL class is not currently associated with the lymphoblastic leukaemia and therefore this is likely to be a new discovery in this area.

Discussion

The results from this algorithm on this gene expression dataset are encouraging. It discovered a ruleset with the same accuracy as See5, but with a different set of genes. This highlights an attractive property of the algorithm: that it can find information not present in rulesets generated by traditional algorithms such as See5.

3.3   Experiment 3 – Myeloma Dataset

This gene expression classification dataset is taken from Page et al (2002) and consists of 70 gene expression values for each of 104 individuals, of whom 73 suffer from myeloma and the remaining 31 are diagnosed as normal. The task is for the classification algorithm to correctly separate the individuals into the appropriate group. A random test set of 40 individuals was used and the remaining 64 used for training.

The hybrid algorithm was run over this shortened dataset and yielded these rules with an increased K-Value of 5 genes:

```
L18972_at_AC = P -2              M63928_at_AC = P -2
X16416_at_AC = P -3              X16416_at_AC = P 3
X16832_at_AC = P 2               U40490_at_AC = A 1
X57129_at_AC = A 3               M33195_at_AC = P -1
L36033_at_AC = A -3              L36033_at_AC = A 2
      ->    normal                     ->    myeloma
 0/22                             0/42
TestError: 1:9                   TestError: 1:31
```

Note that X16416 has strong negative influence on 'normal' and strong positive causal influence on 'myeloma'. The training and test figures represent an outstanding result for the algorithm, misclassifying only 2 examples over the test data. The rules are larger than those generated by See5, but the error is also much smaller. See5 incurred 6 misclassifications over the training and test data (3 in each) compared to 2 for the neural-genetic algorithm with this ruleset:

```
Rule 1: (31/2, lift 3.0)        Rule 3: (58, lift 1.4)
    L36033_at_AC = P                L36033_at_AC = A
    M63928_at_AC = P            ->  class myeloma   [0.983]
->  class normal   [0.909]
                                Rule 4: (59/1, lift 1.4)
                                    M63928_at_AC = A
Rule 2: (1, lift 2.2)           ->  class myeloma   [0.967]
    L36033_at_AC = M
->  class normal   [0.667]      Default class: myeloma
```

Biological Plausibility

The gene most strongly involved with classifying the normal and myeloma states, according to our hybrid method, was X16416. A search on the web reveals that the gene X16416 is known as a proto-oncogene (Fainstein *et al* 1989). Proto-oncogenes are pre-cursors of oncogenes which require a mutation to become carcinogenic. These carcinogenic oncogenes are then implicated in the uncontrolled replication of cells associated with cancer.

Discussion

The development of a ruleset which has considerably less error than that discovered by See5 is an important discovery in its own right. This is enhanced, however, when the gene discovered by the neural-genetic system, but not by See5, appears to have biological significance as it has been identified as a pre-cursor to cancer.

4   Conclusion

No method has as yet become the standard method to be adopted in the domain of gene expression analysis and modelling, mainly because current approaches, given their dependence on statistical techniques, often fail to distinguish the critical genes which contribute significantly to particular sample displaying the phenotype (class). Associations and correlations are often found (as for instance with clustering algorithms), but such clusters do not identify the importance of specific genes for classification or the specific contribution they make to determining or predicting whether a sample falls within a certain class. Such clusters can also be very large, even if they contain just 1% of genes actually measured on a microarray (e.g. a cluster of 1% of 30,000 genes is still 300 genes). An important consideration when processing this type of data is the complexity the algorithm incurs whilst evaluating these combinations. Even with 7000 genes, the total number of gene combinations is 2^{7000} combinations assuming binary attributes. The running times for See5 and the hybrid algorithm were a maximum of a few seconds on a modern 2.6 GHz PC for the datasets seen here. This lack of complexity is vital in this domain where datasets can consist of up to 30,000 genes. In fact the hybrid algorithm may reduce this time further with code optimisation and a more streamlined interface.

The research described here provides a radically different approach to identifying, out of thousands of gene measurements on a microarray, the relatively small number (typically well below 1%) of genes contributing to a disease or specified organism state/phenotype. Those genes which are potentially the most important in terms of a classification are highlighted through high value weights (positive and negative), allowing biologists to easily interpret information yielded during the knowledge discovery process. In short this algorithm provides a solution to discovering classifications from gene expression data which could prove vital to the medical industry. The ability of doctors to correctly diagnose cancer (both ALL and AML have very similar symptoms but respond differently to treatment) is paramount and the applications to other medical conditions is limited only by the data which can be collected. It is hoped that these computer models represent a step towards achieving these types of medical breakthrough.

References

1. Ando, S., Iba H.,. (2001a) "Inference of Gene Regulatory Model by Genetic Algorithms", Proceedings of Conference on Evolutionary Computation 2001 pp712–719
2. Ando, S., Iba H., (2001b) "The Matrix Modeling of Gene Regulatory Networks -Reverse Engineering by Genetic Algorithms-", Proceedings of Atlantic Symposium on Computational Biology, and Genome Information Systems & Technology 2001.
3. Fainstein E, Einat M, Gokkel E, Marcelle C, Croce CM, Gale RP, Canaani E (1989) "Nucleotide sequence analysis of human abl and bcr-abl cDNAs." Oncogene 1989 Dec;4(12):1477–81
4. Golub, T.R., Slonim D.K., Tamayo P., Huard C., Gaasenbeek M., Mesirov J.P., Coller H., Loh M.L., Downing J.R., .Caligiuri M.R., Bloomfield C.D., Lander, E.S. (1999) "Molecular Classification of Cancer: Class Discovery and Class Prediction by Gene Expression Monitoring" Science Vol 286 pp 531–536
5. Narayanan, A., Keedwell, E. C. and Olsson, B. (2002). Artificial Intelligence techniques for Bioinformatics. Paper submitted to *Applied Bioinformatics* and available from http://www.dcs.ex.ac.uk/~anarayan/research.htm.

6. Page, D., Zhan, F., Cussens, J., Waddell, W., Hardin,J., Barlogie,B., Shaughnessy, J. Comparative Data Mining for Microarrays: A Case Study Based on Multiple Myeloma. Poster presentation at International Conference on Intelligent Systems for Molecular Biology August 3–7, Edmonton, Canada.   Technical report available from mwaddell@biostat.wisc.edu
7. Quinlan, J.R. (1993) C4.5: Programs for Machine Learning   Morgan Kaufmann Publishers.
8. Ryu, J., Sung-Bae, C., (2002) "Gene expression classification using optimal feature/classifier ensemble with negative correlation" Proceedings of the International Joint Conference on Neural Networks (IJCNN'02), Honolulu, Hawaii, pp 198–203, ISBN 0-7803-7279-4
9. Shilatifard A, Duan DR, Haque D, Florence C, Schubach WH, Conaway JW, Conaway RC. (1997) "ELL2, a new member of an ELL family of RNA polymerase II elongation factors." Proc Natl Acad Sci U S A 1997 Apr 15;94(8):3639–43
10. Simmons D, Seed B. (1988) "Isolation of a cDNA encoding CD33, a differentiation antigen of myeloid progenitor cells." Journal of Immunology 1988 Oct 15;141(8):2797–800
11. Su, T., Basu, M., Toure, A., (2002) "Multi-Domain Gating Network for Classification of Cancer Cells using Gene Expression Data" Proceedings of the International Joint Conference on Neural Networks (IJCNN'02), Honolulu, Hawaii, pp 286–289, ISBN 0-7803-7279-4
12. Winston, P. H. (1992). Artificial Intelligence (3rd Edition). Addison Wesley.
13. Xu R., Anagnostopoulos G., Wunsch II D.C., "Tissue Classification Through Analysis of Gene Expression Data Using A New Family of ART Architectures", Proceedings on International Joint Conference on Neural Networks, Honolulu, Hawaii, Vol. 1, pp. 300–304, May 2002.

Comparison of AdaBoost and Genetic Programming for Combining Neural Networks for Drug Discovery

W.B. Langdon[1], S.J. Barrett[1], and B.F. Buxton[2]

[1] Data Exploration Sciences, GlaxoSmithKline, Research and Development,
Greenford, Middlesex, UK
[2] Computer Science, University College, Gower Street, London, WC1E 6BT, UK
http://www.cs.ucl.ac.uk/staff/W.Langdon, /staff/B.Buxton

Abstract. Genetic programming (GP) based data fusion and AdaBoost can both improve *in vitro* prediction of Cytochrome P450 activity by combining artificial neural networks (ANN). Pharmaceutical drug design data provided by high throughput screening (HTS) is used to train many base ANN classifiers. In data mining (KDD) we must avoid over fitting. The ensembles do extrapolate from the training data to other unseen molecules. I.e. they predict inhibition of a P450 enzyme by compounds unlike the chemicals used to train them. Thus the models might provide *in silico* screens of virtual chemicals as well as physical ones from Glaxo SmithKline (GSK)'s cheminformatics database. The receiver operating characteristics (ROC) of boosted and evolved ensemble are given.

1 Introduction

Pharmaceuticals discovery research has evolved to the point of critical dependence upon computerised systems, databases and newer disciplines related to biological and chemical information processing and analysis. For instance, bioinformatics has enabled the discovery and characterisation of many more potential disease-related biological targets for screening, whilst cheminformatics concerns the capture and processing of chemical and biological information required to manage and optimise the overall screening process and to support decision-making for chemical lead selection. Machine learning can contribute to discovery processes in a variety of ways:

1. High-Throughput Screening (HTS), and increasingly Ultra-HTS, are resource-intensive. Models developed from diverse sets of previously-screened molecules may lead, via *in silico* screening, to smaller (or prioritised) screening-sets, targeted to specific biological activities.
2. Better ways to choose which of the vast numbers of "virtual molecules" should be synthesis and included in new chemical libraries.
3. After activity has been confirmed in primary screening, there are many additional tests of key properties which are required before forwarding compounds towards development. Traditionally, due to expense or the inability to screen the initialy vast numbers of molecules available, these have

S. Cagnoni et al. (Eds.): EvoWorkshops 2003, LNCS 2611, pp. 87–98, 2003.

been conducted on an *as needed* basis. This is expensive. Time and effort could be saved by computer based screening out of unsuitable chemicals earlier. Machine learning could produce predictions for other key properties such as target-selectivity, toxicity, tissue-permeability, solubility and drug metabolism (i.e. by P450 enzymes).

Here and in some other domains machine learning techniques based on a single paradigm have not been sufficient and so researchers have investigated mechanisms for combining them [Kittler and Roli, 2001; Gunatilaka and Baertlein, 2001]. Existing classifier fusion techniques, such as committees of experts [Jacobs *et al.*, 1991], bagging [Breiman, 1996] and boosting [Freund and Schapire, 1996], typically combine experts of the same type using a fixed way of combining their predictions. E.g. all the experts might be feed forward neural networks whose outputs are: simply summed, a weighted sum might be calculated, or a majority vote taken, to give the collective view of the classifier ensemble. That is, the fusion technique optimises the individual experts (e.g. using back propagation) while keeping the combination rule fixed. Genetic programming offers an alternative, which is to pre-train the experts and optimise the non-linear combination rule. Binary GAs have been used to find good committees of experts [Opitz and Shavlik, 1996; Kupinski and Anastasio, 1999; Kupinski *et al.*, 2000]. However genetic programming gives us the ability not only of deciding which experts to use in the ensemble but also how their predictions are to be combined. I.e. to simultaneously solve both the feature selection problem (at the individual expert level of granularity) and the combination rule. Because the individual experts are pretrained the GP does not need to know how they were trained and so has the ability to form superior ensembles of heterogeneous classifiers [Langdon and Buxton, 2001b; Langdon *et al.*, 2002].

Intelligent classification techniques, such as artificial neural networks (ANN), have had some success at predicting potential drug activity and we have shown genetic programming is able to fuse different neural networks to obtain better predictions [Langdon *et al.*, 2001]. We shall further demonstrate our system and also compare results with a popular boosting technique, AdaBoost [Schwenk and Bengio, 2000].

2   Receiver Operating Characteristics

The Receiver Operating Characteristics (ROC) of a classifier provide a helpful way of illustrating the trade off it makes between catching positive examples and raising false alarms [Swets *et al.*, 2000]. Figures 4 and 5 show ROC curves.

[Scott *et al.*, 1998] suggest the "Maximum Realisable Receiver Operating Characteristics" for a combination of classifiers is the convex hull of their individual ROCs, cf. also [Provost and Fawcett, 2001]. ("Lotteries" in game theory [Binmore, 1990] are somewhat similar.) However we have already shown GP can in some cases do better, including on Scott's own benchmarks [Langdon and Buxton, 2001a] and real world pharmaceutical classification tasks [Langdon *et al.*, 2001].

3 The Pharmaceutical Data

The pharmaceutical data are similar to [Langdon *et al.*, 2001] and [Langdon *et al.*, 2002]. Thousands of chemicals from a chemical library have been tested (using 2 triplicated HTS runs) to see if they inhibit one of the P450 enzymes involved in metabolism. This is an important screen in early drug discovery since P450 inhibition could be expected to lead to problems (when a compound is first evaluated in humans).

The chemicals are a very diverse set, covering the most common types of drug or drug-like compounds, such as would be found in the big pharmaceutical company compound banks. Hence they probably have a range of inhibition mechanisms. Some "primary" enzyme inhibition mechanisms are likely to be much more frequent within the tested set of compounds than others. This is precisely the kind of situation which can defeat individual classifiers.

Chemicals which gave inconsistent screening results (i.e. more than 15% variation between readings) were discarded. The mean of the 6 measurements taken was compared against an activity threshold. Those below the threshold are said to be inactive, while chemicals whose mean exceeded the threshold were classified as inhibitory, i.e. active against the P450 target.

These noise free chemicals were then hierarchically clustered using Ward's linkage in combination with Tanimoto similarity, computed from Daylight 2 Kbit string chemical fingerprint data[1]. Clusters were defined at 0.8 tan. Note unlike our previous work, active and inactive were not separated prior to clustering. This leads to three types of cluster. 1) Pure clusters, i.e. clusters containing either all active or all inactive compounds. 2) Impure, or mixed, clusters. 3) Singleton clusters, i.e. clusters consisting of a single chemical compound.

A total of 699, numerical and categorical, chemical features from a diverse array of families (electronic, structural, topological/shape, physico-chemical, etc.) were computed for each chemical, starting from a SMILES[2] representation of it's primary chemical structure (2-d chemical formula).

The chemicals selected for screening are naturally a highly biased sample. It consists of those chemicals which were considered interesting and necessarily were available. There are two things we would like to know about any predictive classifier; how well it will work on chemicals like those on which it was trained and secondly (and much more difficult), how well will it extrapolate outside the training domain. Naturally we know the distribution of the first but, we do not know the distribution of the second.

The classifiers are trained on chemicals selected from the clean data (i.e. compounds with little HTS noise) at random, ensuring the training data has the same proportion of inhibitory and inactive chemicals and the same proportions of pure, mixed and singleton clusters, as the the whole dataset. The generalisation performance of classifiers is estimated by measuring how well it predicts unseen chemicals, drawn at random from the same distribution.

---

[1] http://www.daylight.com/dayhtml/doc/theory/theory.finger.html
[2] http://www.daylight.com/dayhtml/smiles/

We keep back a number of the chemicals from the singleton clusters. Since these are likely to be the most different from the training set. We use the classifiers' performance on this "extrapolation set" to estimate how well they will perform on novel chemicals.

4   Training the Neural Networks

As before, the 699 features were divided into 15 functionally related groups of about 50 each. Clementine was used to train 5 feed forward neural networks on each of the 15 groups of features. Clementine default parameters (including a 50/50 training set/stop set split, to enable early stopping in order to limit over fitting) were used. (Following disappointing performance (due to over fitting) with C4.5 decision trees we decided to only to use neural networks.)

An imbalance between positives and negatives is common in many data mining tasks. However many machine learning techniques work best when trained on "balanced data sets", i.e. data sets containing an equal mix of inhibitory and inactive examples. [Chawla et al., 2002]. The 1299 compounds were used to create five data sets. Each contained the same 279 inhibitory chemicals and approximately 200 different inactive chemicals. That is, each data set was approximately balanced. Each neural network was trained on one of the 15 groups of attributes selected from one of the five balanced data sets. Making a total of 75 weak classifiers.

5   Genetic Programming Configuration

5.1   Function and Terminal Sets

The genetic programming data fusion system is deliberately identical (except for the choice of random constants) to that described in [Langdon and Buxton, 2001c], cf. Table 1.

In order to use the neural networks within GP they are presented to GP as 75 problem specific functions. Each returns the classification and associated confidence of the corresponding neural network for the current chemical. The Clementine neural networks yield a single floating point number (between 0 and 1). Values below 0.5 are treated as class 0, while those above 0.5 are regarded as predicting class 1. We treat the magnitude of the difference from 0.5 as the indicating the network's confidence.

Normally the output of a neural network is converted into a binary classification (i.e. the chemical is inhibitory or is inactive) by testing to see if the value is greater or less than 0.5. This gives a single point in the ROC square. However by replacing the fixed value of 0.5 by a tunable threshold $(0 \ldots 1)$ we can vary this trade off, so that we get a complete curve in the ROC square. By making the threshold the argument of the function we leave the choice of suitable operating point to the GP. These arguments are treated like any other by the GP and so can be any valid arithmetic operation, including the base classifiers

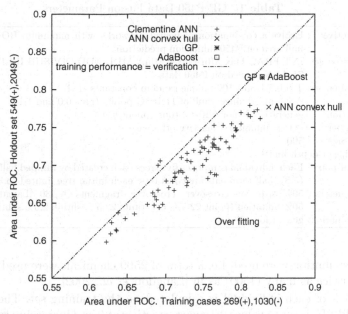

Fig. 1. Area under ROC curve of P450 Clementine neural networks. Points below the diagonal indicate possible over training. Their convex hull is indicated, as are the corresponding points for the boosted and evolved classifiers (which lie almost on top of each other).

themselves. Note GP simultaneously adapts how the non-linear combination of the base classifiers and their operating points.

The terminals or leaves of the trees being evolved by the GP are either constants or the adjustable threshold "T" (see Table 1).

5.2 GP Representation

We continue to create each individual in the initial population with five random trees [Jacobs *et al.*, 1991; Soule, 1999; Langdon, 1998]. Each tree within an individual returns a signed real number. The classification of the individual is the sum of the answers given by the five trees. Note the GP can combine the supplied classifiers in an almost totally arbitrary, non-linear way.

Following [Angeline, 1998] and others, we use a high mutation rate and a mixture of different mutation operators. To avoid bloat, we also use size fair crossover [Langdon, 2000], see Table 1.

5.3 GP Fitness Function

Unlike in previous work and in an effort to reduce over fitting, the chemicals used to train the base level classifiers (ANN) were not used to train the GP. Instead approximately the same number (1300) of chemicals, drawn from the

Table 1. GP P450 Data Fusion Parameters

Objective:	Evolve a combination of neural networks with maximum ROC convex hull area on P450 inhibition prediction
Function set:	INT FRAC Max Min MaxA MinA MUL ADD DIV SUB IFLTE 75 ANN trained on P450 data
Terminal set:	T 0 0.5 1 plus 100 unique random constants -1..1
Fitness:	Area under convex hull of 11 ROC points (plus 0,0 and 1,1)
Selection:	generational (non elitist), tournament size 7
Wrapper:	$\geq 0 \Rightarrow$ inhibitory, inactive otherwise
Pop Size:	500
No size or depth limits	
Initial pop:	Each individual comprises five trees each created by ramped half-and-half (5:8) (half terminals are constants, each initial tree limited to 300)
Parameters:	50% size fair crossover, crossover fragments $\leq$ 30 [Langdon, 2000] 50% mutation (point 22.5%, constants 22.5%, shrink 2.5% subtree 2.5%)
Termination:	generation 10

same distribution were used. I.e. a total of 2599 chemicals were used in training. Almost twice as many (1500) as in [Langdon et al., 2002].

Fitness of each individual is calculated on the training set. The adjustable threshold "T" is set to values 0.1 apart, starting at 0 and increasing to 1. For each setting, the evolved program is used to predict the activity of each chemical in the training set and true positive (TP) and false positive (FP) rates are calculated. Each TP,FP pair gives a point on an ROC curve. The fitness of the classifier is the area under the convex hull of these (plus the fixed points 0,0 and 1,1).

6   Forming a Composite Using AdaBoost

Boosting [Freund and Schapire, 1996] is a deterministic algorithm whereby a series of weak (i.e. poorly performing) but different classifiers are trained. A single composite classifier is formed by the weighted sum of the individual classifiers. The composite should be better than the initial weak classifier.

At each boosting round, a new classifier is trained on weighted training data. At each round the classifier produced is different because at each round the weight associated with each training example is adjusted. The weights of examples on which the last trained classifier did badly are increased. The idea is to encourage the next classifier to be trained to pay more attention to hard examples. Cf. "Dynamic Subset Selection" [Gathercole and Ross, 1994]. As well as being used to adjust the weights of training examples, the performance of each weak classifier is used to determine the strength of its contribution to the final composite classifier.

Note AdaBoost uses the accuracy of the classifier, rather than its ROC, and so implicitly assumes false positives and false negatives have equal costs.

6.1 Matlab Neural Networks Ensembles

The Matlab neural network tool box was used to train single hidden layer, fully connected feed forward networks. Each input unit is connected to one of the networks trained by Clementine (75). The output layer consists of two output units (one for each class). In preliminary experiments, the number of units in the hidden layer was set to 20 and then progressively reduced. Little performance variation was seen until it was reduced to a single node. In the hope of avoiding over fitting and reducing run time the hidden layer was minimised to just two units.

Approximately half the training data was used by Matlab as a stopping set. The training and stop sets do not contain the same chemicals. Matlab used between 7 and 55 training epochs (see Fig. 2).

The default parameters provided by Matlab were used (e.g. back propagation, minimise the sum of square of the difference between output units and the actual class and momentum constant of 0.95). Due to randomised initialisation training a network on the same data will not necessarily produce the same final network each time.

6.2 Boosting Matlab Neural Networks

The first neural network is trained using all the training data (half is used for the stopping set) and initialy each training chemical has the same weight. After the first network has been successfully trained the weights are adjusted, using the AdaBoost algorithm [Schwenk and Bengio, 2000]. At each subsequent iteration, a new training set is created by randomly sampling from training examples in proportion to their weights. The training set used by Matlab is the same size on each boosting cycle. Each time the training set is split approximately equally into actual training and stopping sets.

[Schwenk and Bengio, 2000] suggests boosting should stop once the error exceeds 50%. However, in this application, low performing networks were produced very early in the boosting process, even in the first round. We discard such networks, create a new random sample of training chemicals using the current distribution of weights and then train again. Boosting is stopped after training 20 networks (successful or otherwise). The composite is based only on the successful networks. Approximately 10% of trained networks have to be ignored.

AdaBoost specifies that the class with the highest (weighted) vote of all the weak classifiers will be taken as the predicted class. To produce a complete ROC curve, we use the normalised difference between the weighted votes for each output neuron (i.e. class) as the ensemble's confidence.

7 Results

Figure 3 plots the evolution of fitness (on the GP training set). For the best in the population, the area under ROC on the verification set was also measured. In

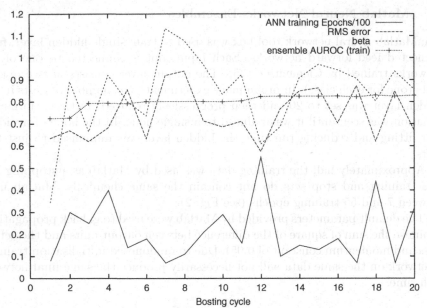

Fig. 2. Performance of individual Matlab neural networks (RMS training error and the AdaBoost weighted pseudoloss function, beta) and of complete AdaBoost ensemble AUROC during boosting. Note beta = error/(1-error) does not converge to zero.

contrast to previous work, the gap in performance on the training and verification sets is modest and grows only slowly. There are two causes for this.

Firstly GP is stopped after generation 10. Earlier work indicated that on problems of this type most of the useful learning was accomplished at the start of the run and most of the apparent performance gain achieved later was illusory, caused by over fitting. (Stopping early also has the advantage of shortening run time to 27 minutes.)

The second potential cause for reduced over fitting, is that the GP is now trained on data that the lower level classifiers have not seen. This appears to have had the beneficial affect of countering any lack of generality which has slipped into the neural networks as they were being trained (see Fig. 1). However the evidence for this is not overwhelming. Certainly in Fig. 3 we see strong correlation between the performance of different GP individuals on the different datasets and so we could reach the conclusion that some parts of the P450 HTS data are simply harder than others.

Figure 4 shows the classifiers produced by AdaBoost and genetic programming have essentially the same performance. (Earlier experiments indicated a small, but significant advantage for a classifier evolved by GP.) However they are significantly better than, not only each classifier in their function set, but also the convex hull of these base classifiers

The use of size fair crossover [Langdon, 2000] and mutation means we do not see explosive increase in program size (bloat [Langdon et al., 1999]) and preliminary experiments suggest over fitting is more closely related to number of

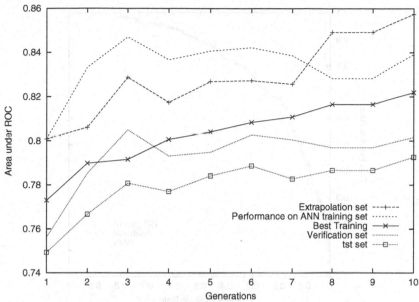

Fig. 3. Evolution of performance of best of generation evolved classifier on other datasets.

generations over which the population has been exposed to the same environment than to the size of the programs.

While earlier experiments using different training data for the base and evolved classifiers were not encouraging, the results for both GP and boosting indicate that using separate chemicals to train base level classifiers and composites of them can indeed be successful.

The performance of the composite classifiers on an "extrapolation" set (cf. Fig. 5) is good. This is encouraging, since we really wish to make predictions for untested chemicals.

8   Conclusions

AdaBoost [Schwenk and Bengio, 2000], like genetic programming (GP), can be used to form non-linear combinations of lower level classifiers. Again like, GP, these exceed Scott's "Maximum Realisable Receiver Operating Characteristics" [Scott *et al.*, 1998] on a non-trivial industrial problem (cf. Figs. 1 and 4).

It is especially encouraging that both methods of automatically forming classifier ensembles are able to extrapolate away from their training data and make predictions on new chemicals, cf. Fig. 5.

Acknowledgements. We would like to thank Sunny Hung, George Seibel, David Corney and Matthew Trotter.

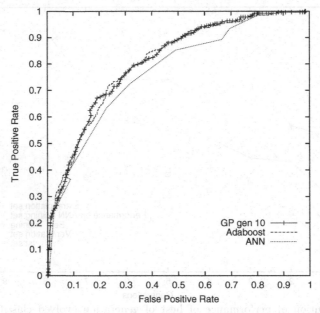

Fig. 4. Receiver Operating Characteristics of evolved and boosted composite classifiers (1298 verification chemicals). In both cases the classifier lies outside the convex hull of their base classifiers (lines without crosses). Note the convex hull classifiers are determined on the training set and so need no longer be convex when used to classify the holdout data.

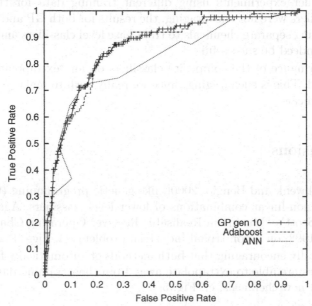

Fig. 5. Receiver Operating Characteristics of evolved and boosted composite classifiers on extrapolation set (779 singleton chemicals).

Source Code

C++ and AdaBoost.M2 Matlab source code can be obtained from
ftp://cs. ucl.ac.uk/genetic/gp-code/ and
ftp://cs.ucl.ac.uk/genetic/boosting/ respectively.

References

Angeline, 1998. Peter J. Angeline. Multiple interacting programs: A representation for evolving complex behaviors. *Cybernetics and Systems*, 29(8):779–806, November 1998.

Binmore, 1990. Ken Binmore. *Fun and Games*. D. C. Heath, Lexington, MA, USA, 1990.

Breiman, 1996. Leo Breiman. Bagging predictors. *Machine Learning*, 24:123–140, 1996.

Chawla *et al.*, 2002. N. V. Chawla, K. W. Bowyer, L. O. Hall, and W. P. Kegelmeyer. SMOTE: synthetic minority over-sampling technique. *Journal of Artificial Intelligence Research*, 16:321–357, 2002.

Freund and Schapire, 1996. Yoav Freund and Robert E. Schapire. Experiments with a new boosting algorithm. In *Machine Learning: Proceedings of the thirteenth International Conference*, pages 148–156. Morgan Kaufmann, 1996.

Gathercole and Ross, 1994. Chris Gathercole and Peter Ross. Dynamic training subset selection for supervised learning in genetic programming. In Yuval Davidor, Hans-Paul Schwefel, and Reinhard Männer, editors, *Parallel Problem Solving from Nature III*, volume 866 of *LNCS*, pages 312–321, Jerusalem, 9-14 October 1994. Springer-Verlag.

Gunatilaka and Baertlein, 2001. Ajith H. Gunatilaka and Brian A. Baertlein. Feature-level and decision level fusion of noncoincidently sampled sensors for land mine detection. *IEEE Transactions on Pattern Analysis and Machine Intelligence*, 23(6):577–589, June 2001.

Jacobs *et al.*, 1991. Robert A. Jacobs, Michael I. Jordon, Steven J. Nowlan, and Geoffrey E. Hinton. Adaptive mixtures of local experts. *Neural Computation*, 3:79–87, 1991.

Jones, 1998. Gareth Jones. Genetic and evolutionary algorithms. In Paul von Rague, editor, *Encyclopedia of Computational Chemistry*. John Wiley and Sons, 1998.

Kittler and Roli, 2001. Josef Kittler and Fabio Roli, editors. *Second International Conference on Multiple Classifier Systems*, volume 2096 of *LNCS*, Cambridge, 2-4 July 2001. Springer Verlag.

Kordon and Smits, 2001. Arthur K. Kordon and Guido F. Smits. Soft sensor development using genetic programming. In Lee Spector *et al.*, editors, *Proceedings of the Genetic and Evolutionary Computation Conference (GECCO-2001)*, pages 1346–1351, San Francisco, California, USA, 7-11 July 2001. Morgan Kaufmann.

Kupinski and Anastasio, 1999. M. A. Kupinski and M. A. Anastasio. Multiobjective genetic optimization of diagnostic classifiers with implications for generating receiver operating characteristic curves. *IEEE Transactions on Medical Imaging*, 18(8):675–685, Aug 1999.

Kupinski *et al.*, 2000. Matthew A. Kupinski, Mark A. Anastasio, and Maryellem L. Giger. Multiobjective genetic optimization of diagnostic classifiers used in the computerized detection of mass lesions in mammography. In Kenneth M. Hanson, editor, *SPIE Medical Imaging Conference*, volume 3979, San Diego, California, 2000.

Langdon and Buxton, 2001a. W. B. Langdon and B. F. Buxton. Genetic programming for combining classifiers. In Lee Spector *et al.*, editors, *Proceedings of the Genetic and Evolutionary Computation Conference (GECCO-2001)*, pages 66–73, San Francisco, California, USA, 7-11 July 2001. Morgan Kaufmann.

Langdon and Buxton, 2001b. W. B. Langdon and B. F. Buxton. Genetic programming for improved receiver operating characteristics. In Josef Kittler and Fabio Roli, editors, *Second International Conference on Multiple Classifier System*, volume 2096 of *LNCS*, pages 68–77, Cambridge, 2-4 July 2001. Springer Verlag.

Langdon and Buxton, 2001c. William B. Langdon and Bernard F. Buxton. Evolving receiver operating characteristics for data fusion. In Julian F. Miller *et al.*, editors, *Genetic Programming, Proceedings of EuroGP'2001*, volume 2038 of *LNCS*, pages 87–96, Lake Como, Italy, 18-20 April 2001. Springer-Verlag.

Langdon *et al.*, 1999. William B. Langdon, Terry Soule, Riccardo Poli, and James A. Foster. The evolution of size and shape. In Lee Spector, William B. Langdon, Una-May O'Reilly, and Peter J. Angeline, editors, *Advances in Genetic Programming 3*, chapter 8, pages 163–190. MIT Press, 1999.

Langdon *et al.*, 2001. W. B. Langdon, S. J. Barrett, and B. F. Buxton. Genetic programming for combining neural networks for drug discovery. In Rajkumar Roy *et al.*, editors, *Soft Computing and Industry Recent Applications*, pages 597–608. Springer-Verlag, 10–24 September 2001. Published 2002.

Langdon *et al.*, 2002. William B. Langdon, S. J. Barrett, and B. F. Buxton. Combining decision trees and neural networks for drug discovery. In James A. Foster *et al.*, editors, *Genetic Programming, Proceedings of the 5th European Conference, EuroGP 2002*, volume 2278 of *LNCS*, pages 60–70, Kinsale, Ireland, 3-5 April 2002. Springer-Verlag.

Langdon, 1998. William B. Langdon. *Genetic Programming and Data Structures*. Kluwer, 1998.

Langdon, 2000. William B. Langdon. Size fair and homologous tree genetic programming crossovers. *Genetic Programming and Evolvable Machines*, 1(1/2):95–119, April 2000.

Opitz and Shavlik, 1996. David W. Opitz and Jude W. Shavlik. Actively searching for an effective neural-network ensemble. *Connection Science*, 8(3-4):337–353, 1996.

Provost and Fawcett, 2001. Foster Provost and Tom Fawcett. Robust classification for imprecise environments. *Machine Learning*, 42(3):203–231, March 2001.

Schwenk and Bengio, 2000. Holger Schwenk and Yoshua Bengio. Boosting neural networks. *Neural Computation*, 12(8):1869–1887, 2000.

Scott *et al.*, 1998. M. J. J. Scott, M. Niranjan, and R. W. Prager. Realisable classifiers: Improving operating performance on variable cost problems. In Paul H. Lewis and Mark S. Nixon, editors, *Proceedings of the Ninth British Machine Vision Conference*, volume 1, pages 304–315, University of Southampton, UK, 14-17 September 1998.

Soule, 1999. Terence Soule. Voting teams: A cooperative approach to non-typical problems using genetic programming. In Wolfgang Banzhaf *et al.*, editors, *Proceedings of the Genetic and Evolutionary Computation Conference*, volume 1, pages 916–922, Orlando, Florida, USA, 13-17 July 1999. Morgan Kaufmann.

Swets *et al.*, 2000. John A. Swets, Robyn M. Dawes, and John Monahan. Better decisions through science. *Scientific American*, 283(4):70–75, October 2000.

Turney, 1995. Peter D. Turney. Cost-sensitive classification: Empirical evaluation of a hybrid genetic decision tree induction algorithm. *Journal of Artificial Intelligence Research*, 2:369–409, 1995.

Cross Validation Consistency for the Assessment of Genetic Programming Results in Microarray Studies

Jason H. Moore

Program in Human Genetics, Department of Molecular Physiology and Biophysics, 519
Light Hall, Vanderbilt University, Nashville, TN, USA 37232-0700
Moore@phg.mc.Vanderbilt.edu

Abstract. DNA microarray technology has made it possible to measure the expression levels of thousands of genes simultaneously in a particular cell or tissue. The challenge for computational biologists and bioinformaticists will be to develop methods that are able to identify subsets of gene expression variables and features that classify cells and tissues into meaningful biological and clinical groups. Genetic programming (GP) has emerged as a machine learning tool for variable and feature selection in microarray data analysis. However, a limitation of GP is a lack of cross validation strategies for the assessment of GP results. This is partly due to the inherent complexity of GP due to its stochastic properties. Here, we introduce and review cross validation consistency (CVC) as a new modeling strategy for use with GP. We review the application of CVC to symbolic discriminant analysis (SDA), a GP-based analytical strategy for mining gene expression patterns in DNA microarray data.

1 Introduction

Biology and biomedicine are in the midst of an information explosion and an understanding implosion. Our ability to generate biological data is far outpacing our ability to understand and interpret its relationship with biological and clinical endpoints. At the heart of this duality are new technologies such as DNA microarrays [1], the serial analysis of gene expression (SAGE) [2], and protein mass spectrometry [3]. Each of these technologies has made it cost-effective and efficient to measure the relative expression levels of thousands of different genes in cells and tissues. The availability of massive amounts of gene expression information afforded by such technologies presents certain statistical and computational challenges to those hoping to use this information to improve our understanding of the initiation, progression, and severity of human diseases.

The first challenge that needs to be addressed is the selection of optimal subsets of gene expression variables from among as many as 10,000 or more candidates. For

S. Cagnoni et al. (Eds.): EvoWorkshops 2003, LNCS 2611, pp. 99–106, 2003.
© Springer-Verlag Berlin Heidelberg 2003

example, with 10,000 variables there are approximately $5 * 10^7$ possible subsets of size two, $1.7 * 10^{11}$ possible subsets of size three, and $4.2 * 10^{14}$ possible subsets of size four. Clearly, the combinatorial magnitude of variable selection precludes an exhaustive search of all possible variable subsets. One solution to this problem is to apply a filter by selecting subsets of variables that each have a statistically significant independent main or marginal effect as evaluated by a univariate statistic such as the Wilcoxon rank sum test [4]. Although the filter approach may reduce the size of the search space, it has the potential of missing gene expression variables whose effects are solely or partially through interactions. This is a well-characterized problem in data mining [5] and in genomics and genetics [6-8].

The second challenge that needs to be addressed is the selection of features of the gene expression variables that can be used to accurately classify and predict biological and clinical endpoints. Current statistical and computational methods such as linear discriminant analysis are simplistic and inadequate. Linear discriminant analysis [9] is a multivariate statistical classification procedure that linearly combines measurements on multiple explanatory variables into a single value or discriminant score that can be used to classify observations. This method is popular because there is a solid theoretical foundation [10] and it is easy to implement and interpret [11]. However, an important limitation is that the linear discriminant functions need to be pre-specified and only the coefficients for each linear predictor are estimated from the data. This limitation is not unique to linear discriminant analysis. Linear, polynomial, and logistic regression also require a pre-specified model [12]. In addition to requiring a pre-specified model, these methods tend to have low power for identifying interactions [6-8].

Symbolic discriminant analysis (SDA) was developed in response to these two challenges [13]. With SDA, both variable and feature selection are carried out with the aid of parallel genetic programming [14, 15]. Here, variables are selected from the complete list of all variables and are combined with optimal mathematical operators to build symbolic discriminant functions that may be linear or nonlinear. An important component of the SDA method is a general cross validation framework that allows prediction error to be estimated. However, due to the stochastic nature of genetic programming (GP), it is common to identify different models within each cross validation split of the data. This makes both hypothesis testing and the identification of a single 'best' model difficult. In response to this challenge, Moore et al. [16] developed cross validation consistency (CVC) as a metric that can be used to facilitate hypothesis testing and model selection with the SDA approach. This general approach can be used for any GP-based modeling procedure in a cross validation framework. We begin with a review of linear discriminant analysis in Section 2. In section 3, we review the SDA approach and its application to real microarray data from a human leukemia study. In Section 4, we review the CVC approach to SDA and its application to real microarray data from a human autoimmune disease study. We conclude in Section 5 with a discussion of the new CVC approach and its general application to GP-based modeling strategies.

2 A Review of Linear Discriminant Analysis

Sir Ronald Fisher developed linear discriminant analysis (LDA) as tool for classifying observations using information about multiple variables [9]. Consider the case in which there are two groups with n_1 and n_2 observations and k variables measured per observation. Fisher suggested forming linear combinations of measurements from multiple variables to generate a linear discriminant score (l) that takes the form

$$l_{ij} = \alpha_1 x_{ij1} + \alpha_2 x_{ij2} + \ldots + \alpha_k x_{ijk} \tag{1}$$

for the ith group and the jth observation in that group where each α is a coefficient and each x is an explanatory variable (e.g. gene expression variable). The goal of LDA is to find a linear combination of explanatory variables and coefficient estimates such that the difference between the distributions of linear discriminant scores for each group is maximized.

Classification of observations into one of the two groups requires a decision rule that is based on the linear discriminant score. For example, if $l_{ij} > l_o$ then assign the observation to one group and if $l_{ij} \leq l_o$ then assign the observation to the other group. When the prior probability that an observation belongs to one group is equal to the probability that it belongs to the other group, l_o can be defined as the median of the linear discriminant scores for both groups. When the prior probabilities are not equal, l_o is adjusted appropriately. Using this decision rule, the classification error for a particular discriminant function can be estimated from the observed data. When combined with cross-validation or independent data, the prediction error can be estimated as well.

3 An Overview of Symbolic Discriminant Analysis (SDA)

An obvious limitation of LDA is the need to pre-specify the linear discriminant function. Additionally, optimal classification of observations into groups may not be possible with a linear combination of explanatory variables. To address these limitations, Moore et al. [13] developed SDA for automatically identifying the optimal functional form and coefficients of discriminant functions that may be linear or nonlinear. This is accomplished by providing a list of mathematical functions and a list of explanatory variables that can be used to build discriminant scores. Similar to symbolic regression [14], genetic programming (GP) is utilized to perform a parallel search for a combination of functions and variables that optimally discriminate between two endpoint groups. GP permits the automatic discovery of symbolic discriminant functions that can take any form defined by the mathematical operators provided. GP builds symbolic discriminant functions using binary expression trees. Each binary expression tree has a mathematical function at the root node and each additional node. Terminals in the binary expression tree are comprised of variables and constants. The primary advantage of this approach is that the functional form of the statistical model does not need to be pre-specified. This is important for the identification of combinations of expressed genes whose relationship with the endpoint of interest may be non-additive or nonlinear [6-8].

In its first implementation, SDA used leave one out cross-validation (LOOCV) to estimate the classification and prediction error of SDA models [13]. With LOOCV, each subject is systematically left out of the SDA analysis as an independent data point (i.e. the testing set) used to assess the predictive accuracy of the SDA model. Thus, SDA is run on a subset of the data (i.e. the training set) comprised of n-1 subjects. The model that classifies subjects in the training set with minimum error is selected and then used to predict the group membership of the single independent testing subject. This is repeated for each of the possible training sets yielding n SDA models. Moore et al. [13] selected LOOCV because it is an unbiased estimator of model error [17]. However, it should be noted that LOOCV may have a large variance due to similarity of the training datasets [17, 18] and the relatively small sample sizes that are common with microarray experiments. It is possible to reduce the variance using perhaps five-fold or 10-fold cross-validation. However, these procedures may lead to biased estimates and may not be practical when the sample size is small [17].

The first application of SDA [13] was to a human leukemia dataset described by Golub et al. [19]. This dataset was selected because previous class prediction methods had been applied with marginal success and two independent datasets were available for analysis. This presented the opportunity to develop symbolic discriminant functions using LOOCV in the first dataset and then validate the predictive ability of these models in the second dataset. The first dataset consisted of 38 acute myeloid leukemia (AML) and acute lymphoblastic leukemia (ALL) samples and was used to develop symbolic discriminant functions of the gene expression variables. The second dataset consisted of an independent set of 34 AML and ALL samples. Approximately 7100 gene expression variables were available for both datasets. For each 37/38 of the data from the first set, Moore et al. [13] ran the parallel GP and selected the resulting models that minimized the misclassification rate and correctly predicted the class membership of the single observation left out of the training step. Optimal symbolic discriminant functions were then evaluated for their predictive ability using the independent dataset consisting of 34 AML and ALL samples. All of the approximately 7100 gene expression variables were available for possible inclusion in a model.

Moore et al. [13] identified two 'near-perfect' symbolic discriminant functions that correctly classified 38 out of 38 (100%) leukemia samples in the training dataset and correctly predicted 33 out of 34 (97.1%) leukemia samples in the independent second dataset. The first near-perfect symbolic discriminant function had four different gene expression variables while the second had just two. Each had different combinations of gene expression variables and different mathematical functions suggesting that there may be many subsets of gene expression variables that define leukemia type. For example, the first discriminant function had the form

$$X_1 * (X_2 + X_3 + X_4) \tag{2}$$

while the second had the form

$$X_1 + X_2. \tag{3}$$

Most of the genes identified by these models are either directly of indirectly involved in human leukemia. For example, variable X_1 in (3) is the *CD33* gene that encodes a differentiation antigen of AML progenitor cells and is a very well-know pathological marker of AML. Additionally, X_3 in (2) is the *adipsin* gene is part of a chromosomal cluster of genes that is expressed during myeloid cell differentiation that is associated with AML. Thus, from approximately 7100 different gene expression variables, SDA identified several genes with direct biological relevance to human leukemia. Further, the level of prediction accuracy obtained with these two symbolic discriminant functions was nearly perfect and significantly better than the class prediction methods of Golub et al. [19].

4   A Cross Validation Consistency Strategy for Symbolic Discriminant Analysis

In the first implementation of SDA [13], models were selected that had low classification and prediction errors as evaluated using LOOCV. The end result of this initial implementation of SDA is an ensemble of models from separate cross validation divisions of the data. In fact, over k runs of n-fold cross validation, there are a maximum of kn possible models generated assuming a 'best' model for each interval is identifiable. In the leukemia example above, 106 final models were selected from a total number of 190 models. In this example, there existed a second independent dataset with which each of the 106 final models could be evaluated for their ability to generalize. Although cross validation and a second independent dataset were used to select models, no formal hypothesis testing was carried out. Further, in many studies, independent second datasets are not available. This motivated the development of a cross validation consistency (CVC) approach to hypothesis testing and model selection [16].

The CVC approach of Moore et al. [16] involves evaluating the consistency with which each gene was identified across each of the LOOCV trials. This is similar to the approach taken by Ritchie et al. [7] for the identification of gene-gene interactions in epidemiological study designs. The idea here is that genes that are important for discriminating between biological or clinical endpoint groups should consistently be identified regardless of the LOOCV dataset. The number of times each gene is identified is counted and this value compared to the value expected 5% of the time by chance were the null hypothesis of no association true. This empirical decision rule is established by permuting the data 1,000 times and repeating SDA analysis on each permuted dataset as described above. In this manner, a list of statistically significant genes derived from SDA can be compiled.

Once a list of statistically significant genes or variables is compiled, a symbolic discriminant function that maximally describes the entire dataset can then be derived. This is accomplished by rerunning SDA multiple times on the entire dataset using only the list of statistically significant candidate genes identified in the CVC analysis. A symbolic discriminant function that maximizes the distance between symbolic discriminant scores among the two endpoint groups is selected. In this manner, a single 'best' symbolic discriminant function can be identified and used prediction in independent datasets.

Moore et al. [16] applied the CVC approach to SDA to the analysis of microarray data from a study of autoimmune disease [20]. In this dataset, expression levels of approximately 4,000 genes were measured in peripheral blood mononuclear cells from 12 normal individuals, seven patients with rheumatoid arthritis (RA), and nine with systemic lupus erythematosus (SLE). Moore et al. [16] identified a total of eight statistically significant genes that differentiate RA subjects from normal and six statistically significant genes that differentiate SLE subjects from normal. Moore et al. [16] reran SDA a total of 100 times on the entire dataset using only the eight statistically significant genes for the RA comparison and the six statistically significant genes for the SLE comparison. A single model of RA and a single model of SLE that maximally described the data were selected. The model of RA had the form

$$[X_1 / (X_4 - X_5)] / (X_3 * X_6) \qquad (4)$$

while the simplified model of SLE had the form

$$-0.89 / [(X_5 - X_3) * -0.14 * (X_2 + X_1 - 1.03)] \qquad (5)$$

The best RA model (4) consisted of five of eight statistically significant genes with arithmetic functions of subtraction, multiplication, and division. The best SLE model (5) consisted of four of six statistically significant genes with all four arithmetic functions present. The classification accuracy of these models was 100%.

Several of the genes identified are either directly or indirectly involved in human immune response and may play a role in either RA or SLE. For example, one of the RA genes encodes defensin 1, a protein that is involved in processes of innate and adaptive immunity. One of the SLE genes is similar to the human proliferating-cell nucleolar antigen gene that may play a role in mechanisms of autoantibody production in lupus patients. Thus, as in the leukemia analysis, SDA identified biologically interesting genes from among thousands of candidates.

5 Discussion and Summary

Symbolic discriminant analysis (SDA) was developed as an attempt to deal with the challenges of selecting subsets of gene expression variables and features that facilitate the classification and prediction of biological and clinical endpoints [14, 16]. Motivation for the development of SDA came from the limitations of traditional parametric approaches such as linear discriminant analysis and logistic regression. Application of SDA to high-dimensional microarray data from several human diseases has demonstrated that this approach is capable of identifying biologically relevant genes from among thousands of candidates.

Moore et al. [16] introduced cross validation consistency (CVC) as a metric to facilitate the identification of a 'best' SDA model that can be used to make predictions in independent datasets. The CVC approach addresses the problem of identifying different models in each cross validation division of the data that results from the stochastic aspects of GP. In combination with permutation testing, CVC allows a subset of statistically significant variables (e.g. genes) to be identified. The statistically significant variables can then be used to model the entire dataset. We suggest that this general approach can be incorporated into any GP-based modeling approach that uses a cross validation framework.

SDA is not the only GP-based approach that has been developed for the analysis of microarray data [15]. For example, Gilbert et al. [21] developed an approach using GP to select gene expression variables and to build classifier rules. Here, the building blocks of the model were gene expression variables, arithmetic functions (+, -, *, /), and a Boolean 'if greater than or equal to' function that returns a 1 if the first argument is greater than or equal to the second, and 0 otherwise. These building blocks were used to construct linear classifier rules of the form "if $(X_1 * X_2) \geq (X_3 - X_4)$ then group=1, else group=0" where the Xs are gene expression variables, for example. Gilbert et al. [21] applied this GP approach to classifying genes as belonging to one of six functional groups using the yeast microarray data of Eisen et al. [22]. A five-subpopulation parallel GP was run 2,000 generations with migration of the top 5% of models to each population every 10 generations. Using this approach, the authors were able to classify and predict group membership with a high degree of accuracy and even discovered some patterns not previously in the literature. Even though a cross validation strategy was not used in this study, it would be possible to integrate the CVC approach described here into the Gilbert et al. [21] approach in exactly the same way it was integrated into SDA.

In summary, the CVC approach provides a framework for hypothesis testing and model building for GP-based approaches that utilize cross validation. Specifically, CVC addresses the problem of identifying different models in different divisions of the data due to the stochastic properties of GP. The CVC approach was developed for the SDA approach but is easily integrated into any GP-based modeling procedure that is performing variable and feature selection for classification and prediction. We anticipate the use of cross validation with GP-based modeling procedures will increase as methods such as CVC become available.

Acknowledgements. This work was supported by National Institutes of Health grants HL68744, DK58749, CA90949, and CA95103.

References

1. Schena, M., Shalon, D., Davis, R.W., Brown, P.O.: Quantitative monitoring of gene expression patterns with a complementary DNA microarray. Science 270 (1995) 467–470
2. Velculesco, V.E., Zhang, L., Vogelstein, B., Kinzler, K.W.: Serial analysis of gene expression. Science 270 (1995) 484–487
3. Caprioli, R.M., Farmer, T.B., Gile, J.: Molecular imaging of biological samples: Localization of peptides and proteins using MALDI-TOF MS. Analyt. Chem. 69 (1997) 4751–4760

4. Bradley, J.V.: Distribution-free statistical tests. Prentice-Hall, Englewood Cliffs (1968)
5. Freitas, A.A.: Understanding the crucial role of attribute interaction in data mining. Artificial Intelligence Reviews 16 (2001) 177–199
6. Moore, J.H., Williams, S.M.: New strategies for identifying gene-gene interactions in hypertension. Annals of Medicine 34 (2002) 88–95
7. Ritchie, M.D., Hahn, L.W., Roodi, N., Bailey, L.R., Dupont, W.D., Plummer, W.D., Parl, F.F. and Moore, J.H.: Multifactor dimensionality reduction reveals high-order interactions among estrogen metabolism genes in sporadic breast cancer. American Journal of Human Genetics 69 (2001) 138–147
8. Templeton, A.R.: Epistasis and complex traits. In: Wade, M., Brodie III, B., Wolf, J. (eds.): Epistasis and Evolutionary Process. Oxford University Press, New York (2000)
9. Fisher, R.A.: The Use of Multiple Measurements in Taxonomic Problems. Ann. Eugen. 7 (1936) 179–188
10. Johnson, R.A., Wichern, D.W.: Applied Multivariate Statistical Analysis. Prentice Hall, Upper Saddle River (1998)
11. Huberty, C.J.: Applied Discriminant Analysis. John Wiley & Sons, Inc., New York Chichester Bisbane Toronto Singapore (1994)
12. Neter, J., Wasserman, W., Kutner, M.H.: Applied Linear Statistical Models, Regression, Analysis of Variance, and Experimental Designs. 3rd edn. Irwin, Homewood (1990)
13. Moore, J.H., Parker, J.S., Hahn, L.W.: Symbolic discriminant analysis for mining gene expression patterns. In: De Raedt, L., Flach, P. (eds) Lecture Notes in Artificial Intelligence 2167, pp 372–81, Springer-Verlag, Berlin (2001)
14. Koza, J.R.: Genetic Programming: On the Programming of Computers by Means of Natural Selection. The MIT Press, Cambridge London (1992)
15. Moore, J.H., Parker, J.S.: Evolutionary computation in microarray data analysis. In: Lin, S. and Johnson, K. (eds): Methods of Microarray Data Analysis. Kluwer Academic Publishers, Boston (2001)
16. Moore, J.H., Parker, J.S., Olsen, N., Aune, T. Symbolic discriminant analysis of microarray data in autoimmune disease. Genetic Epidemiology 23 (2002) 57–69
17. Hastie, T., Tibshirani, R., Friedman, J.: The Elements of Statistical Learning: Data Mining, Inference, and Prediction. Springer, New York (2001)
18. Devroye, L., Gyorfi, L., Lugosi, G.: A Probabilistic Theory of Pattern Recognition. Springer-Verlag, New York (1996)
19. Golub, T.R., Slonim, D.K., Tamayo, P., Huard, C., Gaasenbeek, M., Mesirov, J.P., Coller, H., Loh, M.L., Downing, J.R., Caligiuri, M.A., Bloomfield, C.D., Lander, E.S.: Molecular classification of cancer: class discovery and class prediction by gene expression monitoring. Science 286 (1999) 531–537
20. Maas, K., Chan, S., Parker, J., Slater, A., Moore, J.H., Olsen, N., and Aune, T.M.: Cutting edge: molecular portrait of human autoimmunity. Journal of Immunology 169 (2002) 5–9
21. Gilbert, R.J., Rowland, J.J., Kell, D.B.: Genomic computing: explanatory modelling for functional genomics. In: Whitley, D., Goldberg, D., Cantu-Paz, E., Spector, L., Parmee, I., Beyer, H.-G. (eds): Proceedings of the Genetic and Evolutionary Computation Conference. Morgan Kaufmann Publishers, San Francisco (2000)
22. Eisen, M.B., Spellman, P.T., Brown, P.O., Botstein, D.: Cluster analysis and display of genome-wide expression patterns. Proceedings of the National Academy of Sciences USA 95 (1998) 14863–68

Algorithms for Identification Key Generation and Optimization with Application to Yeast Identification

Alan P. Reynolds[1], Jo L. Dicks[2], Ian N. Roberts[3], Jan-Jap Wesselink[1],
Beatriz de la Iglesia[1], Vincent Robert[4], Teun Boekhout[4], and
Victor J. Rayward-Smith[1]

[1] School of Information Systems, University of East Anglia, Norwich, UK
[2] Computational Biology Group, John Innes Centre, Norwich, UK
[3] National Collection of Yeast Cultures, Institute of Food Research, Norwich, UK
[4] Centraalbureau voor Schimmelcultures, Utrecht, The Netherlands

Abstract. Algorithms for the automated creation of low cost identification keys are described and theoretical and empirical justifications are provided. The algorithms are shown to handle differing test costs, prior probabilities for each potential diagnosis and tests that produce uncertain results. The approach is then extended to cover situations where more than one measure of cost is of importance, by allowing tests to be performed in batches. Experiments are performed on a real-world case study involving the identification of yeasts.

1 Introduction

Identification problems occur in a wide range of disciplines. Biological applications include the identification of plant or animal species from observable specimen characteristics [1], the diagnosis of disease in plants, animals and humans and the identification of bacteria from the results of laboratory tests [2]. Each situation is different, and so a range of identification tools are used.

One such tool is the identification (or diagnostic) key. Given a database of test results for known species, an identification key provides a sequence of tests to be performed, with the choice of test depending upon the results of previous tests. This paper describes the generation and optimization of such keys. The algorithms described handle both variable test costs and uncertain and unknown data, creating keys where tests are performed individually or in batches.

The construction of identification keys is introduced in sections 2 and 3. Section 4 describes and compares greedy key construction heuristics, with results in section 6. Section 7 describes the randomization of the greedy algorithm, to reduce further the cost of the keys produced. Finally, section 8 describes the adaptation of these algorithms to handle batches of tests.

2 Testing Approaches and Key Evaluation

When performing tests to identify specimens, one of three different approaches may be appropriate.

S. Cagnoni et al. (Eds.): EvoWorkshops 2003, LNCS 2611, pp. 107–118, 2003.

Perform all tests at once. This approach is suitable if the speed of identification is more important than the material costs, provided tests may be performed in parallel. In this case, it is desirable to find a minimal cost subset of the tests that maximizes the likelihood of identification [3].

Perform tests individually. If tests cannot be performed in parallel, or the material costs of the tests are high, tests should be performed individually. Each test is chosen after analysing the results of the previous tests.

Perform tests in batches. In some situations, it may be appropriate to perform the tests in batches. For example, if specimens must be sent off to a distant lab for the tests, performing all tests at once may be expensive in terms of the costs of the tests, but performing tests individually will certainly be expensive in terms of time.

This paper focuses on the production of identification keys for the cases where tests are either performed individually or in batches.

A number of different measures may be used to evaluate a key. Some considerations are as follows:

- The expected total test cost should preferably be minimized.
- It may be desirable to complete the identification quickly.
- It may be important to discover specimens of a certain kind quickly. For example, if medical samples are being tested, it may be important to discover dangerous, fast acting but rare diseases quickly.

The expected total test cost is the most commonly used measure of key efficiency [4,5,6], and is used throughout this paper, with the exception of section 8.

3 A Simple Example

Figure 1 shows test data for four yeast species, and the corresponding optimum key, assuming that each species is equally likely. Two tests are needed, regardless of the species the test sample happens to be, so the expected test cost is 20.

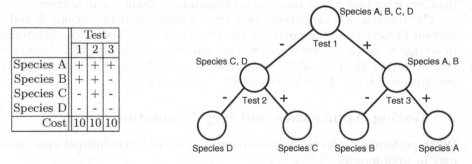

	Test		
	1	2	3
Species A	+	+	+
Species B	+	+	-
Species C	-	+	-
Species D	-	-	-
Cost	10	10	10

Fig. 1. Simple species data with the corresponding optimum identification key.

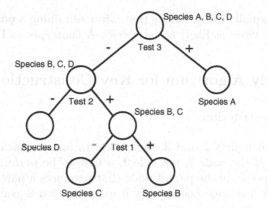

Fig. 2. Alternative identification key

An alternative key is shown in figure 2. Here species A is identified with just one test, but if the sample is equally likely to be any of the species, the expected test cost is now $(10 + 20 + 30 + 30)/4 = 22.5$. However, if species A is more common than the other species, the resulting expected test cost may be less than 20, making the resulting key preferable to that shown in figure 1.

Suppose an extra test is added, that produces a positive result for species A, C and D, but produces an uncertain result for species B. Let each outcome for species B be equally likely. If the test is cheap, it may be present in the optimum key, despite the fact that a positive result does not eliminate any of the potential yeast species. This is shown in figure 3. The cost of identification, if the sample

	Test			
	1	2	3	4
Species A	+	+	+	+
Species B	+	+	-	?
Species C	-	+	-	+
Species D	-	-	-	+
Cost	10	10	10	1

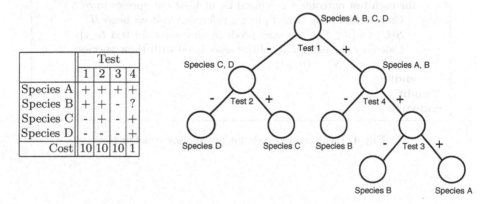

Fig. 3. The optimum key, upon the addition of an extra test.

is species A has increased to 21, but the cost of identification of species B has changed to either 11 or 21, depending on the result obtained for test 4. This leads to an expected test cost of $(21 + (11 + 21)/2 + 20 + 20)/4 = 19.25$, provided each species is equally likely. Notice that, after obtaining a positive result to test 4, the sample is twice as likely to be species A than species B.

4   A Greedy Algorithm for Key Construction

4.1   Key Construction

As illustrated in figures 1 and 3, each node in an identification key has a set of species and, if the node is not a leaf, a test to be performed. A child node contains each species of the parent node that produces a particular result to the parent's test. A leaf node occurs only if identification is complete or no useful tests remain.

A greedy key construction heuristic starts at the top node. At each node, the heuristic either determines that the node is a leaf node, or finds the 'best' test, according to some measure, and creates child nodes associated with each test outcome. The algorithm stops when all nodes have been examined. The three greedy heuristics discussed in this section differ only in the measure used to select tests.

Create node *top* and place a reference to it on empty heap H;
$S(top) :=$ set of all yeast species in database, with associated probabilities;
$T(top) :=$ set of all tests;
while (H not empty)
    Remove a node, N, from the heap;
    if $|S(N)| > 1$ and there exists a useful test in $T(N)$ **then**
        Find the 'best' test $t_{best} \in T(N)$;
        for each test outcome r produced by at least one species in $S(N)$
            Create child node C and place a reference to it on heap H;
            $S(C) := \{s \in S(N) : s$ may produce outcome r for test $t_{best}\}$;
            Calculate the new probabilities associated with these species;
            $T(C) := T(N) - \{t_{best}\}$;
        endfor
    endif
endwhile

Fig. 4. Basic pseudo-code for greedy key construction

4.2   Test Selection

Payne and Preece's review of identification tables and diagnostic keys [7] reports several test selection criteria. Those that are suitable for use with tests with unequal costs are compared in a later paper by Payne and Thompson [4].

Define:

$S = \{(s_1, p_1), \ldots, (s_n, p_n)\}$: Set of potential species and associated
probabilities of occurrence;

S_{ij} : Similar set containing potential species after
test i produces result j;

$P(i, j)$: Probability that test i produces result j;

$E(S)$: Estimate of the cost of completing
identification from set S;

c_i : Cost of test i;

m_i : Number of possible outcomes of test i.

The functions used in Payne and Thompson [4] as selection criteria are all of
the form

$$PT_i = c_i + \sum_{j=1}^{m_i} P(i, j) E(S_{ij}).$$

Here PT_i is the cost of test i plus an estimate of the expected remaining test
costs after test i is performed, and is therefore an estimate of the total test costs.
The test that minimizes this function is considered to be the best test.

A problem with this approach is that of finding a good estimate, $E(S)$. Payne
and Thompson suggest a number of possibilities. However, each of the estimates
suggested can be shown to be either optimistic or pessimistic, depending on the
number and quality of the available tests.

A second problem is that such criteria tend to favour poorly performing
cheap tests over more useful expensive tests. Consider the following example. A
sample specimen may be one of only two species, each equally likely. $E(S)$ is the
estimate of the remaining test costs. Test one has a cost of one, but produces the
same result for both species. Test two has cost five and completes identification.
Then $PT_1 = 1 + E(S)$ and $PT_2 = 5$. If $E(S)$ equals three, the average test cost,
then the useless cheap test is chosen. This choice of test can only increase the
expected test costs.

The approach used in this paper is to calculate a measure of the expected
amount of work performed by the test. If the work performed, upon application
of a test, is defined to be the reduction in the estimate of the remaining test
costs, then the expected amount of work performed is given by

$$W_i = E(S) - \sum_{j=1}^{m_i} P(i, j) E(S_{ij}).$$

This may then be divided by the test cost, to give an estimate of the work
performed per unit cost. The test that performs the most work per unit cost is
selected. This selection criterion has neither of the aforementioned problems.

Of the selection criteria discussed in Payne and Thompson [4], those that use
Shannon entropy [8] are of particular interest. The entropy of set S is given by

$$H(S) = -\sum_{k=1}^{n} p_k \log_2 p_k.$$

$H(S)/\log_2 m$ can be shown to be a very good approximation to the number of tests required for identification if all conceivable tests that produce m results are available. After test i is performed, the expected entropy is given by

$$H(S,i) = \sum_{j=1}^{m_i} P(i,j)H(S_{ij}).$$

Payne and Thompson [4] give the following two functions that may be used as test selection criterion.

$$(PT1)_i = c_i + (c_{min}/\log_2 \bar{m})H(S,i)$$

$$(PT2)_i = c_i + (\bar{c}/\log_2 \bar{m})H(S,i)$$

Here c_{min} is the cost of the cheapest test, $\bar{c}$ is the mean test cost and $\bar{m}$ is the mean number of outcomes of a test.

In the approach used in this paper, Shannon entropy is used as a measure of how much work is required to complete identification. Setting the estimate function $E(S)$ to be equal to the entropy $H(S)$, the expected amount of work performed per unit cost, upon application of test i, is given by the information gain ratio

$$GR_i = \frac{W_i}{c_i} = \frac{H(S) - H(S,i)}{c_i}.$$

The test selected is the one that maximizes this value.

Note that Shannon entropy is commonly used in the closely related research area of decision tree induction in machine learning and data mining [9].

5   The Case Studies

5.1   The Data Used

Data from the Centraalbureau voor Schimmelcultures (http://www.cbs.knaw.nl) was used for the experiments. Results of 96 different tests on 742 yeast species are provided. The possible responses are as shown in table 1. Test costs and the prior probabilities of each specimen are not given.

The data was manipulated in a number of ways in order to provide more than one case study.

Table 1. Possible responses

Meaning	Symbol	Meaning	Symbol
Positive	+	Negative, weak and/or delayed	-,w,d
Negative	-	Positive, weak and/or delayed	+,w,d
Weak	w	Negative and/or positive	-,+
Delayed	d	Unknown	-,+,w,d

Simplification of the results range. Either the full results range of table 1 was used, or this range was simplified. In the latter case, any response that was not 'positive' or 'negative' was replaced by the response 'other'.

Certain and uncertain results. The reported test outcomes were interpreted as either certain or uncertain outcomes. In the first case, the 8 different result values in table 1 are interpreted as distinct test outcomes. In the second, more realistic case, a test produces one of four different outcomes: 'negative', 'weak', 'positive' or 'delayed'. The other four possibilities in table 1 are interpreted as being uncertain outcomes, with each outcome in the symbol list occurring with equal probability. If the simplified results range is used, the result 'other' is interpreted as an uncertain outcome.

Test costs. Tests were either assigned unit costs or random integer costs from 1 to 5.

Prior probabilities. Each species was either assumed to be equally likely, or assigned a random prior probability.

The resulting 16 case studies provided the test bed for the algorithms of this paper. The randomly generated test costs and prior species probabilities are listed at http://www.sys.uea.ac.uk/biodm/.

6   Results: Greedy Heuristics

When each test has unit cost, each of the selection criteria discussed produce the same result, since each selects the test that minimizes the Shannon entropy after the test is performed. The results in this case are shown in table 2.

By using the randomly generated test costs, it is possible to compare the three selection criteria. Table 3 shows the results of this comparison. When creating the test trees, useless tests were discarded, preventing selection criteria $PT1$ and $PT2$ from choosing cheap but useless tests.

When the tests have uncertain outcomes, full identification cannot be guaranteed, as it is possible for two or more species to produce the same result for each of the tests. In this case, tests were applied until no test remains that can possibly produce an identification.

Notice that in all but one of the experiments, selection criterion GR produces keys with lower expected test costs than both $PT1$ and $PT2$. Keys produced with selection criterion GR also tend to be smaller, although this is only really an issue if the key needs to be produced in printed form. Since the keys produced when there are uncertain results and varying test costs are large, the rest of this paper concentrates on the case where test outcomes are certain.

7   Randomizing the Algorithm

When solving an optimization problem, a metaheuristic can often be expected to outperform a greedy algorithm in terms of solution quality. Unfortunately, in this case it is difficult to find a suitable neighbourhood structure for use in local search techniques. However, a simplified GRASP algorithm [10] may be applied

Table 2. Unit test costs

Outcome type	Results range	Species probabilities	Expected #tests	#Nodes
Certain	Simplified	Equal	6.248	1194
		Random	6.176	1209
	Full	Equal	4.624	1073
		Random	4.554	1071
Uncertain	Simplified	Equal	12.887	30223
		Random	12.912	27289
	Full	Equal	8.027	182750
		Random	8.048	231858

Table 3. Random test costs

Outcome type	Results range	Species prob.	Expected cost			#nodes		
			$PT1$	$PT2$	GR	$PT1$	$PT2$	GR
Certain	Simplified	Equal	8.222	8.011	8.015	1311	1245	1300
		Random	10.132	8.136	7.863	1707	1332	1294
	Full	Equal	6.453	6.282	6.240	1200	1170	1165
		Random	8.386	7.542	6.137	1582	1498	1155
Uncertain	Simplified	Equal	37.353	33.824	29.076	893811	535953	205169
		Random	36.986	33.900	29.229	912325	571107	213185
	Full	Equal	20.883	19.078	16.958	5717075	3692803	2165164
		Random	21.059	19.380	17.083	5630820	4167501	2086172

with ease. Instead of choosing the best test, according to selection criterion GR, one of the best three tests is selected at random. Many trees are created and the best is selected.

7.1   Results

Experiments were performed to determine the best selection probabilities for the tests. The selection probability for the best test varied between 0.5 and 0.9. The selection probability for the second best test varied from zero to 0.5, provided that the probability of selecting one of the best two tests did not exceed one. Whenever neither of the best two tests was selected, the third was chosen. Each run created 1000 trees. Ten runs were performed for each case.

Figure 5 shows the mean result of the ten runs for each of the probability settings. In this case the simplified results range was used, test costs were equal to one and each species was assumed to be equally likely. Notice that the best results obtained were found by selecting only the best two tests and setting the probability of choosing the best test to its highest value (0.9). Results for the other case studies were remarkably similar, suggesting that the information gain ratio is a remarkably good choice of test selection criterion.

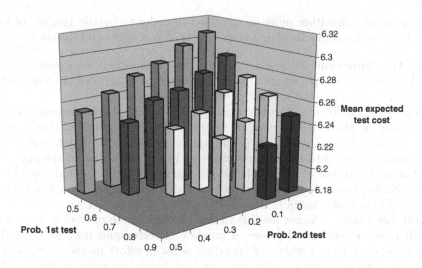

Fig. 5. Mean expected test costs for ten runs

The mean and best results of the ten runs, with test selection probabilities set to 0.9, 0.1 and 0.0 are shown in table 4. The results obtained are not a great deal better than those obtained using the greedy algorithm alone, further suggesting that information gain ratio is a highly effective test selection criterion.

8   Time and Cost Minimization: Performing Tests in Batches

As already discussed, the three approaches to testing — performing all tests at once, performing tests individually and performing tests in batches — are each suited to different situations. When it is important to perform the identification quickly and the material costs are irrelevant, all tests are performed in parallel. When material costs are the overriding factor, each test should be performed individually. It is reasonable to suspect that, when both test duration and material costs are important, testing should be performed in batches.

Table 4. Expected test costs with selection probabilities of 0.9, 0.1, 0.0.

Results	Test outcomes	Equal probabilities		Random probabilities	
		Unit costs	Random costs	Unit costs	Random costs
Mean	Simplified	6.227	8.007	6.152	7.835
	Original	4.606	6.221	4.514	6.109
Best	Simplified	6.221	8.000	6.148	7.825
	Original	4.597	6.214	4.506	6.099

The greedy algorithm must be modified in order to handle batches of tests and objective functions involving both expected test costs and duration.

Objective Function. Functions used for key evaluation are weighted sums, $W = \lambda C + (1 - \lambda)T$, where C is the expected material cost of identification and T is the expected amount of time required.

Batch Quality. To measure the quality of a batch, the information gain is divided by the weighted cost, w, of the batch. This weighted cost is set to $w = \lambda c + (1 - \lambda)t$, where c is the material cost of the batch and t is its duration. Here λ takes the same value as in the evaluation of the key.

The material cost of a batch is simply the sum of the material costs of the individual tests. However, as tests in a batch may be performed in parallel, the duration of a batch is the duration of the longest test.

Batch Selection. It is not feasible to find the best batch of tests by evaluating all possible batches. However, a number of techniques may be applied to this optimization problem. Expending a lot of effort in each optimization is likely to be counter-productive; the best batch according to the selection criterion need not be the best batch for the creation of the key. A simple search algorithm, such as a hillclimber, is likely to be sufficient.

8.1   Results

The following results were obtained by running a stochastic first-found hill-climber at each node of the tree. Neighbourhood moves either added a test to the batch, removed a test, or did both, allowing batch sizes to be determined automatically. The material costs weight, λ, took values between 0.1 and 1. Ten runs were performed for each value of λ and mean results are reported.

Figure 6 shows results for the case study with the full results range, unit material test costs and durations and equal species probabilities. The expected material cost of identification, the expected testing duration and the weighted cost are plotted against the material cost weight. As material test costs become less significant and testing duration becomes more so, the material cost of the key produced by the algorithm increases and the testing duration decreases. Figure 7 illustrates how the average batch size increases as material test costs become less significant compared to testing duration.

9   Further Research

The research outlined in this paper may be extended in a number of ways.

Improved Estimates of Work Performed. Entropy provides a good measure of the work remaining when each species is equally likely, but figure 8 shows that as one species becomes more likely, entropy performs less well.

A possible solution is to use the Huffman algorithm to create optimal trees as shown in Payne and Preece [7], under the assumption that all conceivable binary tests are available. The expected test cost of the trees created may then be used as an alternative to Shannon entropy.

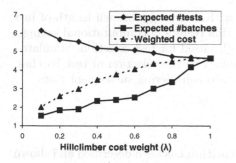

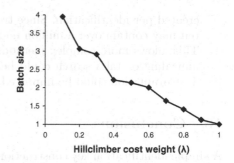

Fig. 6. Expected material costs and time required for identification, with the weighted costs of the keys.

Fig. 7. Batch size.

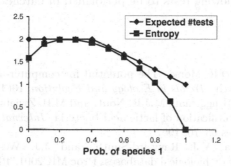

Fig. 8. Expected test costs, provided all conceivable binary tests are available, and entropy for four species, with three equally likely.

Identification of Yeast Class. Sometimes it may only be necessary to determine the class of species to which a specimen belongs, e.g. whether a yeast specimen is one of a set of known food spoilage yeasts. In this case, entropy is not always a reliable measure of the work required to complete identification. Further work to produce a reliable test selection criterion is required.

Although the greedy algorithm described in this paper is less reliable in this situation, the simplified GRASP algorithm has been used to produce useful keys for determining if a yeast specimen is one of the food spoilage species.

Non-independence of Tests. The algorithms described in this paper work under the assumption that tests are independent of each other. In reality this may not be the case. Further work is necessary to account for this.

Practical Application of Greedy Key Construction. In order to *evaluate* a key construction algorithm, it is necessary to create complete identification keys in order to calculate the expected test costs. If a greedy algorithm is used, its *application* to specimen identification does not require the creation of the full key. Instead, the selection criterion is used to determine the tests to be performed as required. The result is that only one branch of the tree is created per identification. Since trees with an average branch length of just ten may contain over a million nodes, this results in computational savings. This allows more complex methods to be used for test selection. Simulated annealing or tabu search could be applied to the selection of test batches. Test evaluation could be improved by also considering subsequent tests.

10   Conclusions

A simple identification key construction algorithm has been described and shown to perform well on a real-world identification problem. This algorithm can handle variable test costs, uncertain test outcomes and species with differing prior probabilities. Furthermore, it has been shown that the algorithm can be extended to handle situations where both material costs and identification time are important, by allowing tests to be performed in batches.

References

1. M. Edwards and D.R. Morse. The potential for computer-aided identification in biodiversity research. *Trends in Ecology and Evolution*, 10(4):153–158, 1995.
2. T. Wijtzes, M.R. Bruggeman, M.J.R. Nout, and M.H. Zwietering. A computerised system for the identification of lactic acid bacteria. *International Journal of Food Microbiology*, pages 65–70, 1997.
3. B. De la Iglesia, V.J. Rayward-Smith, and J.J. Wesselink. Classification/identification on biological databases. Proc MIC2001, 4th International Metaheuristics Conference, ed. J.P. de Souza, Porto, Portugal, 2001.
4. R.W. Payne and C.J. Thompson. A study of criteria for constructing identification keys containing tests with unequal costs. *Comp. Stats. Quarterly*, 1:43–52, 1989.
5. R.W. Payne and T.J. Dixon. A study of selection criteria for constructing identification keys. In T. Havranek, Z. Sidak, and M. Novak, editors, *COMPSTAT 1984: Proceedings in Computational Statistics*, pages 148–153. Physica-Verlag, 1984.
6. R.W. Payne. Genkey: A program for constructing and printing identification keys and diagnostic tables. Technical Report m00/42529, Rothamsted Experimental Station, Harpenden, Hertfordshire, 1993.
7. R.W. Payne and D.A. Preece. Identification keys and diagnostic tables: a review. *Journal of the Royal Statistical Society, Series A*, 143(3):253–292, 1980.
8. C.E. Shannon. A mathematical theory of communication. *Bell Systems Technical Journal*, 27:379–423 and 623–656, 1949.
9. J. R. Quinlan. *C4.5: Programs for Machine Learning*. Morgan Kaufmann, 1993.
10. T. A. Feo and M. G. C. Resende. Greedy randomized adaptive search procedures. *Journal of Global Optimization*, 6:109–133, 1995.

This research was funded by the BBSRC, Grant Ref. No. 83/BIO 12037

Generalisation and Model Selection in Supervised Learning with Evolutionary Computation

Jem J. Rowland

Dept. of Computer Science, University of Wales Aberystwyth,
SY23 3DB, Wales, U.K.
jjr@aber.ac.uk
http://users.aber.ac.uk/jjr/

Abstract. EC-based supervised learning has been demonstrated to be an effective approach to forming predictive models in genomics, spectral interpretation, and other problems in modern biology. Longer-established methods such as PLS and ANN are also often successful. In supervised learning, overtraining is always a potential problem. The literature reports numerous methods of validating predictive models in order to avoid overtraining. Some of these approaches can be applied to EC-based methods of supervised learning, though the characteristics of EC learning are different from those obtained with PLS and ANN and selecting a suitably general model can be more difficult. This paper reviews the issues and various approaches, illustrating salient points with examples taken from applications in bioinformatics.

1 Introduction

Bioinformatics concerns the use of computer-based methods to discover new knowledge about the structure and function of biological organisms. Numerous computational approaches can be applied to the variety of problem areas encompassed by the field. Most of the approaches currently used rely on the formation of an empirical model that is formed on data whose meaning is known and which is then used to interpret new data. Evolutionary computation is particularly successful in many areas of bioinformatics, and for numerous examples of EC-based modelling applied to problems in bioinformatics, see the book by Fogel and Corne [1]. These include sequence alignment, structure prediction, drug design, gene expression analysis, proteomics, metabolomics and more. A single run of an EC-based system produces numerous candidate models; correct model selection and validation is vital to the correctness or otherwise of any results that are subsequently derived from use of that model or any knowledge that is derived from the model itself.

This paper is concerned with EC-based modelling using supervised methods. Here, the aim is to form a model that relates Y data, the known value or values, with the X variables that represent the corresponding measurements or observed

S. Cagnoni et al. (Eds.): EvoWorkshops 2003, LNCS 2611, pp. 119–130, 2003.

attributes. A successful model, once validated, can be used to predict unknown Y values from new sets of X measurements. The nature of the model can also reveal important information about the data and the problem domain. Noise is always present in data sets and originates from instrument variability, sample variability, human errors etc. To produce a predictive model that is generally applicable to similar data, in which the noise is of course different, the model needs to be trained in an appropriate way and properly validated. This should be normal practice.

1.1   The Purpose of Data Modelling

Very broadly, there are two reasons for attempting to form models that capture the characteristics embodied in a data set:

- to understand the structure of the data by discovering explicit relationships between X and Y, or ...
- to form a model that can be used to interpret new X data by predicting the corresponding Y.

The need to ensure generality in the model is of considerable importance in both. It is obviously vital where the model is to be used on new, unseen, data. When we are forming a model to discover an explicit relationship between X and Y in a given data set, unless we take steps to ensure the generality of the model we produce, there are dangers of overfitting or fitting to chance correlations, either of which is likely to render the newly discovered 'knowledge' useless.

1.2   What Do We Mean by Model Selection?

Supervised methods commonly used for interpreting analytical data in biological applications include Partial Least Squares Regression ('PLS') [2], Artificial Neural Networks ('ANN') (e.g.[3]), Genetic Programming [4], numerous variants of Genetic Algorithms [5] and other evolutionary paradigms.

PLS is a deterministic method and we only have to decide how many latent variables, or factors, to use in forming the best model. With ANN a new model is produced each epoch, or with an EC-based method we take the fittest model from each generation, and then choose which of them to use. In each case we need to select a model that generalises sufficiently that it provides a meaningful prediction on previously unseen data.

2   Generalisation and Overtraining

It is normally essential that models are tested on data not used in model formation (e.g. [6]). In this way it is, at least in principle, possible to select a model that embodies the underlying effect but has not 'overtrained' so as to learn also the detail of the noise content of the training data. There are several different approaches to this, not all of equal efficacy and not all straightforwardly applicable to evolutionary methods.

'Leave-k-out' cross validation is often used with methods such as Partial Least Squares Regression [2], and is particularly attractive where the number of available data objects is relatively small. A succession of models is formed, in each case omitting k data objects from the training set. Each successive model is used to predict the Y values of the missing objects from their X variables. The errors calculated by repeating this over the whole data set are averaged and used as a measure of the quality of the model. It relies on the modelling method producing models of similar form for each of the succession of partitions and is commonly used with methods where successive models differ relatively slightly. This is not the case with evolutionary methods (and some others) where successive models may differ drastically and further discussion of 'leave-k-out' cross validation is therefore outside the scope of this paper. There is, however a significant literature on the subject, and section 3.5 below will refer briefly to an evolutionary approach that makes novel use of this method of validation.

Another commonly used method involves dividing the original data set into separate, fixed, partitions. Training is always done on the first partition of the data and performance of a model is assessed on the basis of its performance on a separate partition[1]. During training the learning curves for the training and test sets are monitored. The performance on both sets is expected to to improve initially. Then performance on the test partition can be expected to deteriorate in comparison with the training set, as the model tunes more and more closely to the specific noise characteristics of the training set. A justified criticism of this approach is that in selecting the model for use on the basis of the divergence of these two learning curves the second partition is, in reality, used as part of the training process and is not independent. To counter this, a third partition should be used to assess generality. An ideal model would give equal error measures on all three sets.

'Generality' is measured in terms of 'generalisation error', which is usually considered to be the prediction error on the third set, which has not been used in any way in training or in model selection. We will revisit this metric later.

3   Model Selection and Evolutionary Computation

As noted above, leave-k-out cross-validation is not straightforwardly applicable to EC-based supervised learning methods. With an EC-based supervised learning method such as Genetic Programming [4] or related techniques based on Genetic Algorithms, e.g. [7] there are two particular issues:

---

[1] In this paper the following terminology is used for the three data partitions: the 'training set' is the set on which the training is performed directly, the 'test set' is the partition whose learning curve is also monitored during the training process, and 'validation set' is the final set used for assessing the generality of the model.

- The k models are likely to be significantly different from each other in structure and in the X variables used. This clearly makes it difficult to draw conclusions as to which of the models should be chosen[8].
- What termination criterion should be used for each of the k runs?

Therefore, for EC techniques, the use of partitioned data is generally more practicable. The ANN community commonly uses different partitions to determine the stopping point during training. However, in relation to EC methods, Eiben and Jelasity [9], in criticising many aspects of current EC methodology, point out that:

"[model] evaluation done on the test set and not the training set is not common practice in the EA community."

The clear implication is that use of three sets is rare, and detailed consideration of model selection criteria is less common still.

Of course, there are many cases where multiple sets have been used. One example is provided by an investigation using ANN and GP by Brameier and Banzhaf [10]. They studied the set of medical and bioinformatic classification data sets in the PROBEN1 collection [11]. They used three partitions of their data and selected models using the 'minimum validation error' (in our current terminology this refers to test set) during training of their ANN. For GP they looked at the test at the end of each generation and then used the third set ('validation' set in our terminology) to check the overall best individual when training was complete. Another example is provided by Landavazo and Fogel [12] who describe an investigation aimed at modelling the relationship between the activity of various anti-HIV compounds and their molecular structure. This type of relationship is known as a Quantitative Structure-Activity Relationship (QSAR). Their aim was to achieve a model that could be used in the future to predict the anti-HIV activity of new compounds on the basis of their molecular structure. They judged the onset of overtraining on the basis of a pair of learning curves in which there is a clear minimum in the test set error.

3.1   Illustrative Data Set

In the next subsection, illustrative examples are used that relate to a spectral regression problem. They relate to a study using Fourier Transform Infra Red (FT-IR) Spectroscopy to monitor production by fermentation of a specific metabolite (gibberelic acid) on an industrial scale, with the aim of demonstrating that predictive modelling could be used to determine the metabolite concentration from FT-IR spectra. Data collection and preprocessing, along with all other aspects of the wide-ranging study are described by McGovern [13]. Each spectrum collected was the result of co-adding 256 times during collection and the stored spectra were therefore relatively noise-free, though still subject to drift and baseline variation. McGovern used three data partitions formed using a version of Snee's Duplex algorithm [14] with the aim of ensuring that the test set did not extrapolate the data space occupied by the training set, and the validation set did not extrapolate that of the test set. Measurements were made on triplicated samples and corresponding replicate measurements were retained together within a single data partition.

For illustrative use in this paper these three sets were re-combined and then re-partitioned randomly but again retaining sets of replicate measurements within partitions. A second set of partitions was also produced in the same way, in which each of the variables was artificially contaminated with noise of amplitude equal to 20% of the mean amplitude of all the spectra in the data set. There were 180 individual spectra in all, each consisting of 882 variables. The partitions were of equal size and each of them therefore contained 60 data objects.

3.2 Model Selection Criteria

As in the case of the two examples mentioned above (Brameier & Banzhaf [10] and Landavazo & Fogel [12]), the model that minimises the test set error is often taken as the point at which the training is optimal, yet it is not uncommon for the test set error to continue to decrease, with the training set error decreasing faster. Fig. 1 is such an example, in this case taken from a neural network being trained on the noisy spectral regression data described in 3.1 above. Here the minimum of the test set error is clearly an inappropriate criterion.

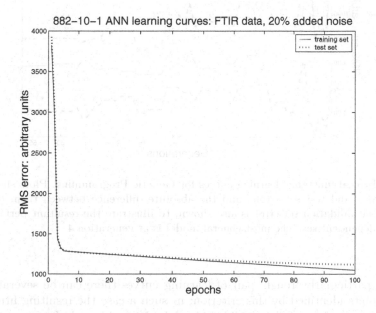

Fig. 1. Training and test set learning curves for a neural network training on spectroscopic data. The divergence of the curves demonstrates that the network is beginning to learn the noise in the data. Their continuing downward trend indicates that the 'minimum of the test set error' is not always the appropriate model selection criterion

Another heuristic often used is the *divergence* of the training and test set curves. Given that a (hypothetical) model with 'perfect' generalisation capability would give equal error measures on all three data partitions, divergence certainly appears to be an appropriate criterion. However, it is commonly interpreted less than rigorously and is sometimes taken simply as the point as which the slopes of the two curves differ, rather than a point at which the errors on the two sets are closest in value. Where learning curves are smooth as in Fig. 1 there is an unambiguous divergence point. However, EC learning curves are rarely smooth. For an example of learning curves from a GP used on the spectral data described in 3.1 see Fig. 2. Which is the 'correct' divergence point in this situation? The answer is provided by taking the model that minimises the absolute difference between the training and test set errors.

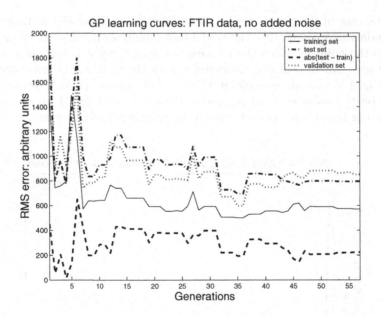

Fig. 2. Typical 'unstable' learning curves for Genetic Programming. Plots shown are the training and test set errors, and the absolute difference between them (bottom trace). The validation set error is also shown, to illustrate the resultant performance on an independent set. The most general model is at generation 4

In a particularly 'rough' pair of learning curves there can be several divergence points identified by this criterion; in such a case the resulting fitness, or predictive accuracy, of the candidate models determines which to use. An example using a variant of the EC-based system described by [7] on the data set described in 3.1 is shown in Fig. 3, where the best generalisation is achieved in a region between 150 and 200 generations. Having identified the 'best' model in this way, its performance on unseen data can then be estimated from the error produced by applying this model to the third (validation) set.

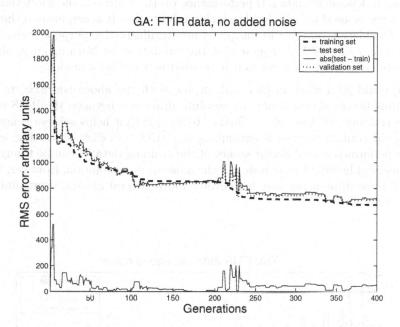

Fig. 3. GA-based learning on spectroscopic data. Examination of the training and test set learning curves suggests multiple candidate models. Plotting their difference (bottom trace) shows that both curves coincide in a region between 150 and 200 generations

While this use of the absolute difference between the training and test set errors is a simple heuristic that provides some improvement on other perhaps less well specified but nevertheless commonly used heuristics, there is still a strong dependence on the characteristics of the data and on the particular random split of the data into three sets. For example, note that in Fig. 4 a different estimate of the generalisation error would be obtained if the test and validation sets were interchanged (which is, of course, an equally likely outcome of splitting the data randomly). A safer estimate is therefore obtained by taking the greatest of the training set error, test error and validation error (of course, the errors on the test and validation sets would usually be greater than the training set error).

Thus, this simple but useful heuristic can be summarised as:

1. Use the model that minimises the absolute difference between the training set and test set errors.
2. For that model, an estimate of its generalisation performance is given by the greatest of the RMS errors of the training set error, the test set error and the validation set error.

This may give a pessimistic estimate but it is usually safer than relying on an estimate that may be over-optimistic. Of course, using the validation set error to measure generalisation performance can only be done once, otherwise it ceases to be an independent data set; performance on the third set, the validation set, must never be used to help decide which model to use. It is emphasised that the plots of validation set error in this paper are for illustrative purposes only. In no way are they intended to suggest that the validation set learning curve should be plotted routinely, even less that it be used in *selecting* a model.

We could go a stage further and, in line with the above heuristic, to help determine fitness we could use the absolute difference between the RMS errors on the training and test sets. A further technique that helps achieve generality during the training process is resampling (e.g.[15,6]); in each generation, evolution is performed on a different subset of the training data. It can be thought of as a means of helping to integrate out the noise during evolution. However, while each of these approaches may help evolve more general models, overtraining is still a possibility.

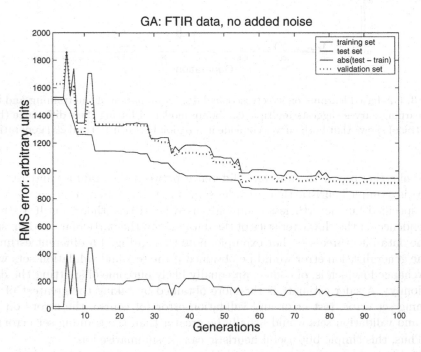

Fig. 4. Illustration of the need to consider both test and validation sets when determining the generalisation error of the model that has been selected. The minimum of the absolute difference between training and test set errors occurs in the first four generations. This gives a generalisation error of about 1620. If the random split of the data had reversed the test and validation sets, the 'best' model would have been at generations 8-10 and estimated generalisation error about 1350

3.3   The 'Generalisation Error' Metric Revisited

Because the 'generalisation error', defined above as the maximum of the RMS errors on the three sets, tends to reduce as the number of generations increases (as in Fig. 4), there may be a temptation to argue that the 'minimum of the test set error' is, after all, the appropriate criterion for model selection, replacing (1) in the heuristic above. However, in selecting a 'useful' model, we are primarily concerned with how well that model is likely to perform on new X data in terms of the degree of uncertainty associated with the resulting Y prediction. An ideal model would produce equal error measures on all three partitions. Thus a model that minimises the absolute difference between the RMS errors on the training and test sets might be expected to lead to reduced uncertainty in the Y value predicted from new X data.

We can consider there to be two objectives in selecting a model:

- Minimize the prediction error (as measured by the RMS error on the test set)
- Minimize the generalisation error (as measured by the greater of the RMS errors on the training and test partitions)

(Then apply the selected model to the validation set to assess the corresponding metrics on independent data.)

Although on the basis of our discussion it would appear that the second of these criteria should always dominate, it is conceivable that there could be circumstances where some tradeoff between them is acceptable and a multiobjective approach to data modelling could be valuable.

It is intuitively appealing, and often quoted, that parsimonious models have greater generality and constraining the expressive power of a modelling method should help achieve more general models. Cavaretta & Chelapilla [15] considered this issue from a theoretical viewpoint and also undertook an empirical investigation using genetic programming. They concluded that highly complex models could exhibit good generality, although their fitness function was based on a resampling scheme that therefore to some extent helped select for generality. Work by Keijzer and Babovic [16], also on genetic programming, provides further theoretical and empirical evidence that parsimony and generality are not necessarily related. Human-interpretability is obviously easier for simple models, and minimization of model complexity could be added as a further objective in the multiobjective approach suggested above.

Llorà, Goldberg et. al. [17] have taken a multiobjective approach in which they seek to optimise predictive accuracy and generality in Learning Systems. However, they equate generality with model parsimony, contrary to [15] and [16], and so the actual objectives they used were accuracy and parsimony. Nevertheless their work provides useful insight into the possibilities offered by a multiobjective approach to model optimisation.

3.4   Bootstrapping, Bagging, and Boosting

Bootstrapping [18] is often quoted as a means of estimating model performance on unseen data and as such it could be used to select between different candidate models identified using as above on the basis of the absolute difference between the RMS errors on the training and test sets. However Kohavi, in an extensive empirical investigation [19] casts doubt on its reliability as an estimator. Bagging [20], or "bootstrap aggregating", combines models, formed by repeatedly resampling the data, by voting. Apart from the consequences of Kohavi's findings, in combining models bagging can obscure the model structure, thereby removing one of the main potential benefits of EC-based methods. Boosting [21] also combines models but by weighting them. However, Quinlan [22] in relation to an investigation based on C4.5 reports that 'boosting sometimes leads to a deterioration in generalization performance'.

3.5   Small Data Sets

In analytical biology it is often expensive and time consuming to generate experimental data. In many instances this leads to a situation where there are insufficient data examples to form properly validated models. In such cases it may be appropriate only to attempt to determine which are the important variables, but model validation is nevertheless important. Moore et. al. [23] report a systematic approach to the analysis of a microarray data set with a typically small number of training examples. They used 'SDA', a GP-based technique with leave-one-out cross validation, so that each run through the data set produced n models, where n is the number of training examples. This was then repeated a number of times to obtain a large number of GP models with good prediction accuracies. The statistically significant genes of all those used in the set of models were then identified by means of their technique of cross-validation consistency (see [24]) and a new model formed using only those genes.

4   Conclusions

This paper has aimed to clarify some of the issues concerning model selection and generality when forming predictive models using evolutionary methods. In doing so it has also presented a simple heuristic that can be used with the 'rugged' learning curves often seen in EC. Various other candidate approaches that are presented in the literature have also been reviewed. Selection of a model that has been suitably validated is essential if the knowledge gained, by subsequent use of the model or from the content and structure of the model, is to be of value.

Models related to biological systems can be treated as hypotheses that can be tested in the laboratory by 'wet' methods. Confirming a model hypothesis in this way is the ultimate validation. However, such work is expensive and prior *in silico* validation is essential so that the laboratory work is suitably focused. For an example of such 'wet' validation, in metabolomics, see Johnson [25]. Moore

[23] detected genes implicated in autoimmune disease whose function could then be investigated further. Where models are to be relied on without confirmation in the laboratory, data of suitable quantity and quality are required, along with particularly careful adherance to appropriate methodologies for model selection and validation.

Acknowledgement. The author would like to thank the anonymous referees for some excellent suggestions. It has not been possible to incorporate all of them into this version of the paper.

References

1. Fogel, G., Corne, D., eds.: Evolutionary Computation in Bioinformatics. Morgan Kauffmann, San Francisco, CA (2003)
2. Martens, H., Naes, T.: Multivariate calibration. John Wiley, Chichester (1989)
3. Bishop, C.: Neural Networks in Pattern Recognition. Oxford University Press, Oxford, U.K. (1995)
4. Koza, J.: Genetic programming: on the programming of computers by means of natural selection. MIT Press, Cambridge, Mass (1992)
5. Holland, J.: Adaptation in Natural and Artificial Systems. University of Michigan Press (1975)
6. Freitas, A.: Data Mining and Knowledge Discovery with Evolutionary Algorithms. Springer Verlag (2002)
7. Taylor, J., Rowland, J.J., Kell, D.B.: Spectral analysis via supervised genetic search with application-specific mutations. In: IEEE Congress on Evolutionary Computation (CEC), Seoul, Korea, IEEE (2001) 481–486
8. Hand, D., Mannila, H., Smyth, P.: Data Mining. MIT Press (2001)
9. Eiben, A., Jelasity, M.: A critical note on experimental research methodology in EC. In: IEEE Congress on Evolutionary Computation (part of WCCI), Hawaii, USA, IEEE (2002) 582–587
10. Brameier, M., Banzhaf, W.: A comparison of linear genetic programming and neural networks in medical data mining. IEEE Transactions on Evolutionary Computation **5** (2001) 17–26
11. Prechelt, L.: PROBEN1 – a set of neural network benchmark problems and benchmarking rules. Technical Report 21/94, Univ. Karlsruhe, Karlsruhe, Germany (1994)
12. Landavazo, D., Fogel, G.: Evolved neural networks for quantitative structure-activity relationships of anti-HIV compounds. In: IEEE Congress on Evolutionary Computation (part of WCCI), Hawaii, USA, IEEE (2002) 199–204
13. McGovern, A., Broadhurst, D., Taylor, J., Gilbert, R., Kaderbhai, N., Winson, M., Small, D., Rowland, J., Kell, D., Goodacre, R.: Monitoring of complex industrial bioprocesses for metabolite concentrations using modern spectroscopies and machine learning: application to gibberellic acid production. Biotechnology & Bioengineering **78** (2002) 527–538
14. Snee, R.: Validation of regression models. Technometrics **19** (1977) 415–428
15. Cavaretta, M.J., Chellapilla, K.: Data mining using genetic programming: The implications of parsimony on generalization error. In: Proc. IEEE Congress on Evolutionary Computation, Washington, DC (1999) 1330–1337

16. Keijzer, M., Babovic, V.: Genetic programming, ensemble methods and the bias/variance tradeoff – introductory investigations. In: Proc. EuroGP 2000. Volume 1802 of LNCS., Springer-Verlag (2000) 76–90

17. Llorà, X., Goldberg, D., Traus, I., Bernadó, E.: Accuracy, parsimony, and generality in evolutionary learning systems via multiobjective selection. Technical Report 2002016, Illinois Genetic Algorithms Laboratory (2002) Also in IWLCS 2002.

18. Efron, B., Tibshirani, R.: An Introduction to the Bootstrap. Chapman & Hall (1993)

19. Kohavi, R.: A study of cross-validation and bootstrap for accuracy estimation and model selection. Intl. Joint Conf. on Artificial Intelligence 14 (1995) 1137–1145

20. Breiman, L.: Bagging predictors. Technical Report 421, Department of Statistics, University of California, Berkeley (1994)

21. Freund, Y., Schapire, R.: Experiments with a new boosting algorithm. In: Machine Learning: Proc. Thirteenth Intl. Conference, Morgan Kauffmann (1996) 148–156

22. Quinlan, J.R.: Bagging, boosting, and C4.5. Proceedings of the National Conference on Artificial Intelligence (1996) 725–730

23. Moore, J.H., Parker, J.S., Olsen, N.J., Aune, T.M.: Symbolic discriminant analysis of microarray data in autoimmune disease. Genetic Epidemiology 23 (2002) 57–69

24. Ritchie, M.D., Hahn, L.W., Roodi, N., Bailey, L.R., Dupont, W.D., Parl, F.F., Moore, J.H.: Multifactor-dimensionality reduction reveals high-order interactions among estrogen-metabolism genes in sporadic breast cancer. American Journal of Human Genetics 69 (2001) 138–147

25. Johnson, H., Gilbert, R., Winson, M., Goodacre, R., Smith, A., Rowland, J., Hall, M., Kell, D.: Explanatory analysis of the metabolome using genetic programming of simple interpretable rules. Genetic Programming and Evolvable Machines 1 (2000) 243–258

Genetic Algorithms on NK-Landscapes:
Effects of Selection, Drift, Mutation, and Recombination

Hernán E. Aguirre and Kiyoshi Tanaka

Faculty of Engineering, Shinshu University
4-17-1 Wakasato, Nagano, 380-8553 JAPAN
{ahernan, ktanaka}@gipwc.shinshu-u.ac.jp

Abstract. Empirical studies have shown that the overall performance of random bit climbers on NK-Landscapes is superior to the performance of some simple and enhanced GAs. Analytical studies have also lead to suggest that NK-Landscapes may not be appropriate for testing the performance of GAs. In this work we study the effect of selection, drift, mutation, and recombination on NK-Landscapes for $N = 96$. We take a model of generational parallel varying mutation GA (GA-SRM) and switch on and off its major components to emphasize each of the four processes mentioned above. We observe that using an appropriate selection pressure and postponing drift make GAs quite robust on NK-Landscapes; different to previous studies, even simple GAs with these two features perform better than a random bit climber (RBC+) for a broad range of classes of problems ($K \geq 4$). We also observe that the interaction of parallel varying mutation with crossover improves further the reliability of the GA, especially for $12 < K < 32$. Contrary to intuition, we find that for small K a mutation only EA is very effective and crossover may be omitted; but the relative importance of crossover interacting with varying mutation increases with K performing better than mutation alone ($K > 12$). We conclude that NK-Landscapes are useful for testing the GA's overall behavior and performance and also for testing each one of the major processes involved in a GA.

1 Introduction

Test function generators[1] for broad classes of problems are seen as the correct approach for testing the performance of genetic algorithms (GAs). Kauffman's NK -Landscapes model[2] is a well known example of a class of test function generator and has been the center of several theoretical and empirical studies both for the statistical properties of the generated landscapes and for their *GA-hardness*[1,3,4,5,6]. Previous works that investigate properties of GAs with NK-Landscapes have mostly limited their study to small landscapes, typically $10 \leq N \leq 48$, and observed the behavior of GAs only for few generations. Recently, studies are being conducted on larger landscapes expending more evaluations[7,8,9,10] and the performance of GAs is being benchmarked against hill climbers[8,9].

Heckendorn et al.[8] analyzed the epistatic features of *embedded landscapes* showing that for NK-Landscapes all the schema information is random if K is sufficiently large predicting that, since "standard genetic algorithms theoretically only perform well when

S. Cagnoni et al. (Eds.): EvoWorkshops 2003, LNCS 2611, pp. 131–142, 2003.

the relationships between schemata can be effectively exploited by the algorithm "[8], a standard GA would have no advantage over a strictly local search algorithm. The authors empirically compared the performance of a random bit climber (RBC+), a simple GA, and an enhanced GA (CHC)[11] known to be robust in a wide variety of problems. Experiments were conducted for $N = 100$ varying K from 0 to 65. A striking result of this study was the overall better performance of the random bit climber RBC+. The authors encourage test generators for broad classes of problems but suggest that NK-Landscapes (and kSAT) seem not to be appropriate for testing the performance of genetic algorithms. Motivated by [8], Mathias et al.[9] provided an exhaustive experimental examination of the performance of similar algorithms including also Davis' RBC[12]. A main conclusion of this study is that over the range $19 \leq N \leq 100$ there is a niche for the enhanced CHC in the region of $N > 30$ for $K = 1$ and $N > 60$ for $1 \leq K < 12$. Yet, this niche is very narrow compared to the broad region where RBC and RBC+ show superiority.

Adaptive evolution is a search process driven by selection, drift, mutation, and recombination over fitness landscapes[2]. All of these are important processes within a GA. In this work we study the effect of each one of them on the performance of GAs. As a methodology, we take a model of generational varying mutation GA (GA-SRM) and switch on and off its major components to emphasize each of the four processes mentioned above. Experiments are conducted with NK-Landscapes for $N = 96$ varying K from 0 to 48. Results by the random bit climber RBC+ are also included. We show that using an appropriate selection pressure and postponing drift make GAs quite robust on NK-Landscapes. With regards to strictly local search algorithms, we see that even simple GAs with these two features perform better than RBC+ for a broad range of classes of problems ($K \geq 4$). We also show that the interaction of parallel varying mutation with crossover improves further the reliability of the GA for $12 < K < 32$. Contrary to intuition, we find that for small K a mutation only EA is very effective and crossover may be omitted; but the relative importance of crossover interacting with varying mutation increases with K performing better than mutation alone for $K > 12$.

We conclude that NK-Landscapes are useful for testing the GA's overall behavior and performance on a broad range of classes of problems and also for testing each one of the major processes involved in a GA, which gives valuable insights to improve GAs by understanding better the complex and interesting behavior that arises from the interaction of such processes.

2   NK-Landscapes

An NK-Landscape is a function $f : \mathcal{B}^N \to \Re$ where $\mathcal{B} = \{0, 1\}$, N is the bit string length, and K is the number of bits in the string that epistatically interact with each bit. Kauffman's original NK-Landscape[2] can be expressed as an average of N functions as follows

$$f(x) = \frac{1}{N} \sum_{i=1}^{N} f_i(x_i, z_1^{(i)}, z_2^{(i)}, \cdots, z_K^{(i)}) \tag{1}$$

where $f_i : \mathcal{B}^{K+1} \to \Re$ gives the fitness contribution of bit x_i, and $z_1^{(i)}, z_2^{(i)}, \cdots, z_K^{(i)}$ are the K bits interacting with bit x_i in the string x. That is, there is one fitness function

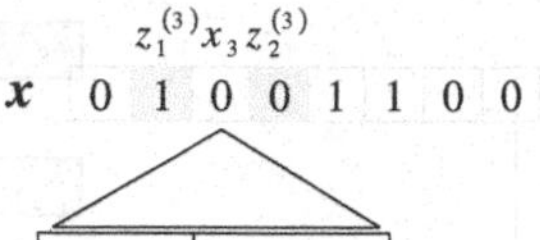

$$z_1^{(3)} x_3 z_2^{(3)}$$

$$x \quad 0 \quad 1 \quad 0 \quad 0 \quad 1 \quad 1 \quad 0 \quad 0$$

$z_1 x_3 z_2$	f_3
000	0.83
001	0.34
010	0.68
011	0.10
101	0.24
110	0.60
111	0.64

Fig. 1. An example of the fitness function $f_3(x_3, z_1^{(3)}, z_2^{(3)})$ associated to bit x_3 in which x_3 epistatically interacts with its left and right neighboring bits, $z_1^{(3)} = x_2$ and $z_2^{(3)} = x_4$ ($N = 8$, $K = 2$)

associated to each bit in the string. NK-Landscapes are stochastically generated and usually the fitness contribution f_i of bit x_i is a number between [0.0, 1.0] drawn from a uniform distribution. Fig. 1 shows an example of the fitness function $f_3(x_3, z_1^{(3)}, z_2^{(3)})$ associated to bit x_3 for $N = 8, K = 2$, in which x_3 epistatically interacts with its left and right neighboring bits, $z_1^{(3)} = x_2$ and $z_2^{(3)} = x_4$, respectively.

For a giving N, we can tune the ruggedness of the fitness function by varying K. In the limits, $K = 0$ corresponds to a model in which there are no epistatic interactions and the fitness contribution from each bit value is simply additive, which yields a single peaked smooth fitness landscape. On the opposite extreme, $K = N - 1$ corresponds to a model in which each bit value is epistatically affected by all the remaining bit values yielding a maximally rugged fully random fitness landscape. Varying K from 0 to $N - 1$ gives a family of increasingly rugged multi- peaked landscapes.

Besides defining N and K, it is also possible to arrange the epistatic pattern between bit x_i and K other interacting bits. That is, the distribution of the K bits among the N. Kauffman investigated NK-Landscapes with two kinds of epistatic patterns: (i) *nearest neighbor*, in which a bit interacts with its $K/2$ left and right adjacent bits, and (ii) *random*, in which a bit interacts with K other randomly chosen bits in the chromosome. By varying N, K, and the distribution of K among the N, we can study the effects of the size of the search space, intensity of epistatic interactions, and epistatic pattern on the performance of genetic algorithms.

The works of Altenberg[13], Heckendorn et al. [8], and Smith and Smith[14,15] are important contributions extending NK-Landscapes to a more general framework of tunable random landscapes and on their analysis. The simplest of the generalized NK-Landscapes[13,8,14] can be expressed as the average of P functions as follows

$$f(\boldsymbol{x}) = \frac{1}{P} \sum_{j=1}^{P} f_j(x_i^{(j)}, z_1^{(i,j)}, z_2^{(i,j)}, \cdots, z_K^{(i,j)}) \tag{2}$$

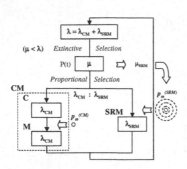

Fig. 2. Block diagram of GA-SRM

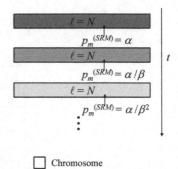

▢ Chromosome

Fig. 3. ADP, Adaptive Dynamic-Probability, mutation strategy in SRM

where $f_j : \mathcal{B}^{K+1} \to \Re$ gives the fitness contribution of the $K + 1$ interacting bits. That is, in this model there could be zero, one, or more than one fitness function associated to each bit in the string.

3 The Algorithms

3.1 GA-SRM

GA-SRM[16,17] applies higher varying mutations parallel to standard crossover & background mutation. The varying mutation operator, called Self-Reproduction with Mutation (SRM), creates λ_{SRM} offspring by mutation alone with a varying mutation rate. The crossover & background mutation operator (CM) creates λ_{CM} offspring by conventional crossover and successive mutation with small mutation probability $p_m^{(CM)}$ per bit. SRM parallel to CM avoids interferences between high mutations and crossover increasing the levels of cooperation to introduce beneficial mutations and create beneficial recombinations, respectively.

Offspring created by SRM and CM coexist and compete for survival and reproduction through (μ, λ) proportional selection[18]. This selection method is extinctive in nature. That is, the number of parents μ is smaller than the total number of created offspring λ. Selection pressure can be easily adjusted by changing the values of μ[18].

GA-SRM enhances selection by preventing fitness duplicates. If several individuals have exactly the same fitness then one of them is chosen at random and kept. The other equal fitness individuals are eliminated from the population. The fitness duplicates elimination is carried out before extinctive selection is performed. Note that under this approach it is possible that two individuals with the same fitness may actually be different at the genotype level. Preventing duplicates removes an unwanted source of selective bias[19], postpones drift, and allows a fair competition between offspring created by CM and SRM.

Mutation rate within SRM is dynamically adjusted every time a normalized mutants survival ratio γ falls under a threshold τ. The normalized mutant survival ratio is specified by

$$\gamma = \frac{\mu_{SRM}}{\lambda_{SRM}} \cdot \frac{\lambda}{\mu} \tag{3}$$

where μ_{SRM} is the number of individuals created by SRM present in the parent population $P(t)$ after extinctive selection. The block diagram of GA-SRM is depicted in Fig. 2. Higher mutations imply that more than one bit will be mutated in each individual and several strategies are possible for choosing the bits that will undergo mutation[16,10]. The mutation strategy used for SRM in this work is ADP (Adaptive Dynamic-Probability). ADP subjects to mutation every bit of the chromosome with probability $p_m^{(SRM)}$ per bit, reducing it by a coefficient β each time γ falls under τ as shown in Fig. 3. Mutation rate $p_m^{(SRM)}$ varies from an initial high rate α to $1/N$ per bit, where N is the bit string length.

3.2 RBC+

RBC+[8] is a variant of a random bit climber (RBC) defined by Davis[12]. Both are local search algorithms that use a single bit neighborhood. We implement RBC+ following indications given in [8,9,12].

RBC+ begins with a random string of length N. A random permutation of the string positions is created and bits are complemented (i.e. flipped) one at the time following the order indicated by the random permutation. Each time a bit is complemented the string is re-evaluated. All changes that results in equally good or better solutions are kept and accounted as an accepted change. After testing all N positions indicated by the random permutation, if accepted changes were detected a new permutation of string positions is generated and testing continues. If no accepted changes were detected a local optimum has been found, in which case RBC+ opts for a "soft-restart". That is, a random bit is complemented, the change is accepted regardless of the resulting fitness, a new permutation of string positions is generated, and testing continues. These soft-restarts are allowed until $5 \times N$ changes are accepted (including the bit changes that constituted the soft restarts). When RBC+ has exhausted the possibility of a soft-restart it opts for a "hard-restart" generating a new random bit string. This process continues until a given total number of evaluations have been expended. The difference between RBC and RBC+ is the inclusion of soft-restarts in the latter.

4 Experimental Setup

In order to observe and compare the effect on performance of selection pressure, parallel varying mutation, and recombination, we use the following algorithms: (i) A simple canonical GA with proportional selection and crossover & background mutation (CM), denoted as cGA; (ii) a simple GA with (μ, λ) proportional selection and CM, denoted as GA(μ, λ); (iii) a GA that uses (μ, λ) proportional selection, CM, and parallel varying mutation SRM, denoted as GA-SRM(μ, λ); and (iv) a version of GA-SRM(μ, λ) with no crossover ($p_c = 0.0$), denoted as M-SRM(μ, λ). To observe the effect of drift the algorithms are used with the fitness duplicates elimination feature either on or off. The superscript ^{ed} attached to the name of a GA indicates that the elimination of duplicates feature is on, i.e cGA^{ed}, GA$^{ed}(\mu, \lambda)$, GA-SRM$^{ed}(\mu, \lambda)$, and M-SRM$^{ed}(\mu, \lambda)$.

Table 1. GAs Parameters

Parameter	cGA	$GA(\mu, \lambda)$	$GA\text{-}SRM(\mu, \lambda)\ ADP$	$M\text{-}SRM(\mu, \lambda)\ ADP$
Selection	Proportional	(μ, λ) Prop.	(μ, λ) Prop.	(μ, λ) Prop.
p_c	0.6	0.6	0.6	0.0
$p_m^{(CM)}$	$1/N$	$1/N$	$1/N$	$1/N$
$p_m^{(SRM)}$	–	–	$[\alpha = 0.5, 1/N]$	$[\alpha = 0.5, 1/N]$
			$\beta = 2, \ell = N$	$\beta = 2, \ell = N$
$\lambda_{CM} : \lambda_{SRM}$	–	–	$1 : 1$	$1 : 1$

The offspring population is set to 200 and all GAs use linear scaling, two-point crossover as the recombination operator and the standard bit-flipping method as the mutation operator after crossover. Linear scaling is implemented as indicated in [20] (p.79), where the coefficients a and b that implement the linear transformation $f' = af + b$ are chosen to enforce equality of the raw (f) and scaled (f') average fitness values and cause the maximum scaled fitness to be twice the average fitness. The number of evaluations is set to 2×10^6 for both GAs and RBC+. The genetic operators parameters used for the GAs are summarized in Table 1. In the case of GA-SRM, from our experiments we observed that values of β in the range $1.1 < \beta \leq 2$ produce similar high results. Setting $\beta > 2$ reduces the mutation segment too fast and poorer results where achieved. Fixing $\beta = 2$ we sampled the threshold τ for all combinations of N and K. Results presented here use the best sampled τ.

We conduct our study on NK Landscapes with $N = 96$ bits varying the number of epistatic interactions from $K = 0$ to $K = 48$ in increments of 4. We use landscapes with *random* epistatic patterns among bits and consider circular genotypes to avoid boundary effects. The figures plot the mean value of the best solution observed over 50 different randomly generated problems (for each value of K) starting each time with the same initial population. The vertical bars overlaying the mean curves represent 95% confidence intervals. Note that we maximize all problems.

5   Simulation Results and Discussion

5.1   Selection Pressure

First, we observe the effect of a higher selection pressure on the performance of a simple GA. Fig. 4 plots results by cGA(200) and GA with $(\mu, \lambda)=\{(100,200), (60,200), (30,200)\}$. Results by the random bit climber RBC+ are also included for comparison. From this figure some important observations are as follows. (i) cGA does relatively well for very low epistasis ($K \leq 8$) but its performance falls sharply for medium and high epistasis ($K \geq 12$). Similar behavior by another simple GA has been observed in [8] and [9]. (ii) GA(μ, λ) that includes an stronger selection pressure performs worse than cGA for low epistasis but it outperforms cGA(200) for medium and high epistasis ($K \geq 12$). The behaviors of GA(100,200), GA(60,200), and GA(30,200) are similar. Results by cGA and GA(μ, λ) indicate the importance of an appropriate selection pressure to pull

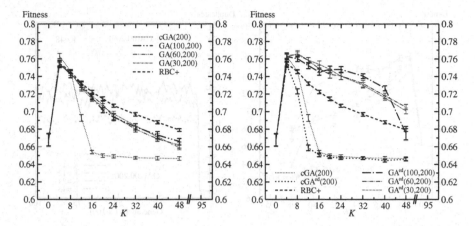

Fig. 4. Higher selection pressure.   **Fig. 5.** Duplicates elimination.

the population to fittest regions of the search space (note that the genetic operators are the same in both algorithms). The selection pressure induced by proportional selection seems to be appropriate only for very small K. As K increases a stronger selection pressure works better. (iii) The overall performance of RBC+ is better than both cGA and GA(μ, λ).

5.2   Duplicates Elimination

Second, we observe the effect of genetic drift by setting on the fitness duplicates elimination feature. Fig. 5 plots results by cGA^{ed}(200) and GA^{ed}(μ, λ) with (μ, λ)= {(100, 200), (60,200), (30,200)}. Results by cGA(200) and RBC+ are also included for comparison. From this figure we can see that eliminating duplicates affects differently the performance of the GAs. (i) It deteriorates even more the performance of cGA. (ii) On the other hand, the combination of higher selection pressure with duplicates elimination produces a striking increase on performance. Note that GA^{ed}(100,200) achieves higher optima than RBC+ for $4 \leq K \leq 40$ while GA^{ed}(60,200) and GA^{ed}(30,200) does it for $4 \leq K \leq 48$.

As shown in 5.1, as K increases a higher selection pressure improves the performance of the simple GA. However, it would also increase the likelihood of duplicates. Preventing duplicates distributes more fairly selective pressure in the population, removes an unwanted source of selective bias in the algorithm[19] and postpones genetic drift[21],[22]. For example, duplicates with lower fitness can end up with more selective advantage than higher fitness individuals. On the other hand, duplicates with higher fitness may lead to premature convergence. GA^{ed}(μ, λ) takes advantage of the higher selection pressure avoiding the unwanted selective bias. The drop in performance by cGA^{ed} compared to cGA suggests that the latter uses the duplicates as a way to increase its selection pressure.

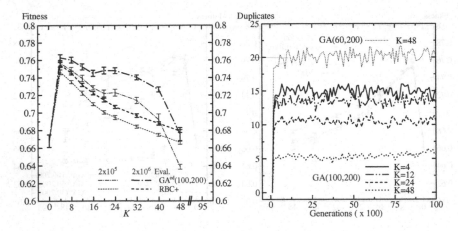

Fig. 6. Mean fitness after 2×10^5 and 2×10^6 evaluations.

Fig. 7. Number of eliminated duplicates.

The optima achieved by $GA^{ed}(100,200)$ is lower than RBC+ for $K = 48$. However, note that $GA^{ed}(60,200)$ and $GA^{ed}(30,200)$ achieved higher optima than RBC+. This suggest that for $K = 48$ even the pressure imposed by $(\mu, \lambda)=(100,200)$ is not enough.

In our study we run the algorithms for 2×10^6 evaluations whereas only 2×10^5 evaluations have been used in previous studies[8,9]. Fig. 6 illustrates the optima achieved by $GA^{ed}(100,200)$ and RBC+ after 2×10^5 and 2×10^6 evaluations. From this figure we can see that allocating more evaluations allows both algorithms to find higher optima, being the rate of improvement by the GA greater than the random bit climber RBC+. Note that for most values of K, even after 2×10^6 evaluations, RBC+ still does not reach the optima achieved by $GA^{ed}(100,200)$ after 2×10^5 evaluations.

Fig. 6 illustrates the number of fitness duplicates[1] eliminated in $GA^{ed}(100,200)$ for landscapes with values of $K = \{4, 12, 24, 48\}$ and the number of duplicates eliminated in $GA^{ed}(60,200)$ for $K = 48$. Most of this fitness duplicates were actual clones (i.e. individuals with the same genotype besides having the same fitness). For example, 148993 fitness duplicates were created during the 10000 generations for $K = 4$ (average in the 50 runs). Out of these, 99.88% corresponded to actual clones. Similar percentages were observed for other values of K. This indicates that the approach is quite effective eliminating clones while being computationally efficient (there is no need to calculate and check hamming distances). Note that (μ,λ) proportional selection is a kind of truncation selection and sorting of the whole population is necessary. Once sorting has been done, the non-duplicates policy requires at most $\mathcal{O}(\lambda)$ to eliminate fitness duplicates. From Fig. 6 it is also important to note that in the case of GA^{ed} (μ,λ) the number of fitness duplicates remains relatively constant throughout the generations for a given K. For example, for $K = 12$ the number of duplicates by GA^{ed} (100,200) is about 14, 7% of

---

[1] Fitness duplicates are counted and eliminated while the offspring population is being truncated from λ to μ. Thus, the number of duplicates of Fig. 6 indicate only those counted until the best μ are found. The number of duplicates in λ can be a little higher.

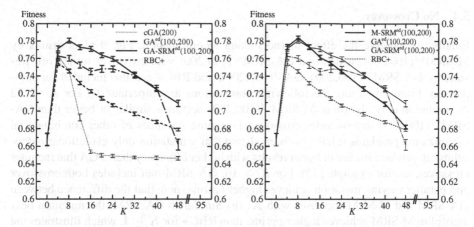

Fig. 8. Parallel varying mutation.   **Fig. 9.** No recombination, M-SRM^{ed}(100,200)

the total offspring population. On the other hand, if the non-duplicates policy is off, the number of duplicates increases considerably with the generations. For the same $K = 12$, the number of duplicates by GA(100,200) at generation 100 is already about 62 and at generation 2000 it has increased to 80, 31% and 40% of the total offspring population, respectively. In addition to avoiding the unwanted selective bias, as mentioned above, by eliminating duplicates we can also increase substantially the likelihood that the evolutionary algorithm will explore a larger number of candidate solutions, which increases the possibility of finding higher optima.

5.3 Parallel Varying Mutation

Third, we observe the effect of parallel varying mutation. Fig. 8 plots results by GA-SRM^{ed}(100,200). Results by cGA(200), GA^{ed}(100,200) and RBC+ are also included for comparison. From this figure the following observations are relevant. (i) The inclusion of parallel varying mutation can improve further convergence reliability. Note that GA-SRM^{ed}(100,200) with ADP achieves higher optima than GA^{ed}(100,200) for $4 \le K < 32$. For $K = 32$ and $K = 40$, however, GA-SRM^{ed}(100,200) is not better than GA^{ed}(100,200), which indicates that varying mutation is not working properly at this values of K. (ii) The optima achieved by GA^{ed}(100,200) is lower than RBC+ for $K = 48$, which seems to be caused by a lack of appropriate selection pressure as mentioned in 5.2. However, for the same selection pressure, GA-SRM^{ed}(100,200) achieved higher optima than RBC+. This nicely shows that the effectiveness of a given selection pressure is not only correlated to the complexity of the landscape, but also to the effectiveness of the operators searching in that landscape. In the case of GA-SRM^{ed}(100,200), the inclusion of varying mutation increases the effectiveness of the operators, hiding selection deficiencies. It would be better to correct selection pressure and try to use mutation to improve further the search.

5.4   No Crossover

Finally, we observe the effect of (not) using recombination. Fig. 9 plots results by M-SRM^{ed}(100,200), which is a GA-SRM^{ed}(100,200) with crossover turned off. Results by GA-SRM^{ed}(100,200), GA^{ed}(100,200), and RBC+ are also included for comparison. From this figure the following observations are important. (i) For $K \leq 12$ the mutation only algorithm M-SRM^{ed}(100,200) performs similar or better than GA-SRM^{ed}(100,200) that includes crossover. For some instances of other combinatorial optimization problems it has also been shown that a mutation only evolutionary algorithm can produce similar or better results with higher efficiency than a GA that includes crossover, see for example [23]. For $K \geq 16$, GA-SRM that includes both crossover and parallel varying mutation achieves higher optima; note that the difference between GA-SRM and M-SRM increases with K. (ii) Similar to GA-SRM, the mutation only algorithm M-SRM achieves higher optima than RBC+ for $K \geq 4$, which illustrates the potential of evolutionary algorithms, population based, with or without recombination, over strictly local search algorithms in a broad range of classes of problems. This is in accordance with theoretical studies of first hitting time of population-based evolutionary algorithms. He and Yao[24] have shown that in some cases the average computation time and the first hitting probability of an evolutionary algorithm can be improved by introducing a population.

The behavior of M-SRM compared with GA-SRM is at first glance counterintuitive and deserves further explanation. Results in [8] imply that the usefulness of recombination would decrease with K. However, Fig. 9 seems to imply exactly the opposite. A sensible explanation for these apparently opposing results comes from the structure of the underlying fitness landscape. Kauffman[2] has shown that clustering of high peaks can arise as a feature of landscapes structure without obvious modularity, as is the case of NK landscapes with K epistatic inputs to each site chosen at random among sites. In this case, recombination would be a useful search strategy because the location of high peaks carry information about the location of the other high peaks[2].

In the case of small K the problems are the easier and a mutation only EA proves to be very effective, although the landscape is more structured than for high K. As K increases the structure of the landscape fades away decreasing also the effectiveness of recombination. However, what figure Fig. 9 is showing is that the decay of mutation alone seems to be faster than the decay of recombination interacting with varying mutation as the complexity of the landscape increases with K and its structure fades. In other words, the relative importance of recombination interacting with varying mutation increases with K respect to mutation alone.

It should be mentioned that recent developments and discussions on the status of the schema theorem[25] might give new insights to better understand the behavior of GAs than the traditional interpretation of the theorem.

6   Conclusions

We have examined the effect of selection, drift, mutation, and recombination on NK-Landscapes for $N = 96$ varying K from 0 to 48. We showed that GAs can be robust algorithms on NK-Landscapes postponing drift by eliminating fitness duplicates and using selection pressure higher than a canonical GA. Even simple GAs with these two features performed better than the single bit climber RBC+ for a broad range of classes of problems ($K \geq 4$). We also showed that the interaction of parallel varying mutation with crossover (GA-SRM) improves further the reliability of the GA. Contrary to intuition we found that a mutation only EA can perform as well as GA-SRM that includes crossover for small values of K, where crossover is supposed to be advantageous. However, we found that the decay of mutation alone seems to be faster than the decay of crossover interacting with varying mutation as the complexity of the landscape increases with K and its structure fades. In other words, the relative importance of recombination interacting with varying mutation increased with K respect to mutation alone. We conclude that NK-Landscapes are useful for testing the GA's overall behavior and performance on a broad range of classes of problems and also for testing each one of the major processes involved in a GA, which gives valuable insights to improve GAs by understanding better the complex and interesting behavior that arises from the interaction of such processes. In the future we would like to look deeper into the relationship among selection pressure, the range for varying mutation, and K. This could give important insights to incorporate heuristics that can increase the adaptability of GAs according to the complexity of the problem.

References

1. K. A. De Jong and M. A. Potter and W. M. Spears, "Using Problem Generators to Explore the Effects of Epistasis", *Proc. 7th Int'l Conf. Genetic Algorithms*, Morgan Kauffman, pp.338–345, 1997.
2. S. A. Kauffman, *The Origins of Order: Self-Organization and Selection in Evolution*, pp.33–67, Oxford Univ. Press, 1993.
3. Y. Davidor, "Epistasis Variance: A Viewpoint of GA-Hardness", *Proc. First Foundations of Genetic Algorithms Workshop*, Morgan Kauffman, pp.23–25, 1990.
4. B. Manderick, M. de Weger, and P. Spiessens, "The Genetic Algorithm and the Structure of the Fitness Landscape", *Proc. 4th Int'l Conf. on Genetic Algorithms*, Morgan Kaufmann, pp.143–150, 1991.
5. L. Altenberg, "Evolving Better Representations through Selective Genome Growth", *Proc. 1st IEEE Conf. on Evolutionary Computation*, Pistcaway, NJ:IEEE, pp.182–187, 1994.
6. J. E. Smith, *Self Adaptation in Evolutionary Algorithms*, Doctoral Dissertation, University of the West of England, Bristol, 1998.
7. P. Merz and B. Freisleben, "On the Effectiveness of Evolutionary Search in High-Dimensional NK-Landscapes", *Proceedings of the 1998 IEEE International Conference on Evolutionary Computation*, IEEE Press, pp. 741–745, 1998.
8. R. Heckendorn, S. Rana, and D. Whitley, "Test Function Generators as Embedded Landscapes", *Foundations of Genetic Algorithms 5*, Morgan Kaufmann, pp.183–198, 1999.
9. K. E. Mathias, L. J. Eshelman, and D. Schaffer, "Niches in NK-Landscapes", *Foundations of Genetic Algorithms 6*, Morgan Kaufmann, pp.27–46, 2000.
10. M. Shinkai, H. Aguirre, and K. Tanaka, "Mutation Strategy Improves GA's Performance on Epistatic Problems", *Proc. 2002 IEEE World Congress on Computational Intelligence*, pp.795–800, 2002.

11. L. J. Eshelman, "The CHC Adaptive Search Algorithm: How to Have a Save Search When Engaging in Nontraditional Genetic Recombination", *Foundations of Genetic Algorithms* , Morgan Kaufmann, pp.265–283, 1991.
12. L. Davis, "Bit-Climbing, Representational Bias, and Test Suite Design", *Proc. 4th Int'l Conf. on Genetic Algorithms*, Morgan Kaufmann, pp.18–23, 1991.
13. L. Altenberg, "Fitness Landscapes: NK Landscapes", *Handbook of Evolutionary Computation*, Institute of Physics Publishing & Oxford Univ. Press, pp. B2.7:5–10, 1997.
14. R. E. Smith and J. E. Smith, "An examination of Tunable, Random Search Landscapes", *Foundations of Genetic Algorithms 5*, Morgan Kaufmann, pp.165–182, 1999.
15. R. E. Smith and J. E. Smith, "New Methods for Tunable, Random Landscapes", *Foundations of Genetic Algorithms 6*, Morgan Kaufmann, pp.47–67, 2001.
16. H. Aguirre, K. Tanaka, and T. Sugimura. Cooperative Model for Genetic Operators to Improve GAs. In *Proc. IEEE Int'l Conf. on Information, Intelligence, and Systems*, pp.98–106, 1999.
17. H. Aguirre, K. Tanaka, T. Sugimura, and S. Oshita. Cooperative-Competitive Model for Genetic Operators: Contributions of Extinctive Selection and Parallel Genetic Operators. In *Proc. Late Breaking Papers Genetic and Evolutionary Computation Conference*, Morgan Kaufmann, pp.6–14, 2000.
18. T. Bäck, *Evolutionary Algorithms in Theory and Practice*, Oxford Univ. Press, 1996.
19. D. Whitley, "The GENITOR Algorithm and Selection Pressure: Why Rank-Based Allocation of Reproductive Trials is Best", *Proc. Third Intl. Conf. on Genetic Algorithms*, Morgan Kauffman, pp. 116–121, 1989.
20. D. E. Goldberg, *Genetic Algorithms in Search, Optimization and Machine Learning*, Addison-Wesley, 1989.
21. L. Eshelman and D. Schaffer, "Preventing Premature Convergence in Genetic Algorithms by Preventing Incest", *Fourth Intl. Conf. on Genetic Algorithms*, Morgan Kauffman, pp.115–122, 1991.
22. D. Schaffer, M. Mani, L. Eshelman, and K. Mathias, "The Effect of Incest Prevention on Genetic Drift", *Foundations of Genetic Algorithms 5*, Morgan Kaufmann, pp.235–243, 1999.
23. K.-H. Liang, X. Yao, C. Newton and D. Hoffman, "Solving Cutting Stock Problems by Evolutionary Programming", *Evolutionary Programming VII: Proc. of the Seventh Annual Conference on Evolutionary Programming (EP98)*, Lecture Notes in Computer Science, Springer-Verlag, vol. 1447, pp.291–300, 1998.
24. J. He and X. Yao, "From an Individual to a Population: An Analysis of the First Hitting Time of Population-Based Evolutionary Algorithms," *IEEE Transactions on Evolutionary Computation*, 6(5):495–511, 2002.
25. C.R. Reeves and J.E. Rowe, *Genetic Algorithms – Principles and Perspectives*, Kluwer, Norwell, MA, 2002.

Multilevel Heuristic Algorithm for Graph Partitioning

Raul Baños[1], Consolación Gil[1], Julio Ortega[2], and Francisco G. Montoya[3]

[1] Dept. Arquitectura de Computadores y Electrónica, Universidad de Almería,
La Cañada de San Urbano s/n, 04120 Almería (Spain),
{rbanos,cgil}@ace.ual.es,
[2] Dept. Arquitectura y Tecnologia de Computadores, Universidad de Granada
Campus de Fuentenueva, Granada (Spain),
julio@atc.ugr.es,
[3] Dept. Ingenieria Civil, Universidad de Granada
Campus de Fuentenueva, Granada (Spain),
pgil@ugr.es

Abstract. Many real applications involve optimisation problems where more than one objective has to be optimised at the same time. One of these kinds of problems is graph partitioning, that appears in applications such as VLSI design, data-mining, efficient disc storage of databases, etc. The problem of graph partitioning consists of dividing a graph into a given number of balanced and non-overlapping partitions while the cuts are minimised. Although different algorithms to solve this problem have been proposed, since this is an NP-complete problem, to get more efficient algorithms for increasing complex graphs still remains as an open question. In this paper, we present a new multilevel algorithm including a hybrid heuristic that is applied along the searching process. We also provide experimental results to demonstrate the efficiency of the new algorithm and compare our approach with other previously proposed efficient algorithms.

1 Introduction

The problem of graph partitioning appears in many different applications, such as VLSI design [1,2], test patterns generation [3-5], data-mining [6], efficient storage of great data bases [7], etc. It consists of dividing a given graph in a set of partitions containing a balanced number of nodes, while the total number of edge cuts is minimised. There is a trade-off between these two objectives as an improvement in one of them usually implies to worsen the other. Moreover, as the problem is NP-complete [8], efficient procedures that provide solutions with a required quality in a reasonable amount of time are very useful. Different strategies have been proposed to solve this graph partitioning problem. These can be classified as approaches based on movements [9,10], based on geometric representations [11], hierarchic multilevel and clustering algorithms [12,13], multilevel recursive bisection algorithms [14] and genetic algorithms [15,16]. There are also hybrid schemes [17] that combine different strategies.

S. Cagnoni et al. (Eds.): EvoWorkshops 2003, LNCS 2611, pp. 143–153, 2003.
© Springer-Verlag Berlin Heidelberg 2003

Recently, multilevel approaches have been considered as a way to obtain good solutions in a reduced amount of time. In this work, we propose a new multilevel algorithm that includes a heuristic that we have previously proposed [17]. As the run times for the proposed procedure go from seconds to several hours, its usefulness is higher as the quality of the solution to find is more relevant than the time required to obtain it (under reasonable time bounds). There are many applications, such as test pattern generation [3], that can be included in this class of applications.

Section 2 gives a more precise definition of the graph partitioning problem and describes the cost function used in the optimisation process. Section 3 describes the proposed algorithm, while Section 4 provides and analyses the experimental results obtained when it is applied to several benchmarks graphs. Finally, Section 5 gives the conclusions of the paper and the future work to do.

2   Graph Partitioning Problem

Given a graph $G=(V, E)$, where V is the set of nodes, with $|V| = n$, and E the set of edges that determines the connectivity among the nodes, the graph partitioning problem consists of dividing G into k parts, $G_1, G_2, ..., G_k$, with the requirement that $Vi \cap Vj = \phi$ ($i \neq j$) for each $V_1, V_2, ..., V_k$; each partition has a number of nodes determined, $n/k - |max\_imbalance| \leq |Vi| \leq n/k + |max\_imbalance|$; and $\sum_{k=1}^{K} Vi = V$. Whenever the nodes and edges have weights, $|v|$ denotes the weight of node v, and $|e|$ denotes the weight for the edge e. All the benchmarks graphs used to evaluate the quality of our algorithm have nodes and edges with weight equal to one ($|v|=1$ and $|e|=1$). However, the implemented procedure is able to process graphs with any weight values.

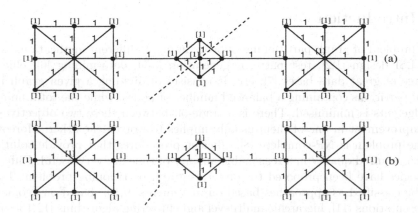

Fig. 1. Two ways of divide the graph.

In figure 1 a graph and two possible partitions of it are shown. In figure 1(a) there are two balanced partitions of the graph, although the number of cuts is not minimized. On the other hand, in figure 1(b) a partition with the optimal number of cuts is given. Nevertheless this partition breaks sensibly the requirement of load balancing. This is a clear example showing the opposition of both objectives. In order to choose one or another solution we have considered a cost function that can be represented as:

$$c(s) = \alpha \cdot ncuts(s) + \beta \cdot \sum_{k=1}^{K} 2^{deviation(k)}$$

In this function both objectives are included. The first term is associated to the cuts of the partition. On the other hand, the second term corresponds to the penalty associated to a deviation with respect to a balanced partition: the maximum difference in the number of nodes of a partition with respect to the state of absolute balance (n/k).

3   Multilevel Procedure Proposed

3.1   Multilevel Paradigm

Recently, the popularity of the multilevel algorithms [12,13] has increased remarkably as a tool for solving optimisation problems. The general scheme of this strategy is shown in figure 2.

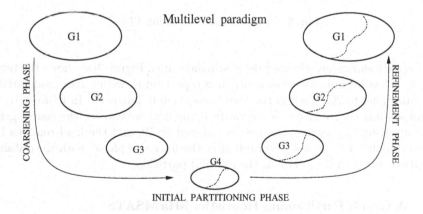

Fig. 2. Multilevel Paradigm.

The idea consists on taking the graph to be divided and to group the nodes building clusters, that will became the nodes of a new graph, called coarse graph. By repeating this process, graphs with less nodes are created, until a graph is obtained, on which it will be able to apply some of the existing procedures to obtain a first partition of the graph. However, this partition is usually not very good, since the coarsening phase intrinsically implies a loss of accuracy. This is the reason to apply a final phase in which the graph is expanded until its original configuration, by applying a refinement phase through a search algorithm in each step.

The main potential of the multilevel paradigm arises in the coarsening phase, as this phase makes it possible to hide a great number of edges within the grouped nodes. This implies a bigger manageability of the coarser graphs during the search process. The selection of the nodes to be grouped can be carried out by different heuristics.

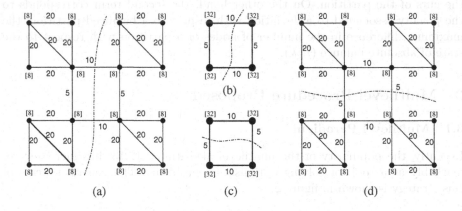

Fig. 3. Utility of the Coarsening Phase.

Figure 3 shows how the multilevel scheme works. Figure 3(a) shows the target graph. It has been divided randomly in two partitions, with a cost associated to the cuts equal to 20, whereas the load is completely balanced. In figure 3(b) the corresponding coarse graph is shown. In figure 3(c) we can see the coarse graph after applying the search. The cost is reduced to 10, and the load remains balanced. Finally, 3(d) shows the graph after the uncoarse phase, with the obtained partition shown in 3(c). This is the optimal partition.

3.2   A Graph Partitioning Heuristic: MLrMSATS

In this section a detailed description of the proposed multilevel heuristic algorithm is given. As we commented previously, the multilevel paradigm consists of three phases. Next, we give a detailed description of each one of these phases.

Coarsening phase. The creation of a coarse graph G_{i+1} (V_{i+1}, E_{i+1}) from the graph obtained in the previous level, G_i (V_i, E_i), consists of grouping pairs of nodes of G_i that are connected by an edge, obtaining a new graph G_{i+1}. Repeating this process, graphs with less nodes are generated, until finally a coarse graph of few hundreds or thousands of nodes is obtained. We have used two different strategies to determine the way to group the nodes. The first strategy selects the nodes randomly. A node not grouped in this level is taken, and another bordering to it and not visited either in this level is chosen randomly to carry out the union. The second strategy consists of grouping the nodes in function of the weight of the edges that connect them. A node not grouped in this level is taken and it is grouped with that one not visited whose common edge has the higher weight. The advantage of this alternative arises in hide the heaviest edges during this phase, it is, the resulting graph is going to be built by edges with less weights. So, the cuts obtained after the graph partitioning will tend to decrease. In both alternatives there are nodes that do not have any free neighbour to be grouped. These nodes will pass directly like node of the inferior level. This causes the weight of the nodes of the new graph to be imbalanced. As the union of nodes is made in pairs, this problem will be worst in the inferior levels. Thus the weights of the nodes in G_i (level i) take values in the interval $[1,2i]$. Supossing that all the nodes of the initial graph have weight equal to one, if we have a graph that descends 10 levels, in the graph G_{10} we can have a node u with weight $|u|=1$, and another v with weight $|v|=2^{10}$. This can cause important problems to get balanced partitions, mainly in the inferior levels, since the movement of very heavy nodes between neighboring partitions could not be accepted by the objective function. In order to solve this problem, we propose to visit the nodes in ascending order of weights. So, first of all the algorithm will visit the nodes with lower weights, i.e., those that could have been isolated in previous levels.

Initial Partitioning phase. After the coarsening phase, the next step consists of making an initial partition of the obtained graph. In this phase of the multilevel proposed algorithm, we have raised the use of the procedure called Levelized Nested Dissection (LND). This procedure, also called Graph Growing Algorithm, starts in a randomly selected node. This node is assigned to the first partition, the same as its adjacent nodes. This recursive process is repeated until the partition reaches n/k nodes. From then, the following visited nodes will be assigned to the following partition, repeating the process until all the nodes are assigned to a certain partition. The position of the initial node will determine the structure of the initial partitions. Thus the random selection of nodes offers a very interesting diversity in the case of treating the problem by evolutionary computation.

Uncoarsening phase. The coarsening phase implies loss of resolution. Thus the application of an algorithm as LND is not enough to obtain a quality partition. So, it is necessary a refinement phase, in which is carried out the inverse process to the coarsening phase. The idea consists of going from level to level, uncoarsening the nodes the same way they had been grouped in the coarsening phase, and applying some of the existing search techniques simultaneously to optimize the quality of the solution. In the upper levels the quality of the solution tends to be improved, since there is more freedom in the uncoarsening phase, although the number of possible movements increases, and therefore the complexity of the searching process. We propose a mixed heuristic based on Simulated Annealing (SA) [18] and Tabu Search (TS) [19]. The use of both procedures define a hybrid strategy that makes it possible not only to escape from local minimums but also to avoid cycles in the search process. SA uses a variable called temperature, t, whose value is reduced based on a factor, $tfactor$, as the number of iterations grows. The variable t acts simultaneously as a control variable for the number of iterations of the algorithm, and to determine the probability of accepting a given solution, according to the Metropolis function. With this function, the reduction of t implies a reduction in the probability of accepting movements that worsen the cost function. On the other hand, the use of a reduced neighborhood at the time of exploring the search space causes the appearance of cycles. To avoid this problem, we use TS as a complement to SA. In [17] a comparison of results of both techniques is presented. Some executions indicate that the use of both techniques improve the results obtained when only SA is applied.

```
begin_MSATS
  generate an initial solution s₀;
  s=s₀; s*=s₀;
  select a initial temperature t₀¿0;
  select the temperature reduction factor, tfactor(α);
  n_failures=0;
  determine the set of possible movements in the reduced neighbor;
  while ((iteration¡max_iteration)and(t¿0)and(n_failures¡max_failures))
  begin
    best_movement=T;
    for each (movement_candidate) do
    begin
      get a neighbor solution s⁻ making a movement of s;
      movement_cost=c(s⁻)-c(s);
      if (metropolis(movement_cost,t)) then
      begin
        accept_movement();
        if (movement_cost ¿ 0) then
          update tabu list;
        s=s⁻;
      end_if
      if (movement_cost ¡ best_movement) then
      begin
        best_movement=movement_cost;
        s=s⁻; s'=s⁻;
      end_if
      n_failures=0;
      s*=s';
    end_if
    else n_failures=n_failures+1;
    end_for
    update the set of candidates movements in the neighborhood;
    t=tfactor·t;
    iteration=iteration+1;
  end_while
  s* is the solution;
end_MSATS

  where :
    s₀ : initial solution;
    s : best solution of the last generation;
    s* : best solution found;
    s⁻ : neighbor solution obtained from s solution;
    s' : temporal best solution;
```

4   Experimental Results

Experimental tests have been performed by using a set of graphs of different size and topology. The set of graphs is a public domain set that is frecuently used to compare graph partitioning algorithms. The experiments have been performed using a 500 MHz Pentium III with 128 Mb of memory. Table 1 gives a brief description of some characteristics of these graphs: the number of vertex, edges, maximun connectivity *(max)* (number of neighbors of the node with the highest neighborhood), minimun connectivity *(min)*, average connectivity *(avg)* and file size.

Table 1. Set of benchmarks used to evaluate our procedure.

Graph	|V|	|E|	max	min	avg	File Size (KB)
add20	2395	7462	123	1	6.23	63
add32	4960	9462	31	1	3.82	90
wingnod al	10937	75488	28	5	13.80	768
fe4elt2	11143	32818	12	3	5.89	341
vibrobox	12328	165250	120	8	26.81	1679
memplus	17758	54196	573	1	6.10	536
cs4	22499	43858	4	2	3.90	506
bcsstk32	44609	985046	215	1	44.16	11368
brack2	62631	366559	32	3	11.71	4358
fetooth	78136	452591	39	3	11.58	5413
ferotor	99617	662431	125	5	13.3	7894
feocean	143437	409593	6	1	5.71	5242
wave	156317	1059331	44	3	13.55	13479

Best solutions for these benchmarks are located and updated at URL : http://www.gre.ac.uk/ c.walshaw/partition/. These solutions indicate the number of cuts classified by levels of imbalance (0%, 1%, 3% and 5%). This is the reason why we use as objective the number of cuts in the objective function described in section 2, and a 5% as restriction in the degree of imbalance. Under these conditions, the objective function described in Section 2, will have as parameters $\alpha =1$, and $\beta =0$; with deviation(k)¡5 for k in the interval [1,K];

We compare the results obtained by our algorithm with METIS [11], a public domain graph partitioning software, also based on the multilevel paradigm. We have used METIS.4.0., which can be downloaded from URL : http://www-users.cs.umn.edu/ karypis/metis/metis/files/metis-4.0.tar.gz. This package has two different implementations, *pMetis*, based in a multilevel recursive bisection, and *kMetis*, based in a multilevel k-way partitioning. Results obtained by *pMetis* are often balanced, while *kMetis* obtains partitions weakly imbalanced.

Table 2 shows the results obtained by MLrMSATS and compared with *pMetis* and *kMetis* (level of imbalance less than 5%). The values of the parameters used are *t0=100* and *tfactor=0.995*. With these parameters the run time of MLrMSATS is two or three magnitudes superior to METIS, although as we can

see the results obtained by MLrMSATS improve (or are very similar to) those
obtained by *pMetis* and *kMetis* in most cases. Average imbalances of *pMetis*
and *kMetis* are lower than the values obtained by MLrMSATS, although with
similar imbalances MLrMSATS obtains better results. For instance, in *add32* or
memplus, MLrMSATS obtains a lower number of cuts than *kMetis* with a similar
imbalance degree.

Figure 4 describes the behaviour of the algorithm when the SA parameters
are modified. More specifically the variation of *tfactor* considerably affects the
behaviour of the algorithm. With *tfactor=0.995* the algorithm obtains better
results, although a greater number of iterations is neccesary and therefore more
computation time is required. By using *tfactor=0.95*, the searching process fi-
nalizes before, but the quality of the solutions found is worse.

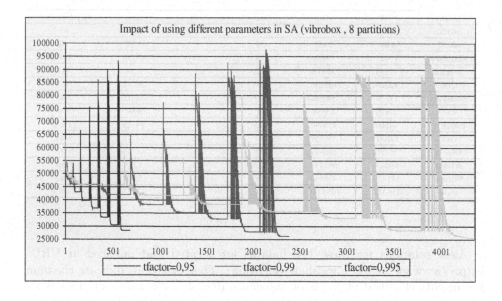

Fig. 4. Refinement phase using Simulated Annealing and Tabu Search.

Table 3 compares rMSATS [17], which applies the hybrid technique directly
in the original graph without using multilevel paradigm, and MLrMSATS. The
same parameters, *t0=100* and *tfactor=0.995*, are used in both cases. As it can
be shown in most cases MLrMSATS improves the results obtained by rMSATS.
These results, along with the conclusions obtained from table 2, demonstrate
the good behaviour of the algorithm here proposed.

5   Conclusions

In this paper we have described a new technique to solve the problem of graph
partitioning. Using an hybrid heuristic that mixes Simulated Annealing and

Table 2. Comparison between MLrMSATS and METIS.

Graph	algorithm	K=2		K=4		K=8		K=16		K=32	
		cuts	imb.	cuts	imb.	cuts	imb.	cuts	imb.	cuts	imb.
add20	pMetis	725	0	1292	0	1907	0	2504	1	3008	3
	kMetis	719	3	1257	3	1857	5	2442	5	3073	74
	MLrMSATS	**696**	5	**1193**	5	**1757**	5	**2186**	5	**2793**	4
add32	pMetis	21	0	42	0	81	0	**128**	0	**288**	1
	kMetis	28	2	44	3	102	3	206	3	352	15
	MLrMSATS	**10**	0	**35**	3	133	5	184	4	294	5
wing nodal	pMetis	1820	0	4000	0	6070	0	9290	0	13237	0
	kMetis	1855	0	4355	2	6337	3	9465	3	12678	3
	MLrMSATS	**1670**	5	**3596**	5	**5387**	5	**8425**	5	**11812**	6
fe4elt2	pMetis	**130**	0	359	0	654	0	1152	0	1787	0
	kMetis	132	0	398	3	684	3	1149	3	1770	3
	MLrMSATS	**130**	0	**350**	2	**632**	5	**1054**	5	**1662**	6
vibro box	pMetis	12427	0	21471	0	28177	0	37441	0	46112	0
	kMetis	**11952**	1	23141	2	29640	3	38673	3	45613	3
	MLrMSATS	12096	3	**20391**	5	**26147**	5	**34791**	5	**42530**	5
mem plus	pMetis	6337	0	10559	0	13110	0	14942	0	17303	0
	kMetis	6453	3	10483	3	12615	3	14604	6	16821	6
	MLrMSATS	**5662**	5	**9427**	5	**12283**	5	**14176**	5	**15661**	5
cs4	pMetis	414	0	1154	0	1746	0	2538	0	3579	0
	kMetis	**410**	0	1173	3	**1677**	3	2521	3	3396	3
	MLrMSATS	418	4	**1103**	4	1696	5	**2430**	5	**3371**	5
bcsstk 32	pMetis	5672	0	12225	0	**23601**	0	43371	0	**70020**	0
	kMetis	5374	0	**11561**	3	27311	3	**42581**	3	82864	3
	MLrMSATS	**5193**	5	18821	5	34596	5	51341	5	78260	7
brack2	pMetis	738	0	3250	0	**7844**	0	**12655**	0	19786	0
	kMetis	769	0	3458	1	8776	3	13652	3	19888	3
	MLrMSATS	**668**	4	**2808**	5	8582	5	13770	5	**18245**	5
fetooth	pMetis	4292	0	8577	0	13653	0	19346	0	29215	0
	kMetis	**4262**	2	7835	3	**13544**	3	20455	3	28572	3
	MLrMSATS	4436	5	**7152**	4	14081	5	**19201**	5	**26016**	5
ferotor	pMetis	2190	0	8564	0	15712	0	23863	0	36225	0
	kMetis	2294	0	8829	3	14929	3	24477	3	36159	3
	MLrMSATS	**1974**	3	**8143**	5	**14495**	4	**23807**	5	**33900**	5
feocean	pMetis	505	0	**2039**	0	**4516**	0	9613	0	14613	0
	kMetis	536	0	2194	2	5627	2	10253	3	16604	3
	MLrMSATS	**317**	3	3035	5	6023	5	**8699**	5	**13954**	5
wave	pMetis	9493	0	23032	0	34795	0	48106	0	72404	0
	kMetis	9655	0	21682	2	33146	3	48183	3	67860	3
	MLrMSATS	**8868**	5	**18246**	5	**30583**	5	**44625**	5	**63809**	5

Tabu Search in the refinement phase of a multilevel algorithm, our procedure obtains partitions of high quality in most of the benchmarks. Although the run time is approximately two or three magnitudes higher than those of METIS

Table 3. Comparison between MLrMSATS and rMSATS.

Graph	algorithm	K=2		K=4		K=8		K=16		K=32	
		cuts	time	cuts	time	cuts	time	cuts	time	cuts	time
add20	rMSATS	843	10	1349	14	1791	29	**2159**	71	**2637**	174
	MLrMSATS	**696**	29	**1193**	31	**1757**	45	2186	104	2793	329
add32	rMSATS	51	9	84	16	154	27	**165**	49	**279**	188
	MLrMSATS	**10**	5	**35**	13	**133**	29	184	83	294	293
wing nodal	rMSATS	1674	99	3626	124	5485	148	**8370**	218	11905	469
	MLrMSATS	**1670**	163	**3596**	196	**5387**	254	8425	447	**11812**	1174
fe4elt2	rMSATS	206	20	385	43	752	65	1223	141	1805	488
	MLrMSATS	**130**	34	**350**	68	**632**	104	**1054**	208	**1662**	693
vibro box	rMSATS	12229	192	20777	246	31092	294	**34460**	396	**42118**	964
	MLrMSATS	**12096**	439	**20391**	434	**26147**	543	34791	851	42530	1864
mem plus	rMSATS	7600	91	10795	114	12644	226	**14158**	259	**15492**	2211
	MLrMSATS	**5662**	171	**9427**	247	**12283**	554	14176	1371	15661	2315
cs4	rMSATS	861	70	1554	111	2261	186	2807	483	**3355**	1898
	MLrMSATS	**418**	151	**1103**	216	**1696**	319	**2430**	598	3371	1706
bcsstk 32	rMSATS	10422	534	29685	708	52697	826	68424	1132	86580	1905
	MLrMSATS	**5193**	418	**18821**	550	**34596**	836	**51431**	1465	**78260**	4407
brack2	rMSATS	711	194	3874	357	**8028**	522	14816	932	20676	2830
	MLrMSATS	**668**	334	**2808**	507	8582	911	**13770**	1626	**18245**	3933
fetooth	rMSATS	3966	386	10486	588	15920	862	23375	1629	32624	4300
	MLrMSATS	4436	725	**7152**	793	**14081**	1206	**19201**	2056	**26016**	4044
ferotor	rMSATS	2058	358	9445	898	17735	1251	28694	1971	39102	4308
	MLrMSATS	**1974**	654	**8143**	1148	**14495**	1687	**23807**	2650	**33900**	5776
feocean	rMSATS	406	598	**2670**	727	6583	1058	12232	2220	18482	8415
	MLrMSATS	**317**	1172	3035	1124	**6023**	1606	**8699**	2979	**13954**	10039
wave	rMSATS	15496	1166	30899	1642	47786	2409	58013	4209	74920	7767
	MLrMSATS	**8868**	1493	**18246**	1595	**30583**	2420	**44625**	4022	**63809**	10127

(depending on the parameters of the SA in the refinement phase), it reaches and improves the results of METIS in most cases. As future researching work, we plan to explore two directions: the use of multiobjective evolutionary techniques to improve the performance of the procedure, and the use of parallel processing to reduce the time required by the algorithm to converge.

References

1. Alpert, C.J., and Kahng, A., Recent Developments in Netlist Partitioning: A Survey. Integration: the VLSI Journal, 19/ 1–2 (1995) 1–81.
2. Banerjee, P.: Parallel Algorithms for VLSI Computer Aided Design. Prentice Hall, Englewoods Cliffs, NJ, 1994.
3. Gil, C.; Ortega, J., and Montoya, M.G., Parallel VLSI Test in a Shared Memory Multiprocessors. Concurrency: Practice and Experience, 12/5 (2000) 311–326.

4. Klenke, R.H., Williams, R.D., and Aylor, J.H., Parallel-Processing Techniques for Automatic Test Pattern Generation, IEEE Computer, (1992) 71–84.
5. Gil, C. and Ortega, J., A Parallel Test Pattern Generator based on Reed-Muller Spectrum, Euromicro Workshop on Parallel and Distributed Processing, IEEE (1997) 199–204.
6. Mobasher, B., Jain, N., Han, E.H., Srivastava J. Web mining : Pattern discovery from world wide web transactions. Technical Report TR-96-050, Department of computer science, University of Minnesota, Minneapolis, 1996.
7. Shekhar S. and DLiu D.R.. Partitioning similarity graphs: A framework for declustering problems.Information Systems Journal, 21/4, (1996)
8. Garey, M.R., and Johnson, D.S, Computers and Intractability: A Guide to the Theory of NP-Completeness, W.H. Freeman & Company. San Francisco, 1979.
9. Kernighan, B.W., and Lin, S., An Efficient Heuristic Procedure for Partitioning Graphics, The Bell Sys. Tech. Journal, (1970) 291–307.
10. Fiduccia, C., and Mattheyses, R., A Linear Time Heuristic for Improving Network Partitions, In Proc. 19th IEEE Design Automation Conference, (1982) 175–181.
11. Gilbert, J., Miller, G., and Teng, S., Geometric Mesh Partitioning: Implementation and Experiments, In Proceedings of International Parallel Processing Symposium, (1995).
12. Karypis, G. and Kumar V.: Multilevel K-way Partitioning Scheme for Irregular Graphs. Journal of Parallel and Distributed Computing, 48/1 (1998) 96–129.
13. Cong, J., and Smith, M., A Parallel Bottom-up Clustering Algorithm with Applications to Circuit Partitioning in VLSI Design, In Proc. ACM/IEEE Design Automation Conference, (1993) 755–760.
14. Schloegel, K., Karypis, G.; and Kumar, V., Graph Partitioning for High Performance Scientific Simulations, CRPC Parallel Computing Handbook, Morgan Kaufmann 2000.
15. Reeves, C.R., Genetic Algorithms, in: C.R. Reeves (eds.), Modern Heuristic Techniques for Combinatorial Problems, Blackwell, London, 1993, 151–196.
16. Soper, A.J., Walshaw, C., and Cross, M., A Combined Evolutionary Search and Multilevel Optimisation Approach to Graph Partitioning, Mathematics Research Report 00/IM/58, University of Greenwich, 2000.
17. Gil, C., Ortega, J., Montoya, M.G., and Banos R., A Mixed Heuristic for Circuit Partitioning. Computational Optimization and Applications Journal. 23/3 (2002) 321–340.
18. Dowsland, K.A, Simulated Annealing, in: C.R. Reeves (eds.), Modern Heuristic Techniques for Combinatorial Problems, Blackwell, London, 1993, 20–69.
19. Glover, F., and Laguna, M., Tabu Search, in: C.R. Reeves (eds.), Modern Heuristic Techniques for Combinatorial Problems, Blackwell, London, 1993, 70–150.

Experimental Comparison of Two Evolutionary Algorithms for the Independent Set Problem

Pavel A. Borisovsky[1] and Marina S. Zavolovskaya[2]

[1] Omsk Branch of Sobolev Institute of Mathematics,
13 Pevtsov str. 644099, Omsk, Russia.
borisovsky@iitam.omsk.net.ru
[2] Omsk State University, Mathemetical Department,
55 Mira str. 644077, Omsk, Russia
zavolovskaja@mail.ru

Abstract. This work presents an experimental comparison of the steady-state genetic algorithm to the (1+1)-evolutionary algorithm applied to the maximum vertex independent set problem. The penalty approach is used for both algorithms and tuning of the penalty function is considered in the first part of the paper. In the second part we give some reasons why one could expect the competitive performance of the (1+1)-EA. The results of computational experiment are presented.

1 Introduction

There is a number of papers devoted to investigation of properties of a simple evolutionary algorithm (1+1)-EA. From a formal view point many randomized algorithms successfully used in complexity theory may be considered as the (1+1)-EA optimizing a needle in the haystack function of specific structure, e.g. the algorithm of C. Papadimitriou for 2-Satisfiability problem with expected running time $O(n^2)$ (see e.g. [1]), the algorithm of U. Schöning [2] for k-Satisfiability problem etc. However, none of these methods actually exploits the full power of the (1+1)-EA since the search process is independent of fitness evaluations until the optimum is found. Many theoretical results show some encouraging properties of the (1+1)-EA viewed as a black-box optimization algorithm. In [3] it was shown that the (1+1)-EA is superior to other algorithms in case of monotone mutation. Also, it was demonstrated there that for some families of instances the (1+1)-EA is relatively competitive compared to some exact algorithms. However, these families of problems may be easily solved using simple local search heuristics, so the random nature of the (1+1)-EA was not so important. In [4] it was shown that the (1+1)-EA is a near-optimal search strategy in the class of needle in the haystack functions in black-box scenario, but at the same time the optimality was established for the trivial random search routine as well.

The common weakness of many theoretical results is that it is difficult to apply them directly to most interesting combinatorial optimization problems. In this situation a very popular approach is the computational experiment. In some papers the statistical approach is used to compare different search operators

S. Cagnoni et al. (Eds.): EvoWorkshops 2003, LNCS 2611, pp. 154–164, 2003.
© Springer-Verlag Berlin Heidelberg 2003

and choose the most appropriate ones or tune their internal parameters, see e.g. [5,6]. Here we consider the statistical investigation of crossover and mutation operators which may help to understand why one evolutionary scheme is better than another.

In this work we compare the (1+1)-EA with a very common version of genetic algorithm (GA) applied to the classical optimization problem of finding the maximum vertex independent set in graph. In Sect. 2 we consider how to choose the appropriate penalty function for this problem. Sect. 3 gives some reasons why one could expect the satisfactory performance of the (1+1)-EA and the results of computational experiment are presented in Sect. 4.

1.1   Notations and Definitions

Let $G = (V, E)$ denote a graph, where $V = \{v_1, ..., v_n\}$ is the vertex set, and $E \subset V \times V$ is the edge set. An *independent set* S is a set of pairwise nonadjacent vertices, i.e. $S \subseteq V$ such that $(v_i, v_j) \notin E$ for all $v_i, v_j \in S$. The vertex independent set problem (VISP) consists in finding the independent set of maximal cardinality. The complement of any independent set $C = V \setminus S$ is a *vertex cover*, each edge of G has at least one endpoint in C. Also, any independent set is a *clique* (complete subgraph) in the complement graph $G' = (V, E')$, where $E' = \{(v_i, v_j) \in V \times V \mid (v_i, v_j) \notin E\}$. Thus the problems of finding maximum independent set, minimum cover and maximum clique are closely related to each other. For these problems a number of heuristics including evolutionary and hybrid algorithms have been developed, see [7,8,9].

In our work we use the VISP as a convenient test problem in order to understand the behavior of two simple evolutionary algorithms being applied in the area of NP-hard optimization. Both of the algorithms use binary representation of solutions, i.e. the binary string (*genotype*) $x \in \{0,1\}^n$ corresponds to the vertex set (*phenotype*) $S(x) = \{v_k \mid x_k = 1\}$. The fitness function used here is based on a penalty term for infeasible solutions like in [7]. The detailed description will be given in Sect. 2.

The first of the algorithms we consider is the genetic algorithm (GA) based on crossover and mutation operators. *Two-point* crossover *Cross* used here chooses randomly two breakpoints and exchanges all bits of the parent individuals between the breakpoints to produce two offspring. Standard mutation operator *Mut* randomly changes every bit in string x with fixed probability $p_{mut} = 1/n$. The initial individuals are built by the procedure *Init* that for each $i = 1, ..., n$ with probability $1/2$ sets $x_i = 1$ and with probability $1/2$ sets $x_i = 0$. Instead of the usual proportional selection we use the *s-tournament* selection which randomly chooses s individuals from the population and selects the best one of them. Some works, for example [8,10], show the advantages of such selection against the proportional one. Furthermore, we use so called *steady-state* replacement method [8,11], which is known to be advantageous to generational replacement. After the child is produced it replaces some individual from the population. Generally, the worst individual in the population is chosen for the replacement. The scheme of the steady-state GA is the following:

1. Initialize population $\Pi^{(0)}$
2. For $t := 1$ to t_{max} do
   2.1 $\Pi^{(t)} := \Pi^{(t-1)}$
   2.2 Choose parent individuals p_1, p_2 from $\Pi^{(t)}$ using s-tournament selection
   2.3 Produce two offspring c_1 and c_2 applying $Cross(p1, p2)$
   2.4 Apply mutation operator $c'_1 := Mut(c_1)$ and $c'_2 := Mut(c_2)$
   2.5 Choose individuals q_1, q_2 from $\Pi^{(t)}$ for replacement
   2.6 Replace q_1 and q_2 by c'_1 and c'_2

The output of the algorithm is the best found individual.

In the algorithm (1+1)-EA, also known as *hillclimbing*, the population consists of only one individual $g^{(t)}$, and on each iteration the mutation operator is applied to $g^{(t)}$ once. If $g' = Mut(g^{(t)})$ is such that $\Phi(g') > \Phi(g^{(t)})$, then we set $g^{(t+1)} := g'$. If $\Phi(g') = \Phi(g^{(t)})$, then $g^{(t+1)} := g'$ with probability $1/2$, and $g^{(t+1)} := g^{(t)}$ otherwise.

As the test instances for experiments we use randomly generated graphs and the problems from DIMACS benchmark set [12] transformed from maximum clique problem.

2   Fitness Function

We will use the same fitness function as it was proposed in [7] but in more general way. This function incorporates penalty for every edge that has two endpoints in S:

$$\Phi_\alpha(x) = \sum_{k=1}^{n} x_k - \sum_{(v_i,v_j)\in E} \alpha x_i x_j \tag{1}$$

The coefficient α defines the "strength" of a penalty. In [7] the value of α was fixed to $2n$. In this section we want to investigate how the choice of the value α influences on the overall performance of the algorithms.

First note that the penalty coefficient α may be chosen from the set $[1, \infty)$. Indeed let for example $\alpha = 1$. Then the algorithm may return infeasible solution but it is possible to correct it removing arbitrary vertex of each edge included in the solution set. The resulting independent set will have cardinality not less than the value of $\Phi_\alpha(x)$.

We have tested the GA and the (1+1)-EA with $\alpha = 1, 2, 3, ...,$ $10, 20, 30, ..., 100$ on random graphs and DIMACS benchmark problems. Each test was repeated 200 times to obtain an average value of the output fitness with sufficient confidence level. In most cases the value $\alpha = 1$ gives better performance for both the GA and the (1+1)-EA.

The most common situation is reflected on Fig. 1, where the computational results for the GA on problem C125.9 are given. The number of generations was $t_{max} = 5000$, the population size was $|\Pi^{(t)}| = 150$, and the selection size was $s = 5$. The thick line shows the average size of the independent set over 200 runs. Two dotted lines show the 95%-confidence interval. As we can see all values $\alpha \neq 1$ give the similar output, but the case $\alpha \neq 1$ significantly differs

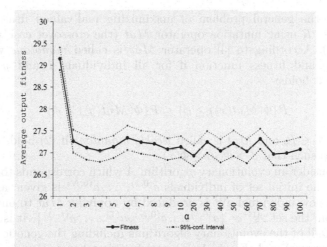

Fig. 1. An average overall performance of the GA on problem C125.9 with different α

from others. It would be very interesting to give an explanation of this effect. Although we are not able to answer this question completely, we would like to consider one example: suppose there is a maximal independent set S, i.e. such set that adding any extra vertex makes it non-independent. In the search space $\{0,1\}^n$ it corresponds to the locally optimal solution. Suppose that there is a vertex $v \notin S$ which has only one adjacent vertex in S. In this case two binary strings correspondent to S and $S \cup \{v\}$ will have the same fitness iff $\alpha = 1$. This means that some "peaks" of the fitness function with $\alpha > 1$ may turn to "plateaus" when $\alpha = 1$ and thus it is easier to reach them. Another way to describe the influence of penalty based on the study of some properties of mutation will be presented in Sect. 3.

These results are well connected with so-called *minimum penalty rule* [13]: the penalty should be kept as low as possible just above the limit below which infeasible solutions are optimal. The intuitive explanation of this rule is that to move from one hill to another the search process has to pass through a valley, and the greater the penalty term is the deeper this valley will be.

Note that besides this the penalty function should take into account some measure of infeasibility of the solution. For example, function (1) may be negative on some solution x. Instead, one can set $\Phi(x) = 1$ for such x, which would be more suitable with respect to minimum penalty rule. But this gives an extremely bad output because the search process more likely will be straying on the flat plateau of values $\Phi(x) = 1$.

3   Connections of Mutation Monotonicity with (1+1)-EA Performance

In this section we consider some properties of mutation operator and fitness function and show how these properties may be used in the analysis of algorithms.

Consider the general problem of maximizing real-valued fitness function $\Phi : \{0,1\}^n \to R$ using mutation operator Mut (the crossover case will be discussed later). According to [3] operator Mut is called *monotone* w.r.t. given search space and fitness function if for all individuals x and y such that $\Phi(x) \leq \Phi(y)$ it holds:

$$P\{\Phi(Mut(x)) \geq \phi\} \leq P\{\Phi(Mut(y)) \geq \phi\} \qquad (2)$$

for arbitrary $\phi \in R$, and Mut is called *weakly monotone* if (2) holds at least for all x, y and ϕ such that $\phi > \Phi(y) \geq \Phi(x)$.

Let us consider an evolutionary algorithm A which corresponds to the following scheme: the initial set of individuals $a^{(0,1)}, \ldots, a^{(0,N)}$ is given, and on each iteration t a new individual $a^{(t)}$ is produced by applying Mut to one of the individuals from the set $\mathcal{A}^{(t)} = \{a^{(0,1)}, ..., a^{(0,N)}, a^{(1)}, \ldots, a^{(t-1)}\}$. It is easy to see that most of all of the evolutionary algorithms including the genetic algorithms (without crossover) and the (1+1)-EA satisfy this scheme. Denote by $\tilde{a}^{(t)}$ the best individual in set $\mathcal{A}^{(t)}$.

Here we formulate the result from [3] which shows that if Mut is weakly monotone then the (1+1)-EA is the best search technique among all algorithms based on mutation operator. Recall that according to Sect. 1.1 by $g^{(t)}$ we denote the current individual on iteration t of the (1+1)-EA.

Theorem 1. *Suppose that the same weakly monotone mutation operator is used in the (1+1)-EA and in the algorithm A. Let the algorithm (1+1)-EA start from the best individual among* $a^{(0,1)}, ..., a^{(0,N)}$. *Then for all* $\phi \in R$ *and* $t > 0$ $P\{\Phi(\tilde{a}^{(t)}) \geq \phi\} \leq P\{\Phi(g^{(t)}) \geq \phi\}$.

For most real-world optimization problems the monotone mutation seems to be unrealistic. Indeed, if $\Phi(x) = \Phi(y)$ then from (2) it follows that $P\{\Phi(Mut(x)) \geq \phi\} = P\{\Phi(Mut(y)) \geq \phi\}$. This is a very strong conjecture and, for example, in the case of the VISP it could be easily disproved by giving an appropriate counterexample. On the other hand, condition (2) expresses in a strict form that the offspring of the more fit parent must have more chances to be sufficiently good than the chances of the offspring of a relatively poor parent. If one can express and prove this in some weaker but similar form then, although it does not necessary mean that the (1+1)-EA is the best possible algorithm, but one can expect that it will show relatively good performance.

In case of the VISP we could consider the probabilities $\gamma_{ij}(\alpha) = P\{\Phi_\alpha(Mut(x)) \geq j\}$, where $x \in \{0,1\}^n$ is such that $\Phi_\alpha(x) = i$. The monotonicity will now mean that $\gamma_{i-1,j} \leq \gamma_{i,j}$ for all i, j. The problem is that the definition of $\gamma_{ij}(\alpha)$ is correct only if $P\{\Phi_\alpha(Mut(x)) \geq j\}$ does not depend on particular x that has $\Phi_\alpha(x) = i$ and, as was already mentioned, this may not be guaranteed for the VISP. Therefore, instead of $\gamma_{ij}(\alpha)$ we consider the estimations $\hat{\gamma}_{ij}(\alpha)$ of conditional probabilities $P\{\Phi_\alpha(Mut(x)) \geq j | \Phi_\alpha(x) = i\}$, where x is a random solution. The estimations $\hat{\Gamma}(\alpha) = \{\hat{\gamma}_{ij}(\alpha)\}$ were obtained statistically by repeatedly generating random solution x, applying mutation operator and computing $\Phi_\alpha(x)$ and $\Phi_\alpha(Mut(x))$. To generate solution x we use procedure *Init* described in Sect. 1.1. Clearly, this procedure gives very poor solutions.

As we would like to try relatively good parents too, we apply to x the repair procedure that removes one endpoint of some edges included in the solution set.

As an illustration we give the fragments of $\hat{\Gamma}(1)$ and $\hat{\Gamma}(100)$, where $i = 10, ..., 13$ and $j = 11, ..., 14$, for problem hamming6-2. Only upper triangles of the matrix are shown because only these parts reflect the "evolvability" of mutation operator, i.e. the ability to produce offspring fitter than the parents:

$$\hat{\Gamma}(1) = 10^{-2} \begin{pmatrix} 12.57 & 1.19 & 0.08 & 0.00 \\ & 10.51 & 0.80 & 0.04 \\ & & 8.60 & 0.68 \\ & & & 8.66 \end{pmatrix}, \hat{\Gamma}(100) = 10^{-2} \begin{pmatrix} 9.11 & 0.89 & 0.06 & 0.00 \\ & 7.82 & 0.62 & 0.04 \\ & & 6.78 & 0.48 \\ & & & 7.09 \end{pmatrix}.$$

As we can see both matrices are monotone in the sense that $\hat{\gamma}_{i-1,j} \leq \hat{\gamma}_{i,j}$. So, if we accept a conjecture that all solutions x such that $\Phi(x) = i$ have almost equal probabilities to be improved after mutation, then we may consider the mutation as "almost monotone" and the coefficients $\hat{\gamma}_{ij}(\alpha)$ may be viewed as good approximations for $\gamma_{ij}(\alpha)$. Thus, due to Theorem 1 one can expect a competitive performance of the $(1+1)$-EA. The actual behavior of the $(1+1)$-EA in comparison with other algorithms depends on how close this isotropic assumption corresponds to the real situation.

The other important issue is that all coefficients of $\hat{\Gamma}(1)$ are greater than the coefficients of $\hat{\Gamma}(100)$ which forecasts higher velocity of the search and this was confirmed by the simulation results presented in Sect.4.

Algorithms with Crossover Operator. Here we briefly show the way the crossover may be analyzed in this study. Consider a *reproduction* operator $\mathcal{R}(x, y)$, where $x, y \in \{0, 1\}^n$, in which the crossover is applied to x and y and then the mutation is applied to both offspring (so $\mathcal{R}(x, y)$ is just a combination of crossover and mutation). Denote by $\tilde{a}_{\mathcal{R}}(x, y)$ the best of two new individuals produced by $\mathcal{R}(x, y)$. Operator $\mathcal{R}$ will be called *monotone* if for arbitrary pairs x^1, y^1 and x^2, y^2 such that $\Phi(x^1) \leq \Phi(x^2)$ and $\Phi(y^1) \leq \Phi(y^2)$ the following condition holds for arbitrary $\phi \in R$:

$$P\left\{\Phi(\tilde{a}_{\mathcal{R}}(x^1, y^1)) \geq \phi\right\} \leq P\left\{\Phi(\tilde{a}_{\mathcal{R}}(x^2, y^2)) \geq \phi\right\}, \tag{3}$$

and $\mathcal{R}$ is called *weakly monotone* if (3) holds for all $\phi > \max\{\Phi(x^2), \Phi(y^2)\}$. According to Theorem 1 in [3] any algorithm based on weakly monotone $\mathcal{R}$ is not superior to the $(1+1)$-EA that uses an "artificial" mutation $\mathcal{M}(x) = \tilde{a}_{\mathcal{R}}(x, x)$. One can see that this variant corresponds to the algorithm $(1+2)$-EA with the mutation operator Mut (on each iteration one parent produces two offspring and the best of the parent and both offspring becomes a new parent, see e.g. [14] for details). It is easy to show that if $\mathcal{R}$ is weakly monotone then $\mathcal{M}$ and Mut are weakly monotone too. Now, the second application of Theorem 1 yields that the $(1+2)$-EA is not superior to the $(1+1)$-EA.

Again we may consider the estimations $\hat{\gamma}_{ijk}(\alpha)$ of the conditional probabilities $P\{\Phi_\alpha(\tilde{a}_{\mathcal{R}}(x, y)) \geq k | \Phi_\alpha(x) = i, \Phi_\alpha(y) = j\}$. The 3-dimensional matrices $\hat{\Gamma}(1)$ and $\hat{\Gamma}(100)$ were obtained and again both the monotonicity and the relationship

$\hat{\gamma}_{ijk}(1) > \hat{\gamma}_{ijk}(100)$ for all i,j and k were observed (we do not present this here because of space limitation). So if the isotropic assumption may be accepted then one can expect the competitive performance of the (1+1)-EA.

4   Computational Experiments

In this part we apply both heuristics to some problems from the DIMACS benchmark set [12]. Experiments on random graphs were also made and the results obtained do not differ significantly from the results of benchmark problems, so they are not reported here. It is interesting to compare both heuristics on different fitness functions, so that the algorithms were tested on each problem with $\alpha = 1$ and $\alpha = 100$. The population size of the GA was set to 150.

Table 1. Comparison of the GA with the (1+1)-EA on DIMACS benchmark set

Problem	t_{max}	$\alpha = 1$		$\alpha = 100$	
		GA	(1+1)-EA	GA	(1+1)-EA
hamming6-2	2500	28.4, **32**(113-10)	31.3 **32**(185-1)	24.1, **32**(33-5)	22.2, **32**(18-2)
c125_9	5000	29.2, 33(7-9)	32.1, **34**(30-64)	27.1, 32(2-2)	27.4, 33(2-3)
c250_9	5000	33.2, 38(3-8)	37.8, 42(2-6)	32.9, 37(8-9)	32.6, 37(3-14)
gen200_9_44	5000	31, 36(2-2)	35.3, 39(6-9)	30.3, 35(1-4)	30.4, 34(3-15)
gen400_9_55	10000	37.6, 42(9-10)	43.9, 49(2-3)	37.2, 42(6-10)	37.2, 43(1-4)
gen400_9_65	10000	37.8, 43(2-5)	43.5, 51(2-1)	37.8, 43(3-6)	37.9, 44(1-1)
gen400_9_75	10000	40.1, 44(4-12)	45.4, 51(2-2)	39.5, 44(2-9)	39.7, 45(1-5)
br200_2	5000	7.1, 10(1-10)	8.6, 10(17-97)	6.9, 9(18-73)	7.3, 9(15-68)
br200_4	5000	11.4, 14(4-19)	13, 16(2-5)	11.2, 14(3-18)	11.2, 14(2-17)
br400_2	10000	16.9, 21(2-3)	20, 24(1-14)	16.9, 20(2-21)	17.1, 21(1-4)
br400_4	10000	16.9, 20(7-20)	19.9, 23(1-17)	17.1, 20(5-21)	17, 20(9-18)
p_hat300-1	10000	5.8, **8**(1-28)	6.6, **8**(12-97)	5.4, 7(2-17)	5.5, 7(17-78)
p_hat300-2	10000	18.9, 23(4-11)	22.7 **25**(35-35)	17.5, 21(10-26)	16.6, 24(1-2)
p_hat300-3	10000	27.5, 32(6-6)	32, **36**(1-2)	25.6, 30(4-12)	25.2, 31(2-4)
keller4	5000	8.1, **11**(5-6)	9.3, **11**(41-12)	7.9 **11**(3-5)	8.1, **11**(2-7)
MANN_a27	10000	117.4, 120(2-15)	118, 120(11-50)	116.9, 119(6-40)	118.3, 121(4-19)

The results are given in Table 1. To provide a fair competition we start the (1+1)-EA from the best individual of the initial population Π^0 of the GA. The number of generations for the GA is shown in column t_{max}. Since each iteration of the GA calls the fitness evaluation twice, we set the maximal number of generations for the (1+1)-EA to $2t_{max}$ (the running time of black-box optimization algorithms is measured as the number of the goal function evaluations, see [4]).

The first number in each cell shows the average fitness value returned by the algorithm over 200 independent runs. The number after comma is the best solution found over all runs. The first number inside the parenthesis shows the number of times that the best solution was found and the second one shows the number of times of finding a solution of size one less.

As one can see, in many cases the (1+1)-EA gives better performance than the GA, but for $\alpha = 100$ the difference between them is very little (there are also some problems where the GA shows better results than the (1+1)-EA for $\alpha = 100$). For $\alpha = 1$ the difference becomes notably sensible. In this case the (1+1)-EA outperforms the GA on all problems both in terms of average output fitness and probabilities of finding near optimal solution.

Generally, the value $\alpha = 1$ provides better results than $\alpha = 100$ especially for the (1+1)-EA. In fact, the only occasion, where the value $\alpha = 100$ happens to be better for the (1+1)-EA, was observed for problem MANN_a27. As it is a special case we would like to discuss it in more details. First we tried the same experiment on some randomly generated graphs with the same number of vertices and edges. The results were similar to other problems, so we suppose that the unusual effect is caused by the special structure of the graph MANN_a27. Secondly we try to observe the dynamics of the search process for this problem.

On Fig. 2 the evolution of fitness value $\Phi_\alpha(g^{(t)})$ is shown on average over 200 runs for $\alpha = 1$ and $\alpha = 100$. The results for the GA are also presented to complete the picture (recall that t here denotes the number of fitness evaluations, so the GA actually makes $t/2$ generations).

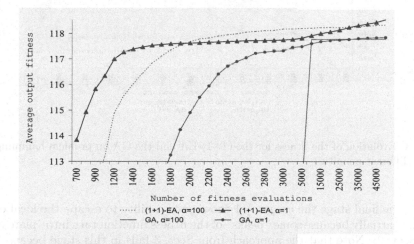

Fig. 2. Evolution of the fitness for the (1+1)-EA and the GA on problem MANN_a27 with different penalties

Note that when t is not large enough, in most cases the infeasible solutions are produced but for $\alpha = 1$ they have higher fitness than for $\alpha = 100$. This feature hinders the adequate understanding of the performance of the algorithm because it is estimated in different measures. To prevent this, when we study the case $\alpha = 100$, for all x we store both fitness functions: $\Phi_{100}(x)$ for using in the search process and $\Phi_1(x)$ for the output. For large numbers t the solutions become feasible and this modification makes no influence (note that in the experiments presented in Table 1 number t_{max} is large enough).

From what is shown on Fig. 2, we can divide the process into two stages: the stage of rapid climbing up on the way to local optimum, and the final stage, when the algorithms try to jump from one local optimum to another one with better fitness (note that such picture is usual for evolutionary algorithms, see e.g. [7]). As we can see on the first stage the value $\alpha = 1$ provides a more rapid ascent. This effect is well explained by the theoretical approach described in Sect. 3 because it is a very realistic assumption that the relatively poor parents have good probabilities to be improved and these probabilities do not differ very much from one parent to another, so we may accept the isotropic assumption.

For problem MANN_a27 the small penalty has led to relatively poor local optima (for both the GA and the (1+1)-EA). We consider this as an exceptional case because the experiments on other problems show the opposite. For a comparison the analogous picture for problem hamming6-2 is shown on Fig. 3.

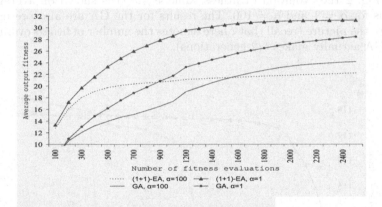

Fig. 3. Evolution of the fitness for the (1+1)-EA and the GA on problem hamming6-2 with different penalties

In the final stage the case $\alpha = 1$ shows more ability to escape the local optimum (partially because some "peaks" of the fitness function turn into "plateaus", see Sect. 2). Note that the approach from Sect. 3 fails in this stage because the isotropic assumption seems to be very unrealistic here. Indeed, for some local optimum x the probability to move to a better solution y is much less than the same probability of any solution z that is a neighbor of y and $\Phi(x) = \Phi(z)$. The final stage is the most difficult for the (1+1)-EA and here it may be inferior to some other algorithms, we can observe this on Fig. 3, where the GA outperforms (1+1)-EA for the case $\alpha = 100$.

We also suppose that some modifications of the (1+1)-EA, such as simulated annealing methods, random restarts or relaxation of the (1+1)-EA, where the worse offspring is allowed to replace the parent, may show much better results in this stage, but we leave it for further study.

5   Conclusion

Penalty functions are the most usual way to adapt some real-world problem for black-box optimization algorithms. Note that it is a very important approach in such areas like allocation and timetabling problems, where even constructing a feasible solution may be NP-hard. In this work the influence of the choice of penalty on performance of the algorithm is considered.

We show from the empirical point of view that the (1+1)-EA may significantly outperform the population based GA on one classic NP-hard combina torial problem. The existence of the connection between penalty coefficient and evolvability of mutation was considered and confirmed by the experimental results. In further studies it would be very interesting to investigate the isotropic assumption in more details and to develop a more strict mathematical technique to describe the effects reported in this work.

Acknowledgements. The research was supported in part by the INTAS grant 00-217. The authors would like to thank Anton V. Eremeev for his very helpful comments and suggestions.

References

1. Motwani, R., Raghavan, P.: Randomized algorithms. Cambridge University Press (1995)
2. Schöning, U.: A probabilistic algorithm for k-SAT and constraint satisfaction problems. Proc. 40th IEEE Symposium on Foundations of Computer Science (1999) 410–414
3. Borisovsky, P.A., Eremeev, A.V.: A Study on Performance of the (1+1)-Evolutionary Algorithm. Proc. of Foundations of Genetic Algorithms-2002 (FOGA-2002) (2002) 367–383
4. Droste, S., Jansen, T., Tinnefeld K., Wegener, I.: A new framework for the valuation of algorithms for black-box optimisation. Proc. of Foundations of Genetic Algorithms-2002 (FOGA-2002) (2002) 197–214
5. Fogel, D. B., Ghozeil, A.: Using fitness distributions to design more efficient evolutionary computations. In Fukuda, T (ed.): Proc. Third IEEE International Conference on Evolutionary Computation, IEEE (1996) 11–19
6. Kallel, L., Schoenauer, M.: A priori comparison of binary crossover operators: No universal statistical measure, but a set of hints. Proc. Artificial Evolution'97, LNCS Num. 1363. Springer Verlag (1997) 287–299
7. Bäck, Th., Khuri, S.: An Evolutionary Heuristic for the Maximum Independent Set Problem. Proc. First IEEE Conference on Evolutionary Computation, IEEE (1994) 531–535
8. Balas, E., Niehaus, W.: Optimized Crossover-Based Genetic Algorithms for the Maximum Cardinality and Maximum Weight Clique Problems. Journ. of Heuristics, 4 (4) (1998) 107–122
9. Bomze, I. M., Budinich M., Pardalos P. M., Pelillo, M.: The maximum clique problem. In Du, D.-Z. and Pardalos, P. M. (eds): Handbook of Combinatorial Optimization, Vol. 4. Kluwer Academic Publishers (1999)

10. Goldberg, D. E., Deb, K.: A comparative study of selection schemes used in genetic algorithms. In G. Rawlins (ed.): Foundations of Genetic Algorithms(FOGA). Morgan Kaufmann (1991) 69–93
11. Syswerda, G.: A study of reproduction in generational and steady state genetic algorithm. In G. Rawlins (ed.): Foundations of Genetic Algorithms (FOGA). Morgan Kaufmann (1991) 94–101
12. See the FTP site:
    ftp://dimacs.rutgers.edu/pub/challenge/graph/benchmarks/clique/
13. Richardson, J.T., Palmer, M.R., Liepins, G., Hilliard M.: Some guidelines for genetic algorithms with penalty functions. In J. David Schaffer (ed.): Proc. of the 3rd International Conference on Genetic Algorithms. Morgan Kaufmann (1989) 191–197.
14. Bäck, Th., Schwefel H.-P.: An overview of evolutionary algorithms for parameter optimization. Evolutionary Computation, **1** (1) (1993) 1–23

New Ideas for Applying Ant Colony Optimization to the Probabilistic TSP

Jürgen Branke and Michael Guntsch

Institute AIFB, University of Karlsruhe (TH), Germany
{branke,guntsch}@aifb.uni-karlsruhe.de

Abstract. The Probabilistic Traveling Salesperson Problem (PTSP) is a stochastic variant of the Traveling Salesperson Problem (TSP); each customer has to be serviced only with a given probability. The goal is to find an a priori tour with shortest expected tour-length, with the customers being served in the specified order and customers not requiring service being skipped. In this paper, we use the Ant Colony Optimization (ACO) metaheuristic to construct solutions for PTSP. We propose two new heuristic guidance schemes for this problem, and examine the idea of using approximations to calculate the expected tour length. This allows to find better solutions or use less time than the standard ACO approach.

1 Introduction

The Probabilistic Traveling Salesperson Problem (PTSP) is a stochastic variant of the well-known deterministic Traveling Salesperson Problem (TSP). In addition to a set of customers with pairwise distances, each customer i also has a probability p_i of actually requiring a visit. The goal in PTSP is to find an a priori tour which minimizes the expected length of an actual tour, where customers that do not require a visit are simply skipped, i.e. no re-optimization takes place. This problem has practical relevance, e.g. if a number of customers require regular visits, but re-optimizing the tour every time would be too expensive. Additionally, a simple scheme has the benefit of the customer and the driver getting used to some routine which usually results in an increased efficiency. Like TSP, PTSP is NP-hard [1].

To find good solutions for PTSP, several researchers have suggested adapting simple TSP heuristics [2,3,4]. Bertsimas et al. [4] provide a thorough analysis of the problem properties, including several bounds and comparisons of a number of simple heuristics. Laporte et al. [5] solve instances with up to 50 cities optimally using a branch-and-cut approach.

Recently, Bianchi et al. [6,7] have applied Ant Colony Optimization (ACO) to the PTSP. ACO is an iterative probabilistic optimization heuristic inspired by the way real ants find short paths between their nest and a food source. The fundamental principle used by the ants for communication is stigmergy, i.e. the ants use pheromone to mark their trails, with a higher pheromone intensity suggesting a better path and inclining more ants to take a similar path. In ACO

S. Cagnoni et al. (Eds.): EvoWorkshops 2003, LNCS 2611, pp. 165–175, 2003.

algorithms, in every iteration, a number of m (artificial) ants construct solutions based on pheromone values and heuristic information, see [8,9,10]. At the end of each iteration, the best ant(s) update the pheromone values corresponding to the choices that were made when constructing the tour.

In this paper, we will propose two ways of improving the performance of ACO when applied to PTSP:

- We examine the effect of approximate evaluation of the expected tour length on the performance of the ACO algorithm, showing that a relatively coarse and quick approximation is sufficient to avoid errors, thus leaving more time for generating solutions.
- We show that it is possible to improve performance by using more problem-specific heuristics to guide the ants during tour construction.

The remainder of this paper is structured as follows. In Section 2, we will take a more in-depth look at the evaluation function for PTSP, i.e. the function used to calculate the expected tour-length, and discuss possibilities for approximation. In Section 3, the basic ACO metaheuristic is explained, and the new heuristic guidance schemes are introduced. Empirical results are presented in Section 4, the paper concludes in Section 5 with a summary.

2   The PTSP Evaluation Function

In this section, we will take a closer look at the evaluation function for PTSP, i.e. the function for calculating the expected tour-length f of a specific tour π, where π is a permutation over $[1, n]$. In the first subsection, formulas for the computation of the complete PTSP evaluation function are given, while the second subsection deals with approximation.

2.1   Full Evaluation

For an n city PTSP instance, there are 2^n possible realizations R_i, $i = 1, \ldots, 2^n$ (each city can either be included or left out of the later instance), each of which can be assigned a probability $p(R_i) \in [0, 1]$, with $\sum_{i=1}^{2^n} p(R_i) = 1$. Let $l_{R_i}(\pi)$ describe the tour-length of π for realization R_i. The expected tour-length can then be formally described as

$$f(\pi) = \sum_{i=1}^{2^n} p(R_i) \cdot l_{R_i}(\pi). \tag{1}$$

Jaillet has shown in [11] that the expected length of a PTSP-tour can be calculated in $O(n^2)$ time. The idea is to consider each possible edge e_{ij} from city i to city j, with the probability that city i is included, city j is included, and all cities between i and j are omitted. Note that for symmetric PTSP problems, where $d_{ij} = d_{ji}$, e_{ij} is also included in the tour if all cities from city j to city i are omitted since the tour is a circle.

Let $q_i = 1 - p_i$ be the probability that city i is not included, $\pi(i)$ denote the city that has been assigned the i-th place in tour π, and d_{ij} denote the distance from city i to city j. Then, the expected tour length can be calculated as

$$f(\pi) = \sum_{i=1}^{n} \sum_{j=1}^{n-1} d_{\pi(i)\pi((i+j) \bmod n)} \cdot p_{\pi(i)} \cdot p_{\pi((i+j) \bmod n)} \cdot \prod_{k=i+1}^{i+j-1} q_{\pi(k \bmod n)}. \quad (2)$$

2.2 Approximate Evaluation

Even when using the above evaluation scheme, the resulting $O(n^2)$ time for calculating the expected tour-length is very long compared to the standard $O(n)$ evaluation for the deterministic TSP. In this subsection, we will consider the possibility of approximating the true PTSP-evaluation to a certain degree, with different trade-offs between evaluation accuracy and evaluation time.

We say that an edge e_{ij} has a depth $\theta \in [0, n-2]$ with respect to a given tour π if there are exactly θ cities on the tour between i and j. A high θ for an edge e_{ij} implies a small probability for the edge to be actually part of a realized tour (as a large number of consecutive cities would have to be canceled), and thus a small impact on the resulting value for f. For the homogeneous case where $\forall i \in [1, n] : p_i = p \in [0, 1]$, this probability has the value $p^2(1 - p)^\theta$, decreasing exponentially in θ. Therefore, it might make sense to simply stop the evaluation once a certain depth has been reached. We define

$$f^\theta(\pi) = \sum_{i=1}^{n} \sum_{j=1}^{1+\theta} d_{\pi(i)\pi((i+j) \bmod n)} \cdot p_{\pi(i)} \cdot p_{\pi((i+j) \bmod n)} \cdot \prod_{k=i+1}^{i+j-1} q_{\pi(k \bmod n)} \quad (3)$$

as the value of the evaluation up to depth θ. Since the expected tour-length will only be needed to compare solutions to one another, the approximate value for the evaluation function can be sufficient even though it might still be substantially smaller than the exact evaluation value.

3 ACO Approaches

This section covers the basic approach used by the ACO metaheuristic to construct a solution. Also, two new heuristic guidance schemes are proposed: one that is based more accurately on increase to overall solution quality, and another that takes the angle between adjacent edges of the tour into consideration.

3.1 Principle of ACO

The general approach of our algorithm for the PTSP follows the ant algorithm of Dorigo et al. [9] used for TSP. In every generation, each of m ants constructs one tour through all n cities of the instance. Starting at a random city, an ant iteratively selects the next city based on heuristic information as well as pheromone information. The heuristic information, denoted by η_{ij}, represents a priori heuristic knowledge about how good it is to go from city i to city j. The pheromone values, denoted by τ_{ij}, serve as a form of memory, adapted over time to indicate which choices were good in the recent past. During tour construction an ant located at city i chooses the next city j probabilistically according to

$$ p_{ij} = \frac{\tau_{ij}^{\alpha} \cdot \eta_{ij}^{\beta}}{\sum_{h \in S} \tau_{ih}^{\alpha} \eta_{ih}^{\beta}}, \tag{4} $$

where S is the set of cities that have not been visited yet, and α and β are constants that determine the relative influence of the heuristic values and the pheromone values on the decisions of the ant.

After each of the m ants have constructed a solution, the pheromone information is updated. First, some of the old pheromone is evaporated on all edges according to $\tau_{ij} \mapsto (1 - \rho) \cdot \tau_{ij}$, where parameter ρ denotes the evaporation rate. Afterwards, a positive update is performed along the tour of the best of the m ants of the iteration and the tour of the elitist ant (representing the best solution found so far). Each of these two positive updates has the form $\tau_{ij} \mapsto \tau_{ij} + \rho/4$ for all cities i and j connected by an edge of the respective tour. For initialization we set $\tau_{ij} = 1$ for every edge e_{ij} $(i \neq j)$.

3.2   PTSP-Specific Heuristics

So far, only the TSP heuristic $\eta_{ij}^{T} = 1/d_{ij}$, where d_{ij} is the distance between cities i and j, has been applied to PTSP in conjunction with ACO[6,7]. This ignores the fact that a customer requires a visit only with some probability. Furthermore, as has been shown in [4], the nearest neighbor heuristic (which is the underlying idea of the standard η-values) behaves poorly for PTSP. Therefore, in this section, we will explore two new heuristics which the ants can use to build solutions.

Depth-based Heuristic. The first idea to improve the heuristic is to more closely estimate the impact on solution quality caused by choosing city j as successor of city i. Instead of only using the distance, we propose using the sum of distances to all cities in the partial tour constructed so far, weighted by their probability of occurrence. Although this only takes into account the tour constructed so far, it should provide a much better approximation of the true additional cost and thus guide the ACO towards more promising solutions.

The expected distances d^P and the corresponding heuristic values η^P can be calculated as

$$ d_{ij}^P = \sum_{k=1}^{n-|S|} d_{\pi(k)j} \cdot p_{\pi(k)} \cdot p_j \cdot \prod_{h=k+1}^{n-|S|} q_{\pi(h)}, \quad \eta_{ij}^P = \frac{1}{d_{ij}^P}. \tag{5} $$

Note that $\pi(n - |S|) = i$. Unfortunately, using this heuristic increases the complexity of constructing a solution to $O(n^3)$ since an ant computes the probabilities in Equation 4 $O(n^2)$ times to build a solution, and an evaluation of

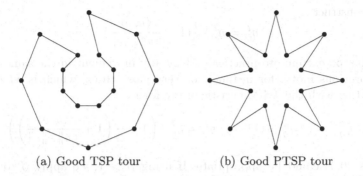

(a) Good TSP tour   (b) Good PTSP tour

Fig. 1. Comparison of the characteristics of TSP and PTSP tours.

Equation 5 takes $O(n)$ time. However, similarly to a bounded depth for the evaluation function as proposed in Section 2.2, we can define a maximum depth ζ for this heuristic by modifying d^P to $d^{P,\zeta}$ via

$$d_{ij}^{P,\zeta} = \sum_{k=\max(1,n-|S|-\zeta)}^{n-|S|} d_{\pi(k)j} \cdot p_{\pi(k)} \cdot p_j \cdot \prod_{h=k+1}^{n-|S|} q_{\pi(h)}. \tag{6}$$

Using this distance for the heuristic $\eta_{ij}^{P,\zeta}$, the time to construct a solution remains at $O(n^2)$ for a constant ζ, albeit the runtime of the ACO algorithm does increase by a constant strongly correlated to ζ.

Angle-based Heuristic. Another heuristic which attempts to give a better indication as to which city to choose next is an angle based heuristic. This heuristic is different from radial sort [4], which sorts all nodes in a single pass. Instead, when deciding which city to visit next, the ant takes into account the angle between the last traversed edge and the one implied by the choice of the next city.

To explain why the angle between adjacent edges of a tour could be exploited heuristically, consider the example shown in Figure 1. The tour on the left is a good TSP tour, while the tour on the right shows a good PTSP tour for the same problem, assuming e.g. $\forall i \in [1, 16] : p_i = 0.5$. Since any city i with $p_i < 1$ might be skipped in a PTSP instance, edges of depth larger than 0 also play a role, in particular edges of depth 1. These tend to be small when the tour has many sharp angles, as is the case for the right tour of Figure 1. A sharp angle tends to send the ant back toward the city it just came from. Thus, letting the ant prefer sharp angles may guide it toward significantly better solutions. If the angle between two edges is γ, we want the ants to make decisions influenced by $\cos \gamma$, which is defined for vectors u, v as $\cos \gamma = \langle u, v \rangle / (\|u\| \cdot \|v\|)$, with $\langle u, v \rangle$ being the scalar product of u and v, and $\|u\| = \langle u, u \rangle$. Then, for u and v parallel, $\cos \gamma = 1$, $\cos \gamma = 0$ for u perpendicular to v, and $\cos \gamma = -1$ for u and v in opposite directions..

We construct

$$\eta_{ij}^{\angle} = \eta_{ij}^T \cdot \frac{1}{2}(1 - \frac{\langle u, v \rangle}{\|u\| \cdot \|v\|}). \tag{7}$$

It may be too extreme to practically deny any movement in the same direction as the previous move, for instance for very close cities, which is why a linear combination with the TSP-heuristic of the form

$$\eta_{ij}^{\angle,c} = c \cdot \eta_{ij}^{\angle} + (1 - c) \cdot \eta_{ij}^T = \eta_{ij}^T \cdot \left(1 - \frac{c}{2}\left(1 + \frac{\langle u, v \rangle}{\|u\| \cdot \|v\|}\right)\right) \tag{8}$$

with $c \in [0,1]$ could be appropriate. If using $\beta > 1$, we apply β to η^T only, since $(\eta^T)^\beta$ can be computed efficiently in a pre-processing step and the result of applying β to the rest of Equation 7 would just lead to a similar result as using a higher value of c.

4   Empirical Evaluation

4.1   Test Setup

To assess the effect of our modifications on the performance of the ACO meta-heuristic on PTSP, we used the `eil101` TSP instance from the TSPLIB [12] and assigned a probability to each city. For simplicity, we restricted our tests to homogenous PTSP, with $p \in \{0.25, 0.5, 0.75\}$, since none of our improvements rely on the question whether the probabilities are homogenous or heterogenous. Due to space limitations, in the following we will focus on $p = 0.25$ and $p = 0.75$.

For the ant algorithm, we used standard settings for the TSP, namely $\alpha = 1$, $\beta = 5$ and $\rho = 0.001$, and let the algorithm construct solutions for 25000 iterations with $m = 10$ ants per iteration. In preliminary tests $\beta = 1$ was also tested but yielded inferior results for each of the heuristics. All results we present are the averages over 50 runs with different random seeds.

4.2   Approximate Evaluation

To determine the effect of using an approximate evaluation function of limited depth instead of the complete one, we tested evaluation depths of $\theta \in \{0, 1, 2, 4, 8, 16, 32, n - 2\}$. Note that for $\theta = 0$, the resulting tour-length corresponds to the TSP tour-length scaled by p^2, and $\theta = n - 2$ is full evaluation.

As can be expected, the higher the approximation depth, the closer the ant algorithm will perform to using the full evaluation function. The lower the approximation depth, the greater the likelihood of the algorithm to make wrong decisions, i.e. to judge one tour to be better than another even though the opposite it true. When comparing evaluation depths, we used the TSP-heuristic for all algorithms. In Figure 2, the relative quality and runtime compared to full evaluation, by definition located at (1,1), after 25000 iterations is presented. The use of an approximative evaluation function lets the algorithm perform between 18.65% and 11.27% faster, for $\theta = 0$ and $\theta = 32$ respectively. The quality of the

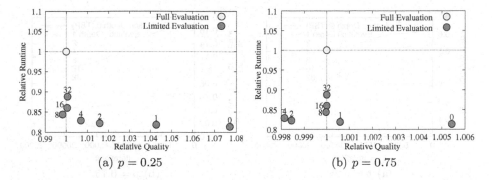

Fig. 2. Deviation in speed vs. quality by approximate evaluation compared to full evaluation for (a) $p = 0.25$ and (b) $p = 0.75$. Grey circles are numbered by approximation depth.

approximation, relative to the full PTSP evaluation function with full depth, depends on the depth of the approximation θ and the probability p. For a low value of p, using a depth of $\theta = 0$ or the TSP function for evaluation leeds to to results that are about 7.7% worse (with a standard error of 0.55%) than those achieved by the full PTSP evaluation, while a depth of $\theta \geq 8$ is sufficient for the quality found to be virtually identical. For a higher value of p, the edges of greater depth have only a marginal impact on solution quality, which is why TSP-like evaluation is only 0.5% worse (with a standard error of 0.17%) than complete PTSP evaluation, and an approximation depth of $\theta = 1$ is sufficient.

We presume the reason for the lesser performance to be the algorithm pursuing wrong paths which it erronously judges to be better than all others found by the ants of an iteration. We therefore tested how often an ant algorithm using an approximate evaluation function promotes a wrong ant to best of an iteration, letting it update the matrix instead of the truly best one. The results of this study are presented in Figure 3 for $p = 0.25$ and $p = 0.75$.

As we can see, depending on the approximation depth θ, the algorithm quite often promotes the wrong ant for updating, more than half the time for the instance $p = 0.25$ for $\theta = 0$ (i.e. TSP). For $\theta = 32$, the best ant is always correctly identified and hence the performance is indentical to full evaluation even for $p = 0.25$. For $p = 0.5$, a depth of $\theta = 8$ is sufficient to virtually eliminate the promotion error, and for $p = 0.75$ this is already true for $\theta = 4$.

It is interesting to note that even though for $p = 0.25$ and $\theta = 8$ or $p = 0.75$ and $\theta = 2$ the algorithm promotes the wrong individual in about 10% resp. 2% of all updates, the performance is not effected by these mistakes. Possible explanations could be that the tour promoted instead of the best tour is only marginally inferior, or that the approximated fitness landscape is easier to search, compensating the promotion errors.

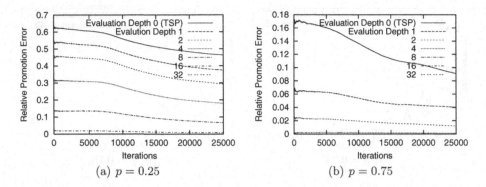

(a) $p = 0.25$ (b) $p = 0.75$

Fig. 3. Percentage of wrong selections up to iteration x, depending on evaluation depth θ for (a) $p = 0.25$, and (b) $p = 0.75$.

4.3   Improved Heuristic Information

For comparing the new heuristics proposed in Subsection 3.2, we always used the full evaluation in order to eliminate the possibility of wrong evaluation being the cause for poor performance. As a basis for comparison, we used the ant algorithm with the TSP heuristic (i.e. $\zeta = 0$).

The depth based heuristic was applied with $\zeta \in \{0, 1, 2, 4, 8, 16, n - 2\}$. Similarly to evaluation depth, a heuristic depth of $\zeta = 0$ yields identical results to using the TSP heuristic, as all heuristic values are again simply scaled by p^2, while $\zeta = n - 2$ is the same as using the complete tour up to the current city for measuring the increase to solution quality. The results are displayed in Figure 4. To avoid clutter, only the results for $\zeta \in \{0, 1, 4, 16\}$ in case of the depth-based heuristic are shown; $\zeta \in \{2, 8\}$ can be interpolated from these results, and the performance of $\zeta = n - 2$ was virtually ($p = 0.25$) or truly ($p \in \{0.5; 0.75\}$) identical to $\zeta = 16$. The figure shows that even using a depth of $\zeta = 1$ instead of the standard TSP heuristic improves the solution quality found by the ant algorithm given the same number of iterations. For $p = 0.25$, moving from $\zeta = 1$ to $\zeta = 4$ leads to a further increase in solution quality, while continuing to move from $\zeta = 4$ to $\zeta = 16$ actually leads to a slight decrease. This seems to be due to the unexpectedly slow convergence behaviour which the algorithm exhibits for $\zeta = 16$. For $p = 0.75$, going beyond $\zeta = 1$ does not yield significantly different results and is therefore not necessary. For $p = 0.5$ (not shown), the results are somewhere in between those for $p = 0.25$ and $p = 0.75$.

The other new heuristic proposed is the angle-based heuristic. Here, we chose the angle influence $c \in \{0.25, 0.5, 0.75, 0.875, 0.999\}$. Again, for the sake of clarity, in Figure 5 we only show the results for some of the parameter settings tested, namely $c \in \{0.5, 0.75, 0.875\}$; the performance of $c = 0.25$ lies in between that of the pure TSP heuristic and $p = 0.5$, while $p = 0.999$ performed so poorly that its results have been omitted entirely. As can be seen, using the ant algorithm in conjunction with this heuristic also leads to a better solution quality for

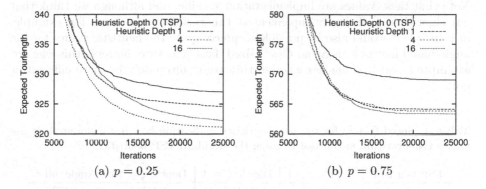

Fig. 4. Comparison of solution quality for depth-based heuristic, for (a) $p = 0.25$, and (b) $p = 0.75$.

$c \leq 0.75$ than using the (purely) distance-based heuristic from TSP. The degree of improvement depends on the emphasis on the angle heuristic compared to the distances. For all values of p using a relatively high value of $c = 0.75$ achieved best results. However choosing an even higher value for c will actually guide the algorithm to worse solutions than the TSP heuristic, as is documented by the performance of $c = 0.875$.

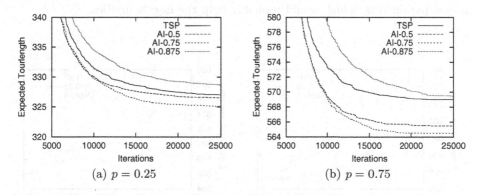

Fig. 5. Comparison of solution quality for angle-based heuristic, for (a) $p = 0.25$, and (b) $p = 0.75$. AI stands for the angle influence c.

So far, we have neglected to mention that these improved heuristics, while resulting in a better solution quality in the same number of iterations, take longer to calculate than the TSP heuristic, whose values are constructed completely during preprocessing at the beginning of the algorithm starts. Hence, we determined the speed factor indicating how much slower the ant algorithm runs when using the new heuristics with the parameters shown above (cf. Table 1).

Note that these values are implementation specific, and although we think that our algorithm is efficiently implemented, this is difficult to prove. For example, in principle it would also be possible to precompute the heuristic values for the angle-based heuristic using an $O(n^3)$ sized data structure, however this was not attempted, and it is unclear whether this would ultimately be more efficient or not.

Table 1. Speed factors for the new heuristics, indicating how much slower their use makes the algorithm in contrast to using the standard TSP heuristic.

TSP heuristic	Depth, $\zeta = 1$	Depth, $\zeta = 4$	Depth, $\zeta = 16$	Angle, all c
1	2.00916	2.61306	4.44108	1.84833

In Figure 6, we took the computation time into account by scaling the curves appropriately. As can be seen, when taking the runtime into account, the proposed new heuristics only beat the TSP heuristic after 25000 "ticks" for $p = 0.25$ and after approximately 18000 "ticks" for $p = 0.75$ for the chosen parameters. However, note that all algorithms are run with identical parameter settings, which work well for 25000 iterations of 10 ants each. For a fair comparison, it would be necessary to take the different number of iterations each of these methods is given into account and run the algorithm with individually optimized parameters, which would probably help the new heuristics.

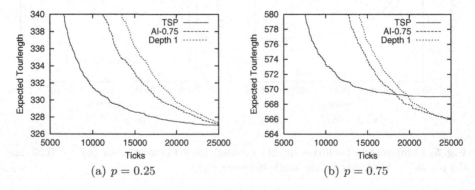

(a) $p = 0.25$ (b) $p = 0.75$

Fig. 6. Comparison of solution quality for angle-based heuristic, for (a) $p = 0.25$, and (b) $p = 0.75$. AI stands for the angle influence c.

5   Conclusion

We have introduced new ideas for improving the performance of Ant Colony Optimization (ACO) when applied to the Probabilistic Travelling Salesperson Problem (PTSP). An approximation of the evaluation function was shown to provide identical results while being only 4% more costly to evaluate than the TSP evaluation function and 14% less costly than the PTSP evaluation function. Also, two new heuristics were proposed which increase performance, leading to better solutions than the TSP heuristic previously used in the same time.

In the future, we plan to investigate whether the proposed heuristics need to be modified for the heterogenous PTSP, and if a combination of depth- and angle-based heuristic is opportune. An adaptive approximation which compares solutions only to such a depth as is necessary until the best solution of an iteration is certain might also be investigated.

References

1. D. J. Berstimas, P. Jaillet, and A. Odoni. A priori optimization. *Operations Research*, 38:1019–1033, 1990.
2. A. Jezequel. Probabilistic vehicle routing problems. Master's thesis, Massachusetts Institute of Technology, Cambridge, MA, 1985.
3. F. Rossi and I. Gavioli. Aspects of heuristic methods in the probabilistic traveling salesman problem. Andreatta, G., F. Mason, and P. Serani (eds.), Advanced School on Stochastics in Combinatorial Optimization, World Scientic, Singapore, pp. 214–227, 1987.
4. D. Bertsimas and L. H. Howell. Further results on the probabilistic traveling salesman problem. *European Journal of Operational Research*, 65:68–95, 1993.
5. G. Laporte, F. V. Louveaux, and H. Mercure. A priori optimization of the probabilistic traveling salesman problem. *Operations Research*, 42(3):543–549, 1994.
6. L. Bianchi, L. M. Gambardella, and M. Dorigo. An ant colony optimization approach to the probabilistic traveling salesman problem. In J. J. Merelo Guervos et al., editor, *Parallel Problem Solving from Nature*, volume 2439 of *LNCS*, pages 883–892. Springer, 2002.
7. L. Bianchi, L. M. Gambardella, and M. Dorigo. Solving the homogeneous probabilistic traveling salesman problem by the aco metaheuristic. In M. Dorigo, G. Di Caro, and M. Sampels, editors, *Ant Algorithms*, volume 2463 of *LNCS*, pages 176–187. Springer, 2002.
8. M. Dorigo. *Optimization, Learning and Natural Algorithms* (in Italian). PhD thesis, Dipartimento di Elettronica , Politecnico di Milano, Italy, 1992. pp. 140.
9. A. Colorni M. Dorigo, V. Maniezzo. The ant system: Optimization by a colony of cooperating agents. *IEEE Transactions on Systems, Man, and Cybernetics*, B(26):29–41, 1996.
10. M. Dorigo and G. Di Caro. The ant colony optimization meta-heuristic. In D. Corne, M. Dorigo, and F. Glover, editors, *New Ideas in Optimization*, pages 11–32. McGraw-Hill, 1999.
11. P. Jaillet. *Probabilistic traveling salesman problems*. PhD thesis, MIT, Cambridge, MA, 1985.
12. http://www.iwr.uni-heidelberg.de/groups/comopt/software/TSPLIB95/index.html, 2002.

An Experimental Comparison of Two Different Encoding Schemes for the Location of Base Stations in Cellular Networks

Carlos A. Brizuela and Everardo Gutiérrez

Computer Science Department, CICESE Research Center
Km 107 Carr. Tijuana-Ensenada, Ensenada, B.C., México
{cbrizuel, egutierr}@cicese.mx, +52-646-1750500

Abstract. This paper presents preliminary results on the comparison of binary and integer based representations for the Base Stations Location (BSL) problem. The simplest model of this problem, which is already NP-complete, is dealt with to compare also different crossover operators. Experimental results support the hypothesis that the integer based representation with a specific crossover operator outperforms the more traditional binary one for a very specific set of instances.

1 Introduction

The planning process of cellular radio networks provides a vast class of combinatorial problems. All these problems have a great importance from the application point of view. Most of these problems are of extreme complexity. One of the most studied problem has been the frequency assignment problem (FAP), for a review see [14]. Other problems are the assignment of cells to switches (ACS) in order to reduce the handoff and wiring costs [12], optimal location of base stations (BSL) [11,8,5,4,10], among others. These problems (FAP, ACS, and BSL) are equivalent to the following traditional combinatorial optimization problems: graph coloring, bin packing, and minimum dominating set, respectively.

There have been many proposed methodologies to find good approximation algorithms for this kind of problem. Some of these methodologies try to find provably good approximations [2], but with very conservative results. Another successful approach is to use meta-heuristics like Genetic Algorithms (GA), Tabu Search (TS), Simulated Annealing (SA), and others (see [1]). The main drawback (by definition of heuristics) of these approaches as opposed to approximation algorithms [2,9], is that they cannot guarantee a given performance for any instance of the problem.

In this paper we deal with the Base Station Location (BSL) problem which is a variant of the Minimum Dominating Set (MDS) problem. A problem whose decision version belongs to the NP-complete class.

Previous works using GA's to deal with this problem considers a binary encoding [4,10]. Here, we propose and conjecture that an integer based representation could be more appropriate for this problem.

S. Cagnoni et al. (Eds.): EvoWorkshops 2003, LNCS 2611, pp. 176–186, 2003.

The remainder of this paper is organized as follows. Section 2 states the problem we are trying to solve. Section 3 proposes the methodology. Section 4 presents the experimental setup and results. Finally, section 5 gives the conclusions and ideas for future research.

2 Problem Statement

The BSL problem in Cellular Radio Networks consist of a set of demand nodes **DN** and a set of possible base station locations **B**. The problem is to decide where to build the BS's such that the maximum possible number of demand nodes are covered, subject to some restrictions related to interference, limited budget, maximum number of built BS's in a given area, among others.

A model for this problem is proposed in [5] and one of the requirements is:

(R2) Every possible base station location $i \in \mathbf{B}$ has a maximal broadcast range $r_{max}(i)$. The actual broadcast range of the built BS has to be determined and is denoted as $r(i) \leq r_{max}(i)$. The sets $\mathbf{N}_i$ of those demand nodes that are supplied by i are then given by $\mathbf{N}_i = \{j | \text{dist}(i,j) \leq r(i)\}$.

Instead we propose that the range for a given base station $i \in \mathbf{B}$ is defined by a function $f_i(d, \theta, h)$, where d is the distance from the base station i to a point in the space, θ the angle with respect to a given reference line, and h is the relative height between the BS and the point in the space. With this function we let open the possibility to include all restrictions that may appear in real-world problems. If we make θ and h constant then we have the same requirement as in [5]. Once this function is defined we can represent the problem as a graph $\mathbf{G} = (\mathbf{DN}, \mathbf{E})$ where $\mathbf{DN}$, the set of vertices, represents the demand nodes and $\mathbf{E}$, the set of edges, represents the set of demand nodes and their corresponding covering BS's. That is, the presence of the edge $(u, v) \in \mathbf{E}$ means that both vertices (v, u) are demand nodes, and one of them v/u is a possible base station location covering the demand node u/v, respectively.

In real-world applications we need to know the minimum number of base stations needed to provide a given coverage. This problem is a variant of what is called Minimum Dominating Set (MDS) problem (see [11]). In order to state the variant let us first introduce the definition of the MDS problem itself.

Definition 1. Minimum Dominating Set Problem [11]. Given a graph $\mathbf{G} = (\mathbf{V}, \mathbf{E})$ where $\mathbf{V}$ is the set of vertices and $\mathbf{E}$ the set of edges, a subset $\mathbf{V}' \subseteq \mathbf{V}$ is said to be dominating if for every vertex $v \in \mathbf{V} - \mathbf{V}'$ there is an edge connecting to a vertex $u \in \mathbf{V}'$ such that $(u, v) \in \mathbf{E}$. The MDS problem consists of finding such a $\mathbf{V}'$ of minimum cardinality $|\mathbf{V}'|$.

Then the variant can be stated as follows: given a number q, find the MDS of a subset $\mathbf{S} \subseteq \mathbf{V}$ such that $q = |\mathbf{S}|$. If the cardinality of $\mathbf{V}$ is N then we need to solve $\binom{N}{q}$ MDS problems in the worst case.

The MDS problem was proven to be NP-complete ([6], page 190). Furthermore, non-approximability results show that it cannot be approximated within

$(1 - \epsilon)ln|\mathbf{V}|$ for any $\epsilon > 0$ ([2], page 370). This result implies that for the worst case scenario even approximations to this problem are difficult to solve. This situation motivates the design of efficient meta-heuristics for specific instances.

3 Methodology

Approximation algorithms in the concept of Hochbaum [9] and Ausiello et al. [2] and meta-heuristics have been the two main approaches to tackle this problem and its variants. The method we propose here is based on Genetic Algorithms and falls in the second category. Genetic Algorithms to solve this problem were previously proposed by Calégari [4], Krishnamachari and Wicker [10], and Meunier et al. [13].

3.1 Encoding

Previous works based on GA's [4,10] proposed a binary representation for each solution. In this representation, the locus indicates the BS number, a value of 1, at this locus, indicates that the location is selected, and a value of 0 indicates that it is not. The length of the chromosome is given by the number of possible locations. Calégari [4] proposes a parallel GA and its main contribution is in the parallel model he uses. Krishnamachari and Wicker [10] compare different meta-heuristics (GA, Tabu Search (TS), Simulated Annealing) and Random Walk for this problem. They conclude that the GA along with TS provide the most consistent results. Maunier et al. [13] presents a GA to solve the problem and considers three different objective functions: minimize the number of selected sites, maximize the throughput, and minimize the interference with some additional constraints. They apparently use the binary encoding for the location of BS, although it is not explicitly indicated in the document [13].

We propose an integer based representation where each integer value indicates the number of the selected base station. The motivation behind the use of an integer based representation is that we can explicitly fix the maximum number of desired base stations by means of the chromosome length. This is not easily done with the binary encoding since we need a mechanism to control the maximum number of one's in the chromosome. The mechanism implies an extra computational cost.

In our proposed representation each gene in the chromosome, e.g. $\mathbf{C} = [1375]$, indicates the number of selected locations, i.e. locations 1,3,5, and 7 are selected. This representation may contain zeros and its maximal length will be the number of possible locations. By looking at this representation and the binary representation we may see no difference at all. However, in the integer representation, as it was mentioned before, we can fix the length of the chromosome, and this will not be a direct thing to do in the binary representation. Another important point in using this representation is that it is possible to design specific crossover operators that are not easily implemented with the binary representation.

The conjecture is that the neighborhood generated by this representation and its crossover operators efficiently exploit the combinatorial structure of the problem.

3.2   Algorithm

Here we present the pseudocode for the implemented algorithm.

1. **for** $i = 1$ to Pop_Size
2.   **do** Generate_Individual(i)
3. **for** $i = 1$ to Pop_Size
4.   **do** Evaluate Objective_Function(i)
5. **for** $i = 1$ to Pop_Size
6.   **do** Select an Individual by RWS and save it in Temporal_Population
7. Replace Population with Temporal_Population
8. **for** $i = 1$ to Num_Of_Iter
9.   **do** Crossover with probability Pc
10.   Apply Generational Replacement
11.   Apply Mutation with probability Pm
12.   **for** $j = 1$ to Pop_Size
13.   **do** Evaluate Objective_Function(j)
14.   **for** $j = 1$ to Pop_Size
15.   **do** Select by RWS and save it in Temporal_Population
16.   Replace Population with Temporal_Population

In this algorithm Pop_size is the algorithm parameter indicating the population size. The function Generate_Individual(i) randomly generates an individual. Evaluate Objective_Function(i) computes the fitness for each individual i, two different functions are implemented, minimization of $f_A = k\frac{N}{R^\beta}$ [10] and maximization of $f_B = \frac{R^\gamma}{N}$ [4]. Here, N is the number of selected locations, R is the covered area rate, i.e the number of covered nodes over the demand nodes, k, β, γ are constants used to give more or less importance to the minimization/maximization of the number of selected locations or the covered area rate. The selection mechanism is the Roulette Wheel Selection (RWS). The number of iterations is given by Num_Of_Iter. The crossover rate and the mutation rate are given by Pc and Pm, respectively.

Crossover. All crossover operators we use here are standard in principle. The modification occurs when we are trying to copy a gene that has been already copied into the offspring. In this case a zero is copied in the offspring.

   PMX. Partially-Mapped Crossover. In this crossover two random positions are selected. Figure 1 illustrates how this operator works, genes from loci 1 to $s1 - 1$ of parent 1 are copied to loci 1 to $s1 - 1$ in the offspring. Loci $s1$ to $s2$ of parent 2 are copied to loci $s1$ to $s2$ in the offspring, and finally loci $s2 + 1$ to n of parent 1 are copied to loci $s2 + 1$ to n in the offspring. In every situation if we are trying to copy a gene that have been already copied in the offspring

PMX

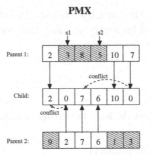

Fig. 1. PMX crossover operator

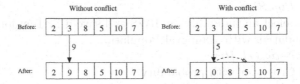

Fig. 2. Mutation operator OP1.

then a zero is copied instead. This situation generates a conflict as it is shown in Fig. 1. This is a modified version of the PMX explained in the book by Gen and Cheng ([7], page 5).

PPX. Precedence-based Crossover. A subset of precedence relations of the parents genes are preserved in the offspring. A binary random mask is generated, the 1's indicate that genes from parent 1 are to be copied and the 0's indicate that genes from parent 2 are to be copied, in the order they appear from left to right. The same conflict resolution as in PMX is applied. This is a modified version of the PPX presented in [3].

TPX. The position of some genes corresponding to parents are preserved in the offspring in the order they appear. As in PPX a random mask is generated, however, for this operator the 1's indicate the selected loci to copy from parent 1, and the 0's indicate the loci to be copied from parent 2. Again the conflict resolution of PMX is used. This is a modified version of the standard two point crossover that can be found elsewhere ([7], page 408).

Mutation. After a set of trials with different mutation operators we have decided to use a single mutation operator which is described in the following.

OP1. The mutation locus is randomly selected, then the gene is replaced by a randomly generated integer in the range 1 to n. If the new value already exists in the chromosome, then the gene is replaced by a zero. Figure 2 illustrates this operator.

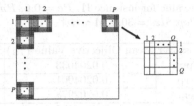

Fig. 3. Grid and cell structure for the artificial problem. MxM grids grouped in PxP cells of QxQ grids

4   Experimental Setup and Results

The instance data are generated in two different ways. The first one is based on the work by Calégari [4] and the second one on the work by Krishnamachari and Wicker [10]. These methods are explained in the following subsections.

4.1   Uniform Generation

This artificial model considers L possible locations distributed in a grid structure of MxM. The grid model is divided into PxP sub-grids, each of size QxQ grids, this structure is shown in Fig. 3. The first P^2 possible locations are fixed in the center of each cell such that these locations have a coverage of 100%, representing the optimal solution. Another $L - P^2$ possible locations are set randomly and uniformly over the whole MxM grid structure. Three different instances were generated, the parameters for each instance are shown in Table 1.

Table 1. Three instances for the artificial problem

Instance	Area (MxM)	Cell Size (QxQ)	Optimal Number of Cells
I1	287 x 287	41 x 41	49
I2	217 x 217	31 x 31	49
I3	147 x 147	21 x 21	49

Two different experiments are performed over these instances. For the first experiment the length of the chromosome (integer encoding) is fixed to a maximum length of 49, which is the optimal number of needed BS's. The second experiment considers a chromosome length equal to the number of possible locations, i.e. equals the number of bits in a binary encoding. We study three different crossover operators for the integer based representation and a single operator (one point crossover) for the binary representation, as it is used in [4]. The mean objective function, the mean covered area, and the mean number of selected base stations over the whole set of runs is computed. This is done for each proposed algorithm. The parameters are fixed according to Calégari [4]

Table 2. Mean objective value for instance I1. $Pc = 0.6$, $Pm = 0.6$, $Num\_Of\_Iter = 1000$, $Num\_runs = 50$, $Pop\_Size = 30$, and $\gamma = 4$

Chromosome Length	Crossover	Mean Objective Value (f_B)	Standard Deviation (%)
	ppx	0.0204033	0.16
K	tp	0.0204001	0.27
	pmx	0.0202978	1.19
	ppx	0.0164076	4.23
K_N	tp	0.0167186	3.70
	pmx	**0.0176148**	**4.86**
	Binary	0.0168669	5.87

Table 3. Mean number of base stations for instance I1. $Pc = 0.6$, $Pm = 0.6$, $Num\_Of\_Iter = 1000$, $Num\_runs = 50$, $Pop\_Size = 30$, and $\gamma = 4$

Chromosome Length	Crossover	Mean Number of BS (N)	Standard Deviation (%)
	ppx	49.0	0.00
K	tp	49.0	0.00
	pmx	49.0	0.00
	ppx	56.8	3.50
K_N	tp	56.0	3.52
	pmx	**53.1**	**3.08**
	Binary	53.7	3.53

where the crossover rate $Pc = 0.6$, the mutation rate $Pm = 0.6$, the maximum number of iterations is set to 1000, the population size is fixed to 30, and $\gamma = 4$.

Table 2 shows the mean objective function for both experiments. K indicates chromosome length of 49, and K_N a chromosome of full length. For the K_N case, we see that the integer encoding with PMX operator outperforms the binary encoding in terms of the objective function mean and standard deviation (maximization problem).

Table 3 shows the mean number of BS, in this case also PMX outperforms the binary representation (minimization).

Table 4 shows the results for mean coverage, again PMX outperforms the binary encoding (maximization).

For all cases the algorithm performance gets worse when changing from K to K_N. This could be due to higher population diversity for K_N (longer chromosome), and the required number of iterations to reach convergence.

4.2   Shadow Fading Model

This approach, for instance generation, was proposed by Krishnamachari and Wicker [10]. The idea here is that the covered area is determined by what is called uncorrelated log-normal shadow fading model. In this model, the power loss in dB at a distance d from the base station is given by the equation [10] $P_{loss} = A + Blog(d) + G$, where G is a zero-mean Gaussian random variable with

Table 4. Mean Coverage for instance I1. $Pc = 0.6$, $Pm = 0.6$, $Num\_Of\_Iter = 1000$, $Num\_runs = 50$, $Pop\_Size = 30$, and $\gamma = 4$

Chromosome Length	Crossover	% Mean Coverage (R)	Standard Deviation (%)
	ppx	99.99	0.04
K	tp	99.99	0.07
	pmx	99.86	0.30
	ppx	98.20	0.77
K_N	tp	98.35	0.53
	pmx	**98.29**	**0.68**
	Binary	97.50	0.78

variance σ^2, A and B are constants, as in [10] $A = 50$, $B = 40$, and $\sigma^2 = 10$. The P_{loss} is computed for every base station and every point in the grid, if $P_{loss} < 100$ dB at a given node then the node is covered by base station i. Notice that the demand nodes are considered to be the whole grid space. This is done without loss of generality since, the specific set of demand nodes can be included at any moment.

In this case the number of possible locations is randomly generated over a whole grid of size MxM. Fig. 4 a) shows a randomly generated pattern for 51 possible locations, Fig. 4 b) shows the covered area for one of the 51 base stations (gray node at the center), and Fig. 4 c) shows the covered area when the 51 locations are selected.

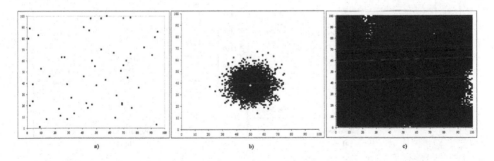

Fig. 4. Possible Base Station Location and Coverage. a) Random Generation of 51 BS. b) Coverage of a single BS. c) Coverage when all 51 BS's are selected

Fig. 5 presents the mean values for the objective function over 50 runs, for different population sizes, and different crossover operators. The parameters values used in each case are indicated in the figure caption. In this case we see a clear superiority of the PMX crossover operator over the others and over the binary representation in terms of objective function optimality (minimization problem). Fig. 6 a) shows the mean coverage, here, the binary shows poorer results than the other operators (maximization). In Fig. 6 b) we see results for

184   C.A. Brizuela and E. Gutiérrez

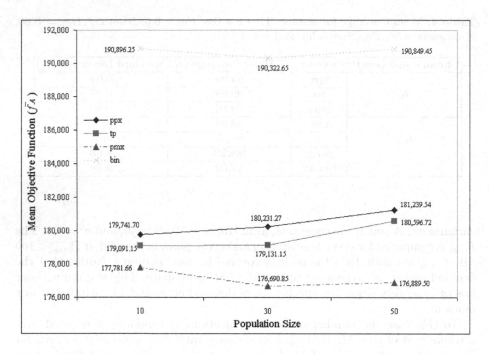

Fig. 5. Mean objective function values. $Pc = 0.7$, $Pm = 0.5$, $Num\_Of\_Iter = 1000$, $Num\_Of\_runs = 50$, $\beta = 3$, $k = 10000$

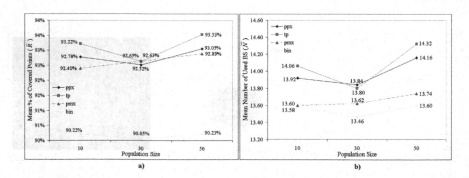

Fig. 6. Mean number of selected base stations, and coverage for different population sizes. a) Mean Coverage and b) Mean Number of BS's. $Pc = 0.7$, $Pm = 0.5$, $Num\_Of\_Iter = 1000$, $Num\_Of\_runs = 50$, $\beta = 3$, $k = 10000$

the mean number of selected base stations, in this case, the binary encoding presents the best result for all population sizes (minimization). Remember that what the algorithm should optimize is the objective function, nor the covered area neither the number of base stations.

Fig. 7 shows a sub-optimal solution of 13 Base Stations and a coverage of 92.33% obtained with the PMX operator.

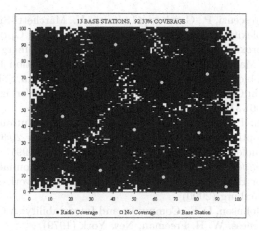

Fig. 7. Sub-optimal solution with 13 selected base stations. Coverage 92.33%. PMX crossover. $Pc = 0.7$, $Pm = 0.5$, $Num\_Of\_Iter = 1000$, $Num\_Of\_runs = 50$, $\beta = 3$, $k = 1000$

5 Conclusions

An integer based representation for the base station location problem has been proposed. The representation along with a particular crossover (PMX) outperforms a binary representation that uses standard operators, over a specific set of instances. In this representation the maximum number of desired base stations can be directly specified by the chromosome length helping to fix the maximum number of base stations to be used.

The representations are compared over two different set of instances. A set of easy instances where the optimal number of base stations is know and easy to compute. The other set, a difficult one, where we do not know the optimal number of base stations and it is not easy to compute. For this case (the difficult one) a clear superiority of the integer representation is observed.

Two main paths are in our interest for future research. The first one is related to the study of how the generated neighborhoods exploit the combinatorial structure of the problem to get better results. The second one is related to the average case performance analysis of the approximation algorithms proposed by Glaβer [8] and their comparison with our proposed procedure.

Acknowledgments. The authors would like to the thank the anonymous referees for their useful comments and interesting ideas for future research.

References

1. Ansari, N., Hou, E.: Computational Intelligence for Optimization. Kluwer Academic Publishers, Boston (1997)

2. Ausiello, G., Crescenzi, P., Gambosi, G., Kann, V., Marchetti-Spaccamela, A., Protasi, M.: Complexity and Approximation – Combinatorial Optimization Problems and Their Approximability. Springer-Verlag, Berlin Heidelberg New York (1999)
3. Bierwirth, C., Mattfeld, D. C., Kopfer, H.: On Permutation Representations for Scheduling Problems. In Proceedings of Parallel Problem Solving from Nature. Lecture Notes in Computer Science, Vol. 1141. Springer-Verlag, Berlin Heidelberg New York (1996) 310–318
4. Calégari, P. R.: Parallelization of population-based evolutionary algorithms for combinatorial optimization problems. PhD thesis, number 2046. Swiss Federal Institute of Technology (EPFL), Lausanne, Switzerland (1999)
5. Galota, M., Glaßer, C., Reith, S., Vollmer, H.: A Polynomial-Time Approximation Scheme for Base Station Positioning in UMTS Networks. Technical Report, Universitat Wurzburg (2000)
6. Garey, M. R., Johnson, D. S.: Computers and Intractability: A Guide to the Theory of NP-completeness. W. H. Freeman, New York (1979)
7. Gen, M., Cheng, R.: Genetic Algorithms and Engineering Optimization. John Wiley & Sons, New York (2000)
8. Glaßer, C., Reith, S., Vollmer, H.: The Complexity of Base Station Positioning in Cellular Networks. ICALP Workshops 2000. Proceedings in Informatics, Vol. 8 (2000) 167–177
9. Hochbaum, D. (ed.): Approximation Algorithms for NP-Hard Problems. PWS Publishing Company, Boston (1997)
10. Krishnamachari, B., Wicker, S. B.: Experimental analysis of local search algorithms for optimal base station location. In Proceedings of International Conference on Evolutionary Computing for Computer, Communication, Control and Power. Chennai, India (2000)
11. Mathar, R., Niessen, T.: Optimum positioning of base stations for cellular radio networks. Wireless Networks, Vol. 6. John Wiley & Sons, New York (2000) 421–428
12. Merchant, A., Sengupta, B.: Assignment of Cells to Switches in PCS Networks. IEEE/ACM Transactions on Networking, Vol. 3 No. 5. IEEE Press (1995) 521–526
13. Meunier, H., Talbi, E., Reininger, P.: A Multiobjective Genetic Algorithm for Radio Network Optimization. In Proceedings of the 2000 Congress on Evolutionary Computation CEC00 (2000) 317–324
14. Murphey, R. A., Pardalos, P., Resende, M. G. C.: Frequency Assignment Problem. In Du, D.-Z., Pardalos, P. M. (eds.). Handbook of Combinatorial Optimization. Kluwer Academic Publishers (1999) 295–377

Landscape State Machines: Tools for Evolutionary Algorithm Performance Analyses and Landscape/Algorithm Mapping

David Corne[1], Martin Oates[1,2], and Douglas Kell[3]

[1] School of Systems Engineering, University of Reading,
Whiteknights, Reading RG6 6AY, United Kingdom
d.w.corne@reading.ac.uk
[2] Evosolve Ltd, Stowmarket, Suffolk, United Kingdom,
moates@btinternet.com
[3] Department of Chemistry, UMIST, Manchester, United Kingdom
dbk@umist.ac.uk

Abstract. Many evolutionary algorithm applications involve either fitness functions with high time complexity or large dimensionality (hence very many fitness evaluations will typically be needed) or both. In such circumstances, there is a dire need to tune various features of the algorithm well so that performance and time savings are optimized. However, these are precisely the circumstances in which prior tuning is very costly in time and resources. There is hence a need for methods which enable fast prior tuning in such cases. We describe a candidate technique for this purpose, in which we model a landscape as a finite state machine, inferred from preliminary sampling runs. In prior algorithm-tuning trials, we can replace the 'real' landscape with the model, enabling extremely fast tuning, saving far more time than was required to infer the model. Preliminary results indicate much promise, though much work needs to be done to establish various aspects of the conditions under which it can be most beneficially used. A main limitation of the method as described here is a restriction to mutation-only algorithms, but there are various ways to address this and other limitations.

1 Introduction

The study of fitness landscapes [14] in the context of evolutionary search strives to understand what properties of fitness landscapes seem correlated with the success of specific evolutionary algorithms (EAs). Much progress has been made, with a number of landscape metrics under investigation as well as sophisticated statistical techniques to estimate landscape properties [1,5,7,9,11,13]. Meanwhile, much effort has also gone into constructing landscapes with well understood properties in attempt to yield hypotheses and guidelines which may apply to 'real' landscapes [4,6,8,11,12]. For the most part, a striking aspect of such investigations has been the ability of EAs consistently to undermine predictions [1,6,10]. Correlation between proposed metrics or landscape features and evolutionary algorithm difficulty tends to be weak, or bothered by the presence of convincing counterexamples. Here we present an alternative ap-

S. Cagnoni et al. (Eds.): EvoWorkshops 2003, LNCS 2611, pp. 187–198, 2003.

proach to exploring landscapes and algorithm performance on them. We model a land-scape as a finite state machine, whose states represent fitness levels and state transition probabilities characterize mutation. Such a model can be approximately inferred from sampling during an EA (or other algorithm) run, and then used for test-driving a range of algorithms under consideration to apply to the 'real' landscape.

This promises to be valuable in cases where good results are at a premium, but fit-ness evaluations on the 'real' landscape are prohibitively expensive, precluding inten-sive *a priori* algorithm comparison and/or parameter tuning; in contrast, a fitness evaluation on a landscape state machine (LSM) is computationally trivial. The success of this technique for algorithm comparison depends on the degree to which the LSM approximation captures those features of the real landscape which are salient in terms of algorithm comparison. Viability also depends on whether sufficiently useful LSMs can be inferred without incurring undue cost in prior search of the real landscape. We start to investigate these questions, and conclude that the technique is promising.

2 Landscape State Machines

What we call a landscape state machine (LSM) is simply a finite state machine (FSM) which models certain aspects of a search landscape. To be specific, given a search space E, an operator M (which takes a point $s \in E$, and returns another point from E) and an associated transition matrix T (such that t_{ij} gives the probability of yielding point $j \in E$ after applying M to $i \in E$). An LSM model of this landscape is a set of states and arcs (S, A), such that S corresponds to a partition of the set E, and A corre-sponds to an abstraction of T. In the extreme, the LSM can model a landscape pre-cisely, with precisely one state for each point in E, and A corresponding precisely to T.

In the general case, one state in the LSM will map onto many points in E. We will normally expect the mapping between S and E to be such that each state s corresponds to a set of points in E with equivalent fitness. More generally, we may define an equivalence relation R, which partitions E into c of equivalence classes $E_1,...,E_c$. The states in the LSM can then correspond precisely to the equivalence classes. In the case that the equivalence relation R forces equality in both fitness and genotype, the LSM model becomes the exact model described above. More generally, and as we later do in section 4, we might define R in such a way that states may be associated with a partition of the fitnesses into bands (e.g. all points with fitness between 0.2 and 0.3 might define an equivalence class).

2.1 Examples and Motivation

Before we discuss the uses of this simple LSM concept, an example will serve to clar-ify and motivate the issues involved. Consider the simple MAX-ONES problem, in which a candidate solution is a binary string of length L, and the fitness of a candidate (which is to be maximised) is the number of 1s it contains. Further, imagine we are interested in addressing this problem with an EA, and will employ the single-gene bit-flip mutation operator (but no other operator). That is, when we mutate a candidate solution, a single bit is chosen at random and its value is flipped.

The LSM in Figure 1 is model for this landscape when $L = 5$. There are six distinct fitnesses in this case, and we can identify each state s_i with the entire set of points whose fitness is i. It should be clear that the arc probabilities are simply derived in this case, and indeed we could easily calculate the corresponding LSMs for this landscape whatever the value of L. For example, if L was 1000 then the arc leading from s_{217} to s_{218} would have probability 1/783, which is the chance of the mutation operator flipping one of the '0' bits of a candidate represented by s_{217}.

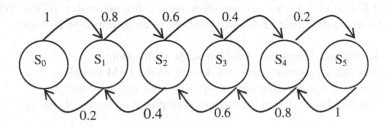

Fig. 1. A landscape state machine for the MAX-ONES function with $L = 5$, assuming single-gene flip mutation

Such a landscape model becomes practically useful when we consider the following scenario. Suppose that one wishes to compare the performance of each of m different EAs on MAX-ONES, using only single-gene bit-flip mutation. Also suppose each algorithm eschews any elements or operations (other than the fitness function) which require genotypic knowledge. E.g. phenotypic crowding may be employed, but not genotypic crowding. The space of algorithms which remain is certainly not overly restricted. E.g. we may wish to compare 1000 algorithms, each of which is a population-based evolution strategy, for all combinations of 10 different population sizes, 10 different selection schemes, and 10 different population-structure strategies (e.g. including islands-based schemes with a variety of migration strategies).

To properly compare these m EAs, we need t runs of each algorithm, and will typically run each trial of each algorithm for a baseline fixed number of fitness evaluations, e. The total number of fitness evaluations will be $m.t.e$, which can easily come to several billions or more on applications of interest. In the case of MAX-ONES the fitness calculation is inexpensive, but consider, for example, the requirement to measure these algorithms' performance when $L=1,000,000$. An intriguing fact is as follows. For algorithms of the type outlined, and landscapes which we can *precisely* model with an LSM, we can replace the landscape by the model, yet predict the resulting comparative algorithm performance with full statistical accuracy. As long as we use a suitable initial distribution of initial fitness labels for the initial population, we can expect the statistical properties of comparative algorithm performance on the described LSM model, of MAX-ONES to precisely (in statistical terms) match performance on the MAX-ONES landscape itself.

It is worth clarifying at this point what we mean by "replace the landscape with the model". To run an EA on an LSM, we simply replace the notion of a candidate solution with that of a state in the LSM. The initial population is simply a state number, and (assuming state numbering is mapped into a simple way onto fitnesses), and selection operates in the normal way. Evaluation is trivial, since the state number corresponds to fitness. Finally, when an individual (state) is mutated, the child state is chosen probabilistically according to the transitions emanating from the parent state.

By using LSM models for differential algorithm performance analysis, it could be possible to significantly reduce development times. The ability to replace a landscape with an LSM model allows, among other things, the possibility of the following method for use in developing EA-based solutions to an optimisation problem:

1.   Derive an LSM model of the landscape in question;
2.   Run many candidate algorithm designs on the LSM model;
3.   Choose a suitable algorithm design given the results of step 2.
4.   Use the chosen design for further development on the real landscape.

Steps 2–3 exploit the fact that an LSM model can be searched far more speedily than the real landscape. To the extent that the dynamics of an algorithm's search on the LSM accurately reflect its dynamics on the real landscape, the information gleaned in steps 2–3 will support an appropriate step 4 choice.

Where precise LSMs can be inferred, the problem is undoubtedly a toy one, but the LSM technique still may uses. E.g., with perfect statistical accuracy we can accurately investigate the relative performance of a wide range of EAs on MAX-ONES with $L=1,000,000,000$, without ever needing to store or evaluate a single 1Gb genome. The thought that this may be possible on more interesting and realistic landscapes is appealing, but the LSM model in such cases will invariably be an *approximate* model. It remains to be seen whether approximate LSMs of interesting landscapes retain the possibility to enable accelerated appropriate choice of algorithm design.

3   Approximate LSMs

Consider the LSM in Figure 2, which models the order 3 deceptive trap function with $L = 6$, again assuming single-gene bit-flip mutation. Note that the use of this particular mutation operator is not a restriction on the method, it just simplifies illustration. In contrast, the restriction to mutation only *is* a limitation which, though not insurmountable, is beyond the scope of a first description of the concept of LSMs and is the topic of ongoing work. Now, Figure 2 is an accurate LSM in the sense that any arc represents the correct proportion of mutations from the sending state which will arrive at the receiving state. State labels correspond precisely to fitnesses, e.g. state s_4 represents all genomes with fitness 4, and the arc to state s_5 indicates that a mutation of a point of fitness 4 is likely to yield a result of fitness 5 with probability 0.33. However, in an important way, it is less helpful than the earlier LSM in capturing the landscape features of importance in exploring search algorithm performance. This is simply because there is high 'profile variance' within some states, which we now explain.

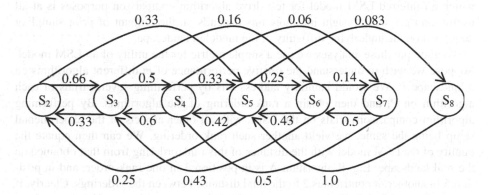

Fig. 2. An LSM for the order 3 deceptive problem with $L = 6$ and single-gene bit-flip mutation

The 'profile' of $s \in E$ is the vector of its transitions t_{sj} for all j. The profile elements correspond precisely to a_{sj} in the case when the LSM has one state for each point in E. In the previous MAX-ONES LSM, the vector of arcs from any state precisely matches the profile of every point 'contained' in that state. But in general the vector of arcs will be a weighted average of these profiles. Consider the state corresponding to a fitness of 6 in Figure 2. Two kinds of points in this landscape have fitness 6. One is exemplified by the point 000110, with one chunk of three genes fully unset, contributing 4, and a second chunk with 2 bits set, contributing 2. The six points of this type each have the profile (0.0, 0.5, 0.0, 0.33, 0.0, 0.16, 0.0) – that, is chance 0.0 of a mutant yielding a fitness of 2, chance 0.5 of a mutant yielding a fitness of 3, and so on. The other kind of point with fitness 6, of which there is just one, is 111111, whose profile is: (0.0, 0.0, 0.0, 1.0, 0.0, 0.0, 0.0). The profile of s_6 in Figure 2 represents these two profiles weighted and combined. A *precise* LSM for this problem would have two different states corresponding to a fitness of 6, one for each of the two corresponding genotypes. The LSM of figure 2, however, is an approximate LSM for this particular problem, since the dynamics of an algorithm running on it will not exactly reflect the dynamics of that algorithm on the real problem.

In general (and which we do in section 4), LSMs will be inferred from data obtained from preliminary sampling runs on the real landscape. Given the general case of no prior landscape knowledge, we will neither know the correct number of states for the LSM to use, nor the number of different profiles that may be shared by points with a given fitness. LSMs inferred by sampling will thus be highly approximated models of the precise LSM for that landscape. In the experiments which follow, we use the simplest kind of basic LSM framework in which (as in figures 1 and 2) each state represents a given fitness level, and there is only one state per fitness level.

4 Empirical Experience

Given that exact LSM models would clearly be very valuable, but approximations are inevitably all we are able to expect in reality, we need to investigate the degree to which an inferred LSM model for test-drive algorithm comparison purposes is at all useful, and get some insight into how this depends on the amount of prior sampling done, and on the underlying difficulty of the modeled landscape.

To underpin these analyses we use a simple metric for the utility of an LSM model. Suppose we wish to understand the relative performance of *m* different algorithms on a landscape *P*. We would naturally address this by performing several trials of each algorithm on *P*, and then obtain a rank-ordering of the algorithms. By performing algorithm comparisons trials on the LSM, with all other aspects of the experimental setup being the same, we yield another such rank ordering. We can then equate the quality of the LSM model with the distance of this rank ordering from that obtained on the real landscape. E.g. if algorithm A is in position 3 in one rank-order, and in position 5 in another, it contributes 2 to the total distance between the orderings. Clearly, if an LSM model scores 0, it precisely predicts the comparative performance of the tested algorithms on the real problem.

We now explain the design of our initial experiments in which we seek to explore LSM utility for approximate LSM models. We first set out, for easy reference, the algorithms and problems used, and later set out the experiments themselves.

4.1 Algorithms, Problems, and Preliminary Sampling

The algorithms tested are the following ten. In each case, the basic algorithm was a standard generational EA with elitism (a single best in the population is automatically entered into the next generation), using tournament selection (with replacement). The mutation operator in all experiments is gene-wise mutation with a probability of 0.001. That is, when a chromosome is mutated, each gene is mutated independently with that probability. This is unlike the single-gene bit-flip mutation operator used previously in illustrations, and leads to a more complicated LSM which is a greater challenge to approximate. The algorithms compared differ in tournament size and population size, and also number of generations (but the latter varied simply to ensure comparable fitness evaluations in each case).

A. Population size 10, tournament size 1 (random selection), 500 generations.
B. Population size 10, tournament size 2, 500 generations.
C–F. Population size 20. 235 generations, tournament sizes 1, 2, 3, 5 respectively.
G–J. Population size 50. 91 generations, tournament sizes 3, 5, 7, 10 respectively.

This choice of algorithms is *ad hoc* and pragmatic, but reflects a suitable collection of potential parameterizations of an EA which might be tested, given the requirement for reasonably fast convergence, and given no prior empirical or theoretical knowledge of the problem in hand.

We use a different EA for preliminary sampling (in order to infer the LSMs used in the algorithm comparison tests). This is the same basic EA, but with a population size of 200 and random selection, run for either 50,000 or 100,000 evaluations. This is again an *ad hoc* choice; later we briefly discuss potentially better, or ways to choose, preliminary sampling techniques. Here we simply note that one salient aspect of this is the number of evaluations. If the 'real landscape' algorithm comparison would consume e evaluations, the number of evaluations consumed in preliminary sampling, p, should be suitably less than e. As we will see, p is 450,000 in the ensuing tests.

The optimization landscapes we explore are MAX-ONES problem with $L = 1000$, and NK landscapes with $N = 1000$, and with K set at 5 or 10. For each landscape the following is done: we run the preliminary sampling EA once on the real landscape, building two LSMs for each, one corresponding to the data gleaned up to 50,000 evaluations and another from the data gleaned up to 100,000 evaluations.

We infer the LSM as follows. Every (parent fitness, child fitness) pair corresponding to a mutation is recorded. With general real-world application in mind (in which there is no prior knowledge of the number of fitnesses), we chose instead to fix the number of states in the LSM at 200, and thus map the information gleaned, in every experiment, onto a 200-state LSM. The (parent) fitnesses sampled were simply scaled uniformly between 0 and 200; then, a state in the LSM was created to correspond to each interval $[x, x+1]$ for x from 0 to 199. Each state therefore corresponded to a number (in some cases 0) of fitnesses, and arc labels were calculated accordingly. A final point worth noting is that we also remembered the fitnesses of the initial population of the sampling run, and scaled them to yield a list of the corresponding states. The LSM trials then used this list of states to build its initial populations. E.g. for a trial run with a population of size 10, the initial population of the algorithm trial on the LSM would be generated by sampling uniformly 10 times from this list.

4.2 Experiments

The aim of these experiments was to gain some preliminary insight into how useful an inferred LSM might be in discriminating the performance of several algorithms. The experimental protocol for a given problem P was as follows:

1. Run the preliminary sampling EA on P in order to harvest data for the LSM.
2. Infer an approximate LSM for P from the data gathered in step 1.
3. Run an algorithm comparison study on P, with the aim of discriminating the relative performances of several chosen EAs.
4. Run the same algorithm comparison study on the LSM inferred for P, establishing the algorithms' relative performance on the LSM.
5. Compare the ordering of algorithms given by step 4 with that given by step 5.

In step 3, mirroring the typical shape of a real-world study which desires enough trial runs for a chance of statistical confidence, yet still needs to be parsimonious in time and resources, we run ten trials of each of the algorithms A–J described in section 4.1. We record the result (best fitness found) in each trial, and a rank ordering is then determined by running t-tests. The same is done in step 4, however this time relative performance is determined by speed of convergence. As we will further discuss later, for various understandable reasons the algorithms would typically converge to the optimum of the LSM (i.e. state 200) before their evaluation limits were reached, and so we could not discriminate in step 4 solely in terms of highest state attained.

A reminder of the object of all this is as follows: given conditions under which such experiments might be successful (i.e. the rank ordering found in step 4 is significantly close to that found in step 3), and supposing we have the task of choosing an algorithm to use in anger on a real problem, step 3 would not be necessary. We could skip to step 4, discover the algorithm which performs best on the LSM, and take this to be a suitably accurate prediction of which algorithm would perform best on the real landscape. Alternatively, we might use the results of step 4 to home in on a smaller selection of choices for tests on the real landscape. The conditions under which this scheme may be successful are easy to state: having an LSM which perfectly captures the landscape. What is to be determined is whether an approximate LSM can capture enough of the salient details of the landscape to produce accurate predictions in step 4.

4.3 Results

Table 1 shows the rank orderings identified in step 3 by running algorithms A–J on each of the three problems addressed here. Table 2 gives the rank ordering of algorithms identified by experiments on the appropriate inferred 200-state LSMs following 50,000 evaluations of the preliminary sampling EA. This table also provides measures of the distance between the LSM-inferred ordering and the real (Table 1) ordering. Associated with each algorithm is its distance d from its position in the corresponding Table 1 ordering, and the sum of these distances is also given. Table 3 provides the same information for the 100,000 evaluation LSMs. A brief summary of the main findings of these preliminary experiments is as follows.

Accuracy of the 50,000-evals LSM for MAX-ONES. The rank order of algorithms obtained by experiments on the 50,000-evals approximate LSM for MAX-ONES is nearly identical to that obtained from experiments on the real MAX-ONES landscape. This augurs very well for the method, since it seems that with 50,000 evaluations' worth of effort, we have predicted with near-perfect accuracy, and in a fraction of the time (since running the LSM experiments is extremely fast) a rank ordering which needed approximately 10 times that amount of effort on the real landscape. Even though MAX-ONES is a simple landscape, this initial result is promising since it shows that an approximate LSM can be used for this purpose, and predicting comparative algorithm performance remains very difficult, even on MAX-ONES, owing the complexity of EA dynamics.

Accuracy of the Approximate LSMs on NK with $K = 5$. The total distance between the 50,000-evals LSM-obtained ordering and the Table 1 ordering is statistically significant, indicating that we can say with confidence that the predictive ability of the 50,000-evals LSM experiments was far better than random. In the 100,000-evals case, the ordering is even better, and the LSM seems particularly good at distinguishing which algorithms are particularly unsuited to this landscape, and well able to identify a collection which should perform well. The improvement in distance as we go from 50,000 to 100,000 evaluations is expected, since more real data has been used in inferring the LSM.

Table 1. Rank-orderings of algorithms *A–J* on each of the three problems studied, identified via *t*-tests on ten trial runs of each algorithm. '1' is best and '10' is worst.

Problem	1	2	3	4	5	6	7	8	9	10
MAX-ONES	B	F	E	D	A	J	I	H	G	C
NK, K = 5	F	B	E	D	J	A	I	H	G	C
NK, K = 10	F	B	E	J	D	I	H	A	G	C

Table 2. Rank-orderings of algorithms *A–J* on each of the three LSMs inferred from 50,000 evaluation preliminary sampling runs. Rankings identified and annotated as in Table 1, but in this case, algorithms are compared in terms of speed of attaining the final state (state 200), using highest state attained where necessary. The column below each rank position gives the distance of the corresponding algorithm from is rank position on that problem in Table 1. The final column provides the total distance from the Table 1 ordering

Problem	1	2	3	4	5	6	7	8	9	10	Distance
MAX-ONES	B	F	E	D	J	A	I	H	G	C	
	0	0	0	0	1	1	0	0	0	0	2
NK, K = 5	B	E	D	F	A	J	H	I	G	C	
	1	1	1	3	1	1	1	1	0	0	10
NK, K = 10	F	J	I	H	G	D	C	B	A	E	
	0	2	3	3	4	1	3	6	1	7	30

Table 3. Rank-orderings of algorithms *A–J* on each of the three LSMs inferred from 100,000 evaluation preliminary sampling runs. Rankings identified and annotated as in Table 2

Problem	1	2	3	4	5	6	7	8	9	10	Distance
MAX-ONES	B	F	E	D	A	J	I	H	G	C	
	0	0	0	0	0	0	0	0	0	0	0
NK, K = 5	B	E	F	D	A	J	I	H	G	C	
	1	1	2	0	1	1	0	0	0	0	6
NK, K = 10	B	E	I	D	F	J	H	A	G	C	
	1	1	3	1	4	2	0	0	0	0	12

Accuracy of the Approximate LSMs on *NK* with *K* = 10. The total distance between the 50,000-evals LSM-obtained ordering and the Table 1 ordering is better than the mean of a random distribution of orderings (33), but is not statistically significant, indicating that we cannot say that the LSM in this case has been able to distinguish between the algorithms better than a random ordering would have done. However, the more accurate 100,000-evals LSM is able to obtain an ordering significantly close to the Table 1 ordering, and certainly provides usefully accurate predictions of the relative performance of the ten test algorithms on this highly epistatic problem.

Trends against Landscape Ruggedness and Sample Size. In partial summary of the above observations, it is worth noting that the accuracy of predicted relative performance decreases with ruggedness of the landscape and increases with the amount of data obtained from (i.e. the length of) the preliminary sampling run. Both of these trends were of course very much expected, however what we did not know *a priori* is whether significant accuracy could be obtained at all without an unreasonably large sample size, and on problems with nontrivial ruggedness. These preliminary experiments have shown that significant accuracy *can* be obtained with a reasonable sample size and on at least one highly rugged problem.

5 Discussion: Prospects and Research Issues

LSMs have a variety of potentially fruitful uses. For example, in the application of Directed Evolution [2,3], in which evolutionary algorithms are applied directly to the generation of novel proteins, one generation can take about 24 hours, and with considerable cost in molecular biology consumables. Similarly, structural design, especially regarding novel structures, often requires highly costly evaluation functions which may include fine-grained airflow or stress simulation. In such cases the need for fast and accurate methods to help design the algorithm is crystal clear.

We have started to explore the viability of inferring them from preliminary landscape data, with a view to obviating the need for algorithm choice and/or tuning runs. The results so far are encouraging, particularly given the following points. First, an arbitrary and simple LSM structure was used in each case, with no preliminary attempt to, for example, find the most suitable number of states for each problem. Second, the preliminary sampling EA was designed only as one of many possible ways to obtain parent/child fitness pairs for a good spread of fitnesses. Further investigation may reveal that alternative sampling approaches may be more effective, such as a Markov Chain Monte Carlo search, or an adaptive EA which explicitly attempts to sample evenly (or in some other suitably defined way) throughout the landscape. As it turned out, the prior sampling EA in each case performed rather less well than the better few of the test EAs in the real experiments; this means that the LSM contained no information about a considerable portion of the landscape traversed by these algorithms; it is therefore encouraging that the LSM was able to do a fair job in each case despite this fact. Third, there is no reason to expect that some useful accuracy of the LSM-based predictions would not be maintained if further (e.g. 20, or 50) different suitable algorithms were tested, and hence the 'true' potential savings in evaluations are potentially vary large. Fourth, LSMs can always be augmented by data from further runs on the real problem, If it turns out that a series of instances in the same problem class can be suitably modeled by the same approximate LSM (which seems intuitively reasonable), then each real algorithm run on the problem yields data which can augment the LSM and hence increase its accuracy for further algorithm choice/tuning experiments. Finally, an exciting prospect is the augmentation of inferred LSMs with prior knowledge or reasonable assumptions about the landscape.

There are many research issues here, ranging through the theory and practice of choosing a sampling method, how best to construct an LSM from given data, and establishing what confidence to bestow on the predictions of an applied LSM model.

A quite separate use of LSMs arisies from the fact that they can be artificially contrived (i.e. we can predefine a transition matrix), yielding a class of 'toy' landscapes which can be tuned in ways interestingly (and in some ways potentially usefully) different from other tunable models such as the NK family. We can construct landscape state machines, and hence models of real landscapes (although see below), at will. In so doing, it is not necessarily easy to directly and precisely control factors such as the degree of epistasis or the sizes of basins of attraction, however we *can* precisely control the number of fitness levels, the density of local optima, and how this density varies as we move around the landscape. The most attractive feature of LSMs in this sense is that we can describe an extraordinarily wide range of landscapes with them. A potential application of this is to search through spaces of landscape state machines to find, for example, landscapes which are particularly suited to one EA rather than another, or to a standard EA rather than hillclimbing, or to simulated annealing rather than hillclimbing, and so forth. This is an area we plan to explore in future work, however it is worth pointing out that one important research issue here is whether arbitrary LSMs correspond to any real landscapes. As it turns out, certain constraints are needed on the profiles of states in the LSM, but we have not yet fully resolved how to ensure an arbitrary LSM which meets these constraints is valid (i.e. corresponds to a realizable landscape over k-ary strings using a standard mutation operator).

6 Conclusions

In this preliminary article we have described a straightforward way to model a search landscape as a finite state machine, which we call a Landscape State Machine. An attractive aspect of such a model is that, for certain landscapes at least, the features relevant to the search dynamics of certain EAs can be fully captured in a very compact model, and this model can be 'run' in place of the 'real' landscape (perhaps saving immense time in fitness computations). Although this is no substitute for locating optima on the real landscape, it does substitute, with full confidence in the case of certain pairings of landscapes and algorithms, in the business of determining the relative performance of algorithms.

The real advantages of LSMs come, however, if the approximations necessary to model more interesting landscapes (in particular, modeling real landscapes via sampled data) are such that the salient aspects of algorithm performance remain captured in the models. In some cases LSMs may be accurate enough to enable correct (and very fast) differentiation between algorithm performances, thus correctly informing how resources should be applied in the case of the real landscape. Preliminary experiments on MAX-ONES and *NK* landscapes have demonstrated that the idea is promising. Further, by searching through spaces of LSMs, in future work we may be able to take a new look at the question of what kind of problem is easy/hard for an EA.

Acknowledgements. We are pleased to acknowledge the support of the BBSRC (UK Research Council for Biological Sciences), and Evosolve (UK registered charity no. 1086384). Thanks to its handy and efficient implementation of *NK* problems, Peter Ross' pga code was used for preliminary sampling and 'real landscape' EAs.

References

1. Altenberg, L. *Fitness distance correlation analysis: an instructive counterexample*. In Th. Bäck, editor, Proceedings of the 7th International Conference on Genetic Algorithms, pages 57–64. Morgan Kaufmann Publishers, 1997.
2. Arnold, F. (1998). Directed evolution, *Nature Biotechnology*, **16**: 617–618.
3. Arnold, F. (2001). Combinatorial and computational challenges for biocatalyst design. *Nature*, **409**: 253–257.
4. Barnett, L. *Ruggedness and neutrality: the NKp family of fitness landscapes*. In C. Adami, R. K. Belew, H. Kitano, and C. E. Taylor, editors, Alive VI: Sixth International Conference on Articial Life, pages 18–27, Cambridge MA, 1998. MIT Press.
5. Davidor, Y. (1991): "*Epistasis Variance: A Viewpoint on GA-Hardness*". In: Foundations of genetic algorithms, ed. G.J.E. Rawlins, Morgan Kaufmann Publishers, pp. 23–35.
6. Grefenstette, J.J. (1992). *Deception considered harmful*. Foundations of Genetic Algorithms, 2. Whitley, L. D., (ed.), Morgan Kaufmann, 75–91.
7. Jones, T. and S. Forrest. *Fitness Distance Correlation as a Measure of Problem Difficulty for Genetic Algorithms*. In L. J. Eshelman, editor, Proceedings of the 6th Int. Conference on Genetic Algorithms, pages 184–192, Kaufman, 1995.
8. Kauffman, S.A. and S. Levin. *Towards a General Theory of Adaptive Walks on Rugged Landscapes*. Journal of Theoretical Biology, 128:11–45, 1987.
9. Kallel, L., Naudts, B. & Reeves, C. (1998). Properties of fitness functions and search landscapes. In *Theoretical aspects of evolutionary computing* (ed. L. Kallel, B. Naudts and A. Rogers), pp. 175–206. Springer, Berlin.
10. Naudts, B. and L. Kallel (2000). A comparison of predictive measures of problem difficulty in evolutionary algorithms, *IEEE Transactions on Evolutionary Computation*, **4**(1):1–15.
11. Reeves, C. R. (1999). Landscapes, operators and heuristic search. *Annals of Operations Research* **86,** 473–490.
12. Stadler, P.F. Towards a Theory of Landscapes," in Complex Systems and Binary Networks, (R. Lopez-Pena et al, eds.), Berlin, New York, pp. 77–163, Springer Verlag, 1995.
13. Stadler, P. F. (1996). Landscapes and their correlation functions. *J. Math. Chem.* **20,** 1–45.
14. Wright, S. (1932). The roles of mutation, inbreeding, crossbreeding and selection in evolution. In *Proc. Sixth Int. Conf. Genetics*, vol. 1 (ed. D. F. Jones), pp. 356–366.

Constrained Coverage Optimisation for Mobile Cellular Networks

Lin Du and John Bigham

Electronic Engineering Department
Queen Mary,University of London
London E1 4NS, United Kingdom
{lin.du, john.bigham}@elec.qmul.ac.uk

Abstract. This paper explores the use of evolutionary algorithms to optimise cellular coverage so as to balance the traffic load over the whole mobile cellular network. A transformation of the problem space is used to remove the principal power constraint. A problem with the intuitive transformation is shown and a revised transformation with much better performance is presented. This highlights a problem with transformation-based methods in evolutionary algorithms. While the aim of transformation is to speed convergence, a bad transformation can be counter-productive. A criterion that is necessary for successful transformations is explained. Using penalty functions to manage the constraints was investigated but gave poor results. The techniques described can be used as constraint-handling method for a wide range of constrained optimisations.

1 Introduction

Mobile cellular networks are by far the most common of all public wireless communication systems. One of the basic principles is to re-use radio resources after a certain distance. The whole area is divided up into a number of small areas called *cells*, with one *base station* giving radio coverage for each cell by its associated *antenna*. This paper considers a novel geographic traffic load balancing scheme achieved by changing cell size and shapes, to provide dynamic mobile cellular coverage control according to different traffic conditions. Study of dynamic cell-size control has shown that the system performance can be improved for non-uniformly distributed users [1]. However changing both cell size *and* shapes has not been studied so far. The formation of cells is based upon call traffic needs. Capacity in a heavily loaded cell can be increased by contracting the antenna pattern around the source of peak traffic and expanding adjacent antenna patterns to fill in the coverage loss as illustrated in Fig. 1.

Realising such a system requires the capability of approximately locating and tracking mobiles in order to adapt the system parameters to meet the traffic requirements. The existing generation of cellular networks has a limited capability for mobile position location, but the next generation of cellular networks is expected to have much better capabilities. The position location capabilities of the

S. Cagnoni et al. (Eds.): EvoWorkshops 2003, LNCS 2611, pp. 199–210, 2003.
© Springer-Verlag Berlin Heidelberg 2003

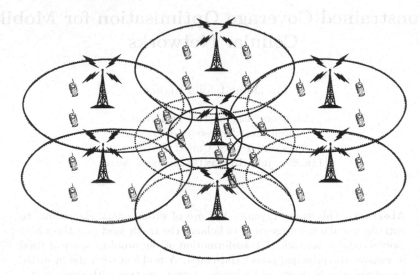

Fig. 1. Cellular coverage control according to geographic traffic distribution.

cellular network can be used to find the desired coverage patterns to serve a current traffic demand pattern. The best-fit antenna pattern from the available antenna resource is then synthesised using a pattern synthesis method, which takes into account the physical constraints of the antenna.

The first step towards intelligent cellular coverage control is to know which combination of antenna patterns is suitable for a traffic distribution. Several optimisation methods can be used. In this work, we explore the use of genetic algorithms, one of the evolutionary algorithms, because of their robustness and efficiency in searching the optimum solution for complicated problems [2], [3], [4]. This is explained in sect. 2.

Section 3 describes the transformations of the coordinate space that remove the central power constraint. While several transformations may be possible and mathematically correct, for rapid convergence care has to be taken to ensure that the transformation does not distort the space in an unsatisfactory way. A criterion that is necessary for successful transformations is also explained.

Computer based simulations are performed to evaluate the system capacity. The simulations are described and results are presented in sect. 4. As a comparison, using penalty function to manage the constraints is also investigated but gave poor results.

Beyond the scope of our application, the techniques described have further applications on constraints handling in optimisations. Attempts of applying this constraint-handling method into other optimisation problems are also investigated, and some of them are presented in sect. 5.

2 Using Genetic Algorithms

Genetic Algorithms are search algorithms based on the mechanics of natural selection and natural genetics. They combine survival of the fittest among string structures. In every generation, a new set of structures is created using parts of the fittest of the previous generation with occasional random variation (mutation). Whilst randomised, genetic algorithms are not simple random walks. They have been shown to efficiently exploit historical information to find new search points with expected improved performance [2]. Since genetic algorithms have already been widely used in many optimisation area, we only focus on how to apply it in our antenna patterns optimisation problem and how to handle the central power constraint.

First, we must design an efficient coding scheme to represent each possible combination of antenna patterns as chromosomes. We use a gain vector, $\vec{G} = [g_1, g_2, g_3, \ldots, g_N]$, in which each gain value is coded as a gene, symbolising antenna gains along N directions. This determines the approximate shape of one antenna pattern. The number of gains N, and the number of bits for each gain, can be determined by the performance and precision requirements. Therefore, the chromosome for a region is formed by combining each set of genes in the region. If we have M base stations, the chromosome will be of the form $[\vec{G}_1, \vec{G}_2, \ldots, \vec{G}_M]$. As many researchers (e.g. [5], [6], [7], [8]) have reported, representing the optimisation parameters as numbers, rather than bit-strings, can speed up the convergence for most real-value optimisation problems. We use a real-coded genetic algorithm, with BLX-α [9] crossover, distortion crossover (intra-crossover within a subset of chromosome, i.e. crossover between two gain values at different directions in one pattern such that it can distort the shapes of individual patterns), and creep mutation operators. The elitism selection is used to prevent losing the best solutions. Since the transformation, which is explained next, handles the central constraint, it is not necessary to use more complicated operators for our optimisation. A cellular network simulator is used to calculate the fitness value for each chromosome.

3 Constraint Handling Method

The genetic algorithms are naturally an unconstrained optimisation techniques. When a new chromosome is created by crossover or mutation in the searching process, it may not be feasible. Many constraints handling approaches have already been proposed [10], [11] and recent survey papers classify them into five categories, namely: use of penalty functions, special representation and operators, separation of objectives and constraints, hybrid methods and other approaches [12]. We investigate a method based on a transformation between search space and feasible space, which ensures that all the products of a crossover or mutation always will be feasible. This falls into the second category mentioned above. It is simpler, and usually better, than the methods that only map unfeasible chromosomes to feasible ones for several reasons. i) A very simple search space can be used, ii) all the feasible or unfeasible chromosomes can be treated without any difference, iii) this transformation can be constructed before starting optimisation, and iv) the transformation can be done as many times as needed, if there are more than one constraint. In their paper [13] the authors propose a general mapping method, which is able to map all the points in searching space into feasible space. It is a numerical method, and involves a lot of computation in searching the boundary points. To avoid this we construct a specific mapping function by analytical means, which is explained next.

The feasible space here is the space that includes all the legitimate subsets (one pattern at N directions) for a chromosome (M patterns for the whole network), $\mathbb{F} = \{\vec{G} = (g_1, \ldots, g_N) : g_i \in \mathbb{R}\}$. A transformation is created such that the search space of the form, $\mathbb{S} = \{\vec{X} = (x_1, \ldots, x_N) : 0 \leq x_i \leq 1, x_i \in \mathbb{R}\}$, can be used. Each time that we need to calculate a fitness value for any chromosome (which is now encoded in terms of the $\vec{X}$), we map the chromosome into the feasible space, calculate the fitness value for new chromosome, and then assign the value to the original chromosome. In this way, we can perform genetic algorithms without constraints.

According to physics theory, the radio frequency (RF) transmitting power at a base station can be expressed as,

$$P_{trans} = \delta \cdot \frac{1}{N} \sum_{i=1}^{N} g_i'^2, \tag{1}$$

where N is the number of gains, g_i' is the $i-th$ gain value along N directions, and δ is a constant.

Then the main constraint of RF power available at the base station is expressed as,

$$P_{min} \leq \delta \cdot \frac{1}{N} \sum_{i=1}^{N} g_i'^2 \leq P_{max}, \tag{2}$$

where P_{min} is the minimum, and P_{max} is the maximum value of RF power for a base station. They are determined by both physical limit and traffic density nearby.

If we choose the same N for all the base stations, (2) can be simplified as,

$$P_{min} \leq \sum_{i=1}^{N} g_i^2 \leq P_{max}, \tag{3}$$

where $g_i = \sqrt{\frac{N}{\delta}} \cdot g_i'$. Therefore, the feasible space $\mathbb{F}$ can be expressed as,

$$\mathbb{F} = \left\{ \vec{G} = (g_1, \ldots, g_N) : P_{min} \leq \sum_{i=1}^{N} g_i^2 \leq P_{max}, g_i \in \mathbb{R} \right\}. \tag{4}$$

Since an N-dimensional cube is used as the search space $\mathbb{S}$,

$$\mathbb{S} = \left\{ \vec{X} = (x_1, \ldots, x_N) : 0 \leq x_i \leq 1, x_i \in \mathbb{R} \right\}, \tag{5}$$

we then need to define a mapping function $f : \mathbb{S} \longrightarrow \mathbb{F}$ to map points from $\mathbb{S}$ into $\mathbb{F}$.

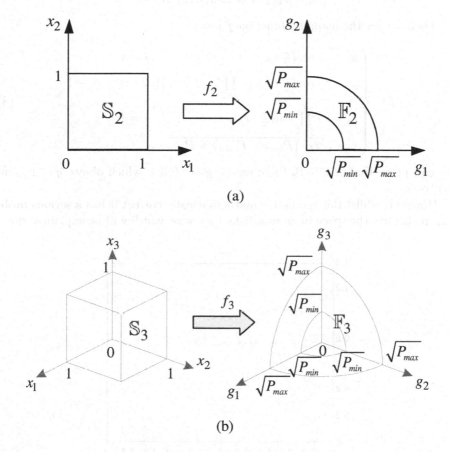

Fig. 2. (a) Mapping from search space $\mathbb{S}_2$ to feasible space $\mathbb{F}_2$, (b) Mapping from search space $\mathbb{S}_3$ to feasible space $\mathbb{F}_3$.

When $N = 2$ and $N = 3$, search space $\mathbb{S}_2$, $\mathbb{S}_3$ and feasible space $\mathbb{F}_2$, $\mathbb{F}_3$ are shown in Fig. 2. Inspired by the transformation from polar coordinates and spherical coordinates to Cartesian coordinates, we can decompose the feasible space $\mathbb{F}$ as,

$$
\begin{cases}
g_1 = r \cdot cos\theta_1 \\
g_2 = r \cdot sin\theta_1 \cdot cos\theta_2 \\
g_3 = r \cdot sin\theta_1 \cdot sin\theta_2 \cdot cos\theta_3 \\
\cdots \\
g_{N-1} = r \cdot sin\theta_1 \cdot sin\theta_2 \cdot \ldots \cdot cos\theta_{N-1} \\
g_N = r \cdot sin\theta_1 \cdot sin\theta_2 \cdot \ldots \cdot sin\theta_{N-1}
\end{cases}
\tag{6}
$$

Then we get $\sum g_i^2 \equiv r^2$.

If we let

$$\begin{cases} r^2 = x_1 \cdot (P_{max} - P_{min}) + P_{min} \\ \theta_i = x_{i+1},\ i = 2,\ 3,\ \ldots,\ N-1 \end{cases}.$$ (7)

Then we get the mapping function f as,

$$f = \begin{cases} g_i = r \cdot cos(\frac{\pi}{2} \cdot x_2), & i = 1 \\ g_i = r \cdot cos(\frac{\pi}{2} \cdot x_{i+1}) \prod_{j=2}^{i} sin(\frac{\pi}{2} \cdot x_j),\ 2 \le i \le N-1 \\ g_i = r \cdot \prod_{j=2}^{N} sin(\frac{\pi}{2} \cdot x_j), & i = N \\ r = \sqrt{x_1 \cdot (P_{max} - P_{min}) + P_{min}} \end{cases}.$$ (8)

We can prove that, $\forall \overrightarrow{x} \in \mathbb{S}$, there exists $\overrightarrow{g} = f(\overrightarrow{x})$, which obeys $\overrightarrow{g} \in \mathbb{F}$, and vice versa.

However, whilst this method is mathematically correct it has a serious problem: it distorts the space in an unsatisfactory way, which will be explained next.

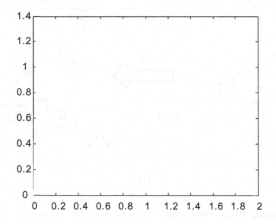

Fig. 3. The marginal PDF of g_i uniformly distributed in feasible space, where $N = 12$, $P_{max} = 5.0$, and $P_{min} = 0.01$

To evaluate how much distortion the transformation causes, we use a PDF (Probability Density Function) based method, which checks the probability of output with uniformly distributed, independent input. Since the possible locations of the optimum solution are unknown in our case, the ideal transformation should generate output points uniformly distributed in feasible space. Because of the symmetry of our feasible space, the marginal PDFs of g_i are identical with these of g_j, where $i \ne j$ and g_i is independent of g_j. One of them is shown in Fig. 3, and will be used as the criterion for evaluating transformation distortion. Whilst this method is not very fine, it works well in our cases, since a little distortion does not affect the performance of a randomised optimisation like GA.

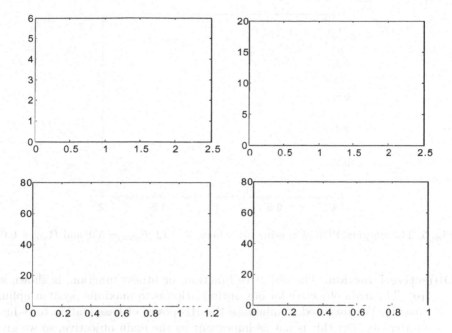

Fig. 4. The marginal PDF of g_1, g_4, g_9, and g_{12}, using (8), where $N = 12$, $P_{max} = 5.0$, and $P_{min} = 0.01$

The marginal PDFs of g_1, g_4, g_9, and g_{12}, calculated by for (8) are shown in Fig. 4. It shows that the PDFs for different g_i are different. g_1 always has the highest probability of being the largest value, while g_N has the lowest probability. This transformation causes too much distortion, and results in slow convergence as seen in Fig. 6 in sect. 4.

To solve this, we chose another mapping function in a similar way described before, which is inspired by the fact of $\sum (\frac{x_i}{\sqrt{\sum x_i^2}})^2 \equiv 1$, as shown in (9),

$$f = \begin{cases} g_i = r \cdot \dfrac{x_i}{\sqrt{\sum\limits_{j=1}^{N} x_j^2}}, \quad i = 1, 2, \ldots, N \\ r = \sqrt{\dfrac{\sum_{j=1}^{N} x_j^\rho}{N}} (P_{max} - P_{min}) + P_{min} \end{cases}, \tag{9}$$

where ρ is a factor used to control the distortion of transformation. Here $\rho = 8.0$ is chosen to get the smaller distortion as seen in Fig. 5. In our experiments, this results in a much faster convergence as seen in Fig. 6 in sect. 4.

4   Optimisation Results

Simulations were performed to test the efficacy of the approach and the potential of the method. The following is the list of simulation specifications.

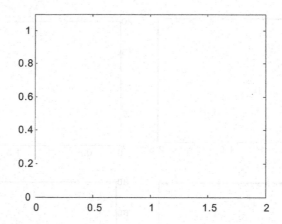

Fig. 5. The marginal PDF of g_i using (9), where $N = 12$, $P_{max} = 5.0$, and $P_{min} = 0.01$

Objective Function. The objective function, or fitness function, is shown as (10). The main objective for our optimisation is to maximise system uplink capacity. We also need to minimise the RF power of base station to reduce interference, but this is not as important as the main objective, so we give it less weight, which is found by experiments.

$$Fitness = Capacity - 0.2 \cdot (RF\ Power\ for\ all\ BSs). \qquad (10)$$

Penalty function. We compare our transformation-based method with a widely used constraints handling method, using penalty function.The penalty function is calculated by (11) and subtracted from the objective function.

$$Penalty = \sum [Penalty\ value\ for\ j - th\ pattern,\ P_j], \qquad (11)$$

where $P_j = \begin{cases} exp(\sum g_i^2 - P_{max}) , & \sum g_i^2 > P_{max} \\ 0 , & \sum g_i^2 \leq P_{max} \end{cases}$.

Simulation Configuration. They are listed as follows:
- 3000 traffic units and 100 base stations. Each base station has the capacity to serve 36 traffic units, i.e. *Total Capacity* = 120% · *Total Demand*;
- 20% of traffic is uniformly distributed in the whole area;
- Other 80% of traffic is distributed in 40 hot spots with normal distribution (the mean value for each hot spots, μ, is uniformly distributed over the whole area, and the standard deviation, $\sigma = 0.2R$).

Results. The optimisation was performed with three different methods. Some optimisation parameters are listed in Table 1.

The optimisation results for two traffic scenarios are shown in Fig. 6 and Fig. 7. The horizontal lines in Fig. 6 indicate the system capacity of a conventional network, i.e. circular shapes, for those traffic scenarios. The results show

Table 1. Optimisation Parameters

		Penalty method	First mapping	Second mapping
Population Size		1500	1500	1500
Generation		1000	1000	1000
Elite Rate		0.1	0.1	0.1
Mapping Function		N/A	(8)	(9)
	BLX-0.5 Weight	2.0	2.0	2.0
	Distortion Weight	0.5	0.5	0.5
RCGA	Creep Weight	0.05	0.05	0.05
Operator	BLX-0.5 Rate	0.2	0.2	0.2
	Distortion Rate	0.3	0.3	0.3
	Creep Rate	0.005	0.005	0.005

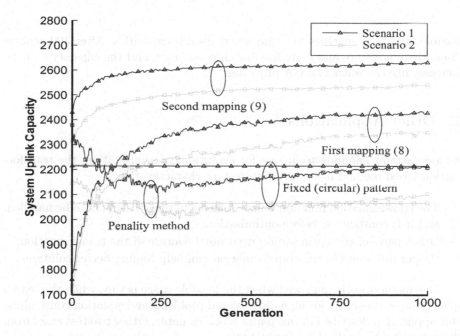

Fig. 6. System uplink capacity given by different optimisation methods

that using the first mapping method (8) is much worse than using the second one (9). It is sometimes even worse than not optimising. This highlights the importance of choosing a proper mapping function in transformation-based constraints handling methods.

The results of using the penalty method are very bad. We believe this to be due to the small size of the feasible space in a huge search space. Most chromosomes at the first 300 generations are infeasible. Initially the higher capacity is obtained by covering more area than antennas can physically do, and so are not

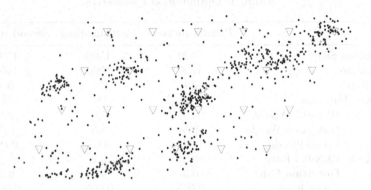

Fig. 7. Part of the cellular coverage optimisation results

feasible, and the capacities here are not realisable capacities. After 300 generations, there are more and more feasible chromosomes, and the capacity start to increase, like the other two GA processes.

5   Other Applications

As a constraint handling technique for evolutionary computation, the transformation-based method has several interesting characteristics:

- The transformation function is only relative to the constraints to be handled, and it is constructed before optimisation;
- Other parts of the optimisation need not be aware of the transformation;
- Proper distortion in the transformation can help finding better solutions.

This method works very well when the feasible space is tiny, while the search space is huge, especially for high dimensional problems and equation constraints. Our application described in this paper is one example. Other two test cases from the paper [14] have also been evaluated.

The first case is to maximise

$$G3(\overrightarrow{x}) = \left(\sqrt{n}\right)^n \cdot \prod_{i=1}^{n} x_i, \tag{12}$$

where $\sum_{i=1}^{n} x_i^2 = 1$, and $0 \leq x_i \leq 1$ for $1 \leq i \leq n$.

The function $G3$ has a global optimum solution at $\overrightarrow{x} = (\frac{1}{\sqrt{n}}, ..., \frac{1}{\sqrt{n}})$ and the value of the function in this point is 1. This case has been tested with an algorithm that redefined the crossover and mutation operators with the specific initialisation in their paper.

To apply our transformation-based method, one possible mapping function for this case is shown as (13).

$$f = \begin{cases} x_i = r \cdot \frac{s_i}{\sqrt{\sum_{i=1}^{n} s_i^2}}, & i = 1, 2, ..., n \\ r = 1.0 \end{cases}, \tag{13}$$

where $\vec{s} \in \mathbb{S}$ and $\vec{x} \in \mathbb{F}$.

The optimisation results showed similar performance to those in their paper. For the case $n = 20$, the system reached the value of 0.99 in less than 5000 generations (with population size of 30, probability of uniform crossover $p_c = 1.0$, and probability of mutation $p_m = 0.06$). However, our method is better at several aspects: 1) The specific initialisation is not necessary; 2) Standard genetic operators, such as uniform crossover and mutation, can be used without any modifications; 3) Usually, it requires more effort to redefine genetic operators than to find proper mapping functions.

Another case is to minimise

$$G11(\vec{x}) = x_1^2 + (x_2 - 1)^2, \tag{14}$$

where $x_2 - x_1^2 = 0$, and $-1 \leq x_i \leq 1$, $i = 1, 2$.

This case has been tested using hybrid methods in their paper, whilst in our test, it became quite simple. One possible mapping function, shown as (15), compresses the two-dimensional space into one-dimensional space.

$$f = \begin{cases} x_1 = (2s_1 - 1) \\ x_2 = (2s_1 - 1)^2 \end{cases}, \tag{15}$$

where $\vec{s} \in \mathbb{S}$ and $\vec{x} \in \mathbb{F}$.

The results present better performance than those in their paper. The system quickly reached the 0.75000455 in less than 30 generations in all runs.

However, when the feasible space is rather large in the search space, or it is too difficult to find a proper mapping function, the transformation-based method is not worthwhile.

6   Conclusion

This paper has proposed and investigated the optimisation of schemes for smart antenna based dynamic cell size and shape control. Real-coded genetic algorithms are used to find the optimum cell size and shapes in the context of the whole cellular network and determine the potential of dynamic geographic load balancing scheme. This has proved important as it has allowed us to construct a benchmark for other real-time and distributed methods.

We also describe an efficient methods to handle the power constraints in GA optimisation. The results show that a proper transformation is very important, as distortion in the mapping can slow down, rather than speed up convergence. The results from using the penalty method reflect the difficulty of using penalties when the solution space is very small relative to the unconstrained search space. The transformation-based methods can be applied into some of the constrained optimisations with a set of relative simple constraints.

References

1. T. Togo, I. Yoshii, and R. Kohno. Dynamic cell-size control according to geo-graphical mobile distribution in a ds/cdma cellular system. In *The Proceedings of the Ninth IEEE International Symposium on Personal, Indoor and Mobile Radio Communications*, volume 2, pages 677–681, 1998.
2. D.E. Goldberg. *Genetic Algorithms in search, optimization, and machine learning.* Addison-Wesley, Reading, Massachusetts, 1989.
3. Raphael Dorne and Jin-Kao Hao. An evolutionary approach for frequency assignment in cellular radio networks. In *The Proceedings of IEEE International Conference on Evolutionary Computation*, volume 2, pages 539–544, Perth, WA, Australia, 1995.
4. F.J. Ares-Pena, J.A. Rodriguez-Gonzalez, E. Villanueva-Lopez, and S.R. Rengarajan. Genetic algorithms in the design and optimization of antenna array patterns. *IEEE Transactions on Antennas and Propagation*, 47(3):506–510, March 1999.
5. D.E. Goldberg. The theory of virtual alphabets. In H.P. Schwefel and R. Manner, editors, *Parallel Problem Solving from Nature*, pages 13–22. Springer-Verlag, 1990.
6. Alden H. Wright. Genetic algorithms for real parameter optimization. In Gregory J. E. Rawlins, editor, *Foundations of Genetic Algorithms*, pages 205–218. Morgan Kaufman, 1991.
7. C.Z. Janikow and Z. Michalewicz. An experimental comparison of binary and floating point representations in genetic algorithms. In R.K. Belew and L.B. Booker, editors, *Proceedings of the Fourth International Conference on Genetic Algorithms*, pages 31–36, 1991.
8. L. Davis, editor. *Handbook of Genetic Algorithms.* Van Nostrand Reinhold, 1991.
9. L.J. Eshelman and J.D. Schaffer. Real-coded algorithms and interval schemata. In L.D. Whitley, editor, *Foundations of Genetic Algorithms*, pages 187–202, 1993.
10. Z. Michalewicz and C.Z. Janikow. Handling constraints in genetic algorithms. In R.K. Belew and L.B. Booker, editors, *Proceedings of the Fourth International Conference on Genetic Algorithms*, pages 151–157, 1991.
11. Carlos A. Coello Coello. A survey of constraint handling techniques used with evolutionary algorithms. Technical report, Lania-RI-99-04, Laboratorio Nacional de Informatica Avanzada, Veracruz, Mexico, 2000.
12. Z. Michalewicz. A survey of constraint handling techniques in evolutionary computation methods. In *Proceedings of the Fourth Annual Conference on Evolutionary Programming*, pages 135–155, Cambridge, Massachusetts, 1995. The MIT Press.
13. S. Koziel and Z. Michalewicz. Evolutionary algorithms, homomorphous mappings, and constrained parameter optimization. *Evolutionary Computation*, 7(1):19–44, 1999.
14. Z. Michalewicz and M. Schoenauer. Evolutionary computation for constrained parameter optimization problems. *Evolutionary Computation*, 4(1):1–32, 1996.

Combinations of Local Search and Exact Algorithms

Irina Dumitrescu and Thomas Stützle

Darmstadt University of Technology, CS Department, Intellectics Group
Alexanderstr. 10, 64283 Darmstadt, Germany
{irina, tom}@intellektik.informatik.tu-darmstadt.de

Abstract. In this paper we describe the advantadges and disadvantages of local search and exact methods of solving NP-hard problems and see why combining the two approaches is highly desirable. We review some of the papers existent in the literature that create new algorithms from such combinations. In this paper we focus on local search approaches that are strengthened by the use of exact algorithms.

1 Introduction

Integer and combinatorial optimisation problems maximise or minimise functions of many variables subject to some problem specific constraints and integrality restrictions imposed on all or some of the variables. This class of problems comprises many real-life problems arising in airline crew scheduling, production planning, Internet routing, packing and cutting, and many other areas. Often, a combinatorial optimisation problem can be modelled as an integer program [25]. However, these problems are often very difficult to solve, which is captured by the fact that many such problems are NP-hard [13].

Because of their difficulty and enormous practical importance, a large number of solution techniques for attacking NP-hard integer and combinatorial optimisation problems have been proposed. The available algorithms can be classified into two main classes: *exact* and *approximate* algorithms. Exact algorithms are guaranteed to find the optimal solution and to prove its optimality for every finite size instance of a combinatorial optimisation problem within an instance-dependent, finite run-time. If optimal solutions cannot be computed efficiently in practice, the only possibility is to trade optimality for efficiency. In other words, the guarantee of finding optimal solutions can be sacrificed for the sake of getting very good solutions in polynomial time. A class of approximate algorithms is that of *heuristic methods*, or simply *heuristics*, and seek to obtain this goal.

Two techniques from each class that have had significant success are Integer Programming (IP), as an exact approach, and local search and extensions thereof called metaheuristics, as an approximate approach. IP is a class of methods that rely on the characteristic of the decision variables of being integers. Some well known IP methods are Branch-and-Bound, Branch-and-Cut, Branch-and-Price, Lagrangean Relaxation, and Dynamic Programming [25]. In recent years remarkable improvements have been reported for IP when applied to some difficult problems (see for example [3] for the TSP). However, for most of the available IP algorithms the size of the instances solved is relatively small, and the computational time increases strongly with increasing instance size. Additional problems are often due to the facts that (i) the memory consumption of

S. Cagnoni et al. (Eds.): EvoWorkshops 2003, LNCS 2611, pp. 211–223, 2003.

exact algorithms may lead to the early abortion of a programme, (ii) high performing exact algorithms for one problem are often difficult to extend if some details of the problem formulation change, and (iii) for many combinatorial problems the best performing algorithms are highly problem specific and that they require large development times by experts on integer programming. Nevertheless, important advantages of exact methods from IP are that (i) *proven optimal* solutions can be obtained if the algorithm succeeds, (ii) valuable information on the upper/lower bounds to the optimal solution are obtained even if the algorithm is stopped before completion, and (iii) IP methods allow to prune parts of the search space in which optimal solutions cannot be located. A more practical advantage of IP methods is that powerful, general-purpose tools like CPLEX are available that often reach astonishingly good performance.

Local search has been shown to be the most successful class of approximate algorithms. It yields high-quality solutions by iteratively applying small modifications (local moves) to a solution in the hope of finding a better one. Embedded into metaheuristics designed to escape local optima like Simulated Annealing, Tabu Search, or Iterated Local Search, this approach has been shown to be very successful in achieving near-optimal (and sometimes optimal) solutions to a number of difficult problems [1]. Advantages of local search methods are that (i) in practice they are found to be the best performing algorithms for a large number of problems, (ii) they can examine an enormous number of possible solutions in short computation time, (iii) they are often more easily adapted to variants of problems and, thus, are more flexible, and (iv) they are typically easier to understand and implement than exact methods. However, disadvantages of local search algorithms are that typically (i) they cannot prove optimality, (ii) they cannot provably reduce the search space, (iii) they do not have well defined stopping criteria (this is particularly true for metaheuristics), and (iv) they often have problems with highly constrained problems where feasible areas of the solution space are disconnected. A drawback from a practical point of view is that there are no efficient general-purpose local search solvers available. Hence, most local search algorithms often require considerable programming efforts, although usually less than for exact algorithms.

It should be clear by now that IP and local search approaches have their particular advantages and disadvantages and can be seen as complementary. Therefore, an obvious idea is to try to combine these two techniques into more powerful algorithms. In this article we are looking at approaches that strive for an integration of these two worlds. While the notion of *integration* is difficult to define, we are thinking of approaches that, although rooted in one of the worlds, take a significant portion of the other world. For example, here we exclude obvious combinations like those that use preprocessing and bound propagation. In fact, only a few articles strive for a real integration of the two approaches. In what follows we will briefly describe methods that have been proposed so far. Our paper does not claim to be an extensive survey of the area, but rather a personal selection of papers that, in our opinion, open new directions of research and put forward ideas that can easily be applied to new problems.

The link between the approaches we consider in this paper is that given an integer problem, a local search technique is applied to the problem to solve and an exact algorithm to some subproblems. The subproblems are either solved in order to define the neighbourhood to explore, to permit a systematic exploration of a neighbourhood, to

exploit certain characteristics of some feasible solutions already discovered, or to find
good bounds on the optimal solutions. We will describe these situations in more detail in
the following sections. For the sake of brevity, in each section we will present in detail
only one method.

2   Defining the Neighbourhood to Explore

The first method we discuss in this paper combines a local search approach with an
exact algorithm that solves some linear programming subproblems defined in order to
reduce the search space and to define neighbourhoods for the local search algorithm.
The subproblems are relaxations of some kind of an integer programming model of the
initial problem, which are strengthened by the addition of some extra constraints. The
optimal solutions of these subproblems are then used to define the search space and the
neighbourhoods for the local search algorithm.

An example of such a method is given in the paper of Vasquez and Hao [33], proposed
for the 0-1 multidimensional knapsack problem. Given n objects, each with a volume a_i
and a value c_i, and m knapsacks, each of capacity b_i, the 0-1 multidimensional knapsack
seeks to maximise the total value of the objects put in the knapsacks such that the capacity
of each knapsack is not exceeded. This problem can be formulated as:

$$\min \sum cx$$
$$\text{s.t. } Ax \leq b \tag{1}$$
$$x \in \{0,1\}^n, \tag{2}$$

where $c \in \mathbf{N}^n$, $A \in \mathbf{N}^{m \times n}$, and $b \in \mathbf{N}^m$, $x \in \mathbf{B}^n$, with $x_i = 1$ if the object i is in the
knapsack and $x_i = 0$, otherwise.

The method proposed by Vasquez and Hao starts by solving an LP relaxation of the
integrality constraint (2) plus an extra constraint that ensures that the solution will be a
good initial solution for the local search procedure. The extra constraint is based on the
fact that the LP relaxation of an IP problem can be far away from the integer optimal
solution, while it is clear that the integer optimal solution can have only a certain number
of components that are not zero. This indeed is the extra constraint used by Vasquez and
Hao: $\sum_{i=1}^{n} x_i = k$, where $k = 1, \ldots, n$. Clearly, the solutions of these problems will
not be integer, however it is hoped that the optimal solution of the original problem
is close to one of the optimal solutions obtained after solving the relaxed problems. In
order not to miss good search areas, every possible k needs to be considered, and
therefore a series of linear programming subproblems need to be solved. Vasquez and
Hao propose a reduction of the number of these problems by calculating some bounds
on k. These bounds are the optimal solutions of two new LP problems solved by the
simplex method. The two problems seek to minimise, respectively maximise, the sum of
the components of x, $(\sum_{i=1}^{n} x_i)$, where x is a feasible solution for the LP relaxation of
the original problem, $(Ax \leq b)$, such that the value of the objective function evaluated
for x is greater than or equal to the value of a known integer lower bound LB for the
original problem plus one, $(cx \geq LB + 1)$. While Vasquez and Hao mention that such a

lower bound can be found by using a different heuristic, they do not provide any details about how they obtained such a bound.

The LP subproblems that remain to be solved after the reduction are solved by the simplex algorithm. Finally, Tabu Search based on the reverse elimination method [15] is used to search the reduced solution space defined as a collection of spheres of specified radii centred at optimal solutions of the LP subproblems solved previously. Vasquez and Hao propose a formula for calculating the radius of such a sphere, which depends on the number of integer and fractional non-zero components of the solution that is the centre of the sphere. The search space is further reduced by limiting the number of objects taken into configurations; for each k, the number of non-zero components of a candidate solution is considered to be *exactly* k, and only the integer candidate solutions are kept. In addition to this, only candidate solutions that improve the current best value of the objective function are kept. The neighbourhoods are defined as add/drop neighbourhoods; the neighbourhood of a given configuration is the set of configurations that differ by exactly two elements from the initial configuration, while keeping the number of non-zero elements constant. An extensive computational study could identify many improved results compared to the best known results at the time of publication; however, Vasquez and Hao acknowledge large computational time, especially for large instances.

In conclusion, the paper of Vasquez and Hao is an example of how easily solvable problems are used to reduce the search space explored by a local search for a difficult problem. While a branch-and-bound approach would also reduce the search space, the advantage of the method presented is that the number of subproblems solved is limited by $n + 2$, as opposed to being possibly exponential in the case of branch-and-bound. However, a drawback of this method is that after reducing the number of subproblems that need to be solved, no further reduction is possible. While in the case of a branch-and-bound approach the computation can be stopped early and information obtained up to that point can still be used, the method of Vasquez and Hao requires finding a solution for all the subproblems identified.

3   Exploring the Neighbourhoods

In this section we present an approach which is useful when the neighbourhoods defined are *very* large and efficient ways of exploring them are needed. When the local search procedure needs to move from the current solution to a neighbouring one, a subproblem is solved. The solution of the subproblem will determine the neighbour that will be used by the local search.

Hyperopt Neighbourhoods.   Burke et al. give such a method in the context of the symmetric and asymmetric Travelling Salesman Problem (TSP) [5,6] and of the Asymmetric Travelling Salesman Problem [6]. We recall that the symmetric TSP is defined on a undirected graph $G = (V, A)$, where $V = \{1, \ldots, n\}$ is the set of nodes, with the nodes representing cities, and A the set of edges. An edge represents a pair of cities between which travel is possible. Every edge has an associated cost. The TSP consists of finding the cheapest tour that visits all cities exactly once. The Asymmetric Travelling Salesman Problem (ATSP) differs from the TSP simply by being defined on a directed graph. In this case, the edges of the graph represent *ordered* pairs of cities between which travel is possible. The oriented edges are usually called arcs.

Burke et al. [5,6] propose a local search method based on a *hyperopt neighbourhood*, which we will exemplify using its example application to the ATSP. Given a feasible tour of the ATSP [6], a *hyperedge* is a subpath of the current tour, in other words, a sequence of successive arcs of the tour. Let i be the start node and j the end node of the hyperedge. We denote the hyperedge by $\mathcal{H}(i, j)$. The *length* of a hyperedge is given by the number of arcs contained. Let t be a feasible tour for the ATSP, considered to start and end at node 1. We introduce a relation of order $\prec_t$ on the set of nodes V as follows: for any $i, j \in V$, we say that $i \prec_t j$ if i is a predecessor of j on the tour, i.e. $t = (1, \ldots, i, \ldots, j, \ldots, 1)$.

Let t be a feasible tour of the ATSP, and $\mathcal{H}(i_1, i_{k+1})$ and $\mathcal{H}(j_1, j_{k+1})$ two hyperedges of length k such that $i_{k+1} \prec_t j_1$ and $\mathcal{H}(i_1, i_{k+1}) \cap \mathcal{H}(j_1, j_{k+1}) = \emptyset$ with respect to the nodes contained. It is obvious that the tour t can be described completely by the four hyperedges $\mathcal{H}(i_1, i_{k+1})$, $\mathcal{H}(i_{k+1}, j_1)$, $\mathcal{H}(j_1, j_{k+1})$, $\mathcal{H}(j_{k+1}, i_1)$. Burke et al. define a *k-hyperopt move* as being a new tour obtained after performing two steps: first, remove $\mathcal{H}(i_1, i_{k+1})$ and $\mathcal{H}(j_1, j_{k+1})$ from the tour t, then add arcs to $\mathcal{H}(i_{k+1}, j_1)$ and $\mathcal{H}(j_{k+1}, i_1)$ such that a new feasible tour is constructed. The set of all k-hyperopt moves is called a *k-hyperopt neighbourbood*.

Obviously, the size of the k-hyperopt neighbourhood increases exponentially with k. Due to the large size of the neighbourhood, instead of exploring the neighbourhood in a classical manner, Burke et al. propose an "optimal" construction of a k-hyperopt move by solving exactly a subproblem: the ATSP defined on the graph $G = (V', A')$, where V' is the set of nodes included in $\mathcal{H}(i_1, i_{k+1})$ and $\mathcal{H}(j_1, j_{k+1})$ and A' the set of arcs in A that have both ends in V'. However, this approach is bound to be efficient only when k is relatively small. Otherwise, a large size ATSP would have to be solved as a subproblem.

In [6] Burke et al. provide numerical results for $k = 3$ and $k = 4$. They mention that they solve the subproblems to optimality using a dynamic programming algorithm, however they do not provide any further details. In [5] Burke et al. consider only the cases when $k = 2$ and $k = 3$ and use enumeration to solve the subproblems.

For a given hyperedge $\mathcal{H}(i_1, i_{k+1})$ the k-hyperopt move to be performed is determined in [5,6] by evaluating the best k-hyperopt move over the set of *all* possible hyperopt moves. Therefore every hyperedge that does not intersect with $\mathcal{H}(i_1, i_{k+1})$ is considered, and for every such hyperedge a subproblem is solved. It is not clear how $\mathcal{H}(i_1, i_{k+1})$ is chosen in the first place or if that hyperedge also changes. This is clearly an expensive approach from the point of view of the computational time. The authors improve their method by using a tabu list that bans some hyperedges from being considered, based on the observation that only the hyperedges that have been affected by previous moves can provide improving hypermoves.

Burke et al. propose several methods that use the concept of k-hyperopt moves in [6]. They involve either a combination of hyperopt with 3-opt moves or the use of a local search based on hyperopt moves inside of a variable neighbourhood search procedure [16]. A numerical comparison between the pure k-hyperopt approach and methods developed on top of it show the latter to be better. Although the k-hyperopt approach appears not to be fully competitive with other approaches for the ATSP [18], we believe that the k-hyperopt approach with further enhancements is a promising direction of research which deserves to be further investigated.

Very Large Scale Neighbourhoods. A similar idea was developed for the class of set partitioning problems. Thompson et al. [31,32] defined the concept of a cyclic exchange for a local search approach, which transfers single elements between several subsets, in a "cyclic" manner. A two-exchange move can be seen as a cyclic exchange of length 2. Thompson et al. showed that for any current solution of a partitioning problem a new graph can be constructed. Costs are associated with its arcs and the set of nodes is split into subsets according to a partition induced by the current solution of the partitioning problem. Finding a cycle that uses at most one node for each subset in the new graph is equivalent to determining a cyclic exchange move for the original problem. Exact and heuristic methods [2,12] that solve the problem of finding the most negative cost subset disjoint cycle (which corresponds to the best *improving* neighbour of the current solution) have been developed, however no work has been yet done on the integration of the exact algorithms within the local search framework.

Dynasearch. Dynasearch is another example where exponentially large neighbourhood are explored. The neighbourhood searched consists of all possible combinations of mutually independent simple search steps and one dynasearch move consists of a set of independent moves that are executed in parallel in a single local search iteration. Independence in the context of dynasearch means that the individual moves do not interfere with each other; this means that the gain incurred by a combined move must be the sum of the gains of the individual moves. In the case of independent moves, a dynamic programming algorithm is used to find the best combination of independent moves. So far, Dynasearch has only been applied to problems, where solutions are represented as permutations; current applications include the TSP [8], the single machine total weighted tardiness problem (SMTWTP) [8,9], and the linear ordering problem (LOP) [8]. Very good performance is reported when dynasearch is the local search routine inside an iterated local search (Section 4, [20]) or guided local search algorithm [34].

4   Iterated Local Search

A successful metaheuristic approach for solving difficult combinatorial problems is Iterated Local Search (ILS) [20], which consists of running a local search procedure many times to perturbations of previously seen local optima. An outline of a simple ILS algorithm is given in Algorithm 1.

Algorithm 1 *Iterated Local Search*

Step 0: Let S be an initial solution.
Step 1: **while** stopping_criterion is not met **do**
(i)   Let S' be the solution obtained from S after a perturbation.
(ii)   Call local_search(S') to produce the solution S''.
(iii)   **if** S'' is better than S **then** $S = S''$.
         enddo
Step 2: Return S.

ILS can be applied in many different ways by, for example, changing the manner in which information about good solutions is collected (Step 1 (iii)), or the way the perturbation is performed. In the rest of this section we will show how different types of subproblems arise when these variations of the ILS are considered.

4.1   Collected Information

The first method focuses on the way the information is collected. It runs an approximate method several times, records the best solutions obtained and uses these to define a subproblem that is solved by an exact algorithm. The solution of the subproblem will then be a solution of the original problem. This idea can be applied to a large number of problems and here we illustrate it by two examples.

Tour Merging. The paper we present next is the report of Applegate, Bixby, Chvàtal, and Cook [4], which is dedicated to the problem of finding near-optimal solutions of the TSP. Applegate et al. propose the use of an exact algorithm to solve a TSP subproblem defined by the outcome of multiple runs of a particular ILS algorithm applied to the original TSP. Applegate et al. propose an approach that makes use of the information obtained after running ILS several times. Their method is based on the observations that (i) high quality tours typically have many edges in common and, in fact, the number of shared edges increases with tour quality and (ii) that very high quality tours share many edges with optimal tours. Their method, called *tour merging*, is a two step procedure.

The first step consists of running an ILS like Chained Lin-Kernighan (CLK) [24,4] or Iterated Helsgaun (IH) [17] a number of times, keeping the best solution obtained at each of the runs (both, CLK and IH are currently among the top performing algorithms for the TSP). In what follows we will denote the set of solutions recorded by T. The second step consists of solving the TSP on the graph induced by this set of tours T on the original graph. In other words, a TSP is solved on a restricted graph that has the same set of nodes as the original graph, but its set of edges consists of the edges that appear at least once in any of the tours collected. Formally, if the original graph is $G = (V, A)$, where V is the set of nodes and A the set of arcs, the reduced graph can be described as $G' = (V, A')$, where $A' = \{a \in A : \exists t \in T, a \in t\}$.

It is hoped that this graph will be sparse with low *branch-width* [26] and also that it will be a graph that contains at least one (near-)optimal tour for the original TSP. Clearly, the TSP subproblem will be easier to solve than the original one. The quality of the tours obtained after running Chained Lin-Kernighan has a direct impact on the cardinality of the set of edges of the reduced graph. If the reduced graph is not very sparse, then the TSP subproblem may be computationally expensive to solve. Note that in tour merging, the subproblem generated is of the same type as the original problem. Here, the advantage of solving a subproblem comes from the reduced size of the subproblem, and therefore a natural direction of research is finding strategies that generate small subproblems of good quality.

In [4] two possibilities for the second step of the method are considered. The first is to use a general TSP optimisation code to find the optimal tour with respect to the graph G'. However, experimental results in [4] suggest that alternative methods could provide solutions much quicker. In particular, they suggest finding (heuristically) a branch-width decomposition of the graph G' and exploiting this decomposition by find optimal tours via a dynamic programming algorithm.

The tour merging approach was tested on a large number of TSPLIB benchmark problems [4]. The results obtained were very good, optimal solutions being identified for all problem tested. However the computational time was large.

Heuristic Concentration. A similar idea is put forward by Rosing et al. in [27] for the p-median problem. Given p, a given positive integer, and a graph that has weights associated with every node and distances associated with every arc, the p-median problem consists of finding p nodes in the graph, called *facilities*, such that the sum of the weighted distances between every node and its closest facility is minimised. The nodes that are not facilities are called *demand nodes*. Rosing et al. propose repeatedly running a heuristic, with different random starts, while collecting the solutions obtained at each iteration. In a second step, they propose obtaining a solution for the p-median problem by solving a subproblem, in fact a restricted p-median problem. This method is called *Heuristic Concentration* (HC). The second step of HC starts with a subset of the best solutions found in the first phase. The facility locations used in these "best" solutions form a subset of the set of all nodes, and they will be collected in what the authors call the *concentration set* (CS). Finally, a restricted p-median problem, with the facility locations restricted to those contained into the concentration set, is solved.

Rosing et al. use CPLEX to solve the LP relaxation of the binary integer model of the reduced p-median problem. In most cases the solution obtained is integer; if not integer, it is easy to find using branch-and-bound. Numerical results are provided in both [27] and subsequent papers [28,29].

We have seen that the two methods presented so far in this section act on the way information is collected at each iteration of the ILS. Then, *after* running ILS, an exact algorithm is used to solve a subproblem determined by that information. Therefore an exact algorithm is applied only once, to exactly one problem. Next, we will talk about a paper that proposes several methods that generate many subproblems.

4.2   Modifying the Perturbation Step

The role of the perturbation in an ILS approach is to introduce relatively large changes in the current solution and to allow the local search to leave local optima, while still conserving the good characteristics of a current solution. A good perturbation step is problem dependent, and should therefore be object of investigation. An idea of obtaining good perturbations for the ILS is defining a subproblem, usually of similar type with the original problem, and solve it using an exact algorithm or a good heuristic.

An example of determining a perturbation by using exact algorithms that solve a subproblem is the paper of Lourenço [22] given in the context of the job-shop scheduling problem. The job-shop scheduling problem is defined for machines and jobs. Each job consists of a sequence of operations, each with a given processing time, that have to be performed in a given order on different machines. The goal is to minimise the completion time of the last job subject to precedence constraints on the operations and additional capacity constraints saying that no machine can work on two operations at the same time. The job-shop scheduling problem can be modelled using the well-known disjunctive graph model [30].

Lourenço describes several variations of the ILS and provides numerical results for all of them. Firstly, she tests several methods of finding an initial feasible schedule. She then proposes the use of local improvement and simulated annealing as local search procedures. Finally, for the perturbation step, several perturbation ideas are put forward and tested for each of the two local search procedures. Since this is where the subproblems we are interested in are generated, we will discuss in some detail how the perturbation is performed.

The first perturbation procedure proposed by Lourenço involves the modification of the disjunctive graph corresponding to the current solution of the job-shop scheduling problem, by removing all the directions given to the edges associated with two random machines in the disjunctive graph. Then Carlier's algorithm [7] (a branch-and-bound method) is applied to one of the machines and the one-machine scheduling problem is solved. This problem can be seen as a very simple version of the job-shop scheduling problem: a number of operations need to be scheduled on one machine in the presence of temporal constraints. The edges corresponding to that machine are then oriented according to the optimal solution obtained. The same treatment is then applied to the second machine. Lourenço mentions that this perturbation idea can create cycles in the disjunctive graph, and suggests a way of obtaining a feasible schedule from the graph with cycles (see [22] for details). In conclusion, at each iteration of the ILS, two subproblems are solved in order to construct a new initial solution for the local search procedure. The subproblems are of a different type than the original problem and of reduced size. However they belong to the same class of job scheduling problems.

A similar perturbation proposed by Lourenço is making use of the early-late algorithm [21] as an exact method. In order to do that preemption is allowed and two one-machine with lags on a chain are solved. Lourenço also gives a simple technique for eliminating cycles. We note that in this case too the subproblems solved are of a different type compared to the original problem.

5   Lower Bounds

Another possibility of combaining exact and local methods is by using exact algorithms to determine lower bounds while iteratively applying construction heuristics. Obtaining good bounds is important, because the bounds can be used by a construction heuristic to determine how to extend a partial solution and to eliminate possible extensions if they lead to solutions worse than the best one found so far. Here, we give an example that exploits these ideas inside an Ant Colony Optimization algorithm.

Ant Colony. Ant Colony Optimisation (ACO) [10] is a recent metaheuristic approach for solving hard combinatorial optimisation problems, which is loosely inspired by the pheromone trail laying and following behavior of real ants. Artificial ants in ACO are stochastic solution construction procedures that probabilistically build a solution by iteratively adding solution components to partial solutions by taking into account (i) heuristic information on the problem instance being solved, if available, and (ii) (artificial) pheromone trails which change dynamically at run-time to reflect the agents' acquired search experience. Of the available ACO algorithms (see [11] for an overview), the Approximate Nondeterministic Tree Search (ANTS) algorithm [23] is of particular interest for us here. Since ANTS was first applied to the quadratic assignment problem (QAP), we present the essential part of the algorithm using this example application.

The QAP can best be described as the problem of assigning a set of objects to a set of locations with given distances between the locations and given flows between the objects. The objective is to place the objects on locations in such a way that the sum of the product between flows and distances is minimal. More formally, in the QAP one is given n objects and n locations, two $n \times n$ matrices $A = [a_{ij}]$ and $B = [b_{rs}]$, where a_{ij} is the distance between locations i and j and b_{rs} is the flow between objects r and s. Let x_{ij} be a binary variable which takes value 1 if object i is assigned to location j and 0 otherwise. Then the problem can be formulated as:

$$\min \sum_{i=1}^{n} \sum_{j=1}^{n} \sum_{l=1}^{n} \sum_{k=1}^{n} a_{ij} b_{kl} x_{ik} x_{jl}$$

subject to the standard assignment constraints

$$\sum_{i=1}^{n} x_{ij} = 1, \ j = 1, .., n; \quad \sum_{j=1}^{n} x_{ij} = 1, \ i = 1, .., n; \quad x_{ij} \in \{0, 1\} \ i, j = 1, .., n$$

When applied to the QAP, in ANTS each ant constructs a solution by iteratively assigning objects to a free location. Given a location j, an ant assigns a still unassigned object i to this location with a probability that is proportional to $\alpha \cdot \tau_{ij}(t) + (1 - \alpha) \cdot \eta_{ij}$, where $\tau_{ij}(t)$ is the pheromone trail associated to the assignment of object i to a location j (pheromone trails give the "learned" desirability of choosing an assignment), η_{ij} is the heuristic desirability of this assignment, and α is a weighting factor between pheromone and heuristic. Lower bound computations are exploited at various places in ANTS. Before starting the actual solution process, ANTS first computes the Gilmore-Lawler lower bound (GLB) [14,19], which amounts to solving a linear assignment problem based on its linear programming relaxation. Along with the lower bound computation one gets the values of the dual variables $u_i, i = 1, \ldots, n$ and $v_i, i = 1, \ldots, n$ corresponding to the assignment constraints. The dual variables v_i are used to define a pre-ordering on the locations: The higher the value of the dual variable associated to a location, the higher is assumed to be the location's impact on the QAP solution cost and, hence, the earlier it is tried to assign an object to that location. The main idea of ANTS is to use at each construction step lower bound computations to define the heuristic information of the attractiveness of a adding a specific assignment of object i to location j. This is achieved by tentatively adding the specific assignment (i, j) to the current partial solution and by estimating the cost of a complete solution containing that assignment by means of a lower bound. This estimate is used as the heuristic information η_{ij} during the solution construction: the lower the estimate the more attractive is the addition of a specific assignment. Using lower bounds computations also presents several additional advantages like the elimination of possible moves if the cost estimation is larger than the so far best found solution. Additionally, tight bounds give a strong indication of how good a move is. However, the lower bound is to be computed at each construction step; hence, the lower bounds should be efficiently computable. Therefore, Maniezzo did not use GLB during the ants' construction steps, but exploits the weaker LBD lower bound, which can be computed in $\mathcal{O}(n)$. For details on the lower bound computation we refer to [23]. Experimental results have shown that ANTS is currently one of the best available algorithms for the QAP. The good performance of ANTS algorithm has also been confirmed in a variety of further applications.

6 Conclusions

The main conclusion of our paper is that there are many research opportunities to develop algorithms that integrate local search and exact techniques and that not much has been done so far in this area. We have presented a number of approaches that use both exact and local search methods in a rather complex way, and that, in our oppinion, can be further improved and extended to a number of different applications than the ones for which they have originally been developed.

Acknowledgments. This work was supported by the "Metaheuristics Network", a Research Training Network funded by the Improving Human Potential programme of the CEC, grant HPRN-CT-1999-00106, and by a European Community Marie Curie Fellowship, contract HPMF-CT-2001-01419. The information provided is the sole responsibility of the authors and does not reflect the Community's opinion. The Community is not responsible for any use that might be made of data appearing in this publication.

References

1. E. H. L. Aarts and J. K. Lenstra, editors. *Local Search in Combinatorial Optimization.* John Wiley & Sons, Chichester, 1997.
2. R.K. Ahuja, T.L. Magnanti, and J.B. Orlin. *Network Flows: Theory, Algorithms, and Applications.* Prentice Hall, Inc. Englewood Cliffs, NJ, 1993.
3. D. Applegate, R. Bixby, V. Chvátal, and W. Cook. On the solution of traveling salesman problem. *Documenta Mathematica*, Extra Volume ICM III:645–656, 1998.
4. D. Applegate, R. Bixby, V. Chvátal, and W. Cook. Finding Tours in the TSP. Technical Report 99885, Forschungsinstitut für Diskrete Mathematik, University of Bonn, Germany, 1999.
5. E.K. Burke, P. Cowling, and R. Keuthen. Embedded local search and variable neighbourhood search heuristics applied to the travelling salesman problem. Technical report, University of Nottingham, 2000.
6. E.K. Burke, P.I. Cowling, and R. Keuthen. Effective local and guided variable neighbourhood search methods for the asymmetric travelling salesman problem. *EvoWorkshop 2001, LNCS*, 2037:203–212, 2001.
7. J. Carlier. The one-machine sequencing problem. *European Journal of Operational Research*, 11:42–47, 1982.
8. R. K. Congram. *Polynomially Searchable Exponential Neighbourhoods for Sequencing Problems in Combinatorial Optimisation.* PhD thesis, University of Southampton, Faculty of Mathematical Studies, UK, 2000.
9. R. K. Congram, C. N. Potts, and S. van de Velde. An iterated dynasearch algorithm for the single-machine total weighted tardiness scheduling problem. *INFORMS Journal on Computing*, 14(1):52–67, 2002.

10. M. Dorigo and G. Di Caro. The Ant Colony Optimization meta-heuristic. In D. Corne, M. Dorigo, and F. Glover, editors, *New Ideas in Optimization*, pages 11–32. McGraw Hill, London, UK, 1999.

11. M. Dorigo and T. Stützle. The ant colony optimization metaheuristic: Algorithms, applications and advances. In F. Glover and G. Kochenberger, editors, *Handbook of Metaheuristics*. Kluwer Academic Publishers, To appear in 2002.

12. I. Dumitrescu. *Constrained Shortest Path and Cycle Problems*. PhD thesis, The University of Melbourne, 2002. http://www.intellektik.informatik.tu-darmstadt.de/~irina.

13. M. R. Garey and D. S. Johnson. *Computers and Intractability: A Guide to the Theory of $\mathcal{NP}$-Completeness*. Freeman, San Francisco, CA, 1979.

14. P. C. Gilmore. Optimal and suboptimal algorithms for the quadratic assignment problem. *Journal of the SIAM*, 10:305–313, 1962.

15. F. Glover and M. Laguna. *Tabu Search*. Kluwer Academic Publishers, Boston, MA, 1997.

16. P. Hansen and N. Mladenovic. *Meta-Heuristics: Advances and Trends in Local Search Paradigms for Optimization*, chapter An Introduction to Variable Neighborhood Search, pages 433–458. Kluwer Academic Publishers, Boston, MA, 1999.

17. K. Helsgaun. An effective implementation of the lin-kernighan traveling salesman heuristic. *European Journal of Operational Research*, 126:106–130, 2000.

18. D. S. Johnson, G. Gutin, L. A. McGeoch, A. Yeo, W. Zhang, and A. Zverovitch. Experimental analysis of heuristics for the ATSP. In G. Gutin and A. Punnen, editors, *The Traveling Salesman Problem and its Variations*, pages 445–487. Kluwer Academic Publishers, 2002.

19. E. L Lawler. The quadratic assignment problem. *Management Science*, 9:586–599, 1963.

20. H. R. Lourenço, O. Martin, and T. Stützle. Iterated local search. In F. Glover and G. Kochenberger, editors, *Handbook of Metaheuristics*, volume 57 of *International Series in Operations Research & Management Science*, pages 321–353. Kluwer Academic Publishers, Norwell, MA, 2002.

21. H.R. Lourenço. *A Computational Study of the Job-Shop and the Flow-Shop Scheduling Problems*. PhD thesis, School of Or & IE, Cornell University, Ithaca, NY, 1993.

22. H.R. Lourenço. Job-shop scheduling: Computational study of local search and large-step optimization methods. *European Journal of Operational Research*, 83:347–367, 1995.

23. V. Maniezzo. Exact and approximate nondeterministic tree-search procedures for the quadratic assignment problem. *INFORMS Journal on Computing*, 11(4):358–369, 1999.

24. O. Martin, S.W. Otto, and E.W. Felten. Combining simulated annealing with local search heuristics. *Annals of Operations Research*, 63:57–75, 1996.

25. G. Nemhauser and L. Wolsey. *Integer and Combinatorial Optimization*. John Wiley & Sons, 1988.

26. N. Robertson and P.D. Seymour. Graph minors. X. Obstructions to tree-decomposition. *Journal of Combinatorial Theory*, 52:153–190, 1991.

27. K.E. Rosing and C.S. ReVelle. Heuristic concentration: Two stage solution construction. *European Journal of Operational Research*, pages 955–961, 1997.

28. K.E. Rosing and C.S. ReVelle. Heuristic concentration and tabu search: A head to head comparison. *European Journal of Operational Research*, 117(3):522–532, 1998.

29. K.E. Rossing. Heuristic concentration: a study of stage one. *ENVIRON PLANN B*, 27(1):137–150, 2000.

30. B. Roy and B. Sussmann. Les problemes d'ordonnancement avec constraintes disjonctives. Notes DS no. 9 bis, SEMA.

31. P.M. Thompson and J.B. Orlin. The theory of cycle transfers. Working Paper No. OR 200-89, 1989.

32. P.M. Thompson and H.N. Psaraftis. Cyclic transfer algorithm for multivehicle routing and scheduling problems. *Operations Research*, 41:935–946, 1993.

33. M. Vasquez and J-K. Hao. A hybrid approach for the 0-1 multidimensional knapsack problem. In *Proceedings of the IJCAI-01*, pages 328–333, 2001.
34. C. Voudouris. *Guided Local Search for Combinatorial Optimization Problems*. PhD thesis, Department of Computer Science, University of Essex, Colchester, UK, 1997.

On Confidence Intervals for the Number of Local Optima

Anton V. Eremeev[1] and Colin R. Reeves[2]

[1] Omsk Branch of Sobolev Institute of Mathematics,
13 Pevtsov str. 644099, Omsk, Russia.
eremeev@iitam.omsk.net.ru
[2] Coventry University, School of Mathematical & Information Sciences,
Priory Street. Coventry CV1 5FB, UK
C.Reeves@coventry.ac.uk

Abstract. The number of local optima is an important indicator of optimization problem difficulty for local search algorithms. Here we will discuss some methods of finding the confidence intervals for this parameter in problems where the large cardinality of the search space does not allow exhaustive investigation of solutions. First results are reported that were obtained by using these methods for NK landscapes, and for the low autocorrelation binary sequence and vertex cover problems.

1 Introduction

Many heuristic search methods for combinatorial optimization problems (COPs) are based on local search. Typically, search is focused on a *neighbourhood* of the current point in the search space, and various strategies can be used for deciding on whether or not to move to one of these neighbours, and if so, to which. In practice, such methods quickly converge to a local optimum, and the search must either begin again by starting from a new point, or use some 'metaheuristic' (e.g., tabu search) to guide the search into new areas. While such local search methods can be arbitrarily bad in the worst case [1], experimental evidence shows that they tend to perform rather well across a large range of COPs. However, experience suggests that the type and size of the neighbourhood can make a big difference to practical performance of the local search methods. Some recent theoretical results also suggest that the local optima are tightly connected with the attractors of the simple genetic algorithm [2].

The neighbourhood structure can be described by a graph, or by its related adjacency matrix, which enables us to give a precise meaning to the idea of a *landscape* (which results from the interaction of this structure with a particular problem instance), and to its characteristics, such as local (and global) optima, and to related properties of an optimum such as its basin of attraction. Formally, we have an optimization problem: given a function (the objective or goal function) of a set of decision variables, denoted by the vector x,

$$g : \mathcal{X} \mapsto \mathbb{R},$$

find

S. Cagnoni et al. (Eds.): EvoWorkshops 2003, LNCS 2611, pp. 224–235, 2003.

$$\arg \max_{x \in \mathcal{X}} g.$$

We define a neighbourhood function

$$N : \mathcal{X} \mapsto 2^{\mathcal{X}},$$

whence the induced landscape $\mathcal{L}$ is the triple $\mathcal{L} = (\mathcal{X}, g, d)$ where d denotes a distance measure $d : \mathcal{X} \times \mathcal{X} \to I\!\!R^{+} \cup \{\infty\}$ for which is required that

$$\left. \begin{array}{l} d(x, y) \geq 0 \\ d(x, y) = 0 \Leftrightarrow x = y \\ d(x, u) \leq d(x, y) + d(y, u) \end{array} \right\} \quad \forall x, y, u \in \mathcal{X}.$$

Usually, the distance measure d is defined by the neighbourhood structure:

$$y \in N(x) \Leftrightarrow d(x, y) = 1.$$

A *locally optimal* vector $x^{o} \in \mathcal{X}$ is then one such that

$$g(x^{o}) \geq g(y) \quad \forall y \in N(x^{o}).$$

We shall denote the set of such optima as $\mathcal{X}^{o}$. (For a fuller discussion on landscapes, see [3] or [4]; for some illustrations of the way landscapes are formed, and how they depend on neighbourhood structure, see [5].)

Many things make an instance difficult, but one of them is the *number* of optima induced in the landscape $\nu = |\mathcal{X}^{o}|$. Besides this an estimate of ν might be useful in defining stopping criteria for local search based metaheuristics, and in experimental testing of the 'big valley' conjecture [6] (for such an estimate may show how representative is the sample of the local optima found).

A number of approaches have been proposed for estimation of ν: one may apply an assumption that the distribution of basin sizes is isotropic [7], or fit certain type of parametric distribution of the basin sizes (exponential, gamma, lognormal etc.) [8,9]. Nonparametric estimates, such as the bootstrap or the jackknife, can also be employed [9,10]. Assuming a particular type of distribution of basin sizes one can obtain the maximal likelihood estimate for ν, or a confidence interval for it. It is interesting that some of these approaches are actually based on the same methods as used by ecologists to estimate the number of animals in a population (see e.g. [11]).

It would be interesting to see how ν depends on the problem size: theoretical results show that many optimization problems have an exponential number of local optima in the worst-case (see [1] and references therein), but experimental results have usually been limited to low-dimensional problems where complete enumeration of solutions is possible. At present, information on experimental evaluations of the parameter ν is scarce even for widely used benchmarks.

In this paper we suggest several methods of statistical evaluation of this parameter in problems where the large cardinality of the search space does not allow exhaustive investigation of solutions. The focus is on the theoretical basis and experimental testing of the methods on some well-known benchmarks.

The paper is organized as follows. Sect. 2 is devoted to the derivation of the computational formulae for the confidence intervals. After some general observations we analyse three methods of collecting the statistical information on an instance: the first repetition time, the maximum first repetition time and the Schnabel census procedure. The computational results obtained using these methods are reported in Sect. 3 for NK landscapes, the low autocorrelation binary sequence problem and the vertex cover problem. Finally the results obtained are discussed in Sect. 4.

2   Confidence Intervals for the Number of Optima

First let us consider the problem of numerical estimation of a confidence interval for the unknown distribution parameter ν, when the distribution function is easily computable for any value of an observable random variable T and any value of ν. Let the distribution function of T be $F_\mu(t) = P\{T \leq t | \nu = \mu\}$ and assume that the observed value $\tau \in Z$ is sampled from this distribution.

In the case of real-valued ν, the $(1 - \alpha_1 - \alpha_2) \cdot 100\%$ confidence interval $[\nu_1(\tau), \nu_2(\tau)]$ for any given α_1, α_2 is found by solving the equations

$$1 - F_{\nu_1}(\tau) = \alpha_1, \quad F_{\nu_2}(\tau) = \alpha_2 \tag{1}$$

for ν_1 and ν_2 (see e.g.[12]). For our purposes to estimate the number of optima we need to modify this estimator for the situation where ν is integer.

Proposition 1. *A conservative* $(1 - \alpha_1 - \alpha_2) \cdot 100\%$ *confidence interval* $[\nu_1(\tau), \nu_2(\tau)]$ *for integer parameter* ν *is given by*

$$\nu_1(\tau) = \min\{\mu : 1 - F_\mu(\tau - 1) \geq \alpha_1\}, \quad \nu_2(\tau) = \max\{\mu : F_\mu(\tau) \geq \alpha_2\},$$

and then $P\{\nu < \nu_1(T)\} \leq \alpha_1$, $P\{\nu_2(T) < \nu\} \leq \alpha_2$.

Proof. Suppose that for some value ν the proposition is false. Then either $\alpha_1 < P\{\nu < \nu_1(\tau)\}$ or $\alpha_2 < P\{\nu_2(\tau) < \nu\}$ or both. In the first case we have:

$$\alpha_1 < P\{\nu < \min\{\mu : 1 - F_\mu(T - 1) \geq \alpha_1\}\} \leq P\{1 - F_\nu(T - 1) < \alpha_1\} =$$

$$P\{T - 1 \geq \min\{\Theta : 1 - F_\nu(\Theta) < \alpha_1\}\} = P\{T \geq \Theta_{\min}(\nu, \alpha_1)\},$$

where $\Theta_{\min}(\nu, \alpha_1) = \min\{\Theta' : 1 - F_\nu(\Theta' - 1) < \alpha_1\}$. But $P\{T \geq \Theta_{\min}(\nu, \alpha_1)\} = 1 - F\{\Theta_{\min}(\nu, \alpha_1) - 1\} < \alpha_1$ by definition of $\Theta_{\min}$, which is a contradiction. In the second case

$$\alpha_2 < P\{\max\{\mu : F_\mu(T) \geq \alpha_2\} < \nu\} \leq P\{F_\nu(T) < \alpha_2\} = P\{T < \theta_{\min}(\nu, \alpha_2)\},$$

where $\theta_{\min}(\nu, \alpha_2) = \min\{\theta : F_\nu(\theta) \geq \alpha_2\}$. But $P\{T < \theta_{\min}(\nu, \alpha_2)\} < \alpha_2$. $\square$

Note that if $F_\nu(t)$ is monotone in ν, it is easy to compute the confidence interval suggested in Prop.1 using (e.g.) the bisection procedure. In what follows we will see that this monotonicity property holds for the estimators of ν we use.

2.1   The First Repetition Time

We consider the case where there are multiple independently initiated restarts of the local search procedure. Let $p_1, ..., p_\nu$ denote the probabilities of finding the corresponding local optima, $p = (p_1, ..., p_\nu)$ and let T denote the first repetition time, i.e. the index number of the restart which yields the first re-occurrence of a previously encountered optimum. In what follows the distribution function of T corresponding to the distribution vector p will be denoted by $F_p(t)$ and by the isotropic distribution vector we will mean $\bar{p} = (1/\nu, ..., 1/\nu)$.

The next proposition shows that the isotropic distribution is extremal in some sense.

Proposition 2. $F_p(t)$ *is minimal only at* $p = \bar{p}$ *for any* $t \geq 2$.

Proof. Let S be the set of all t-element sequences of the elements from $\{1, ..., \nu\}$ without repetitions, and $\sigma(i)$ is the i-th element of sequence σ. Then

$$F_p(t) = 1 - P\{T > t\} = 1 - \sum_{\sigma \in S} \prod_{i=1}^{t} p_{\sigma(i)}.$$

It is easy to see that $F_p(t)$ is a continuous function on a compact set $\Lambda = \{(p_1, p_2, ..., p_\nu) \geq 0 : \sum_{i=1}^{\nu} p_i = 1\}$ when t is fixed. Thus there exists a probability distribution yielding the minimum of $F_p(t)$ on Λ. We claim that except for $\bar{p}$ no other vector is minimal. Suppose the contrary: in the minimal vector p we have $p_i \neq p_j$ for some i, j. Without loss of generality we can assume that $i = 1, j = 2, p_2 = \gamma p_1$ and $\gamma < 1$. Let us consider a new distribution:

$$p'_1 = p'_2 = \frac{p_1 + p_2}{2}, p'_3 = p_3, ..., p'_\nu = p_\nu,$$

and its distribution function $F_{p'}(t)$. We shall now show that $F_{p'}(t) < F_p(t)$, which leads to a contradiction with the supposed optimality of p. Denote the subset of S consisting of the sequences containing element j by $S(j), j = 1, 2$. Then $F_p(t) = 1 - a(p) - b(p) - c(p) - d(p)$ where

$$a(p) = \sum_{\sigma \in S(1) \cap S(2)} \prod_{i=1}^{t} p_{\sigma(i)}, \qquad b(p) = \sum_{\sigma \in S(1) \backslash S(2)} \prod_{i=1}^{t} p_{\sigma(i)},$$

$$c(p) = \sum_{\sigma \in S(2) \backslash S(1)} \prod_{i=1}^{t} p_{\sigma(i)}, \qquad d(p) = \sum_{\sigma \in S \backslash (S(1) \cup S(2))} \prod_{i=1}^{t} p_{\sigma(i)}.$$

Analogously $F_{p'}(t) = 1 - a(p') - b(p') - c(p') - d(p')$ and $d(p') = d(p)$. Note that if the element 2 in all sequences of $S(2) \backslash S(1)$ was substituted by 1, this would yield the set of sequences $S(1) \backslash S(2)$. Thus, on the one hand

$$b(p) + c(p) = \sum_{\sigma \in S(1) \backslash S(2)} \prod_{i=1}^{t} p_{\sigma(i)} + \gamma \left(\sum_{\sigma \in S(1) \backslash S(2)} \prod_{i=1}^{t} p_{\sigma(i)} \right)$$

$$= (1 + \gamma) p_1 \cdot \sum_{\sigma \in S(1) \backslash S(2)} \prod_{i:\sigma(i) \neq 1} p_{\sigma(i)}.$$

On the other hand $p_1' = p_2' = (p_1 + p_2)/2$, so that

$$b(p') + c(p') = (p_1 + p_2) \sum_{\sigma \in S(1) \setminus S(2)} \prod_{i:\sigma(i) \neq 1} p_{\sigma(i)},$$

and hence $b(p) + c(p) = b(p') + c(p')$. Now

$$a(p) = \gamma p_1^2 \cdot \sum_{\sigma \in S(1) \cap S(2)} \prod_{i:\sigma(i) \neq 1,2}^{t} p_{\sigma(i)},$$

$$a(p') = \left(\frac{p_1(1+\gamma)}{2} \right)^2 \sum_{\sigma \in S(1) \cap S(2)} \prod_{i:\sigma(i) \neq 1,2}^{t} p_{\sigma(i)}.$$

But $(1+\gamma)^2 > 4\gamma$ since $\gamma < 1$, so $a(p') > a(p)$ and $F_{p'}(t) < F_p(t)$. $\square$

This proposition is important on its own right but it also yields the monotonicity property of $F_{\bar{p}}(t)$ mentioned above:

Corollary 1. $F_{\bar{p}}(t)$ *is a strictly monotonically decreasing function on ν.*

Proof.
Apply the above result to the case $p = (1/(\nu-1), 1/(\nu-1), \ldots, 1/(\nu-1), 0)$. $\square$

2.2   The Maximal First Repetition Time

If the average repetition time turns out to be short compared to the number of restarts, the sequence of solutions obtained in a multiple restart procedure may be split into a series of subsequences, where each subsequence ends with its first re-occurrence. Let s denote the number of such subsequences and let the random variable T_j denote the length of j-th subsequence (including the re-occurring element). In this case the maximal first repetition time $T^{(s)} = \max_j T_j$ appears to be a more 'stable' estimate for ν compared to the simple first repetition time procedure discussed above. Denote $F_p^{(s)}(t) = P\{T^{(s)} \leq t\}$, then

$$F_p^{(s)}(t) = P\{T_j \leq t \ \forall j = 1, \ldots, s\} = \left[1 - \sum_{\sigma \in S} \prod_{i=1}^{t} p_{\sigma(i)} \right]^s,$$

where S is the set of sequences defined the same way as in Prop.2. Obviously, for a fixed s the isotropic distribution $\bar{p}$ will yield the minimum of $F_p^{(s)}(t)$ again, and it is easy to see that $F_{\bar{p}}^{(s)}(t) = (1 - \binom{\nu}{t} t!/\nu^t)^s$.

Owing to the monotonicity of $F_{\bar{p}}(t)$ (and $F_{\bar{p}}^{(s)}(t)$) in ν, in both cases the confidence interval for the unknown parameter ν can be estimated using Prop.1. Note that if α_2 is set to 0, in these confidence intervals ν_1 becomes a lower bound for ν with a confidence level $100(1 - \alpha_1)\%$ and in view of Prop.2 this lower bound is valid for any probability distribution p on the set of local optima. In what follows we will use $L(t, s, \alpha)$ for the $100(1-\alpha)\%$-confidence lower bound constructed by Prop.1 with $\alpha_1 = \alpha, \alpha_2 = 0$ given the observed value $T^{(s)} = t$ and $F_\nu(t) = (1 - \binom{\nu}{t}t!/\nu^t)^s$.

If the computational resources are limited to r restarts, which all produce distinct outcomes, a valid $100(1 - \alpha)\%$-confidence lower bound for ν could be obtained assuming that the next (hypothetical) restart will yield one of the previously visited optima. It is interesting to see how fast this lower bound $L(r + 1, 1, \alpha)$ grows with the number of restarts. Note that computing $L(r + 1, 1, \alpha)$ we seek for the minimal ν such that $\binom{\nu}{r}r!/\nu^r \geq \alpha$. But

$$\binom{\nu}{r}\frac{r!}{\nu^r} = \frac{\nu!}{(\nu - r)!\nu^r} = \exp\left\{\sum_{j=1}^{r}\ln\left(1 - \frac{r - j}{\nu}\right)\right\} \leq$$

$$\exp\left\{-\sum_{j=1}^{r}\frac{r - j}{\nu}\right\} = \exp\left\{-\frac{r(r + 1)}{2\nu}\right\} \leq \exp\left\{-\frac{r^2}{2\nu}\right\},$$

so $L(r+1, 1, \alpha) \geq 0.5r^2/\ln(\alpha^{-1})$, and e.g. for $\alpha = 0.05$ we have $L(r+1, 1, 0.05) \approx 0.167r^2$. For large values of t this simple estimate is a good approximation to $L(r + 1, 1, \alpha)$.

2.3   Schnabel Census Procedure

A common problem in ecology is to provide an estimate of the size of a population of animals by using the results of repeated samples. One such procedure is known as the *Schnabel census* (see e.g. [11]): we take repeated samples of size n (at suitable intervals of time) and count the number of *distinct* animals seen. The usual assumption is that the probability of catching any particular animal is the same. (Note that this is not the same as the subsequence idea discussed in the previous section: although it would be simple to map the subsequence approach onto the Schnabel census, the probability distribution is different.) We can readily apply this idea to the estimation of ν by repeated restarts of a local search procedure where we take $n = 1$.

Let r be the number of restarts, and K be the random variable for the number of distinct local optima found, while $F(k, \nu, r) = P\{K \leq k\}$ is the distribution function for K in case of the isotropic distribution $\bar{p}$.

For Schnabel census an *approximate* approach to computing the confidence interval for ν via a Normal approximation of the underlying distribution is well-studied (see e.g. [7,11]). The following proposition provides a basis for an *exact* computation of the confidence intervals for ν from the observed value of K in a way suggested by Prop.1.

Proposition 3. $F(k, \nu, r)$ *is a non-increasing function on ν for any $k \in \mathbf{Z}$.*

Proof. By definition, $F(k, \nu, r) = \sum_{i=1}^{k}p(i, \nu, r)$, where

$$p(i, \nu, r) = P\{K = i\} = \frac{\nu!S(r, i)}{(\nu - i)!\nu^r}$$

(see e.g. [13]). $S(r,i)$ stands here for the Stirling number of the second kind [14], i.e. the number of all possible partitions of r-element set into i nonempty subsets. Let us consider the difference $\Delta_\nu(k) = F(k, \nu+1, r) - F(k, \nu, r)$. Obviously $\Delta_\nu(0) = \Delta_\nu(r) = 0$. We shall prove that $\Delta_\nu(k) \le 0$ for $k = 0, ..., r$ by showing that $\Delta_\nu(k)$ is a non-increasing function on k when $0 \le k \le k^*$, and non-decreasing when $k^* \le k \le r$, where $k^* = \nu(1 - \nu^{r-1})/(\nu+1)^{r-1}$. We will carry the proof by induction in two steps: a) for $k \le k^*$ and b) for $k^* \le k$ (note that these ranges overlap at $k = k^*$).

a) The basis of induction is with $k = 0$. Now

$$\Delta_\nu(k) = \Delta_\nu(k-1) + p(k, \nu+1, r) - p(k, \nu, r) < 0,$$

since $\Delta_\nu(k-1) \le 0$ by induction and

$$p(k, \nu+1, r) - p(k, \nu, r) = \frac{k-1}{\nu}\left(\frac{\nu+1}{\nu}\right)^{r-1}\frac{(\nu+1)!S(r,k)}{(\nu+1-k)!(\nu+1)^r} < 0.$$

b) For $k \ge k^*$ we use the basis $k = r$. Here

$$\Delta_\nu(k) = \Delta_\nu(k+1) - p(k+1, \nu+1, r) + p(k+1, \nu, r) \le 0,$$

since $\Delta_\nu(k+1) \le 0$ by induction and

$$p(k+1, \nu+1, r) - p(k+1, \nu, r) = \frac{k}{\nu}\left(\frac{\nu+1}{\nu}\right)^{r-1}\frac{(\nu+1)!S(r,k+1)}{(\nu-k)!(\nu+1)^r} \ge 0.$$

□

For the Schnabel census we also conjecture a statement similar to Prop.2. This claim is supported by experiments, but its proof is not known to us.

3   Computational Experiments

The estimators presented above have been tested in experiments with three sets of optimization problems of different kinds. Among them there are some NK landscapes (see e.g. [15]), the low autocorrelation binary sequence (LABS) problems [16] and some vertex cover problems (VCP) converted from the maximum clique instances suggested in [17]. In all cases we used a 'bit-flip' neighbourhood, i.e. $N(x) = \{y : d_H(x,y) = 1\}$ where $d_H(x,y)$ is Hamming distance. For each instance of NK landscapes and the LABS problems, $r = 1000$ restarts of local search were made, while the VCP instances demanded more of trials. Here we set $r = 10000$ where the computational load was not too great. The data from some of the experiments discussed below have been already used in [10]. For some low-dimensional problems it was possible to perform a complete enumeration of solutions—in these cases the actual ν-values are reported as well.

3.1 NK Landscapes

The experiments with NK landscapes of increasing ruggedness were performed for K ranging from 2 to 12 while keeping $N = 15$. The graph in Fig. 1 shows the total number of local optima and the two-sided confidence intervals for the Schnabel census with the isotropic assumption $p = \bar{p}$. These are displayed around the maximum likelihood estimate 'Sch.ML' [7,10]. This graph also contains the first repetition time lower bounds $L(t_1, 1, \alpha)$, $L(\max_{j=1,\dots,s} t_j, s, \alpha)$ computed on the basis of the same data from $r = 1000$ restarts (where s is the number of subsequences observed in r restarts). In all cases the confidence level is 95%.

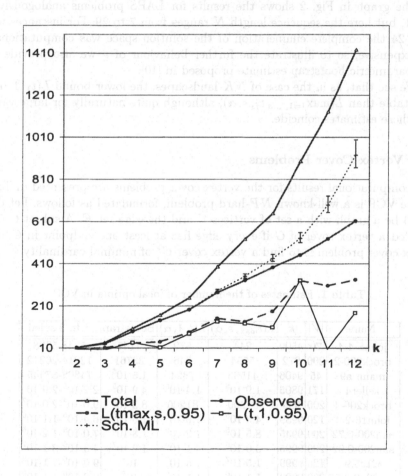

Fig. 1. Number of local optima in NK landscapes

The curve of true values of ν in this figure diverges significantly from the Schnabel census estimate, which clearly shows that the isotropic assumption is invalid for these NK landscapes. The lower bounds based on the first repetition time computed from our data go below the curve of number of observed optima. We shall see that in situations with much greater values of ν, lower bounds based on first repetition time may be more informative.

3.2 Low Autocorrelation Binary Sequences

The LABS problem consists in finding a vector $y \in \{-1, +1\}^N$ that minimizes $E = \sum_{j=1}^{N-1} c_j^2$, where c_j is the autocorrelation of the sequence $(y_1, ..., y_N)$ at lag j, i.e. $c_j = \sum_{i=1}^{N-j} y_i y_{i+j}$.

The graph in Fig. 2 shows the results for LABS problems analogously to Fig. 1, but here the sequence length N ranges from 7 to 29. For instances with $N > 24$ the complete enumeration of the solution space was computationally too expensive, so to illustrate the further behaviour of ν we also provide the non-parametric bootstrap estimate proposed in [10].

We see that, as in the case of NK landscapes, the lower bound $L(t_1, 1, \alpha)$ is less stable than $L(\max_{j=1,...,s} t_j, s, \alpha)$, although quite naturally for large values of ν these estimates coincide.

3.3 Vertex Cover Problems

The computational results for the vertex cover problems are presented in Table 1. The VCP is a well-known NP-hard problem, formulated as follows. Let $G = (V, E)$ be a graph with a set of vertices V and the edge set E. A subset $C \subseteq V$ is called a vertex cover of G if every edge has at least one endpoint in C. The vertex cover problem is to find a vertex cover C^* of minimal cardinality.

Table 1. Estimates of the number of local optima in VCP

Name	$|V|$	k	$L(t_{\max}, s, \alpha)$	$L(t, 1, \alpha)$	Bootstrap	Sch. 2-sided
john8-4-4	70	4509	244	572	7145	5248-5398
brock200-2	200	972	5554	346	33761	12214-26212
mann a9	45	9406	41535	7824	$1.5 \cdot 10^5$	74808-87445
keller4	171	9805	$1.9 \cdot 10^5$	$1.4 \cdot 10^5$	$4.9 \cdot 10^5$	$2.2 \cdot 10^5$-$2.9 \cdot 10^5$
brock200-4	200	999	81200	81200	$8.3 \cdot 10^5$	$9.0 \cdot 10^4$-$2.0 \cdot 10^7$
john16-2-4	120	9939	$4.3 \cdot 10^5$	$1.6 \cdot 10^5$	$1.6 \cdot 10^6$	$6.4 \cdot 10^5$-$1.1 \cdot 10^6$
san200-0.72	200	9945	$8.5 \cdot 10^5$	$1.2 \cdot 10^6$	$1.8 \cdot 10^6$	$7.0 \cdot 10^5$-$1.2 \cdot 10^6$
san200-0.92	200	9995	$3.0 \cdot 10^6$	$4.1 \cdot 10^6$	$1.2 \cdot 10^7$	$4.3 \cdot 10^6$-$3.1 \cdot 10^7$
C125.9	125	9999	$1.5 \cdot 10^7$	$1.5 \cdot 10^7$	10^8	$9.0 \cdot 10^6$-$1.1 \cdot 10^9$
gn200p0.9-55	200	9999	$1.1 \cdot 10^6$	$1.1 \cdot 10^6$	10^8	$9.0 \cdot 10^6$-$1.1 \cdot 10^9$

In order to include this problem into the unconstrained optimization framework accepted throughout this paper we will assume that the solutions set is the

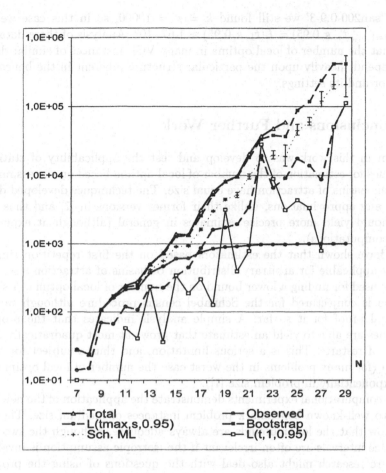

Fig. 2. Number of local optima in LABS problems

set of all $|V|$-element binary strings where ones indicate the vertices belonging to the corresponding subset C. The infeasible solutions are penalized by summing their cardinality with the number of uncovered edges in E multiplied by $|V|$ (by uncovered we mean those edges with both endpoints outside the set C). In our study we take some widely used instances of maximum clique problem from the DIMACS benchmark set [17] and transform them into VCP instances using the standard reduction (see e.g. [18]).

It turned out that large-scale instances in this set are characterised by large values of ν, so in order to provide at least some repetitions in the visited local optima we limited our consideration to graphs with not more than 200 vertices. In the case $r = 1000$ even the smaller problems often yielded no repetitions at all, thus r was set equal to 10000 for all instances except for 'brock200-2' and 'brock200-4'. For the problems 'gn200p0.9-44', 'mann a27', 'san200-0.9-

1' and 'san200-0.9-3' we still found $k = r = 10000$, so in this case we have $L(\max_{j=1,\dots,s} t_j, s, 0.95) = L(t_1, s, 0.95) \approx 1.67 \cdot 10^6$. Analysis of these data indicates that the number of local optima in many VCP instances of similar dimensions depends heavily upon the particular structure inherent in the benchmark generator and its settings.

4   Conclusions and Further Work

Our aim in this work was to develop and test the applicability of statistical techniques for estimation of the number of local optima based on the assumption that their basins of attraction have equal size. The techniques developed do not employ any approximations, unlike their former versions in [7] and thus these tools should yield more precise estimates in general (although at expense of longer computations).

We have shown that the estimators based on the first repetition time are actually applicable for arbitrary distribution of basins of attraction size, when they are used for finding a lower bound on the number of local optima. A similar situation is conjectured for the Schnabel census procedure although we have no formal proof for it so far. A simple analysis indicates that the proposed techniques are able to yield an estimate that grows at most quadratically in the number of restarts. This is a serious limitation, and thus a subject for future research (for many problems in the worst case the number of local optima may grow exponentially in problem size [1]).

The computational experiments demonstrate the application of the new techniques to well-known benchmark problem instances of medium size. These results show that the lower bounds are always adequate, although the two-sided confidence intervals are often irrelevant if the isotropic assumption is invalid.

Future research might also deal with the questions of using the proposed techniques in stopping rules for local search heuristics with multiple restarts. In particular, the lower bounds could give the necessary termination conditions, and if all local optima have an equal chance to be found, sufficient conditions could be formulated as well. Also these techniques may be used for the classification of instances of an optimization problem in terms of their difficulty for local search algorithms when test problems are being generated. The open issues mentioned in this paper are also the subject of future research.

Acknowledgment. Part of this research was carried out when the first author was visiting the UK on a grant from the Royal Society. The authors thank Sergey Klokov for the helpful comments on this paper.

References

1. Tovey, C.A.: Local Improvement on Discrete Structures. In: Aarts, E., Lenstra, J. K. (eds.): Local Search in Combinatorial Optimization. John Wiley & Sons (1997) 57–90

2. Reeves, C.R.: The Crossover Landscape and the Hamming Landscape for Binary Search Spaces. To appear in Proc. of Foundations of Genetic Algorithms – VII. (2003)
3. Reeves C.R.: Fitness Landscapes and Evolutionary Algorithms. In: Fonlupt, C., Hao, J-K., Lutton, E., Ronald, E. and Schoenauer, M. (eds.): Artificial Evolution: 4th European Conference. Lecture Notes in Computer Science, Vol. 1829. Springer-Verlag, Berlin (2000) 3–20
4. Stadler, P.F.: Towards a Theory of Landscapes. In: Lopéz-Peña, R., Capovilla, R., García-Pelayo, R., Waelbroeck, H. and Zertuche, F. (eds.): Complex Systems and Binary Networks. Springer-Verlag, Berlin (1995) 77–163
5. Reeves, C.R.: Landscapes, Operators and Heuristic Search. Annals of Operations Research. 86 (1999) 473–490
6. Boese, K.D., Kahng, A.B., Muddu S.: A New Adaptive Multi-Start Technique for Combinatorial Global Optimizations. Operations Research Letters. 16 (1994) 101–113
7. Reeves, C.R.: Estimating the Number of Optima in a Landscape, Part I: Statistical Principles. Coventry University Technical Report SOR#01–03 (2001)
8. Garnier, J., Kallel, L.: How to Detect All Maxima of a Function? In: Proceedings of the Second EVONET Summer School on Theoretical Aspects of Evolutionary Computing (Anvers, 1999). Springer, Berlin (2001) 343–370
9. Reeves, C.R.: Estimating the Number of Optima in a Landscape, Part II: Experimental Investigations. Coventry University Technical Report SOR#01–04 (2001)
10. Eremeev, A.V., Reeves C.R.: Non-Parametric Estimation of Properties of Combinatorial Landscapes. In: Cagnoni, S., Gottlieb, J., Hart, E., Middendorf, M. and Raidl, G. (eds.): Applications of Evolutionary Computing: Proceedings of EvoWorkshops 2002. Lecture Notes in Computer Science, Vol. 2279. Springer-Verlag, Berlin Heidelberg (2002) 31–40
11. Seber G.A.F.: The Estimation of Animal Abundance. Charles Griffin, London (1982)
12. Mood, A. M., Graybill, F.A., Boes, D.C.: Introduction to the Theory of Statistics. 3rd edn. McGraw-Hill, New York (1973)
13. Johnson, N.L., Kotz, S.: Discrete Distributions. Wiley, New York (1969)
14. Liu, C.L.: Introduction to Combinatorial Mathematics. McGraw-Hill, New York (1968)
15. Kauffman, S.A.: Adaptation on Rugged Fitness Landscapes. In: Lectures in the Sciences of Complexity, Vol. I of SFI studies, Addison-Wesley (1989) 619–712
16. Golay, M.J. E.: Series for Low-Autocorrelation Binary Sequences. IEEE Trans. Inform. Theory. 23 (1977) 43–51
17. Johnson, D.S., Trick, M.A.: Introduction to the Second DIMACS Challenge: Cliques, Coloring, and Satisfiability. In: Johnson, D.S., Trick, M.A. (eds.): Cliques, Coloring, and Satisfiability. DIMACS Series in Discrete Mathematics and Theoretical Computer Science, Vol. 26. AMS (1996) 1–10
18. Garey, M.R., Johnson, D.S.: Computers and Intractability. A Guide to the Theory of NP-Completeness. W.H. Freeman and Company, San Francisco (1979)

Searching for Maximum Cliques with Ant Colony Optimization

Serge Fenet and Christine Solnon

LIRIS, Nautibus, University Lyon I
43 Bd du 11 novembre, 69622 Villeurbanne cedex, France
{sfenet,csolnon}@bat710.univ-lyon1.fr

Abstract. In this paper, we investigate the capabilities of Ant Colony Optimization (ACO) for solving the maximum clique problem. We describe Ant-Clique, an algorithm that successively generates maximal cliques through the repeated addition of vertices into partial cliques. ACO is used to choose, at each step, the vertex to add. We illustrate the behaviour of this algorithm on two representative benchmark instances and we study the impact of pheromone on the solution process. We also experimentally compare Ant-Clique with GLS, a Genetic Local Search approach, and we show that Ant-Clique finds larger cliques, on average, on a majority of DIMACS benchmark instances, even though it does not reach the best known results on some instances.

1 Introduction

The maximum clique problem is a classical combinatorial optimization problem that has important applications in different domains, such as coding theory, fault diagnosis or computer vision. Given a non-oriented graph $G = (V, E)$, such that V is a set of vertices, and $E \subseteq V \times V$ is a set of edges, a *clique* is a set of vertices $C \subseteq V$ such that every couple of distinct vertices of C is connected with an edge in G, i.e., the subgraph induced by C is complete. A clique is *partial* if it is strictly included in another clique; otherwise it is *maximal*. The goal of the *maximum clique problem* is to find a clique of maximum cardinality.

This problem is one of the first problems shown to be NP-complete, and moreover, it does not admit a polynomial-time approximation algorithm (unless P=NP) [2]. Hence, complete approaches —usually based on a branch-and-bound tree search— become intractable when the number of vertices increases, and much effort has recently been directed on heuristic incomplete approaches. These approaches leave out exhaustivity and use heuristics to guide the search towards promising areas of the search space. In particular, [6] describes three heuristic algorithms obtained as instances of an evolutionary algorithm scheme, called GLS, that combines a genetic approach with local search. In this paper, we investigate the capabilities of another bio-inspired metaheuristic —Ant Colony Optimization (ACO)— for solving the maximum clique problem. In Section 2, we describe Ant-Clique, an ACO algorithm for the maximum clique problem.

S. Cagnoni et al. (Eds.): EvoWorkshops 2003, LNCS 2611, pp. 236–245, 2003.

Basically, this algorithm uses a sequential greedy heuristic, and generates maximal cliques through the repeated addition of vertices into partial cliques. ACO is introduced as a heuristic for choosing, at each step, the vertex to enter the clique: this vertex is chosen with respect to a probability that depends on pheromone trails laying between it and the clique under construction, while pheromone trails are deposited by ants proportionally to the quality of the previously computed cliques. In Section 3, we illustrate the behavior of this algorithm on two representative benchmark instances: we study the impact of pheromone, and we show that the solution process is strongly improved by ACO, without using any other local heuristic nor local search. In Section 4, we experimentally compare Ant-Clique with GLS [6], a genetic local search approach, and we show that Ant-Clique finds larger cliques, on average, on a wide majority of benchmark instances of the DIMACS library, eventhough it does not reach the best known results on some benchmark instances.

2   Description of Ant-Clique

The Ant Colony Optimization (ACO) metaheuristic is a bio-inspired approach that has been used to solve different hard combinatorial optimization problems [3,4]. The main idea of ACO is to model the problem as the search for a minimum cost path in a graph. Artificial ants walk through this graph, looking for good paths. Each ant has a rather simple behaviour so that it will typically only find rather poor quality paths on its own. Better paths are found as the emergent result of the global cooperation among ants in the colony. This cooperation is performed in an indirect way through pheromone laying.

The proposed ACO algorithm for searching for maximum cliques, called Ant-Clique is sketched below:

> **procedure** Ant-Clique
>   Initialize pheromone trails
>   **repeat**
>   **for** k **in** $1..nbAnts$ **do**: construct a clique C_k
>   Update pheromone trails w.r.t. $\{C_1, \dots, C_{nbAnts}\}$
>   **until** max cycles reached **or** optimal solution found

Pheromone trail initialization: Ants communicate by laying pheromone on the graph edges. The amount of pheromone on edge (v_i, v_j) is noted $\tau(v_i, v_j)$ and represents the learnt desirability for v_i and v_j to belong to a same clique.

As proposed in [9], we explicitly impose lower and upper bounds τ_{min} and τ_{max} on pheromone trails (with $0 < \tau_{min} < \tau_{max}$). The goal is to favor a larger exploration of the search space by preventing the relative differences between pheromone trails from becoming too extreme during processing.

We have considered two different ways of initializing pheromone trails:

1. Constant initialization to τ_{max}: such an initialization achieves a higher exploration of the search space during the first cycles, as all edges nearly have the same amount of pheromone duringthe first cycles [9].

2. Initialization with respect to a preprocessing step: the idea is to first compute a representative set of maximal cliques without using pheromone —thus constituting a kind of sampling of the search space— and then to select from this sample set the best cliques and use them to initialize pheromone trails. This preprocessing step introduces two new parameters: ϵ, the limit rate on the quality improvement of the sample set of cliques, and n_{Best}, the number of best cliques that are selected from the sample set in order to initialize pheromone trails. More details can be found in [8].

Construction of cliques by ants: ants first randomly select an initial vertex, and then iteratively choose vertices to be added to the clique —within a set of candidates that contains all vertices that are connected to every clique vertex:

> choose randomly a first vertex $v_f \in V$
> $C \leftarrow \{v_f\}$
> $Candidates \leftarrow \{v_i/(v_f, v_i) \in E\}$
> **while** $Candidates \neq \emptyset$ **do**
>> Choose a vertex $v_i \in Candidates$ with probability $p(v_i) = \frac{[\tau_C(v_i)]^\alpha}{\sum_{v_j \in Candidates}[\tau_C(v_j)]^\alpha}$
>> $C \leftarrow C \cup \{v_i\}$
>> $Candidates \leftarrow Candidates \cap \{v_j/(v_i, v_j) \in E\}$
> **end while**
> return C

The choice of a vertex v_i within the set of candidates is made with respect to a probability that depends on a pheromone factor $\tau_C(v_i)$. A main difference with many ACO algorithms is that this factor does not only depend on the pheromone trail between the last added vertex in C and the candidate vertex v_i, but on all pheromone trails between C and $v_i{}^1$, i.e., $\tau_C(v_i) = \sum_{v_j \in C} \tau(v_i, v_j)$.

One should remark that the probability of choosing a vertex v_i only depends on a pheromone factor, and not on some heuristic factor that locally evaluates the quality of the candidate vertex, as usually in ACO algorithms. Actually, we have experimented a heuristic proposed in [1], the idea of which being to favor vertices with larger degrees in the subgraph induced by the partial clique under construction. The underlying motivation is that a larger degree implies a larger number of candidates after the vertex is added to the current clique. When using no pheromone (or at the beginning of the search, when all pheromone trails have the same value), this heuristic actually allows ants to construct larger cliques than a random choice. However, when combining it with pheromone learning, we have noticed that after a hundred or so cycles, we obtain larger cliques without using the heuristic than when using it.

Updating pheromone trails: After each ant has constructed a clique, pheromone trails are updated according to ACO: first, all pheromone trails are decreased

---

[1] This pheromone factor $\tau_C(v_i)$ is computed in an incremental way: at the beginning, when the clique C only contains one vertex v_f, $\tau_C(v_i)$ is initialized to $\tau(v_f, v_i)$; then, each time a new vertex v_k is added to the clique, $\tau_C(v_i)$ is incremented with $\tau(v_k, v_i)$.

in order to simulate evaporation, i.e., for each edge $(v_i, v_j) \in E$, the quantity of pheromone laying on it, $\tau(v_i, v_j)$, is multiplied by an evaporation parameter ρ such that $0 \leq \rho \leq 1$; then, the best ant of the cycle deposits pheromone, i.e., let C_k be the largest clique built during the cycle, and C_{best} be the largest clique built since the beginning of the run, for each couple of vertices $(v_i, v_j) \in C_k$, we increment the quantity of pheromone laying on it, $\tau(v_i, v_j)$, by $1/(1+|C_{best}|-|C_k|)$.

3   Experimental Study of Ant-Clique

As usually in ACO algorithms, the behaviour of Ant-Clique depends on its parameters, and more particularly on α, the pheromone factor weight, and ρ, the evaporation rate. Indeed, diversification can be emphasized both by decreasing α, so that ants become less sensitive to pheromone trails, and by increasing ρ, so that pheromone evaporates more slowly. When increasing the exploratory ability of ants in this way, one usually finds better solutions, but as a counterpart it takes longer time to find them. This is illustrated in figure 1 on two DIMACS benchmark instances.

On this figure, one can first note that, for both instances, when $\alpha=0$ and $\rho=1$, cliques are much smaller: in this case, pheromone is totally ignored and the resulting search process performs as a random one so that after 500 or so cycles, the size of the largest clique nearly stops increasing. This shows that pheromone actually improves the solution process with respect to a pure random algorithm.

When considering instance $gen\_400\_P0.9\_75$, we can observe that, when α increases or ρ decreases, ants converge quicker towards the optimal solution: it is found around cycle 650 when $\alpha=1$ and $\rho=0.995$, around cycle 400 when $\alpha=2$ and $\rho=0.995$, around cycle 260 when $\alpha=2$ and $\rho=0.99$, and around cycle 200 when $\alpha=3$ and $\rho=0.985$. Actually, this instance is relatively easy to solve, regardless of the parameter tuning. In this case, one has better choose parameter values that favor a quick convergence such as $\alpha=3$ and $\rho=0.985$.

However, when considering instance $C500.9$, which is more difficult, one can see that the setting of α and ρ let us balance between two main tendencies. On one hand, when favoring exploration, the quality of the final solution is better, but the time needed to converge on this value is also higher. On the other hand, when favoring convergence, ants find better solutions during the first cycles, but after 500 or so cycles, too many of them have converged on over-marked pheromone trails and the exploration behavior is inhibited. To summarize, one has to choose parameters depending on the availability of time for solving: for 300 cycles, one has to choose $\alpha=3$ and $\rho=0.985$; for 800 cycles one has to choose $\alpha=2$ and $\rho=0.990$; and for more available time one has to choose $\alpha=2$ and $\rho=0.995$. Note that with $\alpha=1$ and $\rho=0.995$, pheromone's influence is too weak and convergence is too slow: the system won't reach optimal solution in acceptable time.

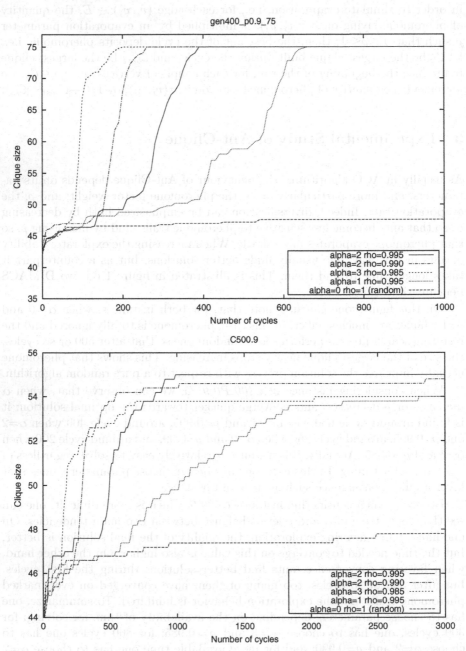

Fig. 1. Influence of pheromone on the solution process: each curve plots the evolution of the size of the largest clique (average on 5 runs), when the number of cycles increases, for a given setting of α and ρ. The other parameters have been set to nbAnts=15, $\tau_{min} = 0.01$, $\tau_{max} = 4$, and no preprocessing. The upper curves correspond to the DIMACS instance $gen\_400\_P0.9\_75$ —that is a rather easy one— and the lower curves to $C500.9$ — that is more difficult.

4   Experimental Comparison of Ant-Clique with GLS

Genetic Local Search: We compare Ant-Clique with genetic local search (GLS) [6], an evolutionary approach that combines a genetic approach with local search. GLS generates successive populations of maximal cliques from an initial one by repeatedly selecting two parent cliques from the current population, recombining them to generate two children cliques, applying local search on children to obtain maximal cliques, and adding to the new population the best two cliques of parents and children. GLS can be instanciated to different algorithms by modifying its parameters. In particular, [6] compares results obtained by the three following instances of GLS: GENE performs genetic local search (the population size is set to 10, the number of generations to 2000, and mutation and crossover rates to 0.1 and 0.9); ITER performs iterated local search, starting from one random point (the population size is set to 1, the number of generations to 20000, and mutation and crossover rates are null); and MULT performs multi-start local search, starting from a new random point at each time (the population size is set to 20000, the number of generations, mutation and crossover rates to 0).

Experimental Setup: Ant-Clique has been implemented in C and run on a 350 MHz G4 processor. In all experiments, we have set α to 2, ρ to 0.995, τ_{min} to 0.01, τ_{max} to 4, and the maximum number of cycles to 3000. We first report results obtained with 7 ants and no preprocessing. In this case, Ant-Clique constructs 21 000 cliques so that this allow a fair comparison with GLS that constructs 20 000 cliques. We also report results obtained when increasing the number of ants to 30, and when using a preprocessing step to initialize pheromone trails (as described in section 2). When a preprocessing step is used, we have set ϵ to 0.001 and n_{Best} to 300.

Test Suite: We consider the 37 benchmark instances proposed by the DIMACS organisers and available at http://dimacs.rutgers.edu/Challenge. Cn.p and DSJCn.p graphs have n vertices; Mann_a 27, 45 and 81 graphs respectively have 378, 1035 and 3321 vertices; brock$n$_m graphs have n vertices; genxxx_p0.9yy graphs have xxx vertices and maximum cliques of size yy; *hamming8-4* and *hamming10-4* respectively have 256 and 1024 vertices; *keller* 4, 5 and 6 graphs respectively have 171, 776 and 3361 vertices; and p_hat$n$−m graphs have n vertices.

Comparison of Clique Sizes: Table 1 compares the sizes of the cliques found by Ant-Clique and GLS (note that for GLS, we report the results given in [6] for the GLS instance that obtained the best average results). On this table, one can first note that GLS clearly outperforms Ant-Clique on the three MANN instances: on these instances, ACO deteriorates the solution process, and the average size of the constructed cliques at each cycle decreases when the number of cycles increases. Further work will concern the study of the fitness landscape of these instances, in order to understand these bad results.

When comparing average results of Ant-Clique with 7 ants and no preprocessing with GLS, we note that Ant-Clique outperforms GLS on 23 instances,

Table 1. Results for DIMACS benchmarks. For each instance, the best known result is presented between parenthesis in first column. The table then displays the results obtained by Ant-Clique (with preprocessing and 7 ants, without preprocessing and 7 ants, and without preprocessing and 30 ants) and reports the best, average and standard deviation of the maximum size cliques constructed over 25 runs. Finally, the table reports results given for GLS in [6]: we display the best, average and standard deviation for the GLS algorithm that obtained the best average results (I for Iter, G for Gene and M for Mult). For each instance, best average results are in bold font.

Graph	Ant-Clique						GLS	
	with preproc 7 ants		without preprocessing					
			7 ants		30 ants			
	best	avg(sdv)	best	avg(sdv)	best	avg(sdv)	best	avg(sdv)
C125.9(34)	34	**34.0**(0.0)	34	**34.0**(0.0)	34	**34.0**(0.0)	34	**34.0**(0.0) I
C250.9(44)	44	**44.0**(0.0)	44	44.0(0.2)	44	**44.0**(0.0)	44	43.0(0.6) I
C500.9(57)	56	54.3(1.1)	56	55.0(0.8)	56	**55.5**(0.7)	55	52.7(1.4) I
C1000.9(68)	65	63.5(1.2)	66	63.6(1.4)	67	**65.0**(1.0)	66	61.6(2.1) G
C2000.9(78)	70	68.5(1.2)	72	68.5(1.9)	73	**70.0**(1.7)	70	68.7(1.2) I
DSJC500.5(14)	13	**12.9**(0.3)	13	12.7(0.5)	13	**12.9**(0.3)	13	12.2(0.4) G
DSJC1000.5(15)	15	**14.0**(0.5)	15	13.9(0.6)	15	**14.0**(0.2)	14	13.5(0.5) I
C2000.5(16)	15	14.9(0.3)	16	14.6(0.6)	16	**15.0**(0.3)	15	14.2(0.4) I
C4000.5(18)	17	15.8(0.6)	16	15.5(0.5)	17	**15.9**(0.4)	16	15.6(0.5) I
MANN_a27(126)	126	125.6(0.3)	126	124.8(0.4)	126	125.5(0.5)	126	**126.0**(0.0) I
MANN_a45(345)	340	338.9(0.7)	338	336.4(0.6)	341	339.2(0.7)	345	**343.1**(0.8) I
MANN_a81(1098)	1092	1090.6(0.7)	1087	1086.3(0.5)	1092	1089.4(0.7)	1098	**1097.0**(0.4) I
brock200_2(12)	12	11.9(0.3)	12	11.9(0.3)	12	**12.0**(0.0)	12	**12.0**(0.0) M
brock200_4(17)	17	16.4(0.5)	16	16.0(0.0)	17	**16.6**(0.5)	17	15.7(0.9) M
brock400_2(29)	25	24.5(0.5)	25	24.4(0.5)	29	**24.7**(1.0)	25	23.2(0.7) I
brock400_4(33)	33	25.3(3.0)	25	24.1(0.3)	33	**25.6**(3.3)	25	23.6(0.8) G
brock800_2(21)	21	19.7(0.6)	21	19.6(0.6)	21	**19.8**(0.5)	20	19.3(0.6) G
brock800_4(21)	20	19.3(0.5)	21	19.6(0.6)	20	**19.7**(0.5)	20	19.0(0.4) I
gen200...44(44)	44	**44.0**(0.0)	44	42.6(1.9)	44	43.8(0.8)	44	39.7(1.6) G
gen200...55(55)	55	**55.0**(0.0)	55	**55.0**(0.0)	55	**55.0**(0.0)	55	50.8(6.4) G
gen400...55(55)	52	51.6(0.5)	52	51.5(0.5)	52	**51.7**(0.5)	55	49.7(1.2) G
gen400...65(65)	65	**65.0**(0.0)	65	**65.0**(0.0)	65	**65.0**(0.0)	65	53.7(7.4) G
gen400...75(75)	75	**75.0**(0.0)	75	**75.0**(0.0)	75	**75.0**(0.0)	75	62.7(12.3) I
hamming8-4(16)	16	**16.0**(0.0)	16	**16.0**(0.0)	16	**16.0**(0.0)	16	**16.0**(0.0) G
hamming10-4(40)	40	38.3(1.0)	40	**38.4**(1.5)	40	38.3(1.7)	40	38.2(1.2) I
keller4(11)	11	**11.0**(0.0)	11	**11.0**(0.0)	11	**11.0**(0.0)	11	**11.0**(0.0) G
keller5(27)	27	**26.9**(0.3)	27	26.6(0.5)	27	**26.9**(0.3)	27	26.3(0.6) I
keller6(59)	54	52.0(1.2)	53	51.3(1.2)	55	**53.1**(1.2)	56	52.7(1.8) I
p_hat300-1(8)	8	**8.0**(0.0)	8	**8.0**(0.0)	8	**8.0**(0.0)	8	**8.0**(0.0) G
p_hat300-2(25)	25	**25.0**(0.0)	25	**25.0**(0.0)	25	**25.0**(0.0)	25	**25.0**(0.0) I
p_hat300-3(36)	36	**36.0**(0.0)	36	36.0(0.2)	36	**36.0**(0.0)	36	35.1(0.8) I
p_hat700-1(11)	11	10.8(0.4)	11	10.8(0.4)	11	**10.9**(0.3)	11	9.9(0.7) I
p_hat700-2(44)	44	**44.0**(0.0)	44	**44.0**(0.0)	44	**44.0**(0.0)	44	43.6(0.7) I
p_hat700-3(62)	62	61.6(0.5)	62	61.5(0.5)	62	**62.0**(0.0)	62	61.8(0.6) I
p_hat1500-1(12)	11	10.9(0.3)	12	**11.2**(0.4)	12	11.0(0.2)	11	10.8(0.4) G
p_hat1500-2(65)	65	**64.9**(0.3)	65	64.8(0.4)	65	**64.9**(0.3)	65	63.9(2.0) I
p_hat1500-3(94)	94	92.9(0.4)	94	92.8(0.5)	94	**93.1**(0.3)	94	93.0(0.8) I

whereas GLS outperforms Ant-Clique on 8 instances. Then, when increasing the number of ants from 7 to 30, the average results found by Ant-Clique are improved for 24 instances, whereas they are deteriorated for 2 instances. When further increasing the number of ants, Ant-Clique can still improve its performances on the hardest instances. For example, for *keller6* and with 100 ants, Ant-Clique found a clique of size 59, and in average found cliques of size 54.1 instead of 53.1 with 30 ants and 51.3 with 7 ants. Finally, one can also note that, when fixing the number of ants to 7, the preprocessing step improves average solutions' quality for 18 instances, whereas it deteriorates it for 4 instances.

CPU times: Table 2 reports the number of cycles and the time spent by Ant-Clique on a 350MHz G4 processor to find best solutions. The table also displays the time spent by the best of the 3 GLS algorithm on a Sun UltraSPARC-II 400MHz as reported in [6][2].

First, note that, within a same set of instances, time spent by Ant-Clique grows quadratically with respect to the number of vertices of the graph, whereas time spent by GLS increases more slowly (even though CPU times are not exactly comparable, as we used different processors). For instance, when considering *C125.9*, *C500.9* and *C2000.9*, that respectively have 125, 500 and 2000 vertices, Ant-Clique with 7 ants and no preprocessing respectively spent 0.4, 34.9 and 689.3 seconds whereas ITER respectively spent 0.5, 2.7 and 24.8 seconds.

Note also that when increasing the number of ants from 7 to 30, the number of cycles is nearly always decreased (for all instances but 3) as the quality of the best computed clique at each cycle is improved. However, as the time needed to compute one cycle also increases, Ant-Clique with 30 ants is quicker than Ant-Clique with 7 ants for 15 instances only (those for which the number of cycles is strongly decreased), whereas it is slower for 21 instances.

Finally, note that when initializing pheromone trails with a preprocessing step, instead of a constant initialization to τ_{max}, the number of cycles is nearly always strongly decreased (for all instances but 2). Indeed, this preprocessing step shorten the exploration phase, so that ants start converging much sooner. One can notice that on very easy instances, the best solution is found during preprocessing, or just after, during the first cycles of the pheromone-oriented solving step. Also, on the 3 *MANN* instances, the best solution is always found during preprocessing as, on these instances, ACO deteriorates solutions quality. However, if the preprocessing step reduces the number of cycles, it is time consuming. Hence, Ant-Clique with preprocessing is quicker than without preprocessing for 24 instances, whereas it is slower for 13. This result must be considered with the fact that preprocessing also allowed to improve solutions' quality for 18 instances, whereas it deteriorates them for 4 instances. Hence, on the maximum clique problem, preprocessing often boosts the resolution.

---

[2] When considering the SPEC benchmarks (http://www.specbench.org/), these two computers exhibit approximately the same computing power: for our 350MHz G4 processor, SPECint95=21.4 and SPECfp95=20.4; for the UltraSPARC-II 400MHz, SPECint95=17.2 and SPECfp95=25.9. Hence, our computer is more powerful for integer processing, but it is less powerful for floating point processing.

Table 2. Time results for DIMACS benchmarks. For each instance, the table displays the results obtained by Ant-Clique (with preprocessing and 7 ants, without preprocessing and 7 ants, and without preprocessing and 30 ants) and reports the number of cycles (average over 25 runs and standard deviation) and the average CPU time needed to find the best solution (on a 350MHz G4 processor). For ant-Clique with preprocessing, we successively report the time spent for the preprocessing step, the time spent for ACO solving, and the total time. The last column displays the time spent by the best GLS approach (on a Sun UltraSPARC-II 400MHz) .

Graph	Ant-Clique							GLS
	with preprocessing			without preprocessing				
	7 ants			7 ants		30 ants		
	nbCy(sdv)	t_P+t_S=	time	nbCy(sdv)	time	nbCy(sdv)	time	time
C125.9	93(61)	1.0+0.2=	1.2	219(63)	0.4	140(49)	0.7	0.5
C250.9	356(144)	2.2+2.5=	4.7	888(523)	6.1	467(98)	6.7	2.4
C500.9	974(614)	4.7+24.5=	29.3	1488(654)	34.9	1211(603)	49.9	2.7
C1000.9	1730(672)	12.9+154.6=	167.6	1972(571)	161.2	1712(602)	202.2	12.4
C2000.9	2065(648)	23.8+672.9=	696.7	2281(498)	689.3	2041(487)	749.5	24.8
DSJC500.5	325(614)	0.9+4.5=	5.4	1049(952)	12.9	605(375)	8.8	2.1
DSJC1000.5	495(612)	1.8+24.4=	26.2	1084(910)	50.6	990(645)	51.3	2.3
C2000.5	576(741)	4.2+108.3=	112.5	1130(919)	203.1	869(615)	159.5	2.3
C4000.5	505(745)	7.6+377.2=	384.8	783(826)	551.0	1097(728)	772.3	15.7
MANN_a27	0(0)	10.1+0.0=	10.1	187(630)	7.8	171(455)	21.3	15.6
MANN_a45	0(0)	67.7+0.0=	67.7	14(14)	4.4	233(508)	231.3	54.4
MANN_a81	0(0)	84.6+0.0=	84.6	25(23)	97.0	7(0)	93.6	693.9
brock200_2	322(718)	0.1+0.6=	0.7	265(492)	0.4	80(113)	0.2	2.7
brock200_4	67(197)	0.3+0.2=	0.5	424(124)	0.9	622(748)	2.3	2.8
brock400_2	813(768)	2.2+10.2=	12.3	955(427)	10.6	954(640)	15.1	2.0
brock400_4	208(151)	2.1+2.7=	4.8	945(468)	10.4	806(605)	12.7	1.3
brock800_2	820(986)	1.4+32.3=	33.7	1437(626)	52.1	1227(766)	51.3	5.2
brock800_4	620(512)	1.4+25.4=	26.8	1437(576)	51.9	1270(776)	53.4	4.1
gen200...44	364(165)	1.3+1.7=	3.0	1065(531)	4.7	376(146)	4.3	1.3
gen200...55	4(5)	2.0+0.0=	2.0	203(31)	0.9	134(25)	1.5	1.4
gen400...55	1423(898)	3.3+24.3=	27.7	1382(742)	22.3	1086(503)	35.8	3.8
gen400...65	281(137)	3.8+4.7=	8.5	718(158)	11.1	412(40)	13.2	4.3
gen400...75	49(42)	4.3+0.8=	5.1	555(443)	8.8	289(22)	9.2	4.6
hamming8-4	0(1)	0.7+0.0=	0.7	134(93)	0.5	59(41)	0.3	0.0
hamming10-4	806(493)	7.9+66.8=	74.6	1750(632)	129.9	1167(323)	103.2	5.3
keller4	1(1)	0.5+0.0=	0.5	9(8)	0.0	3(2)	0.0	0.0
keller5	936(775)	4.0+40.1=	44.1	1417(500)	55.7	1063(598)	50.2	4.0
keller6	1489(768)	23.6+1180.0=	1203.6	2242(375)	1682.4	2112(498)	1731.3	36.2
p_hat300-1	4(4)	0.2+0.0=	0.2	173(164)	0.4	34(31)	0.1	0.4
p_hat300-2	25(25)	0.8+0.1=	0.9	300(48)	1.2	190(47)	1.1	0.5
p_hat300-3	402(248)	1.9+3.1=	5.0	718(228)	4.9	486(96)	5.5	1.5
p_hat700-1	206(274)	0.4+3.0=	3.4	995(375)	14.1	619(310)	9.4	2.6
p_hat700-2	206(127)	3.1+5.5=	8.6	551(86)	12.7	452(99)	13.0	1.2
p_hat700-3	919(664)	5.9+35.8=	41.7	1106(438)	38.0	1030(347)	50.3	4.5
p_hat1500-1	461(745)	1.1+31.7=	32.8	1488(640)	97.0	700(374)	46.3	14.2
p_hat1500-2	791(419)	7.0+92.2=	99.2	1181(580)	125.8	790(220)	94.9	12.2
p_hat1500-3	945(776)	14.9+152.8=	167.8	1360(596)	214.7	1071(422)	195.1	7.1

5   Conclusion

We have described Ant-Clique, an ACO algorithm for searching for maximum cliques. We have shown through experiments that this metaheuristic is actually well suited to this problem: without using any local heuristic to guide them neither local search technics to improve solutions quality, artificial ants are able to find optimal solutions on many difficult benchmark instances.

However, Ant-Clique does not reach the best known results on all instances. The best current approach is Reactive Local Search (RLS) [1], an approach that combines local search with prohibition-based diversification techniques where the amount of diversification is determined in an automated way through a feedback scheme. In the experiments reported here, Ant-Clique has been able to find the best solution found by RLS on 29 instances over 37.

Hence, further work will first explore the integration within Ant-Clique of some local search technics such as the one used by RLS. Actually, the best performing ACO algorithms for many combinatorial problems are hybrid algorithms that combine probabilistic solution construction by a colony of ants with local search [5,9,7]. Also, Ant-Clique performances may be enhanced by introducing a local heuristic to guide ants. Indeed, as pointed out in Section 2, we already have experimented a first heuristic, borrowed from [1], that has not given any improvement. However, we shall further investigate other heuristics.

References

1. R. Battiti and M. Protasi. Reactive local search for the maximum clique problem. *Algorithmica*, 29(4):610–637, 2001.
2. I. Bomze, M. Budinich, P. Pardalos, and M. Pelillo. The maximum clique problem. In D.-Z. Du and P. M. Pardalos, editors, *Handbook of Combinatorial Optimization*, volume 4. Kluwer Academic Publishers, Boston, MA, 1999.
3. M. Dorigo. *Optimization, Learning and Natural Algorithms* (in Italian). PhD thesis, Dipartimento di Elettronica, Politecnico di Milano, Italy, 1992.
4. M. Dorigo, G. Di Caro, and L. M. Gambardella. Ant algorithms for discrete optimization. *Artificial Life*, 5(2):137–172, 1999.
5. M. Dorigo and L.M. Gambardella. Ant colony system: A cooperative learning approach to the traveling salesman problem. *IEEE Transactions on Evolutionary Computation*, 1(1):53–66, 1997.
6. E. Marchiori. Genetic, iterated and multistart local search for the maximum clique problem. In *Applications of Evolutionary Computing, Proceedings of EvoWorkshops2002: EvoCOP, EvoIASP, EvoSTim*, volume 2279, pages 112–121. Springer-Verlag, 2002.
7. C. Solnon. Ants can solve constraint satisfaction problems. *IEEE Transactions on Evolutionary Computation*, 6(4):347–357, 2002.
8. C. Solnon. Boosting ACO with a preprocessing step. In *Applications of Evolutionary Computing, Proceedings of EvoWorkshops2002: EvoCOP, EvoIASP, EvoSTim*, volume 2279, pages 161–170. Springer-Verlag, 2002.
9. T. Stützle and H.H. Hoos. $\mathcal{MAX} - \mathcal{MIN}$ Ant System. *Journal of Future Generation Computer Systems*, 16:889–914, 2000.

A Study of Greedy, Local Search, and Ant Colony Optimization Approaches for Car Sequencing Problems

Jens Gottlieb[1], Markus Puchta[1], and Christine Solnon[2]

[1] SAP AG
Neurottstr. 16, 69190 Walldorf, Germany
{jens.gottlieb,markus.puchta}@sap.com,
[2] LIRIS, Nautibus, University Lyon I
43 Bd du 11 novembre, 69 622 Villeurbanne cedex, France
csolnon@bat710.univ-lyon1.fr

Abstract. This paper describes and compares several heuristic approaches for the car sequencing problem. We first study greedy heuristics, and show that dynamic ones clearly outperform their static counterparts. We then describe local search and ant colony optimization (ACO) approaches, that both integrate greedy heuristics, and experimentally compare them on benchmark instances. ACO yields the best solution quality for smaller time limits, and it is comparable to local search for larger limits. Our best algorithms proved one instance being feasible, for which it was formerly unknown whether it is satisfiable or not.

1 Introduction

The car sequencing problem involves scheduling cars along an assembly line, in order to install options (e.g. sun-roof or air-conditioning) on them. Each option is installed by a different station, designed to handle at most a certain percentage of the cars passing along the assembly line, and the cars requiring this option must be spaced such that the capacity of the station is never exceeded.

This problem is NP-complete [5]. It has been formulated as a constraint satisfaction problem (CSP), and is a classical benchmark for constraint solvers [3, 6,13]. Most of these CSP solvers use a complete tree-search approach to explore the search space in a systematic way, until either a solution is found, or the problem is proven to have no solution. In order to reduce the search space, this approach is combined with filtering techniques that narrow the variables' domains. In particular, a dedicated filtering algorithm has been proposed for handling capacity constraints of the car sequencing problem [9]. This filtering algorithm is very effective to solve some hardly constrained feasible instances, or to prove infeasibility of some over-constrained instances. However, on some other instances, it cannot reduce domains enough to make complete search tractable.

Hence, different incomplete approaches have been proposed, that leave out exhaustivity, trying to quickly find approximately optimal solutions in an opportunistic way, e.g., local search [1,7,8], genetic algorithms [14] or ant colony

S. Cagnoni et al. (Eds.): EvoWorkshops 2003, LNCS 2611, pp. 246–257, 2003.

optimization (ACO) approaches [11]. This paper describes and compares three incomplete approaches for the car sequencing problem.

Section 2 introduces the car sequencing problem, for which several greedy heuristics are described and examined in section 3. Section 4 presents a local search approach, adapted from [8], and section 5 describes an ACO approach that is an improved version of [11]. Empirical results and comparisons are given in section 6, followed by the conclusions in section 7.

2   The Car Sequencing Problem

2.1   Formalization

A car sequencing problem is defined by a tuple (C, O, p, q, r), where $C=\{c_1, .., c_n\}$ is the set of cars to be produced and $O=\{o_1, .., o_m\}$ is the set of different options. The two functions $p : O \rightarrow \mathbb{N}$ and $q : O \rightarrow \mathbb{N}$ define the capacity constraint associated with each option $o_i \in O$, i.e. for any sequence of $q(o_i)$ consecutive cars on the line, at most $p(o_i)$ of them may require o_i. The function $r : C \times O \rightarrow \{0, 1\}$ defines options requirements, i.e., for each car $c_i \in C$ and for each option $o_j \in O$, $r(c_i, o_j)$ returns 1 if o_j must be installed on c_i, and 0 otherwise. Note that two cars may require the same configuration of options. To evaluate the *difference* of two cars, we introduce the function $d : C \times C \rightarrow \mathbb{N}$ that returns the number of different options two cars require, i.e., $d(c_i, c_j) = \sum_{o_k \in O} |r(c_i, o_k) - r(c_j, o_k)|$.

Solving a car sequencing problem involves finding an arrangement of the cars in a sequence, defining the order in which they will pass along the assembly line, such that the capacity constraints are met. We shall use the following notations to denote and manipulate sequences:

- a *sequence*, noted $\pi = < c_{i_1}, c_{i_2}, \ldots, c_{i_k} >$, is a succession of cars;
- the *length* of a sequence π, noted $|\pi|$, is the number of cars that it contains;
- the *concatenation* of 2 sequences π_1 and π_2, noted $\pi_1 \cdot \pi_2$, is the sequence composed of the cars of π_1 followed by the cars of π_2;
- a sequence π_f is a *factor* of another sequence π, noted $\pi_f \subseteq \pi$, if there exists two (possibly empty) sequences π_1 and π_2 such that $\pi = \pi_1 \cdot \pi_f \cdot \pi_2$;
- the *number of cars that require an option* o_i within a sequence π (resp. within a set S of cars) is noted $r(\pi, o_i)$ (resp. $r(S, o_i)$) and is defined by $r(\pi, o_i) = \sum_{<c_l> \subseteq \pi} r(c_l, o_i)$ (resp. $r(S, o_i) = \sum_{c_l \in S} r(c_l, o_i)$);
- the *cost* of a sequence π is the number of violated capacity constraints, i.e.,

$$cost(\pi) = \sum_{o_i \in O} \sum_{\substack{\pi_k \subseteq \pi \text{ so that} \\ |\pi_k| = q(o_i)}} violation(\pi_k, o_i)$$

$$\text{where } violation(\pi_k, o_i) = \begin{cases} 0 & \text{if } r(\pi_k, o_i) \leq p(o_i); \\ 1 & \text{otherwise.} \end{cases}$$

We can now define the solution process of a car sequencing problem (C, O, p, q, r) as the search of a minimal cost sequence composed of the cars to be produced.

2.2 Utilization Rate

The difficulty of an instance depends on the number of cars to be produced and the number of different options configurations, but also on the utilization rate of the different options [10]. The utilization rate of an option o_i corresponds to the ratio of the number of cars requiring o_i with respect to the maximum number of cars in a sequence which could have o_i while satisfying its capacity constraint, i.e., $utilRate(o_i) = \frac{r(C,o_i) \cdot q(o_i)}{|C| \cdot p(o_i)}$. An utilization rate greater than 1 indicates that the demand is higher than the capacity, so that the capacity of the station will inevitably be exceeded; an utilization rate close to 0 indicates that the demand is very low with respect to the capacity of the station.

2.3 Test Suites

All considered instances are available in the benchmark library CSPLib [6]. The first test suite contains 70 problem instances, used in [1,7,11] and grouped into 7 sets of 10 instances per utilization rate 0.60, 0.65, 0.70, 0.75, 0.80, 0.85, and 0.90; e.g. the instances within the group 0.70 are named 70-01, ..., 70-10, and we refer to the group as 70-*. All these instances are feasible ones, and have 200 cars to sequence, 5 options, and from 17 to 30 different options configurations. We also consider a second test suite composed of 9 harder instances, some of which were used in [9]. These instances have 100 cars to sequence, 5 options, from 18 to 24 different configurations, and high utilization rates, equal to 1 for some options; some of them are feasible, whereas some others are over-constrained.

3   Greedy Algorithms

3.1   The Basic Idea and Different Heuristics

Given a car sequencing problem, one can build a sequence in a greedy way, by iteratively choosing the next car to sequence with respect to some given heuristic function to optimize. Obviously, one should choose, at each step, a car that introduces the smallest number of new constraint violations, i.e., given a partial sequence π, the next car is selected within those cars c_j that minimize

$$newViolations(\pi, c_j) = \sum_{o_i \in O} r(c_j, o_i) \cdot violation(lastCars(\pi \cdot <c_j>, q(o_i)), o_i)$$

where $lastCars(\pi', k)$ is the sequence composed of the k last cars of π' if $| \pi' | \geq k$, otherwise $lastCars(\pi', k) = \pi'$. The set of candidate cars that minimize this $newViolations$ function usually contains more than one car so that we may use another heuristic to break further ties. We consider different heuristics:

1. Random choice (Rand): choose randomly a car within the set of candidates.
2. Static Highest Utilization Rates (SHU): the idea, introduced in [10], is to choose first the cars that require the option with highest utilization rate,

and to break ties by choosing cars that require the option with second highest utilization rate, and so on. More formally, we choose the car that maximizes the heuristic function $\eta_{SHU}(c_j) = \sum_{o_i \in O} r(c_j, o_i) \cdot \text{weight}(o_i)$ where $\text{weight}(o_i) = 2^k$ if o_i is the option with the kth smallest utilization rate.

3. Dynamic Highest Utilization Rates (DHU): the heuristic function η_{SHU} is static in the sense that it does not depend on the partial sequence π under construction. We can define a similar function η_{DHU} that is dynamic, by updating utilization rates each time a car is added, i.e., the utilization rate considered to compute the weight of an option o_i is defined by:
   $$\text{dynUtilRate}(o_i, \pi) = \frac{[r(C, o_i) - r(\pi, o_i)] \cdot q(o_i)}{p(o_i) \cdot [|C| - |\pi|]}.$$

4. Static Sum of Utilization Rates (SSU): instead of ranking utilization rates, from the highest to the lowest one, we can simply consider the sum of the required options utilization rates. More formally, we choose the car that maximizes the heuristic function $\eta_{SSU}(c_j) = \sum_{o_i \in O} r(c_j, o_i) \cdot \text{utilRate}(o_i)$.

5. Dynamic Sum of utilization rates (DSU): we can also make the SSU heuristic dynamic by considering the sum of dynamic utilization rates, i.e.,
   $\eta_{DSU}(c_j, \pi) = \sum_{o_i \in O} r(cj, o_i) \cdot \text{dynUtilRate}(o_i, \pi)$.

6. Dynamic Even Distribution (DED): the four previous heuristics choose cars that require options with high utilization rates, whereas the idea of this heuristic is to favor an even distribution of the options. If the average number of cars that require an option is lower in the sequence under construction π than in C, the set of cars, then we favor these cars, and vice versa. More formally, we choose the car that maximizes the heuristic function

$$\eta_{DED}(c_j, \pi) = \sum_{o_i \in O} (r(c_j, o_i) = 0) \ \text{xor} \ \left(\frac{r(C, o_i)}{n} > \frac{r(\pi, o_i)}{|\pi|} \right)$$

This heuristic function is not defined for the first car, when $|\pi| = 0$. Therefore, the first car c_j is selected randomly among those cars requiring the maximum number of options.

3.2 Empirical Comparison of Greedy Heuristics

Table 1 displays results obtained on the benchmark instances described in section 2.3. Remark that standard deviations for SHU and DHU are null. Indeed, all runs on a same instance always construct the same sequence as, at each step, there is only one car that maximizes the η function. For SSU and DSU, standard deviations are rather small, and on some instances they are null, as sums of utilization rates often have different values. For DED, standard deviations are higher, as the η_{DED} function can only take $|O|$ different values and in most cases there is more than one car that maximize it.

When comparing the average costs of the sequences built with the different heuristics, we observe the dynamic heuristics DSU and DHU obtaining rather similar results, whereas DSU and DHU always outperform their static counterpart SSU and SHU. DED is better than DSU and DHU on instances with utilization rates lower than 0.90, whereas on the instances 90-* and the second test suite it is worse. On the one hand, we perceive the instances from suite 1 as easy since each of them could be solved by at least one heuristic (see table 2). On the other hand, all heuristics failed to find a feasible solution for each instance from suite 2. Therefore, suite 2 is considered as more difficult.

Table 1. Results for greedy approaches (average costs on 500 runs, standard deviations in brackets).

Suite	No.	Rand	SSU	DSU	SHU	DHU	DED
1	60-*	20.4 (5.1)	10.2 (0.5)	1.6 (0.0)	10.5 (0.0)	1.4 (0.0)	0.8 (0.9)
	65-*	17.0 (5.0)	13.9 (0.3)	2.3 (0.0)	16.4 (0.0)	2.1 (0.0)	0.8 (1.0)
	70-*	13.8 (4.9)	17.4 (0.4)	3.0 (0.4)	18.6 (0.0)	3.6 (0.0)	0.7 (0.9)
	75-*	12.0 (5.0)	18.2 (0.2)	5.6 (0.4)	18.5 (0.0)	6.7 (0.0)	0.9 (1.0)
	80-*	17.8 (6.3)	16.6 (1.1)	4.3 (0.6)	17.2 (0.0)	5.0 (0.0)	1.0 (1.1)
	85-*	29.4 (7.3)	11.5 (0.6)	4.2 (0.4)	12.4 (0.0)	3.3 (0.0)	1.8 (1.2)
	90-*	54.2 (8.3)	5.9 (0.6)	1.2 (0.3)	6.1 (0.0)	1.6 (0.0)	4.3 (1.7)
2	10-93	59.1 (6.5)	11.0 (0.0)	10.4 (1.6)	11.0 (0.0)	12.0 (0.0)	14.1 (2.5)
	16-81	43.3 (5.1)	9.7 (1.2)	7.7 (0.9)	17.0 (0.0)	11.0 (0.0)	9.5 (2.7)
	19-71	56.0 (6.7)	13.6 (1.7)	8.5 (1.1)	13.0 (0.0)	7.0 (0.0)	10.3 (2.4)
	21-90	49.1 (5.6)	8.0 (0.0)	5.5 (0.7)	9.0 (0.0)	7.0 (0.0)	9.1 (1.9)
	26-82	40.6 (5.9)	6.0 (0.0)	3.6 (0.9)	7.0 (0.0)	3.0 (0.0)	8.8 (2.2)
	36-92	50.4 (6.2)	5.0 (0.0)	8.5 (0.5)	8.0 (0.0)	7.0 (0.0)	12.7 (2.3)
	41-66	34.1 (5.8)	4.7 (1.4)	3.7 (0.7)	6.0 (0.0)	2.0 (0.0)	6.5 (2.0)
	4-72	46.6 (5.8)	4.7 (0.6)	5.0 (0.0)	4.0 (0.0)	2.0 (0.0)	13.3 (3.2)
	6-76	28.1 (4.8)	6.0 (0.0)	6.0 (0.0)	6.0 (0.0)	6.0 (0.0)	9.1 (1.4)

Table 2. Success rates for greedy approaches and suite 1 (500 runs per instance).

No.	Rand	SSU	DSU	SHU	DHU	DED
60-*	0.002	0.002	0.400	0.000	0.500	0.547
65-*	0.002	0.000	0.100	0.100	0.200	0.536
70-*	0.004	0.000	0.000	0.000	0.000	0.580
75-*	0.005	0.000	0.200	0.000	0.100	0.507
80-*	0.001	0.000	0.116	0.000	0.100	0.487
85-*	0.000	0.200	0.218	0.100	0.200	0.251
90-*	0.000	0.300	0.500	0.300	0.500	0.114

4   Local Search

We consider the local search (LS) procedure proposed in [8] for general car sequencing problems involving several different types of constraints, including the one that is the subject of this paper. First, the initial solution is generated randomly – like in [8] – or by some greedy heuristic described in section 3. This solution is then modified by LS, where each iteration consists of randomly selecting one move type out of six different types. Then, a randomly chosen move of the selected type is evaluated regarding the current solution candidate, and the move is accepted and applied, if it does not deteriorate the solution quality. Otherwise the move is rejected and another one is tried on the current solution. The search process is terminated if a feasible solution is found or the evaluation limit is reached.

Table 3 summarizes the move types in a formal way. Insert moves one car from its current position to another, and Swap exchanges two cars. A special case of Swap is SwapS, which exchanges two cars that are similar but not identical regarding the options they require. SwapT is a transposition, another special case of Swap exchanging two neighbouring cars. Lin2Opt inverts a factor, which is defined by $invert(<>) =<>$ and $invert(\pi \cdot < c_l >) =< c_l > \cdot invert(\pi)$. The last type, Random, randomly shuffles a factor, formalized by $shuffle(<>) =<>$ and $shuffle(\pi_1 \cdot < c_l > \cdot \pi_2) =< c_l > \cdot shuffle(\pi_1 \cdot \pi_2)$ with probability $\frac{1}{1+|\pi_1|+|\pi_2|}$.

Besides the specific restrictions listed in table 3, the length of a modified factor is bounded by $n/4$ for all move types, in order to allow fast evaluations. Note that in particular SwapS and SwapT can be evaluated quite quickly since only two neighbouring cars and two options are affected, respectively.

Table 3. Formal description of the move types used within LS.

Type	$\pi \to \pi'$	Restriction
Insert	$\pi_1 \cdot < c_l > \cdot \pi_2 \cdot \pi_3 \;\to\; \pi_1 \cdot \pi_2 \cdot < c_l > \cdot \pi_3$ (forward) $\pi_1 \cdot \pi_2 \cdot < c_l > \cdot \pi_3 \;\to\; \pi_1 \cdot < c_l > \cdot \pi_2 \cdot \pi_3$ (backward)	$1 \le |\pi_2|$
Swap	$\pi_1 \cdot < c_{l_1} > \cdot \pi_2 \cdot < c_{l_2} > \cdot \pi_3 \;\to\; \pi_1 \cdot < c_{l_2} > \cdot \pi_2 < c_{l_1} > \cdot \pi_3$	$1 \le d(c_{l_1}, c_{l_2})$
SwapS	$\pi_1 \cdot < c_{l_1} > \cdot \pi_2 \cdot < c_{l_2} > \cdot \pi_3 \to \pi_1 \cdot < c_{l_2} > \cdot \pi_2 < c_{l_1} > \cdot \pi_3$	$1 \le d(c_{l_1}, c_{l_2}) \le 2$
SwapT	$\pi_1 \cdot < c_{l_1} > \cdot < c_{l_2} > \cdot \pi_2 \;\to\; \pi_1 \cdot < c_{l_2} > \cdot < c_{l_1} > \cdot \pi_2$	$1 \le d(c_{l_1}, c_{l_2})$
Lin2Opt	$\pi_1 \cdot \pi_2 \cdot \pi_3 \;\to\; \pi_1 \cdot invert(\pi_2) \cdot \pi_3$	$2 \le |\pi_2|$
Random	$\pi_1 \cdot \pi_2 \cdot \pi_3 \;\to\; \pi_1 \cdot shuffle(\pi_2) \cdot \pi_3$	$2 \le |\pi_2|$

5   Ant Colony Optimization

The Ant Colony Optimization (ACO) metaheuristic is inspired by the collective behaviour of real ant colonies, and it has been used to solve many hard combinatorial optimization problems [4]. In particular, [11] describes an ACO algorithm for solving permutation CSPs, the goal of which is to find a permutation of n known values, to be assigned to n variables, under some constraints. This algorithm has been designed to solve any permutation CSP in a generic way, and it has been illustrated on different problems. We now describe an improved version of it that is more particularly dedicated to the car sequencing problem. This new algorithm mainly introduces three new features: first, it uses an elitist strategy, so that pheromone is used to break the tie between the "best" cars only; second, it integrates features borrowed from [12] in order to favor exploration; finally, it uses the heuristic functions introduced in section 3 to guide ants.

Our new algorithm follows the classical ACO scheme: first, pheromone trails are initialized; then, at each cycle every ant constructs a sequence, and pheromone trails are updated; the algorithm stops iterating either when an ant has found a solution, or when a maximum number of cycles has been performed.

Construction graph and pheromone trails initialization. To build sequences, ants communicate by laying pheromone on the edges of a complete directed graph which associates a vertex with each car $c_i \in C$. There is an extra vertex, denoted by $c_{nest} \notin C$, from which ants start constructing sequences (this extra vertex will be considered as a car that requires no option). The amount of pheromone on an edge (c_i, c_j) is noted $\tau(c_i, c_j)$ and represents the learnt desirability of sequencing c_j just after c_i.

As proposed in [12], and contrary to our previous ACO algorithm, we explicitly impose lower and upper bounds τ_{min} and τ_{max} on pheromone trails (with $0 < \tau_{min} < \tau_{max}$). The goal is to favor a larger exploration of the search space by preventing the relative differences between pheromone trails from becoming too extreme during processing. Also, pheromone trails are initialized to τ_{max}, thus achieving a higher exploration of the search space during the first cycles.

Construction of sequences by ants. The algorithmic scheme of the construction of a sequence π by an ant is sketched below:

$\pi \leftarrow < c_{nest} >$
while $| \pi | \leq | C |$ **do**
    $cand \leftarrow \{c_k \in C{-}\pi \mid \forall c_j \in C{-}\pi, ((d(c_k, c_j) = 0) \Rightarrow (k \leq j))$ and
                      $newViolations(\pi, c_k) \leq newViolations(\pi, c_j)\}$
    let c_i be the last car sequenced in π (i.e., $\pi = \pi' \cdot < c_i >$)
    choose $c_j \in cand$ with probability $p_{c_i c_j} = \frac{[\tau(c_i, c_j)]^\alpha [\eta(c_j, \pi)]^\beta}{\sum_{c_k \in cand} [\tau(c_i, c_k)]^\alpha [\eta(c_k, \pi)]^\beta}$
    $\pi \leftarrow \pi \cdot < c_j >$
end while

Remark that, at each iteration, the choice of a car is done within a restricted set of candidates *cand*, instead of the set of all non sequenced cars $C - \pi$: in order to break symmetries, we only consider cars that require different options configurations; in order to introduce an elitist strategy, we select among the best cars with respect to the number of new constraint violations.

The choice of a car c_j in the candidates' set is done with respect to a probability that both depends on a pheromone factor τ, and a heuristic factor η. This heuristic factor can be any of the heuristic functions described in section 3.

Updating pheromone trails. After every ant has constructed a sequence, pheromone trails are updated according to ACO. First, all pheromone trails are decreased in order to simulate evaporation, i.e., for each edge (c_i, c_j), the quantity of pheromone $\tau(c_i, c_j)$ is multiplied by an evaporation parameter ρ such that $0 \leq \rho \leq 1$. Then, the best ants of the cycle deposit pheromone, i.e., for each sequence π constructed during the cycle, if the cost of π is minimal for this cycle then, for each couple of consecutive cars $< c_j, c_k > \subseteq \pi$, we increment $\tau(c_j, c_k)$ with $1/cost(\pi)$.

6 Experimental Comparison of ACO and LS

6.1 Experimental Setup

We consider two ACO variants obtained by setting parameters to different values. The first variant, called the "Ant" variant, uses $\alpha=1$, $\rho=0.99$, $\tau_{min}=0.01$ and $\tau_{max}=4$. The second variant, referred to as "Iter", ignores pheromone by using $\alpha=0$, $\rho=1$, $\tau_{min}=\tau_{max}=1$. This resembles an iterated greedy algorithm that repeatedly produces solutions by a randomized greedy heuristic. For both variants, we set β to 6 and the number of ants to 15, and we examine combinations with the greedy heuristics DHU, DSU and Rand described in section 3.

Regarding LS, two variants are considered that differ in the algorithm producing the initial solution. The first variant uses a pure random sequence, whereas the second uses the greedy heuristic DED from section 3.

We use different evaluation limits for ACO and LS, because e.g. on instance 10-93 of suite 2, 15 000 solutions produced by ACO and 200 000 solutions generated by LS need roughly the same CPU time, 20 seconds on a 300 MHz PC.

6.2 Results for Test Suite 1

Table 4 reports results for the first test suite, limiting the number of cycles of Ant and Iter to 100, and the number of moves of LS to 20 000. Note that within 100 cycles, pheromone cannot influence ants, so that Iter and Ant obtain similar results. Actually, instances of this first suite are rather easy ones, so that they can be very quickly solved by all algorithms, provided they use a good heuristic (i.e., DSU, DHU, or DED).

Table 4. Results for Iter, Ant, and LS on the first test suite instances (average costs on 100 runs, standard deviations in brackets).

No.	IterDSU	IterDHU	IterRand	AntDSU	AntDHU	AntRand	LSRand	LSDED
60-*	0.0 (0.0)	0.0 (0.0)	5.4 (1.0)	0.0 (0.0)	0.0 (0.0)	5.2 (0.9)	0.0 (0.1)	0.0 (0.0)
65-*	0.0 (0.0)	0.0 (0.0)	3.0 (0.7)	0.0 (0.0)	0.0 (0.0)	3.0 (0.7)	0.1 (0.1)	0.0 (0.0)
70-*	0.0 (0.0)	0.0 (0.0)	1.4 (0.5)	0.0 (0.0)	0.0 (0.0)	1.4 (0.6)	0.0 (0.1)	0.0 (0.0)
75-*	0.0 (0.0)	0.0 (0.0)	0.6 (0.4)	0.0 (0.0)	0.0 (0.0)	0.6 (0.4)	0.1 (0.2)	0.0 (0.0)
80-*	0.0 (0.0)	0.0 (0.0)	2.0 (0.9)	0.0 (0.0)	0.0 (0.0)	2.0 (0.8)	0.2 (0.3)	0.0 (0.0)
85-*	0.0 (0.0)	0.0 (0.0)	8.3 (1.7)	0.0 (0.0)	0.0 (0.0)	8.2 (1.6)	0.4 (0.6)	0.0 (0.1)
90-*	0.0 (0.0)	0.0 (0.0)	28.1 (2.5)	0.0 (0.0)	0.0 (0.0)	28.1 (2.4)	2.5 (1.2)	0.6 (0.6)

When no heuristic is used (Rand variants), LS is still able to solve all instances, for almost all runs, whereas Iter and Ant have much more difficulties, particularly for those with highest utilization rates[1]. The effect of DED on LS is

---

[1] We also made runs for IterRand and AntRand with 1000 cycles. In this case, on 90-*, the average cost decreases to 23.9 for IterRand, and to 12.9 for AntRand. This shows that pheromone actually allows ants to improve the solution process, even though they do not reach LSRand performances.

rather small compared to the effect of DHU and DSU on Ant and Iter, because DED is applied only once, whereas the heuristic in Ant and Iter is used for each generated sequence. Nevertheless, DED is quite helpful because it performs very well on these instances, as already discussed in section 3.

Finally, note that the results for IterDSU and IterDHU are much better than those reported for DSU and DHU in table 1, because Iter produces a large set of solutions and selects the best one, while the greedy approach produces only one solution for each run. Furthermore, Iter diversifies the search by choosing cars w.r.t. probabilities, which is beneficial when lots of solutions are generated.

6.3 Results for Test Suite 2

Table 5 presents the results for the second set of instances, the more difficult ones. The number of cycles of Ant and Iter is limited to 1 000, and the number of moves of LS to 200 000. Comparing Iter and Ant variants reveals that pheromone has a positive impact since Ant variants yield a better solution quality. In particular for Rand, the worst greedy heuristic, the use of pheromone makes a big difference. Thus, the key success factor of ACO is the employed greedy heuristic, demonstrated by the clear superiority of AntDSU and AntDHU variants. In addition to the heuristics, pheromone helps to improve solution quality, too.

Regarding the different greedy heuristics, DSU and DHU are comparable. It is remarkable that IterDSU and IterDHU achieve very good results. Although they are slightly inferior to their Ant counterparts, the obtained solution quality is comparable to LS. Using DED for the initial solution in LS is better than using a pure random sequence, coinciding to the results for the easy instances. However, LS seems to be slightly inferior to AntDHU and AntDSU. We believe that LS suffers from local optima, because we did not use mechanisms to escape from them. In our previous study for car sequencing problems involving several different types of constraints [8], we observed that solution quality can be improved significantly when using threshold accepting to escape from local optima. Thus, we think the gap between the LS variant we consider here and the best Ant variants could be closed by using an acceptance criterion like threshold accepting. Further, we remark that we have used the local search implementation from

Table 5. Results for Iter, Ant, and LS on the second test suite instances (average costs on 100 runs, standard deviations in brackets).

No.	IterDSU	IterDHU	IterRand	AntDSU	AntDHU	AntRand	LSRand	LSDED
10–93	5.5 (0.6)	6.6 (0.7)	35.4 (1.7)	4.7 (0.5)	4.7 (0.5)	21.9 (1.7)	5.9 (1.1)	5.5 (0.9)
16–81	0.8 (0.4)	2.2 (0.5)	23.1 (1.1)	0.2 (0.4)	0.9 (0.4)	13.0 (1.2)	2.0 (0.8)	1.6 (0.8)
19–71	2.8 (0.4)	3.1 (0.4)	29.9 (1.7)	2.8 (0.4)	2.7 (0.4)	18.1 (1.8)	2.2 (0.4)	2.1 (0.3)
21–90	2.6 (0.5)	2.0 (0.0)	27.1 (1.5)	2.3 (0.5)	2.0 (0.1)	13.4 (1.5)	2.2 (0.4)	2.2 (0.4)
26–82	1.0 (0.2)	0.3 (0.5)	19.3 (1.3)	0.0 (0.0)	0.0 (0.0)	8.9 (1.1)	0.4 (0.5)	0.3 (0.5)
36–92	3.4 (0.6)	2.9 (0.4)	25.9 (1.7)	2.1 (0.3)	2.0 (0.0)	15.7 (1.2)	3.0 (0.5)	2.9 (0.6)
41–66	0.0 (0.0)	0.0 (0.0)	13.3 (1.3)	0.0 (0.0)	0.0 (0.0)	5.2 (0.9)	0.0 (0.2)	0.0 (0.1)
4–72	0.2 (0.4)	1.0 (0.4)	24.1 (1.6)	0.0 (0.0)	0.0 (0.0)	16.3 (1.4)	0.8 (0.6)	0.8 (0.6)
6–76	6.0 (0.0)	6.0 (0.0)	13.1 (0.9)	6.0 (0.0)	6.0 (0.0)	7.6 (0.6)	6.0 (0.0)	6.0 (0.0)

[8], which contains some unnecessary functionality for problems involving other constraint types. Thus, a specific implementation for the special car sequencing problem we consider here, is expected to be more efficient.

Three algorithms – IterDHU, AntDHU, and LSDED – are selected for further experiments regarding the impact of different CPU time limits. We successively limit Iter and Ant to 100, 500, 1 000, 2 500 and 10 000 cycles, which correspond to 20 000, 100 000, 200 000, 500 000, 2 000 000 moves for LS, respectively. The evolution of solutions' quality is shown in figure 1 for four representative instances. For small limits, LS is inferior to IterDHU and AntDHU. However, LS and AntDHU achieve quite similar results for the highest limit. Interestingly, IterDHU is less effective for these high limits. Thus, the pure sampling of iterated greedy is inferior to LS and the pheromone-driven AntDHU, supposed the latter variants are given enough time.

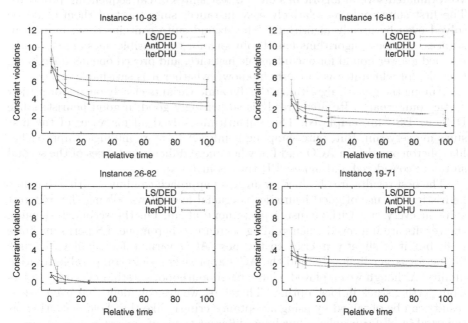

Fig. 1. Results on four representative benchmark instances of the second suite.

During all the runs, we found some new best solutions for instances of the second test suite. Table 6 compares previous results with the outcome of our experiments for the highest CPU time limit. The instances being reported as satisfiable have been solved by AntDHU and LSDED. Further, we found a feasible solution for the instance 26–82, for which it was formerly unknown whether it is satisfiable or not. The lower bound 2 on the number of constraint violations proved by Gent [5] for 19–71 is reached by all three algorithms.

Table 6. Comparison with results reported in [5,6,9]: The best solution is the best value we found in all our experiments, and the success rate (SR) measures for each algorithm the ratio of runs that actually found the best value.

Instance	10-93	16-81	19-71	21-90	26-82	36-92	41-66	4-72	6-76
Satisfiable?	no [9]	yes [6]	no [5,9]	?	?	?	yes [6]	yes [9]	no [9]
Best solution	3	0	2	2	0	2	0	0	6
SR for IterDHU	0.00	0.00	0.43	1.00	1.00	0.70	1.00	0.65	1.00
SR for AntDHU	0.32	0.99	0.99	1.00	1.00	1.00	1.00	1.00	1.00
SR for LSDED	0.85	0.68	1.00	1.00	1.00	1.00	1.00	1.00	1.00

7   Conclusion

We evaluated several algorithms on two test suites of car sequencing problems. The first suite contains relatively easy instances since most of them could be solved even by greedy algorithms. The second suite contains some harder instances. Our best algorithms found solutions for the feasible ones among them, reached a lower bound for one infeasible instance, and proved one instance being feasible, for which it was formerly unknown whether it is satisfiable or not.

Among the greedy algorithms, the dynamic variants clearly outperform their static counterparts. Iterated greedy is surprisingly good, if good heuristics like DHU or DSU are employed. These algorithms solved all instances of the first suite in every run. But, pure sampling is inferior to more intelligent approaches like pheromone-driven ACO and LS, when more difficult instances of the second suite are considered and more CPU time is invested.

The best results are obtained by an ACO approach combined with a dynamic heuristic. The use of good heuristics is crucial and allows solving the first test suite in every run without using pheromone at all. On the second test suite, the results are improved when guiding search by pheromone. LS performs quite well, but it is slightly inferior to the best ACO variant for small CPU time limits; for larger limits, LS and the ACO approaches yield comparable solution quality. Although we employed a large neighbourhood, we still believe LS suffers from getting stuck in local optima. Therefore, we are confident that even better results can be obtained by using acceptance criteria like threshold accepting, as reported in [8] for problems involving different types of constraints.

Although the new results are significantly better than those obtained by a previous ACO algorithm [11], there are other algorithms from literature that deserve a comparison with our algorithms. In particular, we should compare our results with other local search approaches, which are using the Swap neighbourhood, repair heuristics and adaptive mechanisms to escape from local optima [1, 2,7], and with genetic local search proposed in [14]. The difficulty is, however, that different machines, evaluation limits, and benchmarks have been used.

Some other issues are also open and should be the subject of further research. More specific, we want to check the effects of threshold accepting or restarting mechanisms on LS on the benchmarks used in this study. Further, it may be beneficial to integrate local search into the best ACO variant, by applying it to solutions constructed by ants. Since ACO turned out to be very effective, it is challenging to extend it such that it can cope with the real-world problems considered in [8].

References

1. A. Davenport and E. P. K. Tsang. Solving constraint satisfaction sequencing problems by iterative repair. In *Proceedings of the First International Conference on the Practical Applications of Constraint Technologies and Logic Programming*, 345–357, 1999
2. A. Davenport, E. Tsang, K. Zhu and C. Wang. GENET: a connectionist architecture for solving constraint satisfaction problems by iterative improvement. In *Proceedings of AAAI'94*, 325–330, 1994
3. M. Dincbas, H. Simonis and P. van Hentenryck. Solving the car-sequencing problem in constraint logic programming. In *Proceedings of ECAI-88*, 290–295, 1988
4. M. Dorigo and G. Di Caro. The Ant Colony Optimization Meta-Heuristic. In D. Corne, M. Dorigo, and F. Glover (eds.), *New Ideas in Optimization*, 11–32, McGraw Hill, UK, 1999
5. I.P. Gent. Two Results on Car-sequencing Problems. Technical report APES. 1998
6. I.P. Gent and T. Walsh. CSPLib: a benchmark library for constraints. Technical report APES-09-1999, available from http://4c.ucc.ie/~tw/csplib, a shorter version appeared in CP99, 1999
7. J.H.M. Lee, H.F. Leung and H.W. Won. Performance of a Comprehensive and Efficient Constraint Library using Local Search. In *Proceedings of 11th Australian Joint Conference on Artificial Intelligence*, 191–202, LNAI 1502, Springer, 1998
8. M. Puchta and J. Gottlieb. Solving Car Sequencing Problems by Local Optimization. In *Applications of Evolutionary Computing*, 132–142, LNCS 2279, Springer, 2002
9. J.-C. Regin and J.-F. Puget. A Filtering Algorithm for Global Sequencing Constraints. In *Principles and Practice of Constraint Programming*, 32–46, LNCS 1330, Springer, 1997
10. B. Smith. Succeed-first or fail-first: A case study in variable and value ordering heuristics. In *Third Conference on the Practical Applications of Constraint Technology*, 321–330, 1996
11. C. Solnon. Solving Permutation Constraint Satisfaction Problems with Artificial Ants. In *Proceedings of ECAI-2000*, 118–122, IOS Press, 2000
12. T. Stützle and H.H. Hoos. MAX-MIN Ant System. *Journal of Future Generation Computer Systems*, Volume 16, 889–914, 2000
13. E. Tsang. *Foundations of Constraint Satisfaction*. Academic Press, 1993
14. T. Warwick and E. Tsang. Tackling car sequencing problems using a genetic algorithm. *Evolutionary Computation*, Volume 3, Number 3, 267–298, 1995

Evolutionary Computing for the Satisfiability Problem

Jin-Kao Hao, Frédéric Lardeux, and Frédéric Saubion

LERIA, Université d'Angers
2 Bd Lavoisier, F-49045 Angers Cedex 01
{Jin-Kao.Hao,Frederic.Lardeux,Frederic.Saubion}@univ-angers.fr

Abstract. This paper presents GASAT, a hybrid evolutionary algorithm for the satisfiability problem (SAT). A specific crossover operator generates new solutions, that are improved by a tabu search procedure. The performance of GASAT is assessed using a set of well-known benchmarks. Comparisons with state-of-the-art SAT algorithms show that GASAT gives very competitive results. These experiments also allow us to introduce a new SAT benchmark from a coloring problem.

1 Introduction

The satisfiability problem (SAT) [9] consists in finding a truth assignment that satisfies a well-formed Boolean expression E. SAT is one of the six basic core NP-complete problems and has many applications such as VLSI test and verification, consistency maintenance, fault diagnosis, planning ... SAT is originally stated as a decision problem but one may take an interest in other related SAT problems:

- model-finding: to find satisfying truth assignments
- MAX-SAT: to find an assignment which satisfies the maximum number of clauses
- model-counting: to find the number of all the satisfying truth assignments.

During the last decade, several improved solution algorithms have been developed and important progress has been achieved. These algorithms have enlarged considerably our capacity of solving large SAT instances. Recent international challenges organized these years [16,24] continue to boost the worldwide research on SAT.

These algorithms can be divided into two main classes. A complete algorithm is designed to solve the initial decision problem while an incomplete algorithm aims at finding satisfying assignments (model-finding). The most powerful complete algorithms are based on the Davis-Putnam-Loveland procedure [3]. They differ essentially by the underlying heuristic used for the branching rule [5,18,26]. Specific techniques such as symmetry-breaking, backbone detecting or equivalence elimination are also used to reinforce these algorithms [1,17,6].

Existing incomplete algorithms for SAT are mainly based on local search [25,15,20] and evolutionary algorithms [4,13,8,7,11]. The very simple hill-climber

S. Cagnoni et al. (Eds.): EvoWorkshops 2003, LNCS 2611, pp. 258–267, 2003.
© Springer-Verlag Berlin Heidelberg 2003

GSAT [23] and its powerful variant Walksat [22] are famous examples of incomplete algorithms. Though incomplete algorithms are little helpful for unsatisfiable instances, they represent the unique approach for finding models of very large instances. At this time, Unitwalk [14] appears as the best incomplete algorithm. A current trend for designing more powerful algorithms consists in combining in some way the best elements from different approaches, leading to hybrid algorithms [21].

In this paper, we present GASAT, a new hybrid algorithm embedding a tabu search procedure into the evolutionary framework. At a very high level, GASAT shares some similarities with the hybrid algorithm proposed in [8]. GASAT distinguishes itself by the introduction of original and specialized crossover operators, a powerful Tabu Search (TS) algorithm and the interaction between these operators.

In the following sections, we introduce the GASAT algorithm as well as its main components. The performance of GASAT is then assessed on a large range of well-known SAT instances and compared with some best known incomplete algorithms. In this section, we also propose a new benchmark based on a coloration problem. We show thus GASAT competes very favorably with these state-of-the-art algorithms. Future extensions and improvements are discussed in the last section.

2   Basic Components

In this section, we define some basic notions and notations related to evolutionary computation, local search and propositional logic.

2.1   The SAT Problem

An instance of the SAT problem is defined by a set of boolean variables (also called atoms) $\mathcal{X} = \{x_1, ..., x_n\}$ and a boolean formula $\phi \colon \{0,1\}^n \to \{0,1\}$. A literal is a variable (also called atom) or its negation. A clause is a disjunction of literals. The formula ϕ is in conjunctive normal form (CNF) if it is a conjunction of clauses. A (truth) assignment is a function $v \colon \mathcal{X} \to \{0,1\}$. The formula is said to be satisfiable if there exists an assignment satisfying ϕ and unsatisfiable otherwise. In this paper, ϕ is supposed to be in CNF.

2.2   Search Sapce, Fitness, and Selection

Our search method relies on the management of a population of configurations representing potential solutions. Therefore, we present now the basic materials for this population.

Representation. The most obvious way to represent an individual for a SAT instance with n variables is a string of n bits where each variable is associated to one bit. Other representation schemes are discussed in [11]. Therefore the search space is the set $S = \{0,1\}^n$ (i.e. all the possible strings of n bits) and an individual X obviously corresponds to an assignment. $X|i$ denotes the the truth value of the ith atom and $X[i \leftarrow \alpha]$ denotes an individual X where the ith atom has been set to the value α. Given an individual X and a clause c, we use $sat(X,c)$ to denote the fact that the assignment associated to X satisfies the clause c.

Fitness Function. Given a formula ϕ and an individual X, the fitness of X is defined to be the number of clauses which are not satisfied by X:

$$eval\colon S \to I\!N, eval(X) = card(\{c|\neg sat(X,c) \land c \in \phi\})$$

where $card$ denotes as usual the cardinality of a set. This fitness function will be used in the selection process and induce an order $>_{eval}$ on the population. The smallest value of this function equals 0 and an individual having this fitness value corresponds to a solution.

Selection Process. The selection operator is a function which takes as input a given population and extracts some individuals according to a selection criterion. These selected individuals are the chosen parents for the evolution process and will evolve by crossover operations. To insure an efficient search, it is necessary to keep some diversity in the population. Actually, if the selected parents are too similar, some region of the search space S will not be explored. The diversity of the selected population is achieved by introducing the notion of hamming distance ham [12] between strings of bits. This distance corresponds simply to the number of different bits between two strings. Therefore we define the function $select\colon 2^S \times I\!N \times I\!N \to 2^S$ such that $select(P,n,d)$ is the set of the n best X in P according to $eval$ and $\forall X, Y \in select(P,n,d), ham(X,Y) \geq d$. For the sake of simplicity, the parameter d will be automatically adjusted and will not appear in the selection function.

3   The Hybrid Genetic Algorithm: GASAT

3.1   The GASAT Algorithm

The final algorithm is obtained by combining a crossover stage with a local search improvement. Given an initial population, the first step consists in selecting its best elements (i.e. assignments) w.r.t. the function $eval$. Then, crossovers are performed on this selected population. Each child is individually improved using a tabu search. If the child is better than the assignments already in the population then it is added to the current population and the whole process goes on. The general algorithm is described in Fig. 1.

This algorithm can be adjusted by changing the selection function, the crossover operator or the local search method but also by modifying the insertion condition. A variant of the condition used in this algorithm could be to insert the child whatever its evaluation is w.r.t. the whole population. Such a condition would bring more diversity but could also disturb the general convergence of the search.

Data: a set of CNF clauses $\mathcal{F}$, $Maxflip,Maxtries$
Result: a satisfying truth assignment if found

Begin
CreatePopulation(P)
$tries \leftarrow 0$
While no $X \in P$ satisfies ϕ and $tries < Maxtries$
/* Selection */
    $P' \leftarrow Select(P,n)$
    Choose $X, Y \in P'$
/* Crossover */
    $Z \leftarrow crossover(X,Y)$
/* TS improvement */
    $Z \leftarrow Tabu(Z, Maxflip)$
/* Insertion condition of the child */
    If $\forall X \in P, Z >_{eval} X$ then replace the oldest X in P by Z
    $tries \leftarrow tries + 1$
EndWhile
If there exists $X \in P$ satisfying ϕ
    then return the corresponding assignment
    else return the best assignement found
End

Fig. 1. GASAT: general structure

3.2 Basic Stages

Crossover Operator. A crossover or recombination operator has to take into account as much as possible the semantics of the individuals in order to control the general search process and to obtain an efficient algorithm. In the SAT problem, the variables are the atoms and a constraint structure is induced by the clauses. Therefore, while the representation focuses on the atoms, an efficient crossover should take into account the whole constraint structure.

We first define a function $flip: \{0,1\} \rightarrow \{0,1\}$ such that $flip(x) = 1-x$. This function induced a neighborhood relation in S which will be used in the tabu search mechanism when the algorithm changes from a configuration to another one by flipping one of its bits. Then, we define a function $imp: S \times \mathbb{N} \rightarrow \mathbb{N}$ such that $imp(X|i) = card(\{c \mid sat(X[i \leftarrow flip(X|i)], c) \wedge \neg sat(X, c)\}) - card(\{c \mid \neg sat(X[i \leftarrow flip(X|i)], c) \wedge sat(X, c)\})$. This function computes the improvement obtained by the flip of the ith component of X and was previously introduced in GSAT and Walksat [23,22]. It corresponds to the gain of the solution according to the function $eval$ (i.e. the number of false clauses which become true by flipping the atom i minus the number of satisfied clauses which become false). Remark that if this number is negative then the number of false clauses would increase if the flip is performed. This function induces a natural order $>_{imp}$ on the atoms which is extended to the assignments ($X >_{imp} Y$ iff there exists a position i such that $\forall j, X|i >_{imp} Y|j$). This crossover operator produces a child Z from two parents X and Y.

Definition 1. Crossover

For each clause c such that $\neg sat(X, c) \wedge \neg sat(Y, c)$ and for all positions i such that the variable x_i appears in c, we compute $\sigma_i = imp(X|i) + imp(Y|i)$ and we set $Z|k = flip(X|k)$[1] where k is the position such that σ_k is maximum. For all the clauses c such that $sat(X, c) \wedge sat(Y, c)$ and such that $X|i = Y|i = 1$ (resp. $X|i = Y|i = 0$) if the atom x_i appears positively (resp. negatively) in c, we set $Z|i = 1$ (resp. $Z|i = 0$). The positions in Z which have received no value by the previous operations are randomly valued.

Example. The following simple example illustrates the crossover. The problem has five variables, $\{x_1, x_2, x_3, x_4, x_5\}$ and seven clauses.

$$(x_1 \vee x_3 \vee x_5) \wedge (\neg x_2 \vee x_3 \vee \neg x_5) \wedge (\neg x_1 \vee \neg x_2 \vee x_4) \wedge (x_1 \vee \neg x_5 \vee x_4) \wedge (x_2 \vee x_3 \vee x_4) \wedge (\neg x_3 \vee \neg x_4 \vee x_5) \wedge (\neg x_2 \vee x_3 \vee x_4)$$

Let X and Y be two individuals with two false clauses each:

	x_1	x_2	x_3	x_4	x_5
X	1	1	0	0	1
Y	0	1	0	0	1

1) The second and the last clause are false for X and Y. So we compute σ for all the x_i.
For the second clause : $\sigma_2 = 3$, $\sigma_3 = 4$, $\sigma_5 = 2$
For the last clause : $\sigma_2 = 3$, $\sigma_4 = 4$

	x_1	x_2	x_3	x_4	x_5
Z			1	1	

2) With the true clauses for X and Y, x_2 and x_5 can be valued.

	x_1	x_2	x_3	x_4	x_5
Z		1	1	1	1

3) x_1 is randomly valued.

	x_1	x_2	x_3	x_4	x_5
Z	0	1	1	1	1

The main idea of this operator is to keep the structure of the clauses satisfaction induced by both parents by repairing false clauses and maintaining true ones.

At this step, we have defined the evolutionary part of our algorithm which insures the general search with enough diversity to reach some promising regions of $\mathcal{S}$. At this time, a tabu search process will intensify the search process to fully exploit these regions by locally improving the configurations in order to get a solution if possible. Moreover, TS can also explore the search space alone as a diversification process thanks to the tabu list.

---

[1] One should remark that, if a clause is false for both parents, then all the variables appearing in this clause have necessarily the same value in both parents. This comes from the fact that a clause can be false only if all of its literals are false.

Tabu Search Procedure. TS is an advanced local search method using a memory to avoid local optima [10]. The principle is quite simple: it acts as a descent (at each iteration, it makes the best move), but once visited, a configuration is made tabu, that is, the algorithm is not allowed to visit it again for a given number of iterations. Since the memorization of the last configurations could be costly, a common variation consists in making tabu only the moves. Now, a move is done if it is the best one and if it is not tabu. Once executed, the move is included in the tabu list, which acts as a fixed size FIFO queue, and is removed from this list after λ iterations. To consider improvement of the best-to-date solution, the tabu constraint is disabled in the case of moves leading to a better solution than the best-to-date. TS has already been experimented for SAT problem [20].

In our context, it is clear that the moves are possible flips of the value of a given assignment and that the tabu list will be a list of flip index already performed. Here the initial configuration will be given as entry to the TS and not generated randomly.

In order to simplify the tuning of our general algorithm and to guarantee a stable behavior, we propose an automatic adjustment of the tabu list length λ based on some experimental results. Therefore, in the remaining of the paper, we will simplify our TS function as $TS(X, Maxflip)$.

4   Experimental Results

In this section, we evaluate the performance of GASAT on several classes of satisfiable and unsatisfiable benchmark instances and compare it with Walksat [22], one of the well-known incomplete algorithms for SAT, and with UnitWalk [14] (combination of local search and unit propagation), the best up-to-now incomplete randomized solver presented to the SAT2002 competition [24].

4.1   Experimental Conditions

Due to the incomplete and non-deterministic nature of GASAT, Walksat and UnitWalk, each algorithm runs 20 times on each benchmark. This number of runs depends on the time spent by each algorithm.

For the three solvers, we limit the execution time to two hours for each run. We impose a maximum of 10^7 flips for Walksat and GASAT (Walksat with maxtries=10, as suggested in the literature, and maxflip=10^6 and GASAT with 10^3 crossovers, 10^4 flips for each TS step and a tabu list fixed to be 40% of the length of the individuals). UnitWalk runs with the default parameters. Therefore, we can consider that we allow the same power to each of the algorithm in order to get a fair evaluation.

4.2 Benchmark Instances and Evaluation Criterions

Two classes of instances are used : Structured and Random instances.

- Structured instances: `aim-100-1_6-yes1-4`, `aim-100-2_0-yes1-3` (random instances with only one solution), `mat25.shuffled`, `mat26.shuffled` (multiplication of two $n \times n$ matrices using m products [19]), `color-15-4`, `color-22-5` (chessboard coloring instances) `g125.18`, `g250.29` (graph coloring instances) and `dp10u09.suffled`, `dp11u10.suffled` (instances generated from bounded model checking from the well known dining philosophers example).
- Random instances: `glassy-v399-s1069116088`, `glassy-v450-s325799114` (random instances presented to SAT2002 Competition), `f1000`, `f2000` (DIMACS random instances).

The chessboard coloring benchmark `color` is new. It corresponds to the SAT encoding of a problem studied in [2] and that we have generalized. The purpose is to color a chessboard with k colors such that the four corners of every rectangle included in the board are not of the same color.

Two criterions are used to evaluate GASAT and compare it with Walksat and UnitWalk. The first one is the success rate (%) which is the number of successful runs divided by the total number of runs. This criterion is the most important one since it corresponds to the search power of an algorithm. The second criterion is the average running time in second (sec.) for successful runs on a Sun Fire 880 (4 CPU UltraSPARC III 750 Mhz, 8 Go de RAM)[2].

Table 1. Structured instances (if no assignment is found then the best number of false clauses is written between parentheses)

Benchmarks				GASAT		Walksat		UnitWalk	
instances	var	cls	sat	%	sec.	%	sec.	%	sec.
aim-100-1_6-yes1-4	100	160	Y	10	84.53	*(1 clause)*		100	0.006
aim-100-2_0-yes1-3	100	200	Y	100	20.86	*(1 clause)*		100	0.019
mat25.shuffled	588	1968	N	*(3 clauses)*		*(3 clauses)*		*(8 clauses)*	
mat26.shuffled	744	2464	N	*(2 clauses)*		*(2 clauses)*		*(8 clauses)*	
color-15-4	900	45675	Y	100	479.248	*(7 clauses)*		*(16 clauses)*	
color-22-5	2420	272129	?	*(5 clauses)*		*(41 clauses)*		*(51 clauses)*	
g125.18	2250	70163	Y	100	378.660	*(2 clauses)*		*(19 clauses)*	
g250.29	7250	454622	?	*(8 clauses)*		*(34 clauses)*		*(57 clauses)*	
dp10u09.suffled	7611	20770	N	*(39 clauses)*		*(2 clauses)*		*(22 clauses)*	
dp11u10.suffled	9197	25271	N	*(56 clauses)*		*(3 clauses)*		*(20 clauses)*	

---

[2] The version of Walksat is v39 and UnitWalk's is 0.98. GASAT is implemented in C and uses some functions of Walksat.

Table 2. Random instances (if no assignment is found then the best number of false clauses is given between parentheses)

Benchmarks				GASAT		Walksat		UnitWalk	
instances	var	cls	sat	%	sec.	%	sec.	%	sec.
glassy-v399-s069116088	399	1862	Y	*(5 clauses)*		*(5 clauses)*		*(17 clauses)*	
glassy-v450-s325799114	450	2100	Y	*(8 clauses)*		*(9 clauses)*		*(22 clauses)*	
f1000	1000	4250	Y	100	227.649	100	9.634	100	1.091
f2000	2000	8500	Y	*(6 clauses)*		100	21.853	100	17.169

4.3 Comparative Results on Structured Instances

Tables 1 and 2 show respectively the results of GASAT, Walksat and UnitWalk on the chosen structured and random instances.

From these tables, one observes first no clear dominance of one algorithm over the other ones. However, the results show that GASAT performs globally better on the structured instances than random ones. This seems not surprising given that its crossover operator relies on structural information. One observes a particular good performance of GASAT for the four (large) coloring instances compared with Walksat and UnitWalk. At the same time, one observes for random instances that, although GASAT gives competitive results for the glassy instances, it performs less well on the f1000 and f2000.

4.4 Discussions

Note that we report the best number of false clauses when no assignment is found. We could have reported the average performance but the standard deviation is very small for all the solvers.

When the benchmarks are successfully solved, we mention the execution time. From a more abstract point of view, as already mentioned, we approximatively provide the same computation resource in terms of basic operations (such as flips) to each algorithm. Therefore the computation time allowed may be considered to be quite fair.

We do not mention comparisons with complete solvers since, due to the size of the benchmarks, we are more interested in the MAX-SAT problem which is not usually addressed by complete solvers. Moreover, while these solvers can be efficient on the smallest satisfiable structured instances, they do not provide interesting results on larger structured or random instances, especially when these instances are unsatisfiable.

We should mention that we have experimented separately the two main components of our algorithm (crossover and tabu search). TS and crossover alone are able to find solutions but their combination provide the best average results.

5   Conclusion

We have presented the GASAT algorithm, a genetic hybrid algorithm for the SAT problem (and MAX-SAT). This algorithm is built on a specific crossover operator which relies on structural information among clauses and a simple tabu search operator. By their global and local nature, the crossover and tabu search operators act interactively to ensure a good compromise between exploration and exploitation of the search space. Moreover, a selection mechanism based on the Hamming distance is also employed to preserve the diversity of the population.

GASAT was evaluated on both structured and random benchmark instances. Its performance was compared with two state-of-the-art algorithms (Walksat and UnitWalk). The experimentations show that GASAT gives globally very competitive results. In particular, it performs better than the two competing algorithms on the graph coloring (structured) instances. Meanwhile, it seems that GASAT performs less well on some random instances.

The algorithm reported in this paper is a preliminary version of GASAT. Studies are on the way to have a better understanding of its behavior with respect to different classes of problem instances. We are also working on other issues to improve further upon its performance; in particular, a diversification process for tabu search based on unit propagation and criterions for choosing the variables to be flipped in the crossover.

Acknowledgments. The work presented in this paper is partially supported by grants from the LIAMA laboratory and the PRA program. We would like to thank the anonymous referees for their helpful comments and remarks.

References

1. Belaid Benhamou and Lakhdar Sais. Theoretical study of symmetries in propositional calculus and applications. In *CADE'92*, pages 281–294, 1992.
2. May Beresin, Eugene Levine, and John Winn. A chessboard coloring problem. *The College Mathematics Journal*, 20(2):106–114, 1989.
3. Martin Davis, George Logemann, and Donald Loveland. A machine program for theorem-proving. *Communications of the ACM*, 5(7):394–397, Jul 1962.
4. Kenneth A. De Jong and William M. Spears. Using genetic algorithm to solve NP-complete problems. In *Proc. of the Third Int. Conf. on Genetic Algorithms*, pages 124–132, San Mateo, CA, 1989.
5. Olivier Dubois, Pascal André, Yacine Boufkhad, and Jacques Carlier. SAT versus UNSAT. In *Second DIMACS implementation challenge : cliques, coloring and satisfiability*, volume 26 of *DIMACS Series in Discrete Mathematics and Theoretical Computer Science*, pages 415–436, 1996.
6. Olivier Dubois and Gilles Dequen. A backbone-search heuristic for efficient solving of hard 3-SAT formulae. In Bernhard Nebel, editor, *Proc. of the IJCAI'01*, pages 248–253, San Francisco, CA, 2001.
7. Agoston E. Eiben, Jan K. van der Hauw, and Jano I. van Hemert. Graph coloring with adaptive evolutionary algorithms. *Journal of Heuristics*, 4(1):25–46, 1998.

8. Charles Fleurent and Jacques A. Ferland. Genetic and hybrid algorithms for graph coloring. *Annals of Operations Research*, 63:437–461, 1997.

9. Michael R. Garey and David S. Johnson. *Computers and Intractability , A Guide to the Theory of NP-Completeness*. W.H. Freeman & Company, San Francisco, 1978.

10. Fred Glover and Manuel Laguna. *Tabu Search*. Kluwer Academic Publishers, 1998.

11. Jens Gottlieb, Elena Marchiori, and Claudio Rossi. Evolutionary algorithms for the satisfiability problem. *Evolutionary Computation*, 10(1):35–50, 2002.

12. Richard W. Hamming. Error detecting and error correcting codes. *The Bell System Technical Journal*, 29(2):147–160, April 1950.

13. Jin-Kao Hao and Raphael Dorne. A new population-based method for satisfiability problems. In *Proc. of the 11th European Conf. on Artificial Intelligence*, pages 135–139, Amsterdam, 1994.

14. Edward A. Hirsch and Arist Kojevnikov. UnitWalk: A new SAT solver that uses local search guided by unit clause elimination. PDMI preprint 9/2001, Steklov Institute of Mathematics at St.Petersburg, 2001.

15. Wengi Huang and Renchao Jin. Solar, a quasi physical algorithm for sat. *Science in China (Series E)*, 2(27):179–186, 1997.

16. Henry Kautz and Bart Selman. Workshop on theory and applications of satisfiability testing (SAT2001). In *Electronic Notes in Discrete Mathematics*, volume 9, June 2001.

17. Chu Min Li. Integrating equivalency reasoning into davis-putnam procedure. In *Proc. of the AAAI'00*, pages 291–296, 2000.

18. Chu Min Li and Anbulagan. Heuristics based on unit propagation for satisfiability problems. In *Proc. of the IJCAI'97*, pages 366–371, 1997.

19. Chu Min Li, Bernard Jurkowiak, and Paul W. Purdom. Integrating symmetry breaking into a dll procedure. In *Fifth International Symposium on the Theory and Applications of Satisfiability Testing (SAT2002)*, pages 149–155, 2002.

20. Bertrand Mazure, Lakhdar Saïs, and Éric Grégoire. Tabu search for SAT. In *Proc. of the 14th National Conference on Artificial Intelligence and 9th Innovative Applications of Artificial Intelligence Conference (AAAI-97/IAAI-97)*, pages 281–285, Providence, Rhode Island, 1997.

21. Matthew W. Moskewicz, Conor F. Madigan, Ying Zhao, Lintao Zhang, and Sharad Malik. Chaff: Engineering an efficient SAT solver. In *Proc. of the 38th Design Automation Conference (DAC'01)*, Jun 2001.

22. Bart Selman, Henry A. Kautz, and Bram Cohen. Noise strategies for improving local search. In *Proc. of the AAAI, Vol. 1*, pages 337–343, 1994.

23. Bart Selman, Hector J. Levesque, and David G. Mitchell. A new method for solving hard satisfiability problems. In *Proc. of the AAAI'92*, pages 440–446, San Jose, CA, 1992.

24. Laurent Simon, Daniel Le Berre, and Edward A. Hirsch. The sat2002 competition. Technical report, Fifth International Symposium on the Theory and Applications of Satisfiability Testing, May 2002.

25. William M. Spears. Simulated annealing for hard satisfiability problems. In *Second DIMACS implementation challenge : cliques, coloring and satisfiability*, volume 26 of *DIMACS Series in Discrete Mathematics and Theoretical Computer Science*, pages 533–558, 1996.

26. Hantao Zhang. SATO: An efficient propositional prover. In *Proc. of the 14th International Conference on Automated deduction*, volume 1249 of *LNAI*, pages 272–275, Berlin, 1997.

Guiding Single-Objective Optimization Using Multi-objective Methods

Mikkel T. Jensen

EVALife, Department of Computer Science,
University of Aarhus, Ny Munkegade bldg. 540,
DK-8000 Aarhus C, Denmark.
mjensen@daimi.au.dk,
http://www.daimi.au.dk/~mjensen/

Abstract. This paper investigates the possibility of using multi-objective methods to guide the search when solving single-objective optimization problems with genetic algorithms. Using the job shop scheduling problem as an example, experiments demonstrate that by using *helper-objectives* (additional objectives guiding the search), the average performance of a standard GA can be significantly improved. The helper-objectives guide the search towards solutions containing good building blocks and helps the algorithm avoid local optima. The experiments reveal that the approach only works if the number of helper-objectives used simultaneously is low. However, a high number of helper-objectives can be used in the same run by changing the helper-objectives dynamically.

1 Introduction

Over recent years, there has been intense research activity on evolutionary algorithms for multi-objective optimization. A number of different algorithms such as NSGA-II [5], PESA-II [3] and SPEA [16] have been published. Common to these algorithms is that they work by identifying an approximation to the *Pareto optimal set* also known as the *non-dominated set* of solutions. These algorithms are capable of minimizing several different objectives at the same time and identify a set of solutions representing non-dominated trade-offs between conflicting objectives.

Recent research [2,10] indicates that methods from Pareto-based multi-objective optimization may be helpful when solving single-objective optimization problems. Important problems in single-objective optimization include: (*i*) avoiding local optima, (*ii*) keeping diversity at a reasonable level, and (*iii*) making the algorithm identify good building blocks that can later be assembled by crossover. The purpose of the present paper is to further investigate whether multi-objective methods can be used to overcome these difficulties in single-objective optimization. Traditional single-objective optimization methods focus on the one objective exclusively (termed the *primary objective* in the following), but in this paper the notion of *helper-objectives* will be introduced. The idea is that the simultaneous optimization of the primary objective and a number of

S. Cagnoni et al. (Eds.): EvoWorkshops 2003, LNCS 2611, pp. 268–279, 2003.

helper objectives may achieve a better result than if the algorithm focused on the primary objective alone. This can happen if the helper-objective is chosen in such a way that it helps maintain diversity in the population, guides the search away from local optima, or helps in the creation of good building blocks. Most multi-objective algorithms were designed to create diverse non-dominated fronts, so diversity should be increased by most helper-objectives conflicting with the primary objective.

The problem studied in this paper is the job shop scheduling problem using a total flowtime primary objective. A number of helper-objectives and a multi-objective algorithm for minimizing the primary objective will be developed, and the results of this algorithm will be compared to the results of a traditional genetic algorithm working only with the primary objective. On average the total flowtime produced by the multi-objective algorithm is significantly lower than that produced by the traditional algorithm. Furthermore, the computational overhead of using the multi-objective algorithm instead of the standard algorithm is quite small.

The outline of the paper is as follows. The next section discusses related work. The section after that introduces the problem studied in the experiments; the job shop scheduling problem. Section 4 presents the idea of helper-objectives. Section 5 describes algorithms used in the experiments. The experiments are presented and analyzed in section 6. The results and future work are discussed in section 7.

2 Related Work

The work presented in this paper was done independently, but it has come to my attention that the use of Pareto-based based multi-objective methods to achieve better results in single-objective optimization has previously been investigated in the literature. Knowles et al. [10] proposed *multi-objectivization*, a concept closely related to the helper-objectives proposed in this paper. The idea in multi-objectivization is to decompose the optimization problem into subcomponents by considering multiple objectives. The authors study mutation based evolutionary algorithms, and argue that decomposition into several objectives will remove some of the local optima in the search-space, since a solution has to be trapped in all objectives in order to be truly "stuck". The problems investigated are the hierarchical-if-and-only-if function (*HIFF*) and the Travelling Salesperson Problem. For both problems an objective function of the form $A+B$ is decomposed into two separate objectives A and B. For both problems, algorithms based on multi-objectivization outperform comparable algorithms based on single objectives.

The main difference between the study of Knowles et al. [10] and the present one is in the details of how multi-objective methods are used to solve the single-objective problem. Knowles et al. disregard the primary objective in their experiments and use only what is referred to in this paper as helper-objectives[1],

---

[1] They remark that keeping the primary objective is also an option

use only two objectives, and use only static objectives, where the objectives used in this paper are changed dynamically. Besides, the problems studied are completely different.

The idea of decomposing a single-objective problem into several objectives was investigated as early as 1993 by Louis and Rawlins [12]. Their work demonstrated that a deceptive problem could be solved more easily using Pareto-based selection than standard selection in a GA.

An other application of multi-objective methods in single-objective optimisation include the reduction of bloat (uncontrolled growth of solution size) in genetic programming [2,4]. Bleuler et al. [2] and de Jong et al. [4] independently study the effect of using the size of the genetic program as an extra objective in genetic programming. Both studies find that the additional objective both reduces the average size of the solutions found and decreases the processing time needed to find an optimal solution.

Pareto-based selection mechanisms for selection have been investigated in coevolutionary settings by Noble and Watson [13] and Watson and Pollack [15]. Watson and Pollack [15] do experiments on symbiotic combination of partial solutions. Symbiotic combination can be compared to traditional crossover, but the idea is to combine two partial solutions into a new partial (but more complete) solution having all characteristics from both parents. In order to asses the worth of the partial solutions, they have to be turned into full solutions. This is done by constructing templates from other individuals in the population, but since many different templates can be constructed in this way, every individual can be assigned a high number of objectives. Instead of combining these objectives into one "average" objective, the authors use selection based on many objectives and Pareto-dominance. They report that their algorithm performs better than two standard GAs and a previous version of their own algorithm on the shuffled hierarchical-if-and-only-if problem (SHIFF).

A related study is found in [13], in which strategies for poker-playing are evolved in a coevolutionary system. During fitness evaluation the strategies compete against each other, and instead of assigning each strategy a single fitness, selection is based on Pareto-dominance and a concept termed *near-domination*. Noble and Watson find that the Pareto-based algorithm on average evolves better strategies than an equivalent algorithm with a more standard fitness assignment.

In [14] Scharnow et al. present theoretical arguments that the single source shortest path problem can be solved more efficiently by an EA when seen as multi-objective than as a single-objective problem. The authors show that the problem can be solved in expected time $O(n^3)$ when cast as multi-objective, and argue that in some cases seeing the problem as single-objective will be equivalent to a needle-in-a-haystack problem, giving a much higher processing time. Scharnow et al. assume that non-existing edges in the graph and infeasible solutions will be represented as infinite weights. However, non-existing edges and infeasible solutions can also be represented using large (finite) weights. In this case the expected running time of the single-objective algorithm will be identical to that of the multi-objective algorithm.

3   Job Shop Scheduling

The job shop scheduling problem (*JSSP*) [8] is probably the most intensely studied scheduling problem in literature. It is NP-hard. A job shop problem of size $n \times m$ consists of n jobs $J = \{J_i\}$ and m machines $M = \{M_j\}$. For each job J_i a sequence of m operations $(o_{i1}, o_{i2}, \cdots, o_{im})$ describing the processing order of the operations of J_i is given. Each operation o_{ij} is to be processed on a certain machine and has a processing time τ_{ij}. The scheduler has to decide when to process each of the operations, satisfying the constraints that no machine can process more than one operation at a time, and no job can have more than one operation processed at at time. Furthermore, there can be no preemption; once an operation has started processing, it must run until it is completed.

Several performance measures exist for job shop problems. The performance measure used in this paper will be the *total flowtime*

$$F_\Sigma = \sum_{i=1}^{n} F_i, \tag{1}$$

where F_i is the flowtime of job i, the time elapsed from the beginning of processing until the last operation of J_i has completed processing.

4   Helper-Objectives

When selecting a helper-objective for our problem, we should keep in mind that the helper objective *must be in conflict with the primary objective*, at least for some parts of the search-space. If this is not the case, optimizing the primary objective and the helper-objective simultaneously will be equivalent to simply optimizing the primary objective. On the other hand, helper-objectives should reflect some aspect of the problem that we expect could be helpful for the search.

For the example of the total flowtime job shop, it makes sense to use the flowtimes of the individual jobs F_i as helper-objectives. Minimizing the flowtime of a particular job will often be in conflict with minimizing the sum of flowtimes. However, since a decrease in the total flowtime can only happen if the flowtime is decreased for at least one of the jobs, it may help the minimization of the total flowtime if the search focuses on minimizing flowtime for the individual jobs as well. If a schedule with a low flowtime for job J_i but a high total flowtime appears in the population, the crossover operator may be able to combine that schedule with a schedule with a low total flowtime to produce a schedule with an even lower total flowtime.

In a job shop of n jobs, there are n different helper-objectives of this kind. For the benchmarks used in this paper this gives between 10 and 30 potential helper-objectives, depending on the problem instance. Having that many helper-objectives is probably more harmful than beneficial, since most of the individuals in the population could be non-dominated very early in the program run, and the focus of the search will be moved away from the primary objective to an unacceptable degree.

A solution to this problem is to use only a subset of all the helper-objectives at any given time, and sometimes change the helper-objectives. Changing the helper-objectives during the run can easily be accomplished. When the helper-objectives are being changed during the program run (changing helper-objectives will be referred to as *dynamic* helper-objectives), we need to decide how to schedule the changes. Using h helper-objectives at the same time and having a total of t helper-objectives to be used in a run of G generations, each helper-objective can be used for

$$D = \left\lfloor \frac{hG}{t} \right\rfloor$$

generations, assuming all helper-objectives to be used for the same number of generations. Potentially, the helper-objectives could be changed for every generation, but since there will not be time for good building blocks to be formed during such a short period, the opposite approach was taken in the experiments: each helper-objective was used for one period of the maximal length possible. Every D generations the helper-objectives were changed to the next objectives in the *helper-objective sequence*. The sequence in which the helper-objectives are used may have an influence on the search, but since we have no way to know which order is the best one, a random ordering of the helper-objectives was used.

The experiments with the multi-objective algorithm were conducted using one, two, three, and n helper-objectives at the same time, where n is the number of jobs in the problem instance in question. In the following, an algorithm using x helper-objectives simultaneously will be called an *x-helper algorithm*.

5   The Algorithms

The multi-objective algorithm used as a starting point for the algorithm used in the experiments is the *Non-dominated Sorting GA version II*, NSGA-II, published by Deb et al. [5]. This algorithm has been demonstrated to be among the most efficient algorithms for multi-objective optimization on a number of benchmarks. It uses *non-dominated sorting* for fitness assignments.

When selecting a multi-objective algorithm for single-objective optimization, the running time of the algorithm is an important issue. There will be a computational overhead associated with using the helper-objectives, and it is important to minimize this overhead as much as possible. A traditional single-objective GA runs in time $O(GN)$, where G is the number of generations and N is the population size. The traditional implementation of the NSGA-II runs in time $O(GMN^2)$, where M is the number of objectives [5]. However, recent research [9] shows that the algorithm can be modified to run in time $O(GN \log^{M-1} N)$. The vast majority of multi-objective algorithms known today have running times no better than this.

A detailed description of the NSGA-II is outside the scope of this paper. The interested reader is referred to [5,9]. A very brief description of the algorithm follows: Fitness is assigned based on non-dominated sorting: All individuals not dominated by any other individuals are assigned front number 1. All individuals only dominated by individuals in front number 1 are assigned front number 2, etc. Selection is made in size two tournaments. If the two individuals are from different fronts, the individual with the lowest front number wins. If they are from the same front, the individual with the highest crowding distance wins; in this way higher fitness is assigned to individuals located on a sparsely populated part of the front. In every generation N new individuals are generated. These compete with their parents (also N individuals) for inclusion in the next generation.

The author does not expect the results of this paper to be highly dependent on the particular multi-objective algorithm used. Probably other efficient multi-objective algorithms such as SPEA [16] could have been used with similar results. Note however, that some other multi-objective algorithms run in time proportional to N^2, meaning that other choices of multi-objective algorithm may require longer processing times.

5.1   Multi-objective GA for the JSSP

The genetic representation used in the algorithm was the *permutation with repetition* representation widely used for the job shop problem, see e.g. [8]. In this representation, a schedule is represented by a sequence of job numbers e.g. $(1, 3, 1, 2 \ldots)$. If job J_i consists of k operations, then there will be k instances of i in the sequence. Decoding is done using the Giffler-Thompson schedule builder [7]. This decoder builds the schedule from left to right, adding one operation at a time. The sequence of the gene is interpreted as a list of priorities on the operations. The gene $(1, 3, 1, 2 \ldots)$ gives the highest possible priority to the first operation of job 1, followed by the first operation of job 3, the second operation of job 1, etc. The permutation with repetition representation has the advantage that it cannot represent infeasible solutions. All active schedules can be represented, which guarantees that the optimal schedule is in the search-space.

The genetic operators used in the algorithm are the GOX crossover operator and the PBM mutation operator, [8]. The crossoverrate was set to 1.0 and mutation was performed on offspring after crossover. These choices were made after experiments with four different ways of doing crossover: (*i*) GOX followed by PBM, (*ii*) PPX followed by PBM, (*iii*) GOX alone, (*iv*) PPX alone. For each of these possibilities, the algorithm was run 30 times using one dynamic helper-objective on the la26 instance. A population size of 100 and 200 generations were used. This was done for 100 settings of the crossover-rate ranging from 0 to 1. Thus a total of 12000 runs were made in this experiment. Since a crossover-rate of 1.0 doing crossover as GOX followed by PBM showed the best average performance, this parameter setting was used in the rest of the runs. These experiments only used one problem instance, but we assume that the parameter settings are usable on the other instances as well.

Following this experiment, a large number of different population sizes in the range $\{2 \ldots 200\}$ were tried, each run using up to 20000 fitness evaluations. Population sizes in the range 100-120 showed the best performance, so a population size of 100 and a generation limit of 200 were used in the rest of the experiments.

Table 1. The benchmarks used in the experiments. The instances prefixed by `ft` are from [6], while the instances prefixed by `la` are from [11]. In [11], 40 instances divided into eight groups were published, I have used the first two instances from each group.

size	instances	size	instances
10×5	la01,la02	15×10	la21,la22
15×5	la06,la07	20×10	la26,la27
20×5	la11,la12,ft20	30×10	la31,la32
10×10	la16,la17,ft10	15×15	la36,la37

5.2 Traditional GA for the JSSP

A traditional GA for solving the JSSP was created. It used the same representation and decoding as the multi-objective algorithm. The algorithm used an elite of one. As in the multi-objective algorithm, selection was done in two-tournaments. Since the algorithm is fundamentally different from the multi-objective algorithm, it was deemed inadequate to simply use the parameter settings found for that algorithm. Instead, the same effort was put into tuning the parameters of the traditional GA. The four combinations of genetic operators tested for the multi-objective algorithm were tested for the traditional GA, using the exact same method and the same number of program runs. For this algorithm, it turned out that a crossover-rate of 0.75 showed superior performance if the GOX operator was applied alone in crossover. Individuals not created by crossover were generated with the PBM mutation operator.

As for the multi-objective algorithm, a large number of population sizes were tested, and since none of them were superior to a population size of 100, this population size was used in the experiments.

6 Experiments

The multi-objective algorithm was run with 1, 2 and 3 dynamic helper-objectives, and with static helper-objectives using all n helper-objectives at the same time. Experiments were performed on the 18 problems listed in table 1. Each experiment consisted of 500 runs, from which the average performance was calculated. Similar experiments were conducted for the traditional GA, the results are shown in table 2. The table holds one column for every algorithm. The first column reports the average total flowtime F_{Σ} of the traditional algorithm, while the remaining columns report the performance for the multi-objective algorithm. There is a row for every problem instance. For every problem instance, the best performance (lowest number) has been printed bold. The last row holds the total flowtime for each algorithm averaged over all the problems.

The experiments reveal that the multi-objective algorithm performs better than the traditional algorithm when using one dynamic helper-objective. For this parameter setting, the multi-objective algorithm performs better than the traditional algorithm for every single problem instance. On average, the total

Table 2. Average flowtimes found in the experiments.

	traditional	dyn 1	dyn 2	dyn 3	stat n
la01	5029.8	4953.2	**4941.7**	4996.9	5130.3
la02	4653.6	**4600.0**	4604.8	4673.1	4814.9
la06	9365.4	**9205.8**	9316.7	9571.0	9844.1
la07	8850.4	**8663.9**	8804.9	9028.3	9390.7
la11	15843.0	**15562.3**	15900.6	16358.6	16926.4
la12	13437.9	**13267.0**	13512.3	13934.4	14553.5
la16	7851.1	7758.4	**7714.0**	7770.7	7884.5
la17	6828.4	6771.4	**6762.3**	6811.1	6921.5
la21	13784.7	**13728.2**	13837.3	14163.1	14492.2
la22	12816.0	**12750.0**	12814.7	13095.7	13557.1
la26	21779.6	**21643.8**	21946.0	22418.4	22847.7
la27	22425.5	**22309.1**	22597.8	23163.5	23490.7
la31	42344.7	**42131.7**	42720.2	43452.5	44170.0
la32	45683.2	**45542.0**	46238.9	47077.1	47524.2
la36	18064.3	18031.1	**17970.5**	18240.1	18638.7
la37	18941.1	**18939.6**	19022.7	19432.2	19888.2
ft10	8227.5	8077.8	**8016.0**	8064.5	8100.0
ft20	15739.0	**15465.5**	15835.1	16397.5	17488.6
average	16203.5	**16077.8**	16253.1	16591.6	16981.5

flowtime is reduced by 125.7. Assuming a Gaussian distribution of the observations, the performance difference between the these two algorithms is statistically significant on a 95% level for all the problems except la37. Considering the algorithms with more than one helper-objective, the bottom row of the table indicates that averaged over all the problems, these algorithms perform worse than the traditional GA. However, there are a number of problem instances for which the algorithms using two and three helper-objectives outperform the traditional GA. For five of the problem instances, the algorithm using two helper-objectives shows the best average performance.

The performance of the traditional and multi-objective algorithm using one and two dynamic helper-objectives is shown on figure 1. The plots show the average best total flowtime as a function of the generation number for the la01 and la06 problem instances. The errorbars indicate 95% confidence intervals. Comparing the algorithm using one dynamic helper-objective to the traditional algorithm, it is evident that both algorithms start off at the same level of performance, but after a few generations the multi-objective algorithm has a lower average total flowtime. For both problems, the difference between the two algorithms slowly gets larger as time progresses. This behaviour was observed for all of the problems except la37.

Considering the two-helper algorithm, it has a worse performance than both of the other algorithms in the first part of both diagrams. However, in the late stages of the run the two-helper algorithm decreases the flowtime faster than the other algorithms. For the la01 problem, the two-helper algorithm ends up

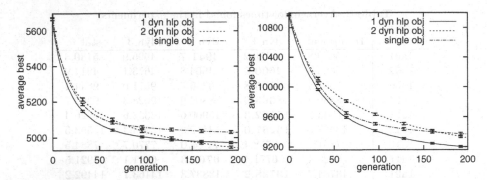

Fig. 1. Average best total flowtime as a function of the generation number for the la01 (left) and la06 (right) problems.

beating both the other algorithms, but for la06 it is unable to catch up with the one-helper algorithm. The plot for la01 is typical for the problems for which the algorithm using two helper-objectives has the best performance. The progression of this algorithm is slow during the early stages of the run, but in the later stages it overtakes the other algorithms.

The running time of the traditional and the one-helper algorithms were investigated. For the one-helper algorithm, the computational overhead over the traditional algorithm ranged from 2% for the largest problems to 13% for the smallest problems. This increase in computational cost seems a low price to pay for the performance improvements found. For the runs using more than one helper-objective, the slow $O(GMN^2)$ version of the NSGA-II was used, and the overhead ranged from 7% to 45%, but this overhead can be expected to drop substantially if a faster implementation of the NSGA-II is used.

6.1   Investigating the Size of the Non-dominated Fronts

The mediocre performance of the multi-objective algorithm for many helper-objectives may be caused by too large non-dominated fronts when the algorithm is running. If most of the population belongs to the first non-dominated front, most of the population will be assigned the same fitness (disregarding crowding), and the selection pressure will be very low. Theoretical investigations reveal that for random solutions the number of non-dominated solutions can be expected to grow with the number of objectives [1]. The average size of the first non-dominated front was investigated for all of the problems. Since the size of the front cannot be expected to be constant during the run, the dependency on the generation number was also investigated. Experiments were made for each combination of problem instance and multi-objective algorithm of table 2.

The average size of the first front is shown in figure 2 for the problems la01 and la06. In each diagram, the average front-size has been plotted for the algorithms using 1, 2 and 3 dynamic helper-objectives. The front-size for the algorithm using all n helper-objectives has not been plotted, since the average

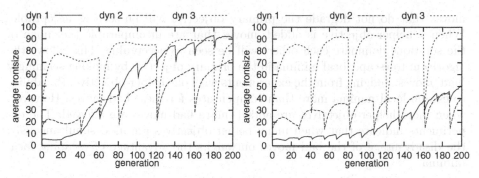

Fig. 2. Average size of the first non-dominated fronts as a function of generation number for the la01 (left) and la06 (right) problems.

front-size was always equal to the population size (100). This was the case for all of the problems, and accounts for the poor performance of this algorithm. Inspecting the plots of figure 2, all of the graphs are quite rugged. The ruggedness is caused by the changes of helper-objectives; every time the helper-objectives are changed, there is a steep decline in the average front-size. The front-size then gradually increases until the helper-objectives are changed again. Comparing the algorithms to each other, it is evident that the average front-size is always quite large for the algorithm using three helper-objectives. This was found to be the case for all of the problems, and the low selection pressure resulting from the large front-size explains the poor performance for this parameter setting.

Considering the algorithm using one helper-objective, in both cases the front-size at the beginning of the run is low. This is the case for all 18 problem instances. For the la06 problem, the front-size stays at a relatively low level during the entire run, while for la01 it increases and almost becomes equal to the population size at the end of the run. Compared to the two-helper algorithm, the front-size starts at a lower level, but it increases much more. For the la01 problem, the front-size of the one-helper algorithm becomes larger than that of the two-helper algorithm around generation 70.

Behaviour similar to this was found in all the problems for which the two-helper algorithm performed better than the one-helper algorithm. For all of these problems, the average front-size of the one-helper algorithm was larger than in the two-helper algorithm from generation 70 or earlier. For the other problems, the front-size of the one-helper algorithm stayed below that of the two-helper algorithm during most of the run. This can be taken as a strong indication that keeping the size of the first non-dominated front small is an important element in tuning this kind of algorithm.

Another reason for the mediocre performance of the algorithms using many helper-objectives could be that as the number of helper-objectives increases, the proportion of non-dominated solutions with a low primary objective decreases. One way to realize this is by considering moves in the search-space as perceived by a search-algorithm. When considering only the primary objective, moves that do not degrade the primary objective will be seen as acceptable. When a helper-objective is added, moves that are incomparable (i.e. none of the solutions dominate each other) will be seen as acceptable. This allows the algorithm to escape local minima in the primary objective by making seemingly 'bad' moves. Judging from the experiments with one helper-objective, the ability to escape local minima more than compensates for these bad moves. However, when more helper-objectives are added, more bad moves are allowed. The experiments indicate that when many helper-objectives are used simultaneously, the disadvantage of the bad moves outweighs the advantage of escaping local minima.

7   Conclusion

This paper has demonstrated that multi-objective methods can be used as a tool for improving performance in single-objective optimization. The notion of *helper-objectives* has been introduced, a helper-objective being an additional objective conflicting with the primary objective, but helpful for diversifying the population and forming good building blocks. Experiments have demonstrated that for the total flowtime job shop scheduling problem, the flow times of the individual jobs can be efficient helper-objectives. For 17 out of 18 problem instances tested, the new approach performed significantly better than a traditional approach.

We have found that using a small number of dynamic helper-objectives is the most promising approach, since using too many helper-objectives at the same time removes the selection pressure in the algorithm. Scheduling the use of dynamic helper-objectives is the subject of future research. From the experiments of this paper it is unclear how many helper-objectives should be used at the same time; the optimal choice is problem dependent. Another possibility for future research is modifying the algorithm to keep a high selection pressure when many helper-objectives are used. This could be accomplished by using a stronger crowding scheme or another method of selection.

Another direction of future research would be applying the methods of this paper to other kinds of problems. In order for the methods to work on a problem, reasonable helper-objectives must be found. Numerical and combinatorial problems in which the objective consists of a sum or product of many terms can probably be solved using the same approach used in this paper; a helper-objective can be defined for each term.

In preliminary experiments, the method was tested on a makespan job shop problem using the same helper-objectives, but the results were not encouraging, probably because minimizing flowtimes of individual jobs does not create good building blocks for that problem. Researching what constitutes a good helper-objective is yet another direction of future research.

Acknowledgement. The author wishes to thank the anonymous reviewers for pointing out relevant literature.

References

1. J. L. Bentley, H. T. Kung, M. Schkolnick, and C. D. Thompson. On the Average Number of Maxima in a Set of Vectors and Applications. *Journal of the ACM*, 25:536–543, 1978.
2. S. Bleuler, M. Brack, L. Thiele, and C. Zitzler. Multiobjective Genetic Programming: Reducing Bloat using SPEA2. In *Proceedings of CEC'2001*, pages 536–543, 2001.
3. D. Corne, N. Jerram, J. Knowles, and M. Oates. PESA-II: Region-based Selection in Evolutionary Multiobjective Optimization. In L. Spector et al., editors, *Proceedings of GECCO-2001: Genetic and Evolutionary Computation Conference*, pages 283–290. Morgan Kaufmann, 2001.
4. E. D. de Jong, R. A. Watson, and J. B. Pollack. Reducing Bloat and Promoting Diversity using Multi-Objective Methods. In L. Spector et al., editors, *Proceedings of GECCO'2001*, pages 11–18. Morgan Kaufmann, 2001.
5. K. Deb, A. Pratab, S. Agarwal, and T. Meyarivan. A Fast and Elitist Multiobjective Genetic Algorithm: NSGA-II. *IEEE Transactions on Evolutionary Computation*, 6(2):182–197, April 2002.
6. H. Fisher and G. L. Thompson. Probabilistic learning combinations of local job-shop scheduling rules. In J. F. Muth and G. L. Thompson, editors, *Industrial Scheduling*, pages 225–251. Prentice Hall, 1963.
7. B. Giffler and G. L. Thompson. Algorithms for solving production scheduling problems. *Operations Research*, 8:487–503, 1960.
8. M. T. Jensen. *Robust and Flexible Scheduling with Evolutionary Computation*. PhD thesis, Department of Computer Science, University of Aarhus, 2001.
9. M. T. Jensen. Reducing the Run-time Complexity of the NSGA-II. *In submission*, 2002. Currently available from http://www.daimi.au.dk/~mjensen/.
10. J. D. Knowles, R. A. Watson, and D. W. Corne. Reducing Local Optima in Single-Objective Problems by Multi-objectivization. In E. Zitzler et al., editors, *Proceedings of the First International Conference on Evolutionary Multi-criterion Optimization (EMO'01)*, pages 269–283. Springer-Verlag, 2001.
11. S. Lawrence. *Resource constrained project scheduling: an experimental investigation of heuristic scheduling techniques (Supplement)*. Graduate School of Industrial Administration, Carnegie-Mellon University, 1984.
12. S. J. Louis and G. J. E. Rawlins. Pareto Optimality, GA-easiness and Deception. In S. Forrest, editor, *Proceedings of ICGA-5*, pages 118–123. Morgan Kaufmann, 1993.
13. J. Noble and R. A. Watson. Pareto coevolution: Using performance against coevolved opponents in a game as dimensions for Pareto selection. In L. Spector et al., editors, *Proceedings of GECCO'2001*, pages 493–500. Morgan Kaufmann, 2001.
14. J. S. Scharnow, K. Tinnefeld, and I. Wegener. Fitness Landscapes Based on Sorting and Shortest Paths Problems. In J. J. Merelo Guervós et al., editors, *Proceedings of PPSN VII*, volume 2439 of *LNCS*, pages 54–63. Springer-Verlag, 2002.
15. R. A. Watson and J. B. Pollack. Symbiotic Combination as an Alternative to Sexual Recombination in Genetic Algorithms. In M. Schoenauer et al., editors, *Proceedings of PPSN VI*, volume 1917 of *LNCS*, pages 425–434. Springer-Verlag, 2000.
16. E. Zitzler and L. Thiele. Multiobjective Evolutionary Algorithms: A Comparative Case study and the Strength Pareto Approach. *IEEE Transactions on Evolutionary Computation*, 3(4):257–271, November 1999.

Analyzing a Unified Ant System for the VRP and Some of Its Variants

Marc Reimann, Karl Doerner, and Richard F. Hartl

Institute of Management Science, University of Vienna, Brünnerstrasse 72,
A-1210 Vienna, Austria
{marc.reimann, karl.doerner, richard.hartl}@univie.ac.at
http://www.bwl.univie.ac.at/bwl/prod/index.html

Abstract. In this paper we analyze the application of an Ant System to different vehicle routing problems. More specifically, we study the robustness of our Unified Ant System by evaluating its performance on four different problem classes within the domain of vehicle routing.

1   Introduction

Vehicle routing and Scheduling has been of major interest to the scientific community for the last 50 years due to the inherent complexity of the associated problems. A large number of different approaches, in recent years mostly meta-heuristics have been proposed for different variants of vehicle routing and scheduling problems, see further Toth and Vigo [1].

On the other hand, goods distribution and more specifically vehicle routing represents an important area of an economy. Basically, in every supply chain transportation occurs between member companies of the chain or from the chain to final customers. Thus, firms have recognized the need to automatize this process and use software to support their distribution process. Part of this software is an optimization tool for routing and scheduling.

However, while it seems that scientific interest and company interest overlap in this regard, this first impression is not completely true. Rather, the level of development of tools in the scientific world is not matched in industry. Most of the tools used in industry are based on very simple optimization techniques developed many years ago.

The reasons for this are manyfold. First, it can not be expected that cutting edge research will instantaneously be transferred to industry. Rather a certain lag has to be accepted. Second, in the quest for ever more sophisticated optimization techniques the scientific community has reached a level of specialization that prevents the tools developed from being applicable to a wide range of problems. Thus, while these tools solve particular problem scenarios much better than the 'old' simple techniques, they lack the flexibility of the latter approaches. However, for practical purposes solution quality is not the only criterion for the choice of an optimization tool.

This gap has been recognized by some researchers. For example Cordeau et al.[2] present four measures of algorithm performance, namely solution quality, speed, flexibility and simplicity. According to their analysis most of the meta-heuristics, while outperforming simple heuristics with respect to solution quality can not compete with simple heuristics when it comes to speed or simplicity. Simplicity is customarily measured in terms of the number of parameters of an algorithm, while flexibility is concerned with the applicability of an algorithm to different variants of a basic problem. This applicabilty covers also issues of robustness, i.e. whether parameter changes are necessary if the problem instance to be solved changes. On the other hand, a research group in Oslo works on a formal representation and algorithms for rich vehicle routing problems, which cover a wide range of problems with different constraints and objectives (c.f. [3]).

In light of these observations, the aim of our paper is to present an Ant System algorithm that is capable of solving different variants of vehicle routing problems reasonably well, with little or no adjustments to the implementation. More specifically, we aim to solve the vehicle routing problem (VRP), the vehicle routing problem with time windows (VRPTW), the vehicle routing problem with backhauls (VRPB) and the vehicle routing problem with backhauls and time windows (VRPBTW). The algorithm is based on the Insertion based AS we presented in Reimann et al. [4].

The remainder of this paper is organized as follows. In the next section we describe the characteristics of the different problems we aim to solve. After that we briefly describe our Unified AS approach. In section 4 we present our rich computational study, before we conclude in Section 5.

2   The Vehicle Routing Problem and Some of Its Variants

The classic VRP (c.f. e.g [1]) aims to find a set of minimum cost routes, each starting and ending at a central depot. These routes have to satisfy the known demands of a number of customers, where each customer must be served by exactly one vehicle, i.e. order splitting is not an option. Of course vehicle capacities and (possibly) maximum tour length restrictions have to be respected. It is assumed that the available fleet is homogeneous, i.e. the vehicles are identical.

This basic setup can be extended in several directions to deal with more realistic scenarios. The most important extensions are those dealing with more complicated restrictions at the customers. More specifically, two characteristics are dominant:

- time windows: Certain or all customers may require that service is performed within a certain time span. In this case, vehicles arriving at a customer before the beginning of the time window have to wait until the customer is willing to accept the service. A vehicle that arrives at a customer location after the end of the time window set by the customer may not serve the customer anymore. However as service is obligatory, this means that any algorithm has to make sure that all vehicles arrive at each customer before the end of the respective time window. The VRPTW has been studied extensively in the last decade; for a recent overview of metaheuristic approaches see (e.g. [5]).

– backhauls: Intermediate companies of a supply chain often deal with suppliers and customers. It has been recognized that a combination of the two distribution areas leads to significant improvements. Thus, a vehicle that delivers goods to customers may also pick up goods from the company's suppliers. This leads to a mixed problem where some locations need delivery while others require pick up of goods. In the backhauling case an additional restriction is that all the pickups be after the deliveries. The intuition for this is that picking up goods before all deliveries are done may lead to the necessity of rearranging goods in the truck. Doing these rearrangements en route is expensive and thus to be avoided. The VRPB has also received a lot of attention recently (see e.g. [6]).

Given these additional problem characteristics, we end up with 4 problem classes. The simplest case is the vehicle routing problem (VRP) without time windows or backhauls. Adding one of the two customer characteristics leads to either the vehicle routing problem with time windows (VRPTW) if time windows are considered, or the vehicle routing problem with backhauls (VRPB) if backhauls are treated. Finally, both types of constraints are added in the vehicle routing problem with backhauls and time windows (VRPBTW).

Compared to the first three problems, the VRPBTW has received only very little attention. The only meta-heuristic approach is due to Duhamel et al. [8] who proposed a Tabu Search algorithm to tackle the problem.

In the next section we will present an Ant System that can be applied to each of the four problems. We will discuss the implementation with respect to the characteristics of the different problems. In closing this section let us note that other possible extensions are the consideration of vehicle heterogeneity, multiple depots or multiple trips per vehicle. These extensions have been studied by different authors (c.f. Gendreau et al. [9], Cordeau et al. [10]). Testing whether our algorithm applies to these problems is left for further research.

3   Ant System Algorithms for the VRP and Some of Its Variants

In this section we briefly describe our Ant System algorithm. In particular, we focus on the constructive heuristics used.

3.1   Ant Systems

The Ant System approach, originally proposed by Colorni et al. (see e.g. [11]) is based on the behavior of real ants searching for food. Real ants communicate with each other using an aromatic essence called pheromone, which they leave on the paths they traverse. In the absence of pheromone trails ants more or less perform a random walk. However, as soon as they sense a pheromone trail on a path in their vicinity, they are likely to follow that path, thus reinforcing this trail. More specifically, if ants at some point sense more than one pheromone trail,

they will choose one of these trails with a probability related to the strenghts
of the existing trails. This idea has first been applied to the TSP, where an ant
located in a city chooses the next city according to the strength of the artificial
trails.

Improved versions of the basic algorithm have been applied to a large number of different combinatorial optimization problems (for an overview see [12]).
Recently, a convergence proof for a generalized Ant System has been developed
by Gutjahr [13]. Generally, the Ant System algorithm consists of the iteration
of three steps:

- Generation of solutions by ants according to private and pheromone infor-
  mation
- Application of a local search to the ants' solutions
- Update of the pheromone information

Below we will discuss the first step in more detail. The second and third step
are reviewed only briefly (c.f. [4] for a more detailed description of these steps).

3.2   Generation of Solutions

As proposed in Reimann et al. [4], we use an Insertion algorithm derived from
the I1 insertion algorithm proposed by Solomon [14] for the VRPTW. This algorithm works as follows: Routes are constructed one by one. First, the unrouted
customer farthest from the depot is selected as a seed customer for the current
route, that is, only this customer is served by the route. Sequentially other customers are inserted into this route according to a cost criterion based on their
distance to the depot as well as the detour and delay caused by inserting them.
Once no more insertions are feasible with respect to time window, capacity or
tour length constraints, another route is initialized with a seed customer and
the insertion procedure is repeated with the remaining unrouted customers. The
algorithm stops when all customers are assigned to routes.

In order to use the algorithm described above within the framework of our
Ant System we need to adapt it to allow for a probabilistic choice in each decision
step. This is done in the following way. To initialize a tour, seed customers are
not chosen deterministically but probabilistically according to their distance
from the depot.

Inserting further customers on the current tour is done using a roulette wheel
selection over all unrouted customers with positive evaluation function κ_i. The
decision rule used can be written as

$$
\mathcal{P}_i = \begin{cases} \dfrac{\kappa_i}{\sum_{h|\kappa_h>0} \kappa_h} & \text{if } \kappa_i > 0 \\[2mm] 0 & \text{otherwise.} \end{cases}
\tag{1}
$$

The chosen customer i is then inserted into the current route at its best feasible
insertion position.

To compute the evaluation function κ_i for inserting an unrouted customer i at its best insertion position on the current tour we first determine for each unrouted customer i the attractiveness of insertion at any feasible insertion position on the current tour. Formally the attractiveness of inserting customer i immediately after customer j can be written as

$$
\begin{aligned}
\eta_{ij} = \max \{0, \\
\alpha \cdot d_{0i} - \beta \cdot (d_{ji} + d_{ik} - d_{jk}) - (1 - \beta) \cdot (b_k^i - b_k) \\
+\delta \cdot (\gamma \cdot type_i + (1 - \gamma) \cdot (1 - type_i))\} \qquad \forall i \in N_u, \forall j \in R_i,
\end{aligned}
$$

where d_{0i} denotes the distance between the depot and customer i and N_u denotes the set of unrouted customers. Further, k is the customer visited immediately after customer j in the current solution, b_k^i is the actual arrival time at customer k, if i is inserted between customers j and k, while b_k is the arrival time at customer k before the insertion of customer i and R_i denotes the set of customers assigned to the current tour after which customer i could feasibly be inserted. Finally, $type_i$ is a binary indicator variable denoting whether customer i is a linehaul ($type_i = 0$) or a backhaul customer ($type_i = 1$). The intuition is that we want to be able to discriminate between linehaul and backhaul customers. Note, that given 'inappropriate' values for the parameters α, β, γ and δ the attractiveness η_{ij} can become zero. Note further, that our attractiveness function is an extension of the function proposed by Solomon [14] for the VRPTW. More precisely, our function reduces to Solomon's function if δ is set to $\delta = 0$.

Given the attractiveness we then compute the evaluation function of the best insertion position for each customer i on the current tour by

$$
\kappa_i = \max \{0, \max_{j \in R_i}[\eta_{ij} \cdot \tfrac{\tau_{ji}+\tau_{ik}}{2 \cdot \tau_{jk}}]\} \qquad \forall i \in N_u,
$$

where τ_{ji} denotes the pheromone concentration on the arc connecting locations (customers or depot) j and i. The pheromone concentration τ_{ji} contains information about how good visiting two customers i and j immediately after each other was in previous iterations. Note, that the same pheromone utilization is done for route initialization, thus augmenting the attractiveness of initializing a route with an unrouted customer i by the search-historic information.

Computing the attractiveness η_{ij} reflects the tradeoff between detour and delay costs associated with inserting a customer. This tradeoff is weighted by the parameter β. A higher value for β puts higher emphasis on the detour and lower emphasis on the delay. The tradeoff is also influenced by customer characteristics such as distance to the depot or type and these characteristics are weighted by the parameters α and γ, δ respectively. More precisely, higher α will favor customers far from the depot, while higher δ will put more emphasis on the customer type. Finally, γ influences the discrimination between linehaul and backhaul customers.

3.3 Local Search

After an ant has constructed its solution, we apply a local search algorithm to improve the solution quality. In particular, we sequentially apply *Move* and *Swap* operators to the solution. Generalized versions of both operators have been proposed by Osman [15] for the VRP.

The *Move* operator tries to eject one or two adjacent customers from their current position and insert it at another position. It is a special case of the 3-opt operator. The elementary *Swap* operator, aims at improving the solution by exchanging a customer i with a customer j. This operator is a special case of the 4-opt operator. Both operators are used to move and exchange customers of the same tour and between tours.

3.4 Pheromone Update

After all ants have constructed their solutions, the pheromone trails are updated on the basis of the solutions found by the ants. According to the rank based scheme proposed in [16] the pheromone level on all arcs is first decreased according to the evaporation factor $(1 - \rho)$. Second, only arcs belonging to either the best solution found so far or to one of the $E - 1$ best solutions found in the current iteration are receiving positive reinforcement. The amount of positive reinforcement depends on the rank of the ant and is proportional to the inverse solution quality found by the ant. That way the search is driven to promising areas of the search space. For more details on this approach we refer to [16].

4 Numerical Analysis

As testing our algorithm on all instances in the four classes would have been computationally too expensive we performed our numerical analysis on selected problem instances. First, to eliminate size effects we only considered problems with approximately 100 customers. In fact, only the VRPB instances have 90 and 94 customers respectively as for this class 100 customer instances are not available. Also, in order not to bias the results by taking 'easy' or 'hard' instances we randomly chose 19 instances. These are:

- VRP (from [17]): vrp3, vrp8, vrp12, vrp14
- VRPTW (from [14]): r101n, r206n, c105n, c207n, rc104n, rc203n
- VRPB (from [18]): lhbh-i1, lhbh-i3, lhbh-j1, lhbh-j3
- VRPBTW (from [7]): bhr101b, bhr102a, bhr103b, bhr104c, bhr105c

4.1 Parameter Settings

Let n be the problem size, i.e. the number of customers to be served, then the Ant System parameters were: $m = \lceil n/2 \rceil$ ants, $\rho = 0.95$ and $E = 4$ elitists. These parameters are standard settings and were not tested systematically as our experience suggests that the rank based Ant System is quite robust.

Generally, the objective for time constrained routing problems is to first minimize the fleet size required to serve all customers and given a minimal fleet size to minimize the total distance travelled. This objective was established by minimization of the following objective function:

$$L = M \cdot FS + TT, \tag{2}$$

where L denotes the total costs of a solution, FS denotes the fleet size found, and TT corresponds to the total travel time (or distance). The parameter M has to be chosen in a way to ensure that a solution that saves a vehicle always outperforms a solution with a higher fleet size. More precisely we set $M = 10000$ in our experiments.

Finally, as we are interested in the interplay of the parameters of our insertion algorithm with the characteristics of the different problems we analyzed the four parameters of the insertion algorithm in the ranges $\alpha \in \{0.5, 1, 1.5, 2\}$, $\beta \in \{0, 0.33, 0.66, 1\}$, $\gamma \in \{0, 0.33, 0.66, 1\}$ and $\delta \in \{0, 0.33, 0.66, 1\}$ on the instances described above.

This means that we tested 256 parameter constellations for each instance. For each constellation we performed 1 run of 10 minutes. All our computations were performed on a Pentium 3 with 900MHz. The code was implemented in C.

4.2   Influence of the Problem Characteristics on the Parameter Values

As stated in the last section, we have tested a large number of constellations for each of the 19 problem instances. As our main interest lies in the robustness of the final approach we have compared the different settings based on averages over all instances. First, we select for each instance the best solution found by our Unified Ant System (over all parameter settings) as a reference value. Each individual setting was then evaluated by relating its performance to the reference values.

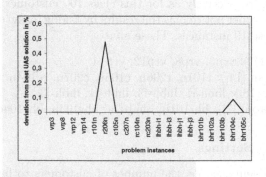

Fig. 1. Robustness of the 'best' setting over the 19 test instances

Using this approach, we found the following 'best' setting:

$$\alpha = 2, \beta = 1, \gamma = 0.33, \delta = 1$$

The average deviation of this setting over the best solution found for each instance is equal to 0.03%. Let us look more precisely at the behavior of this setting over all instances. Figure 1 shows the percentage deviation from the best solution found by our Unified AS.

Clearly, for 17 out of the 19 instances this setting shows excellent behavior with virtually 0% deviation. For the remaining two instances the deviations of 0.1% and 0.5% respectively still are more than reasonable.

Given these encouraging results, let us now perform some sensitivity tests. More precisely, we are interested how changing one parameter from its 'best' setting influences the performance of the Unified AS given that all other parameters remain unchanged. Table 1 shows the deviation in % from the best solution found by our Unified AS, for different settings of the 4 parameters.

Table 1. Sensitivity analysis of the parameter settings

Parameter setting	0.5	1	1.5	2	Parameter setting	0	0.33	0.66	1
α	0.14%	0.07%	0.05%	0.03%	β	0.03%	0.04%	0.04%	0.03%
					γ	0.03%	0.03%	0.03%	0.03%
					δ	0.03%	0.03%	0.03%	0.03%

Clearly, these results show that the parameter α has the strongest influence on solution quality. The other three parameters show basically no sensitivity to their actual setting. To understand this result let us consider the effects of the parameters. Both the parameters α and δ are free parameters, i.e. they can take any values. Moreover, higher levels for these parameters may implicitly increase the number of positive evaluations κ_i. Thus, the number of customers on each truck may increase which in turn leads to a possible reduction in fleet sizes. As this is the main goal, higher values for α and δ should be expected to improve the solution quality. However, both of the free parameters α and δ are at the highest level we tested. This suggests that considering even higher values for these parameters might further improve the performance of the algorithm. This question is not considered in the remainder of this paper, but rather left for future research.

Finally, the parameters β and γ have more indirect effects on the fleet size such that their actual values will influence mainly the second objective namely the total distance travelled.

4.3   Comparison of Our Unified AS with Existing Results

Now that we have analyzed the influence of the actual parameter settings on the solution quality let us evaluate the performance of the 'best' setting we found against state of the art results. More precisely, we consider for each instance the best known solution.[1] Apart from the VRP instances - for those the standard objective is to minimize total distance travelled only - all the other best known results are based on the same lexicographic ordering of objectives we use.

In our comparison, we split the solutions into the fleet size and the total distance travelled. Table 2 shows the results for all of the 19 instances.

Table 2. Comparison of our Unified AS with best known results

Problem instance	Unified AS		Best known results	
	fleet size	total distance	fleet size	total distance
$vrp3$	8	859.68	8	826.14
$vrp8$	9	886.47	9	865.94
$vrp12$	10	848.64	10	819.56
$vrp14$	10	971.82	11	866.37
$r101n$	19	1655.41	19	1650.80
$r206n$	3	979.45	3	912.00
$c105n$	10	828.94	10	828.94
$c207n$	3	589.20	3	588.29
$rc104n$	10	1196.05	10	1135.48
$rc203n$	3	1175.88	3	1060.45
$lhbh - i1$	9	361.08	10	360.00
$lhbh - i3$	5	317.07	5	306.00
$lhbh - j1$	10	352.93	10	352.00
$lhbh - j3$	7	301.67	7	302.00
$bhr101b$	23	1975.07	23	1999.20
$bhr102a$	19	1681.34	19	1677.60
$bhr103b$	15	1405.02	16	1395.88
$bhr104c$	12	1216.32	12	1208.50
$bhr105c$	17	1642.70	17	1633.01
$Total$	202	19244.75	205	18788.16

Our results show that considering the first goal, namely the fleet size, our algorithm improves the best known results in three cases, leading to a reduction in fleet sizes equal to 1.46%. With respect to the second objective, our algorithm deviates by 2.43% from the best known solution. Thus, the reduction in fleet size is achieved through giving up some solution quality associated with the total distance travelled. Overall, our algorithm being the first unified approach for the four classes studied shows very promising behavior.

---

[1] Note, that these best known results are taken from different sources as to date there is no unified approach for the studied problem classes. Additional information about the corresponding sources can be found at
http://www.bwl.univie.ac.at/bwl/prod/reimann/vrpresults.html.

4.4   Flexibility and Simplicity of Our Unified Ant System

The approach presented and evaluated above was developed to deal with the issues of flexibility and simplicity, which become more and more important as additional criteria to compare heuristic algorithms as pointed out in the introduction. The best known results presented above come from a number of researchers using different techniques and (presumably) none of these approaches was designed to be flexible in terms of applicability for different variants of the VRP. Thus, a comparison of flexibility and simplicity is impossible.

However, we believe that our results enable us to make some statements for our approach. First, our algorithm can be applied to (at least) the four types of problems studied without modification. In fact, once the format of the input data is fixed, our algorithm does not make a distinction between the types of problems. Further, while the Ant System features a significant number of parameters, our results suggest that the performance of the algorithm is quite robust over the different types of problems and instances. Finally, the basic principle of the Ant System meta-heuristic is quite simple.

Overall, our algorithm seems to be rather flexible, simple and robust. However, these careful statements show that a much more thorough analysis of these issues is necessary.

5   Conclusions and Future Research

The results presented in this paper suggest that our Unified AS is capable of finding high quality solutions for four important problem classes within the domain of vehicle routing. While some solution quality has to be sacrificed to gain this flexibility, the overall performance of the algorithm is still very reasonable.

However, we clearly have to gather more experimental evidence to support our thesis about the robust performance of our approach. First, we need to extend the parameter ranges for α and δ as we found that these parameters should take the maximum possible values in the domains we currently specified. Second, we have to evaluate our Unified AS on an extended problem set that also includes larger instances. Finally, we aim to extend our model further by giving up the linehaul backhaul precedence relationship and by considering, among others, soft time windows and multiple depots.

Apart from that the tradeoff between the two objectives suggests, that a true multi-objective approach should be applied to further enhance the applicability of the approach to real world problems. An existing approach by Jozefowiez et al.[19] studies this possibility for the basic VRP. Their objectives are the total distance travelled and a load balancing criterion that favors solutions with tours that differ little in their total distance.

Acknowledgments. The authors wish to thank three anonymous referees for valuable comments on the paper and Harald Andree for performing the numerical tests. This work was supported by the Oesterreichische Nationalbank (OeNB) under grant #8630.

References

1. Toth, P. and Vigo, D. (Eds.): The Vehicle Routing Problem. Siam Monographs on Discrete Mathematics and Applications, Philadelphia (2002)
2. Cordeau, J. F., Gendreau, M., Laporte, G., Potvin, J. Y. and Semet, F.: A guide to vehicle routing heuristics. Journal of the Operational Research Society **53** (5) (2002) 512–522
3. http://www.top.sintef.no/
4. Reimann, M., Doerner, K., Hartl, R. F.: Insertion based Ants for Vehicle Routing Problems with Backhauls and Time Windows. In: Dorigo, M. et al. (Eds.): Ant Algorithms, Springer LNCS 2463, Berlin/Heidelberg (2002) 135–147
5. Bräysy, O. and Gendreau, M.: Metaheuristics for the Vehicle Routing Problem with Time Windows. Sintef Technical Report STF42 A01025 (2001)
6. Toth, P. and Vigo, D.: VRP with Backhauls. In Toth, P. and Vigo, D. (Eds.): The Vehicle Routing Problem. Siam Monographs on Discrete Mathematics and Applications, Philadelphia (2002) 195–224
7. Gelinas, S., Desrochers, M., Desrosiers, J. and Solomon, M. M.: A new branching strategy for time constrained routing problems with application to backhauling. Annals of Operations Research. **61** (1995) 91–109
8. Duhamel, C., Potvin, J. Y. and Rousseau, J. M.: A Tabu Search Heuristic for the Vehicle Routing Problem with Backhauls and Time Windows. Transportation Science. **31** (1997) 49–59
9. Gendreau, M., Laporte, G., Musaraganyi, C. Taillard, E. D.: A tabu search heuristic for the heterogeneous fleet vehicle routing problem. Computers and Operation Research **26** (1999) 1153–1173
10. Cordeau, J. F., Gendreau, M., Laporte, G.: A Tabu Search Heuristic for Periodic and Multi-Depot Vehicle Routing Problems. Networks **30** (1997) 105–119
11. Colorni, A., Dorigo, M. and Maniezzo, V.: Distributed Optimization by Ant Colonies. In: Varela, F. and Bourgine, P. (Eds.): Proc. Europ. Conf. Artificial Life. Elsevier, Amsterdam (1991) 134–142
12. Bonabeau, E., Dorigo, M. and Theraulaz, G.: Swarm Intelligence. Oxford University Press, New York (1999)
13. Gutjahr, W. J.: ACO algorithms with guaranteed convergence to the optimal solution. Information Processing Letters. **82** (2002) 145–153
14. Solomon, M. M.: Algorithms for the Vehicle Routing and Scheduling Problems with Time Window Constraints. Operations Research. **35** (1987) 254–265
15. Osman, I. H.: Metastrategy simulated annealing and tabu search algorithms for the vehicle routing problem. Annals of Operations Research. **41** (1993) 421–451
16. Bullnheimer, B., Hartl, R. F. and Strauss, Ch.: A new rank based version of the ant system: a computational study. Central European Journal of Operations Research **7**(1) (1999) 25–38
17. Christofides, N., Mingozzi, A. and Toth, P.: The vehicle routing problem. In: Christofides, N. et al. (Eds.): Combinatorial Optimization. Wiley, Chicester (1979)
18. Jacobs-Blecha, C., Goetschalckx, M.: The vehicle routing problem with backhauls: properties and solution algorithms. Technical Report, School of Industrial and Systems Engineering, Georgia Institute of Technology, Atlanta, Georgia (1992)
19. Jozefowiez, N., Semet, F. and Talbi, E.: Parallel and Hybrid Models for Multiobjective Optimization: Application to the Vehicle Routing Problem. In Guervos, J. M. et al. (Eds.): Parallel Problem Solving from Nature – PPSN VII, Springer LNCS 2439, Berlin/Heidelberg (2002) 271–280

Adapting to Complexity During Search in Combinatorial Landscapes

Taras P. Riopka

Lehigh University, Department of Computer Science and Engineering
200 Packard Lab, 19 Memorial Drive West
Bethlehem PA 18015
riopka@eecs.lehigh.edu

Abstract. Fitness landscape complexity in the context of evolutionary algorithms can be considered to be a relative term due to the complex interaction between search strategy, problem difficulty and problem representation. A new paradigm for genetic search referred to as the Collective Learning Genetic Algorithm (CLGA) has been demonstrated for combinatorial optimization problems which utilizes genotypic learning to do recombination based on a cooperative exchange of knowledge (instead of symbols) between interacting chromosomes. There is evidence to suggest that the CLGA is able to modify its recombinative behavior based on the consistency of the information in its environment, specifically, the observed fitness landscape. By analyzing the structure of the evolving individuals, a landscape-complexity measure is extracted *a posteriori* and then plotted for various types of example problems. This paper presents preliminary results that show that the CLGA appears to adapt its search strategy to the fitness landscape induced by the CLGA itself, and hence relative to the landscape being searched.

1 Introduction

It is well known that problem representation can greatly affect the efficiency of search and can alter the apparent function complexity with respect to the search strategy of the algorithm being used. The application of evolutionary methods often leads to the dilemma of matching the particular search heuristics and problem representation used to the complexity of the problem being solved. Of great benefit would be a method for predicting which algorithm (and its parameters) are best suited to the problem at hand.

One way to approach this problem is try to measure problem complexity using a measure like epistasis variance[1], Walsh sums [2], fitness-distance-correlation (FDC)[3] or density of states [4] and then use the result to classify a given problem for analysis. Unfortunately, problem complexity measures can be misleading as they rely on statistical sampling of the solution space, and except for FDC, completely ignore the question of the search algorithm [5]. As Jones and Forrest have shown [3], the fitness landscape observed for a particular function is an artifact of the algorithm, or more accurately, of the neighborhood induced by the operators the algorithm employs.

S. Cagnoni et al. (Eds.): EvoWorkshops 2003, LNCS 2611, pp. 311–321, 2003.

Another way would be to take an adaptive approach, to devise an algorithm that modifies its own behavior in the process of solving a given problem. Examples of this approach range from the self-adaptive mutation rates used in evolutionary strategies [6] to the evolving representations pioneered by Goldberg [7]. A comprehensive list of such methods is too extensive to be discussed here, but may be found in a recent survey [8]. In the case of self-adaptation, operators are often adjusted based on an analysis of the progress of the algorithm. A frequent criticism is the lag in the time between operator adjustment and algorithm response. In the case of evolving representations, the intent is to discover and analyze variable interactions in order to efficiently deal with them. However, assumptions regarding problem complexity are required to make these methods work.

The Collective Learning Genetic Algorithm (CLGA) [9][10] is an example of the latter approach, but is shown here to also use a type of implicit problem-difficulty measure by virtue of the mechanism used to do recombination. The measure does not ignore the question of the search algorithm because it is part *of* it. In this paper, a **landscape complexity** measure is extracted *a posteriori* by analyzing the structure of the evolving individuals and then plotted for various types of example problems. Preliminary results show that the CLGA appears to respond to problem difficulty naturally, as a consequence of its operation. The CLGA search strategy is reflected in the empirically derived landscape complexity measure, and is shown to adapt to the problem at hand. Adaptation occurs without any explicit operator control and without any assumptions regarding problem complexity.

The paper is organized as follows. Section 2 reviews the main features of the CLGA to provide the necessary background for understanding the landscape complexity measure. Section 3 discusses how the landscape complexity measure is extracted from the CLGA. Section 4 presents and discusses the results, and the paper is summarized and concluded in Section 5.

2   CLGA Description

The following section briefly reviews the main features of the CLGA. For a complete description, the reader is referred to [9][10].

The general structure of the CLGA is the same as that of most evolutionary algorithms, *i.e.,* a population of solutions (or individuals), represented as chromosome strings, evolves through various mechanisms that allow changes to strings; some form of competition then determines which strings survive in subsequent generations. The CLGA differs from these algorithms in one very important way, and that is that recombination involves *not* an exchange of schemata, but an exchange of *information* which is then used by individual chromosomes to guide recombination. Since that information is derived from observations of schemata within the population, recombination is necessarily driven by the quality and consistency of that information.

A CLGA consists of a population of adaptive learning agents called **SmartChromosomes**. Each SmartChromosome consists of two components: an instantiation of a collective learning automaton (CLA) and a chromosome string representing the best solution the SmartChromosome has found so far. The concept of a CLA is derived from collective learning systems theory [11].

A CLA consists of a fixed number of **feature detectors**, each associated with a histogram that contains the accumulated knowledge of the fitness of schemata the SmartChromosome has "observed". Each feature detector monitors a unique set of chromosome sites. Its corresponding histogram contains one bin for every possible permutation of symbols for the chromosome sites, which the feature detector monitors. The number of feature detectors in each SmartChromosome (d) and the cardinality of the set of monitored sites (k) are both parameters of the algorithm. For example, for binary encodings, a feature detector monitoring k chromosome sites will have 2^k bins. Figure 1 shows a graphical depiction of two SmartChromosomes, for two different example configurations. Both monitor the exact same set of chromosome sites, in a binary encoded population with a string length of $N=12$. However, the first uses $d=4$ feature detectors of size $k=1$, while the second uses $d=2$ feature detectors of size $k=2$.

Given a feature detector cardinality of k, the total number of possible combinations of monitored sites is N chromosome sites taken k at a time $C(N,k)$. The number of SmartChromosomes (M) in the CLGA population is determined by the number of feature detectors per SmartChromosome and the **combination ratio** r_c, the fraction of the total number of combinations actually incorporated into the population. Hence,

$$M = \left\lceil \frac{r_c\, C(N,k)}{d} \right\rceil \tag{1}$$

Incorporating all combinations insures problem-representation independence with respect to the recombination process. A population is created by randomly selecting the required number of combinations (without replacement) from the total and distributing feature detectors as randomly as possible among the SmartChromosomes. With large populations and small k, the combination ratio is usually much larger than one and corresponds to the number of copies of each feature detector in the population.

Initial chromosome strings are generated randomly. Each generation consists of five stages: mating, intelligent recombination, directed mutation, evaluation and individual selection, all of which are executed locally by the individual SmartChromosomes.

Mating. Each SmartChromosome mates with m other SmartChromosomes selected randomly from the population.

Intelligent Recombination. Intelligent recombination comprises two processes: an acquisition of knowledge (**inspection**) and application of that knowledge to direct recombination (**interrogation**). A SmartChromosome *inspects* the strings of its mates by noting in each mate's string the particular permutation of symbols at the location of the sites monitored by the SmartChromosome's feature detectors. The bin weight associated with the observed permutation is replaced by the average of the current string fitness and all other string fitnesses that contributed to that bin in the past. A SmartChromosome *interrogates* its mates by superposing the best states of its own feature detectors with those of its mates' feature detectors in order to modify its string.

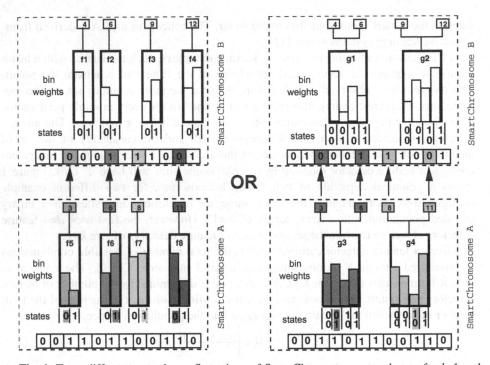

Fig. 1. Two different example configurations of SmartChromosomes are shown, for $k=1$ and $k=2$. During inspection, the feature detectors of SmartChromosome A "observe" the shaded symbols pointed to by arrows. This causes the bin weight for state [0] of feature detector f5 to be replaced by the average of the fitness of SmartChromsome B's string and all other string fitnesses that contributed to that bin in the past. The same occurs for bin weights f6[1], f7[1] and f8[0]. Similarly, for the $k=2$ configuration, the bin weight for state [0 1] of feature detector g3 would be replaced by the average of the fitness of SmartChromsome B's string and all other string fitnesses that contributed to that bin in the past. The same would occur for bin weight g4[1 0]. Note that feature detectors with higher values of k can also be used.

To accomplish this, first all best states are sorted in descending order by weight magnitude. The corresponding permutations of symbols are superposed sequentially state by state beginning with the best state with the largest magnitude overwriting the relevant chromosome sites in the SmartChromosome's string. In order for a state to be **consistent**, none of the monitored sites of the corresponding feature detector may overlap with those of any of the previously integrated states or, in case of overlap, the symbols in the overlapping sites must be identical. States, which are not consistent, are omitted. A more detailed explanation of this process is given in [9] or [10]. The fraction of a chromosome's sites affected by the best state superposition is referred to as the **superposition fraction** (s_f).

Directed Mutation. Standard mutation is not applied in the CLGA, contrary to the experiments presented in [10]. Experiments have since shown that standard mutation is not needed and in fact, actually degrades CLGA performance. Consequently, a *directed* *mutation* operator is applied as follows. Each SmartChromosome maintains a FIFO[1] list of the last *h* unique chromosome strings it has evaluated, where *h* is some small number (to limit time and memory resources) referred to as **history length**. A SmartChromosome first searches its memory to see if the current string has already been evaluated. This will occur only if interrogation results in the creation of a string already in memory. If the string has already been evaluated, a single bit is randomly selected in the string and complemented. A *mutation template* is maintained for each chromosome in memory to keep track of which bits have been changed, so that the next time that same string is obtained, a different bit can be changed. Only those bits that have *not* already been changed are eligible for mutation each time. Once all single bits are exhausted, combinations of two bits are complemented, etc. Due to memory limitations, a reasonable limit on the number of simultaneous bit complements must be enforced. Note that the need for directed mutation is related to the problem of feature detector convergence, a topic currently under research.

Evaluation. The SmartChromosome evaluates its string using the fitness function.

Individual Selection. If a modification results in a string whose fitness is greater than the previous string, the new string is retained, otherwise, the SmartChromosome reverts to the previous one. In effect, the offspring competes with its parent. The Smart-Chromosome inspects its new string regardless of what its relative fitness is, thus learning from both its successes and its failures.

3   A Landscape Complexity Measure

Significant evidence exists to strongly suggest that CLGA performance is independent of the size of the feature detector (*k*). (For a detailed discussion of this interesting result, the reader is referred to [9]). Consequently, it is conceivable that the following analysis for *k*=1 may be extrapolated to higher values of *k* with similar results. Therefore, all of the experiments presented here use a value of *k*=1 for which the following analysis is valid.

Chromosome strings are modified during the interrogation process. If we consider the mating of two SmartChromosomes with *k*=1, feature detectors that they have in common will tend to compete. This case exists in figure 1 for feature detectors f2 and f6 at chromosome site 6. In this example however, the "opinion" of each feature detector as to the best bit value for chromosome site 6 differs. SmartChromosome A (feature detector f6) has a greater bin weight for bit 1 while SmartChromosome B (feature detector f2) has a greater bin weight for bit 0 at that chromosome site. In this particular competition, f6 will prevail because its best state f6[1] has a bin weight that is greater than the bin weight associated with the best state f2[0]. The bit at chromosome site 6 will therefore be set to 1.

In a typical application, reasonable sized populations using *k*=1 give rise to a large number of copies of each feature detector. Consider a single chromosome site. For a given generation, that site will have some number of feature detectors in the popula-

---

[1] FIFO here means that the chromosome that has not been *repeated* in the longest time is replaced first.

tion with one *opinion* as to the best bit state for that chromosome site, while the remainder will have the opposite *opinion*. If we take all those with a best bit state of 0 and compute the average and standard deviation of the bin weight, and do the same for those with a best bit state of 1, we can test the null hypothesis that there is no difference between the means (assuming a t-distribution in a two-tailed test).

Figure 2 shows the calculation of the t-statistic for an example chromosome of length $N=12$. The t-statistic is used to compute the confidence associated with the alternative hypothesis, that there is a difference between the two means. The intuition behind this is that a high confidence should indicate how *sure* the feature detectors have become regarding their *opinion* as to which bit state is correct, regardless of which bit state that is. The idea is that out of the feature detectors that agree with one another, in a competition they will simply set their bit to the agreed value. However, in instances where two feature detectors disagree, a high confidence implies that the difference will most likely be quite stark, and hence the decision as to which bit is set will be quite firm (either all tending to 0 or all tending to 1). On the other hand, a low confidence implies that there is little statistical difference between best states that disagree and hence bits will tend to flip between 0 and 1 more often, inducing a degree of randomness into the decisions being made.

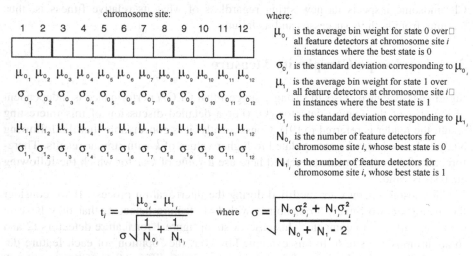

chromosome site:

1 2 3 4 5 6 7 8 9 10 11 12

$\mu_{0_1}\ \mu_{0_2}\ \mu_{0_3}\ \mu_{0_4}\ \mu_{0_5}\ \mu_{0_6}\ \mu_{0_7}\ \mu_{0_8}\ \mu_{0_9}\ \mu_{0_{10}}\ \mu_{0_{11}}\ \mu_{0_{12}}$

$\sigma_{0_1}\ \sigma_{0_2}\ \sigma_{0_3}\ \sigma_{0_4}\ \sigma_{0_5}\ \sigma_{0_6}\ \sigma_{0_7}\ \sigma_{0_8}\ \sigma_{0_9}\ \sigma_{0_{10}}\ \sigma_{0_{11}}\ \sigma_{0_{12}}$

$\mu_{1_1}\ \mu_{1_2}\ \mu_{1_3}\ \mu_{1_4}\ \mu_{1_5}\ \mu_{1_6}\ \mu_{1_7}\ \mu_{1_8}\ \mu_{1_9}\ \mu_{1_{10}}\ \mu_{1_{11}}\ \mu_{1_{12}}$

$\sigma_{1_1}\ \sigma_{1_2}\ \sigma_{1_3}\ \sigma_{1_4}\ \sigma_{1_5}\ \sigma_{1_6}\ \sigma_{1_7}\ \sigma_{1_8}\ \sigma_{1_9}\ \sigma_{1_{10}}\ \sigma_{1_{11}}\ \sigma_{1_{12}}$

where:

μ_{0_i} is the average bin weight for state 0 over all feature detectors at chromosome site i in instances where the best state is 0

σ_{0_i} is the standard deviation corresponding to μ_{0_i}

μ_{1_i} is the average bin weight for state 1 over all feature detectors at chromosome site i in instances where the best state is 1

σ_{1_i} is the standard deviation corresponding to μ_{1_i}

N_{0_i} is the number of feature detectors for chromosome site i, whose best state is 0

N_{1_i} is the number of feature detectors for chromosome site i, whose best state is 1

$$ t_i = \frac{\mu_{0_i} - \mu_{1_i}}{\sigma \sqrt{\dfrac{1}{N_{0_i}} + \dfrac{1}{N_{1_i}}}} \qquad \text{where} \qquad \sigma = \sqrt{\frac{N_{0_i}\sigma_{0_i}^2 + N_{1_i}\sigma_{1_i}^2}{N_{0_i} + N_{1_i} - 2}} $$

Fig. 2. The calculation of the t-statistic used to compute confidence values is shown in relation to an example chromosome of length $N=12$ and the corresponding feature detector bin-weight averages and standard deviations.

A single value was computed by averaging the confidence values over all of the chromosome sites, arriving at a single measure for the entire population for a given generation, referred to as the **average bit certainty**. In the following experiments, the t-score was used to compute confidences due to the small combination ratio.

It is well known that as epistasis increases, the fitness landscape becomes more and more uncorrelated [12]. It would therefore be expected that as epistasis increases the CLGA would find it more and more difficult to learn consistent relationships between bits due to greater variance in solution fitness. Consequently, this measure would be expected to be inversely proportional to the level of epistasis. Results from several experiments support this hypothesis.

4 Experiments

4.1 Fixed Parameters and Performance Metrics

Several problem generators have been developed to facilitate the design of more controlled experiments for testing evolutionary algorithms. Random problems generated by NK-Landscape, LSAT and Multi-modal problem generators were used in the following experiments. The reader is referred to [13] and [14] for details.

The following fixed parameter values for the CLGA were used: $m=1$, $h=4$, $k=1$, $s_f=15\%$, and a population size of $M = 20$. NK-Landscape problems of length $N=30$ were tested using $d=23$, while remaining problems of length $N=100$ were tested using $d=75$, resulting in a combination ratio of approximately $r_c = 15$. Although these parameters were not optimized, they were selected based on heuristics derived from extensive experiments detailed in [9]. A relatively small population size was used for the following experiments, but very similar results were also obtained using larger population sizes as well. Average bit certainty was plotted versus function evaluations in all experiments.

4.2 NK-Landscape Experiments

Fifty random, 30-bit NK-Landscape problems using a neighborhood model were used in the first set of experiments. The CLGA was run for 50000 evaluations on three levels of epistasis, $K=5$, 10 and 15. Results are shown in figure 3.

Note that as epistasis increases, the average bit certainty as measured within the CLGA population decreases. This is consistent with the interpretation given previously, *i.e.*, that the weaker fitness correlations for higher epistasis problems cause the CLGA to behave more randomly. It is interesting to note that NK-Landscape problems with lower values of K seem to have a greater range of average bit certainty values. This is consistent with the intuition that with lower epistasis, there is always some non-zero probability of some problems being "hard" simply by accident.

4.3 L-SAT Experiments

Fifty random, 100-bit 3-SAT problems were used in the next set of experiments. The CLGA was run for 30000 evaluations on three levels of epistasis, with the number of clauses set to $C=430$, 1200, and 2400 for low, medium and high epistasis respectively. Results are shown in figure 4. To our surprise, there did not seem to be much difference between the 3-SAT problems for the three levels of epistasis. This perplexing result however, supports observations made by DeJong, *et al.* [13]. In that paper, the experimenters noted that a low mutation rate (less exploration) and low crossover rate (lower

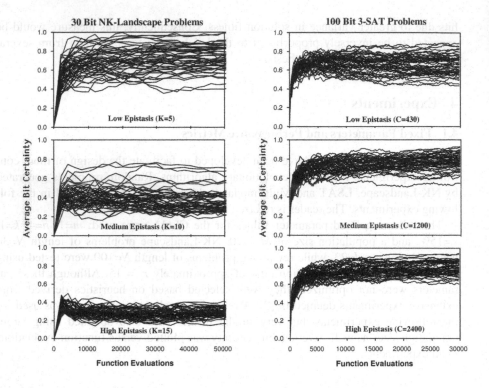

Fig. 3. Average bit certainty plotted for 50 random 30-bit NK-Landscape problems for low, medium, and high epistasis

Fig. 4. Average bit certainty plotted for 50 random 100-bit 3-SAT problems for low, medium and high epistasis

disruption) actually improved results. The results here suggest that the various levels of epistasis for 100-bit 3-SAT problems may contain more structure than previously supposed. The CLGA search strategy, as reflected by the average bit certainty, suggests the need for a less random search over all three levels of epistasis, implying greater problem structure and perhaps explaining why greater preservation of structure (less exploration) resulted in improved performance for DeJong et al.[13].

Note the greater variance in the average bit certainty as epistasis decreases. This is consistent with the fact that the 100 variable 3-SAT is known to have a "phase transition" at approximately 430 clauses [15], resulting in problems that have an almost equal probability of being satisfiable and unsatisfiable and hence computationally most difficult [16]. 3-SAT problems with fewer clauses are almost always satisfiable while those with more are almost always unsatisfiable. The larger variance may reflect the diversity of problems in this critical region. Experiments were also run for 3-SAT problems with fewer clauses, but, as expected, always obtained the optimum faster than an interpretable bit certainty could be extracted.

A second set of experiments was run using a fixed number of clauses ($C = 1200$) but varying the length of the clauses using values $L = 2,3,4$ and 5. Again, the CLGA was run on fifty random problems for 30000 evaluations. Results are shown in figure 5.

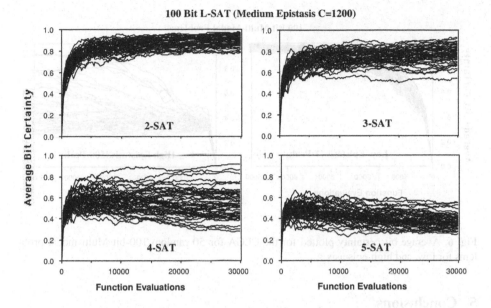

Fig. 5. Average bit certainty plotted for the CLGA for 50 random 100-bit 2-SAT, 3-SAT, 4-SAT and 5-SAT problems using 1200 clauses

In this case, a difference was observed, showing a decrease in bit certainty as clause length increased. This is not surprising if one considers the algorithm used to generate the L-SAT problems. As clause length increases, the probability of satisfying a clause increases (since only a single true value is required to make an entire clause true). A more random strategy is therefore reasonable, since the likelihood of satisfying clauses is high. On the other hand, with short clause length, a more methodical approach may be more efficient (in the long run).

4.4   Multi-modal Experiments

A final set of experiments was run using 50 random, 100-bit Multi-modal problems. The CLGA was run for 30000 evaluations on two levels of epistasis with the number of peaks equal to 1 for low epistasis and 500 for high epistasis. Results are shown in figure 6. Only 5000 evaluations are shown since all optima were found in that time. The results of the last set of experiments were entirely consistent with our expectations. The single peak problems resulted in a less random approach due to the highly correlated fitness landscape. The higher epistasis problems on the other hand, resulted in significantly more random behavior, given the less correlated landscape. Optima, however, were found in both cases in approximately the same time.

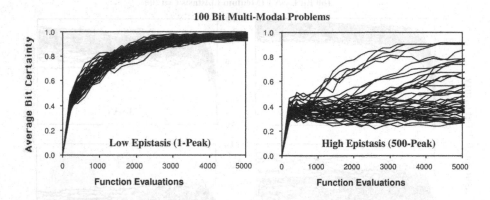

Fig. 6. Average bit certainty plotted for the CLGA for 50 random 100-bit Multi-modal problems for low, and high epistasis

5 Conclusions

How is the computed average bit uncertainty different in principle from any of the measures cited earlier? It too is based on information ultimately obtained from sampling of solutions. However, the difference is that the sampling is done from both successes and failures of the particular *operator* being used (in this case, intelligent recombination). In other words, the information accumulated in the feature detector histograms is based on not only good schemata (gleaned from inspections of a Smart-Chromosome's mates), but on schemata in the neighborhoods of the fitness landscape directly induced by the intelligent recombination. Recall that after modifying its schemata, a SmartChromosome inspects the result, whether it is better or worse than its original string. If the fitness landscape is an artifact of the search algorithm as Jones and Forest [3] suggest, then the CLGA is simply responding to the fitness landscape it is navigating through. If the landscape indicates that its generated solutions are uncorrelated, the CLGA responds by generating trial solutions more randomly. On the other hand, if the landscape indicates its generated solutions are correlated, correlations are strengthened and the CLGA responds by generating more correlated responses. Since correlation structure of fitness landscapes is often considered to be a reliable indicator of fitness landscape complexity [17], this may be a very reasonable search adaptation mechanism.

The empirical landscape complexity measure computed in this paper reflects the strategy chosen by the CLGA to solve the given problem. One cannot make an argument that it is in any way an objective measure of *problem* difficulty. However, if problem difficulty can be considered a relative term with respect to a search algorithm and problem representation, then taken from that perspective the measure is ideal (for the CLGA). The real question is, what kinds of problems evoke which type of behavior? More importantly, is the behavior evoked by a given problem suitable for solving it (efficiently)? The NFL theorem [18] strongly suggests that the answer to this question is no in general, but it is interesting to consider what types of problems respond to the CLGA approach and the scope of its applicability. Future research will investigate information theoretic constraints on the efficacy of the CLGA and hopefully provide more insight into a fascinating new mechanism for evolutionary computation.

References

1. Davidor, Y. (1991). Epistasis Variance. In *Foundations of Genetic Algorithms* Morgan Kaufmann.
2. Heckendorn, R.B., Whitley, D. (1999). Walsh Functions and Predicting Problem Complexity. *Evol. Comp.,* Vol. 7, No. 1, pp. 69–101. MIT Press.
3. Jones, T., Forrest, S. (1995). Fitness Distance Correlation As a Measure of Problem Difficulty for Genetic Algorithms. In *Proc. 6th Int. Conf. on GAs*, pp. 184-192. Morgan Kaufmann.
4. Rose, H., Ebeling,W., Asselmeyer, T. (1996). The Density of States - a Measure of the Difficulty of Optimization Problems. In *PPSN- IV*, Springer-Verlag.
5. Reeves, C. R. (1999). Predictive Measures for Problem Difficulty. In *Proc. 1999 Congress on Evol. Comp.,* pp. 736-743. IEEE Press.
6. Back, T. (1997). Self Adaptation. In *The Handbook of Evolutionary Computation*, pp. 1-23. IOP Publishing and Oxford University Press.
7. Goldberg, Korb, D.E., and Deb, K., (1989). Messy Genetic Algorithms: Motivation, Analysis, and First Results. *Complex Systems*, Vol. 3, pp. 493-530.
8. Hinterding, R., Michalewicz, Z., Eiben, A.E. (1997). Adaptation in Evolutionary Computation: A Survey. In *Proc. 4th Int. Conf. on Evol. Comp.* pp. 65-69.
9. Riopka, T. P. (2002). Intelligent Recombination Using Genotypic Learning in a Collective Learning Genetic Algorithm, *Doctoral dissertation*, GWU, Washington, DC.
10. Riopka, T.P., Bock, P. (2000). Intelligent Recombination Using Individual Learning in a CLGA, In *Proc. Genetic and Evol. Comp. Conf.,* pp. 104-111. Morgan Kaufmann.
11. Bock, P. (1993). *The Emergence of Artificial Cognition*. World Sci. Pub. Co.
12. De Jong, K.A. (1993). Genetic Algorithms are NOT Function Optimizers. In *Foundations of Genetic Algorithms 2*. Morgan Kaufmann.
13. DeJong, K.A., Potter, M.A., Spears, W.M. (1997). Using Problem Generators to Explore the Effects of Epistasis. In *Proc. 7th Int. Conf. on Genetic Algorithms*. Morgan Kaufmann.
14. Heckendorn, R.B., Rana, S. and Whitley, L.D. (1998). Test Function Generators as Embedded Landscapes. In *Foundations of Genetic Algorithms* 5, pp. 183-198. Morgan Kaufmann.
15. Crawford, J.A., Auton, L.D. (1996). Experimental Results on the Crossover Point in Random 3SAT. *Art. Int.* Vol. 81, No. 31.
16. Gomes, C.P., Selman, B. (2002). Satisfied with Physics. *Science.* Vol. 297 No. 5582.
17. Stephens, C.R. (1999). "Effective" Fitness Landscapes for Evolutionary Systems. In *Proc. 1999 Congress on Evol. Comp.,* pp. 703-714. IEEE Press.
18. Wolpert, D.H., Macready, W.G. (1997). No Free Lunch Theorems for Optimization. *IEEE Trans. Evol. Comp.,* vol. 1, no. 1, pp. 67-82.

Search Space Analysis of the Linear Ordering Problem

Tommaso Schiavinotto and Thomas Stützle

Darmstadt University of Technology, Intellectics Group,
Alexanderstr. 10, 64283 Darmstadt, Germany
{schiavin,tom}@intellektik.informatik.tu-darmstadt.de

Abstract. The Linear Ordering Problem (LOP) is an $\mathcal{NP}$-hard combinatorial optimization problem that arises in a variety of applications and several algorithmic approaches to its solution have been proposed. However, few details are known about the search space characteristics of LOP instances. In this article we develop a detailed study of the LOP search space. The results indicate that, in general, LOP instances show high fitness-distance correlations and large autocorrelation length but also that there exist significant differences between real-life and randomly generated LOP instances. Because of the limited size of real-world instances, we propose new, randomly generated large real-life like LOP instances which appear to be much harder than other randomly generated instances. Additionally, we propose a rather straightforward Iterated Local Search algorithm, which shows better performance than several state-of-the-art heuristics.

1 Introduction

The Linear Ordering Problem (LOP) is $\mathcal{NP}$–*hard* and has a large number of applications in such diverse fields as economy, sociology, graph theory, archaeology, and task scheduling [6]. Given an $n \times n$ matrix C, the LOP is the problem of finding a permutation π of column and row indices $\{1, ..., n\}$ such that the function

$$f(\pi) = \sum_{i=1}^{n} \sum_{j=i+1}^{n} c_{\pi(i)\pi(j)}$$

is maximized. In other words, the goal is to find a permutation of the columns and rows of C such that the sum of the elements in the upper triangle is maximized.

Both, exact [6,4,14] and approximate algorithms [1,3,5,11,2] have been proposed for this problem. State-of-the-art exact algorithms can solve fairly large instances with up to a few hundred columns; however, their computation time increases strongly with instance size. Approximate algorithms include constructive algorithms like Becker's greedy algorithm [1], local search algorithms like the $\mathcal{CK}$ heuristic [3] as well as metaheuristics such as tabu search [11], scatter search [2], or iterated dynasearch [5].

These algorithms have typically been tested on real-world as well as randomly generated instances. The LOLIB benchmark library comprises 49 real-world instances that are input-output tables of economical flows in the EU. This is the most widely used set of instances to test algorithms for the LOP and it is available at http://www.iwr.uni-heidelberg.de/iwr/comopt/soft/LOLIB/; for all

S. Cagnoni et al. (Eds.): EvoWorkshops 2003, LNCS 2611, pp. 322–333, 2003.

LOLIB instances optimal solutions are known [7]. Because LOLIB instances are rather small, Mitchell and Borchers [14] generated larger instances in their research on exact algorithms for the LOP. The idea underlying their way of generating instances is that there should be a large number of solutions with costs close to the optimal value in order to obtain hard instances. Thirty of these instances with known optimal solutions are available at http://www.rpi.edu/~mitchj/generators/linord, where also the generator can be found; we will refer to this instances as MBLB (Mitchell-Borchers LOP Benchmarks). Other instances were based on random generators [11], however, the original instances are not publically available and therefore not used in this paper.

One of the contributions of this paper is an analysis of the search space characteristics of the available LOP instances including an autocorrelation analysis [17,19] and a fitness-distance analysis [9]. Our results indicate significant differences between the LOLIB and MBLB instances, surprisingly suggesting that MBLB instances should be, when adjusting for the difference in size, easier to solve for metaheuristics. We generated large real-world like instances and computational results confirm our conjecture. Additionally, our results of the search space analysis suggest that Iterated Local Search (ILS) algorithms [12] are likely to achieve high performance for the LOP. In fact, a rather straightforward ILS algorithm using the $\mathcal{CK}$ local search heuristic appears to be competitive or superior to all previously proposed metaheuristic approaches.

The paper is structured as follows. In the next section we analyze the behavior of different local search algorithms on the two instance classes. Sections 3 and 4 give statistical measures on the instance structure and the results of the landscape analysis. Finally, we give details on the performance of two Iterated Local Search implementations based on two different local searches and give some concluding remarks in Section 6.

2   Local Search

The currently best known constructive algorithm for the LOP is due to Becker [1] and orders the columns (and the rows) in non-decreasing order of the value $q_j, j = 1, \ldots, n$, where q_j is

$$q_j = \frac{\sum_{k=1}^{n} c_{jk}}{\sum_{k=1}^{n} c_{kj}}.$$

The higher q_j the sooner the index j must be in the permutation. Becker's algorithm runs in $\mathcal{O}(n^2 log(n))$ and it returns good solutions; for example, the average deviation from the optimum solutions for LOLIB instances with Becker's heuristic is 9.46% (compared to an average of 30.48% for random permutations) and 2.38% on MBLB instances (random permutations average 40.34% above optimum).

Better solutions are obtained using local search algorithms. We run some with a number of iterative improvement algorithms based on two different neighborhoods, which are defined through the operations applicable to the current solution. The first is the *insert* operation: an element in position i is inserted in another position j. Formally, *Insert* : $\Pi \times \{1, \ldots, n\}^2 \to \Pi$, where Π is the set of all permutations) is defined for $i \neq j$:

$$Insert(\pi, i, j) \triangleq \begin{cases} (\ldots, \pi_{i-1}, \pi_{i+1}, \ldots, \pi_j, \pi_i, \pi_{j+1}, \ldots) & i < j; \\ (\ldots, \pi_{j-1}, \pi_i, \pi_j, \ldots \pi_{i-1}, \pi_{i+1}, \ldots) & i > j; \end{cases}$$

Table 1. Comparison of iterative improvement algorithms based on different neighborhoods on MBLB and LOLIB instances (B.I. indicates that a best-improvement pivoting rule was applied, in the other cases we applied first-improvement). The results are averaged over 100 runs on the 30 instances of MBLB and 49 instances of LOLIB, # Opt. indicates on how many instances at least once a global optimum was found among the 100 runs, and the time given is the total time spent to run the local search on all instances once. If VND is used, this is indicated by two neighborhoods which are separated by; in this case the number of instances for which local search in the second neighborhood led to an improvement is given together with the percentage of the runs in which we observed an improvement.

	Local Search	Avg. Dev. (%)	# Opt.	time (s)	# Inst. Improved	% Improved Runs
	$\mathcal{N}_I$	0.0195	10	9.81	–	–
	$\mathcal{N}_I$ (B.I.)	0.0219	11	187.72	–	–
	$\mathcal{N}_X$	0.3870	0	9.35	–	–
	$\mathcal{CK}$	0.0209	12	**0.22**	–	–
MBLB	$\mathcal{N}_X + \mathcal{N}_I$	0.0182	11	23.33	**30**	**100.00**
	$\mathcal{N}_I + \mathcal{N}_X$	0.0191	**14**	11.44	0	0
	$\mathcal{N}_I + \mathcal{CK}$	**0.0169**	10	9.94	26	26.30
	$\mathcal{CK} + \mathcal{N}_I$	0.0197	**14**	0.90	3	0.40
	$\mathcal{N}_I$	0.1842	42	0.1802	–	–
LOLIB	$\mathcal{CK}$	0.2403	38	**0.0205**	–	–
	$\mathcal{N}_I + \mathcal{CK}$	**0.1819**	**44**	0.1881	33	8.37
	$\mathcal{CK} + \mathcal{N}_I$	0.2360	40	0.0420	2	0.08

We denote this neighborhood by $\mathcal{N}_I$; its size is $|\mathcal{N}_I| = (n-1)^2$.

Another possible neighborhood is $\mathcal{N}_X$, defined by the operation *interchange*; it is given as *Interchange* $: \Pi \times \{1, \dots, n\}^2 \to \Pi$, where we have for $i \neq j$:

$$Interchange(\pi, i, j) \overset{\Delta}{=} (\dots, \pi_{i-1}, \pi_j, \pi_{i+1}, \dots, \pi_{j-1}, \pi_i, \pi_{j+1}, \dots)$$

The size of this neighborhood is $|\mathcal{N}_X| = n(n-1)/2$. We implemented both, a best- and first-improvement version of local search based on the $\mathcal{N}_I$ and the $\mathcal{N}_X$ neighborhoods (the neighborhood is randomly scanned). In addition we also used a *variable neighborhood descent* (VND) style heuristic: if a local minimum with respect to the first neighborhood is met, the search is continued exploiting the second neighborhood; if the localsearch in the second neighborhood finds an improved solution one goes back to the first one.

In addition to these standard neighborhoods, we also implemented the local search algorithm $\mathcal{CK}$ by Chanas and Kobylański [3] that uses two functions *sort* and *reverse*. When applied to a permutation, *sort* returns a new permutation in which the elements are rearranged according to a specific sorting criterion (see [3]), while *reverse* returns simply the reversed permutation. In the LOP case, if a permutation maximizes the objective function, the reversed permutation minimizes the objective function; hence, reversing a good solution leads to a bad solution. The idea of $\mathcal{CK}$ is to alternate sorting and reversing to improve the current solution; in fact, it has been shown that the application of *reverse* and then *sort* to a solution will lead to a solution with a value greater or equal the starting one. The functional description of the algorithm is:

$$(sort^* \circ reverse)^* \circ sort^*$$

where $\circ$ denotes function composition, and the $*$ operator is used to apply any given function iteratively until the objective function does not change. Formally we consider a general function ϕ, and a generic permutation π:

$$\phi^*(\pi) \triangleq \begin{cases} \pi & f(\phi(\pi)) = f(\pi) \\ \phi^*(\phi(\pi)) & \text{otherwise} \end{cases}$$

Unfortunately, the $\mathcal{CK}$ local search induces an ill-defined neighborhood that cannot be used for all the types of search space analysis conducted in section 4.

Table 1 shows the results obtained by applying the various LS algorithms to random initial solutions. It can be noticed that $\mathcal{CK}$ is by more than an order of magnitude faster than the other local search algorithms using a single neighborhood and it obtains a solution quality comparable to the other neighborhoods. Overall, the best solution quality when running local search in a single neighborhood is obtained with the $\mathcal{N}_I$ neighborhood, while local search in the $\mathcal{N}_X$ neighborhood returns poor quality solutions using comparable computation times to *insert*. The VND algorithms show that only by first searching in $\mathcal{N}_I$ and then with $\mathcal{CK}$ yields significant improvements over the single-neighborhood local search, while it is not the case vice versa. In general, these preliminary experiments suggest that good results can be obtained using a single local search and that the improvement through VND is not important. Hence using a single neighborhood in the local search allows for a simpler design of a metaheuristic and a shorter computation time. In the rest of this work we will focus on *insert* and $\mathcal{CK}$, because they are the best performing in terms of solution quality.

3   Structural Analysis of the Instances

The first level of our analysis of LOP instances is based on a high level description of the instances and on the input data, that is, the distribution of the matrix entries.

LOLIB comprises 49 real world instances, of which 39 are of size $n = 44$, 5 of size $n = 50$, 11 of size $n = 56$, and 3 of size $n = 60$. MBLB instances are randomly generated, with the matrix entries generated according to a uniform distribution, and then a number of elements are set to zero. MBLB comprises 30 instances: 5 instances of size 100, 10 of size 150, 10 of size 200 and 5 of size 250.

For all instances we computed the sparsity, the variation coefficient and the skewness of the matrix entries. The sparsity measures the percentage of elements that are equal to zero; it seemed to have a strong influence on algorithm behavior in [14]. The sparsity of the real world examples varies from 11% to 80%, while MBLB instances where generated with fixed sparsity of 0%, 10%, and 20%. Two further measures, which depend to some extent on the sparsity, were computed. The *variation coefficient* (VC) is defined as $\sigma/\bar{X}$, where σ is the standard deviation and $\bar{X}$ is the mean of the matrix entries: it gives an estimate of the variability independent of the size and the range of the matrix entries. The *skewness* is the third moment of the mean normalized by the standard deviation, it indicates the degree of asymmetry of the matrix entries.

Table 2 gives some information about these measures for LOLIB and MBLB instances. We present the results for the two instance classes in two different ways, because of the low variance of these values on the MBLB instances and because these

Table 2. Structural information on LOLIB (left) and MBLB (right) instances; LOLIB instances are grouped according to size, while MBLB instances are grouped according to size and sparsity; 1st Qu. and 3rd Qu. indicate the first and the third quantile, respectively. For MBLB the table entries give the range of the VC and the skewness among the instances in each group.

Size		Min	1st Qu.	Median	3rd Qu.	Max	Mean
44	Sparsity	11.00	31.20	39.36	53.05	80.63	39.36
	VC	4.21	4.59	5.13	5.53	10.34	5.49
	Skewness	9.78	11.51	13.01	15.31	25.70	14.28
50	Sparsity	33.72	33.73	43.20	52.11	58.44	44.64
	VC	5.54	5.72	6.96	10.14	16.17	8.91
	Skewness	13.13	15.16	21.09	29.55	39.21	23.63
56	Sparsity	25.51	25.95	26.91	27.18	28.06	26.67
	VC	4.24	4.38	4.45	5.01	6.45	4.85
	Skewness	10.84	11.48	12.24	16.81	30.52	15.67
60	Sparsity	28.69	28.78	28.86	29.42	29.97	29.17
	VC	5.85	5.98	6.11	6.12	6.14	6.03
	Skewness	23.84	23.96	24.09	24.47	24.85	24.26

Variation Coefficient

Size	Sparsity		
	0%	10%	20%
100	-	-	1.00-1.02
150	0.77-0.78	0.88-0.89	-
200	0.77-0.78	0.88	-
250	0.77-0.78	-	-

Skewness

Size	Sparsity		
	0%	10%	20%
100	-	-	0.98-1.00
150	0.82-0.85	0.87-0.88	-
200	0.82-0.85	0.88-0.89	-
250	0.82-0.84	-	-

instances can easily be grouped on sparsity. As we see, the latter measure determines the skewness and the VC for MBLB instances. This analysis already shows that LOLIB and MBLB instances are very different. While MBLB instances are rather similar (see the low variation in our measures) and generally have low VC and skewness, the real-world instances show much larger differences in all three measures and the VC and skewness are typically much larger.

4 Landscape Analysis

Landscape analysis studies non-trivial features of combinatorial problems. The idea is to "visualize" the search space as a landscape formed by all the solutions (in our case permutations) and a *fitness* value for each solution corresponding in our case to the objective function f [13].

Formally, a landscape for the LOP is described by a triple $\langle \Pi(n), f, d \rangle$, where Π is the set of all permutations of the integers $\{1, \ldots, n\}$, f is the cost function and d is a distance measure, which induces a structure on the landscape. It is natural to define the distance between two permutation π and π' in dependence of the basic operation used by a local search algorithm; typically, the distance then is given by the minimum number of applications of this basic operation needed to transform π into π'. While for the $\mathcal{CK}$ algorithm the *basic operation* is not a clear concept, in the case of the *insert* local search it is clearly the *insert* operation. Unfortunately, as far as we know, there is no efficient way of computing the minimum number of *insert* applications needed to transform one permutation into another one. Therefore, we use for both neighborhoods a surrogate distance that is based on the *precedence metric* [15]: for all pairs of elements j and i we count how often j precedes i in both permutations and then subtract this quantity from $n(n-1)/2$, which is the maximum possible distance.

The first feature of the landscape we studied is its ruggedness: a fitness landscape is said to be rugged if there is a low correlation between neighboring points. To measure

Table 3. Given is the value ℓ/n, that is the correlation length normalized by instance size for LOLIB and MBLB instances.

	Size	Min	1st Qu.	Median	3rd Qu.	Max	Mean
LOLIB	44	0.7536	0.7854	0.7929	0.8031	0.8237	0.7937
	50	0.7642	0.7695	0.7861	0.8043	0.8145	0.7877
	56	0.8004	0.8120	0.8241	0.8311	0.8371	0.8208
	60	0.8383	0.8389	0.8395	0.8399	0.8403	0.8394
MBLB	100	0.9339	0.9350	0.9357	0.9360	0.9371	0.9355
	150	0.9594	0.9595	0.9610	0.9623	0.9645	0.9612
	200	0.9626	0.9690	0.9696	0.9710	0.9744	0.9698
	250	0.9703	0.9742	0.9748	0.9769	0.9775	0.9747

this correlation we can perform a *random walk* in the landscape, to interpret the set of m points $\{f(x_t)\}, t = 1, \ldots, m$ as a time series and to measure the autocorrelation $r(s)$ of points in this time series that are separated by s steps [17,16,19] as

$$r(s) = \frac{1}{\sigma^2(f)(m-s)} \sum_{t=1}^{m-s} (f(x_t) - \bar{f})(f(x_{t+s}) - \bar{f}),$$

where $\sigma^2(f)$ is the variance of the time series, and $\bar{f}$ the mean. Often, the resulting time series can be modeled as an autoregressive process of order one, and the correlation can be summarized by the *landscape correlation length* that is computed as $\ell = -\frac{1}{\ln(|r(1)|)}$ ($r(1) \neq 0$); the lower is the value of ℓ, the more rugged is the landscape. We computed ℓ on the LOLIB and MBLB instances based on the $\mathcal{N}_I$ neighborhood, because $\mathcal{CK}$, as said before, cannot be used for this type of analysis. Table 3 summarizes data collected on all instances grouped by size. The correlation length is always smaller for LOLIB instances than for MBLB instances, which, abstracting from the instance size, indicates that the LOLIB instances are harder for *insert* than the ones of MBLB.

The next step in the our analysis was to generate a large number of local optima for all the instances (13,000 for LOLIB and 1,000 for MBLB instances). Based on these, we first analyzed the number of global optima that were found among the local optima. For *insert* local search, we found on average 0.46% distinct global optima among the distinct local optima for MBLB instances (min. 0%, max. 4.27%) and 14.10% for LOLIB instances (min. 0.47%, max. 85.12%); this indicates that many instances can actually be solved by a random restart algorithm that is run long enough. Furthermore, as Table 1 shows, local optima of MBLB instances have a smaller deviation from the optimum solution value than those of LOLIB instances.

Another standard technique in analyzing the landscape is the study of the fitness-distance correlation ($\rho, -1 \leq \rho \leq 1$) between the quality of the local optima and the distance from the closest global optima [9]. Here we focus on the relationship between the distance to the closest globally optimal solution and the deviation from the global optimum as the fitness function. In this case, a high, positive value of ρ indicates that the solution quality gives good guidance when searching for global optima (the interpretation then is that the better the solution, the closer we get to global optima, on average). Table 4 summarizes the information about fitness/distance correlation and Figure 1 gives some example plots of the fitness-distance relationship.

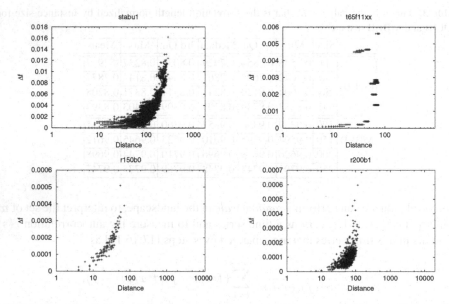

Fig. 1. Plots of the deviation from the optimal value versus the minimum distance of local optima (generated with *insert* LS) from a global optimum, the maximum distance is shown. From LOLIB: stabu1 $\rho = 0.8757$; t65f11xx $\rho = -0.1056$ (lowest). From MBLB: r150b0 $\rho = 0.9395$ (highest); r200b1 $\rho = 0.6144$ (lowest).

All MBLB instances and most LOLIB instances show a very high value for ρ, suggesting that these instances should be relatively easy for restart type algorithms [13]. However, some of the LOLIB instances show even negative fitness-distance correlation; therefore, these instances may pose some problems to metaheuristics despite their small size. In general, the variability of ρ is much higher for LOLIB instances, suggesting that these show a variety of different structures, different from MBLB instances, which are more similar to each other. Additionally, ρ is typically smaller for LOLIB instances than MBLB instances, suggesting that they should be somewhat harder than MBLB instances. Finally, we run a paired t-test for each class to compare the ρ values obtained by the $\mathcal{CK}$ and *insert* local search on each instance; the result was that they are not significantly different both for LOLIB and for MBLB.

5   Iterated Local Search

One main observation of the landscape analysis is the high fitness distance correlation present in many LOP instances, which suggests the usefulness of restart type metaheuristics, that is, metaheuristics that iteratively generate starting points for a local search procedure. Iterated local search (ILS) is such a restart type metaheuristic that despite its simplicity achieved excellent results on several $\mathcal{NP}$-hard problems [12]. ILS iterates in a particular way over the local search process applying three main steps: (i) perturb a locally optimal solution, then (ii) locally optimize it with the local search chosen and

Table 4. Statistical information about fitness/distance correlation (ρ) on local optima returned by $\mathcal{CK}$ (left table) and *insert* (right table) for both classes of instances grouped by size.

	Size	Min	1st Qu.	Median	3rd Qu.	Max	Mean
LOLIB	44	-0.07	0.48	0.67	0.83	1.00	0.59
	50	0.17	0.26	0.49	0.74	0.92	0.52
	56	0.34	0.67	0.72	0.85	0.92	0.73
	60	0.67	0.73	0.79	0.84	0.88	0.78
MBLB	100	0.68	0.75	0.75	0.77	0.83	0.76
	150	0.72	0.84	0.84	0.87	0.90	0.84
	200	0.53	0.66	0.75	0.77	0.91	0.73
	250	0.79	0.79	0.85	0.86	0.87	0.83

	Size	Min	1st Qu.	Median	3rd Qu.	Max	Mean
LOLIB	44	-0.11	0.42	0.64	0.76	1.00	0.58
	50	-0.01	0.40	0.61	0.75	0.89	0.53
	56	0.34	0.65	0.70	0.85	0.94	0.72
	60	0.70	0.71	0.73	0.80	0.88	0.77
MBLB	100	0.71	0.71	0.73	0.81	0.84	0.76
	150	0.70	0.79	0.84	0.86	0.94	0.82
	200	0.61	0.71	0.73	0.78	0.91	0.75
	250	0.79	0.80	0.82	0.84	0.86	0.82

finally (iii) choose, based on some acceptance criterion, the solution that undergoes the next perturbation phase.

We implemented two variants of ILS, both of which apply perturbations based on $\mathcal{N}_X$ and always apply the perturbations to the best solution found so far. Actually, the only difference between the two implementations is that one is based on the $\mathcal{CK}$ local search (ILS$_{\mathcal{CK}}$) and the other on the *insert* local search (ILS$_I$). For both, the parameter tuning has been done on the MBLB instances, for ILS$_{\mathcal{CK}}$ the same settings were found to be good also for LOLIB instances, while ILS$_I$ exhibits some lack of tuning on the latter class of instances due to the bias caused by MBLB instances.

We run both algorithms 100 times on all LOLIB and MBLB instances on an AMD Athlon 1.2GHz machine with 1GB RAM. Table 5 summarizes the maximum time needed to find the known global optima for both instance classes and both ILS algorithms. The results suggest that ILS$_{\mathcal{CK}}$ yields much better performance than ILS$_I$ over the whole benchmark set. For example, ILS$_{\mathcal{CK}}$ was able to find always a global optimum for LOLIB instances in less than 0.34s, except for instance be75np where the maximum time for finding a global optimum was of 1.57s. On the MBLB instances ILS$_{\mathcal{CK}}$ takes 14.27s for the r200e1, and less than the half for all the others. ILS$_I$ requires much larger computation times. For example, the maximum computation times over 100 runs range between 1.29s and 158.33s for the MBLB, and between 0.02s and 45.19s for LOLIB, with a peak of 978s for instance be75np (such a behavior can be in part due to parameter under-tuning). The significantly better performance of ILS$_{\mathcal{CK}}$ over ILS$_I$ is illustrated also by a pairwise comparison between the two algorithms in Figure 2 on the left side, where each point gives the timings of the two different algorithms for the same instance.

Table 5. Summary of the maximum time for finding the optima over 100 runs (the optima has been found in every run) of ILS_{CK} (left side) and ILS_I (right side).

	Size	Min	1st Qu.	Median	3rd Qu.	Max	Mean
LOLIB	44	0.00	0.00	0.01	0.03	0.16	0.03
	50	0.03	0.03	0.09	0.50	1.57	0.44
	56	0.01	0.02	0.03	0.07	0.15	0.05
	60	0.09	0.13	0.16	0.25	0.34	0.20
MBLB	100	0.14	0.20	0.54	0.70	2.41	0.80
	150	0.16	0.21	0.34	0.76	1.81	0.59
	200	0.27	0.94	1.69	4.00	14.27	3.30
	250	0.55	1.44	2.53	3.23	6.25	2.80

	Size	Min	1st Qu.	Median	3rd Qu.	Max	Mean
LOLIB	44	0.02	0.07	0.15	1.19	45.19	3.73
	50	0.16	0.48	6.56	253.90	977.96	247.80
	56	0.160	0.39	0.88	4.66	22.71	4.96
	60	1.40	1.85	2.30	8.12	13.93	5.88
MBLB	100	1.29	1.39	12.37	13.81	34.39	12.65
	150	2.66	3.17	6.94	12.99	27.13	9.27
	200	9.94	21.42	41.60	101.70	158.30	62.51
	250	27.60	49.66	52.21	118.70	133.00	76.23

In addition, we run an algorithm by Mitchell and Borchers (SimpMB) [14], which is based on the Simplex algorithm and that uses a branch and bound procedure to find the integer solution, to compare ILS to exact algorithms. Figure 2 (right side) gives a pairwise comparison between SimpMB and ILS_{CK}. This comparison shows that ILS_{CK} is clearly faster than SimpMB often by several orders of magnitude.

Run-time distribution (RTD) plots depict the probability of finding a global optimum in dependence of computation time [8]. This kind of plots (e.g. Figure 3(a)) point towards an interesting property of the two implementations on MBLB instances: the curves show a similar shape. This suggested to do the same plots but over the number of iterations instead of computation time for the two ILS (Figure 3(b)). As can be observed, by considering only the number of iterations the two algorithms are very similar, and sometimes ILS_I even needs less iterations than ILS_{CK}. In the case shown in Figure 3 ILS_I needs a maximum of 132.99s against 6.25s of ILS_{CK}, but only a maximum of 416 iterations, while ILS_{CK} needs 552. This indicates that, at least on MBLB instances, the two local searches mainly differ on the speed, and they have not an intrinsically different behavior.

We also compared ILS_{CK} to published results for existing LOP algorithms (Table 6). These include a scatter search [2] (SS), elite tabu-search (ETS) [11] and ILS with Dynasearch local search [5]. These algorithms were evaluated on LOLIB instances, instances randomly extracted form the Stanford Graphbase [10], and some additional random instances. Here we focus on LOLIB instances, since we did not have available the other instances. SS and ETS were run on a Pentium 166MHz and Iterated Dynasearch and a SunSparc 5/110. We estimated (using some indirect comparison between cpu95 and cpu2000 benchmarks found in [18]) that our machine is roughly 15 times faster than

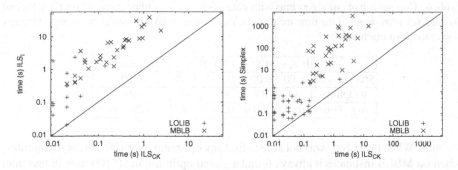

Fig. 2. Pairwise comparison of ILS$_{CK}$ to ILS$_I$ (left side, here the median of the 100 runs is shown) and the SimpMB algorithm (right side, here we use the maximum time of ILS$_{CK}$ over 100 runs).

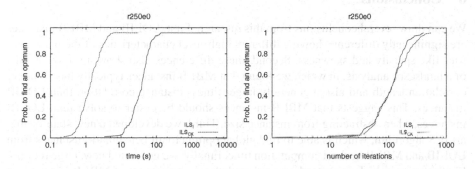

Fig. 3. RTD plots on a MBLB instance: over time (left); over number of iterations (right).

the Pentium 166MHz. In Table 6 we use the original timings given in the papers. There is clearly a ceiling effect due to the small size of the instances, but further experiment on some larger and harder instances would be needed to give a better idea of the quality of the algorithms. In any case, it seems that ILS$_{CK}$ is definitely competitive or may prove to be superior to the best known algorithms for the LOP (notice that ETS and SS did not find optimal solutions for all instances, difference from ILS$_{CK}$).

One additional conclusion from the landscape analysis was that, abstracting from instance size, LOLIB instances appear to be harder than the MBLB ones; therefore we generated new instances of the same size ($n = 250$) as the largest ones in MBLB: a new instance is generated drawing elements from a certain instance of LOLIB (for each instance a corresponding instance of size 250 has been generated), so that the statistical features were the same. As a first step we computed the landscape correlation lengths of the new, randomly generated instances and we observed values ranging between 69% and 75% of the size, a lower value than in the original instances. Then we chose instance be75np-250 (here -250 indicates the size of the new, randomly generated instance); the originalbe75np instance was among the hardest ones from LOLIB. We run SimpMB on be75np-250, which was aborted after 1,140,000 seconds (about 13 days and 4 hours), while on the MBLB instances the maximum time required to run SimpMB was 2987.89

Table 6. Comparison of our algorithms with state-of-the-art algorithms, timings for ETS, SS and Dynasearch correspond to the time measured on a Pentium 166MHz (which is roughly 15 times slower than our machine).

	ILS_{CK}	ILS_I	ETS[11]	SS[2]	Dynasearch [5]
Std. Dev.(%)	0.00	0.00	0.00	0.01	0.00
# Optima	49	49	47	43	49
Avg. Time	0.08	24.06	0.93	3.82	1.22(0.30)

seconds. While the ILS_{CK} was not able to find any optimum over 100 runs of 40 minutes, when on MBLB instances it always found a global optimum in all 100 runs in less than 15 seconds[1].

6 Conclusions

We can draw several conclusions from this research. First, LOLIB and MBLB instances are significantly different, showing different high-level characteristics of the matrix entries like sparsity and skewness. Second, these differences also show up in the results of a landscape analysis, in which we found that MBLB instances typically have a larger correlation length and also a generally larger fitness-distance correlation than LOLIB instances. This suggests that MBLB instances should be easier to solve than LOLIB instances, when abstracting from instance size. Third, we developed a new state-of-the-art ILS algorithm, which is able to find global optima to all benchmark instances from LOLIB and MBLB in short computation time. Finally, we generated new, large real-life like instances, which appear to be significantly harder to solve than MBLB instances of the same size.

In future work we will extend the landscape analysis to a larger number of instances, including all the different types of instances proposed in the literature and examine more closely the relationship of sparsity to instance hardness. A second line of research is to examine the performance of other metaheuristics on the LOP. However, preliminary results indicate that it will be difficult to reach the performance of ILS_{CK}.

Acknowledgments. The authors would wish to thank Prof. John Mitchell and Dr. Brian Borchers for making available the code of the simplex and the referees for some helpful comments. This work was supported by the "Metaheuristics Network", a Research Training Network funded by the Improving Human Potential programme of the CEC, grant HPRN-CT-1999-00106. The information provided is the sole responsibility of the authors and does not reflect the Community's opinion. The Community is not responsible for any use that might be made of data appearing in this publication.

References

1. O. Becker. Das Helmstädtersche Reihenfolgeproblem – die Effizienz verschiedener Näherungsverfahren. In *Computer uses in the Social Science*, Wien, January 1967.

[1] These instances will be made available at the address
http://www.intellektik.informatik.tu-darmstadt.de/~schiavin/lop.

2. V. Campos, M. Laguna, and R. Martí. Scatter search for the linear ordering problem. In D. Corne et al., editor, *New Ideas in Optimization*, pages 331–339. McGraw-Hill, 1999.

3. S. Chanas and P. Kobylanski. A new heuristic algorithm solving the linear ordering problem. *Computational Optimization and Applications*, 6:191–205, 1996.

4. T. Christof and G. Reinelt. Low-dimensional linear ordering polytopes. Technical report, University of Heidelberg, Germany, 1997.

5. R. K. Congram. *Polynomially Searchable Exponential Neighbourhoods for Sequencing Problems in Combinatorial Optimisation*. PhD thesis, University of Southampton, Faculty of Mathematical Studies, UK, 2000.

6. M. Grötschel, M. Jünger, and G. Reinelt. A cutting plane algorithm for the linear ordering problem. *Operations Research*, 32(6):1195–1220, 1984.

7. M. Grötschel, M. Jünger, and G. Reinelt. Optimal triangulation of large real world input–output matrices. *Statistische Hefte*, 25:261–295, 1984.

8. H.H. Hoos and T. Stützle. Evaluating Las Vegas algorithms — pitfalls and remedies. In *Proceedings of the Fourteenth Conference on Uncertainty in Artificial Intelligence (UAI-98)*, pages 238–245. Morgan Kaufmann, San Francisco, 1998.

9. T. Jones and S. Forrest. Fitness distance correlation as a measure of problem difficulty for genetic algorithms. In L.J. Eshelman, editor, *Proc. of the 6th International Conference on Genetic Algorithms*, pages 184–192. Morgan Kaufman, San Francisco, 1995.

10. D. E. Knuth. *The Stanford GraphBase: A Platform for Combinatorial Computing*. Addison Wesley, New York, 1993.

11. M. Laguna, R. Martí, and V. Campos. Intensification and diversification with elite tabu search solutions for the linear ordering problem. *Computers and Operation Research*, 26:1217–1230, 1999.

12. H. R. Lourenço, O. Martin, and T. Stützle. Iterated local search. In F. Glover and G. Kochenberger, editors, *Handbook of Metaheuristics*, volume 57 of *International Series in Operations Research & Management Science*, pages 321–353. Kluwer Academic Publishers, Norwell, MA, 2002.

13. P. Merz and B. Freisleben. Fitness landscapes and memetic algorithm design. In D. Corne, M. Dorigo, and F. Glover, editors, *New Ideas in Optimization*, pages 245–260. McGraw-Hill, London, 1999.

14. J. E. Mitchell and B. Borchers. Solving linear ordering problems with a combined interior point/simplex cutting plane algorithm. In H. L. Frenk *et al.*, editor, *High Performance Optimization*, pages 349–366. Kluwer Academic Publishers, Dordrecht, The Netherlands, 2000.

15. C. R. Reeves. Landscapes, operators and heuristic search. *Annals of Operational Research*, 86:473–490, 1999.

16. P. Stadler. Towards a theory of landscapes. In R. Lopéz-Peña, R. Capovilla, R. García-Pelayo, H. Waelbroeck, and F. Zertuche, editors, *Complex Systems and Binary Networks*, volume 461, pages 77–163, Berlin, New York, 1995. Springer Verlag.

17. P. Stadler. Landscapes and their correlation functions. *J. of Math. Chemistry*, 20:1–45, 1996.

18. Standard Performance Evaluation Corporation. SPEC CPU95 and CPU2000 Benchmarks. http://www.spec.org/, November 2002.

19. E. D. Weinberger. Correlated and uncorrelated fitness landscapes and how to tell the difference. *Biological Cybernetics*, 63:325–336, 1990.

Ant Algorithms for the University Course Timetabling Problem with Regard to the State-of-the-Art

Krzysztof Socha, Michael Sampels, and Max Manfrin

IRIDIA, Université Libre de Bruxelles, CP 194/6,
Av. Franklin D. Roosevelt 50, 1050 Bruxelles, Belgium
{ksocha|msampels|mmanfrin}@ulb.ac.be
http://iridia.ulb.ac.be

Abstract. Two ant algorithms solving a simplified version of a typical university course timetabling problem are presented – Ant Colony System and $\mathcal{MAX}$-$\mathcal{MIN}$ Ant System. The algorithms are tested over a set of instances from three classes of the problem. Results are compared with recent results obtained with several metaheuristics using the same local search routine (or neighborhood definition), and a reference random restart local search algorithm. Further, both ant algorithms are compared on an additional set of instances. Conclusions are drawn about the performance of ant algorithms on timetabling problems in comparison to other metaheuristics. Also the design, implementation, and parameters of ant algorithms solving the university course timetabling problem are discussed. It is shown that the particular implementation of an ant algorithm has significant influence on the observed algorithm performance.

1 Introduction

The work presented here arises out of the Metaheuristics Network[1] (MN) – a European Commission project undertaken jointly by five European institutes – which seeks to compare metaheuristics on different combinatorial optimization problems. In the current phase of the four-year project, a university course timetabling problem is being considered.

The University Course Timetabling Problem (UCTP) is a typical problem faced periodically by every university of the world. The basic definition states that a number of courses must be placed within a given timetable, so that the timetable is feasible (may actually be carried out), and a number of additional preferences being satisfied is maximized. There are also other timetabling problems described in the literature that are similar to UCTP. They include examination timetabling [1], school timetabling [2], employee timetabling, and others [3]. They all share similar properties and are similarly difficult to solve. The general university course timetabling problem is known to be NP-hard, as are many of the subproblems associated with additional constraints [4,5,2]. Even when restricting the interest to UCTP alone, it is difficult to provide a uniform and

---

[1] http://www.metaheuristics.org

S. Cagnoni et al. (Eds.): EvoWorkshops 2003, LNCS 2611, pp. 334–345, 2003.

generic definition of the problem. Due to the fact that course organization as well as additional preferences may vary from case to case, the number and type of soft and hard constraints changes. Hence, the algorithmic solutions proposed for this problem usually concentrate on a particular subproblem.

Recently, in the course of the MN, five metaheuristics were evaluated and compared on instances of a certain reduction of UCTP [6]. The metaheuristics evaluated included: Genetic Algorithm (GA), Simulated Annealing (SA), Tabu Search (TA), Iterated Local Search (ILS), and Ant Colony Optimization (ACO). The $\mathcal{MAX}$-$\mathcal{MIN}$ Ant System ($\mathcal{MMAS}$) [7,8] algorithm for the UCTP was developed later as an approach alternative to the ACO (ACS in fact) developed for the metaheuristics comparison. The purpose of this paper is to present how ant algorithms are performing on such highly constrained problems as UCTP, and analyze the impact of choosing a particular type of ant algorithm.

The remaining part of the paper is organized as follows: Section 2 defines the reduction of the UCTP being solved. Section 3 presents the $\mathcal{MAX}$-$\mathcal{MIN}$ Ant System and the Ant Colony System used for solving the UCTP. Also the major differences and similarities of the algorithms are highlighted. Section 4 presents the experiments that were performed in order to evaluate the algorithms' performance. The results obtained by ant algorithms are also compared to the results previously obtained by other metaheuristics [6]. Finally, Section 5 summarizes the findings and presents the conclusions drawn.

2   UCTP – Problem Definition

For the purpose of evaluating metaheuristics in the course of the MN, a reduction of UCTP has been defined [6,8]. The problem consists of a set of n events E to be scheduled in a set of timeslots $T = \{t_1, \ldots, t_k\}$ ($k = 45$, 5 days of 9 hours each), a set of rooms R in which events can take place (rooms are of a certain capacity), a set of students S who attend the events, and a set of features F satisfied by rooms and required by events. Each student is already preassigned to a subset of events. A feasible timetable is one in which all events have been assigned a timeslot and a room so that the following hard constraints are satisfied:

- no student attends more than one event at the same time;
- the room is big enough for all the attending students and satisfies all the features required by the event;
- only one event is taking place in each room at a given time.

In addition, a feasible candidate timetable is penalized equally for each occurrence of the following soft constraint violations:

- a student has a class in the last slot of the day;
- a student has more than two classes in a row (one penalty for each class above the first two);
- a student has exactly one class during a day.

The infeasible timetables are worthless and are considered equally bad regardless of the actual level of infeasibility. The objective is to minimize the number of soft constraint violations (#scv) in a feasible timetable. The solution to the UCTP is a mapping of events into particular timeslots and rooms.

2.1   Problem Instances

Instances of the UCTP were constructed using a generator written by Paechter[2]. For the comparison being carried out by the MN, three classes of instance have been chosen, reflecting realistic timetabling problems of varying sizes: *small, medium, large*. They differ mostly by number of events being placed (100, 400, 400 respectively) and number of students attending these events (80, 100, 400 respectively). For details on the problem instance generator and exact parameters used to generate the three above classes of the UCTP problem see [6]. For further analysis of the performance of both ant algorithms, 10 additional problem instances were used. Those instances have been proposed as a part of the International Timetabling Competition[3] (*competition instances*). The complexity level of those competition instances may be considered to be between the *medium* and *large* instances. Note, that all the instances used for the tests are known to have a perfect solution, i.e. so that no hard or soft constraints are violated.

3   Ant Algorithms to Be Compared

Ant Colony Optimization (ACO) is a metaheuristic proposed by Dorigo et al. [9]. The inspiration of ACO is the foraging behavior of real ants. The basic ingredient of ACO is the use of a probabilistic solution construction mechanism based on stigmergy. ACO has been applied successfully to numerous combinatorial optimization problems including the quadratic assignment problem, satisfiability problems, scheduling problems etc.

There exist at least two basic variations of the ACO metaheuristic – the $\mathcal{MAX}$-$\mathcal{MIN}$ Ant System initially proposed in [7], and the ACS which is described in detail in [10,11]. While the basic idea of operation is identical for both of those variations, there are some differences. The main difference lies in the way the pheromone is updated, which we will explain below.

Both ant algorithms used for solving the UCTP that are presented here, are based on the general ACO framework [10,12,7]. They both have been shown to be able to produce meaningful results for UCTP instances. The ACS has been compared with other recent metaheuristics in [6], and $\mathcal{MMAS}$ has been shown to significantly outperform the random restart local search algorithm in [8]. However, the performance of the two ant algorithms has never been compared directly in order to assess the impact of the particular design and implementation. This paper provides more insight into the influence of the ant algorithm architecture and its implementation on the performance.

---

[2] http://www.dcs.napier.ac.uk/~benp
[3] http://www.idsia.ch/Files/ttcomp2002/

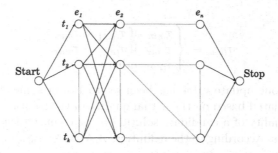

Fig. 1. The construction graph that the ants traverse when building an assignment of events into timeslots.

3.1 Differences and Similarities

The general mode of operation of both ant algorithms is very similar. At each iteration of the algorithm, each of the m ants constructs a complete assignment of events to timeslots. Following a pre-ordered list of events, the ants choose the timeslot for the given event probabilistically, guided by two types of information: heuristic information and stigmergic information. The stigmergic information is in the form of a matrix of *pheromone* values $\tau : E \times T \rightarrow \mathbf{R}^+$, where E is the set of events and T is the set of timeslots. The pheromone values are an estimate of the utility of making the assignment, as judged by previous iterations of the algorithm. Fig. 1 presents the idea of the construction graph that the ants traverse.

After all the events have been assigned to the timeslots, a deterministic matching algorithm assigns the rooms and a candidate solution C is generated. The local search routine [8] is then applied to C. The local search routine, which was provided separately in the course of the MN project, enables the specification of a maximal number of steps and/or maximum running time.

The algorithms compared, differ in the way they use the existing information (both stigmergic and heuristic), and the way they use local search. Also the rules of updating the pheromone matrix are different.

The $\mathcal{MAX\text{-}MIN}$ Ant System introduces upper and lower limits on the pheromone value. If the differences between some pheromone values were too large, all ants would almost always generate the same solutions, which would mean algorithm stagnation. The bounds on pheromone values prevent that. The maximal difference between the highest and the lowest level of pheromone may be controlled, and thus the level of search intensification versus diversification may be balanced. The pheromone update rule becomes then as follows (for the particular case of assigning events e into timeslots t):

$$\tau_{(e,t)} \leftarrow \begin{cases} (1-\rho) \cdot \tau_{(e,t)} + 1 & \text{if } (e,t) \text{ is in } C_{global\_best} \\ (1-\rho) \cdot \tau_{(e,t)} & \text{otherwise,} \end{cases} \tag{1}$$

where $\rho \in [0,1]$ is the evaporation rate. Pheromone update is completed using the following:

$$\tau_{(e,t)} \leftarrow \begin{cases} \tau_{min} & \text{if } \tau_{(e,t)} < \tau_{min}, \\ \tau_{max} & \text{if } \tau_{(e,t)} > \tau_{max}, \\ \tau_{(e,t)} & \text{otherwise.} \end{cases} \tag{2}$$

The pheromone update value has been set to 1 after some experiments with the values calculated based on the actual quality of the solution. The function q measures the quality of a candidate solution C by counting the number of constraint violations. According to the definition of $\mathcal{MMAS}$ $\tau_{max} = \frac{1}{\rho} \cdot \frac{g}{1+q(C_{optimal})}$, where g is a scaling factor. Since it is known that $q(C_{optimal}) = 0$ for the considered test instances, τ_{max} was set to a fixed value $\frac{1}{\rho}$. The proper balance of the pheromone update and the evaporation was needed, and this was controlled by the scaling factor g. We observed that when g was too small, the evaporation was faster than pheromone update, and pheromone levels even on the best paths finally reached τ_{min}. When the values of g were too large, the pheromone values on the best paths grew faster than they evaporated and finally reached τ_{max}, where they were cut-off according to the $\mathcal{MAX}$-$\mathcal{MIN}$ rule. It became apparent that any value of the pheromone update that was close to $\tau_{max} \cdot \rho$ is just as good. Experimental results supported this claim. Hence, we decided that for pheromone update the constant was more efficient than the calculation of the exact value.

In ACS not only the global update rule is used, but also a special local update rule. After each construction step a local update rule is applied to the element of the pheromone matrix corresponding to the chosen timeslot t_{chosen} for the given event e_i:

$$\tau_{(e_i,t_{chosen})} \leftarrow (1 - \alpha) \cdot \tau_{(e_i,t_{chosen})} + \alpha \cdot \tau_0 \tag{3}$$

The parameter $\alpha \in [0,1]$ is the pheromone decay parameter, which controls the diversification of the construction process. The aim of the local update rule is to encourage the subsequent ants to choose different timeslots for the same given event e_i.

At the end of the iteration, the global update rule is applied to all the entries in the pheromone matrix:

$$\tau_{(e,t)} \leftarrow \begin{cases} (1 - \rho) \cdot \tau_{(e,t)} + \rho \cdot \frac{g}{1+q(C_{global\_best})} & \text{if } (e,t) \text{ is in } C_{global\_best} \\ (1 - \rho) \cdot \tau_{(e,t)} & \text{otherwise,} \end{cases} \tag{4}$$

where g is a scaling factor, and the function q has been described above. This global update rule is than very similar to the one used by $\mathcal{MMAS}$ with the exception of not limiting the minimal and maximal pheromone level.

Another important difference between the implementations of the two algorithms, is the way that they use heuristic information. While $\mathcal{MMAS}$ does not use any heuristic information, the ACS attempts to compute it before making every move. In ACS the heuristic information is an evaluation of the constraint violations caused by making the assignment, given the assignments already made. Two parameters β and γ control the weight of the hard and soft constraint violations, respectively.

The last difference between the two ant algorithms concerns the use of the local search. In the case of $\mathcal{MMAS}$, only the solution that causes the fewest number of constraint violations is selected for improvement by the local search routine. Ties are broken randomly. The local search is run until reaching a local minimum or until assigned time for the trial is up – whichever happens first. The local search in case of ACS is run according to a two phase strategy: if the current iteration is lower than a parameter j the routine runs for a number of steps s_1, otherwise it runs for a number of steps s_2. In case of ACS all candidate solutions generated by the ants are further optimized with the use of local search.

Tab. 1 summarizes the parameters used by the two algorithms.

Table 1. Parameters used by the algorithms.

Parameter Name	$\mathcal{MMAS}$	ACS
m	number of ants	
ρ	pheromone evaporation	
$s(j)$	number of steps of the local search	
τ_0	value with which the pheromone matrix is initialized	
τ_{max}	maximal pheromone level	-
τ_{min}	minimal pheromone level	-
α	-	local pheromone decay
β	-	weight of the hard constraints
γ	-	weight of the soft constraints
g	-	scaling factor

4 Performance of the Ant Algorithms

For each class of the problem, a time limit for producing a timetable has been determined. The time limits for the problem classes small, medium, and large are respectively 90, 900, and 9000 seconds. These limits were derived experimentally. All the experiments were conducted on the same computer (AMD Athlon 1100 MHz, 256 MB RAM) under a Linux operating system. The ant algorithms were compared against the best metaheuristics on those instances [6], which were the Iterated Local Search and Simulated Annealing; also against a reference random restart local search algorithm (RRLS) [8], which simply generated a random solution and then tried to improve it by running just the local search. Since all algorithms were run on the same computer, it was easy to compare their performance and a fair comparison could be achieved.

For the remaining **competition instances**, only the ant algorithms were compared against each other and against RRLS. The running time on the same computer was set to 672 seconds. The time limit has been calculated with the use of the benchmark program provided by the organizers of the International Timetabling Competition.

Tab. 2 presents the actual parameters used for running the ant algorithms. The same parameters were used for all runs of both ant algorithms.

Table 2. Parameter settings used by the algorithms.

Parameter	$\mathcal{MMAS}$	ACS
m	10	10
ρ	0.3	0.1
$s(j)$	10 000 000	$\begin{cases} 50\,000 \ j \leq 10 \\ 20\,000 \ j \geq 11 \end{cases}$
τ_0	3.3	10.0
τ_{max}	3.3	-
τ_{min}	0.019	-
α	-	0.1
β	-	3.0
γ	-	2.0
g	-	10^{10}

The ant algorithms were tested on a set of instances of the UCTP as described in Sec. 2.1. The files containing those instances as well as source code of the algorithms and summary of the results may be found on the Internet[4].

In case of the set of **medium** instances, the algorithms were run 40 times on each. For the **large** instances the algorithms were run 10 times, and for the **competition instances**, the algorithms were run for 20 independent trials.

Fig. 2 presents rank comparison of the results obtained for the set of five **medium** instances by the ant algorithms and reference algorithms: Simulated Annealing (SA) and Random Restart Local Search (RRLS). It is clear that the SA performs significantly better than any of the ant algorithms and the reference RRLS algorithm. It is however interesting to see that while $\mathcal{MMAS}$ is performing better than RRLS, the ACS produced solutions significantly inferior to those of RRLS. These differences are significant at least at a p-value of 0.05 in a pairwise Wilcoxon test. Detailed results can also be found on[4].

Fig. 3 presents a similar comparison as Fig. 2, but for two **large** instances. The results are however rather different. It may be said with high statistical significance ($p < 0.01$) that $\mathcal{MMAS}$ is performing best on these instances. ACS is performing worse, and comparably well to the Iterated Local Search (ILS) – the winner of the comparison in [6] on the problem. In this case both ant algorithms beat the performance of RRLS. SA, which was very efficient in case

---

[4] http://iridia.ulb.ac.be/~ksocha/ttantcmp03.html

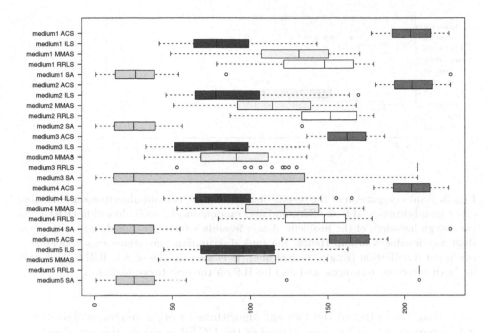

Fig. 2. Rank comparison of the results obtained by the two ant algorithms, leading other metaheuristics (ILS and SA), and also random local search algorithm (RRLS) on five medium instances of the problem. A non feasible solution is considered to be worse than any feasible solution. Hence, the rank distribution may sometimes appear as a one point distribution (single vertical line).

of medium instances failed to provide feasible timetables for the both large instances (similarly to RRLS).

Fig. 4 presents the comparison of the performance of ant algorithms on competition instances. There is no reference data available from other metaheuristics for these instances yet. We run the mentioned earlier RRLS algorithm on these instances, but as it did not provide feasible solutions for any of the instances, we did not include it in the comparison. Hence, the ant algorithms are compared only among themselves. The results show statistically significant better performance of $\mathcal{MMAS}$ in comparison to ACS. Note that Fig. 4 contains also additional results obtained by the modified versions of the ACS and $\mathcal{MMAS}$ algorithms, as described in Sec. 5.1.

5   Conclusions

Based on the results of comparison, it is clear that the two ant algorithms perform differently. The $\mathcal{MMAS}$ performs better than ACS on all instances tested. When

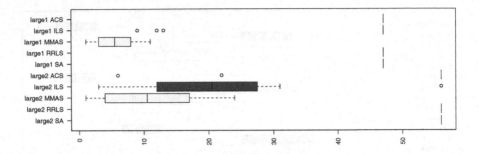

Fig. 3. Rank comparison of the results obtained by the two ant algorithms and leading other metaheuristics (ILS and SA), and also random local search algorithm (RRLS) on two **large** instances of the problem. A non feasible solution is considered to be worse than any feasible solution. Hence, the rank distribution may sometimes appear as a one point distribution (single vertical line). It is seen in case of SA, RRLS, and ACS for both presented instances, and also for ILS on the first **large** instance.

comparing the better of the two ant algorithms to other reference algorithms, it becomes clear that for some classes of the UCTP problem, the ant algorithm proves to be very efficient. While on **medium** instances of the problem, the SA is significantly better than $\mathcal{MMAS}$, while on the **large** instances $\mathcal{MMAS}$ beats any current competitor.

The difference in performance of the two ant algorithms may be due to one or more of the following factors:

- While $\mathcal{MMAS}$ does not use the heuristic information, the ACS uses it extensively. The improvement provided by the heuristic information does not make up for the time lost on its calculation (which in case of the UCTP may be quite high).
- The ACS has a different strategy in using local search than $\mathcal{MMAS}$. While ACS runs the local search for a particular number of steps, the $\mathcal{MMAS}$ tries always to reach the local optimum by specifying extensive number of steps.
- The $\mathcal{MMAS}$ uses local search to improve only one of the solutions generated by the ants, while the ACS tries to improve all the solutions generated. While the approach of $\mathcal{MMAS}$ may lead to discarding some good potential solutions, the approach of ACS may mean that two (or more) very similar solutions will be further optimized by local search, which may be an inefficient use of time.

5.1   Further Investigation

In order to check the hypothesis that due to the design choices made, the ACS actually takes longer to run one iteration, we calculated the number of iterations

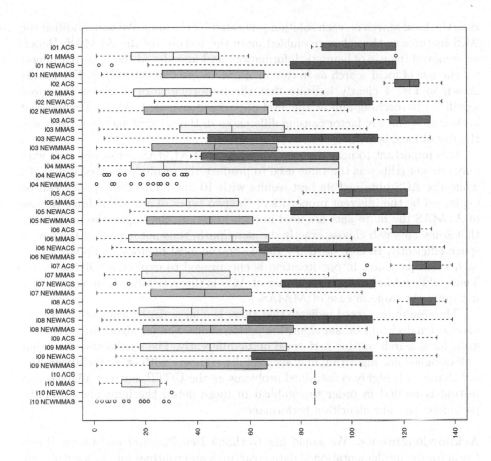

Fig. 4. Rank comparison of the results obtained by the two ant algorithms (ACS and $\mathcal{MMAS}$) on ten `competition instances` of the problem. Also the performance of the versions of those algorithms (NEWACS and NEW$\mathcal{MMAS}$) modified as described in Sec. 5.1 are presented.

done by both algorithms. We counted the number of iterations of both ant algorithms for 5 runs on a single `competition instance`. While the $\mathcal{MMAS}$ performed on average 45 iterations, in case of ACS it was only 21.6. This shows that in fact a single iteration of ACS takes more than twice the amount of time of a single $\mathcal{MMAS}$ iteration. Thus, it is most probable that the first and third of the factors presented above influence the performance of the ant algorithm.

We found it interesting to investigate the topic further. Hence, we decided to run some additional experiments. This time, we tried to make the features of both algorithms as similar as possible, to be able to see which of the factors presented above may be in fact the key issue. We modified the $\mathcal{MMAS}$ so that it runs the local search on each solution generated by the ants. We also modified the ACS features so that they resembled more the features of the $\mathcal{MMAS}$. Hence, we removed the use of heuristic information, and introduced the same parameter for the use of local search as in case of $\mathcal{MMAS}$ (10 000 000 steps). The results shown in Fig. 4 clearly indicate that the performance of ACS has improved significantly reaching almost the level of performance of $\mathcal{MMAS}$. Therefore, it is clear that the key factor causing differences in the original ant algorithms was the use of local search.

It is important to note that the new version of $\mathcal{MMAS}$ performed best with only one ant (this was the value used to produce the results presented in Fig. 4), while the ACS obtained its best results with 10 ants. This discrepancy can be explained by the inherent properties of the two types of ant algorithms. In case of $\mathcal{MMAS}$ the more ants are used in each iteration, the higher the probability that some ants will choose exactly the same path, thus not exploring the search space efficiently. In case of ACS – thanks to the local pheromone update rule – each subsequent ant in one iteration is encouraged to explore a different path. Thus, while adding more ants in case of ACS is theoretically advantageous, it is not quite the same in case of $\mathcal{MMAS}$.

The results presented indicate that there is a large dependency of the particular design decisions on ant algorithm performance. Similar algorithms using the same local search routine performed quite differently. The results also show that well designed ant algorithm may successfully compete with other metaheuristics in solving such highly constrained problems as the UCTP. Further analysis and testing is needed in order to establish in more detail the influence of all the parameters on ant algorithm performance.

Acknowledgments. We would like to thank Ben Paechter and Olivia Rossi-Doria for the implementation of data structures and routines for the local search. Also, we would like to thank Thomas Stützle and Marco Chiarandini for the implementation of the ILS and SA algorithms for UCTP. The information provided is the sole responsibility of the authors and does not reflect the Community's opinion. The Community is not responsible for any use that might be made of data appearing in this publication.

References

1. Gaspero, L.D., Schaerf, A.: Tabu search techniques for examination timetabling. In: Proceedings of the 3rd International Conference on Practice and Theory of Automated Timetabling (PATAT 2000), LNCS 2079, Springer-Verlag (2001) 104–117
2. ten Eikelder, H.M.M., Willemen, R.J.: Some complexity aspects of secondary school timetabling problems. In: Proceedings of the 3rd International Conference on Practice and Theory of Automated Timetabling (PATAT 2000), LNCS 2079, Springer-Verlag (2001) 18–29

3. Cowling, P., Kendall, G., Soubeiga, E.: A hyperheuristic approach to scheduling a sales summit. In: Proceedings of the 3rd International Conference on Practice and Theory of Automated Timetabling (PATAT 2000), LNCS 2079, Springer-Verlag (2001) 176–190

4. Cooper, T.B., Kingston, J.H.: The complexity of timetable construction problems. In: Proceedings of the 1st International Conference on Practice and Theory of Automated Timetabling (PATAT 1995), LNCS 1153, Springer-Verlag (1996) 283–295

5. de Werra, D.: The combinatorics of timetabling. European Journal of Operational Research **96** (1997) 504 513

6. Rossi-Doria, O., Sampels, M., Chiarandini, M., Knowles, J., Manfrin, M., Mastrolilli, M., Paquete, L., Paechter, B.: A comparison of the performance of different metaheuristics on the timetabling problem. In: Proceedings of the 4th International Conference on Practice and Theory of Automated Timetabling (PATAT 2002) (to appear). (2002)

7. Stützle, T., Hoos, H.H.: $\mathcal{MAX}$-$\mathcal{MIN}$ Ant System. Future Generation Computer Systems **16** (2000) 889–914

8. Socha, K., Knowles, J., Sampels, M.: A $\mathcal{MAX}$-$\mathcal{MIN}$ Ant System for the University Timetabling Problem. In Dorigo, M., Di Caro, G., Sampels, M., eds.: Proceedings of ANTS 2002 – Third International Workshop on Ant Algorithms. Lecture Notes in Computer Science, Springer Verlag, Berlin, Germany (2002)

9. Dorigo, M., Maniezzo, V., Colorni, A.: The ant system: Optimization by a colony of cooperating agents. IEEE Transactions on Systems, Man, and Cybernetics **26** (1996) 29–41

10. Bonabeau, E., Dorigo, M., Theraulaz, G. Oxford University Press (1999)

11. Dorigo, M., Gambardella, L.M.: Ant colony system: A cooperative learning approach to the travelling salesman problem. IEEE Transactions On Evolutionary Computation (1997) 53–66

12. Dorigo, M., Di Caro, G., Gambardella, L.M.: Ant algorithms for discrete optimization. Artificial Life **5** (1999) 137–172

Multiple Genetic Snakes for Bone Segmentation

Lucia Ballerini and Leonardo Bocchi

[1] Dept. of Technology, Örebro University
Fakultetsgatan 1, 70182 Örebro, Sweden
lucia@aass.oru.se
[2] Dept. of Electronics and Telecommunications, University of Florence
Via S.Marta 3, 50139 Firenze, Italy
leo@asp.det.unifi.it

Abstract. Clinical assessment of skeletal age is a frequent, but yet difficult and time-consuming task. Automatic methods which estimate the skeletal age from a hand radiogram are currently being studied. This work presents a method to segment each bone complex in the radiogram, using a modified active contour approach. Each bone is modelled by an independent contour, while neighbouring contours are coupled by an elastic force. The optimization of the contour is done using a genetic algorithm. Experimental results, carried out on a portion of the whole radiogram, show that coupling of deformable contours with genetic optimization allows to obtain an accurate segmentation.

1 Introduction

Bone age assessment is a procedure frequently performed in pediatric radiology. A discrepancy between bone age and chronological age indicates the presence of some abnormality in skeletal growth. The assessment of bone age is almost universally performed by examination of the left-hand radiogram. This procedure requires only a minimal exposure, with an high degree of simplicity. Moreover, the hand presents a large number of ossification centers which can be analyzed in order to obtain an accurate evaluation of the skeletal age.

In contrast with such advantages, inspection of the resulting radiogram is a quite complex task. A correct evaluation of the degree of maturation of the bones, requires an high degree of expertise. Several methods have been proposed to perform this evaluation. The most commonly used method is the atlas matching method by Greulich and Pyle [1]. The hand radiogram is visually compared with a series of images reproduced in the atlas, grouped by age and sex. The pattern which appears to be the most similar to the clinical image is selected, and the corresponding age is indicated to assess the skeletal age. The major drawback in this method is its subjectivity, which produces an high degree of variability in the outcoming results, both inter-observer, and intra-observer.

A more complex approach uses the Tanner and Whitehouse (TW2) method [2]. This method involves a detailed analysis of a group of about 20 bones of hand and wrist. Each bone complex is assigned to one of eight classes

S. Cagnoni et al. (Eds.): EvoWorkshops 2003, LNCS 2611, pp. 346–356, 2003.
© Springer-Verlag Berlin Heidelberg 2003

reflecting the various development stages, depending on the degree of calcification and the shape of the complex. In this way, a maturation score is assigned to each bone. A weighted sum of all scores is the used to evaluate the skeletal age. This method yields the most reliable results, but, due to its complexity, it does not present a high application rate (less than 20%).

Several research groups are working to develop automatic methods which can speed up the evaluation process. A complete method can roughly be subdivided in a segmentation step, where the bones are identified and labelled, a feature and shape analysis step which assesses the bone age of each region, and a classification stage which summarizes the partial data to produce the final age assessment.

In this work, we focus on the first stage, the segmentation procedure. The segmentation stage presents an high degree of complexity due to several factors; among those, we face with the presence of several overlapping regions of interest (ROI), the presence of ROI having completely different degree of calcification and overlapping of soft tissue.

We propose the use of Genetic Snakes [3], that are active contour models, also known as snakes [4], with an energy minimization procedure based on Genetic Algorithms (GA) [5]. Snakes optimization through Genetic Algorithms proved to be particularly useful in order to overcome problems of the classical snakes related to initialization, parameter selection and local minima. New internal and external energy functionals have been proposed in our previous works and they have been successfully applied to a variety of images from different domains.

The purpose of this paper is to extend the Genetic Snakes model to handle complex contours, composed of distinct regions, allowing introduction of external knowledge expressed by additional energy terms. Each bone contour is associated to an independent snake, while the anatomical knowledge about relative placement of hand bones is modelled by means of a binding energy which couples together the contours.

The organization of the paper is as follow: in Section 2 we briefly review active contours, the basic notions, their limitations and some improvements proposed in literature. In Section 3 we describe the Genetic Snakes model. In Section 4 we extend our previous formulation by the introduction of the multiple snakes structure and of the binding energy. Experimental results are reported in Section 5.

2   Active Contours

Snakes are planar deformable contours that are useful in several image analysis tasks. They are often used to approximate the locations and shapes of object boundaries on the basis of the reasonable assumption that boundaries are piecewise continuous or smooth.

Representing the position of a *snake* parametrically by $\mathbf{v}(s) = (x(s), y(s))$ with $s \in [0, 1]$, its energy can be written as:

$$E_{snake} = \int_0^1 E_{int} \left[\mathbf{v}(s) \right] ds + \int_0^1 E_{ext} \left[\mathbf{v}(s) \right] ds \qquad (1)$$

where E_{int} represents the internal energy of the snake due to bending and it is associated with *a priori* constraints, E_{ext} is an external potential energy which depends on the image and accounts for *a posteriori* information. The final shape of the contour corresponds to the minimum of this energy.

In the original technique of Kass et al. [4] the internal energy is defined as:

$$E_{int}\left[\mathbf{v}(s)\right] = \frac{1}{2}\left[\alpha(s)\left|\frac{\partial \mathbf{v}(s)}{\partial s}\right|^2 + \beta(s)\left|\frac{\partial^2 \mathbf{v}(s)}{\partial s^2}\right|^2\right]. \tag{2}$$

This energy is composed of a first order term controlled by $\alpha(s)$ and a second order term controlled by $\beta(s)$. The two parameters $\alpha(s)$ and $\beta(s)$ dictate the simulated physical characteristics of the contour: $\alpha(s)$ controls the *tension* of the contour while $\beta(s)$ controls its *rigidity*.

The external energy couples the snake to the image. It is defined as a scalar potential function whose local minima coincide with intensity extrema, edges, and other image features of interest. The external energy, which is commonly used to attract the snake towards edges, is defined as:

$$E_{ext}\left[\mathbf{v}(s)\right] = -\gamma|\nabla G_\sigma * I(x,y)|^2 \tag{3}$$

where $I(x,y)$ is the image intensity, G_σ is a Gaussian of standard deviation σ, ∇ is the gradient operator and γ a weight associated with image energies. Due to the wide and successful application of deformable models, there exist survey papers focusing on different aspects of the model and its variants proposed in the literature [6,7,8,9].

The application of snakes and other similar deformable contour models to segment structures is, however, not without limitations. For example, snakes were designed as interactive models. In non-interactive applications, they must be initialized close to the structure of interest to guarantee good performance. The internal energy constraints of snakes can limit their geometric flexibility and prevent a snake from representing long tube-like shapes or shapes with significant protrusions or bifurcations. Furthermore, the topology of the structure of interest must be known in advance since classical deformable contour models are parametric and are incapable of topological transformations without additional machinery. Due to its own internal energy, the snake tends to shrink in case of lack of image forces, i.e. constant image backgrounds or disconnected object boundaries, and not to move towards the object.

Various methods have been proposed to improve and further automate the deformable contour segmentation process. See the above mentioned surveys for a review of some of them.

As concerns the energy minimization, the original model employs the variational calculus to iteratively minimize the energy. There may be a number of problems associated with this approach such as algorithm initialization, existence of local minima, and selection of model parameters. Simulated annealing [10, 11], dynamic programming [12,13] and greedy algorithm [14,15] have been also proposed for minimization. However they are restricted either by the exhaustive searches of the admissible solutions either by the required accurate initialization.

Some authors propose the application of GA to active contours. Among those, MacEachern and Manku [16] introduce the concept of active contour state and encode the variants of the state in the chromosome of the genetic algorithm. Tanatipanond and Covavisaruch [17] apply GA to contour optimization with a multiscale approach. The fitness function is "trained" from a previously segmented contour. Ooi and Liatsis [18] propose the use of co-evolutionary genetic algorithms. They decompose the contour into subcontours and optimize each subcontour by separate GA working in parallel and co-operating. However, in these approaches the optimization is done in the neighborhood of the snake control points.

In other existing GA-based active contours the optimization is done indirectly, i.e. optimizing the parameters of the contour such as encoding the polygon [19], Point Distribution Models [20,21], Fourier Descriptors [22], Probability Density Functions [23] or edge detector and elastic model parameters [24].

3   Genetic Snakes

In this section we review the genetic snake model, i.e. our model of active contours, where the energy minimization procedure is based on genetic algorithms [3].

The parameters that undergo genetic optimization are the positions of the snake in the image plane $\mathbf{v}_i = (x_i, y_i)$, for $i = 0, ...N$ where N is the total number of snake points. To simplify the implementation we used polar coordinates $\mathbf{v}_i = (r_i, \theta_i)$ with the origin in the center of the contour. Actually, this point must lie inside the object, but its position may be arbitrary. The magnitudes r_i are codified in the chromosomes, while $\theta_i = 2\pi i/N$. The polar representation introduces ordering of the contour points and prevents the snake elements from crossing each other during evolution. The genetic operators can be implemented straightforward on this representation, no additional check is required to ensure that mutation and crossover produce valid individuals.

The fitness function is the total snake energy as previously defined in (1), where E_{int} and E_{ext} are defined in (2) and (3).

The initial population is randomly chosen in a region of interest defined by the user, and each solution lies in this region. This replaces the original initialization with a region-based version, enabling a robust solution to be found by searching the region for a global solution. The region of interest can be the image itself, so the solution can be searched in the whole image, making the initialization fully automatic.

An accurate description of implementation details along with a discussion on the choice of the model coefficients can be found in [3].

The genetic search strategy works against constant image background and overcomes difficulties related to spurious edge-points that can drive the snake to a local minima. To reach an optimal minimum while avoiding local minima, some approaches suggest to consider the whole set of admissible curves and choose the best one.

Other snake optimization methods are for local optimization, where only sub-optimal solutions can be guaranteed. The genetic algorithms are particularly useful in simultaneously handling possible solutions and looking for a global minimum, while avoiding an exhaustive search.

4   Multiple Genetic Snakes

A straightforward extension from a single contour to multiple contours poses a few questions which could prevent convergence. First of all, the spatial relations between facing bones produce sets of parallel edges. Each contour need to be univocally associated to the correct edge to achieve a correct segmentation, while (3) does not allow to discriminate between the edges, because it is sensitive to the gradient modulus, but not to its orientation. We solve this problem with the introduction of a first-order derivative energy, which introduces a directionality in the contours.

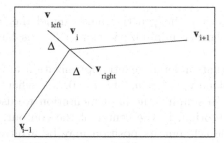

Fig. 1. Points used in the evaluation of derivative energy

For each point $\mathbf{v}_i$ belonging to the snake (see Fig. 1), we define $\mathbf{v}_{left} = (x_{left}, y_{left})$ and $\mathbf{v}_{right} = (x_{right}, y_{right})$, which are placed on a line orthogonal to $\overline{\mathbf{v}_{i-1}\mathbf{v}_i}$, and spaced of a small distance Δ from $\mathbf{v}_i$. To allow a faster computation, the orientation of the line has been constrained to be multiple of $\pi/4$, and the distance Δ is assumed to be one pixel. In this way, the derivative energy is computed as the difference between two pixels belonging to the 8-neighbourhood of $\mathbf{v}_i$, selected accordingly to the direction of the snake in that point. The derivative energy used can then be expressed as:

$$E_{der}[\mathbf{v}_i] = \delta\left[(F * I)(x_{left}, y_{left}) - (F * I)(x_{right}, y_{right})\right] \quad (4)$$

where δ is a weight used to balance the derivative energy with the other terms on external energy and F is a smoothing filter. This definition introduces an energy term that presents a minimum point when the snake is positioned on the image edge, having the brighter region on the left side of the snake, and the darker region on the right side. In our implementation, snakes are running counterclockwise around their center, so the left side corresponds to the internal region, and the right side to the external region. With a positive value of δ, we reach the minimum (negative) energy when the snake encloses a bright region on a darker background, while a null energy corresponds to an uniform image. Using a genetic minimization algorithm, the presence of a negative term does not alter the behaviour of the process, as the sigma scaling method used to evaluate the performance value tunes itself to the average value of the fitness function.

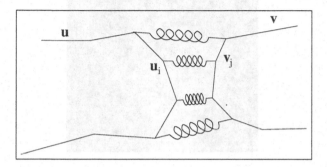

Fig. 2. Binding between adjacent snakes

As introduced before, we also add an additional term to the internal energy, which we call binding energy. This term models the anatomical relationships between adjacent bones, by introducing and elastic force that connects together appropriate points of adjacent snakes, as shown in Fig. 2. The energy associated to the elastic force is assumed to be represented by the relation:

$$E_{bind}[\mathbf{u}(s), \mathbf{v}(t)] = \mu|\mathbf{u}(s) - \mathbf{v}(t)|^2 \tag{5}$$

where μ represents the elastic constant of the spring. The application points of the elastic forces are selected accordingly to the physical relationships which exist between the anatomical regions.

5   Application

The image data set is composed of radiographic images of the left hand and wrists, acquired by means of a conventional radiographic system, and digitized with a spatial resolution of 300 dpi, and a pixel depth of 12 bits. Patient age ranges from 0 to 12 years. Afterward, the images have been downsampled by a factor of three to speed up the segmentation process (see Fig. 3).

In this work we evaluate the application of genetic snakes to the segmentation of a subpart of the hand, the first finger. This allows to develop and test a simpler model, although the task includes most of the segmentation problems. In particular, the three phalanx present a different contrast and mean gray level, due to their different thickness, have different size, and may have an incomplete boundary due to the presence of cartilaginous tissue.

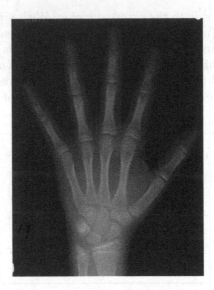

Fig. 3. Radiographic image of the hand

The situation has been modelled using three snakes, which represent the three bones, chained together by means of the binding force. Information about the geometry of the bones are not necessary, but only their relative position.

A region of interest is defined for each snake by setting its center and the minimum and maximum magnitude for each $\mathbf{v}(s)$. Each snake is composed of 36 points, and the binding energy acts on five couples of consecutive points in each junction. In order to simulate the anatomical relationships, binding energy is assumed to act between the upper part ($\theta \simeq \pi/2$) of the first snake and the lower ($\theta \simeq -\pi/2$) part of the second snake. Analogously we connect the upper part of the second snake to the lower one of the third.

The internal energy of the model is given by a weighted sum of E_{int} and E_{bind}, defined respectively in (2) and (5). In our implementation it is possible to use different α and β for each snake.

The image energy is computed as an appropriate combination of E_{ext} and E_{der} ((3) and (4)). Equation (3) has been applied by incorporating three gradient of Gaussians with different σ. The filter F used in (4) is a Diffence of Gaussians, computed for three different couples of σ to obtain three smoothed versions of the image (see Fig. 4)

The fitness function is the total energy of the model i.e. the sum of the internal and image energy:

$$E = E_{int} + E_{bind} + E_{ext} + E_{der} \tag{6}$$

The GA implementation adopted in this work is GAucsd-1.4 [25]. We used most of the default options proposed by the GAucsd package, i.e. Gray-code, fitness sigma scaling, two point crossover, roulette wheel selection. The parameters

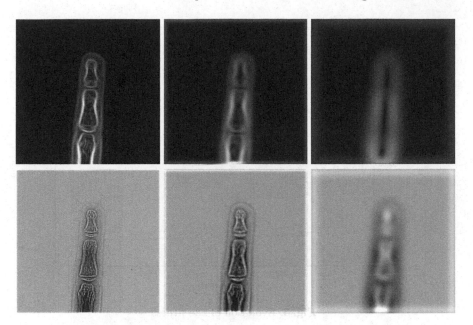

Fig. 4. Gradient of Gaussian images ($\sigma = 0.3, \sigma = 3, \sigma = 10$) and Difference of Gaussian images ($\sigma_1 = 0.5, \sigma_2 = 1.5; \sigma_1 = 1, \sigma_2 = 3; \sigma_1 = 5, \sigma_2 = 8$)

of the GA were: length of the genome = 324, population size = 65000, maximum number of generations = 700, crossover rate = 0.59, mutation rate = 0.00001. Each run takes about 15 min. Figure 5 shows the fitness evolution of the best individual.

We performed several experiments varying the snake energy weighting coefficients. As more internal and external energy terms are considered in our model than in the classical snake, it is hard to determine the appropriate ratio between different forces in the total energy function of the model, due to lack of understanding on the effect of each force on the energy function.

The results showed in Fig. 6 have been obtained with following weights: $\alpha_1 = 0.9$, $\beta_1 = 6$, $\alpha_2 = 0.5$, $\beta_2 = 5$, $\alpha_3 = 0.7$, $\beta_3 = 10$, $\mu = 1.25$, $\delta = -1.2$, $\gamma_1 = 2.5$, $\gamma_2 = 1$. $\gamma_1 = 0.5$. For this setting of weights we did 25 GA runs.

Results have been evaluated by comparing the obtained segmentation with an hand-drawn outline. We can define as TP the portion of bone area correctly enclosed in the automatic segmentation, FP the portion of image not belonging to the bone, but included in the segmented region, and FN as the bone area which is not included in the segmented image. With this assumption, we can evaluate $A_+ = \frac{FP}{TP+FN}$ and $A_- = \frac{FN}{TP+FN}$. In our tests, we found approximately a value of $A_+ = 8\%$, and $A_- = 3.5\%$.

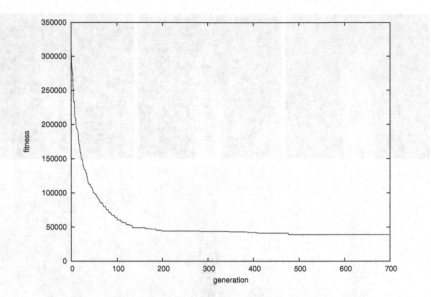

Fig. 5. Evolution of the fitness

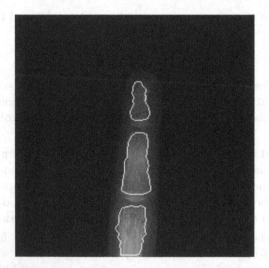

Fig. 6. Segmentation results of the three phalanx

6   Conclusions

In this paper a method for automatic segmentation of hand radiograms is described along with some results obtained with the proposed approach on a subpart of the hand.

The complexity of the skeletal structure in the hand and the variability between different subjects makes very difficult to realize an automatic segmentation of the bones. The proposed method allow to combine the *a-priori* knowledge on the hand structure to the adaptive behaviour of active contours and genetic algorithm. Binding and derivative energy allow to introduce adequate constraints on the geometry of the snake to obtain a satisfactory segmentation.

It is known that the snake model requires either a local minimizer with good initialization or otherwise a global minimizer. Genetic snakes confront and overcome at the same time the two primary problems of initialization and optimization, and provide a global optimization with an automatic initialization.

The encouraging results reported prompt us that the method could be extended and applied to other bone structures as well as to other images. Other extensions could consider the study of the parameters and the functionals governing the snake behaviour. A method to reduce user interaction by automatically assigning snake energy weights is still an open problem. Therefore the evolution of weights could be considered for future studies.

References

1. Greulich, W.W., Pyle, S.I.: Radiographic atlas of skeletal development of the hand and wrist. 2nd edn. Stanford University Press, Palo Alto, CA (1959)
2. Tanner, J.M., Whitehouse, R.H., Marshall, W.A., Healy, M.J.R.: Assessment of skeletal maturity and prediction of adult height (TW2 method). 2nd edn. Academic Press, London (1983)
3. Ballerini, L.: Genetic snakes for medical images segmentation. In: Evolutionary Image Analysis, Signal Processing and Telecommunications. Volume 1596 of Lectures Notes in Computer Science., Springer (1999) 59–73
4. Kass, M., Witkin, A., Terzopoulos, D.: Snakes: Active contour models. International Journal of Computer Vision **1** (1988) 321–331
5. Goldberg, D.E.: Genetic Algorithms in Search, Optimization, and Machine Learning. Addison-Wesley, Reading, MA (1989)
6. McInerney, T., Terzopoulos, D.: Deformable models in medical image analysis: A survey. Medical Image Analysis **1** (1996) 91–108
7. Jain, A.K., Zhong, Y., Dubuisson-Jolly, M.P.: Deformable template models: A review. Signal Processing **71** (1998) 109–129
8. Xu, C., Pham, D.L., Prince, J.L.: Image segmentation using deformable models. In Sonka, M., Fitzpatrick, J.M., eds.: Handbook of Medical Imaging. Volume 2. SPIE Press (2000) 129–174
9. Cheung, K.W., Yeung, D.Y., Chin, R.T.: On deformable models for visual patter recognition. Patter Recognition **35** (2002) 1507–1526
10. Storvik, G.: A bayesian approach to dynamic contours through stochastic sampling and simulated annealing. IEEE Transactions on Pattern Analysis and Machine Intelligence **16** (1994) 976–986
11. Grzeszczuk, R.P., Levin, D.N.: Brownian strings: Segmenting images with stochastically deformable contours. IEEE Transactions on Pattern Analysis and Machine Intelligence **19** (1997) 110–1114
12. Amini, A., Weymouth, T., Jain, R.: Using dynamic programming for solving variational problems in vision. IEEE Transactions on Pattern Analysis and Machine Intelligence **12** (1990) 855–867

13. Geiger, D., Gupta, A., Costa, L., Vlontzos, J.: Dynamic programming for detecting, tracking and matching deformable contours. IEEE Transactions on Pattern Analysis and Machine Intelligence **17** (1995) 294–302
14. Williams, D.J., Shah, M.: A fast algorithms for active contours and curvature estimation. CVGIP: Image Understanding **55** (1992) 14–26
15. Ji, L., Yan, H.: Attractable snakes based on the greedy algorithm for contour extraction. Pattern Recognition **33** (2002) 791–806
16. MacEachern, L.A., Manku, T.: Genetic algorithms for active contour optimization. In: Proc. IEEE International Symposium on Circuits and Systems. Volume 4. (1998) 229–232
17. Tanatipanond, T., Covavisaruch, N.: An improvement of multiscale approach to deformable contour for brain MR images by genetic algorithms. In: Proc. IEEE International Symposium on Intelligent Signal Processing and Communication Systems, Phucket, Thailand (1999) 677–680
18. Ooi, C., Liatsis, P.: Co-evolutionary-based active contour models in tracking of moving obstacles. In: Proc. International Conference on Advanced Driver Assistance Systems. (2001) 58–62
19. Toet, A., Hajema, W.P.: Genetic contour matching. Pattern Recognition Letters **16** (1995) 849–856
20. Cootes, T., Taylor, C.J., Cooper, D.H., Graham, J.: Active shape models – their training and application. Computer Vision and Image Understanding **61** (1995) 38–59
21. Ruff, C.F., Hughes, S.W., Hawkes, D.J.: Volume estimation from sparse planar images using deformable models. Image and Vision Computing **17** (1999) 559–565
22. Undrill, P.E., Delibasis, K., Cameron, G.G.: An application of genetic algorithms to geometric model-guided interpretation of brain anatomy. Pattern Recognition **30** (1997) 217–227
23. Mignotte, M., Collet, C., Pèrez, P., Bouthemy, P.: Hybrid genetic optimization and statistical model-based approach for the classification of shadow shapes in sonar images. IEEE Transactions on Pattern Analysis and Machine Intelligence **22** (2000) 129–141
24. Cagnoni, S., Dobrzeniecki, A.B., Poli, R., Yanch, J.C.: Genetic algorithm-based interactive segmentation of 3D medical images. Image and Vision Computing **17** (1999) 881–895
25. Schraudolph, N.N., Grefenstette, J.J.: A user's guide to GAucsd 1.4. Technical Report CS92-249, Computer Science and Engineering Department, University of California, San Diego, La Jolla, CA (1992)

Mobile Robot Sensor Fusion Using Flies

Amine M. Boumaza[1] and Jean Louchet[1,2]

[1] INRIA, projet FRACTALES, Domaine Voluceau BP 105,
78153 Le Chesnay cedex, France
[2] ENSTA, 32 Boulevard Victor, 75739 Paris cedex 15, France
{Amine.Boumaza|Jean.Louchet}@inria.fr

Abstract. The "Fly algorithm" is a fast artificial evolution-based image processing technique. Previous work has shown how to process stereo image sequences and use the evolving population of "flies" as a continuously updated representation of the scene for obstacle avoidance in a mobile robot. In this paper, we show that it is possible to use several sensors providing independent information sources on the surrounding scene and the robot's position, and fuse them through the introduction of corresponding additional terms into the fitness function. This sensor fusion technique keeps the main properties of the fly algorithm: asynchronous processing. no low-level image pre-processing or costly image segmentation, fast reaction to new events in the scene. Simulation test results are presented.

1   Introduction

The Fly Algorithm [1] is an image processing technique based on evolving a population of points in space (the "flies") using a fitness function designed in such a way that the flies converge onto the physical objects in the scene. The fitness of a fly is calculated using image grey levels. According to the general scheme of "Parisian Evolution" [2], the problem's solution is represented by the whole population rather than by the only best individual. Contrary to the general belief that artificial evolution is slow and not suited for real-time applications, we show that the Fly algorithm is a way to exploit asynchronous data delivery made possible by CMOS camera technology. A matching asynchronous robot path planner using the fly algorithm's output has been proposed in [3]. The evolutionary program's structure is widely application-independent, using problem-specific knowledge expressed in the fitness function, which is not an usual feature in image processing. The aim of this paper is to exploit this property a bit further and show how, without any major alteration of the algorithm's architecture and genetic operators, it is possible to integrate exteroceptive and proprioceptive sensors and use the Fly algorithm as a real-time sensor fusion method.

S. Cagnoni et al. (Eds.): EvoWorkshops 2003, LNCS 2611, pp. 357–367, 2003.

2   Stereovision Using Flies

2.1   Processing Stereo Pairs

Geometry and Fitness Function. A fly is defined as a 3-D point with coordinates (x, y, z). The coordinates of a fly's projections are (x_L, y_L) in the image given by the left camera and (x_R, y_R) in the right camera. The cameras' calibration parameters are known, therefore x_R, y_R, x_L, y_L maybe readily calculated from x, y, z using projective geometry [4] [5]. If the fly is on the surface of an opaque object, then the corresponding pixels in the two images will normally have highly similar neighbourhoods (Fig. 1)[1]. Conversely, if the fly is not on the surface of an object, their close neighbourhoods will be usually poorly correlated. The fitness function exploits this property and evaluates the degree of similarity

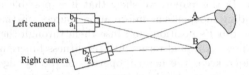

Fig. 1. Pixels b_1 and b_2 , projections of fly B, have identical grey levels, while pixels a_1 and a_2, projections of fly A, which receive their illumination from two different physical points on the object's surface, have different grey levels.

of the pixel neighbourhoods of the projections of the fly onto each image, giving highest fitness values for the flies lying on objects surfaces:

$$fitness(indiv) = \frac{G}{\sum\limits_{colours} \sum\limits_{(i,j)\in N} (L\,(x_L + i,\, y_L + j) - R\,(x_R + i,\, y_R + j))^2} \quad (1)$$

- (x_L, y_L) and (x_R, y_R) are the coordinates of the left and right projections of the current individual (see Fig. 1)
- $L\,(x_L + i,\, y_L + j)$ is the grey value of the left image at pixel $(x_L + i,\, y_L + j)$, similarly with R for the right image.
- N is a neighbourhood introduced to obtain a more discriminant comparison of the fly's projections.

On colour images, square differences are calculated on each colour channel. The numerator G is a normalizing factor designed to reduce the fitness of the flies which project onto uniform regions. It is based on an image gradient norm calculation. The best experimental results were obtained when G is defined as the

---

[1] This may not be completely true if the surfaces differ from Lambert's law, which assumes that for a given illumination, the object's radiance does not depend on the observer's position. Most surface-based stereovision methods are also sensitive to this property.

square root of Sobel's gradient norm [6] [4]: highest fitness values are obtained for flies whose projections have similar and significant pixel surroundings. Additionally, we modified the denominator of the fitness function to reduce its sensitivity to the constant component of the image, which strongly depends on camera sensitivity adjustments.

Artificial Evolution Operators. An individual's chromosome is the triplet (x, y, z) which contains the fly's coordinates. The population is initialised randomly inside the intersection of the cameras' fields of view (Fig. 2). The statistical distribution is chosen in order to obtain uniformly distributed projections in the left image. The values of z^{-1} are uniformly distributed between zero (or $1/d_{max}$) and $1/d_{min}$.

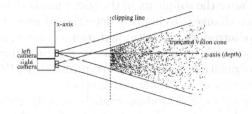

Fig. 2. The fly population is initialised inside the intersection of the cameras 3-D fields of view.

Selection is elitist and deterministic, ranking the flies according to their fitness values and retaining the best individuals (typically 50%).

2-D sharing [7], reduces the fitness values of flies located in crowded areas to prevent them from getting concentrated into a small number of maxima. It reduces each fly's fitness by $K * N$, where K is a "sharing coefficient" and N the number of flies which project into the left image within a distance R ("sharing radius") from the current fly, given by the following formula [8] :

$$R \approx \frac{1}{2} \left(\sqrt{\frac{N_{pixels}}{N_{flies}}} - 1 \right) \tag{2}$$

Mutation allows extensive exploration of the search space. It uses an approximation of a Gaussian random noise added to the flies' chromosome parameters (x, y, z). We chose standard deviations $(\sigma_x, \sigma_y, \sigma_z)$ equal to R, so that they are the same order of magnitude as the mean distance between neighbouring flies.

In order to take into account the frequent straight lines or planar surfaces existing in real-world scenes, we translated this into two *barycentric crossover operators*. The first one builds an offspring randomly located on the line segment between its parents: the offspring of two flies $F_1(x_1, y_1, z_1)$ and $F2(x_2, y_2, z_2)$ is the fly $F(x, y, z)$ defined by $\overrightarrow{OF} = \lambda \overrightarrow{OF_1} + (1 - \lambda) \overrightarrow{OF_2}$. The weight λ is chosen using a uniform random law in $[0, 1]^2$. Similarly, the second operator uses three parents and determines the offspring F such that $\overrightarrow{OF} = \lambda\overrightarrow{OF_1} + \mu\overrightarrow{OF_2} + (1 - \lambda - \mu)\,\overrightarrow{OF_3}$ in the parents' plane, using two random weights λ and μ.

2.2   Processing Stereo Sequences

We have developed several methods to process stereo image sequences. They are described in detail in [8]. The simplest one is the *random approach*, which consists in keeping the same population evolving through frame changes. Thus, only the images (and therefore the parameters used by the fitness function) are modified while the fly population evolves. When motion is slow enough, using the results of the last step speeds up convergence significantly compared to using a totally new random population. This method may be seen as the introduction of a collective memory of space, exploiting the similarity between consecutive frames. It does not alter the simplicity of the static method and does not require significant algorithm changes. To improve detection of new objects appearing in the field of view, we introduced an extra mutation operator, *immigration*, which creates new flies randomly in a way similar to the procedure already used to first initialise the population: used with a low probability (1% to 5%) it favours a convenient exploration of the whole search space.

Dynamical approaches [8] introduce explicit velocity components into each fly's genome. The advantage is a better precision for a given number of generations or evaluations, but this goes with a significantly higher calculation cost at each generation, and therefore a smaller number of generations in a given time interval: the best trade-off will depend on the scene style and complexity of motion.

Unlike in image segmentation-based algorithms, image pixels only need to be read when the fitness of a fly has to be evaluated: therefore the random pixel access allowed by CMOS imagers can be fully exploited and new events in the scene can be processed without usual frame refreshing delays. In the following Sections, as we consider a vision system embedded into a mobile robot, we will use the simple random approach described above, but update the flies' positions at each generation, using the information available about the robot's motion in order to give better initial conditions and allow faster convergence of the fly population. This will be the basis of the proprioceptive fusion described in Section 3.4.

2.3   Path Planning with Obstacle Avoidance

The unusual way the scene is described using the fly algorithm (compared to more classical stereovision algorithms used in mobile robotics), led us to adapt a

---

[2] It is generally accepted that such a crossover operator has contractive properties which may be avoided by using a larger interval. However the goal of this crossover operator is to fill in surfaces whose contours have already been detected, rather than to extend them. It is therefore not desirable to use coefficients allowing the centre of gravity to lie outside of the object's boundary.

classical robot navigation method in order to use the results of the fly algorithm as input data, and build a simulator. The simulator is described in Section 3. To keep the speed advantage of the fly algorithm, we chose to adapt fast potential-based methods from the literature [9]. In former work [3] we defined a force derived from the addition of an attractive (target) and a repulsive (flies) potential acting as a steering command: the blockage situations [10] were resolved using two heuristics creating a new force attracting the robot out of the potential minimum (random walk and wall following methods). In this paper we use a different potential field-based path planner based on harmonic functions [11]. A harmonic function is a function which satisfies Laplace's equation:

$$\Delta U = \frac{\partial^2 U}{\partial x^2} + \frac{\partial^2 U}{\partial y^2} = 0 \qquad (3)$$

One of the interesting properties of harmonic functions is the absence of local minima, which in our case eliminates the blockage situations.

The robot's environment memory consists of a sampled grid which represents a harmonic function. Values of the function at the target and obstacle positions are modelled as Dirichlet boundaries: 1 for obstacles, 0 for the target position [12] [11]. The harmonic function is built iteratively using a finite difference operator, such as the Gauss-Seidel operator which consists in replacing the value of a point with the average value of its four neighbours. After convergence, we end up with a smooth function that has a single global minimum at the target position.

The steering command of the robot is the gradient of the harmonic function. Linear interpolation is used when the robot position falls between grid points. During the movement toward the target, new obstacles detected by the fly algorithm are introduced into the harmonic function according to the local fly density, as high potential Dirichlet boundaries. Since the harmonic function is constantly iterated, there will be a constantly updated obstacle avoiding path for the robot to follow (Fig. 3). The Dirichlet boundaries are set at the positions of the best flies in the robot's field of view.

 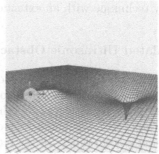

Fig. 3. The robot facing an obstacle (left) and the corresponding harmonic function (right).

2.4   Real-Time Compliance

With the Fly algorithm, each individual evaluation only needs pixel values on a small neighbourhood. CMOS camera technology is well adapted to this requirement as it is able to deliver these values at any time, while conventional image processing techniques do not exploit this feature. Similarly, the Fly algorithm delivers updated scene information at any time, giving the planner faster reaction to scene events (Table 1).

Table 1. Comparison of the Fly algorithm with conventional methods.

	CCD sensor + standard approach	CMOS + flies
image sensor	delay between capture and restitution	asynchronous data reading
image processing (input)	image segmentation must wait end of capture cycle	random access to pixels needed by current fitness calculation
image processing (output)	not available until segmentation process has ended	current state of representation always available
trajectory planner	must wait for image processing output	saves 2 acquisition cycles

3   Data Fusion Using Extra Sensors

Sensor fusion [13] plays an important role in a robot's perception system. It synthesizes the information given by exteroceptive and proprioceptive sensors. Most classical approaches are based on Bayes' theorem. In this Section, we show that in spite of the absence of any formal link between a fly's fitness value and the probability that it lies on an obstacle, similar sensor fusion may be achieved using the fly technique with an extended fitness function.

3.1   Simulated Ultrasonic Obstacle Detectors

Usual ultrasonic obstacle detectors (sonar) are able to detect the presence or absence of an obstacle in their field of view. The view angle depends on the transducer size and acoustic wavelength: we used an array of six simulated sonars with (adjustable) $15°$ angle and a given maximum detection distance, shorter than the vision system range. In the real world, the sonar's detection range may depend on obstacle's size and sound dampening properties. To simulate such a detector, we first use the Zbuffer map from the image synthesis inside the sonar's field of view and calculate the obstacle-robot distances. Their smallest value is returned as the simulated ultrasonic sensor's output.

3.2   Fusion of Exteroceptive Sensors: The Multi-sensor Fitness Function

The new fitness function should integrate information issued by all the exteroceptive sensors. As stated above, it is difficult to give a formal mathematical justification of how to extend the expression of the fitness function to integrate several sensors. In qualitative terms, if a fly's position is in accordance with several independent sensors, its fitness must be increased accordingly. Conversely, a fly confirmed by the visual sensor should still be considered seriously as a real obstacle even if unnoticed by another sensor (e.g. a visible obstacle covered with sound damping material). We defined each sensor's contribution to the fitness function as follows: if a fly is located inside the vision angle of a triggered detector, and its distance to the detector matches the obstacle distance given by the detector accurately enough, then the fly's fitness is increased by a given percentage B :

$$fitness_{new} = (1 + B)\,fitness_{old} \tag{4}$$

If the fly's position does not match any range finder information, then the fitness remains unchanged. We chose a multiplicative rather than additive contribution to the fitness because of the low angular resolution of the ultrasonic sensors: this prevents all flies at the correct distance inside the field of an ultrasonic sensor from getting high fitness values even if they are inconsistent with image data. In addition to the integration of additional sensors into the fitness function, we introduced an additional *immigration* operator: new flies are introduced into the population with a bias in favour of 3-D positions given by the sonars. This helps detecting close obstacles which might otherwise have been overlooked by the system. This method allows the fusion of an arbitrary number of sonars (or any similar exteroceptive sensors) at the fitness level, but it is not appropriate to the integration of proprioceptive sensor information, which we examine in the following Sections.

3.3   Simulated Odometry

On a real robot, odometric data given by wheel rotation coders give an estimation of the robot's position. In most applications, wheel rotation is under control of the trajectory planner. The actual robot trajectory does not exactly match the target trajectory due to "trajectory noise" caused by factors such as wheel slipping, tyre wear or rough ground surface. In the simulator, the robot's actual position is simulated by adding a Gaussian noise to the planner's command, but the sensor fusion algorithm only gets the odometric estimation which is identical to raw trajectory planner data.

3.4   Fusion of Proprioceptive Sensors: Proprioception as an Operator

At each time step, the initial population of flies is based on the preceding generation's results, and positions are updated using the best available knowledge

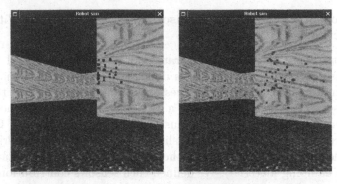

Fig. 4. Frames N and $N + 5$, without proprioception.

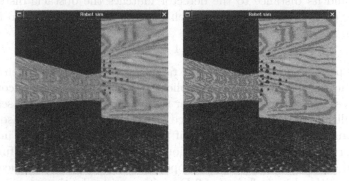

Fig. 5. The same two frames, with proprioception: better obstacle following.

about robot's motion. On the simulator, this information is given by odometric data (Section 3.3), but there could be any other source, e.g. an inertial sensor. The information received by the fly algorithm on the robot's position may not be accurate but remains useful in updating the flies' positions. We noticed that throughout our experiments, after a small number of generations (2 or 3) the flies converge again to the obstacles surfaces.

Figures 4 and 5 show images N and $N + 5$ from a sequence, with a 1 degree right turn of the robot at each frame, and one generation of the fly evolution per frame. On Figure 4, the flies' positions have not been updated using odometric information: the result is a delay in the evolution of flies (the black dots on the images). Figure 5 shows the results on the same sequence but integrating noisy odometric proprioceptive information: even if proprioceptive information is not required for the algorithm to work, it allows better fly stability and more accurate obstacle detection.

4   Simulation Results

Figure 6 shows the functional diagram of a mobile robot simulator using the fly algorithm, including cameras and other sensors (e.g. ultrasonic obstacle detectors). The software implementation uses a shared memory containing the images, sonar data, the fly population and the harmonic function. This memory can be accessed at any time by any process. Figure 7 shows typical examples of robot trajectories produced by the simulator. On the right we may compare the actual robot trajectory (green line) with the robot's internal trajectory representation (yellow line). The purple lines show the obstacles (in perspective).

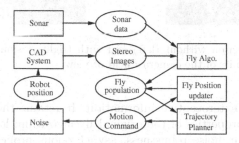

Fig. 6. The complete robot simulator.

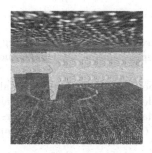

Fig. 7. On the left, an obstacle avoidance trajectory. On the right, the actual robot trajectory (green) and the trajectory given by odometric data (yellow)

Figure 8 shows an example of how the addition of sonar information into artificial evolution improves the detection of a short-range obstacle (on the right side of the robot axis, right image). On the left image, the sonar information has not been used: the flies concentrated onto further obstacles and neglected the closer one.

5   Conclusion

We showed in this paper that the evolutionary strategy underlying the fly algorithm may be extended to perform embedded robot sensor fusion, including

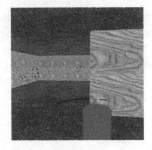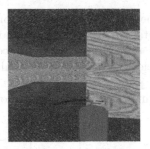

Fig. 8. Obstacle detection without (left) and with (right) fusion of ultrasonic data. Red lines show the limits of the camera's field of view.

proprioceptive and exteroceptive information. It retains the whole structure, original spirit and properties of the original fly algorithm: low calculation cost and fast convergence, noise robustness, asynchronous properties well adapted to state-of-the-art CMOS imagers. The robot simulator presented also includes a non-evolutionary trajectory planner adapted to use the fly population as an input, with matching asynchronous properties.

References

1. J. Louchet. From Hough to Darwin: an individual evolutionary strategy applied to artificial vision. In *Proceedings of Artificial Evolution 99*, Dunkerque, France, November 1999.
2. P. Collet, E. Lutton, F. Raynal, and M. Schoenauer. Individual gp: an alternative viewpoint for the resolution of complex problems. In W. Banzhaf, J. Daida, A. E. Eiben, M. H. Garzon, V. Honovar, M. Jakiela, and R. E. Smith, editors, *Genetic and Evolutionary Computation Conference GECCO99*. Morgan Kaufmann, San Francisco, CA, 1999.
3. Amine M. Boumaza and Jean Louchet. Dynamic flies: Using real-time parisian evolution in robotics. In Egbert J. W. Boers, Stefano Cagnoni, Jens Gottlieb, Emma Hart, Pier Luca Lanzi, Gunther R. Raidl, Robert E. Smith, and Harald Tijink, editors, *Applications of Evolutionary Computing*, volume 2037 of *LNCS*, pages 288–297, Lake Como, Italy, 18 April 2001. Springer-Verlag.
4. R.C. Jain, R. Kasturi, and B.G. Schunck. *Machine Vision*. McGraw-Hill, New York, 1994.
5. R. M. Haralick. Using perspective transformations in scene analysis. *Computer Graphics and Image Processing*, (13):191–221, 1980.
6. R. C. Gonzalez and R. E. Woods. *Digital Image Processing*. Wiley, 1992.
7. J. Holland. *Adaptation in Natural and Artificial Systems*. Univ. of Michigan Press Press, 1975.
8. Jean Louchet, Maud Guyon, Marie-Jeanne Lesot, and Amine Boumaza. Dynamic flies: a new pattern recognition tool applied to stereo sequence processing. *Pattern recognition letters*, 23:335–345, 2002.
9. O. Khatib. Real time obstacle avoidance for manipulators and mobile robots. *The International Journal of Robotics Research*, 5(1):90–99, Spring 1986.

10. Y.Koren and J. Borenstein. Potential field methods and their inherent limitations for mobile robot navigation. In *Procdings of the IEEE Conference On Robotics and Automation, ICRA'91*, pages 1398–1404, Sacramento, California, April 7–12 1991.
11. C. I. Connolly and R. Grupen. On the applications of harmonic functions to robotics. *Journal of Robotic and Systems*, 10(7):931–946, October 1993.
12. C. I. Connolly, J. B. Burns, and R. Weiss. Path planning using laplace's equation. In *The Proceedings of IEEE Internetioal Conference on Robotics and Automation, ICRA'90*, pages 2102–2106, May 1990.
13. B. H. Horn. *Robot Vision*. McGraw Hill, 1986.
14. Cumuli project: Computational understanding of multiple images, http://www.inrialpes.fr/cumuli.
15. J. Holland. *Adaptation in Natural and Artificial Systems*. MIT Press, 1992.
16. D. E. Goldberg. *Genetic Algorithms in Search, Optimisation and Machine Learning*. Addison Wesley, Reading, MA, 1989.
17. D. H. Ballard and C. M. Brown. *Computer Vision*. Prentice Hall, 1982.
18. R. C. Eberhart and J. A. Kennedy. New optimizer using particle swarm. In *Proc. Sixth Int. Symposium on Micro Machine and Human Science, Nagoya*, pages 39–43, Piscataway, NJ, 1995. IEEE Service Center.
19. P. V. C. Hough. Method and means of recognizing complex patterns. U.S. Patent no3, 069 654, 18 December 1962.
20. J. Louchet. Using an individual evolution strategy for stereovision. *Genetic Programming and Evolvable Machines*. Kluwer, to appear, 2001.
21. E. Lutton and P. Martinez. A genetic algorithm for the detection of d geometric primitives in images. In *The Proceedings of the International Conference on Pattern Recognition, ICPR'94*, pages 526–528, Los Alamitos, CA, October 9–13 1994. IEEE Computer Society.
22. M. C. Martins and H. P. Moravec. Robot evidence grid. Technical report, The Robotics Institute, Carnegie Mellon University, March 1996.
23. M. Millonas. Swarms, phase transitions and collective intelligence, artificial life iii. In C.G Langton, editor, *Santa Fe Institute Studies in the Sciences of Complexity*, volume XVII. Addison Wesley, Reading, MA, 1994.
24. I. Rechenberg. Evolution strategy. In J.M. Zurada, R.J. MarksII, and C.J. Robinson, editors, *Computational Intelligence imitating life*, pages 147–159. IEEE Press, Piscataway, NJ, 1994.
25. G. Roth and M. D. Levine. Geometric primitive extraction using genetic algorithm. In *Proceedings of the IEEE Conference on Computer vision ans Pattern Recognition, CVPR'92*, Piscataway, NJ, 1992. IEEE Press.
26. R. Salomon and P. Eggenberger. Adaptation on the evolutionary time scale: a working hypothesis and basic experiments. In *Springer Lecture Notes on Computer Science*, number 1363, pages 251–262. Springer-Verlag, Berlin, 1997.
27. P.K. Ser, S. Clifford, T. Choy, and W.C. Siu. Genetic algorithm for the extraction of nonanalytic objects from multiple dimensional parameter space. *Computer Vision and Image Understanding, vol. 73 no. 1, Academic Press: Orlando, FL*, pages 1–13, 1999.
28. C. K. Tang and G. Medioni. Integrated surface, curve and junction inference from sparse 3-d data sets. In *ICCV98*, pages 818–823, Piscataway, NJ, 1998. IEEE Computer Society Press.
29. John S. Zelek. Complete real-time path planning during sensor-based discovery. In *IEEE/RSJ International Conference on Intelligent Robots and systems*, 1998.

Anticipating Bankruptcy Reorganisation from Raw Financial Data Using Grammatical Evolution

Anthony Brabazon[1] and Michael O'Neill[2]

[1] Dept. of Accountancy, University College Dublin, Ireland.
Anthony.Brabazon@ucd.ie
[2] Dept. of Computer Science and Information Systems,
University of Limerick, Ireland.
Michael.ONeill@ul.ie

Abstract. This study using Grammatical Evolution, constructs a series of models for the prediction of bankruptcy, employing information drawn from financial statements. Unlike prior studies in this domain, the raw financial information is not preprocessed into pre-determined financial ratios. Instead, the ratios to be incorporated into the predictive rule are evolved from the raw financial data. This allows the creation and subsequent evolution of alternative ratio-based representations of the financial data. A sample of 178 publically quoted, US firms, drawn from the period 1991 to 2000 are used to train and test the model. The best evolved model in each time period correctly classified 78 (70)% of the firms in the out-of-sample validation set, one (three) year(s) prior to failure. The utility of a number of different Grammars for the problem domain is also examined.

1 Introduction

The objective of this study is to determine whether the construction of models to predict corporate bankruptcy (failure), using information drawn from financial statements, can be enhanced by allowing the modelling process to evolve different ratio representations from raw financial information. To date, most attempts at developing models for the prediction of corporate failure have utilised a limited set of financial ratios [1]. These ratios are generally selected on an ad-hoc basis by the modeller due to the lack of a strong theoretical framework underlying the failure prediction problem [2]. Unfortunately, the number of ratios which can be calculated from a set of financial statements is large. A set of financial statements could contain several hundred numbers between the primary financial statements and the detailed notes accompanying the primary statements, resulting in many thousands of potential financial ratios. Most studies in this domain utilise similar financial ratios, circularly justifying the choice of ratios by reference to earlier studies. This methodological approach leaves open the possibility that alternative, better, representations of the financial data exist. The selection of quality explanatory variables (ratios) and model form produces

S. Cagnoni et al. (Eds.): EvoWorkshops 2003, LNCS 2611, pp. 368–377, 2003.

a high-dimensional combinatorial problem giving rise to potential for an EAP methodology. Only one previous application of Grammatical Evolution to this domain has been identified [3]. This study extends this, novelly permitting the evolution of the ratios from raw financial data.

This contribution is organised as follows. Section 2 provides a short discussion of prior literature in the corporate failure domain. Section 3 provides an introduction to Grammatical Evolution. Section 4 describes both the data utilised, and the model development process adopted in this paper. Section 5 provides the results of the evolved models. Finally, conclusions and a discussion of the limitations of the contribution are provided in Section 6.

2 Background

Formal research into the prediction of corporate failure has a long history [4] [5]. Early statistical studies such as [6], adopted a univariate methodology, identifying which accounting ratios had greatest classification accuracy when identifying failing and non-failing firms. Although this approach did demonstrate predictive power, it suffers from the shortcoming that a single weak financial ratio may be offset (or exacerbated) by the strength (or weakness) of other financial ratios. Later studies [7] addressed this issue by employing an LDA model, which utilised both financial and market data concerning a firm, and this was found to improve the classification accuracy of the developed models. Other statistical models which have been applied include logit and probit regression models [8] [9] [10]. In recent times, methodologies applied to this problem domain have included neural networks [11] [12], genetic algorithms [13] [14].

2.1 Explanatory Variables Utilised in Prior Literature

Five groupings of explanatory variables, drawn from financial statements, are given prominence in prior literature [15]:

i. Liquidity
ii. Debt
iii. Profitability
iv. Activity / Efficiency
v. Size

Liquidity refers to the availability of cash resources to meet short-term cash requirements. Debt measures focus on the relative mix of funding provided by shareholders and lenders. Profitability considers the rate of return generated by a firm, in relation to its size, as measured by sales revenue and/or asset base. Activity measures consider the operational efficiency of the firm in collecting cash, managing stocks and controlling its production or service process. Firm size provides information on both the sales revenue and asset scale of the firm and also provides a proxy metric on firm history. The groupings of potential explanatory variables can be represented by a wide range of individual financial ratios, each with slightly differing information content. The groupings themselves are interconnected, as weak (or strong) financial performance in one area will impact on another. For example, a firm with a high level of debt, may have lower profitability due to high interest costs. Whatever modelling methodology is applied, the initial problem is to select a quality set of model inputs from a wide array of possible financial ratios, and then to combine these ratios using suitable weightings in order to construct a high quality classifier. Unlike prior work, this study does not prespecify the ratios to be employed in the modelling process, but permits the evolution of a representation based on the raw financial data.

2.2 Definition of Corporate Failure

The definition of corporate failure adopted in this study is the entry of a firm into Bankruptcy proceedings under Chapter 7 or Chapter 11 of the US Bankruptcy code. The selection of this definition provides an objective benchmark as the occurrence, and date of occurrence, of either of these events can be determined through examination of regulatory filings. Chapter 7 covers corporate liquidations and Chapter 11 covers corporate reorganizations, which usually follow a period of financial distress. Under Chapter 11, management is required to file a reorganisation plan in bankruptcy court and seek approval for this plan. When the court grants approval for the plan the firm is released from Chapter 11 bankruptcy and continues to trade. In most cases, Chapter 11 reorganisations involve significant financial losses for both shareholders [16] and creditors [17] of the distressed firm. [18], in a study of the outcomes of Chapter 11 filings, found that 'there were few sucessful reorganisations' (p. 125).

3 Grammatical Evolution

Grammatical Evolution (GE) is an evolutionary algorithm that can evolve computer programs in any language [19] [20] [21]. Rather than representing the programs as parse trees, as in GP [22], a linear genome representation is used. Each individual, a variable length binary string, contains in its codons (groups of 8 bits) the information to select production rules from a Backus Naur Form (BNF) grammar. BNF is a notation that represents a language in the form of production rules. It is comprised of a set of non-terminals that can be mapped to elements of the set of terminals, according to the production rules. An example excerpt from a BNF grammar is given below. These productions state that S can be replaced with either one of the non-terminals *expr*, *if − stmt*, or *loop*.

```
S ::= expr      (0)
    | if-stmt   (1)
    | loop      (2)
```

The grammar is used in a generative process to construct a program by applying production rules, selected by the genome, beginning from the start symbol of the grammar.

In order to select a rule in GE, the next codon value on the genome is generated and placed in the following formula:

$$Rule = Codon\ Value\ MOD\ Num.\ Rules$$

If the next codon integer value was 4, given that we have 3 rules to select from as in the above example, we get $4\ MOD\ 3 = 1$. S will therefore be replaced with the non-terminal $if - stmt$.

Beginning from the the left hand side of the genome codon integer values are generated and used to select rules from the BNF grammar, until one of the following situations arise: (a) A complete program is generated. This occurs when all the non-terminals in the expression being mapped are transformed into elements from the terminal set of the BNF grammar. (b) The end of the genome is reached, in which case the *wrapping* operator is invoked. This results in the return of the genome reading frame to the left hand side of the genome once again. The reading of codons will then continue unless an upper threshold representing the maximum number of wrapping events has occurred during this individuals mapping process. (c) In the event that a threshold on the number of wrapping events has occurred and the individual is still incompletely mapped, the mapping process is halted, and the individual assigned the lowest possible fitness value.

GE uses a steady state replacement mechanism, and the standard genetic operators of mutation (point), and crossover (one point). It also employs a duplication operator, which duplicates a random number of codons and inserts these into the penultimate codon position on the genome. A full description of GE can be found in [20].

4   Problem Domain & Experimental Approach

This section describes both the data utilised by, and the model development process adopted in, this study.

4.1   Sample Definition and Model Data

A total of 178 firms were selected judgementally (89 failed, 89 non-failed), from the Compustat Database[1]. The criteria for selection of the failed firms were:

i. Inclusion in the Compustat database in the period 1991-2000
ii. Existence of required data for a period of three years prior to entry into Chapter 7 or Chapter 11
iii. Sales revenues must exceed $1M

The first criterion limits the study to publicly quoted, US corporations. For every failing firm, a matched non-failing firm is selected. Failed and non-failed firms

---

[1] Firms from the financial sector were excluded on grounds of lack of comparability of their financial ratios with other firms in the sample.

are matched both by industry sector and size (sales revenue three years prior to failure)[2]. The set of 178 matched firms are randomly divided into model building (128 firms) and out-of-sample (50 firms) datasets. The dependant variable is binary (0,1), representing either a non-failed or a failed firm. In this study, rather than pre-specifying financial ratios, GE can create ratios from raw financial data. We have initially restricted our choice of raw financial data to the following twelve items:

   i. Sales
   ii. Net Income
   iii. Gross Profit
   iv. EBIT
   v. EBITDA
   vi. Total Assets
   vii. Total Current Assets
   viii. Total Liabilities
   ix. Total Current Liabilities
   x. Total Long-term Debt
   xi. Cash from Operations
   xii. Free-Cash Flow

This information was collected for each firm for the three years prior to entry, either by it or its matched firm, into Chapter 7 or Chapter 11 (denoted as T3, T2 and T1, where T3 is three years prior to failure)[3]. Three Grammars are employed, in order to examine the impact on predictive accuracy of allowing GE to evolve classification rules of varying complexity. The three Grammars are as follows:

```
Grammar 1

lc ::= output = coeff * ( ( var ) / ( var ) );
var ::= var1[index]  | var2[index]  | var3[index]
      | var4[index]  | var5[index]  | var6[index]
      | var7[index]  | var8[index]  | var9[index]
      | var10[index] | var11[index] | var12[index]
coeff ::= ( coeff ) op ( coeff ) | float
op ::= + | -
float ::= 20 | -20 | 10 | -10 | 5 | -5
        | 4 | -4 | 3 | -3 | 2 | -2 | 1
        | -1 | .1 | -.1
```

```
Grammar 2

lc ::= output = expr ;
expr ::= ( expr ) + ( expr )
       | coeff * (var/var)
var ::= var1[index]
      | .... | var12[index]
coeff ::= (coeff) op ( coeff )
        | float
op ::= + | -
float ::= 20 | -20 | 10 | -10
        | 5 | -5 | 4 | -4 | 3
        | -3 | 2 | -2 | 1 | -1
        | .1 | -.1
```

```
Grammar 3

lc ::= output = expr ;
expr ::= ( expr ) + ( expr ) |  coeff * ( ratio / var )
ratio ::= ratio op ratio | var
var ::= var1[index]  | .... | var12[index]
coeff ::= ( coeff ) op ( coeff ) | float
op ::= + | -
float ::= 20 | -20 | 10 | -10 | 5 | -5 | 4 | -4
        | 3 | -3 | 2 | -2 | 1 | -1 | .1 | -.1
```

---

[2] It is recognised that the use of an equalised, matched sample entails sampling bias and eliminates firm size and industry nature as potential explanatory variables (see [2] for a detailed discussion of these points). It is noted that utilising an unmatched sample imposes its own bias.

[3] The date of entry into Chapter 7 or Chapter 11 was determined by examining US Securities and Exchange Commission (SEC) regulatory filings for each firm.

Grammar one permits the construction of a predicitive rule consisting of a single ratio, formed from any two discrete pieces of raw financial data. This ratio can be rescaled as required by an evolved coefficient parameter. In essence, this Grammar searches for the best univariate predictive model.

Grammar two permits the construction of predictive rules which chain ratios together, producing linear rules of the form:

```
output = coefficient * Ratio X + coefficient * Ratio Y + ...
```

In each of these Grammars, only ratios of the form $\frac{a}{b}$, where a and b are discrete pieces of financial data are permitted.

Grammar three allows the construction of a linear chain of ratios, where the ratios can take the form $\frac{a+b+\cdots}{x}$, greatly increasing the number of possible ratios that can be formed from the raw data.

5   Results

The following results are based on the classification accuracy arising in the out-of-sample datasets. Three series of models were constructed for each Grammar, using raw financial information drawn from one, two and three years (T1, T2 and T3) prior to failure. For each set of models, 30 runs were conducted using population sizes of 500, running for 50 generations, adopting one-point crossover at a probability of 0.9, and bit mutation at 0.01, along with a steady state replacement strategy. Graphs showing average fitness versus generation for each grammar and time period can be seen in Figure 1.

Table 1. The out-of-sample classification accuracies for the best individual (average of best individual over 30 runs) in each Grammar for the three years prior to failure.

Years Prior to Failure	Grammar 1 (%)	Grammar 2 (%)	Grammar 3 (%)
1	74.0 (73.67)	78.0 (70.67)	76.0 (73.53)
2	76.0 (75.93)	76.0 (75.87)	78.0 (77.0)
3	68.0 (63.53)	66.0 (62.93)	70.0 (61.93)

The best individuals evolved for each period (and the average results for the best individual across all 30 runs) are reported in Table 1. Calculation of Press's Q statistic [23] for each of these models rejects a null hypothesis, at the 5% level, for each of the Grammars in periods T1 and T2 that the out-of-sample classification accuracies are not significantly better than chance.

The results are consistent across each of the three Grammars, and do not display degradation between T1 and T2. In order to provide insight into the form of the classifier rules evolved by the Grammars, Table 2 lists the best classifiers[4].

---

[4] In some cases, there was more than one rule producing equivalent classification accuracy. In these cases, the best classifier listed in the table was chosen judgementally.

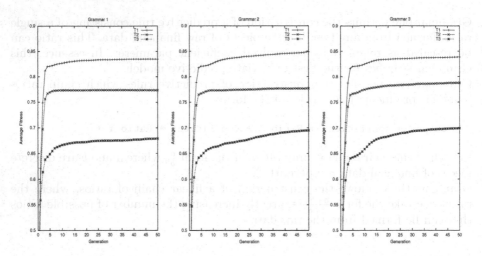

Fig. 1. Average fitness values, plotted for each year prior to failure, for all three grammars on the insample dataset.

5.1 Discussion

Despite using financial data drawn from a wide variety of industrial sectors, the evolved models showed a capability to discriminate between failing and non-failing firms, notably in the two years prior to the event. The risk factors suggested by each model differ somewhat but present plausible findings. In Grammar 1, the best evolved ratios all contain an earnings (profit) component. In all cases, the sign of the ratio coefficient is plausible, with lower earnings indicating greater risk of failure. The specific ratio chosen varies between the periods. In T1, low return on assets is identified as a risk factor, in T2 and T3, profitability relative to the size of the long / short term liabilities of the organisation indicates risk. The classifer rule for T1 under Grammar 2 utilises a combination of ratios which focus on the profitability of sales, the relationship between sales and short-term liabilities, and the relationship between cash generation and sales. The rule for T2 concentrates on measures of profits in relation to the indebtedness of the firm and profitability relative to the sales of the firm. For T3, the rule utilises measures of short-term liabilities and profit, and the rate of cash generation on sales. As for the rules generated under Grammar 1, the coefficient signs are plausible. High levels of firm profitability relative to the sales and level of indebtedness of the firm, are indicative of a non-failing firm, as are high levels of cash generation, relative to sales. The evolved rules do not demonstrate counter-intuitive relationships between the pieces of financial information, and hence there is no evidence of data-mining. The evolved classifiers under Grammar 3 made extensive use of various combinations of net income, cash from operations and short-term liabilities. It would be expected that there would be plausible relationships between these items as firms tend towards failure.

The classification results of the models show considerable consistency across the Grammars. As the Grammars can produce classifiers ranging from univariate

Table 2. The best classifiers evolved for each of the years.

Years Prior to Failure	Best Classifier
Grammar 1	
1	3 - (Total Assets / Net Income)
2	-10 * (Net Income / Total Long Term Debt)
3	5 - 13 * (EBIT / Total Current Liabilities)
Grammar 2	
1	-2*(Sales/Net Income)+3*(Total Current Liabilities/Sales) -20*(Cash from Operations/Sales)+3(Gross Profit/Sales)
2	-20 * (Net Income / Total Long Term Debt) - 3 * (EBIT / Sales)
3	-20*(Net Income/Total Current Liabilities) +3-20*(Cash from Operations/Sales)

models (Grammar 1) to complex linear combinations of financial ratios (Grammars 2 and 3), this suggests that there are a wide variety of classifer models which will produce similar out-of-sample results. The results also suggest that models with the capability to utilise complex combinations of the financial data did not out-perform simpler models. Considering the individual Grammars, it is seen that despite the potential to generate long, complex ratio chains in Grammars 2 and 3, this bloating did not occur and the evolved classifiers are reasonably concise in form.

Finally, it is noted that although GE was required to evolve its own ratio representation of the financial data in developing the classifiers, rather than being supplied with modeller-defined financial ratios (and thereby receiving enhanced domain-knowledge a priori),the classification results are competitive against those of prior studies[5].

6   Conclusions

In assessing the performance of the developed models, a number of caveats must be borne in mind. The premise underlying this paper (and all empirical work on corporate failure prediction) is that corporate failure is a process, commencing with poor management decisions, and that the trajectory of this process can be tracked using accounting ratios. This approach does have inherent limitations. It will not forecast corporate failure which results from a sudden environmental event. Commentators [24] [25] have noted that managers may attempt to utilise creative accounting practices to manage earnings and / or disguise signs of distress. Although not undertaken in this study, the incorporation of

---

[5] A prior study using GE with predefined financial ratios [3], reported best-individual classification accuracies for one and two years prior to failure of 80%

non-financial qualitative explanatory variables or variables related to the firm's share price performance could further improve classification accuracy. Finally, the firms sampled in this study are relatively large and are publically quoted. Thus, the findings of this study may not extend to small businesses.

Despite these limitations, the high economic and social costs of corporate failure imply that models which can indicate declining financial health will have utility. Given the lack of a clear theory underlying corporate failure, empirical modelling usually adopts a combinatorial approach, a task for which GE is well suited. The results of this study support earlier findings that GE has useful potential for the construction of corporate failure models.

References

1. Altman, E. (1993). *Corporate Financial Distress and Bankruptcy*, New York: John Wiley and Sons Inc.
2. Morris, R. (1997). *Early Warning Indicators of Corporate Failure: A critical review of previous research and further empirical evidence*, London: Ashgate Publishing Limited.
3. Brabazon, T., O'Neill, M., Matthews, R., and Ryan, C. (2002). 'Grammatical Evolution and Corporate Failure Prediction', In *Proceedings of the Genetic and Evolutionary Computation Conference (GECCO 2002)*, Spector et. al. Eds., New York, USA, July 9–13, 2002, pp. 1011–1019, Morgan Kaufmann.
4. Fitzpatrick, P. (1932). *A Comparison of the Ratios of Successful Industrial Enterprises with Those of Failed Companies*, Washington: The Accountants' Publishing Company.
5. Smith, R. and Winakor, A. (1935). 'Changes in the Financial Structure of Unsuccessful Corporations', *University of Illinois, Bureau of Business Research, Bulletin No. 51*.
6. Beaver, W. (1968). 'Financial Ratios as Predictors of Failure', *Journal of Accounting Research – Supplement: Empirical Research in Accounting*, 71–102.
7. Altman, E. (1968). 'Financial Ratios, Discriminant Analysis and the Prediction of Corporate Bankruptcy', *Journal of Finance*, 23:589–609.
8. Gentry, J., Newbold, P. and Whitford, D. (1985). 'Classifying Bankrupt Firms with Funds Flow Components', *Journal of Accounting Research*, 23(1):146–160.
9. Zmijewski, M. (1984). 'Methodological Issues Related to the Estimation of Financial Distress Prediction Models', *Journal of Accounting Research – Supplement*, 59–82.
10. Ohlson, J. (1980). 'Financial Ratios and the Probabilistic Prediction of Bankruptcy', *Journal of Accounting Research*, 18:109–131.
11. Shah, J. and Murtaza, M. (2000). 'A Neural Network Based Clustering Procedure for Bankruptcy Prediction', *American Business Review*, 18(2):80–86.
12. Serrano-Cina, C. (1996). 'Self organizing neural networks for financial diagnosis', *Decision Support Systems*, 17:227–238.
13. Varetto, F. (1998). 'Genetic algorithms in the analysis of insolvency risk', *Journal of Banking and Finance*, 22(10):1421–1439.
14. Kumar, N., Krovi, R. and Rajagopalan, B. (1997). 'Financial decision support with hybrid genetic and neural based modelling tools', *European Journal of Operational Research*, 103:339–349.

15. Altman, E. (2000). 'Predicting Financial Distress of Companies: Revisiting the Z-score and Zeta models', *http://www.stern.nyu.edu/ ealtman/Zscores.pdf, October 2001.*
16. Russel, P., Branch, B. and Torbey, V. (1999). 'Market Valuation of Bankrupt Firms: is there an anomaly?', *Quarterly Journal of Business and Economics*, 38:55–76.
17. Ferris, S., Jayaraman, N. and Makhija, A. (1996). 'The Impact of Chapter 11 filings on the Risk and Return of Security Holders, 1979–1989', *Advances in Financial Economics*, 2:93–118.
18. Moulton, W. and Thomas, H. (1993). 'Bankruptcy As a Deliberate Strategy: Theoretical Considerations and Empirical Evidence', *Strategic Management Journal*, 14:125–135.
19. Ryan C., Collins J.J., O'Neill M. (1998). Grammatical Evolution: Evolving Programs for an Arbitrary Language. *Lecture Notes in Computer Science 1391, Proceedings of the First European Workshop on Genetic Programming*, 83–95, Springer-Verlag.
20. O'Neill M., Ryan C. (2001) Grammatical Evolution, *IEEE Trans. Evolutionary Computation.* 2001.
21. O'Neill, M. (2001). Automatic Programming in an Arbitrary Language: Evolving Programs in Grammatical Evolution. PhD thesis, University of Limerick, 2001.
22. Koza, J. (1992). *Genetic Programming.* MIT Press.
23. Hair, J., Anderson, R., Tatham, R. and Black, W. (1998). *Multivariate Data Analysis*, Upper Saddle River, New Jersey: Prentice Hall.
24. Argenti, J. (1976). *Corporate Collapse: The Causes and Symptoms*, London: McGraw-Hill.
25. Smith, T. (1992). *Accounting for Growth.* London: Century Business.

GAME-HDL: Implementation of Evolutionary Algorithms Using Hardware Description Languages

Rolf Drechsler and Nicole Drechsler

Institute of Computer Science
University of Bremen
28359 Bremen, Germany
drechsle@informatik.uni-bremen.de

Abstract. Evolutionary Algorithms (EAs) have been proposed as a very powerful heuristic optimization technique to solve complex problems. Many case studies have shown that they work very efficient on a large set of problems, but in general the high qualities can only be obtained by high run time costs. In the past several approaches based on parallel implementations have been studied to speed up EAs. In this paper we present a technique for the implementation of EAs in hardware based on a the concept of reusable modules. These modules are described in a Hardware Description Language (HDL). The resulting "hardware EA" can be directly synthesized and mapped to Application Specific Integrated Circuits (ASICs) or Field Programmable Gate Arrays (FPGAs). This approach finds direct application in signal processing, where hardware implementations are often needed to meet the run time requirements of a real-time system. In our prototype implementation we used VHDL and synthesized an EA for solving the OneMax problem. Simulation results show the feasibility of the approach. Due to the use of a standard HDL, the components can be reused in the form of a library.

1 Introduction

In the last 10 years evolutionary techniques have been very successful to a large set of problems in very different domains, like e.g. circuit design [1,2,3,4]. Typically, the results obtained by Evolutionary Algorithms (EAs) can be characterized as being of very high quality, but the run time needed to compute the solutions is high compared to other optimization techniques, like hill climbing or gradient descendent. This is mainly due to the simulation-based approach that is used in EAs. Most of the computation time on complex problems is usually consumed by evaluation of the fitness function. The time needed for the overall algorithm, the selection and the application of the evolutionary operators are negligible.

To reduce the run time, in the past mainly two solutions have been proposed. The first one is based on parallel implementations [5]. Since there is usually no close interaction needed between the different elements of a population, the complete problem can easily be distributed over several hardware platforms. Many promising applications have been reported. A speed-up of close to linear (and in some cases

S. Cagnoni et al. (Eds.): EvoWorkshops 2003, LNCS 2611, pp. 378–387, 2003.
© Springer-Verlag Berlin Heidelberg 2003

even super-linear) can be achieved. But still the fitness functions are evaluated in software.

A very promising alternative is to implement the EA directly in hardware [6]. Hardware implementations are often several orders of magnitude faster than an equivalent software realization. The main drawback of this approach is the missing flexibility and the difficulty of the implementation, since this requires detailed knowledge about hardware development. To meet today's requirements to allow for fast prototyping more robust structures are needed that allow an inexpensive development while keeping the advantages of the underlying methods. Especially in the field of signal processing the demands regarding execution speed can often only be met by hardware implementations.

In this paper we present a dynamic hardware concept based on the use of standardized Hardware Description Languages (HDLs). All EA components, like the population (including the initialization), the selection mechanism and the fitness evaluation, are directly implemented as hardware modules that can be synthesized on a target technology. This can be an ASIC or an FPGA. For our implementation we used VHDL [7,8], since it is widely used and a large set of commercial tools exist. The VHDL code describes the different components in a structured way that allows for an easy reuse. This becomes the key of today's successful hardware projects, since without a concise reuse-methodology the complexity cannot be handled [9,10]. The overall structure of the system is closely related to the concept in [11], where GAME (=Genetic Algorithm Managing Environment) has been developed. Since the same modular structure can be found in the hardware counterpart, the library of modules is denoted as GAME-HDL. As a first case study we tested the components on OneMax problem, i.e. maximizing the number of ones in a bit-string. An experimental study including simulation results is given to show the flow of the hardware implementation.

2   Preliminaries

In this section we briefly review the notation and definitions used to make the paper self-contained. First, the concept of EAs is described   (see also [1]) and the components used later are given. Then, an introduction to VHDL – the HDL used throughout the paper – is provided.

2.1   Evolutionary Algorithms

We assume that the reader is familiar with the concepts of simulated evolution for computer optimization problems. Since the late 70s, several concepts using simulated evolution have been proposed. Among them are e.g. *Genetic Algorithms* (GAs) [12], *Evolution Strategy* (ES) [13,14] and *Genetic Programming* (GP) [15]. The different concepts mainly differ in the form of representation of the individuals in a population and in the operators applied, while the overall algorithmic flow is very similar. In the following, we do not distinguish between these "pure" concepts and use the term *Evolutionary Algorithm* as the union of all these techniques.

We do not discuss the components of EAs, i.e. representation, objective function, selection method, initialization of population, and evolutionary operators. (For more details see e.g. [16].) Based on these components the overall structure and information flow of a "classical" EA works as follows:

- Initially a random population is generated.

- The evolutionary operators reproduction, crossover, and mutation are applied to some elements. The elements are chosen according to the selection method.

- The algorithm stops if a termination criterion is fulfilled, e.g. if no improvement is obtained for a given number of iterations.

The EA depends on several parameter settings, e.g. population size, reproduction probability, crossover probability, and mutation probability. A sketch of a "classical" EA is given below.

```
evolutionary_algorithm (problem instance):
  initialize_population ;
  calculate_fitness ;
  do
    apply_operators_with_corresponding_probabilities ;
    calculate_fitness ;
    update_population ;
  while (not terminal case) ;
  return best_element ;
```

2.2 Hardware Description Languages

HDLs are programming languages comparable to languages used in the software domain, like C++ or JAVA, but the compilation process generates a chip instead of an executable program. For this, HDLs contain – in addition to constructs to describe the logic behavior, like if-then-else and loops – also commands to specify implementation specific properties, like timing. The need for standardized HDLs comes from several different aspects. Some of the most important ones are:

- The ever increasing design complexity of today's designs can only be handled by structured languages, since the circuits consist of up to several million transistors and are described by several thousand lines of code.

- Components developed once can be reused several times. In ASIC projects often more than 90% of the modules are reused.

- In design flows several tools have to interact, like the synthesis and simulation tools. The code in the HDL is used as an exchange format.

- The circuits can be developed independent of technology or tool provider.

Typical HDLs used in practice today are VHDL and Verilog. Recently, also languages have been proposed that are closer to software descriptions, since this often allows to describe hardware and software components in the same environment and enables a late decision which parts become software and which are realized by hardware (*hardware-software-co-design*).

For our implementation we used VHDL, since it is the most common language on the European market. Furthermore, several commercial tools are available. Within this article, a complete introduction to VHDL cannot be provided. Instead, a short example is given that explains the main components.

For each module the interface has to be declared. This is done using an ENTITY. It declares the input and output signals of each module and their types.

```
ENTITY AND_OR IS
PORT ( x1, x2, x3 : IN BIT;
                 f : OUT BIT ) ;
END AND_OR ;
```

The functional behaviour is described by the ARCHITECTURE, that has to be provided for each ENTITY.

```
ARCHITECTURE function OF AND_OR IS
BEGIN
   f <= x1 AND (x2 OR x3) ;
END function ;
```

In our example, the module has three inputs and one output. The output f is computed by Boolean operators. Of course, modern HDLs also provide more complex operations and data types, like e.g. addition and multiplication of bit-vectors. For each ENTITY at least one ARCHITECTURE has to exist. If there are several, e.g. some components may be optimised for area while others are better regarding delay, an assignment has to be done using a CONFIGURATION command. Further features are:

- To simplify reuse, corresponding to libraries in other programming languages, frequently used data types and modules can be grouped to build a PACKAGE.
- VHDL furthermore supports different description methods, i.e. structural, behaviour and data flow.
- Signal delays can be modelled.
- Availability of complex data types, like integer, real, or array.
- Structured algorithms can be described by procedures and functions.

For more details on VHDL see e.g. [7,8].

3   EA Implementation in VHDL

For all EA components described in Section 2.1, a VHDL module has been developed. First, the overall structure is outlined. Due to page limitation, a complete description of all modules cannot be given. Instead, we briefly discuss the main features of each element and then show one module, i.e. the realization of the module fitness, in detail. This is justified by the fact that often the fitness evaluation is the most complex operation in EAs regarding execution time. Finally, the components of the package that in sum build the library GAME-HDL are introduced.

The overall flow of the hardware realization of the EA is as follows:

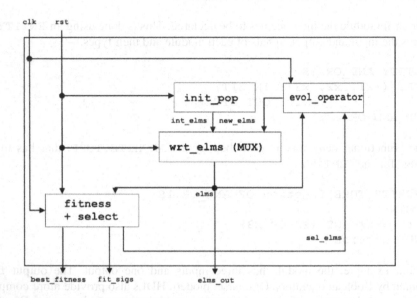

The only external inputs are the clock signal clk and the reset signal rst. The clock is needed, since we consider a synchronous digital design. The reset is used for the initialization of the elements. Of course, it is easy to extend the library to also allow external elements to be added to the population during the initialization, i.e. to use hybrid EAs, where problem specific information can be added to the EA concept. For simplicity, in this application we restrict ourselves to randomly generated elements. The circuit outputs are the best fitness value (best_fitness), the fitness values of all elements in the population (fit_sigs) and the optimized elements (elms_out).

As originally proposed for GAs, we make use of a binary encoding, since there is a one-to-one correspondence to the digital hardware platform. For problems originally not defined over a binary alphabet, the standard encoding techniques known from GAs can be applied.

For the evolutionary operators, crossover and mutation are defined in a classical way on binary strings. A hardware implementation is very easy, since VHDL allows data types such as arrays and control structures as e.g. for-loops. Thus, only a traversal of the strings is needed which can be carried out in linear time.

In our implementation we make use of tournament selection. Here the underlying principle is so simple that a transfer in a HDL is straightforward if the fitness value is available.

In this article it is not possible to describe the full implementation of the EA. Instead, the module for the fitness computation is shown below. (For the reader not very familiar with VHDL this also gives an impression of the code structure.)

First, standard libraries are included:

```
LIBRARY ieee ;
USE ieee.std_logic_1164.all ;
USE ieee.std_logic_arith.all ;
USE ieee.std_logic_signed.all ;
LIBRARY WORK ;
USE WORK.function_pack.all ;
```

Then, the entity is specified:

```
ENTITY fitness IS

PORT( clk             : IN BIT ;          -- clock signal
      rst             : IN BIT ;          -- reset signal
      elements        : IN pop ;          -- population
      best_fitness    : OUT INT_32 ;
      fit_sigs        : OUT INT_32_VECTOR ;
      elms_out        : OUT pop ) ;
END fitness;
```

This gives the input/output behaviour of the module (see previous section and the figure above).

The architecture describes the internal realization of the component. We used the OneMax problem, i.e. in a bit-string the number of ones is maximized, and the procedure below shows the computation of the fitness value for a given element.

```
ARCHITECTURE structure OF fitness IS

PROCEDURE calc_fitness (elements    : pop ;
                        best        : OUT INT_32) IS
          VARIABLE counter  : INT_32;
          VARIABLE max      : INT_32;
BEGIN  -- procedure for calculation of the best fitness
     max := 0 ;
     FOR i IN 0 TO element'right LOOP
          counter := 0 ;
          FOR c IN 0 TO elements(i)'right LOOP
              IF elements(i)(c) = '1' THEN
```

```
                              counter := counter + 1;
                     END IF ;
              END LOOP ;
              IF ( max < counter ) THEN
                     max := counter ;
              END IF ;
        END LOOP ;
        best := max ;
END calc_fitness ;

SIGNAL best : INT_32 := 0 ;

BEGIN
PROCESS (clk)      -- count bits with value 1 in
                   -- elements(i) and return all fitness
                   -- values and best fitness
         VARIABLE var_best : INT_32 ;
         VARIABLE counter : INT_32 ;
    BEGIN
         IF rst = '0' THEN -- initialize fitness values
              best_fitness <= 0 ;
              FOR i IN 0 TO elements'right LOOP
                     fit_sigs(i) <= 0 ;
              END LOOP ;
         ELSIF(clk'EVENT AND clk = '1') THEN
              calc_fitness (elements, var_best) ;
              IF ( best < var_best ) THEN
                     best <= var_best ;
              END IF ;
              FOR i IN 0 TO elements'right LOOP
                     counter := 0 ;
                     FOR c IN 0 TO element(i)'right LOOP
                             IF elements(i)(c) = '1' THEN
                                    counter := counter + 1;
                             END IF ;
                     END LOOP ;
                     fit_sigs(i) <= counter ;
              END LOOP ;
              best_fitness <= best;
         END IF ;
    END PROCESS;
END structure ;
```

The architecture of entity fitness consists of two main components, i.e. a procedure that computes the fitness and a process that describes the overall flow.

3.1 GAME-HDL: Library of Modules

In an analogous way, components for all the other modules, like selection or evolutionary operators, have been defined. They have been bound together in a reusable library. In VHDL these libraries are generated using the PACKAGE command. Following the ideas in [11], where a modular GA library GAME (=Genetic Algorithm Managing Environment) has been developed, the resulting package is called GAME-HDL, since the package considered here has the same underlying philosophy, i.e. a modular development of an EA environment that is motivated by the reuse of components.

4 Case Study: OneMax

The method described above has been implemented in VHDL using the tool MAX+plus II from Altera Corporation under Windows. This system has been chosen since it is available as public domain. Of course, the written code can be directly transferred to other VHDL systems, since only code was written that respects the VHDL standard.

In the following, some screenshots are given from the development environment and from simulation runs. For the initial experiments we considered the OneMax problem, that has been investigated in detail in the EA community. The goal is to maximize the number of ones in a given bit string.

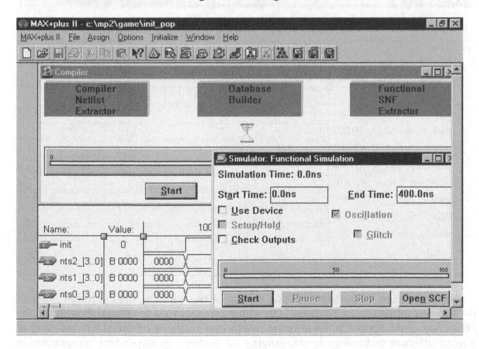

The content.

386   R. Drechsler and N. Drechsler

The desktop is shown above. A context-sensitive editor highlighting the keywords of the language simplifies the development. The code can be compiled and simulated in the same environment.

After the reset, the population has to be initialized. This is done by the module init_pop where all memory elements that are originally set to "all 0" get their values. The corresponding simulation run is shown below.

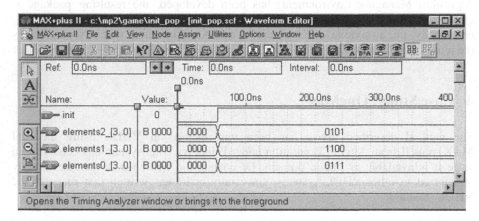

The results of a simulation run for the fitness module described in Section 3 is given in the next figure:

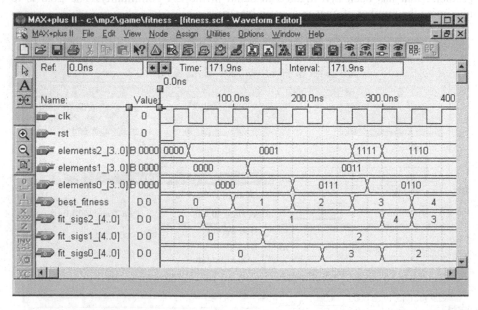

As can be seen on the time scale on the top of the window, the execution times of the fitness evaluation can be carried out within one clock cycle (40ns). This shows the efficiency of the approach. Even better results can easily be obtained by switching to a more efficient technology by exchanging the technology dependent components in the library. Notice that no changes in the VHDL code are required. The signal best_fitness gives the best fitness obtained so far. The fitness function is evaluated with each rising clock edge (see above). The fitness value for best_fitness is available one clock cycle later.

5 Conclusions

In this paper we presented a hardware implementation of Evolutionary Algorithms based on Hardware Description Languages. A tool called GAME-HDL has been implemented in VHDL, one of the most frequently used description languages. GAME-HDL is a library of VHDL modules that can be easily integrated and reused to design hardware-based EAs.

A case study for the OneMax problem has been demonstrated. Simulation runs have been provided that gave an impression of the high execution speed of the resulting circuits.

It is focus of current work to apply GAME-HDL to more complex optimization problems and in this context also experiment with further evolutionary operators and selection principles.

References

1. R. Drechsler, Evolutionary Algorithms for VLSI CAD, Kluwer Academic Publisher, 1998
2. P. Mazumder and E. Rudnick, Genetic Algorithms for VLSI Design, Layout & Test Automation, Prentice Hall, 1998
3. M. Erba, R. Rossi, V. Liberali and A. Tettamanzi, An Evolutionary Approach to Automatic Generation of VHDL Code for Low Power Digital Filters, EuroGP, LNCS 2038, pp. 36–50, 2001
4. R. Drechsler and N. Drechsler, Evolutionary Algorithms for Embedded System Design, Kluwer Academic Publisher, 2002
5. E. Cantu-Paz, Efficient and Accurate Parallel Genetic Algorithms, Kluwer Academic Publisher, 2000
6. S. Koizumi, S. Wakabayashi, T. Koide, K. Fujiwara, N. Imura, A RISC Processor for High-Speed Execution of Genetic Algorithms, GECCO, pp. 1338–1345, 2001
7. R. Lipsett, C.F. Schaefer and C. Ussery, VHDL: Hardware Description and Design, Kluwer Academic Publishers, 1989
8. D. Perry, VHDL, McGraw Hill, 1998
9. M. Keating and P. Bricaud, Reuse Methodology Manual for System-on-a-Chip Designs, Kluwer Academic Publishers, 1999
10. H. Shibata and N. Fujii, Analog Circuit Synthesis Based on Reuse of Topological Features of Prototype Circuits, GECCO, pp. 1205–1212, 2001
11. N. Göckel, R. Drechsler and B. Becker, GAME: A Software Environment for Using Genetic Algorithms in Circuit Design, Applications of Computer Systems, 240–247, 1997
12. J.H. Holland, Adaption in Natural and Artificial Systems, The University of Michigan Press, Ann Arbor, MI, 1975
13. I. Rechenberg, Evolutionsstrategie, Frommann-Holzboog, 1973
14. T. Bäck, Evolutionary Algorithms in Theory and Practice, Oxford University Press, 1996
15. J. Koza, Genetic Programming - On the Programming of Computers by means of Natural Selection, MIT Press, 1992
16. L. Davis, Handbook of Genetic Algorithms, van Nostrand Reinhold, New York, 1991

Evolutionary Approach to Discovery of Classification Rules from Remote Sensing Images

Jerzy Korczak and Arnaud Quirin

Université Louis Pasteur, LSIIT, CNRS, Strasbourg, France
{korczak,quirin}@lsiit.u-strasbg.fr
http://hydria.u-strasbg.fr

Abstract. In this article a new method for classification of remote sensing images is described. For most applications, these images contain voluminous, complex, and sometimes noisy data. For the approach presented herein, image classification rules are discovered by an evolution-based process, rather than by applying an a priori chosen classification algorithm. During the evolution process, classification rules are created using raw remote sensing images, the expertise encoded in classified zones of images, and statistics about related thematic objects. The resultant set of evolved classification rules are simple to interpret, efficient, robust and noise resistant. This evolution-based approach is detailed and validated based on remote sensing images covering not only urban zones of Strasbourg, France, but also vegetation zones of the lagoon of Venice.

1 Introduction

The design of robust and efficient image classification algorithms is one of the most important issues addressed by remote sensing image users. For many years a great deal of effort has been devoted to generating new classification algorithms and to refine methods used to classify statistical data sets [1]. At the time of this writing, relatively few workers in the machine learning community have considered how classification rules might be discovered from raw and expertly classified images. In this paper, a new data-driven approach is proposed to discover classification rules using the idea of evolutionary classifier systems. The unique source of information is a remote sensing image and its corresponding classification furnished by an expert. The images have been registered by various satellites (e.g. SPOT, LANDSAT, DIAS) that use different cameras having various spectral and spatial resolutions [2]. These types of remote sensing images generally contain huge volumes of data. And, sometimes they are very noisy due to coarse spatial resolution or unfavourable atmospheric conditions at the time the images are acquired. In addition, data may be erroneous due to inexperienced operators of the measurement devices.

The aim of this research is to elaborate an evolutionary classification method that, in contrast to classical algorithms, will allow for supervised creation of autonomous classification. In general, classification rules are discovered from the established classifier system ([3], [4]). As we said, the system is data-driven because the formulated classification rules are able to adapt themselves according to the available

S. Cagnoni et al. (Eds.): EvoWorkshops 2003, LNCS 2611, pp. 388–398, 2003.
© Springer-Verlag Berlin Heidelberg 2003

data, environment, and the evolution of classes. In remote sensing, the initial population of classifiers is randomly created from images and given classes, and then evolved by a genetic algorithm until the acceptable solution is found.

In remote sensing literature, several classification approaches are presented, namely:

- pixel-by-pixel: each image pixel is analysed independently of the others according to its spectral characteristic [5],
- zone-by-zone: before classification, the pixels are aggregated into zones, the algorithms detect the borders of the zones, delimit them by their texture, their repetitive motives [6],
- by object: this is the highest recognition, the algorithms classify semantic objects, the algorithms detect their forms, geometrical properties, spatio-temporal relations using the background domain knowledge [7].

Our approach is based on spectral data of pixels; therefore, discovered classification rules are only able to find spectral classes rather than semantic ones. This spectral component of class description is essential to well-recognised thematic classes. It should be noted that the proposed classifier system may be easily adapted to more sophisticated object representation with further research on detailed feature recognition.

The classifier system model has been implemented in the *I See You* (ICU) program, which has been used to validate our approach. In the ICU, we have adapted and extended previously established ideas recognized by classification experts, such as XCS [4], the s-classifiers, and "Fuzzy To Classify System" [8]. We have also been inspired by the works of Riolo [9] on gratification and penalisation, and of Richards [10] on the exploration of the space of classifiers.

The basic notion of our evolutionary classifiers is introduced in Section 2, which follows directly. Section 3 details the discovery process of the classification rules. In this Section, the behaviour of genetic algorithm functions is explained. Finally, three case studies on real remote sensing data are presented in Section 4.

2 Definition of a System of Classifiers in Remote Sensing

Generally speaking, a system of classifiers integrates symbolic learning and evolution-based computing. Classification rules are symbolic expressions and describe conditions to be held and actions to be taken if the conditions are satisfied. Quality of the rules is evaluated according to their classification performance. Here, we must underscore the fact that the rules are not introduced by a programmer or by an expert.

A system of classifiers is called evolutionary if it is able to adapt itself to the environment. This means that it can modify its knowledge and its behaviour according to the situation. For example, in remote sensing, the size of the classes may evolve in one of two ways: (1) if, after an initial classification, there remain non-classified pixels, or (2) if there are pixels belonging to several classes (mixed pixels). When a classifier integrates one of these pixels to one or another class, it is necessary to dynamically adapt classification rules. In this way, certain rules that treat only the simple cases (not mixed pixels) will become useless, and new rules are necessarily created for other cases.

From a functional point of view, a classifier can be defined as a rule representing a piece of knowledge about a class, and may be a conditional expression, such as *if <conditions> then <action>*. In the early classifier systems [4], each part of a rule was a binary message, encoding elementary information such as a value, colour, form, shape, etc. The *"conditions"* part described an entry message in the system, corresponding to conditions that must be fulfilled in order to activate this rule. The *"action"* part defined the action to be carried out when the appropriate conditions were satisfied. This binary encoding scheme is not well adapted to image classification rules.

One of the reasons for this is based on the domain of spectral values that may be assigned to a pixel (from 0 to 255 for 8-bit pixels, or from 0 to 65000 for 16-bit pixels). Of course, binary encoding of rule conditions is possible but the rules would be difficult to understand. Instead, we assert that the evolved rules must be rapidly evaluated and easy to interpret by any user. As a result, condition representation using the concept of an interval could be fully adequate for remote sensing image classification. In terms of machine learning, the rules have to be absolutely specific, meaning that they have to cover the extreme maximum and minimum pixels belonging to any given class.

Before rule specification is explained, recall that a pixel is encoded as a spectral vector, defining a value of reflectance for the n bands of the remote sensing image:

$$<pixel> := [b_1\ b_2\ b_3 ... b_n]$$

In our system, the condition for any rule is built on the concept of spectral intervals defining a given band, corresponding to a given class. Such intervals are a pair of integer numbers, between 0 and the maximum possible value for a pixel of a given band (i.e. 65536 for the pixels defined on 16 bits). This solution allows to partition the space of the spectral values in two ranges: the first containing the pixel values which corresponds to a given class, and the second containing the remainder.

To precisely specify the class definition, a set of intervals is defined for each band of the remote sensing image. Taking onto consideration all bands, one can define the condition part as follows:

$$<condition> := E_1\ and\ E_2\ and\ ...\ and\ E_N$$

where E_i defines a set of intervals for a band i, and N is the total number of bands.

Each E_i is defined as a set of spectral intervals: $E_i := [m_1;\ M_1]\ or\ [m_2;\ M_2]\ or...or\ [m_p;\ M_p]$ where m_j and M_j are, respectively, the minimal and maximum reflectance values authorised for a pixel belonging to a class k for the band i. These intervals are not necessarily disjunctive. By experiments, we have found that if we allow the genetic algorithm to create non-disjunctive intervals, instead of merging them, the results of genetic operators are more interesting. We have noticed that merging intervals significantly diminishes the number of intervals, and in the same time reduces the possibilities to create more efficient rules. The example below illustrates a concept of interval merging: $E = [11; 105]$ or $[138; 209]$ or $[93; 208]$ corresponds after merge operation to $E = [11; 209]$.

To satisfy a rule, a pixel has to match at least one spectral interval for each band. Logically speaking, to associate a pixel to a class, its values have to satisfy the conjunction of disjunctions of intervals that define a condition part of the classification rule. Figure 1 illustrates an example of matching of two pixels against the spectral class. The left figure shows graphically the spectral intervals of the class defined by a given rule. The next two diagrams show spectral signatures of two pixels: the first matches the rule, but the second does not. Hence, only the first pixel of this example may be considered to be an instance of the class.

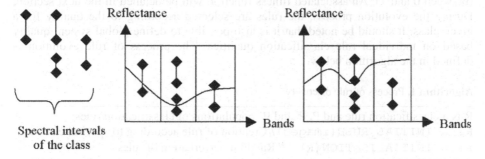

Fig. 1. Matching spectral bands and spectral signature of pixels

This representation of the rule has been chosen mainly because of its simplicity, compactness and uniform encoding of spectral constraints. During experimentation, this representation has demonstrated rapid execution of genetic operators and efficient computing. Of course, one may specify more complex structures using spatial properties of the pixel, with respect to the pixel neighbourhood. Also, one may include features resulting from thematic indices or mathematical operators applied to pixel environment. We may also apply a genetic programming to identify new characteristics. These semantically extended formalisms are interesting, however they not only require more sophisticated genetic operators, but also more powerful computers to perform the calculation in an acceptable amount of time.

3   From the Rule Creation to the Evolution

3.1 Genetic Algorithm

In order to efficiently develop the classification rules, a genetic algorithm initialises interval values according to spectral limits of the classes designated by an expert, for valid zones of the remote sensing image. Initial classification rules are created based on the extreme maximum and minimum values for defined spectral intervals, taking into account every class. It should be noted that by this initialisation, rule searching is considerably reduced, and initial intervals are very close to the final solution. During this process, the initial spectral limits are slightly perturbed by adding a random value to lower and upper spectral limits. Hence, the initial population of classification rules is quite diversified.

This initial pool of classifiers is evolved from a genetic algorithm. Our system searches for a best classifier for each class, independently. A major reason for choosing this procedure is the efficiency of computations; that is, the process of rule discovery is not perturbed by other rules.

The quality of classification rules is based on a comparison of these results with the image classified by an expert. If pixels covered by the classifier perfectly overlap those indicated by an expert, then the system assigns the highest quality value to the classifier; otherwise, in the case of some mismatching, the quality factor is reduced (between 0 and 1). An associated fitness function will be detailed in the next section. During the evolution process, the rules are selected according to the quality for a given class. It should be noted that it is also possible to define global system quality based on individual rule classification qualities. The process of rule evolution is defined in the algorithm below.

Algorithm 1. Process of rule discovery

R is a classification rule and P, P' and P" populations of classification rules.
R: = INITIAL_RULE(images) //Creation of rule according to spectral extremes
P: = INITIALISATION(R)   // Random perturbation of rules
EVALUATION(P)   // Calculation of the fitness function for each rule
do while TERMINATION_CRITERION(P) = false
    P' : = SELECTION_X(P)   // Selection for crossover
    P' : = CROSSOVER(P') U COPY(P)
    P'' : = SELECTION_MUT(P')   // Selection for mutation
    P'' : = MUTATION(P'') U COPY(P')
    EVALUATION(P'')
    P: = REPLACEMENT(P,P'')   // New generation of rules
end_while
Result: R, the classification rule for a given class

As mentioned before, this algorithm must be designed to run independently for each class. This allows for obtaining classifiers according to user requirements without the necessity of carrying out computations for all classes with the same level of quality. This also allows for the maintenance of previously generated classifiers, as well as for the introduction of new ones. Further, the user may define a hierarchy of classes and specialise some classifiers while respecting newly created sub-classes with different levels of classification quality.

3.2 The Evaluation Function

The evaluation function serves to differentiate the quality of generated rules and guide genetic evolution. Usually, this function depends strongly on application domain. In our work, evaluation is based on the classification obtained by the classifier (I_{rule}) and the expertly given classification (I_{expert}). Table 1 defines the values necessary to compute the evaluation function.

Table 1. Characteristics of the evaluation function

		Image classified by the classifier R (I_{rule})	
		No. of pixels activating R	non activating R
Pixel classified by the expert E (I_{expert})	True	$P_{\exp}^{rul}$	$P_{\exp}^{\overline{rul}}$
	False	$\underline{P}_{\exp}^{rul}$	$\underline{P}_{\exp}^{\overline{rul}}$

In any given system, the evaluation function is computed as follows:

$$N_{final} = C_{class} \cdot N_{class} + C_{\overline{class}} \cdot N_{\overline{class}} \tag{1}$$

where $N_{class} = \dfrac{P_{\exp}^{rul}}{P_{\exp}^{rul} + P_{\exp}^{\overline{rul}}}$ and $N_{\overline{class}} = \dfrac{P_{\exp}^{\overline{rul}}}{P_{\exp}^{rul} + P_{\exp}^{\overline{rul}}}$. C_{class} and $C_{\overline{class}}$ are called

the adjusting coefficients, which are used for certain classes that are under- or over-represented. By default the value of these coefficients is equal to ½.

The proposed function has a number of advantages; it is independent of the pixel processing sequence, invariant of the size of classes, and efficient for class discovery with a highly variable number of pixels.

The evolution process converges according to some statistical criteria indicating if the current classifier is near to a global optimum or if the population of rules will not evolve anymore. The termination criterion of the algorithm leans on the statistics of classifier quality evolution. In our system, we take into consideration not only the evolution of quality of the best and the average classifiers, but also the minimum acceptable quality defined by a user and a maximal number of generations to run. If one of these criteria is satisfied, then the process is stopped.

The most difficult determination to make is whether the quality of a classifier is not continuing to evolve. To detect stabilisation of the quality evolution, we have based our heuristics on statistics regarding quality evolution of the best classifier. For example, let Q_k be the quality of the best classifier obtained during the last k generations, and Q_o be the quality of the best classifier of the current generation. The algorithm is interrupted if the following equation is satisfied:

$$\left| \frac{\sum_{k=1}^{P} Q_k}{P} - Q_0 \right| \leq E \tag{2}$$

where P represents the maximum period of quality stabilisation, and E is a maximal variation of this stabilisation compared with the current quality.

It is important to have an initial population of classifiers within the vicinity of the solution to be found. We have proposed two algorithms allowing for the generation of a diversified pool of classifiers close to the expert hidden classification rule. The first, called MinMax, creates maximum intervals covering the expert rule, and the second algorithm, called Spectro, integrates the spectral distribution density and interval partitioning.

3.3 Genetic Operators

One of the most important tasks while designing a genetic algorithm is to invent operators that will create new potential solutions. All of our operators have been specialised on classifier representation, and they have been validated on remote sensing images. With respect to software engineering, the genetic algorithm has been structured into layers corresponding to genetic operations (e.g. selection, mutation, crossover and replacement). The system is viewed as a collection layers with data passed from layer to layer. Layer execution follows from one to another, and genetic operations are invoked in the same sequence. This modular approach makes program maintenance and future extensions much easier.

Selection of classifiers. In general, selection is the operation of allocating reproductive opportunities to each classifier. The reproductive force of a classifier is expressed by a fitness function that measures relative quality of a classifier by comparing it to other classifiers in the population. There are many methods for selecting a classifier [11]. In our system, the selection operator is applied in the following cases:
- choosing the classifier to be reproduced for crossing, or muting;
- repetition of the classifier, depending on whether it completes the genetic pool after having completed the crossover;
- preservation of a classifier from the former genetic pool for the next generation;
- elimination of a classifier in a newly created genetic pool based on an assigned rank.

Selection strategies are well known: the roulette wheel, ranking, elitism, random selection, the so-called tournament, and eugenic selection. Our experiments have shown that roulette wheel selection is most advantageous for the reproductive phase, but the tournament strategy with elitism is best for the generational replacement scheme.

Crossover of classifiers. Crossover requires two classifiers, and cuts their chromosome at some randomly chosen positions to produce two offspring. The two new classifiers inherit some rule conditions from each parent-classifier. A crossover operator is used in order to exploit the qualities of the classifiers.

Each result of the crossover process has to be validated. Validation of the various rule attributes (border limits violation, overpassing, etc) is carried out by a process of interval merging, as shown in Fig. 2.

However, merging not only decreases the number of intervals in the rules, but also generates some information loss. In fact, in order to avoid a premature convergence of rules, it is generally important to preserve for the following generation two distinct intervals instead of a single aggregated one. On the other hand, it is interesting to note that the positive or negative effects of an interval on the quality of the rule can be related to other intervals encoded in the classification rule.

	After crossover	After merging
$[10;13] \vee [24;36] \vee [55;67]$	$[10;13] \vee [48;53] \vee [55;67]$	$[10;13] \vee [48;53] \vee [55;67]$
$[7;27] \vee [48;53] \vee [81;93]$	$[7;27] \vee [24;36] \vee [81;93]$	$[7;36] \vee [81;93]$

Fig. 2. Interval merging after crossover operation

Mutation of classifiers. The mutation operator plays a dual role in the system: it provides and maintains diversity in a population of classifiers, and it can work as a search operator in its own right. The mutation processes a single classification rule and it creates another rule with altered condition structure or variables. The mutation operator to several may be applied on three levels: band level, interval level and border level.

Band mutation consists of a deletion of spectral bandwidth in a chosen classification rule. Its interest is twofold; firstly, the band mutation type allows for simplification and generalization of a rule; secondly, it allows for the elimination of noisy bands that frequently appear in hyper spectral images. The existence of noisy bands significantly perturbs the learning process, as well as the process of evolution convergence.

Interval mutation allows for a chosen band to add, eliminate or cut an interval in two spectral ranges. In case of addition, the new rule is completed by a new interval centred randomly with a user-defined width. The cutting of an interval is done by random selection of a cutting point within the interval (for example, the cutting of *[10;100]* can generate two intervals: *[10;15]* and *[15;100]*). Mutation such as this allows for breakage of continuous spectral ranges. And, this allows for the definition of a spectral tube in which spectral values of the pixels can be assigned to a given class.

Finally, border mutation modifies both boundaries of an interval. This mutation refines the idea of targeting spectral tubes carried out by the other types of mutation. It is worthwhile to note that the mutated rules are systematically validated.

In our system, mutation operators are dynamically adapted. Adjustment is related to the probability of each mutation operator according to its current effectiveness.

Generational replacement. The generational replacement is an operation that determines which of the current classifiers in the population is to be replaced by newly evolved classifiers. According to Algorithm 1, the new generation of classifiers is created from a population of parents (P) and their children after the crossover and the mutation operations (P"). In our system, the following replacement strategies are applied:
- the revolutionary strategy in which only the population of the children completely replaces the parent population (P),
- the steady-state strategy in which new classifiers are inserted in the new population by replacing the worst, oldest member, or the most similar members, or by preserving the best classifiers (elitism).

There exist other replacement strategies integrating for instance the strategy where the best individual of the previous population replaces the worst one of the current population or the strategy where the new individuals having a performance higher than a certain threshold are inserted. However, both these strategies present the risk of having individuals remain in the population, which is not necessarily a problem except in the case of a weak genetic pool in which some individuals of average performances that would profit from immunity.

4   Case Studies and Experiments

In this paper, two case studies involving the remote sensing images of Strasbourg and San Felice (lagoon of Venice) have been chosen. Using these examples, we have addressed the main issues of remote sensing image classification using the evolutionary approach. The first case study, classification of high-resolution SPOT images (3 bands, 8 bits per pixel, resolution 1.3 m), demonstrates a typical problem of classification for urban zones including mixed pixels. The second case demonstrates the performance of classifiers on very noisy hyperspectral images, such as DAIS (80 to 100 bands, 16 bits per pixel, resolution 3m, [12]). Learning was carried out on the lower half of the image, and validation on the whole image. More detailed reports on the experiment can be found in another report by A. Quirin [13].

To illustrate the formalism of rules discovered by our algorithm, an example of a rule which classifies the instances of a *Juncus Maritimus* class follows:

$$(435 \leq B_0 \leq 1647) \wedge ... \wedge ((365 \leq B_{74} \leq 4023) \vee (15643 \leq B_{74} \leq 48409)) \wedge ... \wedge (668 \leq B_{79} \leq 4413)$$

where B_i is the reflectance value for the band i of the considered pixel.

Image of Strasbourg, Stadium Vauban (high resolution image)

Fig. 3. Classification

Comments. For this complex remote sensing image in which 300,000 classifiers have been generated to learn 11 spectral classes, relatively good measures of rule quality have been obtained; the best classifier has global pixel recognition of 90.74%. The classes *Water* and *Shadows* are usually difficult to distinguish. However, rules obtained correctly distinguish between these two classes. The same rules showed a generalization performance of 89 % on whole Strasbourg (an image 23 times larger).

Image of Venice, lagoon San Felice (hyperspectral image)

Fig. 4. Classification

Comments. The initial classified image was very coherent; the performance obtained on the 80 band hyperspectral image is very high (10,000 classifiers have been generated to learn an additional 12 spectral classes with global pixel recognition 96.13%). The algorithm eliminated some noisy or redundant bands, despite the presence of rural classes (a different kind of algae) because the expert was highly relevant.

The two case studies have demonstrated the high capacity of the evolution-based classifiers to interpret and classify heterogeneous and complex images (e.g. high dimension, large number of bands and noisy data that provide a computational complexity of $O(n^3)$, which is quite heavy for a deterministic algorithm). The quality of classification is very high even if there were a high number of noisy bands and mixed pixels. It must be noted that the quality of learning is highly related to the quality of the classified image used for rule discovery. The discovered classification rules are simple and easy to interpret by remote sensing experts. They are also mutually exclusive and maximally specific. The learning time was relatively long due to the large image size and the chosen parameters for the evolution process.

Finally, we must mention the high correlation between obtained results and statistics carried out on the remote sensing image (spectrogram statistics, excluding noisy bands). Classified images by the discovered rules have shown that the evolution-based classifier is able to faithfully reproduce the human expertise.

5   Conclusions and Perspectives

This article has described the evolution-based classifier system applied to remote sensing images. The system has discovered a set of *if … then* classification rules using the fitness function based on image classification quality. These rules, which were proven robust and simple to understand for the user, improve the accuracy of the expert and are sufficiently generic for reusing them on other portions of satellite images. Hence, the rules are of great interest when compared to traditional methods of expertise.

Taking into consideration image complexity and noisy data, the results of our experiments are very encouraging. Case studies have demonstrated that the obtained classifiers are able to reproduce faithfully the terrain reality. The rules are well adapted to recognise large objects on the image (e.g. sport lands), as well as the smaller ones (e.g. trees, shadows, edges of the buildings). The redundant or noisy bands have been successfully identified by our rule representation. The formulation of rule representation has allowed for the modelling of a spectral tube adapted to the granularity of spectral reflectance.

The potential of evolution-based classifiers in remote sensing image classification is just beginning to be explored. Further investigation of ICU's learning efficiency are necessary. Currently, we are starting to work on a more powerful representation of rules including spatial knowledge, temporal relations, and hierarchical representation of objects. We are also trying to optimise system performance, in particular the genetic process implementation and its initial parameters. The classifier system developed by this research work, called ICU, is currently available on our web site, which is http://hydria.u-strasbg.fr.

Acknowledgements. The authors are grateful to Pierre Gançarski, Cédric Wemmert, Christiane Weber, Anne Puissant, Jélila Labed, Massimo Menenti and Mindy Jacobson for their advice, suggestions and comments during the early stages of this research.

References

1. H. H. Bock, E. Diday, (eds.) Analysis of Symbolic Data. Exploratory Methods for Extracting Statistical Information from Complex Data, [in] *Studies in Classification, Data Analysis and Knowledge Organization*, vol. 15, Springer-Verlag, Heidelberg, 1999.
2. C. Weber, *Images satellitaires et milieu urbain*, Hermès, Paris, 1995.
3. K. A. DeJong, Learning with Genetic Algorithms: An Overview, *Machine Learning*, vol. 3, pp. 121–138, 1988.
4. S. W. Wilson, State of XCS Classifier System Research, [in] *Proc. of IWLCS-99*, Orlando, 1999.
5. R. Fjørtoft, P. Marthon, A. Lopes, F. Sery, D. Ducrot-Gambart, E. Cubero-Castan, Region-Based Enhancement and Analysis of SAR Images, [in] *Proc. of ICIP'96*, vol. 3, Lausanne, pp. 879–882, 1996.
6. T. Kurita, N. Otsu, Texture Classification by Higher Order Local Autocorrelation Features, [in] *Proc. of Asian Conf. on Computer Vision*, Osaka, pp. 175–178, 1993.
7. J. Korczak, N. Louis, Synthesis of Conceptual Hierarchies Applied to Remote Sensing, [in] *Proc. of SPIE*, Image and Signal Processing for Remote Sensing IV, Barcelona, pp. 397–406, 1999.
8. M. V. Rendon, *Reinforcement Learning in the Fuzzy Classifier System*, Reporte de Investigaci No. CIA-RI-031, ITESM, Campus Monterrey, Centro de Inteligencia Artificial, 1997.
9. R. L. Riolo, Empirical Studies of Default Hierarchies and Sequences of Rules in Learning Classifier Systems, PhD Dissertation, Comp. Sc. and Eng. Dept, Univ. of Michigan, 1988.
10. R. A. Richards, *Zeroth-Order Shape Optimization Utilizing A Learning Classifier System*, http://www.stanford.edu/~buc/SPHINcsX/book.html, Stanford, 1995.
11. T. Blickle, L. Thiele, *A Comparison of Selection Schemes used in Genetic Algorithms, Computer Engineering and Communication Networks Lab*, TIK-Report Nr. 11, Second Edition, Swiss Federal Institute of Technology, Zurich, 1995.
12. DAIS, M. Wooding, *Proceedings of the Final Results Workshop on DAISEX (Digital AIrborne Spectrometer EXperiment)*, ESTEC, Noordwijk, 2001.
13. A. Quirin, Découverte de règles de classification : classifieurs évolutifs, Mémoire DEA d'Informatique, Université Louis Pasteur, LSIIT UMR-7005 CNRS, Strasbourg, 2002.

Hybrid Evolution Strategy-Downhill Simplex Algorithm for Inverse Light Scattering Problems

Demetrio Macías, Gustavo Olague, and Eugenio R. Méndez

Centro de Investigación Científica y de Educación Superior de Ensenada,
División de Física Aplicada,
Apdo. Postal 2732, 22800 Ensenada, B. C., México.
{dmacias, olague, emendez}@cicese.mx

Abstract. The rough surface inverse scattering problem is approached with a combination of evolutionary strategies and the simplex method. The surface, assumed one-dimensional and perfectly conducting, is represented using spline curves. Starting from rigorously calculated far-field angle-resolved scattered intensity data, we search for the optimum profile using the evolutionary strategies $(\mu/\rho^+\lambda)$. After a fixed number of iterations, the best surface is finally recovered with the downhill simplex method. Aspects of the convergence and lack of uniqueness of the solution are discussed.

1 Introduction

The interaction of optical waves with rough surfaces is a familiar phenomenon. Most of what we see is light that has been scattered by objects (surface and/or volume scattering), and we are always trying to infer information about these objects from the scattered light. Not surprisingly, the scattering of electromagnetic waves from rough surfaces finds applications in many areas of science and technology. Among others, we can mention remote sensing, microscopy, surface metrology, and the characterization of optical surfaces.

In the context of the present paper, the direct scattering problem consists of determining the angular distribution of the scattered intensity from a given surface and conditions of illumination. Although to date there is no general solution to this problem, the subject has been studied extensively [1,2]. The inverse scattering problem deals with the reconstruction of the profile that gave rise to the angular distribution of the scattered intensity. The problem is more complex but, from a practical standpoint, much more important. It is also poorly understood.

Most of the previous works on inverse scattering make use of amplitude data [3,4,5]. Since optical detectors are not phase sensitive, the need to have amplitude, rather than intensity data, constitutes an important drawback that all these methods share. Another important limitation is that these methods are based on approximate models for the interaction between the incident light and the surface, and fail in situation when multiple scattering is important.

S. Cagnoni et al. (Eds.): EvoWorkshops 2003, LNCS 2611, pp. 399–409, 2003.

In a previous paper [6], we have studied the inverse scattering problem considering it as a problem of constrained optimization. We focused our attention on the case of surfaces that constitute realizations of a pre-specified Gaussian random process; that is, the random profiles belong to a well-defined statistical class. This assumption enabled us to use the so-called spectral method [7,8] for generating and representing the surfaces numerically. One of the drawbacks of using that kind of representation, is that its nature precludes the use of recombination and other optimization techniques, such as the downhill simplex method [9]. The other problem is that, as it stands, it cannot deal with deterministic surfaces; the profiles must be random, and we need to know the statistical class they belong to.

In the present paper, we continue along the lines of our previous work, removing some of the limitations. For this, we adopt a representation of the surfaces based on B-spline curves. This, not only permits the treatment of more general surfaces, but also enables us to use a wider range of optimization strategies. As in our previous paper, and unlike other works on the subject, the method described here does not rely on approximate expressions for the field-surface interaction.

The organization of this paper is as follows. In Sect. 2 we introduce the notation, define the physical situation considered, and formulate the inverse scattering problem as an optimization problem. We describe in some detail the mathematical representation of the surfaces. Section 3 is devoted to the description of the algorithms studied. The main results are presented in Sect. 4, together with a discussion and, finally, in Sect. 5 we present our main conclusions.

2 Direct and Inverse Scattering

We consider the scattering of light from a one-dimensional, perfectly conducting rough surface defined by the equation $x_3 = \zeta(x_1)$. The region $x_3 > \zeta(x_1)$ is vacuum, the region $\zeta(x_1) > x_3$ is a perfect conductor, and the plane of incidence is the x_1x_3-plane. With reference to Fig. 1, the surface is illuminated from the vacuum side by a monochromatic s-polarized plane wave making an angle θ_0 with the x_3 axis.

The scattering amplitude $R_s(q|k)$ can be written in the form [8]

$$R_s(q|k) = \frac{-i}{2\alpha_0(k)} \int_{-\infty}^{\infty} dx_1 \left[\exp\{-i\alpha_0(q)\zeta(x_1)\}F(x_1|\omega)\right] \exp\{-iqx_1\}, \quad (1)$$

where

$$F(x_1|\omega) = \left(-\zeta'(x_1)\frac{\partial}{\partial x_1} + \frac{\partial}{\partial x_3}\right) E_2(x_1, x_3)\bigg|_{x_3=\zeta(x_1)} \quad (2)$$

represents the source function, $E_2(x_1, x_3)$ is the only nonzero component of the electric field, $\alpha_0(k) = \sqrt{\omega^2/c^2 - k^2}$, and $\alpha_0(q) = \sqrt{\omega^2/c^2 - q^2}$. The angles of incidence θ_0 and scattering θ_s are related to the components of the incident

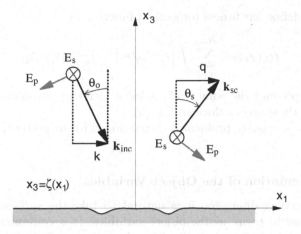

Fig. 1. Geometry of the scattering problem considered.

and scattered wavevectors that are parallel to the mean surface through the expressions

$$k = \frac{\omega}{c} \sin \theta_0 \,, \qquad\qquad q = \frac{\omega}{c} \sin \theta_s \,, \qquad\qquad (3)$$

where c is the speed of light and ω is the frequency of the optical field. The far-field intensity $I_s(q|k)$ is defined as the squared modulus of the scattering amplitude $R_s(q|k)$. The angular distribution of the intensity is then represented by $I(q|k)$ as a function of q.

Due to the lack of general analytical solutions for the direct scattering problem, and the evident complexity of the inverse scattering problem, we approach it as an optimization problem. The goal is to retrieve the unknown surface profile function from scattered intensity data. The angle-resolved scattered intensity used as input data for our algorithms is obtained by rigorous numerical techniques [8]. For this, the source function $F(x_1|\omega)$ is determined by solving numerically an integral equation [8], and the scattering amplitude $R_s(q|k)$ is found through Eq. (1).

As we have mentioned, although the far-field scattered intensity depends on the surface profile function in a complicated way, for one-dimensional surfaces the direct problem can be solved numerically [8]. The closeness of a proposed profile, $z_c(x_1)$, to the original one can be estimated through the difference between the measured angular distribution of intensity $I^{(m)}(q|k)$, and the angular distribution of intensity $I^{(c)}(q|k)$, obtained by solving the direct scattering problem with the trial profile $z_c(x_1)$. The goal then would be to find a surface for which the condition $I^{(c)}(q|k) = I^{(m)}(q|k)$ is satisfied. When this happens, and if the solution to the problem is unique, the original profile has been retrieved.

We, thus, define our fitness (objective) function as:

$$f(\zeta(x_1)) = \sum_{i=1}^{N_{\text{ang}}} \int \left| I_s^{(m)}(q|k_i) - I_s^{(c)}(q|k_i) \right| dq, \tag{4}$$

where N_{ang} represents the number of angles of incidence considered, and the $k_i's$ are related to those angles through Eq. (3).

The inverse scattering problem is then reduced to the problem of minimizing $f(\zeta(x_1))$.

2.1   Representation of the Object Variables

From the preceding discussion it is natural to take the surface heights, or a quantity related to them, as the object variables. It is convenient to choose a representation scheme based on Shoenberg's variation diminishing approxima-tion [10,11]. Assuming that the function $\zeta(x_1)$ consists of planar open curve defined within the interval $[-L/2, L/2]$, it can be represented parametrically in terms of spline curves.

First, the interval $[-L/2, L/2]$ is divided into $(m-1)$ equally-spaced intervals of length $\Delta\tau = L/(m-1)$, where m represents the number of control points. The jth control point, P_j, has coordinates (τ_j, α_j), with $\tau_j = -L/2 + (j-1)\Delta\tau$. The heights α_j of the control points will be chosen as the object variables.

The knots are points at which adjacent spline curves join. The number of knots, ν, is related to the number of control points and the order of the spline, k, by the relation $\nu = m+k$. Throughout this work the knot sequence is assumed uniform but, to force the ends of the curve coincide with the first and last control points, we take

$$t_1 = t_2 = \cdots = t_k = -L/2, \qquad t_{m+1} = t_{m+2} = \cdots = t_{m+k} = L/2. \tag{5}$$

The location of the other (internal) knots is computed through the expression

$$t_{k+j} = \frac{\tau_j + \cdots + \tau_{j+k-1}}{k-1} \qquad j = 1, \ldots, m. \tag{6}$$

The jth B-spline function of order $k = 1$ is defined as

$$B_j^{(1)}(x_1) \triangleq \begin{cases} 1 & \text{for } t_j \leqslant x_1 < t_{j+1}, \\ 0 & \text{otherwise.} \end{cases} \tag{7}$$

For $k > 1$, the B-spline functions are evaluated recursively using the Cox-de Boor Algorithm [12,13],

$$B_j^{(k)}(x_1) \triangleq \omega_j^{(k)}(x_1) B_j^{(k-1)}(x_1) + [1 - \omega_{j+1}^{(k)}(x_1)] B_{j+1}^{(k-1)}(x_1) \tag{8}$$

where

$$\omega_j^{(k)}(x_1) \triangleq \begin{cases} \frac{x_1 - t_j}{t_{j+k-1} - t_j} & \text{for } t_j \neq t_{j+k-1}, \\ 0 & \text{otherwise.} \end{cases} \tag{9}$$

The surface profile function can then be written as

$$\zeta(x_1) = \sum_{j=1}^{m} \alpha_j B_j^{(k)}(x_1). \tag{10}$$

An example of a one-dimensional random profile represented as a spline curve is shown in Fig. 2. The solid line is the random surface contained by the control polygon, shown with a dashed line, and the control points are represented by the diamonds. In this case, the order of the B-spline functions was $k = 4$. Note that the multiplicity of the knots at the beginning and end of the interval expressed by Eq.(5) implies that the profile $\zeta(x_1)$ is tangent to the control polygon at the first and last control points.

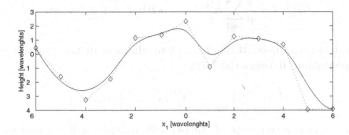

Fig. 2. Random profile represented as a spline curve (solid line). The control polygon is shown with a dashed line and its vertices (control points) with diamonds.

3   Description of the Inversion Algorithms

To retrieve the surface profile, we have studied some hybrid inversion algorithms, combining evolutionary strategies (ES)[14,15] and the simplex method of Nelder and Meade[9]. Using Schwefel's notation,[14] the evolutionary strategies employed are the elitist $(\mu/\rho+\lambda)$ and the non-elitist $(\mu/\rho, \lambda)$ strategies with recombination. Here μ is the number of parents in the initial population, λ is the number of offsprings generated by means of the genetic operators, and ρ is the number of members of the population that generate an intermediate population through recombination[15]. For the case $\rho = 1$, we have the canonical ES. Once the evolution strategies have reached the termination criterion, a local search algorithm (the simplex) is employed to improve the solution.

It is worth mentioning that we have previously used λ to denote the wavelength of the light, which is the usual notation in optical work. It is believed that due to the different context in which the two quantities are employed, the use of the same symbol to denote both should not lead to much confusion.

3.1   The Evolution Strategies

The initial population is a set of μ random surfaces represented by means of Eq. (10) as

$$\zeta^{(i)}(x_1) = \sum_{j=1}^{m} \alpha_j^{(i)} B_j^{(k)}(x_1) \qquad \text{with } i = 1, 2, ..., \mu. \tag{11}$$

The initial ordinates $\alpha_j^{(i)}$ of the control points are generated using zero-mean normally distributed random numbers with standard deviation σ, and the fitness values of all the elements of the population are evaluated through Eq. (4).

When $\rho > 1$, recombination is applied[15], and the newly generated ordinate of the jth control point is given by

$$\langle \alpha_j^{(i)} \rangle_\rho = \frac{1}{\rho} \sum_{i=1}^{\rho} \alpha_j^{(i)} \qquad \text{with } j = 1, 2, \dots, m. \tag{12}$$

Mutation is introduced through random changes in the population [16]. In our implementation, it takes the form

$$\left(\langle \alpha_j^{(i)} \rangle_\rho \right)_{\text{mut}} = \langle \alpha_j^{(i)} \rangle_\rho + N(0, \sigma_{\text{mut}}), \tag{13}$$

where $N(0, \sigma_{\text{mut}})$ is a sequence of statistically independent zero-mean Gaussian random numbers with standard deviation σ_{mut}. No self-adaptation schemes were used in this work.

After mutation, the fitness values of all the elements of the population are evaluated through Eq. (4) and the selection scheme is applied.

In our implementation for the elitist strategies $(\mu + \lambda)$ and $(\mu/\rho + \lambda)$, we chose $\mu = \lambda$. From the union of the initial and the intermediate populations we select a secondary population that consists of the μ surfaces with the lowest associated values of $f(\zeta(x_1))$. This secondary set of surfaces constitutes the starting point for the next iteration of the algorithm. The process continues until the termination criterion is reached, which in our case was set by the maximum number of iterations g.

In the non-elitist strategies (μ, λ) and $(\mu/\rho, \lambda)$, we start with an initial population of μ random surfaces, and keep $\lambda = 10\mu$ during the entire process. Through the recombination and mutation operations, λ new surfaces are generated. This secondary population is evaluated and the best μ random surfaces, those with the lowest associated fitness values, are selected to be the initial population for the next iteration of the algorithm. As with the elitist strategies, the process continues until a termination criterion is reached.

3.2   Multidimensional Local Search Algorithm

At this point, the output of the evolutionary strategy is coupled to a multidimensional local search algorithm, represented by the simplex method [9]. The element with the lowest fitness value is selected and included in the initial population of the simplex. Other elements are generated randomly, employing the representation given by Eq. (10). The dimensions of the simplex, i. e. its number of vertices, is determined by the number of parameters to be optimized. For the present application, this is the m heights α_j of the control points. Thus, the simplex will have $m+1$ vertices, each one of them associated with a surface and a fitness value $f(\zeta(x_1))$. Each element of the population may be visualized as different points in the search space.

The simplex conducts the search towards the optimum by changing its size through a series of defined operations [17]. The aim in each iteration of the algorithm is to find a new point for which the fitness value $f(\zeta(x))$ is lower than the worst fitness value of the previous iteration. The search process ends when a multiple contraction takes place and the simplex collapses to a point. When this happens, the simplex has converged to the optimum and all the members of the population are identical.

4 Results and Discussion

In principle, the data that serves as input to the algorithm should be obtained experimentally. However, in order to evaluate the performance of the proposed algorithms, in these studies we use data obtained through a rigorous numerical solution of the direct scattering problem [8].

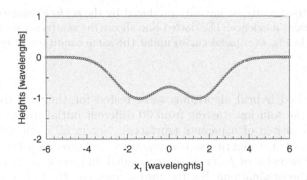

Fig. 3. The profile used in the generation of the scattering data.

For the strategies explored, each element of the initial population consisted of a randomly rough surface. For the $(\mu/\rho + \lambda)$ strategies we chose $\mu = \lambda = 100$, whereas for the $(\mu/\rho, \lambda)$ strategies we set $\mu = 10$ and $\lambda = 100$. The number of elements to be selected from the initial population, depending on whether the recombination operation was performed or not in the search process, was respectively set to $\rho = 2$ or 1. The maximum number of iterations was $g = 300$, which also provided the termination criterion.

In the numerical experiments we considered the retrieval of the deterministic profile shown in Fig. 3. Since the time of computation required to find the optimum increases with the length of the surface, in order to keep the problem to a manageable size, we chose a surface of length $L = 12.1\lambda$. We used 13 control points, splines of order $k = 4$, and 17 knots. The heights α_j of the control points, were generated using normally distributed random numbers with zero mean and a standard deviation of $\sigma = 2\lambda$. For the scattering calculations we used a sampling of $\Delta x = \lambda/10$, so that we had $N = 121$ sampling points on the surface.

The data from which the profile is to be recovered were obtained by illuminating the surface in Fig. 3 from four different directions, defined by the angles of incidence $\theta_0 = -60°$, $\theta_0 = -30°$, $\theta_0 = 0°$, and $\theta_0 = 40°$. In Fig. 4, we show (solid curve) the scattering pattern produced by the surface in Fig. 3 for the case of normal incidence.

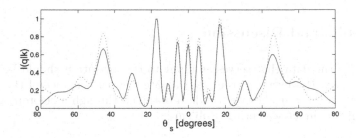

Fig. 4. Normalized scattered intensity produced by the surface depicted in Fig. 3 for the case of normal incidence. The dotted line shows the scattering pattern generated by the surface in Fig. 6(c) (solid curve) under the same conditions of illumination.

The described hybrid algorithms were tested for their relative success by searching for the solution starting from 30 different initial states. Each random initial state consisted of μ parents (surfaces). Not in all of the 30 attempts to recover the profile the algorithms converged to the correct surface. However, we found that a low value of $f(\zeta(x_1))$ corresponded, in most cases, to a profile that was close to the original one. So, the lowest value of $f(\zeta(x_1))$ was used as the criterion to decide whether the function profile had been reconstructed or not.

The convergence behaviour for one run of each of the studied algorithms is presented in Fig. 5. The plots corresponding to the $(\mu/\rho^+_,\lambda)$-ES are shown with solid lines to the left of $g = 300$, whereas a dotted line is used for the canonical strategies $(\mu^+_,\lambda)$. The fitness functions for iterations beyond $g = 300$, obtained by means of the simplex algorithm, are also shown in Fig. 5. The discontinuities in the plots (b) and (c) at $g = 300$ are due to the presence, in the initial populations of the simplex, of the best solutions found by the non-elitist strategy. It is worth pointing out that the highest final fitness values of Fig. 5 correspond to the canonical $(\mu^+_,\lambda)$-ES.

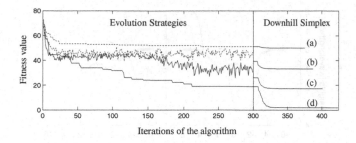

Fig. 5. Convergence behaviour for the best approximation found by the each one of the studied algorithms. The labels correspond, respectively to: (a) the $(\mu + \lambda)$-ES + downhill simplex, (b) the (μ, λ)-ES + downhill simplex, (c) the $(\mu/\rho, \lambda)$-ES + downhill simplex, and (d) the $(\mu/\rho + \lambda)$-ES + downhill simplex.

In Fig. 6 we present typical results obtained with the algorithms considered. The illumination conditions were the same as those used for the generation of the scattering information shown in Fig. 4. Figs. 6(a) and (b) correspond to the case $\rho = 1$, while Figs. 6 (c) and (d) to the case $\rho = 2$. To facilitate the visualization of the results the original profile is shown with circles.

When only the mutation operator was used in the search for the optimum $(\rho = 1)$ we obtain the worst convergence behaviour. The curve shown in Fig. 5(a) corresponds to the reconstruction of Fig. 6(a). It is clear that the retrieved profile does not resemble the original one. The characteristic local search behaviour of the downhill simplex is illustrated by the fact that the improvement of the fitness function after the point $g - 300$ is small. A similar situation can be also seen in Fig. 6(b) for the (μ, λ)-ES.

For the case $\rho = 2$, we obtain a striking result that also illustrates the lack of uniqueness of the solution when intensity data are used. This is shown in Fig. 6(c). The output of the $(\mu/\rho, \lambda)$ evolution strategy is shown with the dotted line, while the solid line corresponds to the solution refined by the downhill simplex. The evolution of the corresponding fitness values is shown as curve (c) in Fig.(5). Although the reconstructions of Fig. 6(c) do not resemble the original profile, the scattering patterns of Fig. 4 demonstrate the fact that different profiles can generate the same, or similar, scattering information.

These results also illustrate a curious property of the scattering problem. If the reconstructions of Fig. 6(c) are reflected with respect to the x_1 and x_3 axis, the resulting profile resembles the sought one. It can be shown [18] that within the Kirchhoff approximation, and if polarization effects are not important, the far-field intensity is invariant under this kind of operation. Such situations lead to multiple solutions of the inverse scattering problem. Interestingly, polarization and multiple scattering effects reduce the number of possible solutions to the inverse problem.

The solution found with the $(\mu/\rho + \lambda)$-ES is depicted with a dotted line in Fig. 6(d). In this case the algorithm retrieved the profile quite well; there are only some subtle differences that are perhaps more noticeable near the ends of the surface. An improvement is obtained when the downhill simplex is used

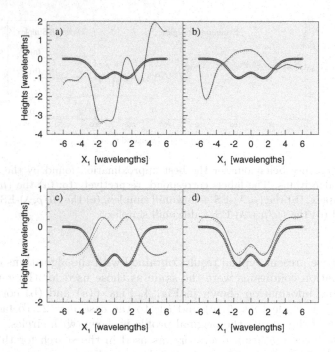

Fig. 6. Reconstruction of the surface profile using: (a) the $(\mu + \lambda)$-ES + downhill simplex, (b) the (μ, λ)-ES + downhill simplex, (c) the $(\mu/\rho, \lambda)$-ES + downhill simplex, (d) the $(\mu/\rho + \lambda)$-ES + downhill simplex. The original profile is plotted with circles and the reconstructions are depicted with a dotted curve for the two evolution strategies, and with a solid curve for the downhill simplex algorithm.

to refine the solution. This can be concluded from the behavior of the fitness function shown in Fig. 5(d). The solution found is shown with the solid line in Fig. 6(d). The vertical displacement of the reconstructed profiles of Fig. 6(d) is quite understandable, as the far-field intensity is insensitive to such shifts. It is also worth pointing out that the displacement is unimportant for practical profilometric applications.

5   Summary and Conclusions

A hybrid evolutionary approach to the inverse scattering problem has been successfully applied to the reconstruction of a one-dimensional, perfectly conducting rough surface. The use of B-splines provides an effective method for the representation of the object variables. In addition, its implementation and evaluation are straightforward. Furthermore, unlike the spectral method used in [6], the representation with B-splines allows the generation of deterministic surfaces.

For the numerical examples presented in this work, the recombination operator played a fundamental role in the convergence to the optimum. Also, the hybridization lead to an additional improvement of the solution found by the evolution strategy.

Acknowledgments. The authors are grateful to CONACyT (México) for financial support.

References

1. P. Beckmann and A. Spizzichino, *The Scattering of Electromagnetic Waves from Rough Surfaces*, (Pergamon Press, London, 1963), p. 29.
2. J. A. Oglivy, *Theory of wave scattering from random rough surfaces*, (Institute of Physics Publishing, Bristol, 1991), p. 277.
3. Wombel, R. J., DeSanto, J. A.: Reconstruction of rough-surface profiles with the Kirchhoff approximation, J. Opt. Soc. Am. A **8**, 1892, (1991).
4. Quartel, J. C., Sheppard, C. J. R.: Surface reconstruction using an algorithm based on confocal imaging, J. Modern Optics **43**, 496, (1996).
5. Macías, D., Méndez, E. R., Ruiz-Cortés, V.: Inverse scattering with a wavefront matching algorithm, J. Opt. Soc. Am. A **19**, 2064, (2002).
6. Macías, D., Olague, G., Méndez, E. R.: Surface profile reconstruction from scattered intensity data using evolutionary strategies, in *Applications of Evolutionary Computing*, S. Cagnoni, *et. el*, eds., (Springer LNCS**2279**, Berlin), 233, (2002).
7. Thorsos, E. I.: The validity of the Kirchhoff approximation for rough surface scattering using a Gaussian roughness spectrum, J. Acoust. Soc. Amer. **83**, 78 (1988).
8. Maradudin, A. A., Michel, T., McGurn, A. R., Méndez, E. R.: Enhanced backscattering of light from a random grating, Ann. Phys. (N. Y.) **203**, 255 (1990).
9. Nelder, J., Mead, R.: A simplex method for function optimization *Computer Journal*, **7**, 308,(1965).
10. Boor, C. de: *A Practical Guide to Splines*, (Springer-Verlag, NY, 1978), p.154.
11. Boor, C. de: Spline Basics, In: Handbook of Computer Aided Geometric Design, G. E. Farin, J. Hoschek and M. Kim (eds.), Elsevier, pp. 141–164, (2002).
12. Cox, M. G.: The numerical evaluation of B-splines, *J. Inst. Math. Applics.*, **10**,134, (1972).
13. Boor, C. de: On calculating with B-splines, *J. Approx. Theory*,**6**, 50,(1972).
14. H. P. Schwefel, *Evolution and Optimum Seeking*, (John Wiley & Sons Inc., NY, 1995), p. 444.
15. H. G. Beyer, *The Theory of Evolution Strategies*, (Springer-Verlag, Berlin, 2001), p.380.
16. Th. Bäck, U. Hammel and H.-P. Schwefel: Evolutionary computation: Comments on the history and current state. IEEE Transactions on Evolutionary Computation, 1(1):3–17, 1997.
17. Press, W. H. (editor) *Numerical Recipes in Fortran*, (Cambridge University Press, Cambridge, 1992) p. 402.
18. Macías, D., Méndez, E. R.: Unpublished work (2001).

Accurate L-Corner Measurement Using USEF Functions and Evolutionary Algorithms

Gustavo Olague[1], Benjamín Hernández[2], and Enrique Dunn[1]

[1] Departamento de Ciencias de la Computación, División de Física Aplicada,
Centro de Investigación Científica y de Estudios Superiores de Ensenada,
Km. 107 carretera Tijuana-Ensenada, 22860, Ensenada, B.C., México
{olague, edunn}@cicese.mx
http://cienciascomp.cicese.mx/Pagina-Olague.htm
[2] Instituto de Astronomía, Ensenada
Universidad Nacional Autónoma de México
Observatorio Astronómico Nacional
km. 103 Carretera Tijuana-Ensenada, Ensenada, B.C., México
{benja}@astrosen.unam.mx

Abstract. Corner feature extraction is studied in this paper as a global optimization problem. We propose a new parametric corner modeling based on a Unit Step Edge Function (USEF) that defines a straight line edge. This USEF function is a distribution function, which models the optical and physical characteristics present in digital photogrammetric systems. We search model parameters characterizing completely single gray-value structures by means of least squares fit of the model to the observed image intensities. As the identification results relies on the initial parameter values and as usual with non-linear cost functions in general we cannot guarantee to find the global minimum. Hence, we introduce an evolutionary algorithm using an affine transformation in order to estimate the model parameters. This transformation encapsulates within a single algebraic form the two main operations, mutation and crossover, of an evolutionary algorithm. Experimental results show the superiority of our L-corner model applying several levels of noise with respect to simplex and simulated annealing.

1 Introduction

Photogrammetry is the science, and art, of determining the size and shape of objects as a consequence of analyzing images recorded on film or electronic media. Close-range photogrammetry as well as computer vision relies on image processing techniques in order to obtain the information required for tasks devoted to perceiving, sensing and measuring the world around a machine vision system. Corners and contours are recognized as a basic characteristic in machine vision, see [1], [2], [3], [4], [5], [6], [7], [8], [9], [10], [11], [12], [13]. High-accurate corner detection is a complex process due to several factors: 1) the attitude, position and orientation, of the camera with respect to the object, 2) the interior orientation of the camera, 3) the fluctuations on illumination and 4) the camera

S. Cagnoni et al. (Eds.): EvoWorkshops 2003, LNCS 2611, pp. 410–421, 2003.

optics, see Olague and Mohr [14]. Several approaches to the problem of detecting feature points have been reported in the literature in the last few years. They can be broadly divided into three groups. The first one consists of boundary based approaches like those proposed by Tsai et al. [15], Medioni and Yasumoto [16], and Sohn et al. [17]. Approaches in the second group involve those based on computing geometric properties directly from the gray-level images, see Beaudet [18], Dreschler and Nagel [19], Kitchen and Rosenfeld [20] and finally Wang and Brady [21]. The third group is composed by approaches based on parametric models, see Deriche and Giraudon [6], Rohr [11] and Olague and Hernández [22], [23].

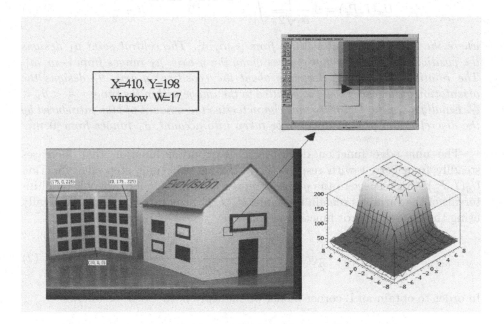

Fig. 1. A zoom of one L-corner of our EvoVisión house is shown as well as its final three dimensional model.

The paper is organized as follows. A first section is devoted to the presentation of our L-corner model. Then we will present the process of modeling L-corners as an optimization problem. Apart from the problem of determining a model that fits the data, there exists the problem of estimating the accuracy of the parameters. Indeed, our approach provides these magnitudes. A novel evolutionary representation is introduced using the concepts of affine transformation. Finally, experimental results are presented in order to show the superiority of our affine evolutionary algorithm against simulated annealing and downhill simplex.

2   L-Corner Model and χ^2 Estimator

Modeling image intensities $(f(x_i, y_i), x_i, y_i)$ of an L-corner to a model that is a linear combination of nonlinear functions of (x_i, y_i) is a subject known as modeling of data, see Figure 1. An L-corner is built of two *Unit Step Edge Functions* (USEF) as follows:

Definition 1 (Unit Step Edge Function) *Let the image coordinates and the set of unknown model parameters be denoted by $I = (x, y)$ and $P = (p_1, \ldots, p_n)$ respectively. The unit step edge function is represented as follows:*

$$U_x(I, P_x) = \pm \frac{1}{\sigma_1 \sqrt{2\pi}} \int_0^x e^{\frac{-(t - y \cdot \tan(\theta_1) - \mu_1)^2}{2\sigma_1^2}} dt + \frac{1}{2} \tag{1}$$

where the image coordinates range from $[-m, m]$. The central point μ_1 designs the position x of the line that crosses along the y-axis. μ_1 ranges from $[-m, m]$. The rotation θ_1 is made clockwise about the (positive) y-axis. θ_1 designs the orientation of the edge model to be fitted to the image within the range $-\frac{\pi}{2} < \theta_1 < \frac{\pi}{2}$. Finally a scaling factor σ_1 that characterizes the amount of blur introduced by the discretization process needs to be taken into account. σ_1 ranges from $[0, m]$.

The unit edge function describes a distribution function that increases steadily from 0 to 1 with respect to the x-axis. The unit step edge function $U_y(I, P_y)$ with respect to the y-axis is represented in a similar way where all intervals of the variables remain the same. $U_y(I, P_y)$ can be evaluated numerically using the Gaussian error function as follows:

$$U_y(I, P_y) = -\frac{1}{2} erf \left(\frac{\sqrt{2}(-y + x \cdot \tan(\theta_2) + \mu_2)}{2\sigma_2^2} \right) + \frac{1}{2} \tag{2}$$

In order to obtain an L-corner model we multiply both USEF as follows:

$$M_L'(x, y, \boldsymbol{P}) = U_x(I, P_x) \cdot U_y(I, P_y) \tag{3}$$

The parameters σ_1, ϑ_1, μ_1, σ_2, ϑ_2, μ_2, A y B, represents the physical and geometrical contours of an L-corner.

Corner's localization was obtained by fitting our parametric model to the image intensities. Estimates for the model parameters $\boldsymbol{P} = (p_1, \ldots, p_n) \in R^n$ are found by minimizing the squared differences between the (nonlinear) model function and the considered gray values:

$$Q = \chi^2 = F(\boldsymbol{P}) = \sum_{i=1}^m \sum_{j=1}^m [I(u_i, v_j) - M_L'(x_i, y_j, \boldsymbol{P})]^2 \tag{4}$$

The intensities and the function values of the model in the considered image area are $I(u_i, v_j)$ and $M_L'(x_i, y_j, \boldsymbol{P})$ respectively. Previous approaches used by Rohr [11] applied the method of Powell utilizing only function values or used the method of Levenberg-Marquardt, see Press et al. [24] incorporating partial derivatives of the model function in order to reduce the computation time. However, a drawback presented on these approaches is that the identification result relies on the initial parameter values and as usual with nonlinear cost functions in general we cannot guarantee to find the global minimum. This problem is overcome in this work using an evolutionary algorithm.

3   Modeling L-Corners as an Optimization Problem

Evolutionary algorithms has raised as a rich paradigm for global optimization. Previous methodologies as the *Down Hill Simplex Method* and *Simulated Annealing* are well known techniques for multidimensional optimization [24]. This section is devoted to our affine evolutionary algorithm for global optimization. Currently, evolutionary algorithms for numerical optimization use real code parameters for which a set of special transformations has been developed. In real coding implementation, each chromosome is encoded as a vector of real numbers of the same length. Several crossover operators have been introduced under the name of arithmetical operators. The arithmetical operators are built by borrowing the concept of linear combination of vectors from the area of convex sets theory. Generally, crossover produces an offspring, which is calculated from the weighted average of two vectors y_1 y y_2 as follows:

$$\begin{aligned} y_1' &= \lambda_1 y_1 + \lambda_2 y_2 \\ y_2' &= \lambda_2 y_1 + \lambda_1 y_2 \end{aligned} \tag{5}$$

if the multipliers are restricted to:

$$\lambda_1 + \lambda_2 = 1, \quad \lambda_1 > 0, \quad \lambda_2 > 0$$

the weighted form is known as convex combination. If the non-negativity condition on the multipliers is dropped, the combination is known as affine combination. Finally, if the multipliers are simply required to be in real space, the combination is known as a linear combination [25]. Another operator is known under the name of dynamic mutation, also called non-uniform mutation, introduced by Janikow and Michalewicz [26]. Dynamic mutation is designed for fine-tuning capabilities aimed at achieving high precision. Given a parent y, if the element y_k of it is selected for mutation, the resulting offspring is $y = [y_1, \ldots, y_k', \ldots, y_n]$, where y_k' is randomly selected from the following two possibilities:

$$\begin{aligned} y_k' &= y_k + \Delta(t, y_k^U - y_k), \; or \\ y_k' &= y_k - \Delta(t, y_k - y_k^L) \end{aligned}$$

where

$$\Delta(t, y) = yr \left(1 - \tfrac{t}{T}\right)^b \tag{6}$$

The function $\Delta(t, y)$ returns a value in the range $[0, y]$ such that the value approaches 0 as t increases. This property causes the operator to search the space uniformly initially, when t is small, and very locally at later stages. t is the generation number, b is a parameter determining the degree of non-uniformity and r is a random number between $[0,1]$. It is possible for the operator to generate an offspring, which is not feasible. In such a case, we can reduce the value of the random number r.

3.1 A Novel Evolutionary Representation

The operations of crossover and mutation can be encapsulated into a single complex transformation as follows. In order to handle affine geometry algebraically, we have to characterize the line i by an invariant equation, and we shall suppose that this equation is $y_0 = 0$. Since, the points of i are now regarded as ideal points, no point with $y_0 = 0$ is actual, and this means that we can represent the actual points of the affine plane by pairs of non-homogeneous coordinates $Y = (Y_1, Y_2)$, where

$$Y_1 = \frac{y_1}{y_0} , \ Y_2 = \frac{y_2}{y_0}$$

The allowable representations $\mathcal{R}_A$ of the affine plane are those representations $\mathcal{R}$ of $\hat{S}_2$ in which the line i has the equation $y_0 = 0$; and this leads at once to the following theorem:

Theorem 1. *If $\mathcal{R}_A$ is any allowable representation of the affine plane, then the whole class $(\mathcal{R}_A)$ of allowable representations consists of all those representations, which can be derived from $\mathcal{R}_A$ by applying a transformation of the form*

$$Y_1' = b_{11}Y_1 + b_{12}Y_2 + C_1$$
$$Y_2' = b_{21}Y_1 + b_{22}Y_2 + C_2$$

where the coefficients are arbitrary real numbers, subject to the condition $|b_{rs}| \neq 0$.

Using Theorem 1, it is possible to transform the n variables of two solutions into a new pair of solutions, according to the following transformation:

$$\begin{pmatrix} Y_{1_1}' & Y_{1_2}' \cdots Y_{1_n}' \\ Y_{2_1}' & Y_{2_2}' \cdots Y_{2_n}' \end{pmatrix} = \begin{bmatrix} \underbrace{b_{11} \ b_{12}}_{crossover} & \underbrace{C_1}_{mutation} \\ b_{21} \ b_{22} & C_2 \end{bmatrix}_n \begin{pmatrix} Y_{1_1} & Y_{1_2} \cdots Y_{1_n} \\ Y_{2_1} & Y_{2_2} \cdots Y_{2_n} \\ 1 & 1 \ \cdots \ 1 \end{pmatrix} \quad (7)$$

Equation 7, can be expanded to the whole population. The advantages of this representation are:

1. Standardized treatment of all transformations.
2. Complex transformations are composed from single transformations by means of matrix multiplication.
3. An n dimensional point can be transformed by applying a set of n transformations.
4. Simple inversion of the transformation by matrix inversion.
5. Extremely fast, hardware supported matrix operations in high-power graphic workstations.

4   Experimental Results

In order to show the robustness of each algorithm and the stability of our L-corner model, we applied Gaussian noise of zero mean and unit variance. This noise is scaled by a constant λ in order to produce perturbations on the synthetic test image. This noise is known as *additive* noise [27]. We decide to use a synthetic image in order to know precisely the location of the corners. The signal-to-noise ratio (SNR) is computed in decibels (DB). Figure 2, shows the structures (a), (b) and (c) with the Gaussian noise scaled by the factors $\lambda = 20, 40, 80$. Table 1, presents final results of corners (a) $\rightarrow$ (f) of Figure 2, using three optimization strategies without noise. It is important to remember that for $SNR \rightarrow 1$ the error on the noisy image is approximately equal to the amount of signal of the synthetic image. If $SNR \leq 1$ the noise is bigger than the original signal and if $SNR \gg 1$ the original synthetic image and the synthetic image with noise are equivalents, the error tends to zero. The test of our L-corner detector considering three optimization algorithms was built as follows:

1. The *Down Hill Simplex*, *Simulated Annealing* and *Evolutionary Algorithm* were applied to the structures (a), (b) and (c) of Figure 2. Those structures show three different corners: straight angle corner, acute angle corner and obtuse angle corner respectively.
2. The size of the window is 13×13 pixels, centered around the pixel:

   a) $(u_0, v_0) = (87, 87)$.
   b) $(u_0, v_0) = (707, 353)$.
   c) $(u_0, v_0) = (396, 397)$.

3. The control parameters of each algorithm are:

   - *Down Hill Simplex*. Maximal number of movements of the simplex $N = 4500$.

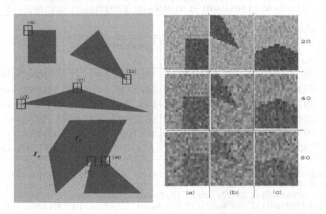

Fig. 2. On the right structures (a), (b) and (c) of the test synthetic image with Gaussian noise scaled by $\lambda = 20, 40, 80$.

Table 1. Exact corner location (u_e, v_e) and parameter values P using the *Down-Hill Simplex*, *Simulated Annealing* and *Evolutionary Algorithm*, considering structures (a) $\rightarrow$ (f) of Figure 2.

Down Hill simplex

Corner	Initial Pixel u_0	v_0	Corner Point u_e	v_e	Aperture α^o	s_{U_1}	s_{U_2}	χ^2
(a)	89	89	87.404142	87.709506	89.76333	+1	-1	1.13215e-15
(b)	706	353	707.028574	353.831573	35.7994	-1	+1	5.44007e-16
(c)	398	397	396.825889	397.466375	149.9239	-1	-1	1.31891e-09
(d)	44	484	40.985292	484.451324	19.1055	+1	+1	4.22482e+04
(e)	570	795	570.573303	792.934090	97.4398	+1	+1	1.22339e-11
(f)	482	795	483.081960	792.272648	37.2723	-1	-1	6.41950e-23

	σ_1	σ_2	μ_1	μ_2	ϑ_1^o	ϑ_2^o	A	B
(a)	6.04357e-02	5.20620e-02	-1.60603e+00	1.33941e+00	-8.74457e-01	6.37787e-01	103.00	76.00
(b)	2.34760e-02	9.25585e-03	4.76497e-01	-4.44535e-01	-3.35800e+01	-2.06206e+01	103.00	76.00
(c)	2.34221e-03	1.72416e-02	-2.75132e+00	-1.82255e-01	-7.35272e+01	1.36033e+01	103.00	76.00
(d)	6.29735e-02	8.53333e-03	-1.25144e+00	-7.01737e-01	7.56428e+01	-4.74830e+00	102.54	76.46
(e)	1.15833e-02	8.79562e-04	-6.67335e-01	2.51901e+00	3.09704e+01	-3.84093e+01	103.00	76.00
(f)	1.33997e-02	4.80644e-03	3.97708e-01	1.86938e+00	1.41271e+01	3.86009e+01	103.00	76.00

Simulated Annealing

Corner	Initial Pixel u_0	v_0	Corner Point u_e	v_e	Aperture α^o	s_{U_1}	s_{U_2}	χ^2
(a)	89	89	87.526970	87.437709	90.01414	+1	-1	1.94364e-10
(b)	706	353	707.017039	353.829675	35.7614	-1	+1	3.56136e-13
(c)	398	397	396.797404	397.427744	149.4558	-1	-1	9.75333e-13
(d)	44	484	40.077467	484.514532	16.977	+1	+1	4.22483e+04
(e)	570	795	570.647257	793.111032	97.7988	+1	+1	6.15267e-01
(f)	482	795	483.046760	792.255049	37.3776	-1	-1	1.54081e-13

	σ_1	σ_2	μ_1	μ_2	ϑ_1^o	ϑ_2^o	A	B
(a)	8.69092e-02	7.67896e-02	-1.47251e+00	1.56314e+00	-1.88830e-02	3.30323e-02	103.00	76.00
(b)	5.01805e-03	7.87605e-03	4.71431e-01	-4.47754e-01	-3.35296e+01	-2.07090e+01	103.00	76.00
(c)	2.11139e-02	2.12681e-02	-2.62621e+00	-1.21410e-01	-7.34947e+01	1.40389e+01	103.00	76.00
(d)	3.16852e-07	3.69060e-04	-1.76589e+00	-7.58448e-01	7.65812e+01	-3.55820e+00	102.56	76.46
(e)	9.50347e-10	7.77463e-10	-4.33799e-01	2.38709e+00	2.97825e+01	-3.75813e+01	102.98	76.03
(f)	1.37867e-02	3.61182e-03	3.66855e-01	1.91193e+00	1.39431e+01	3.86793e+01	103.00	76.00

Evolutionary Algorithm

Corner	Initial Pixel u_0	v_0	Corner Point u_e	v_e	Aperture α^o	s_{U_1}	s_{U_2}	χ^2
(a)	89	89	87.895717	87.644934	89.3094	+1	-1	9.40007e-06
(b)	706	353	706.807636	353.755519	36.2416	-1	+1	1.24802e-06
(c)	398	397	396.943119	397.390921	148.7181	-1	-1	5.69696e-05
(d)	44	484	41.009408	484.473710	18.53073	+1	+1	4.22483e+04
(e)	570	795	570.411461	792.991919	95.7904	+1	+1	4.53057e-04
(f)	482	795	483.170778	792.099214	37.0405	-1	-1	1.17246e-06

	σ_1	σ_2	μ_1	μ_2	ϑ_1^o	ϑ_2^o	A	B
(a)	1.24043e-02	1.41591e-02	-1.11322e+00	1.34753e+00	6.73623e-02	-7.57916e-01	103.00	76.00
(b)	4.32836e-03	9.65062e-03	3.20867e-01	-4.52084e-01	-3.30288e+01	-2.07296e+01	103.00	76.00
(c)	9.80709e-03	8.63958e-03	-2.33377e+00	-1.17649e-01	-7.30867e+01	1.43686e+01	103.00	76.00
(d)	3.31949e-02	2.23349e-03	-1.10955e+00	-7.08228e-01	7.58993e+01	-4.43003e+00	102.59	76.45
(e)	1.22747e-03	1.74839e-03	-8.23944e-01	2.32158e+00	3.15841e+01	-3.73745e+01	103.00	76.00
(f)	1.13199e-02	5.92920e-03	4.34052e-01	1.97009e+00	1.43014e+01	3.86581e+01	103.00	76.00

- *Simulated Annealing.* Initial temperature $T = 1$, size of the equilibrium state $I = 20$, maximal number of iterations $N = 4500$.
- *Evolutionary Algorithm.* Crossover percentage $pc = 0.80$, mutation percentage $pm = 0.05$, convergence percentage $pf = 0.75$, offspring number in the population $P = 22$, maximal number of generations $N = 2000$ approximately equivalent to 4500 movements.

4. 30 samples for each test were performed.
5. Each noisy window $I_r(i, j)$ was normalized to the intensity values $[0, 255]$ considering real numbers through the following function:

$$I_n(j, i) = \frac{255}{max(I_r(j, i)) - min(I_r(j, i))} I_r(j, i)$$

where $I_n(j, i) \mid i, j = 1, \ldots, 2w + 1$ is the normalized studied window including Gaussian noise.

6. Percentage of stop criteria $ftol = 1 \times 10^{-09}$.

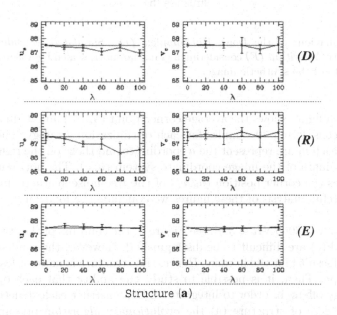

Structure (a)

Fig. 3. Behaviour of the *Down Hill Simplex (D)*, *Simulated Annealing (R)* and *Evolutionary Algorithm (E)* considering a Gaussian noise scaled by a factor λ over the structure (a) of the synthetic image.

As a result of the test the Figures: 3, 4, 5 were built. These figures show the displacement of the corner position (u_e, v_e) of the structures (a), (b) and (c) respectively, considering that a random Gaussian noise was applied over the test

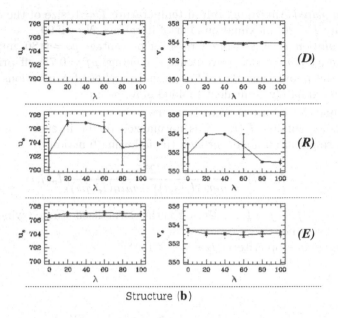

Structure (**b**)

Fig. 4. Behaviour of the *Down Hill Simplex (D)*, *Simulated Annealing (R)* and *Evolutionary Algorithm (E)* considering a Gaussian noise scaled by a factor λ over the structure (b) of the synthetic image.

image. Each figure shows the average corner point $(\overline{u_e}, \overline{v_e})$ and its final standard deviation considering 30 samples for each optimization strategy. The charts on the left of each figure represent the u coordinate and those on the right represent the v coordinate of the image coordinate system (u, v). The horizontal straight line denotes the corner position (u_e, v_e) of the free-noise synthetic image $\lambda = 0$. After a careful analysis of these figures we conclude the following:

1. Beyond $\lambda = 80(SNR \approx 3.5)$, see Figure 3, the contours of the structures (a), (b) and (c) are difficult to be distinguished. However, the random Gaussian noise doesn't blur the borders. Hence, the structure is more or less preserved in shape. Then, it is possible to study the ability that each optimization strategy offers, in order to integrate and reconstruct each structure.

2. In the case of structure (a) the *evolutionary algorithm* presents the best curve of behaviour in presence of noise. Moreover, the maximum standard deviation is obtained for $\lambda = 20$. This value is approximately 0.14 pixels.

3. In the case of structure (b) the *Down Hill Simplex* presents the best curve of behaviour in presence of noise. The maximum standard deviation occurs in $\lambda = 60$ over the u axis. This value is about 0.37 pixels compared to 0.47 pixels obtained by the *evolutionary algorithm* also in $\lambda = 60$. The curve of the *evolutionary algorithm* remains constant around the average value (x_e, y_e) in all cases.

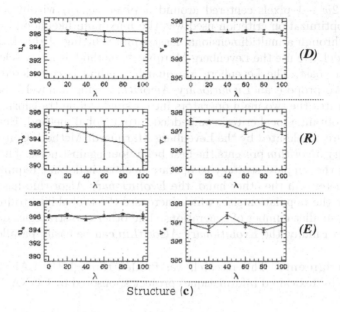

Structure (c)

Fig. 5. Behaviour of the *Down Hill Simplex (D)*, *Simulated Annealing (R)* and *Evolutionary Algorithm (E)* considering a Gaussian noise scaled by a factor λ over the structure (c) of the synthetic image.

4. In the case of structure (c) the *evolutionary algorithm* presents the best curve in presence of noise over the u axis. While, the *Down Hill Simplex* presents the best curve around the v axis.
5. If we observe the error bars that represents the standard deviation of 30 samples for each test considering the three structures. The *Evolutionary Algorithm* presents the best average for each experiment as follows: 1) The average standard deviation of the *Evolutionary Algorithm* is 0.19, 2) The average standard deviation of the *Down Hill Simplex* is 0.25, 3) The average standard deviation of the *Simulated Annealing* is 0.76 pixels.

As a result, the *Evolutionary Algorithm* is less sensitive to noise. Hence, it can be considered more robust. However, the *Down Hill Simplex* offers similar results for the L-corner studied here. Finally, the *Simulated Annealing* shows the worst behaviour in the presence of Gaussian noise.

5   Summary and Conclusions

Accurate L-corner measurement was obtained with a parametric model $M'_L(x, y, \boldsymbol{P})$ using a global optimization approach. The goal was to obtain the best set of parameters $\boldsymbol{P} = (\sigma_1, \mu_1, \vartheta_1, \sigma_2, \mu_2, \vartheta_2, A, B)$, that fits a window of

$2w - 1 \times 2w - 1$ pixels centered around a pixel (u_0, v_0) within a digital image. The optimization criterion used was the maximum likelihood estimator χ^2 obtained through a multidimensional least squares fitting of the data to the L-corner model. We use the Levenberg Marquardt method as many scientists have done in the past. The Levenberg Marquardt method is considered as a local method. We propose an *Evolutionary Algorithm* using a novel representation that integrates the two main operators into a single affine transformation. As a result, we obtain a local criteria embedded into a global method. Hence, our algorithms are accelerated by the Levenberg Marquardt Method. Concluding, the *evolutionary algorithm* presents the best behaviour against noise. The *Down Hill Simplex* is the simplest strategy to operate because it doesn't requires any control parameter. On the other hand, the *Evolutionary Algorithm* has the ability to increase the population size, which increases the probability to find a better result in a smaller number of generations. The other two strategies doesn't have this ability. Finally, the *Evolutionary Algorithm* can be easily parallelized.

Acknowledgments. This research was funded through the LAFMI (Laboratoire Franco-Mexicain d'Informatique) project sponsored by CONACyT-INRIA.

References

1. A. W. Gruen. "Adaptive Least Squares Correlation: A Powerful Image Matching Technique". *S. Afr. Journal of Photogrammetry, Remote Sensing and Cartography.* 14(3), pp. 175–187. 1985.
2. L. Alvarez and F. Morales. "Affine Morphological Multiscale Analysis of Corners and Multiple Junctions". *International Journal of Computer Vision.* 25(2), pp. 95–107, Kluwer Academic Publishers, 1997.
3. Ebner, M., and Zell, A. "Evolving a Task Specific Image Operator." *In Evolutionary Image Analysis, Signal Processing and Telecommunications.* LNCS 1596, Poli et al. (Eds.), EvoIASP. 1999.
4. J. Canny. "A Computational Approach to Edge Detection". *IEEE Trans. on Pattern Analysis and Machine Intelligence.* Vol. 8, No. 6, November, 1986.
5. S. Baker, S. K. Nayar and H. Murase. "Parametric Feature Detection". *International Journal of Computer Vision,* 27(1), pp. 27–50, Kluwer Academic Publishers, 1998.
6. R. Deriche and G. Giraudon. "A Computational Approach for Corner and Vertex Detection". *International Journal of Computer Vision,* 10(2), pp. 101–124, Kluwer Academic Publishers, 1993.
7. T. Lindeberg. "Feature Detection with Automatic Scale Selection". *International Journal of Computer Vision.* 30(2), pp. 79–116, Kluwer Academic Publishers, 1998.
8. D. Marr and E. Hildreth. "Theory of Edge Detection". *Proc. Roy. Soc. London, 207, pp. 187–217,* 1980.
9. R. Mehrotra and S. Nichani. "Corner Detection". *Pattern Recognition.* Vol. 23, No. 11, pp. 1223–1233, 1990.
10. H. P. Moravec. "Towards automatic visual obstacle avoidance". *In Proceedings of the 5th International Joint Conference on Artificial Intelligence,* pp. 584, Cambridge, Massachusetts, USA. 1977.

11. K. Rohr. "Recognizing Corners by Fitting Parametric Models". *International Journal of Computer Vision.* 9(3), pp. 213–230, Kluwer Academic Publishers, 1992.
12. P. L. Rosin. "Augmenting Corner Descriptors". *Graphical Models and Image Processing.* Vol. 58, No. 3, May, pp. 286–294, 1996.
13. Z. Zheng, H. Wang and E. K. Teoh. "Analysis of Gray Level Corner Detection". *Pattern Recognition Letters.* 20, pp. 149–162, Elsevier, 1999.
14. G. Olague and R. Mohr. "Optimal Camera Placement for Accurate Reconstruction". Pattern Recognition, Vol. 35(4), pp. 927–944, 2002.
15. Tsai, D.-M., Hou, H.-T., Su, H.-J. "Boundary-base Corner Detection using Eigenvalues of Covariance Matrices." *Pattern Recognition Letters* 20, 31–40. Elsevier, 1999.
16. Medioni, G., and Yasumoto, Y. " Corner Detection and Curve Representation using cubic B-splines." *Computer Vision, Graphics and Image Processing.* 39, 267–278., 1987.
17. Sohn, K., Kim, J. H., Alexander, W. E. "A Mean Field Annealing Aproach to Robust Corner Detection." *IEEE Transactions on System, Man, and Cybernetics-Part B* 28, 82–90. 1998.
18. Beaudet, P. R. "Rotationally Invariant Image Operators." *In Proc. of the International Conference on Pattern Recognition.* 579–583. 1978.
19. Dreschler, L., and Nagel, H. H. " On the Selection of Critical Points and Local Curvature Extrema of Region Boundaries for Interframe Matching." *In Proc. of the International Conference on Pattern Recognition.* 542–544. 1982
20. L. Kitchen and A. Rosenfeld. "Gray Level Corner Detection". *Pattern Recognition Letters.* No. 1, pp. 95–102, 1982.
21. Wang, H. and Brady M., "Real-time Corner Detection Algorithms for motion Estimation." *Image and Vision Computing.* 13 (9). 1995.
22. G. Olague and B. Hernández. "Autonomous Model Based Corner Detection using Evolutionary Algorithms". *In American Society for Photogrammetry and Remote Sensing.* 12 pages. ASPRS Annual Conference 2001
23. G. Olague and B. Hernández. "Flexible Model-based Multi-corner Detector for Accurate Measurements and Recognition". *16th International Conference on Pattern Recognition.* IEEE Computer Society Press. pp. 578–583, Vol. 2, 11–15 August 2002. Québec, Canada.
24. W. H. Press, B. P. Flanery, S. A. Teukolsky and W. T. Vetterling. "Numerical Recipes in C". Cambridge University Press, Second Edition. 1992.
25. M. Gen and R. Cheng. "Genetic Algorithms and Engineering Design". John Wiley and Sons, Inc. 1997.
26. C. Janikow and Z. Michalewicz. "An Experimental Comparison of Binary and Floating Point Representations in Genetic Algorithms". *In Proceedings of the Fourth International Conference on Genetic Algorithms.* pp. 31–36, San Mateo California, USA. 1991.
27. Dougherty, E. R. Random Processes for Image and Signal Processing. SPIE Optical Engineering Press, and IEEE Press, Inc.1999.

On Two Approaches to Image Processing Algorithm Design for Binary Images Using GP

Marcos I. Quintana[1], Riccardo Poli[2], and Ela Claridge[1]

[1] School of Computer Science, University of Birmingham, Birmingham, B15 2TT,
UK. {M.I.Quintana, E.Claridge}@cs.bham.ac.uk
[2] Department of Computer Science, University of Essex, Colchester, CO4 3SQ, UK.
RPoli@essex.ac.uk

Abstract. In this paper we describe and compare two different approaches to design image processing algorithms for binary images using Genetic Programming (GP). The first approach is based on the use of mathematical morphology primitives. The second is based on Sub-Machine-Code GP: a technique to speed up and extend GP based on the idea of exploiting the internal parallelism of sequential CPUs. In both cases the objective is to find programs which can transform binary images of a certain kind into other binary images containing just a particular characteristic of interest. In particular, here we focus on the extraction of three different features in music sheets.

1 Introduction

In the last few years, a variety of evolutionary approaches have been applied to the problem of discovering algorithms for image processing, so much so that Evolutionary Image Processing can almost be considered a separate field of research. A relatively small subset of these approaches (e.g. [1,2,3,4,5,6]) have been based on Genetic Programming (GP) [7].

Mathematical Morphology (MM) is well known as a powerful tool for various image-processing tasks [8]. It is suitable for shape related processing since morphological operations are directly related to the object shape. To design a MM procedure (i.e. algorithm) some expert knowledge is necessary to properly select the structuring elements and to make an adequate selection of the morphological operators sequence. To date, no one has attempted to evolve MM algorithms using GP (although Yoda *et al.* [9] have suggested using genetic algorithms for this task). In this paper we start by describing the results of our efforts in this direction.

GP is usually quite demanding from the computation load and memory use point of view. This is particularly true when using GP to solve image processing problems, where the number of fitness cases can be extremely large, and finding improvements to the programs in a relatively small population can be difficult (which can easily lead to extensive bloat). Our approach based on GP with MM primitives is not immune from this. So, as a second contribution in this paper we have explored an alternative representation and primitive set which is based

S. Cagnoni et al. (Eds.): EvoWorkshops 2003, LNCS 2611, pp. 422–431, 2003.
© Springer-Verlag Berlin Heidelberg 2003

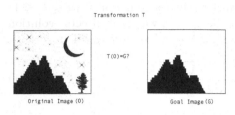

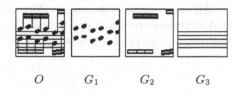

$$O \qquad G_1 \qquad G_2 \qquad G_3$$

Fig. 2. Examples of binary transformations. The original fragment of a music sheet (O) and three possible goals extracted by hand: heads (G_1), hooks (G_2) and lines (G_3).

Fig. 1. Transformation from Original Image (O) to Goal Image (G).

on a technique to speed up GP. Many ideas have been applied to improve the performance of GP. The technique we have used is known as Sub-Machine-Code GP (SMCGP) [10,11]. SMCGP is a GP variant aimed to exploit the intrinsic parallelism of sequential CPUs. SMCGP extends the scope of GP to the evolution of parallel programs running on sequential computers. These programs are faster as, thanks to the parallelism of the CPU, they perform multiple calculations during a single program evaluation.

Adorni *et al.* have suggested using SMCGP for the efficient design of low-level vision algorithms [12,13,14]. They applied SMCGP in a plate detection and character recognition approach. The results obtained show that SMCGP is able to generate programs that are both accurate and efficient.

Here we have used GP (and SMCGP) to evolve transformation algorithms on music sheets. We aim to transform an original image into a goal image containing only a particular feature. The features analyzed are heads, hooks and lines.

The paper has the following structure. In Section 2 we describe the problem at hand and its features. In Section 3 we describe the GP approach for MM algorithm design on binary images. In Section 4 we describe how SMCGP can be used to speed up the evolution. In Section 5 we analyze the obtained results. Finally in Section 6 we draw some conclusions.

2   Test Problem and Fitness Function

Many image processing tasks for binary images can simply be described by providing an original binary image and a goal image (possibly drawn by hand). As shown in Figure 1 the problem then is to find an appropriate computational procedure which can perform that same transformation.

For the experiments presented in this paper we study the extraction of heads, hooks and lines in music sheets. The different features in a music sheet are interesting for the image transformation problem since each feature is easily recognizable for a human but the features involved are easy to miss-detect for a machine vision system. A set of images belonging to the training set is shown in Figure 2.

All our experiments were performed using a fitness function F $(0 \le F \le 1)$ that evaluates the similarity between two images in a way that directs evolution towards a good trade-off between *sensitivity* (SV) and *specificity* (SP):

$$F = 1 - \frac{\sqrt{(1 - SP)^2 + (1 - SV)^2}}{\sqrt{2}}, \tag{1}$$

where $SV = \frac{TP}{TP+FN}$ and $SP = \frac{TN}{FP+TN}$, and TP is the number of true positives, FP is the number of false positives, TN is the number of true negatives and FN is the number of false negatives. For these experiments, the goal is to convert the original image (O) into the goal image (G). If $G(x_i, y_i) = 1$ then we have a true positive if $O(x_i, y_i) = 1$. If, instead, $O(x_i, y_i) = 0$ we have a false positive error. If $G(x_i, y_i) = 0$ then we have a true negative if $O(x_i, y_i) = 0$, and a false negative if $O(x_i, y_i) = 1$. The true/false positive/negative numbers are obtained by integrating over every pair of (x_i, y_i) coordinates in the original image.

3   GP for MM Algorithm Design

Let us start by describing our approach to evolving image processing algorithms based on GP and MM primitives.[1]

3.1   Basic Morphological Operations

The language of MM is *set theory*. As such, morphology offers a unified and powerful approach to numerous image-processing problems. Sets in MM represent the sets of objects in an image. For example, the set of all black pixels in a binary image is a complete description of the image. In binary images, the sets in question are members of the 2–D integer space Z^2, where each element of a set is a tuple (x, y) representing the coordinates of a black pixel in the image. Gray-scale digital images can be represented as sets whose components are in Z^3. Sets in higher dimensional spaces can contain other image attributes, such as color and time varying components.

The two morphological operations most used in MM are *dilation* and *erosion* and most of the algorithms developed by experts to perform a particular task make use of them.

If A and B are sets in Z^2 and $\emptyset$ denotes the empty set, the dilation of A by B, $A \oplus B$, requires performing the reflection of B about its origin, then shifting this reflection $\widehat{B}$ by $z \in Z^2$ to obtain $(\widehat{B})_z$. The dilation of A by B, as shown in Equation 2, is the set of all z displacements such that $(\widehat{B})_z$ and A overlap by at least one nonzero element:

$$A \oplus B = \{z | (\widehat{B})_z \cap A \neq \emptyset\} \tag{2}$$

Figure 3 exemplifies this operation.

---

[1] Additional details on the experimental setup can be found in [15].

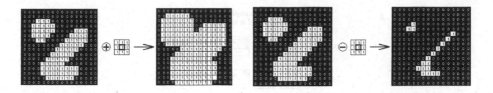

Fig. 3. Dilation.   **Fig. 4.** Erosion.

The erosion of A by B, $A \ominus B$, is defined as the set of all points z such that $\widehat{B}$ translated by z is contained in A. The erosion of A by B, shown in Equation 3, is exemplified in Figure 4:

$$A \ominus B = \{z | (\widehat{B})_z \subseteq A\} \tag{3}$$

These two equations are not the only definitions for dilation and erosion, but they are usually preferred in practical implementations because of their analogy with the operation of convolution for linear filtering.

3.2 Evolution of Morphological Algorithms with GP

The GP approach suggested assumes that it is possible to a find a sequence of morphological operators in the MM algorithm's search space to convert an image into another containing only a particular feature of interest. To show this idea we select musical sheets as examples and extract some features from them.

The process is visualized in Figure 5. The first step is to create some examples of correct feature extraction by hand to be used as training sets in GP. Next, it is necessary to define the type, size and number of structuring elements to be used in the GP search. After the *primitive set* is defined, we start the GP search to obtain a (near) optimum tree representing a MM algorithm sequence (see below). The best sequence evolved, according to the fitness function, is also analyzed visually to decide whether its high fitness value corresponds to a good perceptual image quality.

When a MM algorithm is designed by hand, the programmer usually chooses a regular structuring element to make a sequence of operations. There is really no reason for choosing a regular element except that we, as humans, are more likely to understand *regularities* such as squares, lines, triangles, etc. However, it is entirely possible that one could get much better results by allowing *irregular* structuring elements. So, in our primitive set for GP we also include irregular structuring elements (constructed randomly). We use structuring elements of sizes 3×3, 5×5 and 7×7 such as those shown in Figure 6. In this work, we include 11 regular and 11 irregular structuring elements of each size.

We use a function set including two functions, EVAL1 and EVAL2 of arity 1 and 2, respectively, which are used only for sequencing purposes. The terminal set includes nested primitives of the form $x(yz[w])$ where: $x \in \{e, d\}$, represents a morphological operators (erosion and dilation); $y \in \{R, I\}$, represents the type of

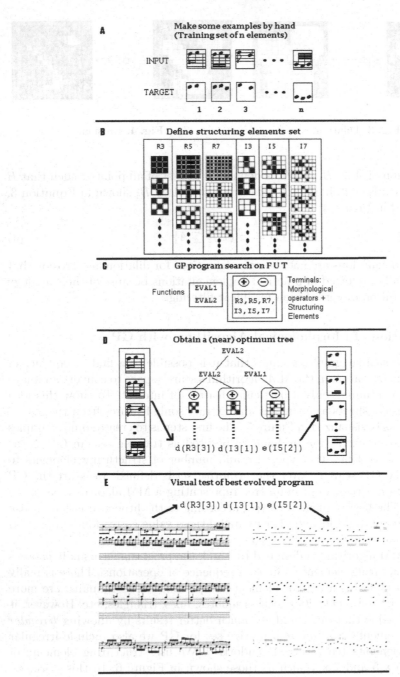

Fig. 5. Process to obtain a MM algorithm using GP. A) Make examples by hand.
B) Decide structuring elements to use. C) Perform GP search. D) Obtain a (near)
optimum tree. E) Evaluate best result.

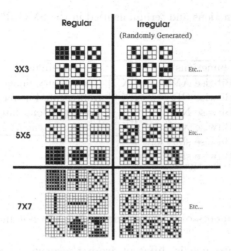

Fig. 6. Examples of regular and irregular structuring elements of various sizes.

structuring element selected (regular and irregular); $z \in \{3, 5, 7\}$, represents the size of structuring element and $w \in \{1..11\}$ represents the structuring element index.

As shown in Figure 5.D, at evaluation time each GP tree is transformed into a linear representation, which, when read from left to right, represents an MM algorithm. This is applied to the training set in order to evaluate the fitness of the corresponding GP tree.

The GP parameters we have used are similar to those suggested in the literature [7]. A crossover rate of 0.9 and 0.1 mutation rate were used. Mutation was based on the *ramped half and half* initialization method.

The sizes of the images in the training set were 16×16, 32×32, 64×64 and 128×128. The number of images in the training set was 1, 5, 15 and 25 in different experiments. Due to the high computation and memory load, all the experiments were performed using a population of 50 individuals run for 100 generations.

4   Using SMCGP to Evolve Image Processing Algorithms for Binary Images

In the MM approach described in Section 3 we were forced to use tiny populations and short runs, while, nowadays in GP usually people use much bigger populations and/or much longer runs.

So, we decided to test an alternative approach: we encoded the images using unsigned long integers to take advantage of the SMCGP paradigm. In this approach each image is represented as a vector of 32 unsigned long integers (32 bits); each element of the vector represents a row in the image. We used the Function and Terminal sets presented in Table 1, where bitwise operations are

Table 1. Functions and terminals used in the SMCGP approach.

Functions		Terminals	
AND	Bitwise AND	X[1]	One binary image
OR	Bitwise OR	X[2]	represented using 32
NOT	Bitwise NOT	...	unsigned long integers
XOR	Bitwise XOR	X[32]	(One row each.)
SL	Bitwise shift left		
SR	Bitwise shift right		

applied to all the elements of a vector (note they are not morphological operations).

Note that our approach is different from those where SMCGP is used to evaluate N fitness cases in parallel [11]. Here we use the parallelism of the CPU to process multiple pixels per program evaluation.

We used populations of 500, 1000 and 5000 programs run for 1000, 500 and 100 generations respectively (to fix the number to fitness evaluations to 500000 in each run) using 20 different random seeds. A 0.9 crossover rate was adopted. Mutation based on the ramped half and half initialization method was applied with a rate of 0.1.

5 Results and Analysis

5.1 GP with MM Approach

GP easily found different algorithms to solve the *heads* extraction problem by using MM primitives. The degree of accuracy is very difficult to evaluate visually and, depending on the evaluator, it does not always match the numerical fitness values obtained. However, we believe that many of the results obtained are as good as those that could be obtained by an expert writing the algorithm by hand. Figure 7 shows an original image and Figure 8 shows the result of an MM algorithm generated by GP.

The *hooks* extraction problem is a difficult task. That feature could be mismatched with lines, heads or other features present in a musical sheet. In spite of that, some GP generated algorithms present good approximations to the desired task. One example is presented in Figure 10.

Contrary to expectation, the *lines* extraction problem was the one presenting the most difficulties for the GP approach proposed. The results obtained were tending either to completely white images or to confuse lines with other features. We show an example in Figure 9 where GP accurately finds the lines, but also includes most of the hooks present in the test image.

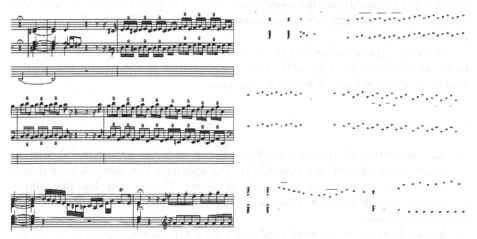

Fig. 7. Example of testing image for visual purposes.

Fig. 8. Example of a good visual result for heads on the image in Figure 7 using GP and MM.

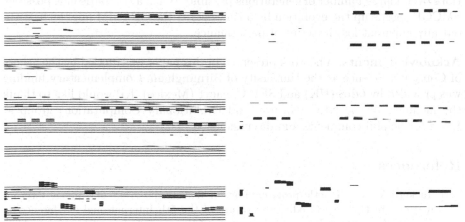

Fig. 9. Example of a good visual result for lines on the image in Figure 7 using GP and MM.

Fig. 10. Example of a good visual result for hooks on the image in Figure 7 using GP and MM..

5.2 SMCGP-Based Approach

We made a total of 720 SMCGP runs. We had: 3 features to extract (heads, hooks and lines), 4 different numbers of examples in the training set (1, 5, 10, 15), 3 different population-size/number-of-generation combinations (500/1000, 1000/500 and 5000/100) and 20 different random seeds.

The SMCGP approach speeds up evolution in an impressive way. When the comparison is possible (recall that the GP and MM approach can't evolve so many programs) the SMCGP approach is up to 5 times faster. It is also able to evolve bigger populations for more generations. Indeed, the SMCGP approach finishes all the runs properly while some of the runs of GP with MM operators ran out of memory.

In terms of performance on the training set, the SMCGP approach appears to be as good as GP with MM. Due to space limitations we cannot provide examples of the output produced in this case.

6   Conclusions

We have described an approach to the evolution of image processing algorithms for binary images based on GP and MM. The approach has shown a good degree of accuracy in experiments with musical sheets. However, it has also shown a computational bottleneck when using big populations and running them for many generations.

As an alternative, we have explored a SMCGP-based approach which was hoped to speed up the evolution of algorithms for binary images. The SMCGP approach has been quite successful in this, making the evolution of big populations over a large number of generations possible. When a comparison is possible, SMCGP speeds up the evolution by 5 times w.r.t. the GP+MM approach, without any apparent loss in terms of performance.

Acknowledgments. The work presented in this paper was funded by the School of Computer Science at the University of Birmingham. Complementary funding was provided by ORS (UK) and SEP-Conacyt (Mexico). RP would like to thank the members of the NEC (Natural and Evolutionary Computation) group at Essex for helpful comments and discussion.

References

[1] Tackett, W.A.: Genetic programming for feature discovery and image discrimination. In Forrest, S., ed.: Proceedings of the 5th International Conference on Genetic Algorithms, ICGA-93, University of Illinois at Urbana-Champaign, Morgan Kaufmann (1993) 303–309

[2] Daida, J.M., Hommes, J.D., Ross, S.J., Vesecky, J.F.: Extracting curvilinear features from SAR images of arctic ice: Algorithm discovery using the genetic programming paradigm. In Stein, T., ed.: Proceedings of IEEE International Geoscience and Remote Sensing, Florence, Italy, IEEE Press (1995) 673–675

[3] Poli, R.: Genetic programming for image analysis. In Koza, J.R., Goldberg, D.E., Fogel, D.B., eds.: Genetic Programming 1996: Proceedings of the First Annual Conference, Stanford University, CA, USA, MIT Press (1996) 363–368

[4] Teller, A.: Evolving programmers: The co-evolution of intelligent recombination operators. In Angeline, P.J., Kinnear, Jr., K.E., eds.: Advances in Genetic Programming 2. MIT Press, Cambridge, MA, USA (1996) 45–68

[5] Howard, D., Roberts, S.C., Brankin, R.: Target detection in SAR imagery by genetic programming. In Koza, J.R., ed.: Late Breaking Papers at the Genetic Programming 1998 Conference, University of Wisconsin, Madison, Wisconsin, USA, Stanford University Bookstore (1998)

[6] Ebner, M., Zell, A.: Evolving a task specific image operator. In Poli, R., Voigt, H.M., Cagnoni, S., Corne, D., Smith, G.D., Fogarty, T.C., eds.: Evolutionary Image Analysis, Signal Processing and Telecommunications: First European Workshop, EvoIASP'99 and EuroEcTel'99. Volume 1596 of LNCS., Goteborg, Sweden, Springer-Verlag (1999) 74–89

[7] Koza, J.R.: Genetic programming: On the programming of computers by natural selection. MIT Press, Cambridge, Mass. (1992)

[8] Serra, J.: Image Analysis and Mathematical Morphology. Academic Press (1982)

[9] Yoda, I., Yamamoto, K., Yamada, H.: Automatic acquisition of hierarchical mathematical morphology procedures by genetic algorithms. Image and Vision Computing **17** (1999) 749–760

[10] Poli, R., Langdon, W.B.: Sub-machine-code genetic programming. In Spector, L., Langdon, W.B., O'Reilly, U.M., Angeline, P.J., eds.: Advances in Genetic Programming 3. MIT Press, Cambridge, MA, USA (1999) 301–323

[11] Poli, R.: Sub-machine-code GP: New results and extensions. In Poli, R., Nordin, P., Langdon, W.B., Fogarty, T.C., eds.: Genetic Programming, Proceedings of EuroGP'99. Volume 1598 of LNCS., Goteborg, Sweden, Springer-Verlag (1999) 65–82

[12] Adorni, G., Cagnoni, S., Gori, M., Mordonini, M.: Efficient low-resolution character recognition using sub-machine-code genetic programming. In: WILF 2001. (2002) In press.

[13] Adorni, G., Cagnoni, S., Mordonini, M.: Efficient low-level vision program design using sub-machine-code genetic programming. Workshop sulla Percezione e Visione nelle Macchine, available at citeseer.nj.nec.com/539182.html (2002)

[14] Adorni, G., Cagnoni, S.: Design of explicitly or implicitly parallel low-resolution character recognition algorithms by means of genetic programming. In R., R., M., K., Ovaska, S., Furuhashi, T., F., H., eds.: Soft Computing and Industry: Recent Applications. (Proc. 6th Online Conference on Soft Computing), Springer (2002) 387–398

[15] Quintana, M.I., Poli, R., Claridge, E.: Genetic programming for mathematical morphology algorithm design on binary images. In Sasikumar, M., Hegde, J.J., Kavitha, M., eds.: Proceedings of the International Conference KBCS-2002, Mumbai, India, Vikas (2002) 161–170

Restoration of Old Documents with Genetic Algorithms

Daniel Rivero, Rafael Vidal, Julián Dorado, Juan R. Rabuñal, and Alejandro Pazos

Univ. A Coruña, Fac. Informatica, Campus Elviña, 15071, A Coruña, Spain
danielrc@mail2.udc.es, infrvr00@mail.ucv.udc.es,
{julian, juanra, ciapazos}@udc.es

Abstract. Image recognition is a problem present in many real-world applications. In this paper we present an application of genetic algorithms (GAs) to solve one of those problems: the recovery of a deteriorated old document from the damages caused by centuries. This problem is particularly hard because these documents are affected by many aggresive agents, mainly by the humidity caused by a wrong storage during many years. This makes this problem unaffordable by other image processing techniques, but results show how GAs can succesfully solve this problem.

1 Introduction

Old documents suffer the effect of time. Manipulation, humidity and wrong storage affect them and make them difficult to be read. Moreover, the more manipulated and read they are, the more deteriorated they get. As they are unique, i.e. they were written when there was no press, this problem makes them even more valuable. Thus, their restoration is needed for the conservation of the ancient knowledge they contain. In many cases much of this knowledge is completely lost, so there is the need to recover their information while it is still available.

These documents are already affected by many agents. These agents have made them in many parts very difficult to be read, reaching the point that only an expert human is able to read them. Some other parts are so damaged that their information is completely lost.

The recovery of these documents is a huge work that involves many steps. These steps include noise filtering, character recognition, etc. The aim of this paper is to make an attempt to cover a first step in the processing of these documents: the discrimination between letters and paper in these damaged documents. Page images of old manuscripts are usually terribly dirty and considerable large in size. To overcome this problem, in this paper we propose a new effective method for separating characters from noisy background, since conventional threshold selection techniques are inadequate to cope with the image where the gray levels of the character parts are overlapped by that of the background.

To solve this problem, we used GAs to develop a filter for character discrimination. The characteristics of GAs make them a powerful tool to solve this particular problem.

S. Cagnoni et al. (Eds.): EvoWorkshops 2003, LNCS 2611, pp. 432–443, 2003.
© Springer-Verlag Berlin Heidelberg 2003

1.1 Image Filtering

Image filtering [1][2] is the transformation from one image into another by means of a convolution transformation. This operation can be done either in the frequency or spatial domain. In this paper, we are interested in spatial domain, where a filter function h is convolved with the input image. After this convolution, a new image appears as a result of a transformation done to the input image. This transformation is usually done for various objectives, e.g. noise clearing or border detection.

Discrete convolutions are done by shifting the kernel h over the image x and multiplying its values with the corresponding pixel values of the image. This kernel will be a matrix of numbers, defining a spatial window of weights. The convolution process is expressed in the following formula:

$$y(i, j) = \sum_{m=-M/2}^{M/2} \sum_{n=-M/2}^{M/2} h(m,n) \cdot x(i - m, j - n) \tag{1}$$

Where $y(i,j)$ denotes the pixel value of the resulting image at position (i,j), $h(m,n)$ is the weigth of the kernel at position (m,n) for a kernel matrix of size M x M, and $x(i-m,j-n)$ is the value of the pixel at position (i-m,j-n) of the original image.

There are several traditional kernels for specific operations. For example, some kernels for noise filtering will make a mean value of the neighbour pixels (including the one being processed). There are also other known kernels for other operations, such as edge detection. For this particular task, there are many different kernels, like the Laplacian convolution kernel [1], the Roberts edge detector [3], the Sobel edge operator [4], the Prewitt Edge operator [5] or the Kirsch edge operator [6].

1.2 Genetic Algorithms

A GA [7][8] is a search technique inspired in the world of biology. More specifically, the Evolution Theory by Charles Darwin [9] is taken as basis for its working. GAs are used to solve optimization problems by copying the evolutionay behaviour of species. From an initial random population of solutions, this population is evolved by means of selection, mutation and crossover operators, inspired in natural evolution. By applying this set of operations, the population goes through an iterative process in which it reaches different states, each one is called *generation*. As a result of this process, the population is expected to reach a generation in which it contains a good solution to the problem. In GAs the solutions of the problem are codified as a string of bits or real numbers.

GAs have been very used in the field of signal processing. Their use for filter design was extensively described in [10] and [11], but they have many other applications, like designing QRS detectors [12]. In image processing they have shown that they are a very useful tool [13], and, as an evolution of GAs, Genetic Programming (GP) has also shown that it can be succesfully applied to many different image processing problems like image analysis [14] or object detection [15].

However, the main application of GAs in the field of image processing is probably image segmentation.

Image segmentation is a low-level image processing task that tries to divide an image into homogeneous regions. This technique is very used in medical applications. In these applications image segmentation is a fundamental pre-processing step in systems that perform tasks such as planning of surgical operations or diagnosis. In these systems images play an important role, whether they are in 2D [16][17] or 3D [18]. One the most extensive work in the area of image segmentation with GAs is [19] and [20].

2   Description of the Problem

In this particular problem we will use GAs to develop a filter. The objective is, from an initial deteriorated image, to get another image that only has the result of the original writing on it. The desired filter must be able to discriminate the original letters from the paper, even in some parts where this document is very damaged and deteriorated.

A representative part of this document is shown on Fig. 1. Here we can see that in some parts it is very damaged, due to humidity or ink, and in some other parts the information is completely lost. This can happen either because bents caused by wrong storage of the document or because it is broken.

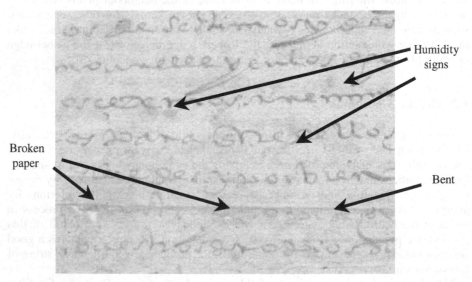

Fig. 1. Part of the document.

There are two ways to detect characters in a text. The first is by a comparison with others taken as examples, like ANNs do. The second way to detect characters is by choosing them taking as a basis some characteristics they have, like colour or shape.

In the first way there are some works done with oriental alphabets. These alphabets have a very high number of characters in comparison with Latin alphabet. In those alphabets it is necessary that characters are extracted more clearly because it is more difficult to recognize them in comparison with Latin ones.

The second solution can be divided into two classes: the ones that detect the border of the characters and select those pixels that are inside the character, and those that try to recognise the pixels from the characters by comparing them with their neighbourhood. The first technique was used in the work by Wen Hwang and Hu Chang titled "Character Extraction from Documents Using Wavelet Maxima" [21]. Some relevant works in the second technique are the work titled "Character Extraction from Noisy Background from an Automatic Reference System" [22] and, more recently, the work titled "Old Text Reconstruction: An Artificial Intelligence Approach" [23]. These two works make a selection of pixels looking at their colour, neighbourhood and the document to determine if they are characters or not.

3   Configuration of the Genetic Algorithm

To solve this problem, we used a real value-based GA. The aim of using a GA is to develop a filter which can do the transformation from an original image into another one that only has the text. As we want to develop a filter, it will be a window with a size of M x M. Thus, the codification of the filter will be a string of M*M real values designing the values of the window.

This window will be applied to an input image that will be a part of the document to be restored. For this purpose, we chose one of the most damaged parts of the document and asked our expert to manually restore it. The resulting window from the GA should be able to return this restored image from the original damaged part, as is shown on Fig. 2.

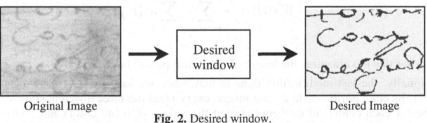

Original Image   Desired Image

Fig. 2. Desired window.

The fitness function will return the mean value of the distances between each value of the pixels of the desired restored image and their correspondent from the output image. Let o(i,j) be the pixel value of the output image at position (i,j); this output image is obtained as was explained, from the original image and following the equation 1. Let O(i,j) be the pixel value of the desired image at position (i,j). Let N_1 be the number of rows and N_2 be the number of columns of the original, output and desired image, the fitness function is the following:

$$f = \frac{1}{N_1 \cdot N_2} \sum_{i=1}^{N_1} \sum_{j=1}^{N_2} |O(i, j) - o(i.j)| \qquad (2)$$

From this initial configuration, several changes have been done.

First, as we work with colour images, we will need to work with three different windows, one for each colour. Colour images are composed by three images: each one for each of the three colours: red, green and blue. Here we will work with these three images separately because of the properties of the original image: the ink used for writing has different color components of red, green and blue, and so it has the paper, humidity spots, etc. So, we expect the GA to take advantage of this feature when it developes the three windows and it makes it easier for the GA to discriminate between text and paper. This is one of the main reasons that took us to use GAs.

As we will evolve three different windows, these will be named as W_r, W_g and W_b, standing for red, green and blue window respectively. These three windows will be convolved with the original image on its three components, and the output image will be the mean value of the three resulting images, having as output a grey-scaled image. The whole fitness function can be seen on Fig. 3.

As we are now working with three windows, the length of the GA value string will be M*M*3.

Second, after many different trials done, we noticed that the system works much better if the middle value of the window (i.e., the value that corresponds to the weight of the pixel being treated) is not present in the AG string. This value will be the opposite of the sum of the rest of the values. This modification can be seen in the following formula, where W_r is the final window and w_r is the window designed by the genotype, with its middle value $w_r(0,0)$ equal to zero:

$$W_r(0,0) = - \sum_{i=-M/2}^{M/2} \sum_{j=-M/2}^{M/2} w_r(i, j) \qquad (3)$$

After this modification, the length of the genotype will be M*M*3 − 3.

Finally, the last modification done is that when we load the original image, we invert its color attributes. In a color image, every pixel has three different color values (one for each color), all of them between the values of 0 (no color) and 255 (most intensity of the color). The inversion we do is to take these values as 255 − original value, so 0 will be the most intensity of color and 255 the absence of color. This process is done due to the fact that the effect of the multiplication convolution process is higher in those pixels with higher values, and if we want to remark the text, we will obtain better results if we make it have the highest values (closer to 255) rather than lower (closer to 0), which would be the normal case. Text pixels, darker, are normally closer to 0 than paper pixels, which are white and then closer to 255. With this transformation we will make dark pixels (text) have a value closer to 255.

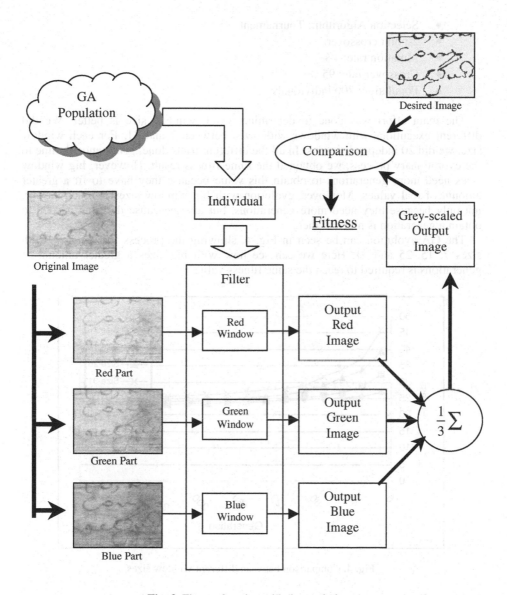

Fig. 3. Fitness function with three windows.

4   Results

With this configuration, we did many different trials with different parameter configurations. The configuration that worked best is the following:

- Selection Algorithm: Tournament.
- 2-point crossover.
- Mutation rate: 4 %.
- Crossover rate: 95 %
- Population: 700 individuals.

The main effort was done to determine which window size is better. We did different executions with windows with sizes between 7 and 31. For each window size, we did 20 independent runs. In all the different trials done, given enough time to the evolutionary process, we obtained the same fitness result. However, big window sizes need more generations to obtain this value because they have to fit a greater amount of real values. Moreover, evolution with big window sizes is much slower, not only because they need more generations, but also beacause the time needed to obtain a generation is much higher.

The GA evolution can be seen in Fig. 4, showing the process for windows with sizes 7, 15, 25 and 31. Here we can see how with big sizes a greater amount of generations is required to reach the same fitness value.

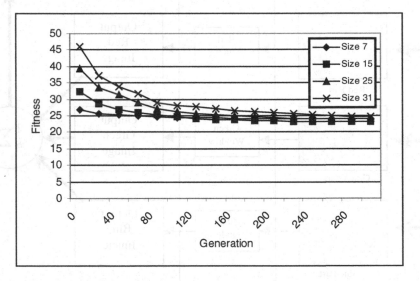

Fig. 4. Comparison between different window sizes

The differences with different window sizes are more visible in the output image. With a small window size (7, 9 or 11) we obtain an output image with very few visible text (the resulting image with size 7 is unreadable).

With a bigger window, we get an image with a more visible text, but the document has some noise. Taking a bigger size of the window (17, 19 and higher) the differences between the resulting images are less visible. In these cases we obtain a document with the text perfectly readable and visible, but with much more noise than with smaller windows. In Fig.5 we can see some examples of the output images corresponding to the input shown in Fig.1 with window sizes of 11, 13, 15 and 17.

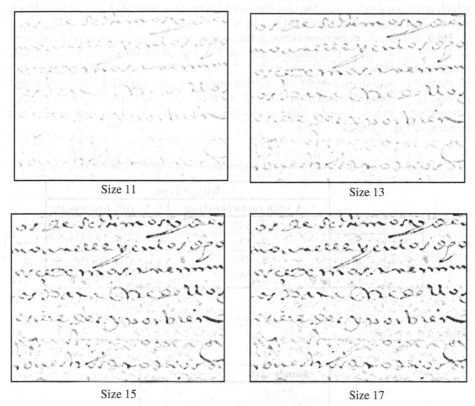

Fig. 5. Resulting images with window sizes of 11, 13, 15 and 17

Once we have obtained these images, we processed them again and did several transformations as described in [23]. The main transformations done are:

- First, we select pixels after comparing them with the mean value of their neghbourhood. Their value will be modified according to the variance value of the neighbour pixels. The neighbourhood size is 9x9.
- Second, we work with groups of pixels, deleting those groups with less than 10 pixels.
- We can now take the mean value of the groups of pixels and delete those pizels that are not dark enough.
- Finally, we calculate the mean value of all the pixels deleted in the whole process. If any group has a value lower than that mean, it is deleted.

After this process, we obtained very clear and readable images. However, even with this second processing, those output images obtained with big window sizes still had a lot of noise. This noise has the shape of dark spots in the document, due to old humidity signs. The images with a window size lower than 11 are completely unreadable. Thus, this second processing of the images is very useful in those images

440   D. Rivero et al.

which have some noise with the text that can be removed, and this has lead us to think that the best window sizes are 11, 13 and 15.

Table 1 draws a comparison between the four images obatained with window sizes of 11, 13, 15 and 17. The measurement used is the mean difference between that image and the desired one, shown on Fig. 2, i.e., the fitness value. These values can be compared with the fitness of the original image and the obtained only with image processing techniques, shown on Table 2. In this table we can see how the image processing techniques improve the results obtained with GAs. These six images can be seen on Fig. 6 and 7.

Table 1. Fitness values with different window sizes and with and without processing.

		Image type	
		GA with no processing	GA with processing
Window Size	11	23.845	21.227
	13	23.534	20.159
	15	23.489	19.538
	17	23.653	19.099

Table 2. Fitness values for the original image and the image obtained only with image processing techniques.

Image type	Fitness
Original	71.165
Only processed	69.129

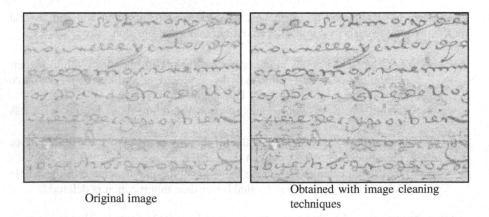
Original image   Obtained with image cleaning techniques

Fig. 6. Resulting images after a cleaning process from GAs and from the original image

The images in Fig. 7 correspond to those of Fig. 5 after processing them. Note that in Table 1 the fitness values of the images obtained with GAs are very similar. However, after processing them the fitness values are lower when we use higher sizes of the windows. This is because the processing operations can eliminate the noise introduced by GAs, and with small windows there is few noise to eliminate. In all the four cases exposed in this table, the fitness values are much better than the one corresponding to the image obtained only by image processing techniques.

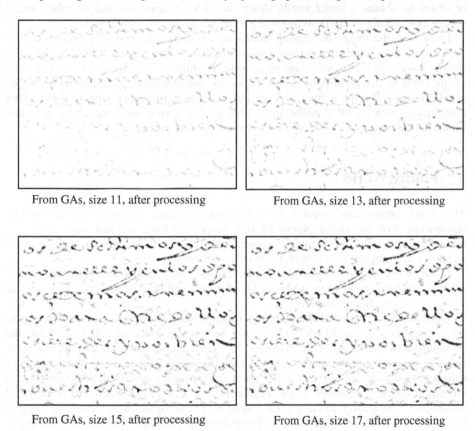

From GAs, size 11, after processing   From GAs, size 13, after processing

From GAs, size 15, after processing   From GAs, size 17, after processing

Fig. 7. Resulting images after a cleaning process from GAs and from the original image

5   Conclusions

In this paper we have presented an application of GAs in the field of image processing. We have shown how we can adapt them to solve a difficult problem: to discriminate between text and paper in a document deteriorated by time. As result, we have obtained the same document, but now it only has the original text. However, some areas were so damaged (in some of them the paper was even broken) that their information could not be recovered.

We have also studied the effect of window size in the result. If we use a small window we will obtain an image with few noise, but the text will be few readable.

With a big size, the text will be much more readable, but we will have much more noise. The possibility of applying other techniques in the resulting image makes possible to choose a window size that leads to obtaining a resulting image with some noise that can be removed. Thus, different image processing techniques can be mixed in order to obtain a better result. After the GA process, the use of other image processing techniques helps to produce a more readable and visible text.

Even with the image processing techniques, some of the resulting images can be seen even more confusing to the human eye than the original ones or the ones obtained with the described image processing techniques. This has made the choosing of the window size an important task. However, the technique presented in this paper is useful not only for a human read, but also as a pre-processing tool for other techniques, such as OCR techniques. This is the reason why we chose a fitness function that produces a high contrast between ink and paper.

6   Future Works

This work opens new research lines that can be done in the field of old text restoration. For the development of this work we have worked only with one document, so a future work could be to study the possibility of obtaining more general windows that can restore a complete set of documents.

In this work we have obtained windows using GAs. It is also interesting the use of other AI techniques, such as ANNs or GP, for the obtention of the document, and so we can compare them with GAs.

Once we have obtained the resulting images, the restoration process goes on in other directions, as could be the detection of when a piece of text is a letter, the recognition of that letter or the translation of text. These tasks are particularly hard, because for their commitment we must recognize old text, and old writing is very different than today's. There is also a great difference in writing in those documents written in different centuries, so this problem becomes really difficult. However, OCR techniques have proved to have good results at text recognition. Recently, techniques based on Sub-machine-code Genetic Programming [24] have shown their ability to recognise characters [25] and seem to be a powerful tool for this particular task.

References

1.   Castleman, K. R.: Digital Image Processing. Prentice-Hall (1996)
2.   Russ, J. C.: The Image Processing Handbook (third edition). CRC Press LLC (1999)
3.   Roberts, L. G.: Machine Perception of Three-Dimensional Solids in J.T. Tippett, ed., Optical and Electro-Optical Information Processing, MIT Press, Cambridge, MA, (1965) 159–197
4.   Davis, L. S.: A Survey of Edge Detection Techniques. CGIP, 4:248–270. (1975)
5.   Prewitt, J.: Object Enhacement and Extraction, in B. Lipkin and A. Rosenfeld, eds., Picture Processing and Psychopictorics, Academic Press, New York (1970)
6.   Kirsch, R. A.: Computer Determination of the Constituent Structure of Biological Images, Computers in Biomedical Research, 4 (1971) 315–328

7.   Holland, J. H.: Adaptation in Natural and Artificial Systems. University of Michigan Press (1975.)
8.   Goldberg, D. E.: Genetic Algorithms in Search, Optimization and Machine Learning. Addison-Wesley Reading, MA (1989)
9.   Darwin, C.: On the Origin of Species by means of Natural Selection or the Preservation of Favoured Races in the Struggle for Life. Cambridge University Press, Cambridge, UK, sixth edition, (1864), originally published in 1859.
10. Suckley, D.: Genetic algorithm in the design of FIR filters, IEE Proceedings-G, vol. 138 (1991) 234–238
11. Nambiar, R., Tang, C.K.K., Mars, P.: Genetic and learning automata algorithms for adaptive digital filters, in Proc. ICASSP-92, vol. 4, New York, NY (1992) 41–44
12. Poli, R., Cagnoni, S., Valli, G.: Genetic design of optimum linear and non-linear QRS detectors, IEEE Trans. On Biomed. Engineering, vol. 42, no. 11 (1995) 1137–1141
13. Bounsaythip, C., Alander, J.T.: Genetic Algorithms in Image Processing – A Review, Proc. Of the 3rd Nordic Workshop on Genetic Algorithms and their Applications, Metsatalo, Univ. Of Helsinki, Helsinki, Finland, (1997) 173–192
14. Howard,D., Roberts, S. C.: A Staged Genetic Programming Strategy for Image Analysis, Proceedings of the Genetic and Evolutionary Computation Conference. Vol. 2. (1999) 1047–1052
15. Howard, D., Roberts, S. C.: The Boru Data Crawler for Object Detection Tasks in Machine Vision, Applications of Evolutionary Computing, Proceedings of EvoWorkshops2002: EvoCOP, EvoIASP, EvoSTim/EvoPLAN, (2002) 222–232
16. Ramos, V., Muge, F.: Image Colour Segmentation by Genetic Algorithms (2000)
17. Ramos, V.: The Biological Concept of Neoteny in Evolutionary Computation – Simple Experiments in Simple Non-Memetic Genetic Algorithms (2001)
18. Cagnoni, S., Dobrzeniecki, A.B., Poli, R., Yanch, J.C.: Genetic Algorithm-based Interactive Segmentation of 3D Medical Images, Image and Vision Computing 17 (1999) 881–895
19. Bhanu, B., Lee, S.: Genetic Learning for Adaptive Segmentation, Kluwer Academic Press (1994)
20. Bhanu, B., Lee, S., Ming, J.: Adaptive Image Segmentation using a Genetic Algorithm, IEEE Transactions on Systems, Man, and Cybernetics 25(12), pp. 1543–1567. (1995)
21. Hwang, W., Chang, H.: Character Extraction from Documents using Wavelet Maxima, Image and Vision Computing. Volume 16, Issue 5 (1998) 307–315
22. Negishi, H., Kato, J., Hase, H., Watanabe, T.: Character Extraction from Noisy Background for an Automatic Reference System, Proceedings of the Fifth International Conference on Document Analysis and Recognition, Bangalore, India, 20–22 September (1999)
23. Vidal, R.: Old Text Reconstruction: An Artificial Intelligence Approach, Graduate Thesis, Facultad de Informática, Universidade da Coruña (1999)
24. Poli, R., Langdon, W.B.: Sub-machine-code Genetic Programming. In L. Spector, U.M.O'Reilly W.B.Langdon and P.J.Angeline, editors, Advances in Genetic Programming 3, MIT Press , chapter 13 (1999) 301–323
25. Adorni, G., Cagnoni, S.: Design of explicitly or implicitly parallel low-resolution character recognition algorithms by means of genetic programming, in Roy, R., Koppen, M., Ovaska, S., Furuhashi, T., Hoffmann, F. (eds.), Soft Computing and Industry: Recent Applications, (Proc. 6th Online Conference on Soft Computing). Springer (2002) 387–398

The Effectiveness of Cost Based Subtree Caching Mechanisms in Typed Genetic Programming for Image Segmentation

Mark E. Roberts

School of Computer Science, University of Birmingham, B15 2TT, UK
M.E.Roberts@cs.bham.ac.uk

Abstract. Genetic programming (GP) has long been known as a computationally expensive optimisation technique. When evolving imaging operations, the processing time increases dramatically. This work describes a system using a caching mechanism which reduces the number of evaluations needed by up to 66 percent, counteracting the effects of increasing tree size. This results in a decrease in elapsed time of up to 52 percent. A cost threshold is introduced which can guarantee a speed increase. This caching technique allows GP to be feasibly applied to problems in computer vision and image processing. The trade-offs involved in caching are analysed, and the use of the technique on a previously time consuming medical segmentation problem is shown.

1 Introduction

The search ability of GP comes at a considerable price. As each individual in the GP population is a program, it needs to be executed to test the solution, and this has to be done for every individual in the population for a number of generations. This introduces a bottleneck, as the programs must be interpreted (the overhead of a full compilation is usually too high), and as each of the functions is used hundreds of thousands of times, even a relatively simple set of functions can result in a long run time. For this reason, GP work is often limited to dealing with numerical operations, or simple symbolic manipulations. Now, however, computing power is on the verge of allowing systems to use more expensive operators in order to accomplish a wider variety of tasks. Surprisingly, very little work has been done on trying to reduce this execution time, with most people opting instead to simplify or modify their problem.

1.1 GP for Imaging Tasks

One such area is the use of GP for computer vision and image processing. A simple real-valued GP system may do around 5 or 6 arithmetic operations per node, whereas an image processing operation on a reasonable size image may perform several million arithmetic operations *per node*. Due to the huge amounts of information processing involved, operations involving images are very expensive, and

S. Cagnoni et al. (Eds.): EvoWorkshops 2003, LNCS 2611, pp. 444–454, 2003.
© Springer-Verlag Berlin Heidelberg 2003

doing them hundreds of thousands of times in a GP run requires an enormous amount of computation. Computing power has increased to a level where these sorts of problems can start to be explored, but until now, any substantial GP imaging tasks have become infeasible due to the time involved, or have been reduced to using low dimensional features extracted from the images rather than the images themselves. Work described in [1,2,3,4,5], are examples of systems that first pre-process the images and extract key features, and then use GP to perform a numerical optimisation using those features as terminals.

The use of images in GP does not require any change to the paradigm. The only change is in the datatypes and functions used. In an image based GP system, the function set would contain image processing operations, and the terminal set would include the input images. Only very few attempts have been made at using images as the main GP datatype in this way; examples include [6] and [7].

The ability to use images in GP makes it possible to attempt to evolve programs to tackle computer vision problems in a manner similar to supervised learning. The system can be given a set of inputs and a set of target outputs, and it has to find the common transformation that will convert inputs to targets. Figure 1 shows a dataset of this type; the system would have to find a way to create the outputs shown from the input images. This type of system would be especially useful to non-programmers, who could supply input and target images, and a program could be evolved for them to perform the task.

It is obvious that anything that can be done to speed up this type of system would be a great advantage. This work presents a subtree evaluation caching scheme that does this by minimising the re-evaluation of duplicated subtrees, both within and across generations. As an example, the use of this system on a non-trivial binary segmentation problem will be shown, where the system has to learn a general segmentation algorithm from a set of user-supplied manual segmentations.

2   Basic Principle

It is a fundamental principle of genetic programming that subtrees present in one generation will be present in subsequent generations. Normally these trees are wastefully re-evaluated in spite of being exact copies of trees that have already been evaluated.

The basic principle of the caching mechanism is to store the results of subtree evaluations so that they do not need to be re-evaluated if they appear again in the same or subsequent generations. This may seem like an obvious thing to do, but the use of this type of system has very rarely been reported in the literature. Systems that did use it were reported by Handley [8] and Ehrenburg [9] who used graphs to represent the entire population. That approach resulted in reduced space requirements as duplicate trees were only stored once, as well as a time reduction as the evaluations were cached at the subtree's root node. Langdon [10] also used a caching method which stored ADF evaluations. However, systems of this type have not been widely used or developed further, and their potential

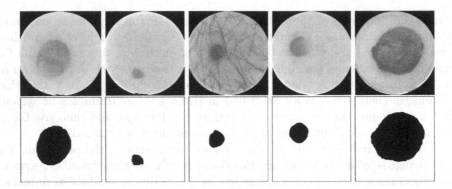

Fig. 1. Sample mole images, showing the wide variety of shapes and spatial variations in grey levels, and their target segmentations which would be used in a supervised learning situation

has not yet been fully realised, especially in areas such as imaging where the benefits are much larger due to the more costly operators involved. Obviously caching can only be applied where evaluations have no side effects.

3   System Description

The basic GP system used is a strongly typed GP system similar to the one described by Montana [11], in order to achieve the flexibility required for imaging operations. In this case a typed system is used in order to allow numerical and coordinate data (from now on called *decimals* and *points* respectively) to be passed to image processing operators. This allows much greater flexibility than could be achieved by just using a standard single-typed GP. The entire system is written in C++ and includes hand optimised image processing operations to try to minimise the cost of these operators.

The individuals are represented in a classical tree-based structure, and are initially generated at random using the ramped half-and-half method [12] up to an initial depth of 7. The root node of each tree is constrained to be a function that produces an image as an output. Individuals are selected to undergo genetic operations using a tournament selection scheme. Genetic operations, type restricted crossover, mutation and direct copying into the next generation, are applied to the selected individuals probabalistically in order to produce the new generation.

3.1   Experimental Setup

The operators used in the function set consisted of both image processing operations and real number arithmetic operations. The image processing functions used included edge filtering and edge strength at given angles, dilation and erosion, merging, thresholding, logical and/or operations, normalisation, and region intensity functions (mean, min, max). The terminal set consisted of decimal constants, coordinate point constants, and image constants (representing the input image). The decimal and point constants are created randomly.

These operators were chosen as they represent a generic set of operators that are often used by humans attempting to hand-segment images. The point and decimal constants are included for the same reason.

3.2 Fitness Evaluation

The fitness evaluation uses a function proposed by Poli [6] for similar segmentation problems, summed over all of the N images in the training set. It effectively implements the well known measures of *sensitivity*, the ability to correctly segment sections of the image containing the feature, and *specificity*, the ability to correctly segment the sections which do not contain the feature. These are formally defined using measures of true positives (TP), true negatives (TN), false positives (FP) and false negatives (FN) which are used in this fitness function.

$$ f = \sum_{i=1}^{N} FP + FN \exp\left(10\left(\frac{FN}{p} - \alpha\right)\right) \tag{1} $$

where p represents the number of pixels present in the target segment(s).

The α parameter in this function allows the relative importance of sensitivity and specificity to be varied. It is important to consider these properties, as a simple pixel difference calculation could easily miss features; consider the case of a very small feature on a uniform background – a very high fitness could be achieved by just matching the background, but totally missing the feature. Also, as in Poli's work, the system is not trying to find solutions whose output matches the target output – instead there is a wrapper function which first thresholds the image. So, in actual fact the system is trying to find the solution that, when thresholded, best matches the target output. This is a much more effective method.

3.3 Caching

A simple caching system would store every subtree evaluation for later use (and this was done in [8]), but this technique is not feasible when working with images for several reasons -

1. In image based GP, the evaluations of the subtree may actually be images which take thousands of times more storage space than a simple single numerically-typed GP system.
2. In some cases it is more efficient to just evaluate the tree rather than find and copy the results from somewhere else.

Given that cache space is limited, the problem arises of making the most efficient use of a limited amount of space. Two important issues arise from this; determining whether it would be worthwhile to cache a tree, both in terms of time and space, and determining when to remove a tree from the cache in order to free space for other trees. These issues are addressed in section 3.4.

3.4   Cache Implementation

The cache is implemented as a simple hash table. Each entry in the table stores a copy of the tree and the result of evaluating that tree. Near-unique hashes are generated by a simple recursive function. Using the caching system, the general node evaluation procedure is that we look for the tree in the cache and return it if it is there. If not, the tree is evaluated, added to the cache and then returned.

If the subtree's root node outputs an image then the result is cached to disk, if it returns any other type it is stored in memory as the storage requirements for these other types are minimal. This disk caching introduces the most significant inefficiency into the system, and will be discussed further in section 5.

The hash lookup functions are very efficient, and the within-bucket tree equality checks are also extremely fast as this particular implementation of the typed GP system stores a lot of information about the subtrees below it which allows very early termination during equality checks.

Evaluation Contexts. This basic scheme will obviously only work if the terminals in the expression have the same value every time the tree is evaluated. However, the same tree is often evaluated with different instantiations of its variables. For instance in the segmentation example, the value of the image constant terminal could be any one of the 5 images in the training set. To solve this, the system instead stores an array of results instead of just a single one, representing each of the different contexts that the tree can be evaluated in.

Removing Cache Entries. Due to the nature of crossover, subtrees in generation n are guaranteed to be found in generation $n + 1$, and this is where the benefit of caching comes from. However, with limited cache space, the question arises as to how long subtrees should kept in the cache. Keeping too many items in the cache will use up valuable space, but keeping too few items will reduce the number of cache hits.

In order to selectively remove less useful items from the cache to free space for newer items, after every generation the system removes all cache items that have not been needed during the evaluation of the last δ generations, where δ is one of the run parameters, and in this case was set to 3. This simple behaviour is what we would intuitively expect – if a subtree has not been seen in the population for a while, then it is unlikely that it will appear again.

Deciding What to Cache. As the cache space is limited, a decision needs to be made as to whether or not it is worthwhile to cache a result. A simple scheme would be to only cache subtrees with more than a given number of nodes, but due to the nature of the operators in the function set this may give highly undesirable results. The great variety of operators present means that there is a difference of several orders of magnitude in the time complexities of the various functions. A 4-node tree performing an orientation filter would do hundreds of times more computation than a 1000 node tree doing simple decimal arithmetic. Clearly tree size alone is not a good measure of the value of caching.

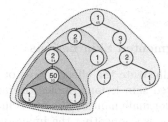

Fig. 2. The overcaching problem illustrated. The large numbers are the node costs and the smaller numbers are the entire subtree costs. The expensive deepest subtree causes all highlighted trees above it to be unnecessarily cached.

3.5   Cost Based Caching

A more sensible scheme for deciding whether or not to cache a subtree is to take into account a measure of how long the tree would actually take to evaluate and how long it would take to cache. This can be easily achieved by timing each of the operators, over thousands of runs, and timing the caching operation.

For any subtree, its estimated evaluation time can be worked out by recursively summing the timing values for each of its subtrees. If this is greater than the time of a cache/uncache operation then a speed increase can be guaranteed. In actual fact setting the threshold as the time of a single cache/uncache operation is not a wise decision, as the tree will hopefully be uncached (i.e. used in evaluations) a number of times. The cost threshold value is therefore set as the time of a cache operation plus the time of c cache retrieval operations, where c is a run parameter, set in this case to 5.

A Problem: Over-caching. Using the method described above, the caching would be very inefficient. Using a recursive sum of subtree costs will give a cost value for the subtree, but if an expensive subtree deep down in the tree is cached and all other operators above it are not expensive, then the inexpensive subtrees will be unnecessarily cached. This is illustrated in Figure 2. If we assume that cost threshold is set as 20, then the first subtree has a cost 52 and will correctly be cached. However, all of the subtrees that take input from this subtree will inherit its cost and be pushed over the threshold even if they themselves are inexpensive. This will lead to a severe inefficiency, as it would be far quicker to just evaluate these inexpensive trees. This means that the program will be slowed down by caching the trees, then slowed down again by retrieving them from the cache, and cache space will also be wasted.

This problem can be avoided by zeroing the cost of a node once it is cached. In the case of Figure 2, as soon as the subtree with cost 52 is evaluated its result is cached and the cost is reset to zero. All trees including it will then see it as a zero cost node and none of them will be cached unless the threshold is exceeded by some other uncached nodes.

4   Results

4.1   Detection of Pigmented Skin Lesions

The problem chosen to illustrate the effectiveness of the caching scheme is the segmentation of pigmented skin lesions (moles) in images. Research into automatic diagnosis systems for malignant melanoma is increasing, and the first step in any system of this type is segmenting the image into lesion and background groups. This is a very difficult task considering the huge variation in size, shape, colour, and lighting conditions that is present in these types of images [13]. Some examples of these are shown in Figure 1 along with their target segmentations.

As stated previously, the performance on the task is not the primary focus of this work, but it is worth presenting the success on the problem as it shows how the caching system can be used to attempt to solve real-world problems which were previously too time consuming. Each tree in the population is evaluated

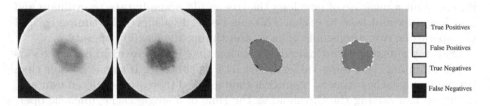

Fig. 3. Two sample segmentations from the unseen test-set

on each of the five images in the training set, shown in Figure 1. The fitness is calculated as described in section 3.2. Performance on this problem is gauged using the measures of sensitivity and specificity. The best solution contained 95 nodes and had a fitness of 419.1 (calculated using Equation 1 on the 5 training images). On the training set, all results had a specificity above 99% and a sensitivity above 90%. The same solution program was run on an unseen test set containing 67 similar images. The performance was highly successful, with an average sensitivity of 92.3 and an average specificity of 97.2. This performance shows a great deal of potential as the system was able to find a solution which generalised from such a small training set. Figure 3 shows two examples from test-set which were successfully segmented in spite of their faded borders.

4.2   Cache Performance

The caching system was tested against an uncached run (with identical random seed) in several different configurations designed to test the caching system in a variety of situations. Four configurations were chosen which give a variety of different behaviours. These are shown below

1. Population size of 100 for 10 generations with a maximum tree depth of 15
2. Population size of 100 for 10 generations with a maximum tree depth of 17

3. Population size of 1000 for 50 generations with a maximum tree depth of 17
4. Population size of 2000 for 50 generations with a maximum tree depth of 17

For each of the tests, the total number of subtree evaluations was counted and the total elapsed time was measured. This measure of time *does* include all time spent on disk access due to caching, and therefore gives us a true indication of the time taken including all the overheads of the caching. The results shown are of the best run (in terms of fitness) from a set of 20 runs.

Reduction in number of evaluations. The number of subtrees evaluated in cached versions of the problems was obviously always lower than in the equivalent uncached version. Table 1(a) shows the percentage reductions in the number of subtree evaluations.

It can be seen from this data that the benefit of caching increases with the size of the problem, in terms of the number of trees that need evaluating. This can be explained due to the fact that as a GP progresses, larger trees are produced, due to the improvement of solutions, and to bloat. This effect is increased as the generations progress and so in experiments 3 and 4 there are many more larger trees present. This means that in the population there will be more trees that exceed the cost threshold, and therefore there is more benefit to be gained from caching them.

The reductions that can be seen are substantial. In experiments 3 and 4, reductions of 18 million and 103 million trees were produced. These represent 60% and 66% reductions in the number of subtrees evaluated. This is significant, as experiments 3 and 4 represent real sets of parameters which can solve this problem whereas, experiments 1 and 2 are toy sets of parameters, introduced to quantify scalability.

One of the biggest implications of this reduction is illustrated well by the graphs shown in Figure 4. These show the number of subtree evaluations needed in experiments 3 and 4, both per generation and cumulatively. As expected, the cumulative plot for the uncached versions rise steeply, confirming the common sense argument that tree size increases non-linearly as the generations progress. The plot showing the cached performance is virtually linear. This implies that the number of subtrees required in each generation in the cached version is roughly the same, and that the caching lessens the impact of increasing tree sizes. This can also be seen in the per-generation plots as the cached lines "smooth-out" the effect of the increasing tree size. Even at points where the uncached plot rises

Table 1. Performance of cached and uncached systems on the 4 problem configurations. (a) shows the number of subtrees evaluated. (b) shows the decrease in time (in seconds).

<table>
<tr><td colspan="4">(a)</td><td colspan="4">(b)</td></tr>
<tr><td>Config.</td><td>Uncached</td><td>Cached</td><td>% Decrease</td><td>Config.</td><td>Uncached</td><td>Cached</td><td>% Decrease</td></tr>
<tr><td>1</td><td>173460</td><td>149225</td><td>14.0</td><td>1</td><td>894</td><td>880</td><td>1.57</td></tr>
<tr><td>2</td><td>639845</td><td>344201</td><td>46.2</td><td>2</td><td>3444</td><td>2695</td><td>21.7</td></tr>
<tr><td>3</td><td>31202930</td><td>12299365</td><td>60.1</td><td>3</td><td>149016</td><td>75496</td><td>49.3</td></tr>
<tr><td>4</td><td>155174805</td><td>51333005</td><td>66.9</td><td>4</td><td>531468</td><td>254260</td><td>52.2</td></tr>
</table>

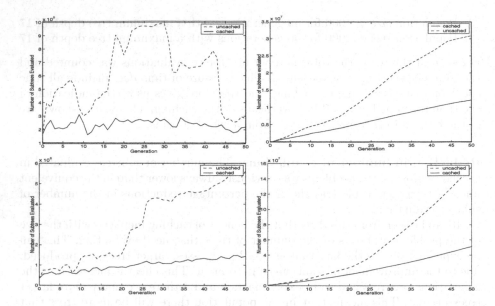

Fig. 4. Left graphs: show the number of individuals evaluated per generation (configs. 3 and 4). Right graphs: show the cumulative number of subtrees evaluated in cached and uncached versions. The cached versions stay linear in spite of the larger trees

very steeply, the cached version still stays largely linear. For any GP system this is a very good result. For an expensive run involving imaging operations it shows the potential for a large increase in the size of problems that could be tackled.

Reduction in time. The reductions in subtree evaluations described in the previous section are very encouraging, but are not very useful unless there is a corresponding decrease in the actual elapsed time i.e. the cost of caching is not greater than the actual evaluation cost. Ideally the percentage decrease in time would be exactly the same as the percentage decrease in the number of subtrees evaluated, but this would require that all of the caching operations happened instantaneously. The difference between the two percentages therefore represents the overheads of the caching operations. Table 1b shows the time in seconds to complete each of the experiments, for both cached and uncached versions, and the percentage decrease that this represents.

5   Discussion

Factors affecting caching. The effectiveness of caching is highly dependent on the problem being tackled, because that problem defines the configuration of the GP system. The biggest factor is probably the number of unique trees that can be created, which is defined by the size and composition of the function and terminal sets, and the maximum allowed depth of tree. For example if there was only one function of arity 2, two terminals, and a maximum depth of 1, then all 4 unique trees could be cached. If however, there were 20 functions, 10 terminals, and a maximum depth of 17, then only a fraction of the unique trees could be stored. Ideally, every unique tree encountered in the run, with a cost that exceeds the threshold, would be stored, but with limited storage space this cannot happen and cache elements need to have an age limit.

In this problem 13 functions were used along with 3 types of random constants, in trees up to 17 deep. This means that there are a huge number of unique trees, and it is a very encouraging result to see the caching mechanism working so well on this problem. On a simpler problem, with fewer unique trees, the caching effect would be much more profound.

The population size chosen has a big impact, as a larger population will have a higher number of subtrees above the cache cost threshold. With limited storage space, this would have to be compensated for by reducing the age limit and/or increasing the cost threshold. Obviously the diversity of the population will also have a bearing in the same way.

The age limit of the cache elements again has an influence on the effectiveness of the cache. With unlimited storage space, every tree could be stored indefinitely in case it was encountered again, thus totally minimising re-evaluations. However, when this is not the case, the age limit needs to be set as high as possible within the constraints of the available storage space.

Cache efficiency. The efficiency of the caching system, i.e. the disparity between the reduction in evaluations and the reduction in time, is the critical factor in the viability of this sort of system. The system presented here has shown a real time reduction in all of the experiments, but not as much as the reduction in evaluations would imply is possible. This is mostly down to the use of hard disk space for the storage of image type caches. The efficiency could be greatly improved by storing the images in memory rather than on disk which is several orders of magnitude faster, but these runs required about 25 gigabytes of disk space for the cache, and this amount of memory will not be available for several years on standard computers.

With the current relative speed of processors and hard disks, this technique provides a significant saving. Obviously, this relative speed may change as new technologies become available and caching may become more or less beneficial depending on which way the balance shifts.

6   Conclusions

Although there are other techniques known to speed up the GP process, they are often limited to more restrictive representations (such as some linear GP systems) which are not as suitable for this type of imaging task. This work has presented a caching system for tree-based genetic programming which, although demonstrated on a specific problem, is a general method which speeds up image based GP runs in a variety of situations. The use of the caching system minimises the effects of normal tree growth and bloat. The 50% reduction in run time is a significant result and brings some GP imaging problems into a timeframe in which they can be properly explored. Also, unlike previous caching systems, this work presents a method which does not blindly cache every subtree, but only those that would result in a performance increase.

References

1. Tackett, W.A.: Genetic programming for feature discovery and image discrimination. In: Proceedings of the 5th International Conference on Genetic Algorithms, ICGA-93, Morgan Kaufmann (1994) 303–309
2. Johnson, M., Maes, P., Darrel, T.: Evolving visual routines. In: Proceedings of the Fourth International Workshop on the Synthesis and Simulation of Living Things, MIT Press (1994) 198–209
3. Ross, B.J., Fueten, F., Yashkir, D.Y.: Edge detection of petrographic images using genetic programming. In Whitley, D., Goldberg, D., Cantu-Paz, E., Spector, L., Parmee, I., Beyer, H.G., eds.: Proceedings of the Genetic and Evolutionary Computation Conference, San Francisco, USA, Morgan Kaufmann (2000) 658–665
4. Agnelli, D., Bollini, A., Lombardi, L.: Image classification: an evolutionary approach. Pattern Recognition Letters 23 (2002) 303–309
5. Roberts, S.C., Howard, D.: Genetic programming for image analysis: Orientation detection. In Whitley, D., Goldberg, D., Cantu-Paz, E., Spector, L., Parmee, I., Beyer, H.G., eds.: Proceedings of the Genetic and Evolutionary Computation Conference, San Francisco, USA, Morgan Kaufmann (2000) 651–657
6. Poli, R.: Genetic programming for feature detection and image segmentation. In Fogarty, T., ed.: Proceedings of the AISB'96 Workshop on Evolutionary Computation. Volume 1143 of Lecture Notes in Computer Science., Springer (1996) 110–125
7. Belpaeme, T.: Evolution of visual feature detectors. In Poli, R., Cagnoni, S., Voigt, H.M., Fogarty, T., Nordin, P., eds.: Late Breaking Papers at EvoIASP'99, University of Birmingham Technical Report CSRP-99-10 (1999)
8. Handley, S.: On the use of a directed acyclic graph to represent a population of computer programs. In: Proceedings of the 1994 IEEE World Congress on Computational Intelligence, Orlando, Florida, USA, IEEE Press (1994) 154–159
9. Ehrenburg, H.: Improved direct acyclic graph handling and the combine operator in genetic programming. In Koza, J.R., Goldberg, D.E., Fogel, D.B., Riolo, R.L., eds.: Genetic Programming 1996: Proceedings of the First Annual Conference, Stanford University, CA, USA, MIT Press (1996) 285–291
10. Langdon, W.B.: Pareto, population partitioning, price and genetic programming. Research Note RN/95/29, University College London, UK (1995)
11. Montana, D.J.: Strongly typed genetic programming. Evolutionary Computation 3 (1995) 199–230
12. Koza, J.R.: Genetic Programming: On the Programming of Computers by Means of Natural Selection. MIT Press (1992)
13. Ganster, H., Pinz, A., Rohrer, R., Wildling, E., Binder, M., Kittler, H.: Automated melanoma recognition. IEEE Transactions on Medical Imaging 20 (2001) 233–239

Pixel Statistics and False Alarm Area in Genetic Programming for Object Detection

Mengjie Zhang, Peter Andreae, and Mark Pritchard

School of Mathematical and Computing Sciences,
Victoria University of Wellington, Wellington, New Zealand
{mengjie,pondy,markp}@mcs.vuw.ac.nz
http://www.mcs.vuw.ac.nz/~mengjie,pondy

Abstract. This paper describes a domain independent approach to the use of genetic programming for object detection problems. Rather than using raw pixels or high level domain specific features, this approach uses domain independent statistical features as terminals in genetic programming. Besides position invariant statistics such as *mean* and *standard deviation*, this approach also uses position dependent pixel statistics such as *moments* and local region statistics as terminals. Based on an existing fitness function which uses linear combination of detection rate and false alarm rate, we introduce a new measure called "false alarm area" to the fitness function. In addition to the standard arithmetic operators, this approach also uses a conditional operator *if* in the function set. This approach is tested on two object detection problems. The experiments suggest that position dependent pixel statistics computed from local (central) regions and nonlinear condition functions are effective to object detection problems. Fitness functions with false alarm area can reflect the smoothness of evolved genetic programs. This approach works well for the detecting small regular multiple class objects on a relatively uncluttered background.

1 Introduction

As more and more images are captured in electronic form, the need for programs which can find objects of interest in a database of images is increasing. For example, it may be necessary to find all tumors in a database of x-ray images, all cyclones in a database of satellite images or a particular face in a database of photographs. The common characteristic of such problems can be phrased as "Given $subimage_1, subimage_2, ..., subimage_n$ which are examples of the objects of interest, find all images which contain this object and its location(s)". Figure 5 shows examples of problems of this kind. In the problem illustrated by figure 5 (b), we want to distinguish and find the centers of all of the New Zealand 5 cent and 10 cent coins. Examples of other problems of this kind include target detection problems [1], [2], [3] where the task is to find, say, all tanks, trucks or helicopters in an image. Unlike most of the current work in the object recognition area, where the task is to detect only objects of one class [1], [2], our objective is to detect objects from a number of classes.

S. Cagnoni et al. (Eds.): EvoWorkshops 2003, LNCS 2611, pp. 455–466, 2003.

Domain independence means that the same method will work unchanged on any problem, or at least on some range of problems. This is very difficult to achieve at the current state of the art in computer vision because most systems require careful analysis of the objects of interest and a determination of which features are likely to be useful for the detection task. Programs for extracting these features must then be coded or found in some feature library. Each new detection/vison system must be hand-crafted in this way. Our approach is to use easily computed pixel statistics such as mean and variance of the pixels in a sub-image and to evolve the programs needed for object detection.

There have been a number of reports on the use of genetic programming in object detection and classification [3], [4], [5], [6], [7], [8], [9]. Typically, simple features such mean and standard deviation or even high level features are used to form terminals, and the four arithmetic operations form the function set. However, genetic programming with pixel statistics of only mean and standard deviation for object detection often results in many false alarms. The main reason is that mean and standard deviation are position and rotation invariant and not effective for object localisation.

The function set consisting of the four standard arithmetic operations is simple, domain independent and easy to use. However, it is not effective at expressing conditional programs — programs with a different "shape" in different parts of feature space. The programs constructed with these functions are also often difficult to understand and interpret.

The main objective of a detection system is to achieve a high detection rate and a low false alarm rate, so that fitness functions in genetic programming for object detection are often based on detection rate and false alarm rate or similar measures [10]. A problem with such fitness functions is that they may not reflect small improvements of evolved genetic programs, which makes the genetic search much harder.

1.1 Goals

This paper investigates three new developments in domain independent object detection using genetic programming.

- The first is to explore the use of position dependent pixel statistics to improve the object detection performance.
- The second is to investigate the effect of adding a conditional *if* operator to the function set on genetic search and on the comprehensibility of the constructed programs.
- The third is to explore the effect on the evolutionary process of an additional measure in the fitness function.

2 The Approach

2.1 Overview of the Approach

Figure 1 shows an overview of this approach, which has a learning process and a testing procedure. In the learning/evolutionary process, the evolved genetic programs use a square input field which is large enough to contain each of the objects of interest. The programs are applied, in a moving window fashion, to the entire images in the training set to detect the objects of interest. In the test procedure, the best evolved genetic program obtained in the learning process is then applied to the entire images in the test set to measure object detection performance.

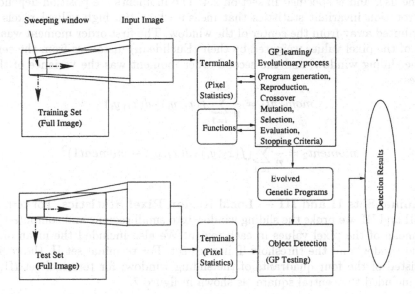

Fig. 1. An overview of the GP approach for multiple class object detection.

In this system, we use tree-structured programs to represent genetic programs. The ramped half-and-half method was used for generating the programs in the initial population and for the mutation operator. The proportional selection mechanism and the reproduction, crossover and mutation operators were used in the learning process.

In the remainder of this section, we address the other aspects of the learning/evolutionary system: (1) Determination of the terminal set; (2) Determination of the function set; (3) Development of a classification strategy; (4) Construction of the fitness measure; (5) Selection of the input parameters and determination of the termination strategy.

2.2 Terminals

For object detection problems, terminals generally correspond to image features. The simplest statistics (mean and variance of the pixel values) are position and rotation invariant statistics, representing overall brightness/intensity and the contrast of a region of the image. We considered two approaches to introducing position dependence. The first was to use statistics that depended on the positions as well as the values of the pixels. The second was to use position invariant pixel statistics on smaller, local regions of the sliding window. We investigated three terminal sets.

Terminal Set I — Whole Window Pixel Statistics. The first terminal set consisted of the *mean, variance, first order moment, second order moment* of the whole window, and a constant threshold value *T*. The threshold T is set by the user and is specified in section 2.4. The moments are position dependent but rotation invariant statistics that measure how the high valued pixels are distributed away from the center of the window. The first order moment was the sum of the pixel values weighted by their (Euclidean) distances from the center of the sliding window, and the second order moment was the variance of these values:

$$moment_1 = \frac{1}{n} \sum_{i=1}^{n} f(x_i, y_i) \cdot d(x_i, y_i) \tag{1}$$

$$moment_2 = \frac{1}{n} \sum_{i=1}^{n} (f(x_i, y_i) \cdot d(x_i, y_i) - moment1)^2 \tag{2}$$

Terminal Sets II and III — Local Region Pixel Statistics. For terminal sets II and III, we broke the sliding window into small regions and computed just the mean of the pixel values in each region. We also included the mean of the whole window and the threshold T in each set. For terminal set II, the regions consisted of the four quadrants of the sliding window; for terminal set III, we also included the central square, as shown in figure 2.

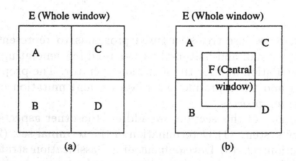

Fig. 2. Local region terminals. (a) Terminal set II; (b) Terminal set III.

2.3   Functions

In the function set, the four standard arithmetic and a conditional operation was used to form the non-terminal nodes:

$$FuncSet = \{+, -, *, /, if\}$$

The $+$, $-$, and $*$ operators have their usual meanings — addition, subtraction and multiplication, while $/$ represents "protected" division which is the usual division operator except that a divide by zero gives a result of zero. Each of these functions takes two arguments. The *if* function takes three arguments. The first argument, which can be any expression, constitutes the condition. If the first argument is positive, the *if* function returns its second argument; otherwise, it returns its third argument. The *if* function allows a program to contain a different expression in different regions of the feature space, and allows discontinuous programs, rather than insisting on smooth functions.

2.4   Object Classification Strategy

The output of a genetic program in a standard GP system is a floating point number. Generally genetic programs can perform one class object detection tasks quite well where the division between positive and negative numbers of a genetic program output corresponds to the separation of the objects of interest (of a single class) from the background (non-objects). However, for multiple class object detection problems described here, where more than two classes of objects of interest are involved, the standard genetic programming classification strategy mentioned above cannot be applied.

In this approach, we used a different strategy called *program classification map*, as shown in figure 3, for the multiple class object detection problems [10]. Based on the output of an evolved genetic program, this map can identify which class of the object located in the current input field belongs to. In this map, m refers to the number of object classes of interest, v is the output value of the evolved program and T is a constant defined by the user, which plays a role of a threshold.

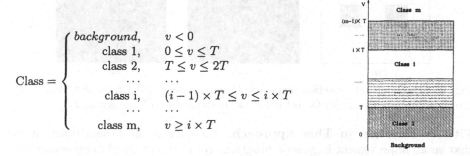

$$
\text{Class} =
\begin{cases}
background, & v < 0 \\
class\ 1, & 0 \le v \le T \\
class\ 2, & T \le v \le 2T \\
\cdots & \cdots \\
class\ i, & (i-1) \times T \le v \le i \times T \\
\cdots & \cdots \\
class\ m, & v \ge i \times T
\end{cases}
$$

Fig. 3. Mapping of program output to an object classification.

2.5   Fitness Function

Fitness functions used for object detection. The goal of object detection is to achieve a high detection rate and a low false alarm rate. In genetic programming, this typically needs a "multi-objective" fitness function. An example fitness function [10] is:

$$fitness(DR, FAR) = W_d * (1 - DR) + W_f * FAR \qquad (3)$$

where DR is the Detection Rate (fraction of objects correctly identified and localised) and FAR is the False Alarm Rate (the number of incorrectly identified or localised objects as a fraction of the number of true objects). The parameters W_d, W_f reflect the relative importance between detection rate and false alarm rate. Note that the fitness is zero for the ideal case.

Although such a fitness function accurately reflects the performance measure of an object detection system, it is not very smooth. In particular, small improvements in an evolved genetic program may not be reflected in any change to the fitness function. The reason is the clustering process that is essential for the object detection — as the sliding window is moved over a true object, the program will generally identify an object at a cluster of window locations where the object is approximately centered in the window. It is important that the set of positions is clustered into the identification of a single object rather than the identification of a set of objects on top of each other.

A poor program may (incorrectly) identify a large cluster of locations as an object. A modified program may identify a smaller cluster of locations (as shown in figures 4 (b) and (c)). Although the second program is better than the first, and is closer to the ideal program, it has exactly the same FAR since both programs have one false positive. A fitness function based solely on DR and FAR cannot correctly rank these two programs, which means that the evolutionary process will have difficulty selecting better programs.

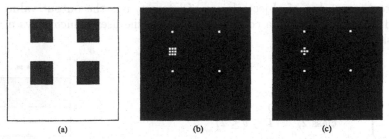

(a)   (b)   (c)

Fig. 4. Sample object detection maps. (a) Original image; (b) Object detection map produced by program 1; (c) Object detection map produced by program 2.

Fitness Function in This Approach. To smooth the fitness function so that small improvement in genetic programs could reflect small improvement in fitness measure while large improvement in genetic programs could reflect large improvement in fitness, we introduced a new measure, *false alarm area*, into the above fitness function.

The new fitness of a genetic program is obtained as follows:

1. Apply the program as a moving $n \times n$ window template (n is the size of the input field) to each of the training images and obtain the output value of the program at each possible window position. Label each window position with the 'detected' object according to the object classification strategy described in figure 3. Call this data structure a detection map.

2. Find the centres of *objects of interest only* by the clustering algorithm:
   - Scan the detection map for an object of interest. When one is found mark this point as the centre of the object and continue the scan. Skip pixels in $n/2 \times n/2$ square to right and below this point.
3. Match these detected objects with the known locations of each of the desired true objects and their classes. A match is considered to occur if the detected object is within *tolerance* pixels of its known true location. Here, the *tolerance* is a constant parameter defined by the user.
4. Calculate the detection rate DR, the false alarm rate FAR, and the false alarm area FAA of the evolved program. The DR refers to the number of small objects correctly reported by a detection system as a percentage of the total number of actual objects in the image(s). The FAR, also called false alarms per object, refers to the number of non-objects incorrectly reported as objects by a detection system as a percentage of the total number of actual objects in the image(s). The FAA is the number of false alarm pixels which are not object centres but are incorrectly reported as object centres before (without) clustering.
5. Compute the fitness of the program according to equation 4.

$$fitness(FAR, DR) = A \cdot (1 - DR) + B \cdot FAA + C \cdot FAR \qquad (4)$$

where A, B, C are constant weights which reflect the relative importance of detection rate versus false alarm rate versus false alarm area.

2.6 Parameters and Termination Criteria

The parameter values used in this approach are shown in table 1.

In this approach, the learning/evolutionary process is terminated when one of the following conditions is met:

- The detection problem has been solved on the training set, that is, all objects in each class of interest in the training set have been correctly detected with no false alarms. In this case, the fitness of the best individual program is zero.
- The number of generations reaches the pre-defined number, *max-generations*.

3   Image Databases

We used two different databases in the experiments. Example images are given in figure 5. Database 1 (Easy) was generated to give well defined objects against a uniform background. The pixels of the objects were generated using a Gaussian generator with different means and variances for each class. There are two classes of small objects of interest in this database: circles and squares. Database 2 (Coins) was intended to be somewhat harder and consists of scanned images of New Zealand coins. There are two object classes of interest: the 5 cent coins and 10 cent coins. The objects in each class have a similar size but are located

Table 1. Parameters used for GP training for the three databases.

Parameter Kinds	Parameter Names	Easy Images	Coin Images
Search	population-size	300	500
	initial-max-depth	4	5
	max-depth	8	12
Parameters	max-generations	150	200
	input-size	16×16	20×20
Genetic	reproduction-rate	10%	5%
	cross-rate	65%	70%
	mutation-rate	25%	25%
Parameters	cross-term	15%	15%
	cross-func	85%	85%
Fitness	T	80	100
	A	50	50
Parameters	B	1000	1000
	C	100	200
	tolerance (pixels)	2	2

at arbitrary positions and with different rotations. Although the task is simple for humans with the original scanned image (Fig 5 (b)), we used a low resolution version (Fig 5 (c))that reduced the computation time, but resulted in a significantly more difficult task.

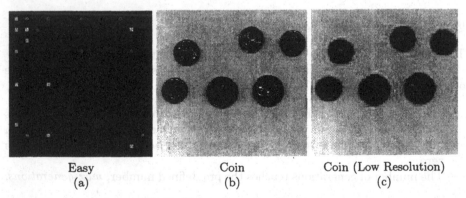

Easy	Coin	Coin (Low Resolution)
(a)	(b)	(c)

Fig. 5. Object Detection Problems

4   Results

This section describes the detection results. For all cases, the experiments were repeated 10 times and the average results on the *test set* were presented.

Table 2. Object detection results for the easy images.

Easy Images			Object Classes	
			circles	squares
Best Detection Rate(%)			100	100
False Alarm Rate (%)	GP1	Term set I	240	0
	GP2	Term set II	0	0
	GP3	Term set III	0	0

4.1 Easy Images

Table 2 shows the best results of the GP approach with the three different terminal sets for the easy images. Detecting squares was relatively straight forward,
where the GP with all the terminal sets achieved a 100% detection rate with no
false alarms. For circles, GP2 and GP3 achieved ideal results, but GP1 resulted
in 240% false alarm rate at a detection rate of 100%. GP2 took 147 generations
to achieve the ideal result but GP3 only used 87 generations. This suggests that
the position invariant statistics on local regions are more effective than the position dependent statistics on the whole window. The addition of the central local
region is also helpful for object localisation.

Sample object detection maps which gives an intuitive view of the detection
results using GP2 is shown in figure 6. In figures 6 (b) and (c), a white point
(pixel) represents a match between a detected object centre and a desired object
centre for circles and squares. This figure clearly shows that GP2 achieved ideal
performances for detecting objects of interest in these images.

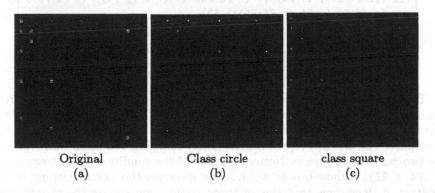

Original	Class circle	class square
(a)	(b)	(c)

Fig. 6. Sample Object detection map for the easy images.

4.2 Coin Images

Experiments with the coin images gave similar results to the easy images, as
shown in table 3. GP1 resulted in a large number of false alarms for both classes.
While GP2 greatly improved the performance, it still produced some false alarms
for detecting the 5 cent coins. GP3 achieved the best (ideal) performance. These

Table 3. Object detection results for the coin images.

Coin Images			Object Classes	
			coin005	coin010
Best Detection Rate(%)			100	100
False Alarm	GP1	Term set I	440	210
Rate (%)	GP2	Term set II	34	0
	GP3	Term set III	0	0

results confirmed that the moment statistics on the whole input window did not provide sufficient position dependence for this object detection problem. Statistics on local regions, especially if the central region was included, could do a better job. The results also demonstrated that the augmented fitness function worked well for the two object detection problems.

4.3 Comprehensibility

To check the role the conditional *if* operator played in the evolutionary process, an evolved genetic program produced by GP2 for the easy images is shown in figure 7. In this figure, F1, F2, F3, F4, F5 represent the mean of local regions A, B, C, D and the whole window (E) on figure 2, respectively. Note that this program has had some manual simplification to remove the large amount of redundancy in the original program.

```
(if (if (- (+ T T) F4) (+ (- F3 F5) (- (+ T T) F4)) (+ F1 F5))
    (+ (/ F2 (- F1 (+ F4 F5))) F3)
    (- (* F3 F4) T)
)
```

Fig. 7. A sample genetic program.

This program is considerably simpler than the programs generated in previous experiments that did not include an *if* operator. The program consists of two parts — two different expressions that will be used in different parts of the feature space. The first line represents the (fairly complex) condition that segments the two parts of the space. Notice the part of the condition that is equivalent to (F4 < 2T). Although it is still hard to interpret the exact meaning of this program, it does show that this program could approximate the classification rule — we have two classes of interest and the thresholds T and 2T occurred in this program which made the program have three ranges (for two classes of interest plus the background).

5   Conclusions

The goal of this paper was to investigate three new developments in multiclass, domain independent object detection using genetic programming. The approach was tested on two object detection problems and achieved good results.

- We developed three different terminal sets based on domain independent, statistical image features. The position dependent pixel statistics on the four quadrants of a sliding window did a better job than the statistics on the whole window only. In particular, statistics containing a central local region seem more effective for these detection problems than without. Our results suggest that, in genetic programming for object detection problems, input terminals should include local region pixel statistics not just the whole input sliding window.
- We augmented the function set to include a conditional *if* function. The evolved genetic programs with an *if* function were more comprehensible than without. In addition these programs could simulate a different "shape" in different parts of feature space. Our results suggest that, for multiclass object detection problems, a function set should include an *if* function.
- We modified the fitness function by including a new measure called false alarm area. A fitness function should be an accurate, smooth measure of the evolved genetic programs. The false alarm area measure was able to identify small improvements to genetic programs that could not be detected without the measure. The new fitness function that included false alarm area was therefore smoother and able to reflect both large and small improvements of genetic programs.

This approach has a number of limitations, which need to be considered and improved in the future:

- The programs generated by the evolutionary process contain a large amount of redundancy. Simplification of these programs should be considered during the evolutionary process so that the search space can be greatly reduced.
- The training process is relatively long (on the order of several days) due to a large number of input images (examples) applied in the window moving procedure. A better method of producing representative training examples should be considered.
- While this approach did very well on the two detection problems here, it should be tested on more difficult object detection problems in the future.

References

1. Paul D. Gader, Joseph R. Miramonti, Yonggwan Won, and Patrick Coffield. Segmentation free shared weight neural networks for automatic vehicle detection. *Neural Networks*, 8(9):1457–1473, 1995.
2. Daniel Howard, Simon C. Roberts, and Richard Brankin. Target detection in SAR imagery by genetic programming. *Advances in Engineering Software*, 30:303–311, 1999.
3. Walter Alden Tackett. Genetic programming for feature discovery and image discrimination. In Stephanie Forrest, editor, *Proceedings of the 5th International Conference on Genetic Algorithms, ICGA-93*, pages 303–309, University of Illinois at Urbana-Champaign, 17–21 July 1993. Morgan Kaufmann.

4. Karl Benson. Evolving finite state machines with embedded genetic programming for automatic target detection within SAR imagery. In *Proceedings of the 2000 Congress on Evolutionary Computation CEC00*, pages 1543–1549, La Jolla Marriott Hotel La Jolla, California, USA, 6–9 July 2000. IEEE Press.
5. Cristopher T. M. Graae, Peter Nordin, and Mats Nordahl. Stereoscopic vision for a humanoid robot using genetic programming. In Stefano Cagnoni, Riccardo Poli, George D. Smith, David Corne, Martin Oates, Emma Hart, Pier Luca Lanzi, Egbert Jan Willem, Yun Li, Ben Paechter, and Terence C. Fogarty, editors, *Real-World Applications of Evolutionary Computing*, volume 1803 of *LNCS*, pages 12–21, Edinburgh, 17 April 2000. Springer-Verlag.
6. Daniel Howard, Simon C. Roberts, and Conor Ryan. The boru data crawler for object detection tasks in machine vision. In Stefano Cagnoni, Jens Gottlieb, Emma Hart, Martin Middendorf, and Günther Raidl, editors, *Applications of Evolutionary Computing, Proceedings of EvoWorkshops2002: EvoCOP, EvoIASP, EvoSTim*, volume 2279 of *LNCS*, pages 220–230, Kinsale, Ireland, 3–4 April 2002. Springer-Verlag.
7. F. Lindblad, P. Nordin, and K. Wolff. Evolving 3d model interpretation of images using graphics hardware. In *Proceedings of the 2002 IEEE Congress on Evolutionary Computation, CEC2002*, Honolulu, Hawaii, 2002.
8. Jamie R. Sherrah, Robert E. Bogner, and Abdesselam Bouzerdoum. The evolutionary pre-processor: Automatic feature extraction for supervised classification using genetic programming. In John R. Koza, Kalyanmoy Deb, Marco Dorigo, David B. Fogel, Max Garzon, Hitoshi Iba, and Rick L. Riolo, editors, *Genetic Programming 1997: Proceedings of the Second Annual Conference*, pages 304–312, Stanford University, CA, USA, 13–16 July 1997. Morgan Kaufmann.
9. Mengjie Zhang and Victor Ciesielski. Genetic programming for multiple class object detection. In Norman Foo, editor, *Proceedings of the 12th Australian Joint Conference on Artificial Intelligence (AI'99)*, pages 180–192, Sydney, Australia, December 1999. Springer-Verlag Berlin Heidelberg. Lecture Notes in Artificial Intelligence (LNAI Volume 1747).
10. Mengjie Zhang, Victor Ciesielski, and Peter Andreae. A domain independent approach to multiclass object detection using genetic programming. Technical Report CS-TR-02-4, Victoria University of Wellington, April 2002.
11. Riccardo Poli. Genetic programming for feature detection and image segmentation. In T. C. Fogarty, editor, *Evolutionary Computing*, number 1143 in Lecture Notes in Computer Science, pages 110–125. Springer-Verlag, University of Sussex, UK, 1–2 April 1996.
12. Riccardo Poli. Genetic programming for image analysis. In John R. Koza, David E. Goldberg, David B. Fogel, and Rick L. Riolo, editors, *Genetic Programming 1996: Proceedings of the First Annual Conference*, pages 363–368, Stanford University, CA, USA, 28–31 July 1996. MIT Press.

The Emergence of Social Learning in Artificial Societies

Mauro Annunziato and Piero Pierucci

Plancton
Lung.ere degli artigiani 28, 00153, Rome, Italy
http://www.plancton.com
plancton@plancton.com

Abstract. The most recent advances of artificial life research are opening up a new frontier: the creation of simulated life environments populated by *autonomous agents*. In several cases a new paradigm for learning is emerging: social learning as a form of self-organization of many individual learning. In this paper two different approaches are presented and discussed: genetic competition and partial emulation. Finally an example of application of these concepts.

1   Introduction

The most recent advances of artificial life research are stimulating the creation of environments populated by evolving *autonomous agents*. In these environments artificial beings can interact, reproduce and evolve [5, 7, 10, 17], and the environment itself can be seen as a laboratory where to explore the emergence of social behaviours like competition, cooperation, relationships and communication [6, 8] . It is still not possible to approach a reasonable simulation of the incredible complexity of human or animal societies, but these environments can be used as tools to explore some basic aspects of the evolution [1, 2, 3, 11, 12, 13, 14, 15, 16, 18] including the development of forms of social learning.

The combination of these concepts with robotics technology or with immersive-interactive 3D environments (virtual reality) are changing quickly well known paradigms like *digital life, man-machine interface, virtual world.* The virtual world metaphor becomes interesting when the artificial beings can develop some form of learning, increasing their performances, adaptation, and developing the ability to exchange information with *human visitors*. In this sense the evolution enhances the creative power and meaningful of these environments, and human visitors experience an emotion of a shift from *a simplified simulation of the reality* to a *real immersion into an imaginary life.* We may think that these realizations are the first sparks of a new form of life: simulated for the *soft-alife* thinkers, real for the *hard-alife* thinkers, or a simple imaginary vision for the artists.

A key aspect distinguishes the learning experiments carried out in the contest of artificial societies in respect to the classic experiments and modelling of the artificial intelligence. This aspect is connected to the contribute of the social sharing and

S. Cagnoni et al. (Eds.): EvoWorkshops 2003, LNCS 2611, pp. 467–478, 2003.
© Springer-Verlag Berlin Heidelberg 2003

interaction in order to increase the learning process. In this case the knowledge is referred to the society rather than the single agent (*collective knowledge*). The social learning process can be considered as the result of the self-organization of the knowledge of its components. This knowledge can be expressed not only in terms of shared information but also in terms of relationships and inherited or emulated behaviours. For the enormous implication in many different science fields (biology, learning engineering, robotics, sociology, economy, arts, networks, etc...), the mechanisms of social learning represents a fashionable challenge to explore both in science and art.

Some reference experiments can be found in the pioneer works of K. Sims [13] and D. Terzopoulos [17]. They developed interesting models for evolving digital creatures. In those experiments, they fix a specific task (swimming in a marine environment or winning a duel for the food) and trained the individuals through genetics selection or optimisation functions. The goal was to obtain creatures for computer graphics applications. Very interesting experiences are described by the Polyword environment of Yager [18] and the relation between individual and social learning of Parisi and Cecconi [10]. In these works the social component of learning is more enhanced through the interactions of individuals in an environment. The individuals are equipped with neural networks that evolve the weights in the time. In the case of Yager [18] very complex creatures with a Hebbian learning approach are utilized. This study is focused on the same pilot problem utilized in this paper: the food tracking.

The final goal of our work is the development of audio-visual interactive installations embedding new scientific approaches in an artistic frame. Our intention is explore the creative content and suggestions that the artificial life (*alife*) environments have in their potentialities. The suggestion we want to communicate is the evocation of an artificial society able to self-develop in the time and interact with the humans. The idea of the *social development* is very ambiguous and wide. This aspect it could be a drawback by a scientific point of view because of the effort to implement a solution for realistic problem is really huge. By an artistic point of view, the main aspect is not the difficulty of the solved problem but the communication of the paradigm of the - *autonomously evolving society* -. The basic idea is a vision of future digital worlds as a way to better explore the mechanisms which are on the base of the formation of our societies, languages, psyche.

Our long term goal is create continuous learning mechanisms to achieve complex tasks in social contexts. Along this direction, we do fix any specific target except the survival. The digital creatures should be able to derive all the living functions (search for the food, competition/co-operation, communication, language) directly as priorities or intermediate goals to reach a better adaptation in the environment under an evolutionary pressure. In this effort we have built a roadmap for the realization of this context in terms of interactive installations: a) to realize a population of socially evolving creatures, b) establish an hybrid world where digital beings and humans can interact, c) create the conditions for the development of an autonomous language in the artificial society exchanging symbols (i.e. sounds and voice) with the humans.

At the current state of development we are applied several experiments. Similarly to the mentioned case of Yager, in these experiments a community of autonomous agents, equipped with a personal neural network, autonomously develop a behaviour to recognize and search for the food to survive. This task is obtained realizing an evolutionary pressure that pushes the individuals to evolve. In this way, adaptation is not an option for the individuals but a survival need. In this paper we explore two different paradigms for the development of this ability: genetic evolution and social learning.

Finally we will trace a brief synthesis of a first realization of the artificial-human interaction and a performative experience of dance and alife in a theatre. We recommend to see the reference [4] for another aspect of this framework regarding the development of an autonomous language in the artificial society and speaking interaction with the humans.

2   The Alife Environment

The alife environment is a three-dimensional space where the artificial individuals (or *autonomous agents*) can move around. During the single iteration (*life cycle*) the individuals move in the space, interact with other individuals, exchange information, and reproduce generating another individuals. The data structure of the individual is composed by parameters regarding specie, reproduction, interaction, dynamics, life, and the current values of the information coded in the individual neural network. A basic variable of the status is the *energy*. The energy is a sort of probability of surviving for the individual. It is gained through the food eaten by the individual at each life cycle, and is needed to move and reproduce. Low energy values causes the death of the individual.

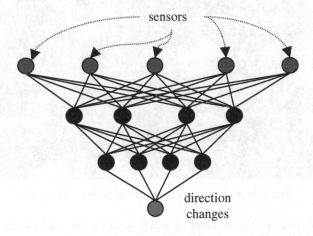

Fig. 1. The neural network to control the movement

The central structures of the individual are the sensors and the movement controller. The sensors have the goal to locate substances in the environment surrounding the individual position in term of presence of food and relative direction. Any individual is endowed with a neural network controlling the movement. This net is composed of four layers of neurons with 4 neurons in the input layer, 4 neurons in two hidden layers and 1 neuron in the output layer (see fig. 1). The net topology is arbitrary and not optimised. The input layer is connected to information coming from the sensors. The output layer defines the change in the movement direction (curvature).

Therefore, the agent movement is the result of the application of the network to the input information in order to decide the new movement direction. Typical reactions are moving towards (or escape from) a substance or show indifference. When the individual enter in a cell with a substance, the individual *eats* the substance that disappears from the environment. The food increases the individual energy.

The reproduction model is aploid: one parent-one child. A probabilistic model for self-reproduction is performed at every life cycle. The *fecundity* probabilistic parameter is recorded in the genotype. Reproduction can occur only if the individual has energy greater than a specific amount. In the reproduction, an amount of energy is transferred from the parent to the child.

In the reproduction, the status of the child-individual is derived from the parent except for random mutations in relation to a *mutation average rate* and *mutation maximum intensity*. In such a way the child will have a similar behaviour but with some little differences in respect to the parent.

Fig. 2. Digital creatures living in the artificial life world

When an individual tries to enter in a cell with another individual an interaction occurs. Several kind of interaction have been developed depending on the experiment: competition, co-operation and indifference as described in the following paragraph.

3   Evolution and Social Learning in an Artificial Life World

In these experiments we put a number of individuals in the environment (typically 256) with the neural network initially filled with random numbers. In the environment we random distribute a fixed rate of food bits. Each bit occupies a single cell and it disappears after a fixed number of life cycles (lifetime, typically 10 cycles). The experiment consists in the autonomous development of the ability of the individuals to recognise the presence of food in the neighbourhood, move toward the bit and eat the bit itself. To obtain this knowledge, the population has to modify progressively the neural networks in order to react to the input information in the best way to survive. Any explicit target for food search is a-priori implemented in the individual behaviour.

We have realised three different experiments corresponding three different adaptation mechanisms. Two of these mechanisms are based on genetic evolution: a) direct competition and b) competition for the resources. The third one is based on c) social learning utilizing the behaviour partial emulation model.

Our interest is not the knowledge level reached by the best individual but a global feature of the society. In order to monitor the adaptation progress of the society, we measure the food bits currently present in the environment. When the individuals are not expert in the food tracking, this number is high. The food disappears for accidental eating (an individual passing randomically over a food cell) or for passed lifetime. When the population develop ability to track the food, the food bits decrease rapidly due to intentional passing of the individuals over a food cell.

3.1 Evolution through Generations: Direct Competition

This first experiment is based on the genetic evolution through the direct competition. The individuals don't change the network weights during their life but only through the genetic mutations in the reproduction. The selection mechanism is a direct competition based on energy. When two individuals meet on the same cell, they fight. The individual with the higher value of energy, wins and survive while the looser dies. For each individual, the energy level is the balance of the energy increased by the food and the one consumed in the life cycle. An increase of the ability to eat food produces an increase of the energy and of the probability to win in the fights.

In the plot of fig. 3 a diagram of the average food density in time is shown for all the experiments. Each time sample corresponds to the average of 100 life cycles. At the beginning, the food presence increases up to reach the maximum corresponding to the equilibrium between the food randomically consumed and the one periodically distributed. After the maximum, a slow decrease of the food presence is exhibited corresponding to the individual learning. After 10000 cycles the curve exhibit a sharp decrease due the a reduction of the amplitude of the mutation on the net weights. Finally a saturation value is reached corresponding to the maximum ability that the individuals can reach trough this mechanism.

The increase of the ability to eat is clearly demonstrated looking to the alife animation. At the beginning the individuals move in a very chaotic pattern. Along the evolution, some individuals succeed to reach the food after some strange trajectories. At the end, when a food bit compares in the environment, immediately many individuals converge towards the food. The one that has the best ability, succeeds to reach the food increasing its energy. The others don't eat and will be filtered out by some more able competitor.

3.2   Evolution through Generations: Competition for the Resources

In the second experiment, the adaptation mechanism is based on the genetic evolution through the competition for the resources. Also in this case, the individuals don't change the network weights during their life but only through the genetic mutations in the reproduction.

The situation is quite similar to the previous one but with two differences:
- when two individuals meet, they ignore the meeting and have any interaction;
- the energy consumed in life cycle is quite higher in respect to the previous case.

In this case the individuals are forced to eat in order to avoid the decrease of the energy under the survival threshold. In few words they compete for the resources instead to compete directly each other. The plot of fig. 3 shows a trend similar to the previous case, but the final value is lower. This mechanism is more efficient than the previous one. It recalls an ecosystem where the animals compete mostly for the resources in the context co-evolution of different species.

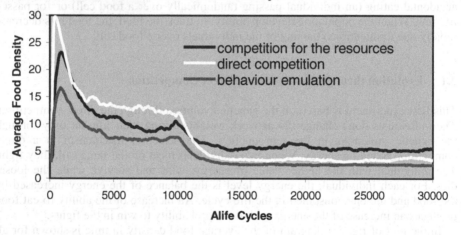

Fig. 3. Comparison of the efficiency of the three different strategies of learning: direct competition, competition for the resources, behaviour emulation.

3.3   Learning through Communication: The Partial Emulation Model

The third mechanism we experimented, is not based on evolution through genetic mutations but it regards the learning during the single individual life and it is connected to the social communication mechanisms. In some sense it is related to the *cultural advancement* of the population: when two individuals meet, they communicate exchanging their information about own developed behaviour.

In this experiment the individuals exchange information about the weights of the neural networks controlling the movement and responding to the presence of food. An emulation mechanism is activated when a meeting between two individuals occurs. In the meeting, the individual with lower energy modifies the weights of its neural networks. In few words, the behaviour could be synthesised by the sentence: *if you have a higher energy respect to me, it could be better for me try to partially emulate your behaviour.* This mechanism represents a sort of translation of the genetic mutation in the cultural domain. In formulas:

$$W_{Ai} = W_{Bi} * \alpha + W_{Ai} * (1-\alpha)$$

Where W_{Ai} is the i-th weight of the network of the moving individual and W_{Bi} is the i-th weight of the network of the met individual; α is the emulation factor typically ranging between 0.1 and 0.5.

The individuals do not die, but when the energy goes to zero, they are forced to apply small changes to their behaviour, that means small changes to the neural network weights.

As the previous case, the plot of fig. 3 shows the same trend, but comparing to the other cases, the values reached with this mechanism are lower and faster reached. This means that this mechanism is the most efficient in respect to the others. This comparison has only a reference value because of in the reality these mechanisms are contemporary present and the real living beings are much more complex.

In this case, the competition is similar to the stock market competition. When an individual becomes quite able to eat, it increases its incoming of energy without any competitor. The other individuals try to emulate and learn from him. When the others reach its level and someone becomes better, the first individual starts to have an attenuation of the incoming energy and then a drastic energy reduction up to finish its energy. At this point it is forced to change its behaviour to come back to a positive energy incoming.

This form of learning is the most intriguing because of its feature of dynamics and *volatility*. In fact, the produced knowledge is still a product of the whole society but it is moved dynamically between the various individuals. Although the knowledge is generated during the life of the individuals, it can be transmitted through the generations. In this sense is the one more similar to the *culture*.

3.4   Some Remarks about the Consciousness Dilemma

To have a visualisation of the ability reached autonomously by the digital creatures, in fig. 4 we report two sequences of life with creatures passing close a food bit. In the first sequence the individuals are at the beginning of the training experiment. They exhibit indifference for the food. The second sequence is related to trained individuals. In this case is quite clear a strong finalisation of the creatures movement to catch the food.

It should be noted that in the described mechanism the individuals achieve the ability to eat but they don't develop any form of *consciousness* of eating or *intentional direction* towards the specific target of eating. Simply they establish a relation between some behaviour (the weights of the neural networks) and the satisfaction of some survival needs (the feeding to increase the energy and to longer survive).

We could apply the same procedure to a higher communication level, like sound messages or the development of a language. Probably, we could allow the development of the complex behaviour relating it to an increase of adaptation. When the selection mechanism is extended to the competition between societies and groups also some behaviour like affect, parent care can be revisited as survival needs. In principle a high level of *adaptive behaviour* and *intelligence* could be reached without any form of consciousness.

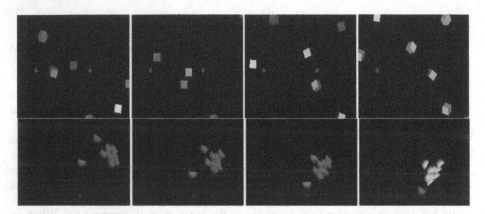

Fig. 4. The creatures at the beginning (top) and end (bottom) at of the training experiment.

For the involved implications about relation between individual and social behaviour, we have decided to use the social learning paradigm of the third experiment, as a fertile platform to generate metaphors, open problems and questions that is the natural environment of an artwork. Some questions remain still open: what is really the consciousness ? How could it be developed in a digital being ? Are intelligence and culture possible without consciousness ?

4   Human-Artificial Interaction in the Hybrid Environment

In the previous sections we have shown the realization of an artificial world where the creature can learn and exchange information. So far all the world is confined in the digital domain. A real jump in the potential of these worlds is to establish a contact between this world and humans. The idea is a sort of cross-fertilization between humans and digital beings. There are many approaches to establish this communication which corresponds different communication metaphors. In the following we describe the paradigm we selected among the many possible ones.

The basic idea is to combine two different channels of interaction: biochemical and symbolic. The biochemical interaction is based on the idea that humans emit substances in the digital environment when they move in an interaction area. Depending by the type of substance, the digital creatures are attracted or not by the person. This channel of interaction is more intended for a primitive interaction and it will discussed in the following. The symbolic communication is the exchange of symbols a) in between the creatures and b) between the humans and the creatures. The set of symbols and their meaning emerge as a society feature as a result of a cultural evolution process and interaction with humans. This channel of communication is intended for specific installations based on voice exchange between humans and creatures. It is not discussed in this paper for space reasons but you can find explanations in [4].

The starting point is the place where the interactions occur. So we have to re-define the borders of the environment. In the interactive installation, the image of the artificial world is projected on a 2D screen. The area for the human interaction consists in the area in front of the screen. To interact, a person has to enter in this area in order to produce modification in the artificial world. In such a way we have extended a dimension of the environment in the real world building an hybrid real-digital ecosystem. The interaction area is observed by video-cameras acquired in the computer. A tracking program detects people presence in terms of change detection in the image. This information is mapped as substances emitted by the real people in the digital dimension of the environment. There is a spatial coherence between the location of the human and the digital environment. A program flag is used to decide if the emitted substance is food for the creatures or poison.

Fig. 5. Playing with digital entities through a biochemical communication.

If a creature is moving in the same location it reacts to the substance. The creatures have been trained with the procedure illustrated in the third experiment. Furthermore their networks have been trained to track the food and avoid the poison. As results the creatures are attracted by humans emitting fooding substances and repulsed by humans emitting poison substances. See fig. 5 for pictures of the installation.

4.1   Alife and Dance: The *Aurora di Venere* Performance

The described installation was used on an alife-dance performance shown at the Theatre of the *Palais de San Vincent* (Italy), in March 2001. *Aurora di Venere* was presumably the first live performance in Italy including alife interacting with the dancers. The performance (about 30 min.) included 8 dancers, 6 computers (SGI and PCs), 6 video-projectors and 8 sound amplifiers for 3D sound rendering around the theatre. Two video-projectors were focused on two large screens (12x8 m.) located at the background and at the front (semi-transparent) of the theatre's stand. The other 4 projectors covered the entire ceiling of the theatre that has a dome shape.

In the performance, the dancers interact with digital creatures projected on the stand screens (see fig. 6). The performers dance in the middle of the screens, and they seem completely immersed inside the digital creature movements. The dancers play with the images of the artificial individuals which move following their own personality: they attract and repel the creatures through the *biochemical communication* mechanism explained before.

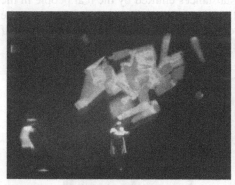 

Fig. 6. Pictures from the alife-dance performance *Aurora di Venere*: the dancers play with the digital creatures projected over the background and over a semi-transparent screen of the theatre's stand.

During the performance the story grows in intensity when the artificial beings (fig. 7) escape from the front screens invading the public and the theatre ceiling. They search for people movements and produce 3D sounds travelling in the theatre. At the end the whole internal pseudo-spherical surface of the theatre is invaded by digital beings.

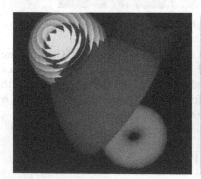 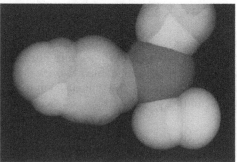

Fig. 7. Creatures for the *Aurora di Venere* alife-dance performance.

5   Conclusions

We have explored several ways to build digital creatures living in an artificial world, able to learn from the sensorial experience and through genetics. Several paradigms of evolution and learning has been experimented in order to achieve autonomously simple tasks like search for food. A basic paradigm of the social learning based on a behaviour partial emulation paradigm has been selected as a fertile platform for development of alife-art contexts. These concepts have been applied in an interactive installation where visitors can interact with the artificial creatures through a mechanism of substance emission-reception. This installation has been involved in an alife-dance performance in a theatre.

Rather than conclusions, this experience opens many questions about
What does digital life means ?
Is it really possible to develop an autonomous culture in alife worlds ?
Is it possible to have knowledge without consciousness ?
How far this knowledge could go?
Maybe the only reasonable conclusion today is to raise these questions. Using imagination and art to find some answer.

References

1.   Annunziato, M.: Emerging Structures in Artificial Societies, in Creative Application Lab CDROM, Siggraph, Los Angeles (CA). 1999. For Artificial Societies see also www.plancton.com.
2.   Annunziato, M., Pierucci, P.: Relazioni Emergenti. Experimenting with Art of Emergence, Leonardo, Volume 35, Issue 2 (April 2002).
3.   Annunziato, M.: Ed. of Special Iussue of YLEM Newsletters on Artificial Societies, Sept.-Oct. 2001. Articles of C. Sommerer, L. Mignonneau, K. Rinaldo, P. Pierucci, J. Prophet.
4.   Annunziato, M., Pierucci, P.: Human-Artificial Ecosystems: searching for a language. In Proc. of Int. Conf. of Generative Art, Milan, Italy, Dec. 2002, www.generativeart.com.
5.   Epstein J., Axtell, R.: Growing Artificial Societies, Brooking Institution Press, The MIT Press, 1996.
6.   Lipson H., Pollack, J.: Evolving Creatures. In: Int. Conf. Alife VII, Portland (OR), 2000.
7.   Langton, C.: Artificial Life. C. Langton Ed. Addison-Wesley. pp. 1–47, 1989.
8.   Kaplan, F.: Semiotic schemata: selection units for linguistic cultural evolution. In: Int. Conf. Alife VII, Portland (OR), 2000.
9.   Monod, J.: Chance and Necessity. New York: Knopf, 1971.
10.   Parisi. D., Lecconi, F., Individual versus social survival strategies, Journal of Artificial Societies and Social Simulation vol. 1, no. 2, 1998.
11.   Ray, T. S.: Evolution as Artist, in Art @ Science, C. Sommerer and L. Mignonneau. Eds., Springer-Verlag. 1998.
12.   Rinaldo, K.: Autopoiesis, In: Int. Conf. Alife VII, Portland (OR), Workshop Artificial life in Art., 2000
13.   Sims K.: Evolving Virtual Creatures, in Computer Graphics, Siggraph Conf. Proc., 1994.

14.   Sommerer, C., Mignonneau, L.: A-Volve - an evolutionary artificial life environment. In: Artificial Life V . C. Langton and C. Shimohara Eds., MIT, pp. 167–175. 1997.
15.   Sommerer, C., Mignonneau, L. : Art as a Living System, in Art @ Science, C. Sommerer and L. Mignonneau Eds., Springer-Verlag. 1998.
16.   Tosa, N. et al.: Network neuro-baby with robotics hand, symbiosis of human and artifact, elsevir Science B.V, pp. 77–82. 1995,
17.   Terzopoulos, D., Rabie, T., Grzeszczuk, R.: Perception and Learning in Artificial Animals, in Proc. of Int. Conf. Artificial Life V, Nara, Japan, May 1996.
18.   Yeager, L.: Computational Genetics, Physiology, Metabolism, Neural Systems, Learning, Vision, and Behaviour on PolyWorld: Life in a New Context, in C. Langton ed., Artificial Life III, 263–298, Addison-Wesley, 1994.

Tabula Rasa: A Case Study in Evolutionary Curation

Jon Bird[1], Joe Faith[2], and Andy Webster[3]

[1] Centre for Computational Neuroscience and Robotics (CCNR),
School of Biological Sciences, University of Sussex, Brighton, BN1 9QG, UK
jonba@cogs.susx.ac.uk
[2] School of Informatics,
University of Northumbria,
Newastle-upon-Tyne, NE1 8ST, UK
joe.faith@unn.ac.uk
[3] Falmouth College of Arts,
Falmouth, Cornwall, TR11 4RH, UK
andy.webster@falmouth.ac.uk

Abstract. This paper describes a novel use of evolutionary techniques to curate a main stream art show, *Tabula Rasa*. This allowed an open-ended approach to curation that was entirely in keeping with the artistic motivations behind the project. We detail a major difficulty with this approach that will be faced by any future automatic curation systems.

1   Introduction

This paper describes a recent art show, *Tabula Rasa*, which was curated using an automatic, self-organizing process implemented with a genetic algorithm. We believe that this project will be of interest to the evolutionary music and art community for the following reasons:

- it is an example of evolutionary methods being used in a mainstream art context;
- it is a novel application of evolutionary methods;
- it has highlighted a major difficulty that will be faced by any future self-organizing curation systems.

The project was the result of a collaboration between an artist and two scientists. In this paper we initially describe the artistic motivations underlying the project in order to ground the technical details that follow.

2   The Fine Art Context of *Tabula Rasa*

In this section we outline some of the art issues that *Tabula Rasa* was exploring, in order to show why the use of evolutionary techniques was conceptually appropriate in this project. Two of the central motivations of the art show were:

S. Cagnoni et al. (Eds.): EvoWorkshops 2003, LNCS 2611, pp. 479–489, 2003.

- curating an art show in a bottom-up, artist-led fashion, rather than the usual top-down, curator-led way;
- taking art out of an institutional context and displaying it in a public arena, where there are fewer expectations or preconceptions that influence its reception.

It was hoped that these two elements would lead to an open-ended art show consisting of a diverse body of works. We viewed our role in the project as creators of a 'blank slate' whose final form would be shaped by the films that were submitted, the automatic process that curated them, and the way that they were received in the public context where they were displayed.

2.1 Submission Process

We advertised for artists to submit moving image works for outdoor projection in the centre of Croydon, UK, and emphasized that there were no thematic or stylistic constraints. Three practical limitations were the format of the entries (mini DV tape); the fact that there would be no sound with the projections; and that the films could not be over 20 minutes in duration. Otherwise, the artists had complete freedom to submit any films they liked. At the outset we also decided that every film submitted would be projected, as long as its content could be legally displayed in public. We had 58 submissions for the project (some of them multiple films) and the entry was international. Each artist was required to send 5 keywords that described the work they were submitting. We asked the artists to use nouns as their keywords, but we did not enforce this requirement strictly. Artists were told that an automatic curation system would use these keywords to organize the submitted works into 4 distinct groups, one for each of the projection sites. The artists who submitted works were prepared to take a risk and give up control over the context in which their art would be displayed: they did not know the buildings it would be displayed on or the other works that would be shown with it; furthermore they were willing to display their work in a non-institutional context where the viewers would be very different from a typical art audience.

2.2 Art in a Public Context

'It is precisely at the legislative frontier between what can be represented and what cannot that the postmodern operation is being staged – not in order to transcend representation, but in order to expose that system of power that authorizes certain representations while blocking, prohibiting or invalidating others.' In [1], page 166.

Public art is displayed at sites that are not generally considered to be conventional art environments. However, it is often commissioned and funded by galleries or museums and has been criticized on the grounds that although outdoors, it is a self-celebration of traditional art institutions. The aim of *Tabula Rasa* was to situate art in an urban and commercial context so that it was shaped by the everyday collective experience, rather than predetermined institutional objectives. The centre of Croydon is ideally suited to this: the squeaky clean, luminous, air conditioned, anaesthetized spaces of galleries, museums and cinemas are replaced by dirty, noisy streets subject to the vagaries of the weather and changing light conditions. In contrast to institutional spaces, in public spaces there are no entry points and no exits, no clear beginnings or ends. There is no invitation to experience the work, nor do the public expect or necessarily even want to see 'art'. None of the institutional apparatus which Owens [1] critiqued is present.

To this end, with the help of Croydon skyline (an organizational arm of Croydon Borough Council) we negotiated permission to project massive images on 4 buildings in the centre of Croydon. 2 of the sites had 60' by 40' projections and 2 had 25' by 16' projections. One site was the side of a Croydon College building, one the side of a theatre, one a wall opposite a bus stop in a side street, and one was the side of an office block on a major road through Croydon.

3   Evolutionary Curation

The video submissions to *Tabula Rasa* were partitioned into 4 distinct groups according to the similarity of meaning of their 5 associated keywords. In this section we describe the metric used to calculate the similarity of 2 films: the Jiang Conrath [2] semantic relatedness measure based on the WordNet semantic network [3]. In this paper, when we refer to the similarity of 2 films, it is always in terms of the semantic relatedness of their associated keywords, and not in terms of their content.

3.1   WordNet

WordNet is an extremely comprehensive semantic network created as an attempt to model the lexical knowledge of a native speaker of English. English nouns, verbs, adjectives, and adverbs are organized into synonym sets (synsets), each representing one underlying concept. These synsets are interlinked into a network by a variety of relations. WordNet can therefore be conceptualized as a kind of super-thesaurus.

The noun portion of WordNet is by far the most developed part of the network, hence our request for the keywords to be nouns. It consists of more than 60,500 synsets, representing over 107,400 noun senses, which are interlinked by over 150,000 arcs. Each arc represents one of the nine relations adopted by WordNet's creators:

Synonymy the relation used to form synsets. For example, 'dog' is a member of the same synset as 'hound'.

Hypernymy the is-a relation: e.g., 'plant' is a hypernym of 'tree' since tree is-a plant.

Hyponymy the subsumes relation (inverse of hypernymy).

Meronymy the set of three relations that can be collectively referred to as part-of:

Component-Object e.g., 'branch' is a meronym of 'tree' since 'branch' is a component of 'tree'.

Member-Collection e.g., 'tree' is a meronym of 'forest' since 'tree' is a member of 'forest'.

Stuff-Object e.g., 'aluminium' is a meronym of 'airplane' since 'aluminium' is the stuff that 'airplane' is made from.

Holonymy the set of three relations that can be collectively referred to as has-a (and that are the respective inverses of meronymy).

Antonymy very roughly, the opposite-of relation: e.g. 'rise' and 'fall' are antonyms, as are 'brother' and 'sister'.

The IS-A hierarchy (hypernymy/hyponymy) constitutes the backbone of the noun subnetwork, accounting for close to 80% of the links in WordNet. At the top of the hierarchy are 11 abstract concepts, termed unique beginners, such as *entity* ('something having concrete existence; living or nonliving'), *psychological feature* ('a feature of the mental life of a living organism'), *abstraction* ('a concept formed by extracting common features from examples'), *shape/form* ('the spatial arrangement of something as distinct from its substance') and *event* ('something that happens at a given place and time').

3.2 Semantic Distance

Given a semantic network such as WordNet, we can define a crude measure of the distance between two words as the minimum number of steps it takes to traverse from a synset containing the first word to a synset containing the second. Thus, for example, the distance between 'brother' and 'comrade' is 0 (as they are part of the same synset) , whereas the distance between 'brother' and 'sister' is 1 (as they are directly linked synsets). There are many possible distance measures that use a variant of this arc-counting method. However they all suffer from the fact that the links between synsets are not uniform: some synsets seem conceptually closer to others [4]. For example, the hypernymy relation between 'safety valve' and 'valve' seems intuitively to cover less ground than that between 'knitting machine' and 'machine'. The intuitive notion of 'covering ground' can be expressed formally by noting the position of words within hypernymy (IS-A) hierarchies. In particular, the higher in a hierarchy that the common synset between two words occurs, then the more common and less specific it will be: 'machine' is more general than 'valve', and so synonyms and hyponyms of 'machine' will be more common than those of 'valve'. Hence the link between valves and safety-valves is closer than that between machine and knitting machines.

The commonness of a word can be measured empirically using a large corpus of example texts, such as Brown's corpus of American English [5]; the measured commonness of a word is then inversely proportional to the frequency with which that word occurs in that corpus. The Jiang-Conrath [2] metric of semantic similarity was used to calculate the distance between keywords for *Tabula Rasa*, since it combines both the number of links between words, *and* the relative frequency of their common synset. This measure also accords well with human judgements of distances between word pairs (with a correlation coefficient of > 0.85 [6]).

Table 1. A table of the submitted keywords

1	Building	City	Street	Derivé	Model
2	Screen	Drawing	Tape	Window	Hand
3	Photography	Landscape	History	Time	Homage
4	Time	Paradise	Butterfly	Death	Black
5	Head	Mask	Mines	Waste	Cage
6	People	Car	Food	Door	Street
7	Cell	Drift	Encounter	Procedure	Endless
8	Man	Lamb	Air	Ewe	Straw
9	Textile	Stitch	Structure	Iterance	Projection
10	Treadmill	Hamster	Wheel	Cage	Revolution
11	Light bulb	Glasses	Cigarette	Wheel	Smoke
12	Flower	Curtain	Net	Cut	Screen
13	Mirror	Television	Puppet	Copycat	Reflection
14	Artificial-life	Sand	Geometry	Suggestion	Confrontation
15	Ground	Window	Door	Root	Building
16	Goodness	Badness	Hopelessness	Prosperity	Life
17	Protest	England	Liberalism	Suburbia	Leisure
18	Memory	Journey	Archive	Experimental	Blankness
19	Skyline	Moon	Phase	Luna	Night
20	Ground	Sea	Sky	Bird	Balloons
21	Build	Play	Bricks	Distract	Pizza
22	Cycle	Urban	Mundane	Hope	Despair
23	Subjectivity	Lyricism	Language	Abstraction	Deconstruction
24	Garden	Crystals	Light	Growth	Mountain
25	Architecture	Cells	Humans	Motion	Planet
26	Food	Television	Body	Reflection	Game
27	Graffiti	Astronomy	Documentary	Fiction	Linear
28	Electron	Microscope	Insect	Animal	Plant
29	Head	Eyes	Face	Body	Eyelids
30	Ornament	Ornament	Ornament	Ornament	Ornament
31	Night	Light	Car	Road	Flash
32	Bunker	Tourist	Identity	Authorship	Technology
33	Offices	Heels	Loops	Fugitives	Indoor spaces
34	Classification	Culture	Masculinity	Language	Power
35	Pig	Hair	Eye	Ear	Nose
36	Shell	Glider	Foot	Vector	Span
37	Sun	Bath	Room	Shadows	Light
38	Play	Building	Control	Absurd	Mechanical
39	Bones	Life	Memory	Plant	Steel
40	Boredom	Humour	Artifice	Contingency	Lo-tech
41	Hand	Journey	Heart	Animation	Landscape
42	Desert	Man	Clouds	Parrott	Tree
43	Road	Land	Car	Factory	Sea
44	Violence	Numb	Narrow	Spam	Vacuum
45	Liquid	Static	Flame	Particles	Light
46	Dust	Surface	Fluid	Landscape	Tentacles
47	Dance	Dervish	Space	Movement	Axis
48	Children	Dictionary	Fruit-machine	School	Billboard
49	Feet	Leaves	Concrete	Sand	Skin
50	Subjectivity	Lyricism	Language	Abstraction	Deconstruction
51	Banana skin	People	Feet	Shoes	Mind
52	Ladder	Paint pot	People	Window	Mind
53	Track	Ears	Shadows	Gravel	Mind
54	Waste	Water	Sea	Cleaners	Brush
55	Bubbles	Glass	Water	Light	Universe
56	Sunlight	Dust	Space	Window	Plant
57	Health	Smoking	America	Sport	Sterile
58	Sheffield	England	World	Galaxy	Universe

The Jiang-Conrath metric actually measures similarity between words, where $d_{jc}(w_i, w_j) = 0$ if w_i and w_j bear no similarity and $d_{jc}(w_i, w_j) = 1$ if they are identical. However for the purposes of *Tabula Rasa* the *distance* between words is required, which is taken to be $1 - d_{jc}(w_i, w_j)$.

4 Film Distance Data

Once the distance between individual words has been calculated, we define the
distance between 2 films $D(F_i, F_j)$ as the mean distance between the individual
keywords that describe those films:

$$D(F_i, F_j) = \frac{\sum_{m=1}^{N} \sum_{n=1}^{N} d(w_{im}, w_{jn})}{N^2}$$

where w_{im} designates the mth keyword of the ith film, and there are N keywords
per film. In the experiment reported here $N = 5$.

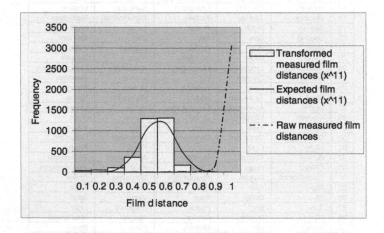

Fig. 1. Frequency plots for the raw film distances, transformed film distances and
expected values of the transformed film distances. After transforming each data point
the variance increases, but is still small (0.227).

A fragment of the resulting distance matrix for submitted keywords is shown
below:

	Building	City	Street	Derivation	Model
Building	0.000	0.907	0.895	0.936	0.906
City	0.907	0.000	0.926	0.920	0.931
Street	0.895	0.926	0.000	0.948	0.925
Derivation	0.936	0.920	0.948	0.000	0.939
Model	0.906	0.931	0.925	0.939	0.000

Note that the word distance matrix is symmetric and has a zero-diagonal,
since the distance from a word to itself is zero. Also note that the range of
distances is small: typically in the range $(0.85, 0.95)$. This means that, given the
set of keywords that were submitted, most word pairs are regarded as almost

maximally distant. Nonetheless there are significant differences that mostly accord with human intuition: $d(Building, Street) < d(Building, Derivation)$, for example. The 58 sets of submitted keywords are shown in Table 1. There was very little variance in the 3364 raw distances between the 58 works: the raw data is heavily left skewed (skewness $= -6.78$) and has a mean of 0.93 and a standard deviation of only 0.031. Essentially, all of the films submitted were *very* different from each other, in terms of the semantic distances between their associated keywords, as measured by the Jiang-Conrath metric. In order to increase the variance of the film distances, we transformed each distance, D, using a power transformation:

$$D = D^x,$$

where $x = 11$ was found to give the biggest increase in the variance of the data. The transformed data had a three fold increased standard deviation of 0.099, a mean of 0.476 and a more normal distribution, with a reduced skewness of -1.45.

In addition it was found that some of the submitted keywords were not recognized by WordNet. If possible, substitute words were found with the agreement of the artists. For example, the word 'derivation' in the distance matrix above was the agreed substitution for the submitted keyword 'derivé'. If this was not possible, then the distance to unrecognized keywords was taken to be the mean of all the other power transformed distances (0.476).

5   Genetic Algorithm

The genetic algorithm was used to cluster similar films together, while keeping a similar number of films at each site. Each genotype had a fixed-length of 58, each locus on the genotype representing one of the submitted works. Each allele was an integer value in the range $(1, 4)$, representing which one of the four projection sites the corresponding film was assigned to. 10 evolutionary runs were completed, each 5,000 generations long and using a population size of 200. The algorithm used a rank-based selection procedure, with the fittest individual being 7 times more likely to be selected than an individual with the mean population fitness value. Elitism was employed, with the 10 fittest individuals being preserved in the next generation. There was a 90% chance of single-point crossover and a mutation probability of 2% per locus. If the best individual fitness did not increase for 500 generations, then the mutation rate was increased by 1%, up to a maximum of 4% per locus. If the individual best fitness increased then the mutation rate was reduced by 1%, down to a lowest level of 2% per locus (see section 6 for details of one of the consequences of this mutation regime). The experiment used these genetic algorithm parameter values as they had been successful in other applications. Besides the dynamic mutation rate, no fine-tuning of these parameter values was carried out.

5.1   Fitness Function

The fitness function had two components, $F_{similarity}$ and F_{spread}. The first component rewarded solutions that minimized the distance between the films clustered at each site; the second component penalised solutions that did not cluster similar numbers of films at each site. The second was necessary to prevent the algorithm clustering a small number of films at a couple of sites, in order to minimize the distance at these sites.

As the distance between two films is symmetrical, and their order is not important, the mean distance between films at a site s, $\overline{D}_s$, was calculated as the average of the combinations (2-subsets) of distances, C_2^n, where n is the number of films at the site. $F_{similarity}$ was calculated as:

$$1 - \frac{\sum_{s=1}^{S} \overline{D}_s}{S},$$

where S is the number of sites, in this case 4.

The acceptable range of numbers of films at each site was calculated as:

$$\frac{N}{S} \pm \frac{N}{S^2},$$

where N is the total number of films and S is the total number of sites. The calculation was rounded to give an integer solution. For the experiment reported here, it meant that the permissible number of films at each site was between 11 and 18. Each site that had a number of films outside of this range incurred a penalty p of $\frac{1}{S}$. F_{spread} was calculated as:

$$1 - \sum_{s=1}^{S} p,$$

where S is the number of sites.

$F_{similarity}$ was weighted by 0.8 and F_{spread} by 0.2, giving a final fitness value in the range $(0, 1)$.

6   Results

As most of the films submitted were very dissimilar, in terms of the semantic relatedness of their associated keywords as measured by the Jiang-Conrath metric, there was very little structure for the genetic algorithm to discover in the data. This is shown graphically in figure 2, where the fitness increases from a mean value of 0.57 in the first generation to a best fitness level of 0.66 by generation 1500, only increasing to 0.662 by generation 5,000. The reason that the mean fitness drops after a peak at generation 1900 is that the mutation rate started to increase once the best individual fitness stopped increasing.

A number of controls were carried out to check the efficacy of the genetic algorithm. Firstly, 1,000,000 genotypes were randomly generated and their fitness calculated. The 10 best random genotypes were selected; the mean fitness of this randomly generated elite was 0.627 and the standard deviation was 0.001.

Secondly, a k-nearest-neighbour algorithm [7] was applied to the distance data, with k fixed at 4 and using the spread constraints that were applied in the genetic algorithm (that is, each site had to contain between 11 and 18 films). The 10 best solutions generated by this algorithm were selected; the mean fitness of this elite was 0.632 and the standard deviation was 0.005. These two control groups were then compared to the best individuals generated by each of the 10 evolutionary runs.

In all 10 of the evolutionary runs, the highest fitness achieved by the randomly generated genotypes and k-nearest-neighbours algorithm was reached by the best individual in each population within 70 generations. Using a two-sample, unequal variance (heteroscedastic) t-test, a pair-wise comparison of these 3 groups was carried out. It was found that the best evolved solutions are significantly different ($p < 0.001$) from the random and k-nearest-neighbour solutions. The genetic algorithm performed better than the other two control techniques. As a final control, a small number of individuals were asked to cluster the films into similar sites, with 11 to 18 films on each site: that is, they were clustering the films under the same constraints as the genetic algorithm. The fitness values achieved by humans were not significantly different from the mean random fitness level and all of the subjects reported that the task was extremely difficult.

The standard deviation of the best fitness value achieved in each of the 10 evolutionary runs was only 0.002: all of the runs reached a plateau at essentially the same fitness level, 0.66, even though all of the solutions found were different. The only common structure shared by the best evolved solutions is that they cluster the films into 2 groups of 11 and 2 groups of 18. It is a good strategy to create two groups containing the smallest number of films allowed by the fitness function and to try and minimize their mean distance. This is because the mean distance of the other two groups does not increase significantly by increasing their size, as most of the films are very different from each other, as measured by the semantic relatedness of their associated keywords. However, there were no groupings of 11 films that were similar enough to increase the fitness level beyond 0.665.

7   Discussion

The automatic curation system in *Tabula Rasa* succeeded in the sense that all of the solutions it generated are significantly better than random ones, those generated by the k-nearest-neighbours algorithm and those constructed by human test subjects solely on the basis of keywords. By using an evolutionary curation system it was possible to organize an art show in a novel, open-ended way.

The rationale was to put groups of films with the smallest mean semantic distance together on the same site, so that any similarities might become apparent to a viewer. However, the submitted films were all very dissimilar, in terms of the semantic relatedness of their associated keywords as measured by the Jiang-Conrath metric, and therefore there was not enough structure in the data to allow the algorithm to discover 4 similar groups. In order for any clustering analysis to succeed there have to be groups in the data such that an object in a group is more like other objects in that group than objects outside of the

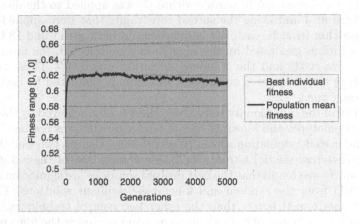

Fig. 2. Graph of the best individual fitness and the mean population fitness over 5,000 generations. Each data point in the graph is the mean of 10 runs.

group. The 10 evolutionary runs produced solutions of almost identical fitness, and yet there is very little similarity between them in terms of the films that were grouped together.

The choice of keywords was very open-ended, in keeping with the motivation underlying the art show. One way to enhance the semantic relatedness of the films' associated keywords would be to use a restricted set of keywords. However, as this set becomes smaller, the process becomes more like conventional, top-down curation. Another possibility is to increase the number of keywords used to describe each individual artwork. However, this approach might require impracticably large numbers of keywords before it provided enough structure to form clusters of similar films. Another solution would be to continue to have an open submission policy, without any thematic or stylistic constraints, but to then use some criteria to tag the films on the basis of their content. This form of analysis would probably be problematic with a very diverse group of artworks, such as those submitted to *Tabula Rasa*, which varied widely in style and content. Determining the appropriate level of top-down control and the best way to categorize artworks is a challenge that will face the designers of any future automatic curation system.

John Cage defended Satie's music by saying, 'to be interested in Satie one must be disinterested to begin with, accept a sound is a sound...give up illusions about ideas of order, expressions of sentiment, and all the rest of our inherited claptrap.' In [8], page 77. Similarly, it is perhaps appropriate, given the artistic motivations behind *Tabula Rasa*, that the open-ended curation process encouraged such a diverse submission of art works that there were no obvious groupings that could be identified by an evolutionary algorithm purely on the basis of the semantic relatedness of the 5 keywords associated with each film.

References

1. Owens, C.: The discourse of others: Feminists and postmodernism. In: Beyond Recognition. University of California (1992)
2. Jiang, J., Conrath, D.: Semantic similarity based on corpus statistics and lexical taxonomy. In: Proceedings of the International Conference on Research in Computational Linguistics. (1997)
3. Miller, G.A., Beckwith, R., Felbaum, C., Gross, D., Miller, K.: Five papers on wordnet. CSL Report 43, Princeton University (1990) Revised 1993.
4. Resnick, P.: Using information content to evaluate semantic similarity. (1995)
5. Francis, H., Kuvcera, H.: Frequency Analysis of English Usage: Lexicon and Grammar. Houghton Mifflin, Boston MA (1982)
6. Budanitsky, A.: Lexical semantic relatedness and its application in natural language processing (1999)
7. Bishop, C.M.: Neural Networks for Pattern Recognition. Clarendon Press, Oxford (1995)
8. Cage, J.: Silence: Lectures and Writings. Marion Boyars, London (1968)

MusicBlox: A Real-Time Algorithmic Composition System Incorporating a Distributed Interactive Genetic Algorithm

Andrew Gartland-Jones

Computational Creativity Research Group, The University of Sussex, Brighton, UK.
drew@atgj.org

Abstract. This paper discusses the motivation, design and construction of a generative music system, 'MusicBlox', (by the author) that utilises a domain specific, knowledge rich Genetic Algorithm (GA). The paper begins by describing the functionality and musical aims of the project, and goes on to detail the implementation of a GA as part of the project's compositional sub-system, including a discussion of the suitability of using a GA for compositional tasks. The paper concludes that the developed GA is able to produce musically successful results, but that significant additional work still needs to be undertaken before it achieves all the aims outlined.

1 Introduction

This paper is a musical and technical report on the development of a real-time composition system: 'MusicBlox'.

Although the aim is for the system to be controlled through real, physical objects, the prototype currently exists as a 3D graphical model. This paper discusses the motivation and current state of the prototype, in particular the development of a Genetic Algorithm as part of the algorithmic composition system.

2 What Is MusicBlox

The MusicBlox project uses blocks, similar in size and shape to children's wooden building blocks, which are combined together in a similar way to make physical structures. However, each block has the ability to play and compose music, so building a physical structure results in creating a piece of music.

Imagine you have a single block. You give it a small musical fragment (its 'home' music), put it on the table and start it off playing. The block then begins repeating its musical phrase, either sporadically or continuously.

Make another block, this time with a different piece of 'home' music. Place it next to the first block and they start playing together, and in synch.

S. Cagnoni et al. (Eds.): EvoWorkshops 2003, LNCS 2611, pp. 490–501, 2003.

Now press a button on block 1 labelled 'send music'. This causes block 1 to send its music to block 2. Block 2 then performs a composition activity, based on its own 'home" music, *and* the music it has just been passed by block 1. The compositional aim for the block is to produce a *new* musical section that has a thematic relationship with both its home music and the music it has just been passed. Block 2 then starts playing its new music.

Now imagine a chain or group of several blocks in any 3D structure. All blocks have a 'send music' button, so the start of the chain does not have to be block 1. If a block is passed some music it recomposes itself, then passes its new music on to *all* of its neighbours.

Block 1 sends its music to Block 2 >> Block 2 re-composes itself.
Block 2 sends its new music to Block 3 >> Block 3 re-composes itself.
Block 3 sends its new music to Block 4 >> Block 4 re-composes itself. Etc....

By sending out music from a starting point all other blocks within a specified range recompose, and the collective music of the structure is transformed. It is important to clarify that each block holds onto its 'home' music throughout, enabling any music composed by it to remain thematically related, despite the constant process of re-composition each block undertakes. In this way the composer of the home music for all blocks maintains a compositional thumbprint on the evolving musical structure.

The development of the overall piece of music will therefore be determined by:

1. The design of the composition system described here
2. The nature of the music imported as 'home' music into each block
3. How the blocks are built into structures
4. How these structures change over time
5. How the user sends music around the structure

In effect the listener/performer is able to shape the overall music by choosing to send musical fragments from blocks they like to *influence* other blocks. One part of the structure may have composed some ideas the user likes. A block from that group could then be placed in another part of the structure to see what effect it has.

2.1 Musical Aims

I am first and foremost interested in composing music, in the sense that the project will succeed or fail on the perceived quality of its musical output. Although I, like most artists/researchers using algorithms for the generation of artistic products, am interested in the nature of 'artificial' creativity, this is not the driver for the project. It is not a computer modelling of a psychological system, a study of the nature of musical style, nor aims to highlight various philosophical points of view. Such perspectives do influence the work, and may be stimulated by the results, but they are not conceptual layers through which I wish to mediate in order to create a piece of music or music composition system. In this sense the project is an extension of my own compositional practice, so I *am* keen to examine how composers go about the task of writing music.

As the aims are subjective (like most other composers), it is especially importance that I outline the *musical* aims of the project.

In terms of my previous work, MusicBlox can be seen as continuing an interest in exploring the shifting relationship between the roles of composer, performer, and listener. Western musical practice in the 20th Century progressed in two opposite directions with regard to such relationships. Many composers, especially those connected with post 1945 avant-garde, (e.g. Webern, Bartok, Boulez), chose to provide increased specification on musical performance, by providing dynamic articulations on every note for example. The effect of this is to subtract from the opportunity for creative interpretation by the performer. Other composers released control, making scores less specific. The chance elements and graphic scores of John Cage (e.g. Fontana Mix, 1960), and the aleatoric music of Lutoslawski (e.g. Venetian Games, 1961), are examples of this relaxation, providing performers with a greater degree of creative freedom. Aside from Cage's 4' 33" however, where the audience are also the performers, this still leaves *listeners* with little or no control.

My work before MusicBlox centred on interactive music. In effect what this usually consisted of was pre-composed musical fragments, sometimes sounds, sometimes synchronised loops, that were combined in real-time under the control of some kind of a user interface, such as ultrasound sensors, light sensors, etc. This undoubtedly shifts the creative balance, allowing the listener a higher degree of involvement in the realised piece, but I was unhappy with the limited depth of musical exploration enabled by common sensing mechanisms, and the high degree of predefinition required when using pre-composed fragments.

By considering the limitations of predefined music I was led to a conception of the compositional act as being the process of defining a search space for exploration by the performer and/or listener. This is an extension of the notion of the act of composition itself being an exploration of the potentially very large search space permitted by the compositional practice of the composer, even considering limitations of instruments, styles, genres etc. I was now regarding this exploration as a two-stage process. Firstly the composer limits the larger search space, potentially every combination of any number of notes, by defining an area within the space. Secondly the performer or listener explores this more limited space and defines his or her own aural realisation within it.

Another way of looking at this is to see the combined input of composer and system designer as mapping out an area within what Boden [1] refers to as a *conceptual space*, which is defined as "an accepted style of thinking in a particular domain". This conceptual space maybe already in existence, or could be a new addition to the domain; in Boden's terms, the input from the composer and system designer then maybe either E-Creative (exploratory creativity) or T-Creative (transformational creativity). The user is asked to *explore* that space and is therefore performing potentially *combinational*, and/or *exploratory-transformational* creativity. The system itself is creative only in the sense that it encodes the musical constraints of composer, system designer, and user.

The degree of control relinquished or retained by the composer can be seen as points on a continuum reflecting the size of the limited search space (or conceptual space) defined for the non-composer. Highly specified scores allow a very restrictive search space for creative interpretation, further along the continuum graphic scores increase the space available to a performer. So what I was looking for was a way to define a search space that is wider than a limited set of pre-composed fragments, perhaps to provide a guide through that space, and to create a mechanism for this space to be explored by the performer/listener.

It would be a composing system, rather than playback system, with an interface that encouraged depth of involvement, and would provide the performer/listener with a high degree of creative influence on the actual realisation, but would also still enable a composer's thumbprint to be present. This obviously relates the work of Cope [2], as the 'thumbprint' I describe could easily be seen as musical 'style', but what I needed was not a way of recreating style, but of playing with parameters of style, and using it as a compositional *influence* on a new piece.

2.2 What Does Each Block Need?

In order to perform the tasks outlined above each block needs two linked but distinct systems: A *bottom-up* composition system, and a *top-down* composition system. Bottom-up refers to the development of small-scale ideas which may combine to form part of an actual piece, such as melodic fragments, or harmonic colours. Top-down refers to the development of form and structures which may exists at a higher level of abstraction in the piece than actual fragments of music.

Bottom-Up Composition System

The bottom-up system is responsible for taking in music passed to it and creating a new fragment of music based on what its 'home' music is, perhaps what it is currently playing, and the nature of the music just passed to it. The reason the system is termed bottom-up is due to the relatively short duration of the musical fragments being passed around and composed.

Top-Down Composition System

The top-down system takes a higher view. Not all blocks will be playing at the same time. As well as music being passed around between blocks, information is also distributed on what is currently sounding. Each block then uses its top-down system to determine whether it is appropriate for it to join in, and is therefore a distributed system responsible for the overall form of the emerging piece.

A complex relationship is possible between the two distributed systems, determining the manner in which each influences the other. This relationship is crucial before any notion of large scale structures are possible with the system, but is beyond the scope of this paper and is currently under development.

The remainder of the paper discusses the bottom-up composition system.

3 Bottom-Up Composition System

Many mechanisms have been employed in the task of algorithmic composition, from the stochastic processes utilised in (possibly) the earliest example of computer generated composition: *The Illiac Suite* 1955 [3], to the more recent application of cellular automata by composers such as Miranda [4]. I decided to focus on the use of a GA model in the bottom-up composition system for two reasons: the increased use of GA's in generative artistic systems, and the resonance the simple GA model created with at least part of the processes used in my own (non-generative) compositional practice. For a useful overview of work on GA's and musical composition see Burton and Vladimirova [5].

3.1 What Aspects of Musical Creativity Are Sympathetic to GA Simulation?

A commonly used compositional process may be described as taking an existing musical idea and changing it in some way. Musicians in various styles and genres may follow the process differently, some through improvisation, others through pencil and paper, but what is most often practiced is taking an existing idea and *mutating* it to provide a new idea. In fact mutation is closely related to notions of development, which lie at the heart of western musical concepts of form and structure. It may even be possible to see development as *directed* mutation.

In this model of [idea ⇒ mutation ⇒ new idea] we see three aspects to the creative act: creating/selecting the initial starting idea, mutation processes, and assessment of the results of mutation. With the core elements of GA's being mutation (including cross-over) and fitness assessment there appears to be a strong correlation with at least some aspects of the 'human' *process* of generating musical material. The degree of any similarity is likely to be in part determined by the *nature* of the mutation and selection processes.

We do not, of course, escape any conceptual difficulties related to artificial creativity by using this model. The similarity of process does suggest however, that GA's might be a good starting point and suitable as a creative support tool.

3.2 Composition as Exploring a Search Space

GA's were developed primarily as an *optimised* search tool. Even if there is a correlation between the mechanisms of a GA, and the (human) compositional process of [idea ⇒ mutation ⇒ new idea], can the aims of composers, who as a group are not usually happy to accept the most efficient solution over the most interesting, be met by mechanisms primarily devised for optimisation? If interesting pathways are to be found, any traversal through a search space should be guided in some way.

"...creativity is not a random walk in a space of interesting possibilities, but...directed". Harold Cohen [6].

"As with most problem-solving activities, musical tasks like composition, arranging and improvisation involve a great deal of search...Certainly, the notion of 'right' is individual, but a typical musician 'knows what she likes', and the aesthetic sense guides the search through the various problem spaces of notes, chords, or voicings". Biles [7].

How then, is the search to be directed? The direction taken by a GA at any given time is determined primarily by the nature of the mutation and crossover operators, and the mechanisms for selection. Due to the clear importance placed on selection I will discuss this aspect in more detail later.

In a standard GA the mutation and crossover operators are random and take place on a simple bit array genotype. The bit array obviously has a task specific representation in some part of the system, but the actual manipulations take place on the *abstraction*, causing them to be more useful in a generic problem space but less useful in specifically *musical* manipulations.

In order to make the GA more 'musical', two important changes are made to the algorithm with regards to mutation and crossover. Firstly, the genotype was made less abstract, and more closely related to the problem space (the actual genotype used in the system is discussed below), and secondly the mutation/crossover operators are designed with musical processes in mind. The algorithm makes *musical* mutations to *musical* phenotypes. This is not a unique perspective, and one of the most notable musical GA systems, and in my mind one of the most successful, GenJam by Al Biles [7] also performs musical operators on two types of musical phenotype. Indeed in the most recent incarnation of GenJam the operators are considered sufficiently 'musical' to dispense with fitness measures entirely [8].

The difference here is that while GenJam has been developed within a specific musical style allowing acceptable restrictions to be placed on the genotype design, such as the limitations of pitch to 14 values per half measure, and a rhythmic resolution of eighth notes, 'MusicBlox' is not *intended* to be style dependent. Biles also makes clear that the limitations on genotype design are deliberately imposed in order to limit the potential search space.

"...allowing greater rhythmic and chromatic diversity would increase the string lengths needed for measures, thereby exploding the size of the space searched"[7].

With operator limitations however, the actual size of the space explored by the system is not necessarily 2^n (with n representing the string length of the genotype), so such considerations become a balance between genotype design and genetic operators.

3.3 The Fitness Function Problem

Much of the work previously undertaken in the area of using GA's for composition focuses on the problem of assessing fitness. Often determining what is 'good' about a particular phenotype is equated to deciding what is 'good' music in a general sense.

There are two common approaches to assessing fitness:

1. Use a human critic to make the selection - an Interactive Genetic Algorithm (IGA). This requires an individual to use all the real world knowledge they possess to make decisions as to which population members should be promoted into the next generation.
2. Automatic Fitness Assessment. Which means encoding sufficient knowledge into the system to make fitness assessment automatic after each mutation and crossover process.

The problem with using an IGA, is the time taken to make assessments, first described by Biles [7] as the "fitness bottleneck". This describes the difficulty in trying to assess many possible musical solutions due to the temporal nature of the medium. The benefit however, is that we are not required to develop a formal rules base. With Automatic Fitness Assessment we escape the fitness bottleneck, but face the daunting task of encoding the essence of the music we want to encourage.

In fact what we often see in examples of GA music are two distinct goals being suggested, if not explicitly articulated:

1. Creating 'original' music, where the aims are more closely aligned with the subjective process of conventional composition. When a composer sits down in front of a blank sheet of paper, they don't normally feel the need to explicitly define 'what music is' before notes are placed on the page, even though what they write may in fact contribute to, or even extend, such a definition.
2. The 'objective' validation of a given rules-set, as well as other aspects of the algorithm, by providing compositional output as a test. In effect asking, 'how good are my rules'. Examples of this work include Wiggins, Papadopoulos, Phon-Amnuaisuk, Tuson, [9], where the output is designed to be assessable 1st year undergraduate harmony exercises.

The primary difference of course is that the first goal is subjective and creative, whilst the second aims to be objective and may be seen as principally a musicological, or music-analytical task. I am not suggesting that the two goals are mutually exclusive, as most musicians know, merely that much previous work tends to emphasise one or the other.

Baring in mind my previously stated creative goals, and the desire to map out a search space, I decided to supply the algorithm with a starting musical fragment to evolve from, and a target musical fragment to evolve towards. In the case of an individual Music Block, the starting musical fragment is most often the 'home' music of the block, and the target musical fragment is any music received from an adjacent block. This has the effect of both directing the search and making fitness assessment significantly easier. The fitness function can then be described as the similarity of a specific phenotype to the provided target music. The nature of the similarity test is described in section 3.6 below.

This approach has similarities to previous work on the use of GA's for thematic bridging by Horner and Goldberg [10]. My aim however, is not to play in sequence all the solutions found on the evolutionary path from starting to target music, in the manner of phase or minimalist music, but to generate a single point solution at a specified location between the two. Another key difference is that Horner and Goldberg see the nature of the path as one of optimisation: "Thematic bridging bears a close resemblance to another GA application, job scheduling". In discussion of their results they describe developing, "An alternative approach…in an attempt to reduce the problem's nonlinearities to a reasonable level and improve linkage". By contrast the evolutionary path in MusicBlox aims to be as loose as possible, whilst still retaining forward motion between starting to target music. In essence the function is to *explore* the space, and find musically interesting solutions, rather than perform a linear morph.

Another related earlier work is that of Ralley [11]. Here the initial population of the GA was "…seeded with melodies similar to the user's input melody", which has obvious similarities to my notion of starting music. There is also a similarity in goal, and Ralley states two aims: to develop a GA as a tool for searching melody space interactively, and to use the system to move towards an explicit goal. The developed system was deemed successful in the first aim, "…in that it produced new and novel musical material derived from the users input", but less successful in the second.

3.4 The Algorithm

The individual steps of the algorithm are detailed below:

1. Create initial population. The home music of the block is used to create a population of identical phenotypes, which comprise the initial population.
2. Select a population member (in turn) and perform mutation operators and crossover. As described above, rather than 'blind' bit flipping, the mutations have been modelled on processes used in composition, from addition and deletion of notes, to transpositions, inversions and reversals. Additionally, the application of any operators that require the selection of new pitches is based on a harmonic evaluation of the musical target.
3. Assess the fitness of the mutated population member. Fitness is defined as how similar the phenotype is to the target, discussed below.
4. Accept into the population or discard. If the fitness value for the mutated population member is higher than the lowest fitness found in the population, it replaces the low fitness population member, and is stored as a musical point on the evolutionary path to the target, otherwise it is discarded.
5. Repeat until the specified point on the evolutionary path towards the target is reached.

3.5 The Genotype

Rather than represent musical attributes as a bit array, a simple musical object model was chosen. The genotype also contains meta-data attributes, which hold fitness values for each phenotype. A simplified object model for the genotype is described below:

Section: The Section object is the outer object representing a fragment of music. It has attributes for target match fitness, and an array of note objects.

Note: The note object represents the values for a single note, and has attributes for *velocity, pitch, duration*, and target *fitness* (0/1).

3.6 Fitness Assessment

Fitness of each phenotype is assessed on how *similar* it is to the target. The notion of *musical* similarity is complex to measure and is the focus of work on current extensions to the algorithm. Currently however, similarity is simply defined as:

Musical Section fitness = # of Notes with a Fitness of 1/ # of Notes in Target,

Where the fitness of a note is 1 if it has the same pitch as a note in the same position in the target musical section, 0 otherwise.

This measure relates to that used by Horner and Goldberg [10], where:

"Fitness=0.5*(*Pcommon*+*Pcorrect*), where *Pcommon* is the proportion of elements in the resulting pattern also in the final pattern, and *Pcorrect* is the proportion of elements in the resulting pattern which match their corresponding element in the final pattern."

In essence then, the current measure of similarity is the same as *Pcorrect* in Horner and Goldberg.

3.7 The Mutation Operators

As discussed above, the system is required to evolve musical fragments towards a target, *and* generate musically interesting results on the way. These are potentially conflicting, multi-fitness functions, and the mechanism chosen for resolving this is to not assess musical interest directly, but include processes through the mutation operators that are more likely to provide a musically successful output.

The mutation operators used (presented in the order applied) include:

1. Add a note
2. Swap two adjacent notes
3. Transpose a note pitch by a random interval
4. Transpose a note pitch by an octave
5. Mutate the velocity (volume) of a note
6. Move a note to a different position
7. Reverse a group of notes within a randomly selected start and end point
8. Invert notes within a randomly selected start and end point. (Inversion around an axis of the starting pitch).

9. Mutate the Duration of a note
10. Delete a note

Each of the mutations listed is applied according to a user specified probability, and only to individual, or groups of notes, if the application of the mutation will not in any way alter notes that already have a targetFitness value of 1. The fitness of each phenotype is recalculated after each mutation. Therefore if a mutation has resulted in a note having a fitness of 1, the note will not be mutated further. This optimisation of mutations enables the evolution process to be efficient enough to withstand interesting musical manipulations and still move towards the target.

A variety of crossover methods were implemented and although the musical results were often interesting the effect was to significantly hamper the algorithm's search for the target. Crossover was therefore removed from the algorithm pending further investigation.

3.8 Selecting New Pitches

When the system is started a harmonic analysis of the target musical section is made. A table is constructed (see fig. 1 below) in which each index location represents the target's 'degree of membership' of that scale type (Major and Minor), for that scale degree (C to B).

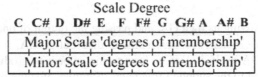

degree of membership = number of pitches present in scale/number of pitches

Fig. 1. This shows the table representing the 'degree of membership' of the target analysed. The table could be extended to include other scale type.

These analysis values, in conjunction with user-supplied weightings, are used to determine the *probability* as to which scale (eg. C# major) a new pitch should be drawn from when transposing a pitch within an octave, or creating a new note.

Once a scale is selected, the final pitch within that scale is chosen using a similar process. Now however, instead of the target music being analysed, the table is populated from an analysis of the other notes in the phenotype, and additional user defined fitness weightings are included to rate certain intervals over others. The degree to which this pitch selection process is applied maybe controlled by the user, giving the system a type of harmonic 'dial' that can be turned up to give tight constraints, or down to produce 'free' harmonies.

This rough estimation of harmonic analysis allows new pitches to be selected that are more likely to be present in the target music, but with enough fluidity that new

500   A. Gartland-Jones

harmonies will often occur. Of course, a more conventional harmonic analysis could
be applied, but this solution offers a more flexible harmonic model than methods associated with historical, functional (motion directed) harmony.

3.9 Rhythm, Note Duration, and Dynamics

In the current model no assessment is made of musical similarity on the basis of note
duration or dynamic. Also, no attempt is made to form 'constructed' rhythms in phenotypes on the evolutionary path. All such aspects are left to chance or are an outcome
of the mutations and assessments described above, for two reasons. Firstly, rhythmic
noise is more acceptable in the musical output than harmonic noise. As a common
characteristic of formal composition systems is a trade off between predictabil-
ity/consistent results, and experimentation/variable results, the effect of any rhythmic
noise encountered is considered acceptable given the benefits of greater experimentation. The second reason is the difficulty in resolving incorporation of additional fitness
measures. The later aspect is currently under investigation as part of an attempt to
achieve a more useful measure of musical similarity.

4 Conclusions and Next Steps

By each block possessing its own GA, and passing the output of one as the target input
of another, the whole system becomes a kind of distributed IGA.

In terms of the GA itself, it does, at least to me, appear to produce 'musical' output
that is clearly related to the supplied fragments. Samples of some output from the GA,
which were used in interactive installation 'nGen.1' exhibited as part of Brighton
Festival 2002, UK, and funded by South East Arts, may be downloaded at
http://www.atgj.org/drew/, following the link to 'Musical Examples', then scrolling
down to 'Example of OriGen Output'.

Other possible next steps for further development include:
- Assessing fitness duration and velocity as well as pitch.
- Fitness evaluation could be extended beyond single yes/no note match, and
  could include an analysis of note patterns, which could be graded as closer or
  further away from the target.
- I have asserted that the mutation operators are based on some of the processes I find myself using when developing musical ideas, but an obvious next
  step would be to research the compositional processes of other composers in
  a more systematic way. Mutation operators could then be made both more
  suitable for an individual composer, and more complex as a combination of
  smaller elements.

Acknowledgements. I would like to thank Prof. Phil Husbands and Dr. Adrian Thompson for ongoing information, advice and general support.

References

1. Boden, M. Computer Models of Creativity, in *Handbook of Creativity*, ed. Sternberg, R. J. Cambridge University Press (1999). ISBN 0-521-57604-0.
2. Cope, D. *Computers and Musical Style* (1991). ISBN 0-89579-256-7.
3. Hiller, L, & Issacson, L. Musical Composition with a High-Speed Digital Computer, Journal of the Audio Engineering Society (1958).
4. Miranda, E, R. On the Origins and Evolution of Music in Virtual Worlds, in *Creative Evolutionary Systems*, ed. Bentley, P, Corne, D, Academic Press (2002).
5. Burton Anthony R, & Vladimirova, T. Generation of Musical Sequences with Genetic Techniques. Computer Music Journal. Volume: 23 Number: 4 pp. 59–73, (1999).
6. Cohen, H. A self-Defining Game for One Player: On the Nature of Creativity and the Possibility of Creative Computer Programs. Leonardo, Vol. 35, No1, pp. 59–64, (2002).
7. Biles, John. A. GenJam: A Genetic Algorithm for Generating Jazz Solos. Proceedings of the International Computer Music Conference 94 (1994).
8. Biles, John. A. GenJam in Transition: from Genetic Jammer to Generative Jammer. Proceeding of Generative Art 2002, Milan 11–13 December (2002).
9. Wiggins, G. Papadopoulos, G. Phon-Amnuaisuk, S, Tuson, A. Evolutionary Methods for Musical Composition. Proceedings of the CASYS98 Workshop on Anticipation, Music & Cognition, Liège (1998).
10. Horner, A. and Goldberg, D. E. Genetic algorithms and computer-assisted music composition. Technical report, University of Illinois (1991).
11. Ralley, D. Genetic Algorithms as a Tool for Melodic Development. Proceedings of the International Computer Music Conference 95 (1995).

Towards a Prehistory of Evolutionary and Adaptive Computation in Music

Colin G. Johnson

Computing Laboratory
University of Kent
Canterbury, Kent, CT2 7NF
England
C.G.Johnson@ukc.ac.uk

Abstract. A number of systems have been created which apply genetic algorithms, cellular automata, artificial life, agents, and other evolutionary and adaptive computation ideas in the creation of music. The aim of this paper is to examine the context in which such systems arose by looking for features of experimental music which prefigure the ideas used in such systems. A number of ideas are explored: the use of randomness in music, the view of compositions as parameterized systems, the idea of emergent structure in music and the idea of musicians performing the role of reactive agents.

1 Introduction

In recent years the use of evolutionary algorithms (genetic algorithms, genetic programming, et cetera) and related ideas of "adaptive systems" (artificial life, cellular automata, agents) have been used in many ways in creating music. These ideas are surveyed in [6,10,2,3], and have been the topic of a number of recent workshops[1]. Typically such systems are concerned with the generation of electronic music; however in some situations these techniques are used to generate scores for acoustic realization (e.g. [14,17]).

The aim of this paper is to investigate how some of the ideas involved in the creation of pieces of music via evolutionary algorithms (e.g. the use of randomness as a source of musical material, the use of ideas of reactive agency as a structural principle in music) can be found in mid-to-late twentieth century experimental music.

The main motivation for this is to understand one of the contexts in which these experiments in automated composition and structuring of musical materials arose. It is interesting to speculate as to whether the various experiments with the components of evolutionary and adaptive algorithms "paved the way"

---

[1] See e.g.
http://www.generativeart.com/, http://www.csse.monash.edu.au/~iterate,
http://galileo.cincom.unical.it/esg/Music/ALMMAII/almmaII.htm and
http://www.tic.udc.es/~jj/gavam/

S. Cagnoni et al. (Eds.): EvoWorkshops 2003, LNCS 2611, pp. 502–509, 2003.
© Springer-Verlag Berlin Heidelberg 2003

for the application of these algorithms in composition. Would musicians have been led to consider these ideas without the "scene setting" which these musical experiments provided? Furthermore are there areas of computing which could lead to the development of interesting musical developments, yet which haven't been explored because the right kind of musical groundwork needed to provide the links between the computational and musical structures has not been layed?

Defining exactly what is meant by "experimental" (or an alternative term such as "avant-garde") music is potentially complex (see e.g. the discussion in [13]). Basically I mean the sort of music discussed e.g. by Griffiths in [9]. I shall use the term "experimental music" as a shorthand for these musics for the remainder of the paper.

2 Randomness

One of the key features of evolutionary computation is *randomness*, which acts as the main generator of novel combinations of material, the creator of new material and the engine driving the route through search-space.

Small-scale randomness and variability is a feature of all musical performance. Nonetheless the use of randomness as a deliberate organizing (or disorganizing) feature of music became an important feature of experimental music from the 1960s onwards. However this move towards randomness does not come from a single motivation.

One motivation is to help performers to move away from cliché and provide new sources of material which would not otherwise be considered. In particular John Cage has emphasized that it is difficult for a performer just to do something radically innovative just by trying to do it; they end up falling into cliché and habit:

> Tom Darter: *[...]* some of the players said, "Given that this is going to appear random to the audience, can't we do anything?" What would be your response to that?
> John Cage: Well, if they just do anything, then they do what they remember or what they like, and it becomes evident that this the case, and the performance and the piece is not the discovery that it could have been had they made a disciplined use of chance operations. [12]

There may be some prefiguring here of the way in which randomness is used in computational search and exploration algorithms. Instead of attempting to encode a set of suggested rules for dealing with certain kinds of situations (*a la* expert systems), which reaches a limiting point, the randomness of search enables new ideas and combinations of ideas to be formed.

A second motivation, which we may also associate with Cage, is to do with separating intention and action.

> Another idea in *Cartridge Music* [7] is not just transforming sound, but that of performers getting in each other's way in order to bring about

nonintention in a group of people, so, if one person is playing a cartridge with an object inserted in it, another one is turning the volume down independently. I never liked the idea of one person being in control. - John Cage [12].

This use of randomness promotes a different relationship between composer and audience than is traditional in music. Instead of the action being focused on creating a particular outcome, the focus is on carrying out the action itself, regardless of consequence. Indeed the 'absence of intention' [12] is a key component of the performance/composition aesthetic in Cage's music.

An related important aspect of Cage's aesthetic is the idea of the score as an indication of the actions to be performed, rather than the desired outcome. Such an idea seems radical in music, yet is natural in computing; computers are typically programmed by giving them instructions for action (which in the case of a stochastic algorithm may give rise to an unpredictable result) rather than by specifying the desired end-result of a procedure. Indeed it is interesting to note that much effort is being expended in computing research in trying to go the other way, and find ways of programming computers in terms of specifications of desired output rather than giving an explicit description of action.

A third aim of randomness is radically different. This is the idea of creating music based on the use of stochastic processes to generate the fine-level details of a piece of music (e.g. the particular pitches used and the placement of sounds in time). Instead of a compositional process being concerned with the placement of those individual sound-events, the composition becomes a matter of manipulating the statistical distributions in the sound-space. This approach to composition was pioneered as pioneered by Iannis Xenakis in the 1950s and 60s, giving rise to pieces such as $ST/4$ [24], and detailed in the book *Formalized Music* [25]. Some related ideas are discussed by Stockhausen (e.g. the essay *Composing Statistically* in [22]).

This approach to composition arises from a line of thought about complexity in music. Around the time that these pieces were created increasingly complex musical superstructures were being created using 'mathematical' schemes for the organization of musical materials. Given the complexity of material produced as an end result of these schemes, the replacement of the direct methods which a stochastic generation of the same 'density' of complexity seems a natural progression. Perhaps there is a similarity here to statistical mechanics; even though we believe that the large scale behaviour of a fluid is governed by a vast number of small deterministic actions of molecules, it is easier to produce a similar effect (in this context for studying the phenomena in question) by making a statistical approximation to the deterministic phenomenon.

A similar kind of thinking seems to underpin the creation of music via methods such as cellular automata (e.g. [15,16,19,20]). In such systems the composition lies in the creation of rules about the relationship between the objects in the system, rather than about the direct placement of individual sounds. Nonetheless there are differences too; in a cellular automaton model the manipulation of the global effect is done in a bottom-up fashion; the music emerges from the effects of the rules. In the Xenakis approach the attitude is much more top-down, concerned with the sculptural manipulation of the space of where the sounds are allowed to be. Perhaps the cellular automata models are closer to a tight version of the total serialist systems, where parameters such as pitch, volume, et cetera, are organized by visiting different values in a fixed order.

3 Music and Parameters

Leigh Landy [13] has suggested that a key characteristic of experimental music in the mid-late twentieth century is the conceptualization of music in terms of *parameters*. Firstly he notes that a trend in musical history has been towards the expansion of the scope of a particular musical parameter (though it is unlikely to have been seen as such until recently). For example pitch expands out from early scales with a small number of notes, to more complex scales involving more pitches, to a point at which pitch can be described as a continuum measured in Hertz. He points out that the idea of regarding music as exploration of parameter-space was first formalized by Joseph Schillinger in the 1940's, and traces the development of the idea first as an *analytic* technique, then examines the role of explicit thought about parameters as a compositional tool. We may want to think of this as the analytic technique being turned around to form the basis of a synthetic technique.

There are a number of ways in which the formalization of music in terms of manipulable parameters can be used directly by composers. For example (again the example is from [13]) Stockhausen makes use of the idea of coupled actions through parameterization, e.g. in matching physical action to sound by realizing both of them from a single underlying parameter. Another example comes from the idea that timbral characteristics can be manipulated directly in the same way as other characteristics of music, e.g. in Schönberg's idea of the *Klangfarbenmelodie*—a melody made up of different timbres ('sound colours') rather than pitches.

A related point has been made by Barrett [1], who suggests that composers such as Boulez, Xenakis and Stockhausen work within *"a general field theory of which particular compositions are offshoots"*. Offshoots (suggesting the idea of moving away from a particular starting point) may not be the best choice of word here; perhaps the concept is more of choosing a particular point within a parameter space (or a particular route through that space), or else one of "collapsing" the parameter space to a particular piece.

Such an approach has been considered by Pierre Boulez:

> I prefer to think of mobile form as a *material form*, i.e to regard it as a basis for one or more "fixed" scores chosen from the multiplicity of possibilities. [5].

Many of these ideas have close parallels with the idea of *search space* in computer science (a concept which was being developed at the same time as many of these ideas were being developed in music) and with the older idea of *phase space* in theoretical physics. The conception of music existing in some kind of space which can be explored seems to be key to many of the applications of evolutionary and adaptive systems in music; after all, it is difficult to apply a search algorithm without first describing the space to be searched.

The granularity of this space varies from system to system. Sometimes an entire composition may exist at each point in the phase space (e.g. consider a space where each point consists of a set of rules for a cellular automaton which in turn gives rise to the ordering of sounds as that automaton runs). Alternatively the act of moving through the space may itself create the temporal structure of the piece (examples are given in [15,18]), with each timestep representing e.g. a time interval, or the triggering of a single action.

A number of interesting perspectives on the shape of sound space (and whether we always want to consider sounds as living in such a space) are given by Wishart [23].

4   Emergence and Interaction

Another concept which is important in many evolutionary and adaptive pieces is the idea of structure as an emergent property from small-scale actions in the music (a bottom-up approach to structure) rather than as a top-down concept where an explicit form is created and various musical components placed within that form.

This relates to another concept which is the idea of musicians (or sound-creating devices) acting as *agents*, reacting to other musicians according to some instructions. The newness here is perhaps in the idea of simultaneously following a score and reacting to other players. Clearly micro-reactions (e.g. adjusting tuning to be "in tune" with other players) have long been a part of music, and improvising alongside and in response to other players is a core part of most musical traditions.

These ideas are important in a number of experimental compositions too. Some examples are given in the work of Cornelius Cardew, in particular in *The Great Learning* [8]. An excerpt which shows this clearly is paragraph 7 from the piece, illustrated (in it entirety) in figure 1. A performance of this piece can easily consist of 60–90 minutes of complex, structured music, yet the entire formal structure can be expressed on a single page. The piece consists of a number of lines of text, each of which are sung independently by each member of a large vocal ensemble. However there is interaction between the various members of the ensemble, because at the end of each line the singer chooses a note which they can hear being sung elsewhere in the ensemble. Therefore the piece gradually converges on a single pitch or pair of pitches. The fact that each performer moves through the piece at their own pace adds to the richness of the texture.

This "convergence" clearly reflects aspects of the text. In particular other parts of the text from other "paragraphs" in the piece make the ideas clearer still, emphasizing the Confucian ideal of good government arising first from personal clarity of thought and self-discipline, through families and communities

The Great Learning, paragraph 7

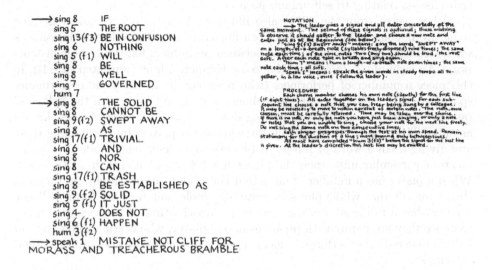

Fig. 1. Paragraph 7 from Cardew's *Great Learning*.

and eventually giving rise to 'equilibrium' throughout the empire. Thus the bottom-up emergence of structure reflects the ideals of the piece.

These aspects have been emphasised more than once in analytical exegeses of this piece. Michael Parsons emphasises the economy of the notation and the way in which it exploits differences and random initial choices (perhaps in contrast to many traditional musics which rely on commonality as the basis for their structure, e.g. keeping to a common time):

> The idea of one and the same activity being done simultaneously by a number of people, so that everyone does it slightly differently, [...] gives one a very economical form of notation—it is only necessary to specify one procedure and the variety comes from the way everyone does it differently. This is an example of making use of "hidden resources" in the sense of natural individual differences (rather than talents or abilities) which is completely neglected in classical concert music. (Michael Parsons, in [21]).

Some related themes are discussed by Dusman:

Paragraph 7 successfully isolates the individual as the only variety-introducing factor in the piece. As in much experimental music, the score is composed in such a way as to produce a range of outputs, valuing its probabilistic properties of generation, not its specific ones. What is perceived by the listener in Paragraph 7 is an evolutionary harmonic procedure.

Other issues relating to self-organization are given by Beyls [4].

Other pieces from the 1970s also illustrate ideas of agency and emerging structure in contrasting ways, and both illustrate the idea of the evolution of a sonic structure through interactions between performers following some instructions. An example is *Telephone Matrix* by Kenneth Maue (reprinted in [11]). In this piece a number of performers (who need not be in the same performance space) are each equipped with a telephone. The piece begins with the "players" (as the performers are termed) writing down a few words and phrases, known as units. These units are then transmitted to other players by telephoning them and giving them a word or phrase; if two attempts have been made to communicate a particular unit, then the player must purge that unit from their list. When a player has a number of units that they can form into a phrase, they can then transmit that whole phrase; eventually whole sentences are formed. When a player has a full sentence, they can stop participating. When everyone has a sentence they are happy with (apart from one player, who is still calling and not being answered) the performers meet together to "compare the results of their activities".

References

1. Richard Barrett. Michael Finnissy—an overview. *Contemporary Music Review*, 13(1):23–43, 1995.
2. Peter J. Bentley, editor. *Evolutionary design by computers*. Morgan Kaufmann, 1999.
3. Peter J. Bentley and David W. Corne, editors. *Creative Evolutionary Systems*. Morgan Kaufmann, 2001.
4. Peter Beyls. Musical morphologies from self-organizing systems. *Interface*, 19:205–218, 1990.
5. Pierre Boulez. *Orientations*. Faber and Faber, 1986.
6. Anthony R. Burton and Tanya Vladimirova. Generation of musical sequences with genetic techniques. *Computer Music Journal*, 23(4):59–73, Winter 1999.
7. John Cage. *Cartridge Music*. Peters Edition, 1960. EP6703.
8. Cornelius Cardew. *The Great Learning*. Matchless Publications, 1971.
9. Paul Griffiths. *Modern Music: the Avant-Garde since 1945*. Dent, 1981.
10. Colin G. Johnson. Genetic algorithms in visual art and music. *Leonardo*, 35(2):175–184, 2002.
11. Rogerr Johnson, editor. *Scores: An Anthology of Modern Music*. Schirmer, 1981.
12. Richard Kostelanetz, editor. *Conversing with Cage*. Omnibus Press, 1998.
13. Leigh Landy. *What's the Matter with Today's Experimental Music?* Harwood Academic Publishers, 1991.

14. K. McAlpine, E. Miranda, and S. Hoggar. Making music with algorithms: A case-study system. *Computer Music Journal*, 23(2):19–30, 1999.
15. Eduardo Reck Miranda. Cellular automata music: An interdisciplinary project. *Interface*, 22:3–21, 1993.
16. Eduardo Reck Miranda. Granular synthesis of sounds by means of a cellular automaton. *Leonardo*, 28(4):297–300, 1995.
17. Eduardo Reck Miranda. On the music of emergent behaviour: What can evolutionary computing bring to the musician? In *Workshop Proceedings of the 2000 Genetic and Evolutionary Computation Conference*, 2000.
18. Artemis Moroni, Jonatas Manzolli, Fernando J. von Zuben, and Ricardo Gudwin. Vox populi: evolutionary computation for music evolution. In Bentley and Corne [3].
19. Gary Lee Nelson. Sonomorphs: An application of genetic algorithms to the growth and development of musical organisms. In *Proceedings of the Fourth Biennial Art and Technology Symposium*, pages 155–169. Connecticut College, March 1993.
20. Gary Lee Nelson. Further adventures of the Sonomorphs. Conservatory of Music Report, Oberlin College, 1995.
21. Michael Nyman. *Experimental Music: Cage and Beyond*. Schirmer, 1974. (second edition by Cambridge University Press, 1999).
22. Karlheinz Stockhausen. *Stockhausen on Music*. Marion Boyars, 2000. Edited by Robin Maconie.
23. Trevor Wishart. *On Sonic Art*. Harwood Academic Publishers, 1996. second edition, revised by Simon Emmerson; first edition 1985.
24. Iannis Xenakis. *ST/4 for string quartet*. Boosey and Hawkes, 1962. 080262.
25. Iannis Xenakis. *Formalized Music: Thought and Mathematics in Composition*. Indiana University Press, 1971.

ArtiE-Fract: The Artist's Viewpoint

Evelyne Lutton, Emmanuel Cayla, and Jonathan Chapuis

INRIA - Rocquencourt, B.P. 105, 78153 LE CHESNAY Cedex, France,
{Evelyne.Lutton,Emmanuel.Cayla,Jonathan.Chapuis}@inria.fr,
http://www-rocq.inria.fr/fractales/ArtiE-Fract

Abstract. ArtiE-Fract is an interactive evolutionary system designed for artistic exploration of the space of fractal 2D shapes. We report in this paper an experiment performed with an artist, the painter Emmanuel Cayla. The benefit of such a collaboration was twofold: first of all, the system itself has evolved in order to better fit the needs of non-computer-scientist users, and second, it has initiated an artistic approach and open up the way to new possible design outputs.

1 Randomness in a Creative Process: Freedom or Uncontrollability?

The programming of a software for artistic purposes is a challenging task: the computer software framework usually locks the user inside many interaction constraints that are sometimes considered as an obstacle to creativity.

Recent advances in interactive evolutionary algorithms (IEA) [2] have initiated many attractive works, mainly based on the idea of "maximising the satisfaction of the user" via a guided random search [1,15,19,20,21,24,25,26,27]. The use of randomness in this particular way yields an additional "creativity component," that may or may not be considered as helpful by the artist, with respect to the way this random component is controllable. The success of such approaches is thus strongly dependent at least on the choice of a convenient representation and of an adequate set of genetic operators.

Additionally, a problem to be considered very carefully when using randomness in an artistic process is linked to the use of randomness itself, which can be considered by the user as an uncontrollable component of the system, triggering a reflex of reject. A disturbing question can indeed be raised: who is the "artist": the user or the machine ? Both positions can be defended, of course, and several fully machine-driven artistic attempts have been performed.

In the ArtiE-Fract experiments [8], we have tried to put the artist in the loop, and worked to limit this negative perception of the use of randomness. The system has been designed and programmed in order to let the user fully drive or partially interact with the evolutionary process at any moment (ArtiE-Fract is not an "artist"). The idea is to give the possibility to the human user to tame and to adapt the random behaviour of the system to his own artistic approach.

This paper is organised as follows: The ArtiE-Fract software is presented in section 2 (details can also be found in [8]) the artistic experiment is then detailed

S. Cagnoni et al. (Eds.): EvoWorkshops 2003, LNCS 2611, pp. 510–521, 2003.
© Springer-Verlag Berlin Heidelberg 2003

in section 3, and figures 4 to 9 present a sample of the work of Emmanuel Cayla. Main influences on the design of ArtiE-Fract and conclusions are presented in section 4.

2   ArtiE-Fract: Interactive Evolution of Fractals

Fractal pictures have always been considered as attractive artistic objects as they combine complexity and "hierarchical" structure [3,13]. Going further into their mathematical structure (for example with iterated function systems attractors, [4]) one has a more or less direct access to their characteristics and therefore, shape manipulation and exploration is possible [11,22].

In ArtiE-Fract, an Evolutionary Algorithm (EA) is used as a controlled random generator of fractal pictures. The appropriate tool is interactive EA, i.e. an EA where the function to be optimised is partly set by the user in order to optimise something related to "user satisfaction." This interactive approach is not new in computer graphics [27,24], we extended it to the exploration of a fractal pictures space and carefully considered flexibility with the help of advanced interactive tools related to the specific fractal model that is used.

2.1   Man-Machine Interaction

Interaction with humans usually raises several problems, mainly linked to the "user bottleneck" [20]: human fatigue and slowness[1]. Solutions need to be found in order to avoid systematic and boring interactions. Several solutions have thus been considered [20,26,2]:

- reduce the size of the population and the number of generations,
- choose specific models to constrain the research in a priori "interesting" areas of the search space,
- perform an automatic learning (based on a limited number of characteristic quantities) in order to aid the user and only present him the most interesting individuals of the population, with respect to previous votes of the user.

Allowing direct interactions on the phenotype's level represents a further step toward efficient use of IEA as a creative tool for artists. The idea is to make use of the guided random search capabilities of an EA to aid the creative process. This is why in ArtiE-Fract, the user has the opportunity to interfere in the evolution at each of the following levels:

- initialisation: various models and parameters ranges are available, with some "basic" internal fitness functions;
- fitness function: at each generation, a classical manual rating of individuals may be assisted by an automatic learning, based on a set of image characteristic measurements (may be turned on or off);

---

[1] The work of the artist Steven Rooke [21] shows the extraordinary amount of work that is necessary in order to evolve aesthetic images from a "primordial soup" of primitive components.

- *direct interaction with the genome*: images can be directly manipulated via a specialized window and modified individuals can be added or replaced in the current population (a sort of interactive "local" deterministic optimisation). A large set of geometric, colorimetric and structural modifications are available. Moreover, due to the specific image model, some control points can be defined on the images that help distort the shape in a convenient, but non trivial manner;
- *parameter setting and strategy choices* are tunable at any moment during the run.

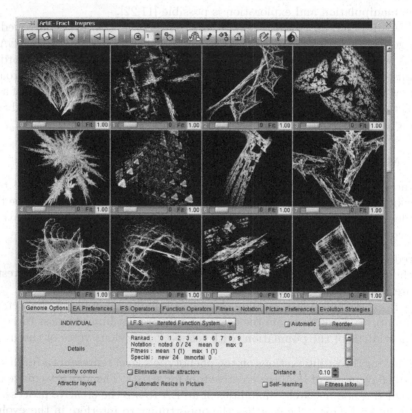

Fig. 1. Main Window of ArtiE-Fract

2.2   Advanced Evolutionary Strategies

Another specific component of ArtiE-Fract is the Parisian approach implementation, which also can be turned on or off at any moment of the evolution. This component has been designed to favour exploration and genetic diversity.

The Parisian approach has been designed relatively recently [10] and is actually a generalisation of the classifier systems approaches [12]. It is based on the capability of an EA not only to push its best individual toward the global optimum, but also to drive its whole population in attractive areas of the search space. The idea is then to design a fitness landscape where the solution to the problem is given by the whole population or at least by a set of individuals, and not anymore by a single individual. Individuals do not encode a complete solution but a part of a solution, the solution to the problem being then built from several individuals that "collaborate."

This approach is to be related to the spirit of co-evolution: a population is a "society" that builds in common the solution that is search for, but on the contrary to co-evolution, the species are not specifically identified and separated. Of course the design of such algorithms becomes more complex than for a direct –standard– EA approach, and the diversity of the population is a crucial factor in the success of a Parisian approach. Moreover, splitting the problem into interconnected sub-problems is not always possible. However, when it is possible to do so, the benefit is great: a Parisian approach limits the computational waste that occurs in classical EA implementations, when at the end of the evolution, the whole final population is dumped except the best individual only. Experiences and theoretical developments have proved that the EA gains more information about its environment than the only knowledge of the position of the global optimum. The Parisian approach tries to use this important feature of EAs.

A Parisian EA may have all the usual components of an EA, plus the following additional ones:

- *two* fitness functions : a "global" one that is calculated on the whole population or on a major portion of it (after a clustering process or elimination of the very bad individuals, for example), and a "local" one for each individual, that measures how much this individual contributes to the global solution.
- a distribution process at each generation that shares the global fitness on the individuals that contributed to the solution,
- a diversity mechanism, in order to avoid degenerated solutions where all individuals are concentrated on the same area of the search space.

Developing a Parisian EA for interactive creative design tools is based on observation of the creative process. Creation cannot be reduced to an optimisation process: artists or creative people usually do not have precisely in mind what they are looking for. Their aim may fluctuate and they sometimes gradually build their work from an exploration. "User satisfaction" is a very peculiar quantity, very difficult to measure, and to embed in a fitness function. This is the reason why ArtiE-Fract has been equipped with a Parisian approach mode that can be activated at any time during the run of the system using a translation module between classical and Parisian populations.

2.3   Evolution of Attractors of Iterated Function Systems

Another important aspect of ArtiE-Fract is the choice of the search space. As we have told before, a way to limit "user fatigue" is to reduce the size of the search space in order to navigate in a space of *a priori* interesting shapes. The choice made in ArtiE-Fract is the space of 2D fractal shapes encoded as iterated function systems (IFS). This gives access to a wide variety of shapes, that may appear more or less as "fractals."

ArtiE-Fract thus evolves a population of IFS attractor pictures, and displays it via an interface, see figure 1. More precisely, these IFS attractor pictures are encoded as sets of contractive non-linear 2D functions (affine and non-affine), defined either in cartesian or polar coordinates. A set of contractive functions represents an IFS, i.e. a dynamical system whose attractor can be represented as a 2D picture. These mathematical objects were considered as interesting as they allow to encode rather complex 2D shapes with a reduced number of parameters.

IFS were extensively studied in the framework of image and signal compression [11,14,23,4], however all IFS models explored in fractal compression were based on affine sets of contractive functions.

From an artistic standpoint, affine IFS give access to an interesting variety of shapes (the "self-affine" fractals). But the use of non-affine functions, beside the scientific interest of exploring this rather unknown space, yields a variety of shapes that may look "less directly" fractal. This is another of the specifics of ArtiE-Fract: three models of IFS are used (affine, mixed and polar), separately or in combination. Each of them induces a slightly different topology on the search space, which gives privileged access to various image types.

Figures 1 and 2 present a set of images created by several users (non-necessarily artists !), that suggest biological images or vegetation, as well as some very "geometric" ones.

This additional freedom, based on the use of non-linear functions seems to be experimentally attractive to artists, as it allows the expression of various inspirations.

Fig. 2. Sample images obtained with ArtiE-Fract

3 Example of an Artistic Process: How Emmanuel Cayla Uses **ArtiE-Fract**

Emmanuel Cayla's first approach to ArtiE-Fract is precisely this flexible access to fractals. Actually the artist advocates that a new intimacy is built between mathematics and painting as it happened during the "Renaissance" with geometry, proportion and perspective. This new relationship doesn't imply only fractals but those are certainly meant to play a key role in this reunion.

To start his exploration of the fractals' universe on the ArtiE-Fract software, the painter first decided to set the graphical parameters of a set functions and to stick with those. He had indeed the feeling that it is too difficult to properly browse such a search domain without defining a static reference framework. The first of these decisions was to put colours aside and to work only with black ink on a white background. The second parameter had to do with noise, also known as "grain" or "distorsion" whose level was uniquely set over the whole initial population.

Indeed, the "noisify" operator of ArtiE-Fract, is an important component of this artistic approach. This operator adds a random noise to the value returned by the function during the drawing of the attractor image, see figure 3. This is an important factor of visibility of scattered attractors, as it conditions the thickness of the simulated "paintbrush."

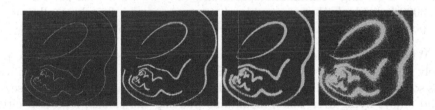

Fig. 3. Original IFS and 3 mutated IFS with various noise levels.

With these two settings, the painter produced the first generations of images having approximately (as it is however produced by a random process, even if it is strongly limited in this particular case) the desired characteristics.

The third of these *a priori* constraints parameters is the image format, in other words the graphical proportion and size of the generated images. For the moment, the default setting (square images) has been used. This point has however been considered as a limitation by the artist, and has to be considered in future evolutions of the software.

Experience shows that extraction of individuals that are interesting from the pictural standpoint is indeed a directed process. All the individuals that are acknowledged as "picturally effective" by the artist, i.e. all the individuals he identifies as matching his artistic visions, are the outcome of a process based on selection and gradual construction. Selection of individuals during the successive generations is a bit more sophisticated of a process than a simple "good"/"bad" categorisation. The painter is familiar with images and the criteria used for conservation and "reproduction" of individuals in the creation process of ArtiE-Fract are going to be based on observations such as : "This individual is interesting for the feeling of movement it produces, I will mate it with that one whose use of blacks and blurs is appealing; I will also add this other one, it is a bit less interesting but the other two could greatly benefit from the way it plays with lines..." As one can see, the artist indeed plays the role of constructor. ArtiE-Fract might bring a great deal of randomness in the process but painters have always worked with randomness and taken advantage of "artistically interesting incidents."

Another interesting aspect is the tuning of the various "usual" genetic parameters, for instance, population size. It appeared pointless for the painter to work with generations rich of hundreds or thousands of individuals for the five following reasons:

– Computation time increases with the number of individuals.
– In any case the population will need sorting.
– Whatever happens, one will remain within the same family of shapes.
– There are enough individuals exhibiting originality, even in smaller sized populations.
– One is dealing with infinite spaces and may we work with 20, 40, 100 or 3000 individuals that we would only encompass an infinitely small space of possibilities with respect to the potential creativity of the system.

What the artist is looking for, before anything else, is expression, more precisely poetic expression. And with ArtiE-Fract, it works. This means that with this software, one is able to browse and discover meaningful shapes like those our mind enjoys flying over: lakes, mountains piercing the clouds, forgotten cities, trees taking over old walls, people standing on icebergs near the North Pole... And it is essentially here, although not only, that this approach of numeric arts moves us.

4   Conclusions and Future Developments

Of course this fruitful collaboration has deeply influenced the design of many components of ArtiE-Fract, as an artist has sometimes a completely different viewpoint on the software tools.

For example, besides the specific way he selects and notates the images, strong control tools were considered as crucial. Direct interactions were thus designed[2], such as simply "killing" an individual and controlling the "reproduction-elimination" step of the evolutionary process. The artist also stressed on the fact that the visual evaluation of a picture is strongly dependent on the surrounding

---

[2] A "strongly controlled evolution mode" is now available in ArtiE-Fract.

Fig. 4. DANCING WOMAN *(Danseuse)* – Emmanuel Cayla

Fig. 5. SWIMMING FOSSILS *(Fossiles nageant)* – Emmanuel Cayla

images and background, point that was neglected by the ArtiE-Fract developers so far.

Another point has to do with the creation of repetitive sequences such as borders: The "modulo," "mirror" and "symmetries" effects have been for instance programmed to produce shapes that can be continuously juxtaposed.

Fig. 6. HARSH HORIZON *(Horizon violent)* – Emmanuel Cayla

Fig. 7. FUNNY CIRCLES *(Drôles de cercles)* – Emmanuel Cayla

Until now, Emmanuel Cayla based his work mainly on the "global evolution" tools of ArtiE-Fract, i.e. the Parisian evolution modes were used only as harsh exploration tools, to open new research directions at some stages, when the population becomes too uniform, for instance. He however produced unexpected shapes, in comparison to what was produced before by inexperienced designers.

Fig. 8. FOX *(Renard)* – Emmanuel Cayla

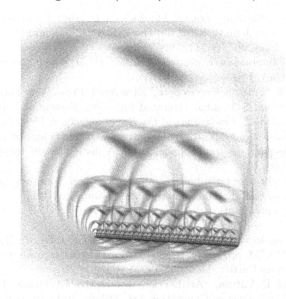

Fig. 9. WATER WHEEL *(Roue d'eau)* – Emmanuel Cayla

His use of black and white was also noticeable, see figures 4 to 9. It stressed on the importance of predefined simple evolution modes:

– that concentrate the search on some specific aspects of the design, such as "evolution of color only," "evolution of shape only,"

− that tune the degree of randomness in the evolution process, like "strong control" or "weak control."

Natural follow-ups to this project include working with colours, new operators, and getting together with the world of applied fine arts (e.g. decorated ceramics or fabric along with a lot of other potential opportunities for original applications of the system). This will demand new specifications for new functionalities in ArtiE-Fract. These improvements will focus on structure properties and design of the shapes where fractals grow as well as on composition of colour palettes and their expression in the system. And finally, as Emmanuel Cayla says, "ArtiE-Fract is a tool that should help us come up with new shapes in the world of artistic drawing."

Acknowledgments. The authors would like to thank Yann Semet for his kind help during the writing of this paper.

References

1. P. J. Angeline, "Evolving Fractal Movies", *GP 1996: Proceedings of the First Annual Conference*, John R. Koza and David E. Goldberg and David B. Fogel and Rick L. Riolo (Eds), pp 503–511, 1996.
2. W. Banzhaf, "Interactive Evolution", *Handbook of Evolutionary Computation*, 1997, Oxford Univ. Press.
3. M. F. Barnsley, "Fractals Everywhere," *Academic Press*, New-York, 1988.
4. M. F. Barnsley and S. Demko, "Iterated Functions System and the Global Construction of Fractals," *Proceedings of the Royal Society*, A 399:243–245, 1985.
5. E. Cayla, "Tapis-bulles 1," prototype d'une approche multi-agents des formes et des couleurs, 3èmes journees francophones sur l'IA distribuees et les systemes multi-agents. Chambéry 1995
6. E. Cayla, "L'atelier des tapis-bulles', interprétation et "compression" des poèmes graphiques", Journees d'etudes et d'echanges "Nouvelles techniques pour la compression et la représentation des signaux audiovisuels", CNET-CCETT, Cesson-Sévigné, 1995
7. E. Cayla, "D'un logiciel de CAO aux variations Goldberg de Bach", Le micro bulletin, CNRS n°53, pp. 162–182, 1994 (Communication produite aux entretiens du Centre Jacques Cartier, Montréal 1992)
8. J. Chapuis and E. Lutton, "ArtiE-Fract: Interactive Evolution of Fractals", 4th International Conference on Generative Art, Milano, Italy, December 12–14, 2001.
9. P. Collet, E. Lutton, F. Raynal, and M. Schoenauer, "Individual GP: an Alternative Viewpoint for the Resolution of Complex Problems," GECCO 99, *Genetic and Evolutionary Computation Conference*, Orlando, Florida, USA, July 13–17, 1999.
10. P. Collet, E. Lutton, F. Raynal, M. Schoenauer, "Polar IFS + Parisian Genetic Programming = Efficient IFS Inverse Problem Solving", *Genetic Programming and Evolvable Machines Journal*, Volume 1, Issue 4, pp. 339–361, October 2000.
11. B. Forte, F. Mendivil and E.R. Vrscay, 'Chaos Games' for Iterated Function Systems with Grey Level Maps, SIAM Journal of Mathematical Analysis, vol 29, No 4, pp 878–890, 1998. http://citeseer.nj.nec.com/146020.html

12. D. A. Goldberg, "Genetic Algorithms in Search, optimisation, and Machine Learning," *Addison-Wesley Publishing Company, inc.*, Reading, MA, January 1989.
13. J. Hutchinson, "Fractals and Self-Similarity," *Indiana University Journal of Mathematics*, 30:713–747, 1981.
14. A.E. Jacquin, "Fractal Image Coding: a Review," *Proc. of the IEEE*, 81(10), 1993.
15. S. Kamohara, H. Takagi and T. Takeda, "Control Rule Acquisition for an Arm Wrestling Robot", in IEEE Int. Conf. on System, Man and Cybernetics (SMC'97), vol 5, Orlando, FL, USA, 1997.
16. J. Lévy Véhel, F. Paycha and J.-M. Rocchisani Multifractal Measurements of Lung Diseases In First International Conference on Fractals in Biology and Medicine, 1993.
17. E. Lutton, J. Lévy Véhel, G. Cretin, P. Glevarec, and C. Roll, "Mixed IFS: resolution of the inverse problem using genetic programming," *Complex Systems*, 9:375–398, 1995.
18. P. Machado, A. Cardoso, All the truth about NEvAr. Applied Intelligence, Special issue on Creative Systems, Bentley, P. Corne, D. (eds), Vol. 16, Nr. 2, pp. 101–119, Kluwer Academic Publishers, 2002.
19. N. Monmarche, G. Nocent, G. Venturini, and P. Santini. "On Generating HTML Style Sheets with an Interactive Genetic Algorithm Based on Gene Frequencies", AE 99, Dunkerque, France, November 1999, Selected papers, Springer Verlag, LNCS 1829, 1999.
20. R. Poli and S. Cagnoni, "Genetic Programming with User-Driven Selection : Experiments on the Evolution of Algorithms for Image Enhancement", in 2nd Annual Conf. on GP, pp 269–277, 1997.
21. S. Rooke, "The Evolutionary Art of Steven Rooke" http://www.azstarnet.com/~srooke/
22. F. Raynal, E. Lutton, P. Collet, "Manipulation of Non-Linear IFS Attractors Using Genetic Programming" CEC99, *Congress on Evolutionary Computation*, Washington DC, USA, July 6–9, 1999.
23. D. Saupe and R. Hamzaoui, A bibliography for fractal image compression. Institut für Informatik, Universität Freiburg, Germany. 1996. Available as ftp://www.informatik.uni-freiburg.de/documents/papers/fractal/biblio.ps.gz.
24. K. Sims, "Artificial Evolution for Computer Graphics," *Computer Graphics*, 25(4):319–328, July 1991.
25. H. Takagi, M. Ohsaki, "IEC-based Hearing Aids Fitting", IEEE Int. Conf. on System, Man and Cybernetics (SMC'99), Tokyo, Japan, vol 3, pp 657–662, Oct. 12–15, 1999.
26. H. Takagi, " Interactive Evolutionary Computation : System Optimisation Based on Human Subjective Evaluation", IEEE Int. Conf. on Intelligent Engineering Systems (INES'98), Vienna, Austria, pp 1–6, Sept 17–19, 1998.
27. S.J.P. Todd and W. Latham, Evolutionary Art and Computers, Academic Press, 1992.
28. Some related web sites: http://www.genarts.com/karl
http://www.cosy.sbg.ac.at/rec/ifs/
http://www.accad.ohio-state.edu/~mlewis/AED/Faces/
http://draves.org/flame/

Evolutionary Music and the Zipf-Mandelbrot Law:
Developing Fitness Functions for Pleasant Music

Bill Manaris[1], Dallas Vaughan[1], Christopher Wagner[1],
Juan Romero[2], and Robert B. Davis[3]

[1] Computer Science Department, College of Charleston,
Charleston, SC 29424, USA
{manaris, neil.vaughan, wagner}@cs.cofc.edu

[2] Creative Computer Group - RNASA Lab - Faculty of Computer Science,
University of A Coruña. Spain
jj@udc.es

[3] Department of Mathematics and Statistics, Miami University
Hamilton, OH 45011, USA
davisrb@muohio.edu

Abstract. A study on a 220-piece corpus (baroque, classical, romantic, 12-tone, jazz, rock, DNA strings, and random music) reveals that aesthetically pleasing music may be describable under the Zipf-Mandelbrot law. Various Zipf-based metrics have been developed and evaluated. Some focus on music-theoretic attributes such as pitch, pitch and duration, melodic intervals, and harmonic intervals. Others focus on higher-order attributes and fractal aspects of musical balance. Zipf distributions across certain dimensions appear to be a necessary, but not sufficient condition for pleasant music. Statistical analyses suggest that combinations of Zipf-based metrics might be used to identify genre and/or composer. This is supported by a preliminary experiment with a neural network classifier. We describe an evolutionary music framework under development, which utilizes Zipf-based metrics as fitness functions.

1 Introduction

Webster's New World Dictionary (1981) defines beauty as *the quality attributed to whatever pleases or satisfies the senses or mind, as by line, color, form, texture, proportion, rhythmic motion, tone, etc., or by behavior, attitude, etc.* Since computers are ultimately manipulators of quantitative representations, any attempt to model qualitative information is inherently problematic. In the case of beauty, an additional problem is that it is affected by subjective (cultural, educational, physical) biases of an individual – that is, beauty is in the eye (ear, etc.) of the beholder. Or is it?

S. Cagnoni et al. (Eds.): EvoWorkshops 2003, LNCS 2611, pp. 522–534, 2003.
© Springer-Verlag Berlin Heidelberg 2003

Musicologists generally agree that music communicates meaning [12]. Some attempt to understand this meaning and its effect on the listener by dissecting the aesthetic experience in terms of separable, discrete sounds. Others attempt to find it in terms of grouping stimuli into patterns and studying their hierarchical organization [4], [10], [11], [15], [16]. Meyer [13, p. 342] suggests that emotional states in music (sad, angry, happy, etc.) are delineated by statistical parameters such as dynamic level, register, speed, and continuity. Although such state-defining parameters fluctuate locally within a music piece, they remain relatively constant globally.

In his seminal book, Zipf [25] discusses the language of art and the meaning communicated between artists and their audiences. He demonstrates that phenomena generated by complex social or natural systems, such as socially sanctioned art, tend to follow a statistically predictable structure. Specifically, the frequencies of words in a book, such as Homer's Iliad, plotted against their statistical rank on logarithmic scale, produce a straight line with a slope of approximately −1.0. In other words, the probability of occurrence of words starts high and decreases rapidly. A few words, such as 'a' and 'the', occur very often, whereas most words, such as 'unconditionally', occur rarely. Formally, the frequency of occurrence of the nth ranked word is $1/n^a$, where a is close to 1.

Similar laws have been developed independently by Pareto, Lotka, and Bendford [1], [6], [17]. These laws have inspired and contributed to other fields studying the complexity of nature. In particular, Zipf's law inspired and was extended by Benoit Mandelbrot to account for a wider range of natural phenomena [9]. Such phenomena may generate lines with slopes ranging between 0 (random phenomena) and negative infinity (monotonous phenomena). These distributions are also known as *power–law* distributions [19].

Zipf distributions are exhibited by words in human languages, computer languages, operating system calls, colors in images, city sizes, incomes, music, earthquake magnitudes, thickness of sediment depositions, extinctions of species, traffic jams, and visits of websites, among others.

Research in fractals and chaos theory suggests that the design and assembly of aesthetically pleasing objects – artificial or natural – is guided by hidden rules that impose constraints on how structures are put together [5], [18]. Voss and Clarke [23], [24] have suggested that music might also be viewed as a complex system whose design and assembly is partially guided by rules subconscious to the composer. They have also demonstrated that listeners may be guided by similar rules in their aesthetic response to music.

1.1 Zipf Distribution in Music

Zipf mentions several occurrences of his distribution in musical pieces. His examples were derived manually, since computers were not yet available. His study focused on the length of intervals between repetitions of notes, and the number of melodic intervals; it included Mozart's 'Bassoon Concerto in Bb', Chopin's 'Etude in F minor, Op.

25, No. 2,' Irving Berlin's 'Doing What Comes Naturally,' and Jerome Kern's 'Who' [25, pp. 336-337].

Voss and Clarke [23], [24] conducted a large-scale study of music from classical, jazz, blues, and rock radio stations collected continuously over 24 hours. They measured several fluctuating physical variables, including output voltage of an audio amplifier, loudness fluctuations of music, and pitch fluctuations of music. They discovered that pitch and loudness fluctuations in music follow Zipf's distribution.

Voss and Clark also developed a computer program to generate music using a Zipf-distribution noise source (aka 1/f or pink noise). The results were remarkable: *The music obtained by this method was judged by most listeners to be much more pleasing than that obtained using either a white noise source (which produced music that was 'too random') or a 1/f² noise source (which produced music that was 'too correlated'). Indeed the sophistication of this '1/f music' (which was 'just right') extends far beyond what one might expect from such a simple algorithm, suggesting that a '1/f noise' (perhaps that in nerve membranes?) may have an essential role in the creative process.* [23, p. 318]

2 Zipf-Based Metrics for Evolutionary Music

We have developed several Zipf-based metrics that attempt to identify and describe such balance along specific attributes of music [8]. These musical attributes may include the following: pitch, rests, duration, harmonic intervals, melodic intervals, chords, movements, volume, timbre, tempo, dynamics. Some of these can be used independently, e.g., pitch; others can be used only in combinations, e.g., duration. Some attributes are more straightforward to derive metrics from, such as melodic intervals; others are more difficult, such as timbre. These attributes were selected because they (a) have been used in earlier research, (b) have traditionally been used to express musical artistic expression and creativity, and/or (c) have been used in the analysis of composition. They are all studied extensively in music theory and composition. Obviously, this list of metrics is not complete.

We have automated several of these metrics using Visual Basic and C++. This allowed us to quickly test our hypothesis on hundreds of musical pieces encoded in MIDI. The following is a brief definition of selected metrics:

- **Pitch:** The relative balance of pitch of music events, in a given piece of music. (There are 128 possible pitches in MIDI.)
- **Pitch mod 12:** The relative balance of pitch on the 12-note chromatic scale, in a given piece of music.
- **Duration:** The relative balance of durations of music events, independent of pitch, in a given piece of music.
- **Pitch & Duration:** The relative balance of pitch-duration combinations, in a given piece of music.

- **Melodic Intervals:** The relative balance of melodic intervals, in a given piece of music. (This metric was devised by Zipf.)
- **Harmonic Intervals:** The relative balance of harmonic intervals, in a given piece of music. (This metric was devised by Zipf.)
- **Harmonic Bigrams:** The relative balance of specific pairs of harmonic intervals in a piece. (This metric captures the balance of harmonic-triad (chord) structures.)
- **Melodic Bigrams:** The relative balance of specific pairs of melodic intervals in a piece. (This metric captures the balance of melodic structure and arpeggiated chords.)
- **Melodic Trigrams:** The relative balance of specific triplets of melodic intervals in a piece. (This metric also captures the balance of melodic structure.)
- **Higher-Order Melodic Intervals:** Given that melodic intervals capture *the change of pitches* over time, we also capture higher orders of change. This includes the *changes in melodic intervals*, to the *changes of the changes in melodic intervals*, and so on. These higher-order metrics correspond to the notion of derivative in mathematics. Although a human listener may not be able to consciously hear such high-order changes in a piece of music, there may be some subconscious understanding taking place. [1]

2.1 Evaluation

We evaluated the effectiveness of our Zipf metrics by testing them on a large corpus of quality MIDI renderings of musical pieces. Additionally, we included a set of DNA-generated pieces and a set of random pieces (white and pink noise) for comparison purposes. Most MIDI renderings of classical pieces came from the Classical Archives [22].

Our corpus consisted of 220 MIDI pieces. Due to space limitations, we summarize them below by genre and composer.

1 *Baroque:* Bach, Buxtehude, Corelli, Handel, Purcell, Telemann, and Vivaldi (38 pieces).
2 *Classical:* Beethoven, Haydn, and Mozart (18 pieces).
3 *Early Romantic:* Hummel, Rossini, and Schubert (14 pieces).
4 *Romantic:* Chopin, Mendelssohn, Tarrega, Verdi, and Wagner (29 pieces).
5 *Late Romantic:* Mussorgsky, Saint-Saens, and Tchaikovsky (13 pieces).
6 *Post Romantic:* Dvorák and Rimsky-Korsakov (13 pieces).

---

[1] Surprisingly, our study revealed that, once a Zipfian distribution is encountered in a lower-order metric, the higher-order metrics continue to exhibit such distributions. However, the higher-order slopes progressively move towards zero (high entropy – purely random). This result suggests that balance introduced at a certain level of assembly may influence the perceived structural aspects of an artifact many levels removed from the original. If generalizable, this observation may have significant philosophical implications.

7 *Modern Romantic:* Rachmaninov (2 pieces).

8 *Impressionist:* Ravel (1 piece).

9 *Modern (12 Tone):* Berg, Schönberg, and Webern (15 pieces).

10 *Jazz:* Charlie Parker, Chick Corea, Cole Porter, Dizzy Gillespie, Django Reinhardt, Duke Ellington, John Coltrane, Miles Davis, Sonny Rollins, and Thelonius Monk (33 pieces).

11 *Rock:* Black Sabbath, Led Zeppelin, and Nirvana (12 pieces).

12 *Pop:* Beatles, Bee Gees, Madonna, Mamas and Papas, Michael Jackson, and Spice Girls (18 pieces).

13 *Punk Rock:* The Ramones (3 pieces).

14 *DNA Encoded Music:* actual DNA sequences encoded into MIDI, and simulated DNA sequences (12 pieces).

15 *Random (White Noise):* music consisting of random note pitches, note start times, and note durations "composed" by a uniformly-distributed random number generator (6 pieces).

16 *Random (Pink Noise):* music consisting of random note pitches, note start times, and note durations "composed" by a random number generator exhibiting a Zipf distribution (6 pieces).

2.2 Results

Each metric produces a pair of numbers per piece. The first number, *Slope*, is the slope of the trendline of the data values. Slopes may range from 0 (high entropy – purely random) to negative infinity (low entropy – monotone). Slopes near −1.0 correspond to Zipf distribution. The second number, R^2, is an indication of how closely the trendline fits the data values – the closer the fit, the more meaningful (reliable) the slope value. R^2 may range from 0.0 (extremely bad fit – data is all over the graph) to 1.0 (perfect fit – data is already in a straight line). We considered R^2 values larger than 0.7 to be a good fit.

Every celebrated piece in our corpus exhibited several Zipf distributions. Random pieces (white noise) and DNA pieces very few (if any) Zipf distributions. Table 1 shows average results for each genre in terms of slope, R^2, and corresponding standard deviations. The average for all musical pieces (excluding DNA, pink noise, and white noise pieces) across all metrics is −1.2004, a near-Zipf distribution; the corresponding R^2 (fit) across all metrics is 0.8213.

Additionally, we performed statistical analyses on these results to identify patterns across genres. First we drew side-by-side boxplots of each metric for each genre. In the following results, genres are numbered as 1 - Baroque, 2 - Classical, 3 - Early Romantic, 4 - Romantic, 5 - Late Romantic, 6 - Post Romantic, 7 - Modern Romantic, 9 - Twelve-tone, 10 - Jazz, 11 - Hard Rock, 12 - Pop, 13 - Punk, 14 - DNA, 15 - Pink noise, and 16 - White noise. Although boxplots are not formal inference tools, they are useful in data analysis because any substantial differences in genres should be evident from visual inspection. These side-by-side boxplots revealed several interesting patterns. For instance, genres 14 (DNA) and 16 (random music – uniformly dis-

Table 1. Average results across metrics for each genre

Genre	Slope	R^2	Slope Std	$R^{2 \, Std}$
Baroque	-1.1784	0.8114	0.2688	0.0679
Classical	-1.2639	0.8357	0.1915	0.0526
Early Romantic	-1.3299	0.8215	0.2006	0.0551
Romantic	-1.2107	0.8168	0.2951	0.0609
Late Romantic	-1.1892	0.8443	0.2613	0.0667
Post Romantic	-1.2387	0.8295	0.1577	0.0550
Modern Romantic	-1.3528	0.8594	0.0818	0.0294
Impressionist	-0.9186	0.8372	N/A	N/A
Twelve-Tone	-0.8193	0.7887	0.2461	0.0964
Jazz	-1.0510	0.7864	0.2119	0.0796
Rock	-1.2780	0.8168	0.2967	0.0844
Pop	-1.2689	0.8194	0.2441	0.0645
Punk Rock	-1.5288	0.8356	0.5719	0.0954
DNA	-0.7126	0.7158	0.2657	0.1617
Random (Pink)	-0.8714	0.8264	0.3077	0.0852
Random (White)	-0.4430	0.6297	0.2036	0.1184

tributed pitch) are easily identifiable by their pitch metric. We also discovered that the first seven genres (baroque, classical, and the five genres with the word "romantic" in them) appear to have significant overlap on all of the metrics. This is not surprising, as these genres are relatively similar – people refer to such pieces as "classical music." However, further examination suggests that it may be still be possible to identify styles and composers by using combinations of metrics.

We also performed analyses of variance (ANOVAs) on the data to determine whether or not the various metrics had significantly different averages across genres. We found that all of the p-values are significant, meaning there are differences among genres within our corpus. To better visualize the ANOVA results we generated confidence interval graphs. When displayed using side-by-side boxplots, genre samples may overlap due to simple natural variation and/or some unusual individual pieces within the genres. However, the confidence intervals given in the ANOVA output characterize where the mean slope for each metric within each genre is located. When these intervals do not overlap, there is a statistically significant difference between genres. Figure 1 shows the results for the harmonic interval metric.

Overall, twelve-tone and DNA differ from the other genres significantly. In terms of the pitch metric, Late Romantic appears to differ significantly from Hard Rock and Pop. In terms of pitch mod 12, there are several genres that differ significantly from Jazz. Jazz and Baroque appear to differ in duration. Jazz definitely differs from most other genres in terms of pitch & duration.

Moreover, several other interesting patterns emerge. For instance, in the pitch-mod-12 metric, twelve-tone music exhibits slopes suggesting uniform distribution – average slope is −0.3168 with a standard deviation of 0.1801. In particular, Schönberg's pieces averaged −0.2801 with a standard deviation of −0.1549. This was comparable to the average for random (white noise) pieces, namely −0.1535.

to the average for random (white noise) pieces, namely –0.1535. Obviously, this metric is very reliable in identifying twelve-tone music. For comparison purposes, the next closest average slope for musical pieces was exhibited by Jazz (–0.8770), followed by Late Romantic (–1.0741).

One-way ANOVA: Harmonic versus fam

```
Analysis of Variance for Harmonic
Source      DF        SS        MS         F         P
fam          9   13.4172    1.4908     29.80     0.000
Error      187    9.3550    0.0500
Total      196   22.7721
                                    Individual 95% CIs For Mean
                                    Based on Pooled StDev
Level        N      Mean     StDev  -+---------+---------+---------+-----
  1         38   -1.3261    0.1947          (-*-)
  2         18   -1.4124    0.1501       (--*-)
  3         14   -1.4938    0.1693      (--*--)
  4         24   -1.3821    0.2315         (-*--)
  5         13   -1.3707    0.2026        (--*--)
  9         15   -0.9219    0.1696                    (--*--)
 10         33   -1.2233    0.1631            (-*-)
 11         12   -1.3099    0.1494         (--*--)
 12         18   -1.3709    0.1811        (--*--)
 14         12   -0.3679    0.5572                              (--*--)
                                    -+---------+---------+---------+-----
Pooled StDev =    0.2237              -1.60     -1.20     -0.80     -0.40
```

Fig. 1. ANOVA confidence interval graph for the harmonic metric across all genres. Genres 9 (twelve-tone) and 14 (DNA) are identifiable through this metric alone

Overall, these results suggest that we may have discovered certain **necessary but not sufficient conditions** for aesthetically pleasing music. It should be noted that the original inspiration for this project was how the Zipf-Mandelbrot law is used in many other domains to identify "naturally" occurring phenomena – phenomena with a natural "feel" to them [6]. So, it is not surprising to us that we are finding this correlation between pleasant music and instances of Zipf distribution.

3 Composite Metrics

These results suggest that aesthetically pleasing aspects of music may be algorithmically identifiable and classifiable. Specifically, by combining metrics into a weighted composite (consisting of metrics that capture various aspects throughout the possible space of measurable aesthetic attributes) we may be able to perform various classifications tasks. We have experimented with composite metrics having (a) various weights assigned to individual metrics and (b) conditional combinations of individual metrics.

Fig. 2.a. Sample at resolution 1

Fig. 2.b. Sample at resolution 2

Currently, we are exploring various neural network configurations and classification tasks. In one experiment, we developed an artificial neural network to determine if Zipf metrics contained enough information for authorship attribution. For instance, we have used the Stuttgart Neural Network Simulator (SNNS) [20] to build, train and test an artificial neural network. Our corpus consisted of Zipf metrics for two data sets: (a) Bach pieces BWV500 through BWV599, and (b) Beethoven sonatas 1 through 32. From these we extracted training and test sets. The trained neural network was able to identify the composer of a piece it had not seen before with 95% accuracy. We believe that this can easily be improved with a refined training set and/or the fractal metrics discussed in the next section.

Composite metrics, implemented through neural network classifiers, could be used to identify pieces that have similar aesthetic characteristics to a given piece. Composite metrics may also help derive a statistical signature (identifier) for a piece. Such an identifier may be very useful in data retrieval applications, where one searches for different performances of a given piece among volumes of music. For instance, during an earlier study [8], we discovered a mislabeled MIDI piece by noticing that it had identical Pitch-mod-12 slope and R^2 values with another MIDI piece. The two files contained different performances of Bach's Toccata and Fugue in D minor.

3.1 Fractal Metrics

The Zipf metrics presented so far are very promising. However, they have a significant weakness. They measure the global balance of a piece. For instance, consider the sample shown in figure 2.a. The pitch metric of this sample is perfectly Zipfian (slope $= -1.0$, $R^2 = 1.0$). However, locally, the sample is extremely monotonous.

This problem can be easily addressed using a fractal method of measurement. The metric is applied recursively at different levels of resolution: We measure the whole sample, then split it into two equal phrases and measure each of these, then split it into four equal phrases, and so on. For instance, at resolution 2 (see sample at figure 2.b), the slope of the left side is negative infinity (monotone). The slope of the right side is -0.585. By dividing the sample into two parts, the lack of local balance was quickly exposed.

Preliminary tests with music corpora indicate that aesthetically pleasing music exhibits Zipf distributions at various levels of resolution. Depending on the piece of music, this will go on until the resolution reaches a small number of measures. For instance, in Bach's Two-Part Invention No. 13 in A minor (BWV.784), this balance exists until a resolution of three measures per subdivision.

This recursive measuring process may be used to calculate the fractal dimension of a sample relative to a specific metric. This is known as the *box-counting* method. Taylor [21] has used this approach in visual art to authenticate and date paintings by Jackson Pollock using their fractal dimension, *D*. For instance, he discovered two different fractal dimensions: one attributed to Pollock's dripping process, and the other attributed to his motions around the canvas. Also, he was able to track how Pollock refined his dripping technique – the fractal dimension increased through the years (from $D \approx 1$ in 1943 to $D = 1.72$ in 1952).

Preliminary results suggest that, as observed for simple Zipf metrics, aesthetically pleasing music exhibits several fractal dimensions near 1, as opposed to aesthetically non-pleasing music or non-music. For instance, the pitch fractal dimension for Bach's 2-Part Invention in A minor is 0.9678. The results of these experiments are only preliminary. We believe that these fractal metrics will prove much more powerful than simple metrics for ANN-based classification purposes.

4   Evolutionary Music Framework

We have shown that Zipf-based metrics are capable of capturing aspects of the economy exhibited by socially sanctioned music. This capability should be very useful in computer-assisted composition systems. Such systems are developed using various AI frameworks including formal grammars, probabilistic automata, chaos and fractals, neural networks, and genetic algorithms [2], [3], [14].

We are currently evaluating the promise of Zipf-based metrics for guiding evolutionary experiments. Based on our results, we believe that Zipf-based fitness functions should produce musical samples that resemble socially sanctioned (aesthetically pleasing) music. If nothing else, since Zipf distributions appear to be a necessary, but not sufficient condition for aesthetically pleasing music, such fitness functions could minimally serve as an automatic filtering mechanism to prune unpromising musical samples.

4.1 Genotype Operations

Our system is based loosely on Machado's NEvAR system [7] – a powerful system for evolutionary composition of visual art. In our adaptation of the NEvAR framework, a phenotype is a music score. A genotype is represented as a tree. Leaf nodes are music phrases. Non-leaf nodes are operators that, when interpreted, generate a phenotype (see figure 3).

Genotype operators, such as +, –, and * are related, but not completely analogous, to their mathematical definitions. The following is an overview of low-level genotype operators. We use the word "element" to refer to an arbitrary genotype sub-tree.

+ (addition) takes two elements, A and B, and returns the union of the two preserving their respective start times, end times, and pitches.

− (subtraction) takes two elements, A and B, and returns the set of notes in which
B is NOT enveloped by A.

& (concatenation) takes two elements, A and B, and appends B to the end of A.

* (multiplication) takes two elements, A and B, and replaces each instance of B
with a complete repetition of A, but transposed from A's starting note to B.
Each repetition is appended to the last.

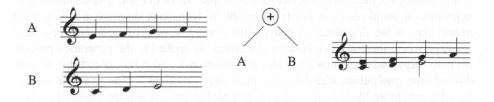

Fig. 3. Example of the addition operation. The tree at the center is the genotype representation. The music scores at left and right are phenotype representations of the two operands
and the result of their addition, respectively.

These low-level operations are used to evolve themes. Once a theme has been
evolved, higher-level operations are applied to evolve other aspects of the notes,
phrases, and piece as a whole. These operations include common compositional
devices such as retrograde, diminution, augmentation, inversion, imitation, harmoni-
zation, temporal quantization, harmonic quantization, and transposition.

Finally, we include genetic operators for evolving sub-trees such as mutate (subtree mutation), and fit (sub-tree evaluation). These allow for introducing improvised
phrases within larger compositions.

We maintain control over the probabilities and complexity restrictions of when
these operations take place. For instance, if within a certain genre an operation is
found to be more prevalent in the first few levels of the tree than the last few, this fact
can be used to weight the corresponding probabilities of that operation taking place
among the various levels of the generations.

In the NEvAr system, elements are composed of collections of pixels that form a
two-dimensional image. Expressions are evaluated on a pixel-by-pixel basis, and the
results can be used as the arguments in the next operation in the expression tree. Any
sub-tree can be mutated in one of five ways: 1) swapping arbitrary sub-trees, 2) replacing arbitrary sub-trees with randomly created ones, 3) inserting a randomly created node at a random insertion point, 4) deleting a randomly selected node, and 5)
randomly selecting an operator and changing it [7].

These sub-tree mutations could prove to be valuable in the context of music, since,
unlike most visual art, music is defined almost exclusively by an abstract (in the sense

of layers building upon other layers) composition of notes to phrases, phrases to melody, melody to section, and sections to piece. Since this method is closer to the actual process of composition, the results of these operations should minimally produce something that resembles a "standard" musical piece, at least in structure.

The most important question to be answered is, "to what degree should we have certain operations, and where?" If the answer to this is, "anywhere, anytime," then there will likely be many more non-standard compositions created than if the answer were based on music theory or probability (dependent on whatever genre of music one is attempting to emulate). Although a fitness test (in the case of NEvAr, a human) would usually decide which generations stayed and which did not, a valid fitness test, in terms of musical beauty, is nearly impossible to formulate, since the goal is so hard to articulate in the first place. A better solution may be somewhere in between these answers, where there are weightings and restrictions applied to the generation process (to both elemental operations and sub-tree operations), and a fitness test that at least discards the generations which are not minimally "musical." The generations produced would more likely be of a structured type, but there is still the possibility that a less structured generation would make it, provided it passed the fitness test. Among the possible fitness tests, combined Zipf metrics are worthy candidates since they do not depend solely on musical tastes or rules of theory, but on the more abstract idea of balance begetting beauty.

5 Conclusion

We have shown the promise of using the Zipf-Mandelbrot law to measure the balance, and to a certain degree, pleasantness of musical pieces by applying this law to various musical attributes, such as pitch, duration, and note distances. The results of the ANN experiment suggest that a neural network is capable of distinguishing pieces based on their Zipf metrics, and so can be used in part or whole as a fitness test for each generation. Using a neural network as a fitness test could also be used to constrain the generation process to create certain types of music, like Classical, Jazz, etc., or to create pieces that are similar to particular composers.

We also discussed the generation of musical pieces through an evolutionary framework comprised of genetic operations similar to those of the NEvAr framework. This will allow the structured formulation of music, either with or without human interaction. Although this system may produce music that is statistically similar to socially sanctioned music, it is not clear if the result will be truly aesthetically pleasing music. Therefore, this tool could at least assist a human composer by enforcing minimal conditions for aesthetically pleasing music and, thus, producing rough musical sketches for inspiration and further refinement.

Acknowledgements. The authors would like to acknowledge Penousal Machado for his suggestion to combine Zipf metrics with an ANN for authorship attribution. They also acknowledge José Santiago, Cernadas Vilas, Mónica Miguélez Rico, and Miguel Penin Álvarez for conducting the ANN experiment. Tarsem Purewal, Charles McCormick, Valerie Sessions, and James Wilkinson helped derive Zipf metrics and MIDI corpora. This work has been supported in part by the College of Charleston through an internal R&D grant.

References

1. Adamic, L.A.: "Zipf, Power-laws, and Pareto – a Ranking Tutorial" (1999) http://ginger.hpl.hp.com/shl/papers/ranking/
2. Balaban, M., Ebcioğlu, K., and Laske, O., (eds.): Understanding Music with AI: Perspectives on Music Cognition. Cambridge: AAAI Press/MIT Press (1992)
3. Cope, D.: Virtual Music, MIT Press Cambridge (2001)
4. Howat, R.: Debussy in Proportion. Cambridge University Press (cited in [16]) Cambridge (1983)
5. Knott, R.: "Fibonacci Numbers and Nature". www.mcs.surrey.ac.uk/Personal/R.Knott/Fibonacci/fibnat.html (2002)
6. Li, W.: "Zipf's Law". http://linkage.rockefeller.edu/wli/zipf/ (2002)
7. Machado, P., Cardoso, A.: "All the truth about NEvAr". Applied Intelligence, Special issue on Creative Systems, Bentley, P., Corne, D. (eds), 16(2), (2002), 101–119
8. Manaris, B., Purewal, T., and McCormick, C.: "Progress Towards Recognizing and Classifying Beautiful Music with Computers". 2002 IEEE SoutheastCon, Columbia, SC (2002) 52–57
9. Mandelbrot, B.B.: The Fractal Geometry of Nature. W.H. Freeman New York (1977)
10. Marillier, C.G.: "Computer Assisted Analysis of Tonal Structure in the Classical Symphony". Haydn Yearbook 14 (1983) 187–199 (cited in [16])
11. May, M.: "Did Mozart Use the Golden Section?". American Scientist 84(2). www.sigmaxi.org/amsci/issues/Sciobs96/Sciobs96-03MM.html (1996)
12. Meyer, L.B.: Emotion and Meaning in Music. University of Chicago Press Chicago (1956)
13. Meyer, L.B.: "Music and Emotion: Distinctions and Uncertainties". In Music and Emotion – Theory and Research, Juslin, P.N., Sloboda, J.A. (eds), Oxford University Press Oxford (2001) 341–360
14. Miranda, E.R.: Composing Music with Computers. Focal Press Oxford (2001)
15. Nettheim, N.: "On the Spectral Analysis of Melody". Interface: Journal of New Music Research 21 (1992) 135–148
16. Nettheim, N.: "A Bibliography of Statistical Applications in Musicology". Musicology Australia 20 (1997) 94–106
17. Pareto, V.: Cours d'Economie Politique, Rouge (Lausanne et Paris) (1897) Cited in [9]
18. Salingaros, N..A., and West, B.J.: "A Universal Rule for the Distribution of Sizes". Environment and Planning B(26) (1999) 909–923
19. Schroeder, M.: Fractals, Chaos, Power Laws. W.H. Freeman New York (1991)
20. Stuttgart Neural Network Simulator (SNNS), http://www-ra.informatik.uni-tuebingen.de/SNNS/).

21. Taylor, R.P., Micolich, A.P., and Jonas, D.: "Fractal Analysis Of Pollock's Drip Paintings". Nature, vol. 399, (1999), 422
    http://materialscience.uoregon.edu/taylor/art/Nature1.pdf
22. The Classical Music Archives. www.classicalarchives.com.
23. Voss, R.F., and Clarke, J.: "1/f Noise in Music and Speech". Nature 258 (1975) 317–318
24. Voss, R.F., and Clarke, J.: "1/f Noise in Music: Music from 1/f Noise". Journal of Acoustical Society of America 63(1) (1978) 258–263
25. Zipf, G.K.: Human Behavior and the Principle of Least Effort. Addison-Wesley New York (1972) (original publication Hafner Publishing Company, 1949)

Genophone: Evolving Sounds and Integral Performance Parameter Mappings

James Mandelis

School of Cognitive and Computing Sciences,
University of Sussex,
United Kingdom
jamesm@cogs.susx.ac.uk

Abstract. This project explores the application of evolutionary techniques to the design of novel sounds and their characteristics during performance. It is based on the "selective breeding" paradigm and as such dispensing with the need for detailed knowledge of the Sound Synthesis Techniques involved, in order to design sounds that are novel and of musical interest. This approach has been used successfully on several SSTs therefore validating it as an Adaptive Sound Meta-synthesis Technique. Additionally, mappings between the control and the parametric space are evolved as part of the sound setup. These mappings are used during performance.

1 Introduction

This paper describes the Genophone [5][6], a hyper instrument developed for sound synthesis and sound performance using the evolutionary paradigm of selective breeding as the driving process. Sound design on most current commercial systems relies heavily on an intimate knowledge of the SST (Sound Synthesis Technique) employed by the sound generator (hardware or software based). This intimate knowledge can only be achieved by investing long periods of time playing around with sounds and experimenting with how parameters change the nature of the sounds produced. Often such experience can be gained after years of interaction with one particular SST. The system presented here attempts to aid the user in designing sounds and control mappings without the necessity for deep knowledge of the SSTs involved. This method of design is inspired by evolutionary techniques, where the user expresses how much particular sounds are liked and then uses these sounds to create new ones through variable mutation and genetic recombination. The aim of the system is to encourage the creation of novel sounds and exploration rather than designing sounds that satisfy specific a priori criteria.

Through the use of "locked" parameter sets (where their values are unchangeable), variable control is exercised on the non-deterministic effect of the evolutionary processes. This feature, on the one hand, exercises some control on the shape evolution takes and on the other, allows a gradual familiarisation with the SST involved (if desired). Manual manipulation of the parameters for the particular SST is also provided, therefore allowing for precise control, if desired.

S. Cagnoni et al. (Eds.): EvoWorkshops 2003, LNCS 2611, pp. 535–546, 2003.
© Springer-Verlag Berlin Heidelberg 2003

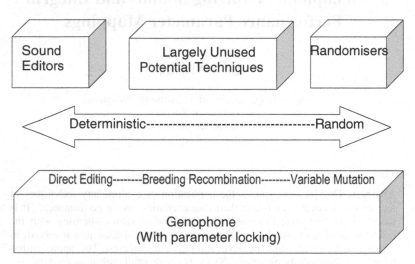

Fig. 1. Variable degrees of determinism in sound design

Genophone [5][6] is a "Hyperinstrument" or "Virtual Musical Instrument" [4][10][8][7], comprising of a dataglove, synthesiser and a PC that runs the evolutionary software. Real-time information from the glove is used to manipulate parameters that affect the sounds produced. The mapping of the finger flex and the sound changes is one-to-many. That is, a single finger flex (one of five) can control multiple parameters. This problem of mapping lower dimensionality performance controllers to higher dimensionality parameters [12][1][10] is also tackled within the same evolutionary framework. The resulting mappings are mainly used during performance by changing sound characteristics in real-time.

The selective breeding process generates System Exclusive MIDI messages (SysEx) for sound definitions, which are then sent to the synthesiser to be rendered. This level of abstraction facilitates the use of different external synthesisers with minimal effort. It also taps into the ability of commercial synthesisers to produce musical sounds by their design, and the existing wealth of sounds available for them. In previous attempts of using AL for sound synthesis a lower level definition was used, therefore limiting the usage of the system to one particular piece of hardware (or software emulation) that employed only a single SST. Also, musical sounds are much harder to evolve when lower level definitions are used [14][13].

This project has also shown that evolutionary methods can be used successfully on several sound-synthesis-techniques, demonstrating the feasibility of a generic approach in sound design for all different sound-synthesis-techniques. This approach is fast; often only a few generations are needed for evolving sounds that are interesting and of good quality. It is also very easy and fun to use, as well as easy to learn.

2.3 Software Description

There are three main structures used in this program;
1. The Instrument Template; describes the parameters used in making up a patch in a particular instrument (in this case the Prophecy). It defines things like SysEx headers, parameter names, type and range of values. It also defines sections of parameters (and sections of sections) that reflect the logical design of the instrument. The state of parameters "Locked" or "Unlocked" is also stored here temporarily as well as the current parameter values (not persistent).
2. The Values (Genotype); is a particular set of values for the parameters defined above, i.e. of the required type in the required range. Each Genotype translates to a SysEx message that defines a sound.
3. The Filter; is a list of "Unlocked" parameters. When a filter is applied to the current patch it has the effect of "Unlocking" any "Locked" parameters.

The state of parameters ("Locked" or "Unlocked") is used for limiting the effects of loading new patches, mutation, and recombination. Filters are used to fill the gap between total knowledge of the SST (knowing how each parameter changes the sound) and total ignorance of the principles underlying the sound production. By enabling or disabling sets of parameters it is possible to influence just parts of the instrument's logical structure, i.e. the effects, mixer or oscillator section, etc.

2.4 Workflow

Patches (individuals) can be loaded into population containers where they can be:
– Previewed, send to the synthesiser for rendering.

– Rearranged by the user, where the vertical position indicates relative preference, after they have been previewed.
– Selected, so genetic operators can be applied to them (one for variable mutation or more for recombination).
– Deleted from the population, saved to file, or copied into another population container.

Any number of individuals in a population can be highlighted as parents for operators to be applied (all possible couples will be used for recombination operators or just one for variable mutation). A mix-and-match approach can be used for the operators; they can be applied at any stage and in any order. Also at any stage, new parents can be brought into the gene-pool, removed from it, or spawn new populations for seeding. This way multiple strains can be evolved and maintained, and also be used for speciation or backtracking from dead-end-alleys.

2.5 Genetic Operators

2.5.1 Preference as Fitness
By arranging individuals within a population at different heights, the user is able to indicate his/her relative preference on the sounds and mappings that those individuals

represent. From this vertical placement the fitness of each individual is derived. This fitness is used by the recombination operators as bias. It is also used by the recombination operators for assigning to each offspring a temporary estimated fitness, which is derived from its parent fitness's weighted by the Euclidian distance of its parameters from each parent. After previewing the sounds, the fitness should be changed later by the user, by simple rearrangement of individuals within the new population window.

2.5.2 Parameter Locking
Only the "Unlocked" parameters are changed by the following operators, "Locked" parameter values just get copied across from the original patch.

2.5.3 Variable Mutation
The variable mutation operation creates a new population of ten mutants based on a previously selected individual. The amount by which the parameters are mutated from their original value is governed by the *"Mutation Factor"*. It can have a value between 0 and 1, when 0 the new value is the same as the old, when 1 the new value can be anything within the valid range of values for that particular parameter.

2.5.4 Recombinations
For each couple of selected individuals, a random crossover point is selected in the range of 1 to the number of *"Unlocked variables"* and an other at a wrapped offset derived by the number of *Unlocked Variables* x *"Depth Factor"*(0<,<1). Originally two offspring are produced as parent clones (one of each parent), then their unlocked parameters are overwritten, but only those defined within the wrapped range between the two Cross-Over-Points. The overriding values are derived for each recombination algorithm as;

Swap Recombination: The values of the other parent's parameters are used.

Probabilistic Recombination: Each value used for overwriting is selected probabilistically from one the parents, where the probability is proportional to the relative parent fitness values.

Interpolating Recombination: The parameter value used for overwriting is derived between the parents' values weighted by their fitness. Consequently, in the two offspring, the region between the Cross-Over-Points is identical.

2.5.5 Discussion of Operators
- *Variable Mutation* is quite effective when using a sensible *"Mutation Factor"*; this parameter has to be found by rule-of-thumb for each particular SST as well as the number of unlocked variables. It is a very subjective choice. *"Mutation Factor"* values usually range from 0.01 to 0.9.
- *Swap Recombination* is the one used most, it preserves clusters of parameters and their values, and its results are slightly more predictable. This is due to the high *epistatic* correlation of parameter genotype locus and parameter function.

- *Probabilistic Recombination* seemed more likely to disrupt clusters of related parameters with the resulting sounds been more likely to be unfit (no sound produced at all, or was too alien from the parents to be usable), sometimes this could be considered to be an advantage since the resulting sounds though alien were quite attractive. One could say that results from this recombination type are more radical.

- *Interpolating Recombination* also seemed likely to produce relatively less suitable sounds for different reasons. By using "in between" values quite often the sounds were indistinct, quiet or just silent. One could say that this recombination type is too conservative or uncommitted in its results.

3 Genophone and VMIs

The system can be viewed as Virtual Musical Instrument and as such can be considered as a new step in the development of musical instruments. Mulder has suggested the following classification and development of musical instruments [7]:

The first step, suggested by Mulder, is traditional acoustic instruments that are manipulated in a certain way in order to produce the sounds. The next development is the use of electronics in order to apply sound effects on acoustic instruments. The manipulations remain essentially the same. His comments on the characteristics of these types of instruments are: "Limited timbral control, gesture set and user adaptivity. Sound source is located at gesture."[7]

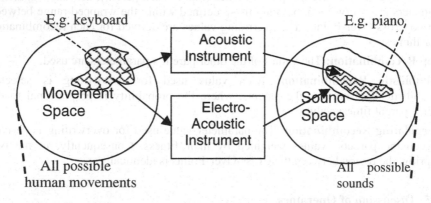

Fig. 3. Steps 1 & 2 of instrument development [7]

The next step, suggested by Mulder, are Electronic Musical Instruments; where the essential manipulations of a piano produce sounds that mimic other acoustic or electronic instruments. His comments on the characteristics of these types of instruments are: "Expanded timbral control, though hardly accessible in real-time and discreetised; gesture set adaptivity still limited. Sound emission can be displaced."[7]

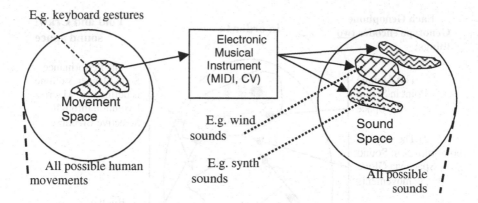

Fig. 4. Step 3 of instrument development [7]

The next step, suggested by Mulder, involves Virtual Musical Instruments where gestures from motion caption devices are used to drive sound engines. His comments on the characteristics of these types of instruments are: "Expanded real-time, continuous timbral control; gesture-set user selectable and adaptive. Any gestures or movements can be mapped to any class of sounds."[7]

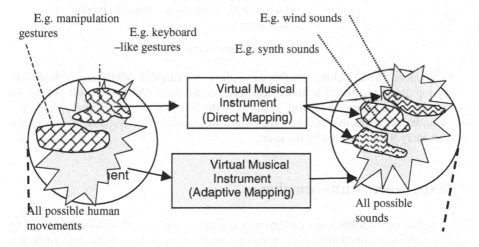

Fig. 5. Steps 4 [7] & 5 [5][6]of instrument development

As an improvement to the last step, and an extension to the overall classification, the author suggests a new class. It involves VMIs that provide a framework for adaptive generation of sounds and their gesture mappings.

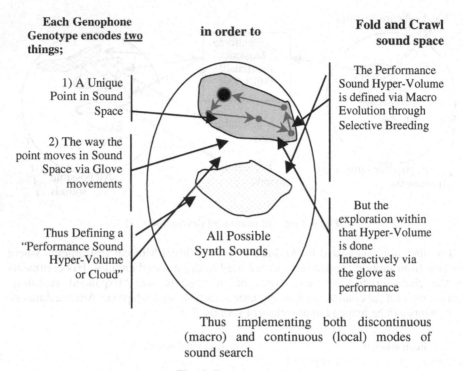

Each Genophone Genotype encodes two things;

in order to

Fold and Crawl sound space

1) A Unique Point in Sound Space

2) The way the point moves in Sound Space via Glove movements

Thus Defining a "Performance Sound Hyper-Volume or Cloud"

All Possible Synth Sounds

The Performance Sound Hyper-Volume is defined via Macro Evolution through Selective Breeding

But the exploration within that Hyper-Volume is done Interactively via the glove as performance

Thus implementing both discontinuous (macro) and continuous (local) modes of sound search

Fig. 6. Genophone operation

Genophone [5][6] belongs to this new class of Adaptive VMIs and exhibits the following characteristics: Expanded real-time, continuous timbral control; gesture-set is user designed via breeding. Any gestures or movements can be mapped to any class of sounds where both the mappings and the sounds are subject to the same evolutionary forces applied by the user.

4 Experiences with Genophone

The preliminary results from this project are encouraging and will be followed by system enhancements that will allow more complex experiments to be performed and move into the next phase. Most of the initial aims for this pilot phase have been satisfied and can be summarized as:

4.1 Usage Modes

The system has been used in three distinct modes of operation:

As a Solo Instrument; where the right hand plays a melody on the keyboard while the left hand is changing the sound via the glove. I found the Solo mode the most difficult to operate, mainly because of lack of keyboard virtuosity on my part. The results were much better when I used a sequencer to play melodies I knew via MIDI, while at the same time I was changing the sound with the glove. This way it is possible take a "sterile" sounding MIDI file and breath a lot of life into it.

As a Single Event / Sound Effect Generator: a single note is played either as drone or until the sound expires. The glove is not often used in this mode with the exception of drones. The Sound Effect Generator mode was much easier to use, it produced a lot of single event sounds that were very rich and dynamics. Quite often they did not play well musically on the keyboard, but had enough structure and complexity to be satisfying into themselves. Often providing their own melodic or rhythmic framework. Drone sounds where also produced in this way, sometimes using the glove to change their characteristics.

As a Pattern Arpeggiator; where an arpeggiated pattern is chorded with the right hand, while the left hand is changing the sound via the glove. The Pattern Arpeggiator mode is the most fun to use; it has an instant appeal due to the responsiveness of the glove and rhythmic structures can be created in a very intuitive way. Also the repetition of the phrase facilitates the perception and prediction of the sound changes within a rhythmic framework. I found it quite difficult having to play the chords with right hand, but this problem can be easily solved with a right hand glove.

It is obvious from the above that there is an appropriate mode of operation for creating each part of a track, whether these parts are rhythmic, melodic, drones or single events.

4.2 Hand Rearing vs. Hand Design

The ease of use of the interface was a surprising outcome. The selective breeding paradigm is an accessible one and users were able to breed complex sounds after only a brief introduction. The sounds produced were of such quality that would take someone with quite a bit of experience in the SST involved if they were to be programmed manually, which would be much slower. The overall process is directional not goal orientated one; it is not designed to satisfy a priori sound specifications, i.e. "I would like to produce a bell sound". It does not preclude the possibility of doing so in indirect ways, though. For instance, if bell or bell-like sounds are used for seeding, then is conceivable that a satisfactory bell sound will be produced within a few generations of selective breeding and variable mutation.

4.3 Meta-SST

Different SSTs can be used without the use of Specific Domain Knowledge. It was an initial requirement that no specific domain knowledge should be used in the system. That is, the parameters are treated as going into a black box, no knowledge of their function is kept in the system. As a result a new SST can be added by just specifying the System Exclusive Implementation Chart of the new synthesiser. As a down side, when sounds are produced that are interesting but very quiet then the there is no evolutionary way in the current framework to address the problem. The only solution is to either selectively breed part of the genotype that is suspected of being responsible, or manually tweak individual values until the desired result is achieved.

4.4 Recombination vs. Mutation

The Evolutionary Paradigm can be successfully applied for the creation of novel sounds often with surprising complexity. It seems that viable (fit) parameter sections are preserved through the genetic recombination, as it is also the case with Genetic Algorithm optimisation. In other words, if the starting sounds are professionally designed ones then the offspring are likely to be of comparable quality. This is also shown by the observation that genetic recombination produces higher quality results than if mutation is used alone. In implementations [14] where no genetic recombination is used, and mutation or a type of "genetic space crawling" is used instead, it is much harder to produce sounds that are complex and of high (subjective) quality.

4.5 MIDI SysEx & Musicality

The level of abstraction (SysEx MIDI) was the right choice. Synthesisers are designed for musical sounds and aspects like keyboard mapping are already implemented in them. If an approach was taken where the parameters were encoding low-level sound production [13], it would have been much harder to produce musically acceptable sounds.

4.6 Expressivity & Mapping

The use of the glove is very responsive and expressive. The original professionally programmed sounds that are used for seeding contain mappings that are relevant to the parameters used by the particular patch. These mappings seemed to be preserved and combined better by the *Probabilistic Recombination* operator (due to the high epistatic correlation). Although there are no widely accepted a priori mappings between hand movements and sound changes (since they are also evolved), they are usually easily internalised by the player [3]. That is, after a few finger flexions the brain seems to be able to assimilate the correspondence of movements and sound changes. This was further facilitated by using arpeggiated phrases, it seems that phrases are processed better by short term memory in order for the brain to notice the sound changes and play with the rhythm. The easy internalisation of the mapping actually came as a surprise; it was thought originally, that some unchanging, directly programmed mapping would have to be used, in order for the brain to learn the movement-to-sound mapping. One example of such mapping is "sound sculpturing" [9] in which the sound is represented by a three-dimensional object where changes in its shape are translated to changes in sound. It was also thought that some kind of structured language would have to be used for describing those (evolved or not) mappings [2][4][10][8][7][11] if no direct one-to-one mapping was used. The use of such formalisation [14] has not been necessary yet, partly because the Prophecy implements its own one-to-many (1 to 4) mapping formalisation through SysEx parameters, and partly because the relatively low dimensionality of the input device (dataglove) which has only five degrees of freedom.

5   Future Directions

It would be interesting to see if the ease of internalising mappings is retained when input devices of more channels are used i.e. more than five. In the future, when different synthesisers and input devices (with more degrees of freedom) are used, the issue of a mapping formalisation will have to be readdressed. Also the two processes for *sound* evolution and *motion-to-sound-mapping* evolution will have to be separated from the same genotype. More operators are currently being developed and tested. Since this is an exploration system each operator has unique properties that can be used appropriately to guide the search.

Acknowledgments. Special thanks to Jon Bird, Andrew Gartland, Jon MacCormak, Alan Montgomery and Sam Woolf for their incisive feedback and stimulating discussions; to Prof. Margaret Boden for inspiring me on computers and creativity during my most impressionable years. Last, but by no means least, to my supervisor Prof. Phil Husbands for his support, patience and often needed clarity of thought, also to his wife Allison for her tips in glove making.

References

1.   Choi, I., Bargar R., Goudeseune C,. "A manifold interface for a high dimensional control interface." Proceedings of the ICM conference (Banff, Canada), 385–392. San Francisco CA, USA: International Computer Music Association, 1995.
2.   Keane, D. & Gross, P., "The MIDI baton," Proceedings International Computer Music Conference, Columbus, Ohio, USA. San Francisco CA, USA: International Computer Music Association, 1989.
3.   Krefeld, V., "The Hand in the Web: An interview with Michel Waisvisz," *Computer Music Journal*, 14 (2), pp. 28–33, 1990.
4.   Machover, T. & Chung, J., "Hyperinstruments: Musically intelligent and interactive performance and creativity systems," Proceedings International Computer Music Conference, Columbus, Ohio, USA. San Fransisco CA, USA: International Computer Music Association, 1989.
5.   Mandelis, J., "Genophone: An Evolutionary Approach to Sound Synthesis and Performance," Proceedings ALMMA 2001: Artificial Life Models for Musical Applications Workshop, Prague, Czech Republic: Editoriale Bios, pp. 37–50, 2001. http://www.cogs.susx.ac.uk/users/jamesm/Papers/ECAL(2001)ALMMAMandelis.ps
6.   Mandelis, J., "Adaptive Hyperinstruments: Applying Evolutionary Techniques to Sound Synthesis and Performance," Proceedings NIME 2002: New Interfaces for Musical Expression, Dublin, Ireland, pp. 192–193, 2002. http://www.cogs.susx.ac.uk/users/jamesm/Papers/NIME(2002)Mandelis.pdf
7.   Mulder, A.G.E., "Virtual Musical Instruments: Accessing the Sound Synthesis Universe as a Performer," Proceedings of the First Brazilian Symposium on Computer Music, pp. 243–250, 1994.
8.   Mulder, A.G.E. Fels, S.S. & Mase, K., "Empty-handed Gesture Analysis in Max/FTS," Proceedings of the AIMI international workshop on Kansei – the technology of emotion, Antonio Camurri (ed.), pp 87–90, 1997.

9.   Mulder, A.G.E. Fels, S.S. & Mase, K., "Mapping virtual object manipulation to sound variation," IPSJ SIG notes Vol. 97, No. 122, 97-MUS-23 (USA/Japan intercollege computer music festival), pp. 63–68, 1997.
10. Pressing, J., "Cybernetic issues in interactive performance systems," *Computer Music Journal*, 14 (1), pp. 12–25, 1990.
11. Rovan, J.B. Wanderley, M.M. Dubnov, S. & Depalle, P., "Instrumental Gestural Mapping Strategies as Expressivity Determinants in Computer Music Performance," presented at "Kansei – The Technology of Emotion" workshop, 1997.
12. Wessel, D. and Wright, M. "Problems and Prospects for Intimate Musical Control of Computers," ACM SIGCHI, CHI '01 Workshop New Interfaces for Musical Expression (NIME'01) ,2000.
13. Woolf, S., "Sound Gallery: An Interactive Artificial Life Artwork," MSc Thesis, School of Cognitive and Computing Sciences, University of Sussex, UK, 1999.
14. Yee-King, M. (2000).,"AudioServe – an online system to evolve modular audio synthesis circuits," *MSc Thesis,* School of Cognitive and Computing Sciences, University of Sussex, UK, 2000.

Genetic Improvisation Model
A Framework for Real-Time Performance Environments

Paul Nemirovsky and Richard Watson

MIT Media Laboratory, 20 Ames St., Cambridge, MA, USA 02139
{pauln, watsonr}@media.mit.edu

This paper presents the current state in an ongoing development of the Genetic Improvisation Model (GIM): a framework for the design of real-time improvisational systems. The aesthetic rationale for the model is presented, followed by a discussion of its general principles. A discussion of the Emonic Environment, a networked system for audiovisual creation built on GIM's principles, follows.

1 Introduction

Increasing numbers of tools for improvisation have appeared over the last few years, mimicking a growing awareness among researchers of the value in an improvisational approach to creativity. Improvisational methods have been successfully employed by human performers for thousands of years, yet, until recently, remained untouched by most engineers and interaction designers. Projects such as Voyager [1], Galapagos [2], ChaOs [3], and Swarm Music [4] ushered in a welcome change – while not explicitly improvisational, they explore media-informational spaces in nearly-improvisational ways, allowing their users to modify the exploratory process in the course of exploration. We have attempted to expand this research direction by creating a playground for building non-idiomatic[1] improvisational environments.

Non-idiomatic improvisation has been traditionally considered a process that does not lend itself to modeling; by its nature it seeks to defy any fixed model as a complete description of its behavior. A good improviser is one who is able to simultaneously avoid algorithmic description, always surprising his audience, and appeal to some aesthetic criteria (local or global[2]) in shaping his performance. In the context of improvisation, the characteristics and success of a performance can be described in terms of global aesthetic criteria; one of the most intuitive and widely used of these is *energy*. Why is it a useful concept? First, energy is a popular analogy with improvising[3] musicians and artists; many of them regard it as a convenient metaphor to think about one's improvisational talents ("Zorn has a great energy"). Second, speaking in terms of energy enables us to think about changes in behavior over time, regarding them in the context of an overall flow rather than as a composition of discrete actions.

---

1 Not following one fixed aesthetic idiom, such as a particular music style.
2 Here, *global criteria* define what constitutes an overall good improvisation, while *local criteria* are those of an immediate context.
3 From here on *improvisation* & any derivative words refer to the non-idiomatic improvisation.

S. Cagnoni et al. (Eds.): EvoWorkshops 2003, LNCS 2611, pp. 547–558, 2003.

Local aesthetic criteria on the other hand are dependent on an endless range of social, cultural, and technical circumstances. No model of improvisation should seek to define these criteria decisively. Models that do end up capturing a specific algorithmic behavior rather than the 'essence' of improvisation: its ability to always generate new criteria and mix the idiomatically unmixable, resulting in a never-ending exploratory process. We therefore strive to design improvisational models that allow both explicit as well as implicit criteria[4], with the latter emergent from the interaction of system agents, both autonomous as well as reactant to the human performer.

The GIM presents a foundation for constructing a vehicle for improvisational exploration. Such vehicle may exercise various degrees of autonomy. Similar to its physical counterpart, a GIM-based system could take the performer 'around the city' all on its own. Our research, however, is directed toward exploring situations of co-improvisation, where the performer and the system are symbiotically contributing to a shared performance. In one conception, the performer's role is to provide a general evaluation of the vehicle's activities and express high-level desires to be carried out by the vehicle. Such a vehicle is neither an autonomous improviser nor a passive assistant. In this model, a lower degree of precise control available to the human performer (in comparison to one who uses traditional instruments) is compensated by a lower responsibility for the output. As the system constantly produces new and modifies old materials and structures, the performer has no assurance as to the output yet is integrally involved in evaluating, transforming, and exchanging it with others. We foresee such co-improvisational approaches becoming manifest in computational tools for all walks of art making, information exchange, education, and entertainment.

How can the relationships between different elements in a co-improvisational system and the system and its users be modeled? We found a directed graph (or *network*) representation most useful. Rather than representing one-to-one relationships between input and output, we model associability of elements within an environment[5]. In this depiction, changing one element or connection affects everything else.

A second question that immediately arises is how to uncover configurations of the network that lead first to interesting connectivity and ultimately to compelling new behavioral modes or trends. Here, we argue that genetic algorithms are uniquely fit to meet this challenge, offering a largely unconstrained ('blind') evaluative process, which, when combined with the mutable non-linear structure of the network, provide a fertile ground for interesting emergent behavior. We discuss these issues further in the first section where we outline the GIM.

The only way to test the assumptions and predictions made in the GIM is by developing a system that implements it. A further question then becomes: what type of evaluation procedures are we to follow in deciding whether the system is taking part in a co-improvisational process or only behaving randomly? To address this question, the second section is dedicated to elaborating a number of guidelines that a system must follow to be considered as GIM-compliant.

Finally, in the third section, functional guidelines in hand, we discuss our implementation, the Emonic Environment (EE). We evaluate just how well we fulfill the principles set forth by the GIM and report on the current development state.

---

4 Meant here as a set of aesthetic principles observed by a performer when creating the media.
5 Here and throughout the paper, *environment* is meant as a workspace for creative activity.

2   GIM Motivations

Motive for the Architecture

It seems to us that the development of particularly interesting computational models of improvisation is hindered largely by adherence to (a) Western ideas of music and art (as a disassociated sphere of "fine art", guided by a set of immovable principles such as tonality, melodic principles, proportion, etc) and (b) the predominant computer science paradigm which concerns itself with predefined operational rules or parameters (what Koza [5] calls the strive for simplicity, convergence, conciseness).

In opposition to the school of thinking that studies channels of human perception and activity in isolation, we believe, drawing our inspiration from Varela [6], Claxton [7], and others that any creative activity is ultimately a social act, tied to the world around and within the improviser. As an implication of that view, creative processes, whatever their output mediums might be, cannot be seen as conceived solely in a particular medium; a musician improvising with sounds employs a much wider non-sonic array of procedural and perceptual memories – images, smells, sounds, etc., tied together and shaped in their perception by the social world of the improviser as well as the external stimuli being perceived.

As a performer improvises, he thinks up new 'things to say'. In order to investigate his thought process, we must ask two important questions: (1) what kind of representation does the improviser employ and (2) how do all the disparate 'things' become unified into one improvisational experience? It has been argued (e.g. in Marcus [8]) that the representations are never distinctly high- or low- level, but rather always a combination. We have adopted this conception as it makes sense and suits our model well. We have named the special combination of levels of representation a *mediated layer*, borrowing from Stafford's [9] concept of a mediating image.[6] Through forming connections in the mediated layer, new meaning is then made.

Creative thoughts rarely come solely in a form of "play a C4, then an E3 flat, then double each note produced with a bass that is always on the distance of a triton from the upper voice". Certainly, such procedural thinking is useful; if we were to never think in precise terms of melody and harmony (or point-of-view and principles of montage in video), we wouldn't be able to produce many sounds or images; our creative drive would never be realized as it would remain solely conceptual.

What such literal thinking lacks, however, is flexibility and abstraction. For in this paradigm, how can we define an algorithm for generating 'cool' sound, exchanging 'sad' melodies, or applying 'fairytale-like' video effects? We see then, that what we need is something that can mediate between the two layers. We must both be able to make manifest an abstract concept or feeling and at the same time be able to react to the very surface level perceptual happenings, realizing their importance contextually and fitting them into an overall conceptual framework. The two layers are interlinked: the sounds/images that the improviser produces are influenced by what is taking place around him (other performers, the crowd, sounds, colors, etc.); at the same time the surrounding world (and thus his perception of it) is influenced by his action of pro-

---

[6] Stafford believes that there is always a mediating image between two entities of an analogy; the image becomes the space of interaction / negotiation between the two entities (objects).

ducing these media artifacts. We refer to these layers as perceptual and structural respectively.

An improviser continuously cycles between the two, evaluating his own performance and attending and responding to the surrounding context. This relationship of feedback brings about a continuous assemblage of the two layers into a mediated one and the resulting creation of one unified experience. This process of controlling the flow of information can be thought of as controlling the energy of improvisation.

From this discussion, it is evident that both the structural and perceptual representations are essential to our understanding of the improvisational processes. Moreover, a dynamic interaction between the two layers appears to be necessary.

Motive for the Methods of Functional Alteration

Too Much Potential? A Case for Evolution

From our description above, it is apparent that any system built on the GIM will be rich in both the number and types of elements it contains as well as in the type of interactions between the elements and the layers to which they belong. A possible problem, then, in adopting this sort of model is that it allows for too many potential states or configurations. That is, in setting up a system to model improvisation, we are forced to relax any constraints on dynamics we might tend to put in place in order to manage the size of the parameter[7] space. Consequently, we will potentially find ourselves drowning in a sea of optional configurations from which we must fish out "just the right" one whose product suits our goals and context. If we spent the rest of our lives trying out all the possible combinations of parameters, we would but just graze the surface of that which is available. Must we conclude, then, that a true model of improvisation is incompatible with an implementation that can use it?

GIM vs. TSP

Those familiar with genetic algorithms know of the traveling salesman (optimization) problem (TSP) [10] – an example that illustrates how GAs may overcome painful levels of combinatorial complexity. Though a GIM parameter space is much larger than that of the TSP, our goal criteria are not as exclusive as those deemed by the salesman. Namely, our goal is not to find *the* best configuration, but to find a configuration whose product pleases us. We may safely presume that several such configurations exist. So, although our parameter space is larger, so too is our solution space[8].

Another feature difference of our problem, which goes hand in hand with that of multiple goal states, is the nature of the fitness criteria. We say TSP is transparent because we know exactly what it takes for one parameter configuration to be better than another. However that is not the case in the creative paradigm addressed by GIM. Instead, we must explore by trial and error. Our fitness criteria are *implicit* and *user specific*, the exact definitions inaccessible to us and to others.

---

[7] Parameters, here, refer to the specifications of elements, element properties and interactions between elements in the system.

[8] A further interesting topic of investigation is just *how* this solution space grows with the parameter space. For lack of room, we postpone this discussion until a later time.

We can furthermore distinguish the two problems by noticing that not only the criteria change between users but also that they change for a given user, even over the course of a single performance.

Genetic Algorithm for Exploration

The distinctions we've drawn out above between our problem and that of the TSP point to a very different picture of the job of a GA in our model. What we've sketched here is not a tool for *optimizing* a set of parameters given fixed goals and constraints toward a 'best' solution. Instead we're employing a GA to suggest possible configurations to the user and traverse the parameter landscape based on feedback from that user regarding those suggestions. We cannot say that this movement on the part of the GA is optimization for three reasons: any 'peaks' in the fitness landscape corresponding to our parameter space are (1) plural, (2) fuzzy and potentially plateau-ed, and (3) dynamic.

Instead, we might say, our GA is being used as a tool for *exploring*. The GA's job is to work with the user to find new directions for movement in the parameter space and pick which shall be followed. Furthermore, there is an inherent feedback process involved. As the GA helps the user to traverse the parameter space, the user generates and refines the criteria by which he is judging the suggestions. The whole procedure leads to a richly dynamic, interesting, and unpredictable improvisational trajectory.

Genetic Advantages

We might imagine that we could adopt other search algorithms for use in conjunction with a system built out of the principles of GIM. It is worth mentioning, then what advantages a GA approach might offer over its competitors.

1. GAs are 'blind'. That is, they do not have a plan for what direction they will search in next. A trajectory is mapped solely in virtue of how well local movements along that trajectory bring about positive (desired) changes. This seems to fit better with the idea of improvisation-as-exploration than might an algorithm that tries to 'figure out' the correct direction to move in.
2. GAs are context sensitive. They take into account a continuously changing context while deciding on new directions of movement within a parameter space. This is something that is 'built in' to the algorithm and thus doesn't require any sort of reconfiguration. It is a characteristic that is well suited to an improvisational setting, which, by definition is responsive to context.
3. GAs can be designed to act solely on the basis of emergent behavior. That is, instead of an algorithm that might consider what changes to any particular parameter might do to the system in virtue of itself alone, GAs may be (and usually are) designed to respond to the overall effect. This is important for improvisers who are concerned with how change of a single element may alter interactions between elements in the totality of the performance.

Principles

To define the principles of the GIM, the following questions must be considered: what is improvisation? What aspects of it does GIM try to model? Who would be the consumer of a system built on the GIM – would he be a composer, performer, audience, or some other new type of participant?

1st: Dynamic nature of improvisational structures. Structural representations used in the course of improvisation are incorporated, modified, and purged dynamically to satisfy the improviser's changing goals and attention. The criteria guiding improviser's behavior (in terms of expectation and evaluation procedures he employs) evolve in the course of a performance; an improviser changes what he considers the "right thing to do" based on the combination of these evolving criteria and the stimuli from his environment that only manifest themselves as the performance develops.

2nd: Changing, multiple-leveled focus. An improviser thinks about what he's doing in many different ways. Continuously switching between macro- and micro-level representations, he attends to the very minute (e.g. a particular sound) at one moment, only to switch and think about general development (e.g. a climax) a second later.

3rd: Diversity of types. Improvisation is a result of interrelating multiple perceptual inputs and memories; an improvisation whose 'output' is audio is nevertheless an improvisation that includes visual, tactile, and other formative content. A sound might be inspired by an image, which in turn is inspired by a text or another sound; this free and proactive interaction of types is integral to the improvisational process.

4th: Relevance of context. Following on the above point, the improviser's decision-making is rooted in the totality of his perception of the moment. Thus medium-specific laws of decision-making should be used cautiously in deciding the subsequent output, for the perception of any media is in itself an act shaped by the context. Indeed, improvisation is not formed in a vacuum or in one medium separated from others; it strives to incorporate or reflect the environment in which it is created.

5th: Process, not artifact production, as the goal. An improviser, unlike a Western composer, feature-film cinematographer, or product designer[9], is not concerned with the production of a final artifact, a sonata, a pop-song, a chair, or a movie. While improvisation might be recorded and as such seen as a fixed construct, the true point of improvisation is the process of exploration, contextualizing and interrelating memories and perceptions[10]. An improviser's job is to weave together an array of 'sketches', which gain their relevance (and meaning) only as the improvisation unfolds.

6th: Absence of a plan. Planning does not seem to be the optimal way to think about the process of improvisational creation. Instead, the act of improvisation is better thought of as one of exploration. Another way of putting it is that an improviser is far less concerned with perfectly playing to a specification than he is with breaking new ground and learning from unintended mistakes and successes.

---

9 This of course is not a binary dichotomy; discussion of near-improvisational compositional movements however is beyond the scope of this paper; see for example Nyman's [12] excellent account of the experimental music movement.

10 This memories-perception inside-outside Descartian dichotomy is questionable to say the least; we use it for the reasons of space, not to endorse it. The idea of the rhizome [13] seems to be much better fit for the description of improvisational processes.

7th: Issues of control/responsibility. In an improvisational performance, no fixed contract specifying responsibilities of control (balance of power) exists between the performers; the criteria that define the degree to which each party assumes creative control over different aspects of the ongoing improvisation are set dynamically, according to implicit and explicit negotiations between the performers. Giving up part of the control also frees the improviser from the preoccupation with creating a perfect finalized product. In other words, improvisation implies a lower cost of experimentation, allowing spontaneous exploration of new, 'unproven', ideas.

8th: Continual feedback. Improvisation is not evaluated at one point in time or space. Over the course of a performance, improvisers provide feedback to each other. This feedback ranges from general and vague to particular and precise; what defines its value is the ability of the recipients to learn from it and move in new directions. The learning is not procedural; it cannot be summarized by a symbolic rule. Instead, it can be described as discovering patterns where one didn't see them before.

9th: Meaning-making through exchange. In an improvisational group action, construction of meaning happens through the exchange of elements. In other words, a sound or an image acquires its meaning only through the details of its history of use – where and how it has been employed before. [11] These details determine how it or similar elements are perceived the next time they are encountered.

10th: Audience as participant. From the passive audience of the linear storytelling to the nearly equally passive audience of the multiple-choice "interactive" environments, a strict giver / taker dichotomy has been enforced between the consumer (the audience) and the producer (the performer)[11]. In improvisation however such distinction is obsolete; anyone can co-improvise, so long as the effect of his activity is heard / seen in one way or another by the other performers. Similarly, even when not actively participating in the act of media creation, the audience is not regarded as passive; it is viewed as a part of the improvisational circle.

3   From Principles to Implementation Guidelines

From the above principles may be extracted a set of guidelines that a system based on the GIM should follow. The guidelines enumerated below are directly derived from and continuous with the list above.

1. The structure of the system must be mutable. It must allow for the addition and removal of modular substructures within the system. The system must also have an alteration process that has continual and consistent access to this structure. Furthermore, the system must be able to accommodate a performer's dynamically changing goals and expressive behavior.

2. The system must be accessible on many levels. Performers must have the option of changing very minute details of particular elements in a performance media stream; at the same time, they must be able to have control of persistent abstract patterns.

3. The system must handle a diverse range of media types, both as input and as elements within the system.

---

11 The few fortunate exceptions from this rule do not do much more than recombine already existent materials without modifying any of their properties.

4. These diverse types of input must be able to 'speak' to one another. That is, the system must have ways to connect various types of media, whether directly through the operators that output the media or through indirect structural relations.

5. The system must have focus not on constructing a polished 'work' but on the process during a performance. This is not to say that an appreciable sequence (that might recur) may not come out of the process, but only that the system must not focus on producing the output as a single whole.

6. Unpredictability must permeate the system; nothing should be 'laid out'. Not only should it be unforeseen where the system will be at the end of a performance but also (to one extent or another) where it will be in the next few seconds. This fits with the idea that improvising involves taking risks.

7. The system should have a configurable 'autonomy level' which determines to what degree it is running independently; specification of what and when the system controls should be open to runtime modification.

8. An ongoing evaluation process from performer to system (and maybe from system to performer) should be in place. Feedback in this process must have the potential to be both specific to particular areas and general, expressing views on the performance of the system as a whole.

9. The system must have the capability to exchange media and system structures with other users (on separate instances of the system), and to employ these shared elements in various ways.

10. Methods of sensing the audience (e.g. using cameras or microphones) should be developed for and within the system.

4 The Emonic Environment

The Emonic Environment (EE) is our attempt at building a system according to the GIM-based guidelines. Using the EE, performers are able to create, modify, and exchange audiovisual content.

The EE is designed with a multi-layered architecture, not unlike an animal nervous system. Interfaces to the system from outside environments are situated in the Input layer, on the bottom. In the middle, the 'brains' of the system exist as a neural network in the Structural layer. Finally, on top, media events are scheduled and processed by a population of media operators known as *emons* which are bonded to one another through a web of connections in the Perceptual layer. This layer is analogous to the motor system of a mobile organism.

This layered design addresses the need for improvisers to access the system on many levels. It affords performers a capability to dynamically refocus their objectives while creating and manipulating media in realtime. The layered design also modularizes the system making it easier to understand from the inside and easier to expand.

Joints between the layers serve to enable flow from outside world influences to inner dynamic structures and finally to media operators which reconnect the system to the outside world. Links within the layers enable richly interactive (largely unpredictable) emergent behavior to arise.

The EE is written in Java and uses various third party components including JSyn for sound synthesis, playback and processing; JMSL for event timing; and custom libraries for audio processing developed in the MIT Media Lab.

Layer Details
The Input layer of the EE consists of tangible and software interfaces that sample information from the outside world (e.g. computer mouse, video camera, gesture controller). The aim of the Input layer is to integrate the world around a performer into the ongoing improvisation. Input received from each interface is used to control functional elements in one or many of the three layers using *transforms*, which map between a particular device's data format and that of a particular element. One category of transform, for instance, maps from a custom gestural controller (called the Emonator) which outputs arrays of 144 short values to a one-dimensional double value. Over time this produces a signal that can be routed to emons on the Perceptual layer to modulate properties such as frequency or amplitude.

The Structural layer is a recurrent neural network, populated with nodes and weighted connections (or *associations*). The network and its elements individually each have a number of properties which can be controlled either explicitly (by the performer) or by a system administered process:

Association: Path, Weight, Time Delay (inner-node stimulus travel time)
Node: Activity, Decay Rate (of activity), Propagation Threshold (above which the node propagates any incoming stimuli), In/Out Stimulus Scalar
Network: Max Simulation Propagations, Auto-Activation Threshold (under which new spontaneous activity will be introduced), Auto Management Features (when operating without any performer input; not yet implemented).

The architecture of a recurrent neural network offers both intricate and largely unpredictable behavioral profiles while at the same time adhering to constraints that generate perceivable patterns of activity.

The Perceptual layer is populated with *emons* – constructs that receive data from other emons or elements in other layers and translate it (if necessary) into directives to modulate the generation, modification and presentation of media that they control. Each emon serves one media function (e.g. sample player or sine-wave generator) with one or more mutable properties (e.g. amplitude). It modifies these properties according to an incoming array of data points (either fresh from another source – emon, input interface, etc. – or 'docked' and reused over a period of time) passed to it directly or through a translation process. Perhaps the most important source that an emon 'listens' to is the collective of nodes on the Structural layer. By attending and responding to these nodes, the layer of emons acquires complex patterns of behavior.

The range of emons employable in the Perceptual layer is diverse. They categories include: audio and video sample playing; audio waveform generating; audio processing (e.g. filtering); video processing; lighting control; and textual generating and playing. Sampled audio and video playing emons have the following properties:

Sampled Audio: Playback, Speed, Direction, Loop Pos. {Start, End}, Amplitude
Sampled Video: Playback, Speed, Direction, RGB channel control, Loop Position {Start, End}, Blend Amount.

These diverse species of emons are able to talk to one another either directly or through a *translator* agent. Emons of a common format (e.g., audio) in most cases may transfer signals directly. (e.g., a sine-wave generator emon can pass a signal directly to an audio sample-playing emon to modulate its amplitude) Otherwise the signals are passed through these translators which make them readable to the receiver.

Elements on all of the different layers as well as all media employed by a performer are sharable. That is, the system is designed to communicate with other instances of the same system, providing means for immediate collaboration as well as co-improvisational performance with other EE performers.

The system is designed to be operated using changeable degrees of autonomy. This variability allows it to be disseminated in a striking range of circumstances. On one end of the spectrum, a musician might employ the system in a performance where he requires full-on immediate influence in the Structural and Perceptual layers. On the other, a visual artist might set up a semi-permanent installation where ongoing network activity in the Structural layer continually and independently modulates room lighting controlled on the Perceptual layer, taking inputs only from ambient sensors.

Evolution Implementation

Evolution engages the EE on all three layers. The process may run singularly, combining the configurations of elements on different layers into one large genetic code; or it may run as distinct processes, allowing different layers in the network to evolve on different time scales.

Evolution modifies the Input layer by controlling just where and how data is mapped between an input interface and another layer. When specified an evolutionary process will run that will modify *transforms* (described above) and destinations of those transforms, shaping the way the outside world affects the system.

Elements in the Structural layer are possibly the prime target of the evolutionary process. Through evolution, the connectivity of the network as well as the properties of individual nodes ripen, developing a complexity as a whole which may never be found manually. Just as the brain of an animal evolves over time to suit its environment, the brain of the EE evolves to suit a performer in a given context.

In the Perceptual layer, the evolutionary process modifies the associated *translator* agents of emons, changing how an emon responds to various forms of input; the connectivity of the web of emons; and its association with the Structural layer's nodes.

We have integrated evolution in our system as three modes of operation. Each mode offers the performer a different way to interact with and change the system and each is appropriate for different goals and contexts in which the system may be used.

1. Browse (offline) In Browse mode, several 'child' networks run in the background. The networks are initialized in corresponding states of activity and each network's output is recorded over a set period of time. Each recording is voted on by the performer, and decisions about breeding, mutating, and killing of the source networks are made based on these votes in a tournament fashion. This mode is designed for a performer who is interested in configuring a network offline in a simple, hassle-free way.

2. Explore (online) For a user who would like to perform with the network, or perhaps just wants a continuous interactive experience with the system, Explore mode offers a realtime option. Here, parameters of a single network are mutated on a consistent periodic basis. Voting is an ongoing process whereby votes are captured and mapped to the appropriate configurations (present at the time of voting) according to user's actions. Direction of mutation is modified continually based on the voting.

3. Navigate (online) Navigate mode offers a more 'hands-off' option for the performer. In this mode, the performer specifies one or more saved network configurations called *magnets* to direct the evolutionary process. The magnets either attract or repel the state of the online network toward or away from that configuration, thus producing interesting phase trajectories in a performance. The mode is made more interactive when the user dynamically (de)emphasizes magnets; or when he combines it with Explore mode, allowing the user to 'drop' magnets as he goes.

Current Status
As of now, emons are strictly dependent on the nodes they listen to (one node per emon, right now). No translator agents have yet been implemented. As such, each emon understands only two types of data: directives from a connected node and the specific media it plays. Data is not yet independent of those emons that use it – while it's possible for emons to exchange media or source nodes, no independent 'data repositories' to which an emon or an agent could connect have been implemented.

Emons currently operate on sampled audio and video, in three forms: (1) prerecorded, (2) as realtime input and (3) streaming (from a URL). The first crop of emons that perform audio synthesis and processing are being developed and should be available over the next couple of months. Emons that operate on text (sampling and generation) are currently being planned. In addition, we are currently developing a bridge between audio emons and DirectX audio plug-ins as well as a built-in sample maker.

Though the functionality of the EE is currently far more limited than the sketch we have given above, the system is able to produce interesting sequences and combinations of sounds and video which enable (through choice of media data and Structural configuration), many various musical and visual patterns. For instance, loading up an array of choral voices yields a satisfying series of choir music arrangements. Combining recordings of a string orchestra with the political orations of Fidel Castro produces an unexpected simultaneous combination of comedy and drama. The environment has been met with good reactions by an initial testbed of non-professional performers. Those working with the EE have been enthusiastic, spending extended amounts of time and generally commenting on how it reminds them of one band or another.

5 Conclusion

This paper is a progress report. We have focused on theory, showing justification for the creation of the GIM, articulating its main principles, and describing our ongoing implementation efforts. It is our hope and plan to continue exploring improvisational landscapes in the context of evolution, bringing about a change in how we regard creation, modification and dissemination of media in computer environments.

Acknowledgements. The authors thank Glorianna Davenport and Ariadna Quattoni for their ideas and help.

References

1.   Lewis, G., *Interacting with latter-day musical automata.* Contemporary Music Review, 1999. **18**(3): p. 99–112.

2.   Sims, K., *Galapagos*. 1997. Information online at: http://www.genarts.com/galapagos/
3.   Miranda, E. R., *On the Origins and Evolution of Music in Virtual Worlds*, in Bentley, P. and D. Corne, *Creative evolutionary systems*. 2002, San Francisco, CA. San Diego, CA: Morgan Kaufmann ; Academic Press.
4.   Blackwell T.M. and Bentley P.J. (2002) "Improvised Music with Swarms", 2002 Congress on Evolutionary Computation, CEC-2002
5.   Koza, J.R., *Genetic programming : on the programming of computers by means of natural selection*. Complex adaptive systems. 1992, Cambridge, Mass.: MIT Press.
6.   Varela, F.J., E. Thompson, and E. Rosch, *The embodied mind : cognitive science and human experience*. 1991, Cambridge, Mass.: MIT Press.
7.   Claxton, G., *Cognitive psychology : new directions*. International library of psychology. 1980, London ; Boston: Routledge & Kegan Paul.
8.   Marcus, G.F., *The algebraic mind : integrating connectionism and cognitive science*. Learning, development, and conceptual change. 2001, Cambridge, Mass.: MIT Press.
9.   Stafford, B.M., *Visual analogy : consciousness as the art of connecting*. 1999, Cambridge, Mass.: MIT Press.
10.   Traveling Salesman Problem. Information online at http://www.math.princeton.edu/tsp/
11.   Mauss, M., The gift; forms and functions of exchange in archaic societies. 1954, Glencoe, Ill.,: Free press. xiv, 130 p.
12.   Nyman, M., *Experimental music : Cage and beyond*. 2nd ed. 1999, Cambridge ; New York: Cambridge University Press.
13.   Deleuze, G. and F.l. Guattari, *A thousand plateaus : capitalism and schizophrenia*. 1987, Minneapolis: University of Minnesota Press.
14.   Minsky, M.L., *The society of mind*. 1st Touchstone ed. Touchstone book. 1988, New York: Simon & Schuster.
15.   Bentley et al, in Bentley, P. and D. Corne, *Creative evolutionary systems*. 2002, San Francisco, CA. San Diego, CA: Morgan Kaufmann ; Academic Press.
16.   Sloboda, J.A., *Generative processes in music : the psychology of performance, improvisation, and composition*. 2000, Oxford [England] New York: Clarendon Press.
17.   Boden, M.A., *The philosophy of artificial life*. Oxford readings in philosophy. 1996, Oxford ; New York: Oxford University Press.
18.   Dawkins, R., *The selfish gene*. New ed. 1989, Oxford; New York: Oxford Univ. Press.
19.   Miranda, E.R., *Readings in music and artificial intelligence*. Contemporary music studies, v. 20. 2000, Amsterdam: Harwood Academic.
20.   Miranda, E.R., *Composing music with computers*. Music technology series. 2001, Oxford ; Boston: Focal Press.
21.   Moroni, J. et al, *Vox Populi: Evolutionary Computation for Music Evolution*.
22.   Rowe, R., *Interactive music systems : machine listening and composing*. 1993, Cambridge, Mass.: MIT Press.
23.   Lotman, I.F.U.M., *Universe of the mind : a semiotic theory of culture*. 1990, Bloomington: Indiana University Press.
24.   Degazio, B., Evolution of Musical Organisms, Leonardo Music Journal, MIT Press, 1997
25.   Hendriks-Jansen, H., *Catching ourselves in the act : situated activity, interactive emergence, evolution, and human thought*. Complex adaptive systems. 1996, Cambridge, Mass.: MIT Press. xii, 367.
26.   Todd, P.; Werner, G.M., Frankensteinian methods for evolutionary music composition in Griffith, N. and P.M. Todd, Musical networks : parallel distributed perception and performance. 1999, Cambridge, Mass.: MIT Press. xv, 385.

On the Development of Critics in Evolutionary Computation Artists

Juan Romero[1], Penousal Machado[2], Antonio Santos[1], and Amilcar Cardoso[3]

[1] Creative Computer Line – RNASA Lab – Fac. of Computer Science – University of Coruña, Spain.
[2] Instituto Superior de Engenharia de Coimbra; Quinta da Nora, 3030 Coimbra, Portugal
[3] CISUC – Centre for Informatics and Systems; University of Coimbra, Pinhal de Marrocos, 3030 Coimbra, Portugal
{jj,nini}@udc.es, {machado,amilcar}@dei.uc.pt

Abstract. One of the problems in the use of evolutionary computer systems in artistic tasks is the lack of artificial models of human critics. In this paper, based on the state of the art and on our previous related work, we propose a general architecture for an artificial art critic, and a strategy for the validation of this type of system. The architecture includes two modules: the *analyser*, which does a pre-processing of the artwork, extracting several measurements and characteristics; and the *evaluator*, which, based on the output of the *analyser*, classifies the artwork according to a certain criteria. The validation procedure consists of several stages, ranging from author and style discrimination to the integration of critic in a dynamic environment together with humans.

1 Introduction

The creation of art with computers is an old dream and there were several systems that create, assess or process art by using all types of computational methods. Over the last few years, many evolutionary computation (EC) systems have been dedicated to the creation of art [1]. Indeed, natural evolution has created some truly beautiful forms, and evolutionary computation techniques have proved effective in fields that require a certain degree of creativity [2].

An EC process consists of two stages, the generation of new individuals and their evaluation. Considering a multi-agent architecture, one can view the evolutionary process as an interplay between two types of agents, creators and critics. The creators are responsible for the generation of the new works; the critics evaluate the generated works, thus determining their survival probabilities. The evaluation plays a key role in EC, since it guides the search procedure. However, in fields that encompass a vast amount of subjective and cultural criteria, the development of an evaluation method presents considerable problems. As such, the main difficulty in the application of EC techniques to the field of the arts lies in the development of an appropriate fitness function.

"We are moving towards creation of machine art, or direct collaboration with machines, in which software makes aesthetic judgments." [3]. The present paper is about the development of artificial art critics (AAC), i.e. agents that evaluate artworks based on aesthetic and cultural criteria. In the real world the "value" of an artwork can

S. Cagnoni et al. (Eds.): EvoWorkshops 2003, LNCS 2611, pp. 559–569, 2003.

only be assessed by taking into account its cultural context. One of the shortcomings of nowadays Computational Artists is precisely their isolation. In order to provide a cultural context to our agents, we propose their integration in a Hybrid Society of artists and critics, both computational and human.

Next we describe the structure of this paper. Section 2 shortly reviews the state of the art in the development of AAC's, classifying current systems into different classes, and making an analysis of the characteristics of these classes. Section 3 presents a brief description of some of our previous research efforts, directly connected with the generation and evaluation of artworks with EC techniques.

Based on the analysis presented in section 2 and on the conclusions derived from our previous work, in section 4, we propose a general architecture that tries to maximize the virtues of the different approaches and minimize their shortcomings. The proposed architecture includes two modules: the *analyser*, which does a pre-processing of the artwork, extracting several measurements and characteristics; and the *evaluator*, which, based on the output of the analyser, classifies the artwork according to a certain criteria. Thus, unlike most of the evaluation systems, the assessment does not deal with the artwork directly; instead, it is based on some of the artwork's characteristics. In our proposal, the *analyser* is static, i.e. the properties taken into account do not change over time. The *evaluator* module is adaptive, which means that the way the characteristics are evaluated, changes through time.

One of the main difficulties on the development of Computational Artist, and more specifically AAC's is their validation. We propose a multi-stage validation methodology. The first steps of this procedure, allow the objective, yet meaningful, assessment of the developed AAC's, providing a solid basis for their development. The later steps consider more dynamic criteria, and include testing the AAC's in a hybrid society of humans and artificial agents.

Finally, in section 5, we draw a series of conclusions and outline future research.

2   State of Art

We can distinguish two roles in any system of artistic creation: the creator (or author) and the critic (or audience). This section briefly analyses systems in which the generation role is played by an evolutionary algorithm, focusing on the classification of the approaches used to evaluate the artworks. Due to lack of space, we cannot include a wide list of references. A more elaborate survey can be found in [1]. In the field of music generation, several works analyse this aspect, such as: Burton [4], Romero [5], and Todd [6].

Taking into account the great number of evolutionary systems devoted to artistic tasks, it is convenient to classify them according to the type of approach used to implement the AAC. Following this criterion we obtain four main classes: interactive; example based; rule based; and EC based.

In the first type of system – the interactive type – the role of the critic is played by a human being, who evaluates the pieces generated by the system, and thus guides the evolutionary process. The critic's role can be played by a single person or by a group; in the latter case the generated works are evaluated simultaneously by several people, and the system uses the average value. Some examples of works in this category are [7-15], in the musical domain, and [16-25], in the visual domain.

The main drawback of this type of system is the time involved in the user's evaluation of the artworks, which severely limits the number of generations [17, 26, 27]. This drawback is particularly severe in the musical domain. Additionally, the lack of consistency of the human evaluation also raises difficulties.

These negative consequences of human evaluation have led researches to develop systems that try to learn the user preferences from a set of examples. The most common approach relies on the use of Artificial Neural Networks (ANN's). Typically, the ANN's are trained using, either, a set of pieces of a particular style and/or author, or, a set resulting from an interactive evolutionary system. Some examples of evolutionary systems that resort to example based evaluation are [28, 29, 30], in the musical domain, and [31], in the visual domain.

This type of systems appears to have great potential, however, the results achieved are often disappointing, particularly in the visual arts domain. The main difficulties in the development of these systems are: the creation of a representative training set; and the huge amount of information present in the training artwork.

In rule-based systems, the AAC is formed by a set of rules that conduct the system. The set of rules tries to express some knowledge of the concerned domain, and typically, results from stylistic requirements of a particular type of artwork. In the musical domain, one can find several examples of this type of system, some examples being [27, 32, 33]. We weren't able to find any system using this type of approach in the field of visual arts, which can't be considered surprising, when we take into account that musical theory is way more developed than visual art theory.

The main problem of this type of approach is their lack of generality. These systems are based around a particular vision of an artistic style or theory. As such, their adaptation to other styles is difficult, if not impossible. Their main advantage lies in the possibility of using a set of formalizations, structures, metrics, and knowledge, which makes the analysis of the pieces of art easier.

A further possibility is to use an evolutionary system as AAC. A typical approach would be to co-evolve two populations: one of creators and one of AAC's. In this type of system the fitness of a given individual (creator or critic) is determined by the interplay of between agents, giving rise to new and isolated aesthetics. Some examples of this type of approach are: [34] that composes music; [35, 36] that uses artificial life techniques to create musical themes and sounds; and [37] that resorts to co-evolution to evolve images.

3 Background Work

This section shortly describes some of our previous work, which is the basis for our ongoing research. This description comprises examples of some of the classes of systems introduced in the previous section. We will focus on the analysis of the critics used.

3.1 Music Generation and Hybrid Society

The Tribe project started in 1998 [12, 13] and aims at building artificial models of human music composers. The underlying idea of this project is to follow the evolution of human music through time, i.e., start by focusing on the most primitive and simple forms of music, and gradually build more complex models.

Tribe is a typical interactive evolutionary system that composes this very primeval music, namely purely percussive. Each *tribe* consist on one percussive pattern. The user assigns the fitness of each tribe, in an interactive way.

The experience with the Tribe project revealed the need to facilitate the design of social artificial beings, able to develop and assess new creative products. Following this goal, a model of an egalitarian society populated by humans and artificial beings was proposed, giving rise to the Hybrid Society Project[1]. Since the performance of the agents acting in Hybrid Society, is evaluated by a dynamic society of both artificial and human agents, this paradigm allows an adequate validation of a social system.

Moreover, this paradigm provides a natural learning approach, intermediary between learning by discovering and learning by reinforcement, as it mimics the dynamics of a human society.

Although this approach presents some logistic difficulties since it needs the participation of several humans. However, it is more flexible than standard interactive evolutionary systems, since it is designed to allow the change of the human participants through time. In standard interactive evolutionary systems, the interaction of one human, or a close group, it is required during the all duration of the experiment. Moreover, Hybrid Society takes into account the "opinion" of each of the humans (and artificial systems) in a group and not just an average opinion.

Hybrid Society allows the incorporation of computer systems that manage several artificial agents, which can have different "genetic code" and distinct behaviour at the same time. In this sense, each computer system can be seen as a species with different individuals. This capacity provides the necessary conditions to allow the use of hybrid society in conjunction with evolutionary computation systems.

In the first experiments this paradigm was applied in the musical domain [38]. Two artificial species were considered: one of artificial creators, based on the previously mentioned Tribe project; and one of AAC called "Oreja" (Spanish word for ear). Additionally, the Hybrid society also included human critics and creators.

The AAC resorted to a set of ANN's, which were trained using evolutionary computation techniques. The weights of the connections between the nodes, and the architecture of the net were included in the genetic code, and, could thus evolve through time. Each rhythmic pattern is codified as an array of 160 binary elements, 10 percussive instruments in 16 slots of time. The short dimension of the rhythmic patterns allows us to use them, directly, as input for the ANN's. In the future we intend to apply Hybrid Society to the evolution of more complex structures, the size of these structures will, eventually, make it unfeasible to use them directly as input. As stated before, our idea is to extract relevant characteristics, properties and measures, of the artworks and use them as input for the ANN's. This idea will be discussed in more detail in section 4.

---

[1] More information about the Hybrid Society Project can be found in http://www.hybridsociety.net

3.2 Visual Art

NEvAr (Neuro Evolutionary Art) is a research project that aims at building an artificial artist in the field of visual arts. In the simplest mode of operation NEvAr is an evolutionary art tool, it allows the evolution of populations of images, which are evaluated by a user that guides evolution.

NEvAr was inspired in the works of Dawkins [39] and Sims [21] and shares several similarities with the latter. It resorts to Genetic Programming [40], and as such the genotypes are trees constructed from a lexicon of functions and terminals. The function set is composed mainly of simple functions such as arithmetic, trigonometric and logic operations. The terminal set is composed of a set of variables x and y and random constants. The phenotype (image) is generated by evaluating the genotype for each (x,y) pair belonging to the image. Thus, the images generated by NEvAr can be seen as graphical portrayals of mathematical expressions.

As usual, the genetic operations (recombination and mutation) are performed at the genotype level. In order to produce colour images, NEvAr resorts to a special kind of terminal that returns a different value depending on the colour channel – Red, Green or Blue – that is being processed.

In [17] we made an assessment of NEvAr as a tool. According to our analysis the artworks evolved with NEvAr reflect the aesthetic and artistic principles of the user[2]. This analysis also revealed the importance of the individual's database. In NEvAr the user has the possibility to store highly fit individuals in a database. Later these individuals can be injected in an ongoing experiment, or used to create a non-random initial population. By resorting to this database, one can significantly decrease the amount of time necessary to create "good" images. Since the recombination possibilities are virtually infinite, the generated images will still be new, and, in most cases, innovative. However, they will share several characteristics with the selected database individuals, which can be considered an inspiring set [17].

With time, the database size has increased drastically, which has led to the development of automatic seeding procedures. The basic idea was to select a set of database images that resemble one supplied by the user, and then make this set as part of the initial population. To achieve this goal we needed to develop a way to compare images. After testing several measures of distance among images, we came to the conclusion that a direct comparison was not appropriate [17]. Instead, we extracted some characteristics of the images that were deemed important (in this case several complexity estimates), and then performed the comparison based on these characteristics [17]. Although limited, this approach gave promising results, especially if we take into account that we used a very small number of characteristics.

As stated in the beginning of this section the ultimate goal of this project is to create a computational artist. This means that NEvAr should be able to work autonomously, and hence, assign fitness to the individuals based on aesthetic criteria. In other words, NEvAr must be able to act, at least to some extent, as its own critic.

The approach used in NEvAr to evaluate images, is based on our personal views about aesthetics, and relies on the idea that the aesthetic value of an image is connected with the sensorial and intellectual pleasure resulting from its perception. Moreover, this pleasure is deeply related with the perceived complexity of the

---

[2] For examples of artworks generated with NEvAr can be found at:
http://www.dei.uc.pt/~machado.

sensorial stimulus, and with the complexity of the percept (the representation of what is perceived).

Based on this notion, and using estimates of image and processing complexity, we developed an evaluation procedure that, basically, attributes high fitness values to images that are, simultaneously, visually complex and easy to process. A full description of this procedure, along with some experimental results can be found in [41], for a discussion of the importance of complexity in art see, e.g., [42] [43].

The evaluation procedure only takes into account the lightness information of the images, discarding the hue and saturation information. Therefore, in this mode of execution, we are limited to greyscale images.

An analysis of the role of colour and the way colour is assigned, particularly in abstract art, leads to the conclusion that artists (certainly not all, but at least a significant proportion) usually work with a limited colour palette, and that the spatial relation between colours usually follows a set of rules. This is consistent with the view that each artist constructs its own artistic language, which complies with an implicit grammar.

The idea of creating a program to give colour to the greyscale images created by NEvAr emerged naturally. Unfortunately the development of such a program poses several problems. We are therefore developing a system that learns to colour images from a set of training ones. This approach has, potentially, several advantages over a built-in colouring procedure, namely: we do not need to code by hand a set of colouring rules; the results of the system are less predictable; and we can use paintings made by well-known artists as training set, thus learning to colour images according to their style.

In [44] we present our current approach in which, we employ Genetic Programming to evolve computer programs that mimic the colourings of the training instances. As before, the experimental results indicate that making a direct comparison between the colourings yields, poor and uninteresting results. By taking into consideration some of their underlying characteristics, we were able to improve these results significantly, obtaining interesting colouring programs [44].

4 Model Proposal

In this section we propose a model for the development of an AAC. The design of this model was based on a set of characteristics that we consider desirable:

- Generality – We want a model that allows the development of AAC's for different domains; the domain specific tasks should be carried out by specialized modules, allowing an "easy" adaptation of the AAC to new domains.
- Independence – We are interested in AAC's that are able to perform autonomously in an egalitarian society of humans and artificial beings. Therefore, the agents should be able to perceive the artworks in a standard and uniform representation (binary files as bitmaps or midi) and evaluate them. In other words, they should be able to "see" (or ear, touch, etc., depending on the type of piece) the artworks and form a judgement base on what they "see". An AAC can build its own internal representation of an artwork, but it cannot access the original artwork representation (assuming that this representation exists).

- Adaptability – The AAC's should evolve and adapt with time. This however is not a strict requirement, a static AAC can also be successful; we are primarily interested in this type of AAC, because they mimic better the behaviour of human critics.

- Sociability – Ideally, the AAC's should be able to adjust their behaviour according to the demands of the society in which they are integrated.

4.1 Architecture

Figure 1 presents a rough outline of the proposed architecture. The AAC is composed by two main modules: an *analyser* and an *evaluator*.

The *analyser* is static and receives as input a direct representation of the artwork (e.g., bitmap file, midi sequence, etc.) producing some sort of analysis. The *evaluator* is a purely adaptive system, based, for instance, on ANN's or Evolutionary Computation techniques, and forming a judgement based on the analysis created by the first module.

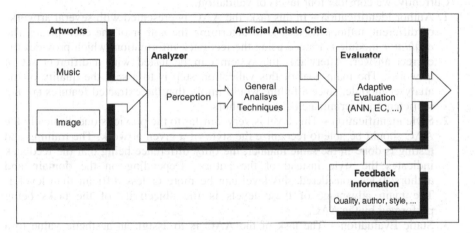

Fig. 1. Outline of the proposed model.

The analysis comprises two stages, called *perception* and *general analysis*. The first step is domain specific, while the second is mostly domain independent. The *perception* stage builds a percept of the artwork, i.e. some sort of internal representation, outputting the percept and information about the perception task (e.g. the complexity of the task). The *general analysis* stage uses generic analysis techniques to extract relevant information from its input.

In some cases this division between stages is more conceptual than real. The main idea is that, the perception acquires information about domain specific parameters, which are then analysed. We do not impose any kind of constraints to the type of internal representation, nor to the range of techniques used on the *general analyser* module. Therefore, the range of techniques that can be used includes statistical, rule based, algorithmic, symbolic, or sub symbolic techniques, etc.

The second module, *evaluator*, takes as input the characterization done by the previous analyser, and outputs an assessment of the artwork. In order to allow adaptation it receives feedback information, which reflects the quality of its appraisal.

The role of this information will become clearer in the next sub-section dedicated to the validation methodology.

The proposed architecture allows a certain degree of independence between the search for new relevant features, analysis, and evaluation. Moreover, the adaptive *evaluator* module can give information about the features that are relevant in the assessment of an artwork.

4.2 Validation

The validation of an AAC involves great difficulties, due mainly to the subjective nature of the task. In this section we propose a multi level validation methodology. In each level the AAC is presented with a different task. The idea is to begin with tasks in which the response of the AAC can be evaluated easily, and then move to task that involve a higher subjectivity and dynamics. In each level it is possible to compare distinct methods and approaches in order to obtain better AAC's, without losing the adaptation capacity.

Currently, we consider four levels of validation:

1. Author Identification – In this task, the AAC is presented with several artworks of different authors. Its task is to determine the author of the each piece. The *evaluator* module is trained using the feedback information, which provides the correct answer. Afterwards the system can be tested with a different set of artworks. The main goal of this validation step is to assess the quality of the analyser module, since a failure would imply that the extracted features are not sufficient to discriminate between authors.

2. Style identification – This level is very similar to the previous one, basically the AAC should be able to recognize the style of a given artwork. The training and testing is done in the same manner, the only difference being that the feedback indicates the style instead of the author. Depending on the domain and authors/styles considered, this level can be more or less difficult than level 1. The main advantage of these levels is the objectivity of the tasks being performed by the AAC.

3. Static Evaluation – The task of the AAC is to assign an aesthetic value to a series of artworks, which were previously evaluated by humans. The main difficulty in performing this test is the construction of a database of consistently evaluated artworks. One option is to use some sort of evolutionary art tool to generate the artworks. Alternatively, we can resort to pieces of classical art. However, we need works of varying quality (which includes "bad" pieces). Depending on the set of positive and negatives examples used, the task of the ACC can be either difficult or easy.

4. Dynamic Evaluation – In this fourth level the AAC is part of a society of agents, which can be artificial or human (see the description of the Hybrid Society in section 3). The success of the AAC depends on the appraisal of his judgements by the other members of the society. This type of test introduces a new, social and dynamic dimension to the validation, since the value of an artwork varies over time and depends on the agents that compose the society. Although challenging, we think that this type of validation is the most natural one, since it tries to mimic the conditions in which human critics and authors perform. Furthermore, it is the only test that takes into consideration that the value of an

artwork depends on its cultural context, and that consequently, the critic must be sensible to this context and adapt to its cultural surroundings.

The validation methodology presented here tries to find a compromise between automated and human-like validation. We are fully aware of the difficulty of the proposed tasks. However, it is our belief that AAC's, which are capable to overcome only some of the levels, can still be interesting and useful.

In the first levels (1-3) of validation it is possible to assess the performance of the *analyser* and *evaluator* module independently, since the output of the *analysis* module (in conjunction with the feedback information), can be seen as a training instance to the evaluator. In the latest level, this is no longer possible since the feedback information does not reflect directly the quality of the artworks, but only an appraisal of the AAC actions by the society, and it changes dynamically in time. So, the system must work in real time.

5 Conclusions and Further Work

This paper has analysed the current state of the art in the development of art critics, focusing on the ones employed in evolutionary art systems. Based on this analysis, and on the experience acquired in the development of previous systems, we have proposed a generic model for the development of artificial art critics. In order to allow an easy adaptation to different domains, the proposed architecture separates generic from domain specific components. Furthermore, it also establishes a boundary between static and adaptive modules.

The validation of an artificial art critic is a complex task. We proposed a multi level validation methodology that is aimed at simplifying and automating this task. Since the proposed architecture is not domain dependent, we would also like to test the possibility of mapping aesthetical principles between different domains, e.g. visual art and music.

The research in the area of artificial art critics and artists is still on an embryonic stage. It is our hope that the ideas presented here may help its development, by supplying a common framework that allows the comparison of different techniques, and facilitates the collaboration between researchers.

Acknowledgements. This work was partially supported by grants from Xunta de Galicia, project "Hybrid Society" from University of A Coruña, and also by the Portuguese Ministry of Education, program PRODEP III, Action 5.3.

References

1. Johnson, C., Romero, J.: "Genetic Algorithms in Visual Art and Music". In Leonardo. MIT Press. Cambridge MA Vol 35, (2). pp. 175–184. 2002
2. Bentley, P. J.: "Is Evolution Creative?". In P. J. Bentley and D. Corne (Eds.), Proceedings of the AISB'99 Symposium on Creative Evolutionary Systems (CES), pp. 28–34. Published by The Society for the Study of Artificial Intelligence and Simulation of Behaviour (AISB). Edinburgh.1999.
3. Karnow, C. E.A.: "In collaboration with machines". Leonardo, (30):4 pp 248. 1997
4. Burton, A. R., Vladimirova, T.: "Generation of musical sequences with genetic techniques". Computer Music Journal, 23(4):59–73, Winter 1999.

5.   Romero, J., Santos, A., Dorado, J., Arcay, B., Rodriguez, J.: "Evolutionary Computation System for Musical Composition". In Mathematics and Computers in Modern Science, pp. 97–102. World Scientific and Engineering Society Press. 2000
6.   Todd, P.M.: "Evolving musical diversity". In Andrew Patrizio, Geraint A. Wiggins, and Helen Pain, Eds. Proceedings of the AISB'99 Symposium on Creative Evolutionary Systems. Society for Artificial Intelligence and the Simulation of Behaviour, pp. 40–48, 1999.
7.   Biles, J. A.: "GenJam in Perspective: A Tentative Taxonomy for Genetic Algorithm Music and Art Systems". In Wu, A.S. (Ed.), Workshop GAVAM 2000 Genetic and Evolutionary Computation Conference, pp. 133–136. Morgan Kaufmann. San Francisco CA. 2000.
8.   Horowitz, D.: "Generating rhythms with genetic algorithms". In Proceedings of the 1994 International Computer Music Conference, 1994.
9.   Jacob, B.L.: "Composing with genetic algorithms". In Proceedings of the 1995 International Computer Music Conference, pages 452–455, 1995.
10. Jacob, B.L.: "Algorithmic composition as a model of creativity". Organized Sound, 1(3):157–165, 1996.
11. Moroni, A., Zuben, Von F., Manzoli, J.: "Arbitration: Human machine interaction in musical domains". In Leonardo. MIT Press. Cambridge MA, Vol. 35, (2). pp. 185–189. 2002.
12. Pazos, A., Santos, A., Dorado, J. and Romero, J.J.: "Genetic music compositor". In P. Angeline, Z. Michalewicz, M. Schoenhauer, X. Yao, and A. Zalzala, Eds, Proceedings of the 1999 Congress on Evolutionary Computation, pp. 885–890, Vol. 2. IEEE Press. 1999.
13. Pazos, A., Santos, A., Dorado, J. and Romero, J.J.: "Adaptive aspects of rhythmic composition: Genetic music". In W. Banzhaf, J. Daida, A. E. Eiben, M. H. Garzon, V. Honavar, M. Jakiela, and R. E. Smith, Eds., Proceedings of the Genetic and Evolutionary Computation Conference, page 1794, Vol. 2. Morgan Kaufmann, 1999.
14. Putnam , J. B.: "Genetic programming of music". Technical Report, New Mexico Institute of Mining and Technology, 1994.
15. Ralley, D.: "Genetic algorithms as a tool for melodic development". In Proceedings of the 1995 International Computer Music Conference, pp. 501–502. 1995.
16. Haggerty, M.: "Evolution by aesthetics, an interview with W. Latham and S. Todd". .IEEE Computer Graphics, 11(2):5–9, March 1991.
17. Machado, P., Cardoso, A.: "NEvAr – The Assessment of an Evolutionary Art Tool". In: Wiggins, G. (Ed.). Proceedings of the AISB'00 Symposium on Creative & Cultural Aspects and Applications of AI & Cognitive Science, Birmingham, UK, 2000.
18. Rosenman, M.A.: "The generation of form using evolutionary approach". In D. Dasgupta and A. Michalewicz, Eds., Evolutionary algorithms in Engineering Applications. Springer, 1997.
19. Rowland, D. A.: "Computer Graphics Control over Human Face and Head Appearance", Genetic Optimisation of Perceptual Characteristics. PhD thesis, University of St. Andrews, 1998.
20. Rowland, D. A, Biocca, F.: "Evolutionary Cooperative Design Methodology: The Genetic Sculpture Park". In Leonardo. MIT Press. Cambridge MA Vol 35, (2). Pp. 193–196. 2002.
21. Sims, K.: "Artificial evolution for computer graphics".Computer Graphics, 25(4):319–328, 1991.
22. Sims, K.: "Interactive evolution of dynamical systems". In F.J. Varela and P. Bourgine, Eds., Towards a Practice of Autonomous Systems: Proceedings of the First European Conference on Artificial Life, pages 171–178. Mit Press, 1992.
23. Soddu, C.: "Recognizability of the idea: the evolutionary process of Argenia". In A. Patrizio, G. A. Wiggins, and H. Pain, Eds., Proceedings of the AISB'99 Symposium on Creative Evolutionary Systems. Society for Artifical Intelligence and the Simulation of Behaviour, pages 18–27. 1999.

24. Todd, S. and Latham, W.: "Mutator, a subjective human interface for evolution of computer sculptures". IBM United Kingdom Scientific Centre Report 248, 1991.
25. Unemi, T.: "SBART 2.4 An IEC Tool for Creating Two-Dimensional Images, Movies and Collages". In Leonardo. MIT Press. Cambridge MA Vol 35, (2). pp. 189–192. 2002
26. Biles, J. A.: "GenJam: A genetic algorithm for generating jazz solos", In Proceedings of the 1994 International Computer Music, 1994.
27. Papadopoulos, G. and Wiggins, G.A.: "A Genetic Algorithm for the Generation of Jazz Melodies", In STeP'98. Proceedings of STeP'98: 8th Finnish Conference on Artificial Intelligence. September 7–9, 1998, Jyväskylä, Finland. 1998.
28. Biles, J. A., Anderson, P. G. and Loggi, L. W., "Neural Network Fitness Functions for an IGA", in Proceedings of the International ICSC Symposium on Intelligent Industrial Automation (ISA'96) and Soft Computing (SOCO'96), 26–28 March 1996 (Reading, U.K.: ICSC Academic Press, 1996) pp. B39-B44. 1996.
29. Burton, A. R. and Vladimirova , T., "Genetic Algorithms Utilising Neural Network Fitness Evaluation for Musical Composition", in G.D. Smith, N.C. Steele and R.F. Albrecht Eds., International Conference on Artificial Neural Networks and Genetic Algorithms Norwich, UK. pp. 220–224, Springer, 1997.
30. Spector, L. and Alpern, A.: "Criticism, Culture and the Automatic Generation of Artworks", In Proceedings Twelfth National Conference on Artificial Intelligence (AAAI-94) August 1-4, pp. 3–8. AAAI Press. 1994.
31. Baluja, S., Pomerleau, D. and Todd, J.: "Towards Automated Artificial Evolution for Computer-Generated Images", In "Connection Science 6, No. 2, pp. 325–354. 1994.
32. Wiggins, G.A., Papadopoulos, G., Phon-Amnuaisuk, S. and Tuson, A.: "Evolutionary Methods for Musical Composition", In International Journal of Computing Anticipatory Systems, 1(1), 1999.
33. McIntyre, R.A.: "Bach in a Box: The Evolution of Four-Part Baroque Harmony Using Genetic Algorithm", In Proceedings of the 1994 IEEE International Conference on Evolutionary Computation, pp. 852–857, IEEE Press, 1994.
34. Todd, P.M. and Werner, G.M.: "Frankensteinian Methods for Evolutionary Music Composition", In N. Griffith and P.M. Todd, Eds., Musical Networks: Parallel Distributed Perception and Performance. Cambridge, MA: MIT Press, 1998.
35. McAlpine, K., Miranda, E. and Hoggar, S.: "Making Music with Algorithms: A Case-Study System", Computer Music Journal, 23, No. 2, 19–30. 1999.
36. Miranda, E.: "Granular Synthesis of Sounds by Means of a Cellular Automaton", Leonardo, No. 4, 297–300. 1995.
37. Greenfield, G. R.: "On the Co-Evolution of Evolving Expressions, International Journal of Computational Intelligence and Applications", Vol. 2, No. 1, pp. 17–31. 2002
38. Romero, J: "Metodología Evolutiva para la construcción de modelos cognitivos complejos. Exploración de la 'creatividad artificial' en composición musical" (In Spanish). Ph.D. Thesis. University of A Coruña. 2002.
39. Richard Dawkins: "The Blind Watchmaker". Penguin, 1990.
40. Koza, J.: "Genetic Programming: On the Programming of Computers by Means of Natural Selection". MIT Press. 1992.
41. Machado, P. and Cardoso, A.: "All the truth about NEvAr". Applied Intelligence, Special issue on Creative Systems, 16(2):101–119. 2002
42. Arnheim, R.: "Entropy and Art", University of California Press, 1971.
43. Arnheim, R.: "Art and Visual Perception: A Psychology of the Creative Eye", University of California Press, 1954.
44. Machado, P., Dias, A., and Cardoso, A.: "Learning to Colour Greyscale Images". The Interdisciplinary Journal of Artificial Intelligence and the Simulation of Behaviour, 1(2), 2002.

Genetic Algorithms for the Generation of Models with Micropopulations

Yago Sáez[1], Oscar Sanjuán[1], Javier Segovia[2], and Pedro Isasi[3]

[1] Lenguajes y Sistemas Department, Faculty of Computer Science, Universidad Pontificia Salamanca, Madrid, {ysaez,osanjuan}@vector-it.es
[2] Lenguajes y Sistemas Department, Faculty of Computer Science, Universidad Politécnica de Madrid, fsegovia@fi.upm.es
[3] Departamento de Informática, Universidad Carlos III Madrid, isasi@ia.uc3m.es

Abstract. The present article puts forward a method for an interactive model generation through the use of Genetic Algorithms applied to small populations. Micropopulations actually worsen the problem of the premature convergence of the algorithm, since genetic diversity is very limited. In addition, some key factors, which modify the changing likelihood of alleles, cause the likelihood of premature convergence to decrease. The present technique has been applied to the design of 3D models, starting from generic and standard pieces, using objective searches and searches with no defined objective.

1 Introduction

Evolutionary computation are learning techniques that use computational models that follow a biological evolution metaphor. In these techniques a population of individuals (solutions) evolves until the convergence criterion is reached, usually until finding a near optimal solution. The success of these techniques are based on the maintenance of the genetic diversity, for which it is necessary to work with large populations. However, not always it is possible to deal with such a large populations. There are two main areas where it becomes necessary to work with micropopulations. The first one refers to those problems where adequacy values must be estimated by a human being; and the second one relates to the problems that involve a high computational cost.

Within the first area, object of study of the present article, interactive evolutionary computation techniques can have various applications, such as the design, both artistic and functional, of bidimensional and tridimensional objects. A parallel search for solutions, allows this kind of techniques to evaluate multiple designs, thus easily generating designs that stimulate and even surpass the creativity of artists and engineers. A typical example of the application of evolutionary computation to simple figures can be seen in Biomorphs (Dawkins, 1986). It is a very simple application, where the user interactively decides which figures resemble most a real insect. In

S. Cagnoni et al. (Eds.): EvoWorkshops 2003, LNCS 2611, pp. 570–580, 2003.

practice, applications have been developed, such as those for the design of coffee tables, (Bentley 1999). The aim in this case is to find a table that meets certain requirements, such as, for instance, working surface, measurements, stability, lack of floating objects, etc.

Applications which make use of this technique for other practical purposes have also been developed, such as the digital treatment of images (Ngo and Marks, 1993) (Sims, 1994 a, b), the composition of musical works (Moore, 1994) and even the design of sculptures (Rowland, 2000), the automatic design of figures (F. J. Vico, 1999), the artistic design (Santos et al. 2000, Unemi, 2000), or the generation of gestures in 3D figures aimed at virtual worlds (Segovia et al. 1999, Berlanga et al. 2000).

In most of the above-mentioned references, the selection criterion is based on a merely artistic and personal point of view. In this kind of applications, problems become more complex, since the criterion used is rather subjective and, furthermore, it varies according to the user's opinions or personal preferences. This paper presents a series of modifications to the usual procedure of interactive evolutionary computation in order to be able to overcome the problem that arises when using micropopulations.

2 Environment of the Problem

The environment of the problem of interactive evolutionary computation that needs to be solved is that of the creation of a 3D design of general-purpose "tables". In this creative process the participation of a human being is needed in order to determine which tables are more creative, more functional, or which ones better fit their needs. In addition, it must be allowed to combine and modify different types of objects within the same design, i. e., the tables are complex objects generated from the genetic mutation of their various pieces.

Care will also be taken that the developed techniques shall be based on a generic design, so that they can be applied to the generation of other types of objects, such as cars, logos, and so on.

In this scheme, each table constitutes an individual of the population to be evolved, and each table is made up of various pieces. Each kind of piece that makes up the table corresponds to a different gene string. There are different types of pieces, namely, the predefined ones, based on real models, hereinafter referred to as 'standard' pieces; and the random ones, generated from cones, cylinders, cubes and spheres, referred to as 'generic' pieces. Each one of them has its own alleles, which correspond to the kind of piece -only if it is standard- material, measurement, angle, position, etc.

3 Mixed Fitness

Within interactive evolutionary computation, when working with small populations, two problems are encountered:

1. As regards individuals, it is necessary to carry out a subjective evaluation and, besides, it is rather complicated to ask the user to allocate numerical values.

2. Due to the fact that there is not a considerable genetic diversity, the algorithm tends to converge towards very different results in few generations.

The solution to these problems means changing the general scheme of the canonical algorithm. See the next figure:

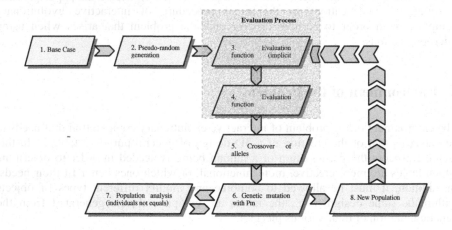

Fig. 1. Flowchart of the Genetic Algorithm that has been used.

The first above-mentioned limitation is imposed by the manner in which the user actually chooses the designs, and also by the complexity that arises when allocating a numerical value to each individual (mostly if the range is wide, or if there is a great variety).

In a genetic algorithm, an evaluation function, which causes the best ones to survive and evolve, is always used. Nevertheless, when this evaluation function depends on the user, he or she cannot be compelled to always having to provide certain fixed values to the generated individuals. It must be taken into account that the user has no knowledge on Evolutionary Computation, and therefore, when asked to supply some adequacy values for the designs, this must be done in the easiest, most straightforward and understandable way.

In this case, it has been chosen not to give a numerical value for the evaluation of individuals, but to choose those more interesting designs, from the point of view of the user (subjetive measure). The selected individuals are going to be used as a seed for the generation of the next population.

In the application developed as an example, in order to facilitate things, the user will receive six individuals as input, from a total population, and he/she will be able to choose and classify the three that better fit his or her personal preferences, needs, and so on. As an initial option (the case base), the user can also suggest the system the rough measurements of the table he/she wants as well as its number of legs.

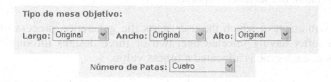

Fig. 2. Case base for the application

After this, the system makes a pseudo-random generation in order to create the first population that will be the seed for the evaluation process (see step 3 and 4, figure 1).

As for the second limitation detected, it has been decided to develop a combined evaluation function, where a 'mixed fitness' value is used. This value of mixed fitness is made up of two other values, an implicit fitness value, also called automated fitness (Machado 2002), which is estimated by applying some automatic evaluation functions to individuals, and an explicit fitness value, which is defined by the user.

The mixed fitness allows the widening of genetic diversity, since the population can be increased without the need to complicate the operation on the part of the user. This one will evaluate a subset of the population, which will have been previously filtered by the automatic evaluation function. The implicit fitness is estimated by a function it adds that calculates stability, the gravity centre and the contact among pieces (it is explained forward). In our example, the implicit fitness function is estimated for ten individuals generated through the use of the algorithm described in figure 1, and among these individuals, the best six ones are presented to the user, in order for he/she may evaluate the explicit fitness.

Therefore, the user will receive six individuals as input, from a total population, and he will be able to choose and classify one, two or three that most fit his or her personal tastes, needs, and so on (explicit fitness value).

If the user does not choose any individual at all, then all new genotypes will be generated through random mutation, and the process is the same as in the first generation. It can also occur that the user does not like how the population has evolved in the last mutation, and desires to return to the previous one, or even start the whole process again.

It was found that when generating designs from standard pieces, the algorithm in very few generations had a tendency towards the same kind of tables, and this is due to the fact that when working with a limited number of standard pieces, designs are restricted to mere combinations. This problem of convergence towards the same attractor was solved firstly through the inclusion of a wide population of standard pieces and secondly making some adjustments on the reproduction mechanism, which allowed us to modify the mutation likelihood and make them be able to inherit from different parents. This measurement was considered necessary as it was observed (see Trace 1) that, during the process of objective searching, there may be situations in which the features of several individuals are of interest; for instance, the type of legs of one individual, the surface of another.

At the very moment in which the user starts the generation process of a new set of individuals, after having selected one, two or three designs through the method of classification, each gene is allocated a fitness value.

The value of mixed fitness is given by the estimation of the implicit fitness value of all elements in the population as well as of the explicit fitness value of each one of them, obtained from the user's judgement on all generated designs.

The implicit fitness value (see step 3, figure 1) is obtained after having performed the following calculations:

1. *Contact Function:* a function that evaluates the existing contact among pieces. If there is no contact among them, then the fitness value is zero.
2. *Stability Function:* a function that calculates the contact between the various pieces and the floor, by drawing a concentric circle starting from the centre of the model. The more stability, the higher the fitness value.
3. *Robustness Function:* a function that calculates the model structure by evaluating the contact that exists among pieces, the higher the contact, the more robustness.
4. *Working Surface Function:* a function that permits taking into account if a table is a practical one, by estimating its working surface.

The implicit fitness value is used to select the "n" more valid individuals and to generate the population that will be showed to the user (showing population).

4 Population Generation

The showing population is composed by six individual, among them, the user can select one, two, or three. The selected individuals(parents) are used for generating the new population. Ten new individuals are generated, less if elitism has been choosen by uniform crossover of the parents (see step 5, figure 1). If only one parent has been selected, all the new individuals are the same. When more than one individual has been selected as a parent, each new individual is built up by selecting, in a uniform way, a gene from one of the parents.

To this last generated individuals two new operations are made. From one hand, an operator is performed in order to eliminate repeated individuals in the population (step 7). By means of this operator, each generated individual is compared with the rest, if a repeated one is founded, the individual is generated again with the same procedure as before. In the other hand, in order to avoid the premature convergence of the algorithm, and infinite cycles, a mutation operator has been included. The mutation operator is in charge of randomly altering each one of the individual's genes and this is done with a mutation likelihood of Pm. Such likelihood is a variable one and it depends on the user. The user can select it and so decide whether it is better to supply individuals in a population with a greater variety or tend towards more common features. We show it like a creativity factor.

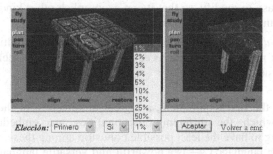

Fig. 3. Mutation likelihood

The aim of mutation is to increase the number of design primitives (Bentley, 1999), apart from avoiding the overuse, after several generations, of only one element of the alphabet, and this one being utilized in the same position for all chromosomes. This would involve that that specific feature would never change, and, therefore, it could happen that an optimal solution was never reached. However, the changing likelihood of the mutation operator must not be high, since that could damage the generation of the designs, see Figure 3. The aim, through the mutation operator, is to create elemental variations in the population and guarantee, in theory, that any point in the searching space can be reached.

During the reproduction process, genes can mutate following this possibilities:
(step 6)

1. The gene does not mutate any value.
2. The value that corresponds to the individual's location mutates.
3. The values responsible for the individual's dimensions and the materials they are made up of, mutate.
4. It mutates that part of the genes that corresponds to dimensions, materials they are made up of and location.
5. The whole gene mutates.

Then simply apply weights and weigh up according to results.

After a playing all this operators, a new population is generated and all the proccess is going to be repeated until the user decides he has founded the right design.

5 Architecture

This application is shown as a useful tool for the design of customized tables, without the need for the user to have any in-depth knowledge on computer-aided design, nor, of course, on genetic algorithms.

This architecture's design has been carried out under the premises of a system that is:

1. *Adaptable:* i.e. flexible in the design of different objects, that is, a system that is easy to modify, so that it can accomplish other kinds of designs, such as cars, motorbikes, and so on.

2. *Universal:* a system is needed that is accessible from multiple languages and different platforms; one that is easily adaptable to new technologies, such as the Internet.

In order to meet the first requirement, a logical architecture of the system has been developed, in which the different functionalities are embedded into various subsystems, trying to have the least possible cohesion among these.

The programming language C++ is a powerful and versatile language, mostly when it is combined with object-oriented methodology. And DCOM components are a standard for the development of applications. In this way it is assured that the work done can be exploited from very different applications, from a remote Web application, up to a local executable.

In the developed application, the User Interface has been carried out using the ASP technology of Microsoft®. VRML designs will be displayed through ASP pages and it will concurrently serve as an interface among users and DCOM components. A future objective would involve developing a Java-based architecture in order to use the benefits of X3D standard, still to be approved. From the ASP pages the COM/DCOM component is instantiated on the Web server where the Web application is being run, without the need for the client to download code, nor compile it, thanks to which we get a flexible, reliable and, most of all, platform-independent solution.

In the last version, among the improvements developed, several record files have been included which allow the system administrator to take a close look to all the events that are happening during the execution of the program and to gather information for further analysis. For the use of this application, the operation is the same as that seen in [18], the improvements have focused on the optimization of generations —through source code debugging (by freeing memory, sorting methods, etc.)— as well as on comparative studies of performance and compatibility among the various VRML interpreters available on the market, and also, as it has been pointed out, on carrying out a permanent registration of events. It has also been developed a straightforward application in Visual Basic which accesses the component and offers some of its fuctionality from ASP pages.

Now some traces, of how the system is actually working, are shown where it is possible to see how the design has evolved and interesting results are easily achieved.

Trace 1. Objective Search with only one parent: a table for the lounge.

Restrictions of the user: Width: 1.5m Length: 1.5m Height: 0.5

After developing twenty different examples we come to the conclusion that an objective search with only one parent always converges to the same model with very few iterations. This feature makes it quite interesting if a model that is liked enough can be found.

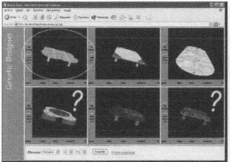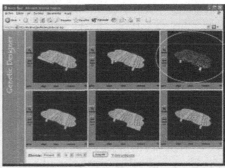

Many cases have been detected where the first iteration has several models which are valid for an ongoing search, that is why the possibility of inheriting features of up to three models was enabled. Once that the search space has been restricted by increasing the mutation factor to 15%, we find interesting variations on the model chosen in the previous evolution stage.

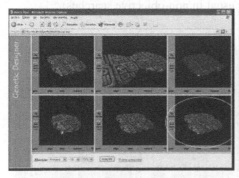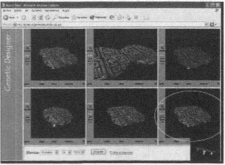

(End of Trace 1)

Conclusion: the system initially puts forward very different options, the user refines the features and restricts the search space with low mutation factors, until focusing on the desired table, and then, according to its features, the user tells the system, by increasing the mutation factor, to offer variations.

Tests were conducted starting from the first iteration with a mutation factor of 15% or more, with the clear drawback that the features that were chosen by the user could get lost due to gene mutation.

Trace 2. Non-objective Search with only one parent
With no restrictions on the part of the user.
At a non-objective search the user expects the system to provide him with models that are useful as an original idea, with an indeterminate purpose. After having conducted several tests, it is proven how the system, with low mutation factors (15% or lower) needs many iterations to provide any new ideas.

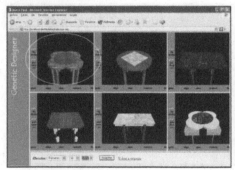
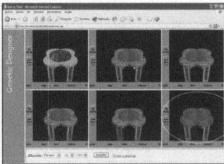

It is observed how, after three iterations, the model has hardly been changed, and the micropopulation demands new features, that is why we set the mutation factor at a 50%.
It is soon seen that the individual variety increases and how one of the proposals further motivates the user (something quite usual in non-objective searches).

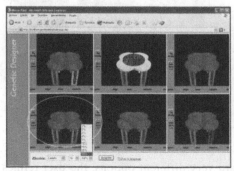

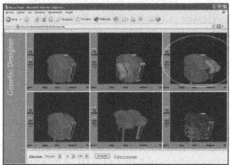 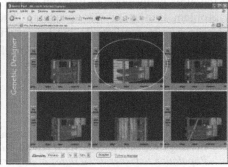

(End of Trace 2)

Conclusion: in non-objective searches it is necessary to increase the mutation factor in order to have new proposals from the system starting from the models chosen by the user (creaativity factor for the user). The originality of final designs is obvious in the following example.

6   Conclusions

The restriction imposed by the number of individuals of the population, along with the fact that the user is the one that decides which individuals evolve, has made it necessary to modify the canonical genetic algorithm. The general operation scheme has been modified in order to make use of a mixed evaluation function. Such function makes a previous selection of individuals and then present them to the user, who eventually chooses after having sorted them. Furthermore, the scheme has been changed so as to enable multiparent generation, to utilize an elitist strategy and to permit the return to previous generations and thus avoiding generations with few chances of surviving. As for the reproduction mechanism it has become necessary to modify the selection and the mutation operators, with the aim of avoiding premature convergence of individuals.

This method has been tested in an application with the aim of attaining functional and creative designs; and thus the real object model as well as the generic object model have been tested; experiments with and without objective models have been conducted, with and without tracks, in the base case. After the study, both models are granted for valid, if we take into account the either creative or functional end, that we want to confer; and even very interesting results have also been obtained with the mixed intermediate model, which can combine both working models. It has been proved how the mixed fitness, combined with that modification of the selection and mutation operators favours an optimal design with the least possible number of iterations.

The improvements carried out with respect to debugging and code cleaning-up result in better response times. As for the VRML viewers tested, it is recommended the use of the already disappeared Blaxxun, as the best one regarding performance (although a problem in texture was detected). The viewer Cortona from ParallelGraphics features a more comfortable, but less efficient interface for the surfer. Anyway, both are seen as valid for the purposes of this application.

References

1. Bentley P. (1999) *From Coffee Tables to Hospitals: Generic Evolutionary Design,* Evolutionary design by computers, Morgan-Kauffman, pp. 405–423.
2. Berlanga A., Isasi P. Segovia J. (2000) *Interactive Evolutionary Computation with Small Population to Generate Gestures in Avatars,* Proceedings of GECCO 2001, Artificial Life, Adaptative Behavior, and agents
3. Chambers L. (1995) *Practical handbook of genetic algorithms.* Vols. 1,2 editado por Lance Chambers, CRC Press.
4. Dawkins R. (1986) *The blind watchmaker,* Longman Scientific and Technical, Harlow.
5. F.J. Vico, F.J. Veredas, J.M. Bravo, J. Almaraz *Automatic design sinthesis with artificial intelligence techniques.* Artificial Intelligence in Engineering 13 (1999) 251–256
6. Holland J.H. (1975) *Adaptation in Natural and Artificial Systems,* University of Michigan Press.
7. Holland J. H. (1991) *The Royal Road for Genetic Algorithms: Fitness Landscapes and GA Performance.* Proceedings of the First European Conference on Artificial Life, Cambridge, MA: MIT Press. pp.1–3, 6–7.
8. Holland J.H. (1995) *Hidden order: how adaptation builds complexity.* Addison Wesley, Reading Massachussets.
9. Moore, J.H. (1994) *GAMusic: Genetic algorithm to evolve musical melodies.* Windows 3.1 Software disponible en:http://www.cs.cmu.edu/afs/cs/project/ai-repository/ai/areas/ genetic/ga/systems/ gamusic/0.html.
10. Ngo J.T. y Marks J., (1993), Spacetime Constraints Revisited. Computer Graphics, Annual Conference Series pp. 335–342.
11. Rowland D. (2000) *Evolutionary Co-operative Design Methodology: The genetic sculpture park.* Proceedings of the Genetic ad Evolutionary Computation Conference Workshop, Las Vegas.
12. Santos A., Dorado J., Romero J., Arcay B., Rodríguez J. (2000) *Artistic Evolutionary Computer Systems,* Proceedings of the Genetic ad Evolutionary Computation Conference Workshop, Las Vegas.
13. Segovia J., Antonio A., Imbert R. Herrero P., Antonini R. (1999) *Evolución de gestos en mundos virtuales,* Proceedings of CAEPIA 99.
14. Sims K., (1991) *Artificial Evolution for Computer Graphics,* Computer Graphics, Vol. 25, (4), pp. 319–328.
15. Sims K., (1994a) Evolving Virtual Creatures. In *Computer Graphics.* Annual Conference Series (SIGGRAPH '94 Proceedings), Julio 1994, pp. 15–22.
16. Sims K., (1994b) Evolving 3D Morphology and Behaviour Schemes. In Fogel, L. J. Angeline, P.J. and Back, T. *Proceedings of the 5th Annual Conference on Evolutionary Programming,* Cambridge, MA: MIT Press, pp. 121–129.
17. Unemi T. (2000) SBART 2.4: an IEC Tool for Creating 2D images, movies and collage, Proceedings of the Genetic and Evolutionary Computation Conference Program, Las Vegas.
18. Y.Sáez, O.Sanjuan, J.Segovia (2002) AEB'02 Algoritmos Genéticos para la Generación de Modelos con Micropoblaciones, Mérida, España.
19. Machado, P., Cardoso, A., All the truth about NEvAr. Applied Intelligence, Special issue on Creative Systems, Bentley, P. Corne, D. (eds), Vol. 16, Nr. 2, pp. 101–119, Kluwer Academic Publishers, 2002.

Evolution of Collective Behavior in a Team of Physically Linked Robots

Gianluca Baldassarre, Stefano Nolfi, and Domenico Parisi

Institute of Cognitive Sciences and Technologies,
National Research Council of Italy (ISTC-CNR),
Viale Marx 15, 00137 Rome, Italy
{baldassarre, nolfi, parisi}@ip.rm.cnr.it

Abstract. In this paper we address the problem of how a group of four assembled simulated robots forming a linear structure can co-ordinate and move as straight and as fast as possible. This problem is solved in a rather simple and effective way by providing the robots with a sensor that detects the direction and intensity of the traction that the turret exerts on the chassis of each robot and by evolving their neural controllers. We also show how the evolved robots are able to generalize their ability in rather different circumstance by: (a) producing co-ordinated movements in teams with varying size, topology, and type of links; (b) displaying individual or collective obstacle avoidance behaviors when placed in an environment with obstacles; (c) displaying object pushing/pulling behavior when connected to or around a given object.

1 Introduction

How can a group of robots that are physically connected to form a single physical structure display coordinated movements? In this paper we consider a group of physically connected robots forming a linear structure that have to move as straight and as fast as possible in the environment. Given that the initial orientation of the tracks of each robot is randomly chosen, the robots should first negotiate a common direction and then move along that direction in a coordinated fashion.

This is one of the research problem we are facing within a project founded by CEC in which we are developing "swarm-bots" [2, 3, 5], i.e. a groups of robots (each called "s-bot") that are able to self-assemble so as to form different physical structures and to cooperate in order to solve problems that cannot be solved individually.

Each s-bot has its own neural-network controller that generates motor outputs in response to sensory inputs. Since the s-bots are physically connected they need to coordinate in order to move together. As we will see, by providing the individual s-bots with traction sensors that detect the direction and intensity of the force that the turret exerts on the chassis and by utilizing an evolutionary technique [4], we were able to find a simple and effective solution: evolved s-bots display the ability to coordinate toward a unique direction that emerges from the negotiation between the individuals.

S. Cagnoni et al. (Eds.): EvoWorkshops 2003, LNCS 2611, pp. 581–592, 2003.

Moreover, we will show how neural controllers evolved for the ability to produce coordinated movements in a swarm-bot which includes four assembled s-bots forming a linear structure generalize to rather different circumstances. In particular, the evolved s-bots are able to: (a) produce coordinated movements in swarm-bots with varying size, topology, and type of links; (b) display individual and collective obstacle avoidance behaviors when placed in an environment with obstacles; (c) spontaneously produce object pushing/pulling behavior when connected to or around a given object.

2 Experiments and Results with a Simple Linear Formation

In order to investigate the problem described, we evolved the control systems of four physically linked s-bots forming a linear structure that were asked to move as fast and as straight as possible (Fig. 1). Experiments were run in simulation by constructing a software based on the rigid body dynamics simulator SDK Vortex^{TM}. This simulator reproduces the dynamics, friction and collisions between physical bodies.

Fig. 1. Four physically linked s-bots forming a linear structure. For each s-bot, the cylinder and the parallelepiped respectively represent the turret and the chassis. The large disks and small spheres respectively represent the motorized and passive wheels. The line between two s-bots represents the link between them. The white line above each s-bot indicates the direction and intensity of the traction (see below)

Each s-bot (Fig. 1) consists of a rectangular chassis of size 3.5×3.5×1.0 cm provided with two motorised and two passive wheels with a width of 0.2 cm and a radius of 0.75 and 0.375 cm, respectively. Each s-bot is also provided with a cylindrical turret with a radius of 2.75 cm and a height of 1.0 cm that is connected to the chassis through a motorised "hinge joint" that can rotate around a vertical axis. Each s-bot has a physical link through which it can be attached to another s-bot along the perimeter of its turret. The link consists of another "hinge joint" that has a rotation axis parallel to the ground plane and is perpendicular to the line formed by the four s-bots. The density of the turret was set at 0.5 and the density of the other components at 1.0.

To speed up the simulations, we used a low gravitational acceleration coefficient (9.8 cm/s^2). This low value, that causes a low friction of the wheels on the ground, was compensated for by setting the maximum torque of the motors at a low value (see below). This allowed us to set the parameter that determines the granularity at which

Vortex approximates the differential equations used to simulate the bodies' dynamics at a rather high value, 0.1 (this parameter is interpreted as 100 ms).

Vortex uses a Coulomb friction model. Coulomb friction causes a force, opposite to the forces that cause the relative motion of objects, with an intensity μ N. μ is the friction coefficient and N is the total force, normal to the plane of contact, that depends on the forces applied to the bodies. μ is set at 0.6 and 0.0 respectively for the motorised and passive wheels. The activation of the motor neurons is used to control the motorised wheels by setting their desired velocity within the range [-10, +10] radians per second, and the maximum torque of the motors at 20 dynes-centimetre. The desired velocity applied to the turret-chassis motor is set on the basis of the difference between the activation of the left and right output neuron. If this difference is positive the chassis tends to rotate rightward (with respect to the turret and when seen from above), otherwise it tends to rotate leftward.

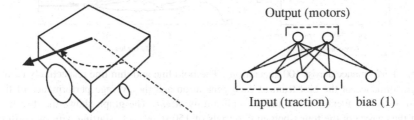

Fig. 2. *Left:* Traction force detected by an s-bot's traction sensor. The large and small circles respectively represent the right active wheel and front passive wheel. The dashed line and the full arrow respectively indicate the s-bot's orientation and the direction and intensity of the traction. The dashed arrow indicates the angle between the chassis' orientation and the traction. *Right:* The neural controller of the s-bots

Each s-bot is provided with a "traction sensor", placed at the turret-chassis junction, that returns the direction (i.e. the angle with respect to the chassis' orientation) and the intensity of the force of traction (henceforth called "traction") that the turret exerts on the chassis (Fig. 2). Traction is caused by the movements of both the connected s-bots and the s-bot's own chassis. Notice that the turret of each s-bot physically integrates the forces that are applied to the s-bot by the other s-bots. As a consequence the traction sensor provides the s-bot with an indication of the average direction toward which the team is trying to move as a whole. More precisely, it measures the mismatch between the directions toward which the entire team and the s-bot's chassis are trying to move. The intensity of the traction measures the size of this mismatch. From the point of view of each s-bot, this type of information is very easy to use for changing the direction of its own chassis in order to follow the rest of the team or to push the team to move toward a different desired direction.

Each s-bot's controller (Fig. 2, right) is a neural network with 4 sensory neurons that encode the traction. These neurons are directly connected with 2 motor neurons that control the two motorized wheels and the turret-chassis motorized joint. The 4 sensory neurons encode the intensity of the traction (normalized in [0.0, 1.0] on the basis of its maximum value registered in an experiment where the s-bots moved randomly) from four different preferential orientations with respect to the chassis (front, right, back and left). For each sensor, this intensity decreases linearly with respect to the absolute value of the difference between the sensor's preferential orientation and the traction's direction, and is 0 when this value is bigger than 90 degrees. The activation state of the motor units is normalized between [−10, +10] and is used to set the desired speed of the two corresponding wheels and the turret-chassis motor.

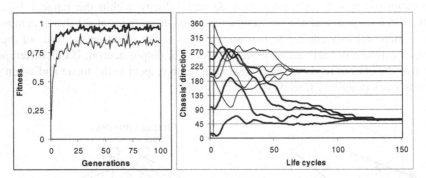

Fig. 3. *Left:* Performance across 100 generations. The bold line and thin line respectively represent the performance of the best team of each generation and the average performance of the population of teams, averaged over the 10 replications. *Right:* The graph shows the direction (angle) of the chassis of the four s-bots in two trials of 150 steps each, starting with two different initial random orientations (the bold and thin lines refer to the two trials)

The connection weights of the neural controller of the s-bots are synthesized by using an evolutionary technique [4]. The initial population consists of 100 randomly generated genotypes that encode the connection weights of 100 corresponding neural controllers. Each connection weight is represented in the genotype by 8 bits that are transformed in a number in the interval [−10, +10]. Therefore, the total length of the genotype is 10 (8 connection weights and two biases) × 8 = 80 bits.

As in the experiments reported in [1] and [6], we used a single-pool-single-genotype selection schema, i.e. we evolved a single population of genotypes each of which encodes the connection weights of a team of identical neural controllers. Each genotype is translated into four identical neural controllers ("clones") that in the experiments reported here correspond to a team of four s-bots forming a linear structure. As shown in [6], neural controllers might differentiate during lifetime as a result of their internal dynamics. However, in the experiments of this paper, as in the case of [1], neural controllers do not adapt online and do not have any internal state that can allow them to respond differently depending on their previous sensory experiences. Therefore, all s-bots always react in the same way to the same sensory states.

The team is allowed to "live" for 5 epochs each including 150 cycles. In each cycle, for each s-bot: (1) the activation state of the sensors is set according to the procedure described above; (2) the activation state of the two motor neurons is computed according to the standard logistic function; (3) the desired speed of the two wheels and of the motor controlling the motorized joint between the chassis and the turret are set according to the activation states of the motor units. At the beginning of each epoch the chassis of the four s-bots are assigned random orientations. However, to have a fair comparisons between different teams, all the teams of the same generation started with the same 5 randomly selected orientations in the 5 epochs. The 20 best genotypes of each generation were allowed to reproduce by generating 5 copies of their genotype with 3% of their bits replaced with a new randomly selected value. The evolutionary process lasted 100 generations. The experiment was replicated 10 times by starting with different randomly generated initial populations.

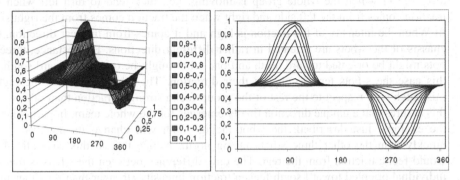

Fig. 4. *Left:* A typical evolved strategy (seed 4). Horizontal axis: angle of traction (0 is front, 90 right, 180 back, and 270 left). Depth axis: intensity of traction, between 0 and 1. Vertical axis: difference of the activation of the two neurons controlling the left and right wheels and the turret-chassis motor. When this value is around 0.5 the two wheels have the same speed (the s-bot is going straight), when it is above 0.5 the left wheel has a speed higher than the right wheel and the chassis turns right, and when it is below 0.5 the chassis turns left. *Right:* The graph shows the sections of the surface reported on the left graph, for different intensity levels

To force the teams of s-bots to move as fast and as straight as possible, we devised a fitness function based on the Euclidean distance between the barycentre of the team at the beginning and at the end of each epoch:

$$\text{Fitness} = ((x_0 - x_{150})^2) * (y_0 - y_{150})^2)^{1/2} / \text{MaxiDist} \tag{1}$$

where x_t and y_t are the coordinates, on the ground plane, of the average position occupied by the four s-bots at cycle t and MaxiDist is the maximum distance that a single s-bot can cover in 150 cycles by moving straight at maximum speed.

Fig. 3 shows how the fitness of the population, averaged over the 10 replications, changes across 100 generations. At the end of the evolution, the best team of each replication was tested for 100 epochs, and their average performance was: .947 .943 .931 .923 .839 .934 .765 .860 .946 .945.

Direct observation of behaviour shows that s-bots start to pull in different directions, orient their chassis in the direction where the majority of the other s-bots are pulling, move straight along this direction that emerges from this negotiation, and compensate successive mismatches in orientation that arise while moving. Fig. 4 shows a compact representation of the typical strategy of an evolved team. In particular, the figure shows the difference between the activation of the two motor neurons corresponding to the left and right wheels of an s-bot (normalized in the range [0.0, 1.0]) plotted against the angle and the intensity of the traction. The analysis of the figure and of the corresponding behaviour displayed by s-bots, indicates that evolved individuals adopt a simple strategy that can be described in the following way:

1) When the chassis of the s-bots are oriented toward the same direction, the intensity of the traction is null and the s-bots move straight with maximum speed.

2) When the traction intensity is low, the chassis of the s-bots are oriented toward similar but non-identical directions. In this case s-bots tend to turn toward the average direction in which the whole group is moving, i.e., they tend to turn left when the traction comes from the left side and right when the traction comes from the right side.

3) When the intensity of the traction is high and it comes from the rear direction, the chassis of the s-bots are oriented in rather different directions. For instance, three s-bots might be oriented toward north and one s-bot might be oriented toward south. In this case the s-bots tend to change their direction. The s-bots that have the higher mismatch with respect to the rest of the group feel a stronger traction than others, and this assures that a unique direction finally emerges for the whole team. In particular, in the example just described, the s-bot facing south will change its direction more quickly than the other three s-bots facing north. Notice that in this case all s-bots would feel a traction from the rear. The only difference between the s-bots is that the individual oriented toward south feels a traction intensity stronger than the other individuals. Aside from this schematic description, note that the non-linearities in how s-bots react to traction coming from different angles and of different intensities seem to play an important functional role that we are still trying to understand.

The right graph of Fig. 3 shows that the team direction that emerges from the s-bots' negotiation can be any possible one. This demonstrates that the strategy does not rely upon any type of alignment between the turret and the chassis. This is the key aspect of the traction sensor that allows the strategy to generalize to the wide variety of situations illustrated in the following sections.

3   Generalization

The strategy evolved with teams of four aligned s-bots illustrated in the previous section, generalises under many different conditions and exhibits a number of interesting emergent properties when placed in different environments. This suggests that this simple control strategy might be useful in a large number of cases.

3.1   Coordination in Swarm-Bots with Varying Size, Topologies, and Links

To verify how this form of behaviour generalizes in new circumstances, we tested the control strategies evolved in the experiments described above on larger or smaller teams of s-bots (from 2 up to 10 s-bots) assembled so as to form a linear structure (recall that the different s-bots of a team have identical neural controllers, hence for each team made up by a given number of s-bots, we had to create the same number of clone controllers to guide them). The results of these tests show that s-bots maintain their ability to negotiate a single direction and to produce a coordinated movement along it independently of the size of the team (results not shown). In the case of two s-bots, in few tests, depending on the initial orientation of their chassis, s-bots fail to converge toward a single direction and circle around the group's centre. This situation is a dynamic equilibrium because the centripetal force caused by the link between the two s-bots causes a traction toward the group's centre that makes the s-bots turn toward it. This dynamic equilibrium has also been observed in other experiments (see Section 3.3). In the majority of cases, however, s-bots generalize successfully to variations of the team size.

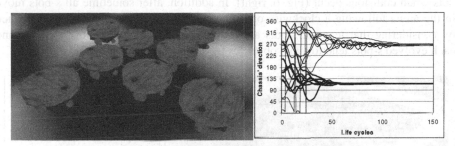

Fig. 5. *Left:* Eight s-bots connected by rigid links so as to form a star formation. *Right:* The orientation angle of the chassis of the eight s-bots of a star formation (bold lines) and snake formation (thin lines) in 150 steps

S-bots also display an ability to produce coordinated movements when assembled so as to form topologies that are different from the linear topology with which they have been evaluated during evolution. By testing a team of eight s-bots connected so as to form the star structure shown in the left part of Fig. 5, we observed that they displayed an ability to negotiate a unique direction and to move toward such emergent direction (see Fig.5, right).

Finally, s-bots also display an ability to produce coordinated movement when assembled by means of flexible instead of rigid links. Flexible links are made up by two segments connected by a hinge joint that allows the connected s-bots to rotate between them on the ground plane. By testing eight s-bots connected by flexible links so as to create a linear structure (snake formation), we observed that s-bots were able to negotiate a unique direction and to produce a coordinated movement along such direction also in this case. At the beginning of each trial the formation deformed as a consequence of the different orientation of the s-bots' chassis, but after some time it settled to a stable configuration and stable direction (Fig. 5, right). Given that in flexible assembled structures, the motor action produced by s-bots affects both the shape of the swarm and the traction perceived by the s-bots, these results seem to indicate that the evolved strategy is extremely robust and allows the s-bots to coordinate even when the traction sensors provide incomplete information about the movements of the team.

3.2 Individual and Collective Obstacle Avoidance

By placing s-bots in an environment with obstacles we observed that they display individual and collective obstacle avoidance behaviour. This can be explained by considering that collision with obstacles, by generating a traction force toward the direction opposite with respect to the direction of movement, might allow s-bots to turn away from obstacles.

In a first test, eight unconnected s-bots were placed in a squared arena surrounded by walls and including 4 large cylindrical obstacles (Fig. 6, left). In this situation, s-bots move straight when far from obstacles and turn avoiding obstacles and other s-bots when collisions occur (Fig. 6, right). In addition, after sometime all s-bots move in the arena following a clockwise or anticlockwise direction. This can be explained by considering that team of s-bots are provided with identical control systems and each evolved s-bots tend to avoid obstacles by turning left or right.

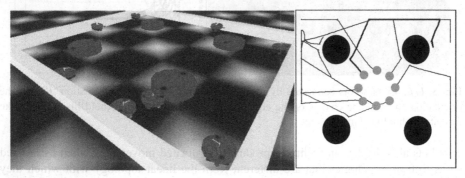

Fig. 6. Individual obstacle avoidance. *Left:* The arena with the eight s-bots, the walls, and the obstacles (larger cylinders). *Right:* Traces left by the s-bots (thin lines) in 300 cycles. The bold line is the trace left by an s-bot that hits an obstacle, the wall, a companion, and the wall again along its path

In a second set of tests, we placed assembled s-bots in the same environment surrounded by walls and including obstacles, and observed that swarm-bots displayed a collective obstacle avoidance behaviour independently from the topology with which they were assembled and the rigidity or flexibility of the connections used. Fig. 7 shows the behaviour of: a star formation connected with rigid links (top); a circular formation connected with flexible links (middle); a snake formation connected with flexible links (bottom). In all cases, the swarm-bots display an ability to coordinate and to collectively avoid walls. In the case of the two formations provided with flexible links, swarm-bot tend to change their shape after colliding with obstacles. However, given that they also tend to persevere in their direction of movement, they are also able to pass narrow passages eventually deforming their shape according to the configuration of the obstacles. In many cases (i.e. the control systems evolved in some of the replications of the experiments described in section 2) they never got stuck during the long time of observation (up to 4 hours: with the computer we used, Vortex performed about 30 cycles per second). This can be explained by considering that swarm-bots assembled through flexible links are dynamical systems that keep changing shape until they disentangle from the obstacles (small variations are also observed in the case of swarm-bot connected through rigid links due the fact that these might also bend of few degrees when subjected to significant forces). By moving s-bots change their relative positions with respect to other s-bots so that the whole swarm-bot always generates new configurations and has extremely reach dynamics.

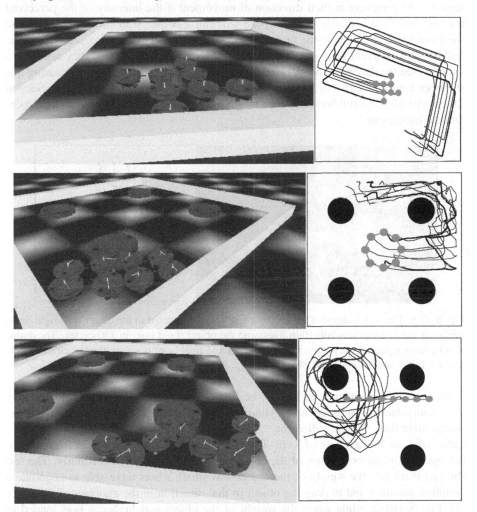

Fig. 7. Swarm-bots displaying collective obstacle avoidance. *Top*: A star formation assembled through rigid links. *Middle*: A circle formation with flexible links. *Bottom*: A snake formation with flexible links. *Left*: The light parallelepipeds are walls while the large cylinders are obstacles inside the arena. *Right:* The small gray circles and lines represent the initial shape of the swarm-bots. The large full circles are the obstacles inside the arena. The lines display the traces left by the s-bots showed in the left part of the figure during 600 cycles (bold lines highlight the traces left by two particular s-bots)

3.3 Collective Pushing and Pulling of Objects

We also observed that s-bots connected to an object, or connected so as to form a closed structure around an object, tend to pull/push the object in a coordinated fashion. This can be explained by considering that evolved s-bots tend to follow the direction of the team (for simplicity we will call this the "conformist tendency") but also have a tendency to persevere in their direction of movement if the intensity of the perceived traction is not too high (for simplicity we will call this the "stubborn tendency"). The stubborn tendency is due to the fact that when the direction of the perceived traction is 180 degrees different from the direction of the motion of the s-bots, they tend to go straight (see the flat area around 180 degrees in Fig. 4; the size of this area causing this stubborn tendency varies in different replication of the experiments reported in Section 2). Incidentally, the stubborn tendency might play a role in the ability to produce coordinated movements.

Fig. 8. *Left:* The ants swarm-bot formed by eight s-bots connected to the object with rigid links. *Right:* Traces left by the s-bots (thin lines) and the object (bold line) in 150 cycles. The dotted small circles represent the s-bots' initial (bottom) and final (top) position, the big circles represent the initial and final position of the object

Two sets of tests have been run. In the first set, the s-bots formed a circle and they were individually connected to a cylindrical object through a rigid link but not between them (this will be called the "ants formation", Fig. 8, left). The object had a radius of 4 cm, and a height of 3 cm. The density of the object was varied from 0.01 to 0.3, and the initial orientation of the chassis of the s-bots was set randomly. The test showed that when the weight of the object was small, s-bots were able to negotiate a common direction and to drag the object in that direction in the majority of the cases (cf. Fig. 8, right), while when the weight of the object was higher, s-bots tended to move in circle around it by displaying the behaviour described in Section 3.1. With a density of 0.3 the group succeeded in dragging the object only if the initial chassis' orientation was set to be the same for all s-bots, and only in 6 out of 10 cases (i.e. in 6 out of the 10 teams obtained by using the control strategies evolved in the 10 different replications of the experiments described in Section 2).

In the second set of tests, s-bots were assembled so as to form a circular structure around the object (Fig. 9, left). The results were similar to those obtained with the ants formation (Fig. 9, right). The only difference was that the formation deformed its shape so that some s-bots pushed the object while other s-bots pulled the other s-bots.

These results show that assembled s-bots evolved to move together as fast as possible also display an ability to coordinate through an external object to which they are attached and to pull/push the object in a coordinated fashion.

4 Conclusions and Directions of Future Research

We described how a group of simulated robots that are physically linked so as to form a single physical structure can display coordinated movements. We showed that the problem can be solved in a rather simple and effective way by providing the robots with a sensor that detects the direction and intensity of the traction that the turret exerts on the chassis and by evolving the neural controllers. The evolved strategy exploits the fact that the body of a swarm-bot (i.e. a robot constituted by a group of autonomous but aggregated individual robots) physically integrates the effects of the movements of the individual robots. Once individual robots are provided with traction sensors that allow them to detect the result of this integration, the problem of producing coordinated movements can be easily solved. In fact this sensors allow individual robots to have direct access to global information about what the entire group is doing.

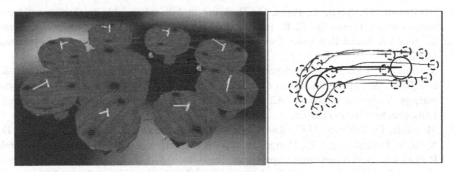

Fig. 9. *Left:* Eight s-bots assembled with flexible links forming a ring around the object. *Right:* Traces left by the s-bots (thin lines) and the object (bold line) in 150 cycles. The dotted small circles represent the swarm-bot's initial (left) and final (right) position, the big circles represent the initial and final position of the object

We also showed how neural controllers evolved for the ability to produce coordinated movement in a swarm-bot of four robots forming a linear structure are able to generalize in rather different circumstances: (a) they produce coordinated movements in swarm-bots with varying size, topology, and type of links; (b) they display individual or collective obstacle avoidance when placed in an environment with obstacles; (c) they spontaneously produce object pushing/pulling behavior when assembled to or around a given object. These results suggest that this strategy might constitute a basic functionality that, complemented with appropriate additional functions, might allow assembled robots to display a large number of interesting behaviors.

In future work we would like to evolve swarm-bots able to move toward a given target and to assemble and disassemble on the basis of their current goal and of the environmental conditions. From this point of view the results reported in this paper on individual and collective obstacle avoidance behavior suggest that the problem of controlling individual robots and teams of assembled robots might be solved with uniform and simple control solutions. Moreover, the results reported in the paper on the ability to generalize to rather different topologies of assembled robots, suggest that these control solutions might scale up to significantly complex setups.

Acknowledgments. This research has been supported by the SWARM-BOT project founded by the Future and Emerging Technologies program (IST-FET) of the European Community under grant IST-2000-31010. The information provided is the sole responsibility of the authors and does not reflect the Community's opinion. The Community is not responsible for any use that may be made of data appearing in this publication. The Swiss Government supported the Swiss project's participants (grant 01.0012).

References

1. Baldassarre, G., Nolfi, S., Parisi, D.: Evolving mobile robots able to display collective behaviour. In: Hemelrijk, C. K. (ed.): International Workshop on Self-Organisation and Evolution of Social Behaviour. Swiss Federal Institute of Technology, Zurich Switzerland (2002)
2. Mondada F., Floreano D., Guignard A., Deneubourg J.-L., Gambardella L.M., Nolfi S., Dorigo M.: Search for rescue: an application for the SWARM-BOT self-assembling robot concept. Technical Report, LSA2 - I2S – STI. Swiss Federal Institute of Technology: Lausanne, Switzerland (2002)
3. Mondada, F., Pettinaro, G.C., Kwee, I., Guignard, A., Gambardella, L.M., Floreano, D., Nolfi, S., Deneubourg, J.-L., Dorigo, M.: SWARM-BOT: A Swarm of Autonomous Mobile Robots with Self-Assembling Capabilities. In: Hemelrijk, C.K. (ed.): International Workshop on Self-Organisation and Evolution of Social Behaviour. Swiss Federal Institute of Technology, Zurich Switzerland (2002)
4. Nolfi, S., Floreano, D.: Evolutionary Robotics. The MIT Press, Cambridge Ma. (2000)
5. Pettinaro, G.C., Kwee, I., Gambardella, L.M., Mondada, F., Floreano, D., Nolfi, S., Deneubourg, J.-L., Dorigo, M.: SWARM Robotics: A Different Approach to Service Robotics. In Proceedings of the 33rd International Symposium on Robotics. International Federation of Robotics, Stockholm Sweden (2002)
6. Quinn, M., Smith, L., Mayley, G., Husband, P.: Evolving teamwork and role allocation with real robots. In Proceedings of the 8th International Conference on The Simulation and Synthesis of Living Systems (Artificial Life VIII) (In press)

Exploring the T-Maze: Evolving Learning-Like Robot Behaviors Using CTRNNs

Jesper Blynel and Dario Floreano

Autonomous Systems Lab
Institute of Systems Engineering
Swiss Federal Institute of Technology (EPFL)
CH-1015, Lausanne, Switzerland
{Jesper.Blynel, Dario.Floreano}@epfl.ch

Abstract. This paper explores the capabilities of continuous time recurrent neural networks (CTRNNs) to display reinforcement learning-like abilities on a set of T-Maze and double T-Maze navigation tasks, where the robot has to locate and "remember" the position of a reward-zone. The "learning" comes about without modifications of synapse strengths, but simply from internal network dynamics, as proposed by [12]. Neural controllers are evolved in simulation and in the simple case evaluated on a real robot. The evolved controllers are analyzed and the results obtained are discussed.

1 Introduction

Learning in neural networks is normally thought of as modifications of synaptic strengths by for example back-propagation or Hebbian learning. This view was in 1994 challenged by Yamauchi and Beer in [12], where the authors described the abilities of fixed synapse continuous time recurrent neural networks (CTRNNs) to display reinforcement learning-like properties by exploiting internal network dynamics. The task studied was generation and learning of short bit sequences. In [11] this work was extended to an artificial agent task where the relationship between the positions of a goal and a landmark in an environment had to be learned. However, the movement of the agent was restricted and it was equipped with artificial high level goal- and landmark-detection sensors. These restrictions were loosened in the recent work by [10] where an extended version the same landmark navigation task was studied. In the present work we apply a similar approach in which a simulated Khepera robot has to navigate in first a simple and then a double T-Maze. The task for the robot is to locate and "remember" the location of a reward-zone in the environment it happens to be evaluated in. In contrast to the above mentioned work the evolved behaviors are verified by testing them on a real robot in a real environment. Previous work on T-Maze navigation in evolutionary robotics includes delayed response tasks where the robots had to perform one or several turns in a maze on the basis of light source cues given to the robot [5][13]. In contrast to these works our focus is how to retain information over successive trials in the same of a different environment.

S. Cagnoni et al. (Eds.): EvoWorkshops 2003, LNCS 2611, pp. 593–604, 2003.

This becomes possible by equipping the robot with a sensor to detect the position of the reward-zone used for fitness evaluation.[1]

A different line of research has studied how agents in a self-organized ways can learn internal models of the environment [9]. The authors successfully trained a hierarchy of recurrent neural networks to predict increasingly complex information about the environment. The high level information which emerged was in which of two rooms the agent was currently navigating. The authors argued that the model learned could later on be used to generate action plans for goal seeking behaviors as in [8]. In the present work no explicit model of the environment exists, but is tightly coupled with both the learning of behaviors and the generation of motor actions. This corresponds with our belief that, as pointed out by [12], a direct distinction between mechanisms responsible for behavior and mechanisms responsible for learning is hard to defend biologically.

2   Neural Architecture and Genetic Encoding

Continuous-time recurrent neural networks (CTRNNs) are utilized for the experiments in this paper. The state of each neuron can be described by the following differential equation:

$$\frac{d\gamma_i}{dt} = \frac{1}{\tau_i}\left(-\gamma_i + \sum_{j=1}^{N} w_{ij}A_j + \sum_{k=1}^{S} w_{ik}I_k \right) \qquad (1)$$

where N is the number of neurons, i $(= 1, 2, ..., N)$ is the index, γ_i describes the neuron state (cell potential), τ_i is the time constant, w_{ij} the strength of the synapse from the presynaptic neuron j to the postsynaptic neuron i, $A_j = \sigma(\gamma_j - \theta_j)$ is the activation of the presynaptic neuron where $\sigma(x) = 1/(1 + e^{-x})$ is the standard logistic function and θ_j is a bias term. Finally, S is the number of sensory receptors, w_{ik} is the strength of the synapse from the presynaptic sensory receptor k to the postsynaptic neuron i and I_k is the activation of the sensory receptor ($I_k \in [0, 1]$). As in [12] the *Forward Euler* numerical integration method with step size $\Delta t = 1$ is applied to equation (1). The range of the rest of the parameters are the following:

$$\tau \in [1, 50], \quad \theta \in [-1, 1] \ and \ w \in [-5, 5]$$

The network architecture is shown in figure 1(b). The network consists of 6 fully interconnected neurons (4 hidden + 2 motor outputs) and 5 sensory receptors. Every neuron has synaptic connections from all neurons and all sensory receptors. The receptors are configured as follows: 4 inputs from the infrared proximity sensors paired two-by-two and 1 additional input from a floor sensor pointing downwards measuring the surface brightness. The 4 proximity values

---

[1] Note that in contrast to traditional reinforcement learning no direct reward is given to the robot. The evolved robots has to discover themselves the relationship between the input of this sensor and the fitness score.

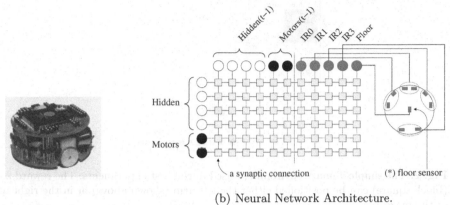

(a) Khepera Robot.

(b) Neural Network Architecture.

Fig. 1. The *Khepera robot* (a) used in the experiments. A standard Khepera has 8 infrared sensors distributed around the body used for object proximity and light intensity measurement. An extra infrared sensor pointing downwards (*) has been added in the center of the robot body in order to measure the surface brightness below the robot. *Neural Architecture* (b): The network consists of 6 fully interconnected neurons (4 hidden + 2 motor outputs) and 5 sensory receptors. Every neuron has synaptic connections from all neurons and all sensory receptors. The *sensory receptors* are wired to the robot as shown (b, right) (motor connections not shown for clarity).

are scaled between 0 and 1. The floor sensor input is set to 1 if the robot is inside a black reward-zone and 0 otherwise. The activation of the 2 output neurons, linearly scaled between -10 and 10, are used to set the wheel-speeds of the robot.

The network parameters are encoded in a bitstring genotype. Each neuron has 13 encoded parameters: A time constant (τ), a bias threshold (θ), and 11 synaptic strengths (w_{ij}). Each of the 78 network parameters is encoded linearly within its range using 5 bits, resulting in a total genotype length of 390 bits.

3   Experiment 1: Simple T-Maze

In the first experiment a robot has to navigate a simple T-maze (fig. 2). The experiment is carried out in a realistic simulation of the Khepera robot (fig. 1(a)) based on sensor sampling [6] and adding 5% uniform noise to the sampled values. Initially the robot is positioned as shown in figure 2, and the task is to find and stay on the black reward-zone which can be positioned in either the left or the right arm of the maze. The position of the reward-zone stays fixed during each epoch. The robot is tested for 4 epochs of 5 trials each - two epochs with the reward-zone in each arm of the maze. The neural network controlling the robot is initialized (by setting the state of each neuron to zero) at the beginning

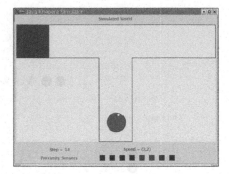

Fig. 2. The simple T-maze environment used in the first experiments. The reward zone (black square) can be positioned either the left arm (shown above) or in the right arm of the maze.

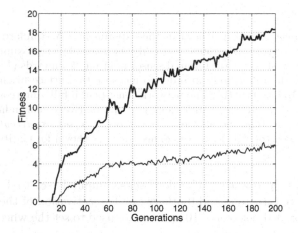

Fig. 3. *T-Maze Task.* Thick line shows best fitness and thin line shows population mean (both are averaged over 10 replications of the experiment).

of each epoch but *not* between trials within the same epoch. This means that the robot can potentially build up and store information in the dynamic state of the network between trials within the same epoch. The optimal behavior of the robot in this environment is to use the first trial of each epoch to locate and "remember" the position of the reward-zone, and thereafter move directly towards it for the remaining trials of the epoch. To put additional evolutionary pressure on this behavior, the number of available sensory-motor steps is 360 in the first trial of each epoch and only 180 in the remaining 4 trials. Given the size of the maze this means that the robot only has time to explore the whole maze during the first trial of each epoch. In addition a poison-zone (white square in figure 4(b)) is positioned opposite of the reward-zone in the last 4 but *not* the first trial of each epoch. An individual is immediately killed if it steps over the poison-zone. The *fitness function* is simply the sum of trials an

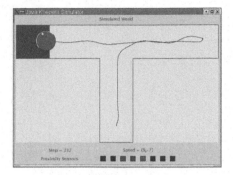 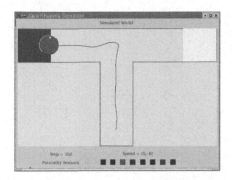

(a) *Trial 1*: The robot explores the environment, and after some time locates and stays on in the reward-zone.

(b) *Trials 2-5*: For the remaining trials the robot exploits the "knowledge" gained in trial 1 and moves directly towards the reward-zone.

Fig. 4. Robot traces of an epoch with the reward-zone to the left

individual ends its life inside the reward-zone. Since each individual is tested for a total of 20 trials the maximal possible fitness is 20. Notice that there is no direct pressure on evolving fast moving robots given this fitness function. This is however compensated by the fact that the number of steps in each trial is limited and has been adjusted to fit the size of the environment.

The experiments are carried out using a standard genetic algorithm with rank-based selection. A population of 200 randomly generated neural controllers is evolved for 200 generations. At every generation the best 40 individuals make 5 copies each. One copy each of the 5 best individuals remains unchanged (elitism). For the rest of the population single-point crossover with a probability of 0.04 and bit-switch mutation with a 0.02 probability per bit is applied. The whole experiment is repeated 10 times using different initializations of the computer's pseudo-random number generator.

The fitness results of the evolutionary runs on this experiment are shown in figure 3. The thick line shows best fitness and the thin line shows population mean, both are averaged over 10 replications of the experiment. The evolutionary process found individuals able to collect the maximal fitness of 20 in 6 out of the 10 replications of the experiment. The maximal fitness in the 4 remaining runs was around 16. The behavior of an individual from the final generation of one of the successful runs is shown in figure 4 and 5. The robot starts out in trial 1 of the first epoch (figure 4(a)) by exploring the maze until it locates the reward-zone where it stays the remaining time of the trial. In the following trials (figure 4(b)), the robot is able to retain the "knowledge" gathered during trial 1 and always turns left at the T-junction in order to move towards and stay on the reward-zone. In the epochs with reward to the right the robot moves directly towards the reward-zone in trial 1 (figure 5(a)), since the default behavior of this

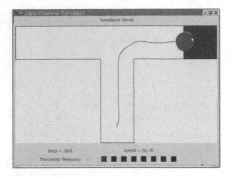 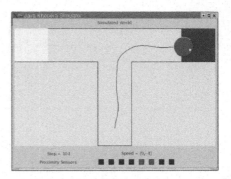

(a) *Trial 1*: The default behavior of this individual of turning right takes it directly to the reward-zone.

(b) *Trials 2-5*: For the remaining trials this behavior is repeated.

Fig. 5. Robot traces of an epoch with the reward-zone to the right.

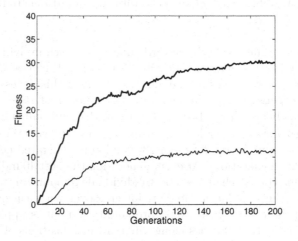

Fig. 6. *T-Maze task with reward switching.* Thick line shows best fitness and thin line shows population mean (both averaged over 10 replications of the experiment).

individual is to turn right at the first junction after an re-initialization of the neural controller. For the remaining trials this successful behavior is repeated (figure 5(b)).

3.1 Analysis

In order to better understand the functioning of the evolved neural controllers, some further analysis on an individual from one of the successful runs was done. The neural activities of each neuron were recorded over two epochs - one with reward to the left and one with reward to the right. It was found that the essential information about the current environment is stored in one of the hidden neurons. The activity of this neuron approaches zero at the end of the trials with the reward to the left and approaches one otherwise. By initializing every other neuron in the network as normal, but setting this neurons activity to either zero or one, it could be controlled which way the robot turns at the T-junction. In other words the state of this neuron "stores" the robots current assumption about the environment and is updated when these assumptions are not met. For more details about the analysis performed please refer to [1].

3.2   Transfer to the Real Robot

A way of verifying the evolutionary robotics results obtained in simulation is to test the evolved neural controllers on a real robot. For this purpose the T-Maze shown in figure 7 was built. The best individual from each of the 10 replications of experiment 1 task was tested. Initially however, the results of the tests were rather poor. None of the 10 controllers were able to reliable navigate the robot. This result indicates that the functioning of the evolved CTRNNs was specific to the sensory-motor conditions encountered in the simulator. This observation confirms our earlier results that CTRNNs, despite of their ability to display learning-likes abilities, lack the sensory-motor adaptability found in e.g. plastic Hebbian synapse networks [2]. However, several techniques for reducing this "reality gap"-problem, by adding noise at different levels of the simulation, have been proposed [5][6]. With this in mind the simulator was changed in the following way: Sensor noise levels were increased from 5% to 10%. In addition, 10% uniform noise was added to the distance traveled by each wheel at each timestep. Furthermore, the initial conditions of each tested individual changed in the following way: The starting position was randomized within a 4 by 4 cm square, and the orientation was randomized within the range forward +/- 15 degrees. With these modifications an incremental evolution lasting 20 generations was launched, seeded with a population from one of the original runs. This time the transfer to the real robot was perfect. The best individual of the last generation was able score a fitness value of 20, i.e. finding the reward-zone in every trial. No significant behavioral differences compared to the simulation were observed.

4   Experiment 2: Simple T-Maze with Reward Switching

In order to further investigate the learning-like capabilities of CTRNNs the task for the robot was now made slightly more complex. In experiment 1 the robots could rely on the fact that the position of the reward-zone remained fixed during a whole epoch. Evolved robots were able to explore the whole environment during trial 1 of each epoch in order to locate this position, but would they also be able to adapt if the reward position was changed later on within the same epoch. This turned out not to be the case. When testing the individual analyzed in section 3.1 by placing the reward-zone to left for 5 trials and then switching

Fig. 7. *The real T-maze environment.* (Note: surface colors have been reversed compared to the simulated environment, but the processed input from the floor-sensor remains 1 inside reward-zone and 0 otherwise) .

the reward position to the right *without* resetting the neural network, the robot would continue to turn left a the T-Junction in the trials after the switch took place.

A new experiment was now set up in order to check if this lack of adaptivity to environmental changes taking place later on in an epoch was due to a limitation in the learning capabilities of the network, or simply given by the fact this condition was never met during evolution. The evolved robots could simply have found a minimalistic solution. In this new evolution the duration of each epoch was increased to 10 trials. The reward-zone position remained fixed in the first 5 trials but was then switched to the other side in the 5 last trials of each epoch. Each individual was still tested for 4 epochs, 2 with the reward initially to the left and 2 with the reward initially to the right. The fitness function remained the same, and since each individual was tested for 40 trials in total the maximum possible fitness was now 40.

The average result of 10 replications of the experiment is shown in figure 6. The resulting best fitness in the 10 replications varied between 22 in the worst case and 38 in the best. In the latter case the best individual did realize that the reward position had changed in trial 6, and was able to locate the new position. However in some of the following trials it would still turn the wrong way thus ending up in a the poison-zone. In order to increase the performance the last bit an incremental evolutionary approach was now applied. The evolutionary conditions remained the same, but instead of seeding the evolution with a random population, it was initially seeded with a population consisting of the best individual of each of the 200 generations from the best replication of the previous evolution. Again 10 replications were performed. In most of the runs the fitness level stayed at 38, and even dropped to 32 in one case (graph not shown). It seemed that level 38 solution was a local optimum which was difficult to escape. In one replication, however, the fitness level reached the maximal value of 40, and when tested afterwards the best individual from this replication could reliably solve the task. When the reward position switched at trial 6 of each epoch, the

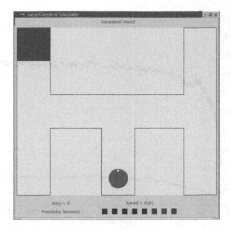

Fig. 8. *Double T-Maze.* The maze now has three T-junctions. The reward-zone is positioned either in the upper left (as shown) or upper right corner. The 3 remaining corners each contains a poison-zone during trials 2-5 (but *not* trial 1) of each epoch.

robot would at first move towards the previous location, but when not finding the reward-zone here anymore it would turn around and initiate a search until the new reward position was located. In the remaining trials of each epoch the robot then again turned directly towards the reward-zone, resulting in the total fitness of 40.

5 Experiment 3: Double T-Maze

In the previous section it was shown that evolved robots were able to completely solve the simple T-Maze even in the case when the reward position was switched during an epoch. However, the robots only had to retain one piece of information based on previous experiences, namely whether to turn left or right at the T-junction. The investigations were now turned towards a double T-maze with several T-junctions thus further complicating the task (see figure 8). To compensate for the increased size of the maze the number of sensory-motor cycles was increased to 400 in trial 1 and 200 in the remaining trials. The reward-zone could appear in either upper-left and upper-right corner of the maze.

During the first trial only the reward-zone was present, and in the following trials poison-zones were placed in the 3 remaining corners. The other parameters remained unchanged. As in the simple T-Maze, the case where the reward position remained unchanged during an entire epoch was tested first. An evolutionary process seeded with populations of random individuals was launched, but the results were very poor under these conditions. Basically the fitness remained at zero all the time, with very few exceptions of fitness 1 coming and going for a couple of generations in one of the replications of the experiment. In fact this result is not that surprising considering the fitness function used. In order for a random initial individual to collect some fitness and kick-off the

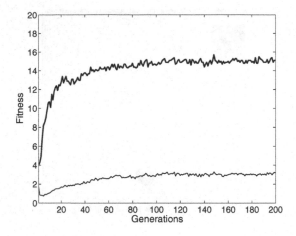

Fig. 9. *Double T-Maze task.* Thick line shows best fitness and thin line shows population mean (both are averaged over 10 replications of the experiment).

evolutionary progress it had to navigate all the way to one of the upper corners. If it could not do that it simply got zero fitness. A solution to this problem could have been to design an incremental fitness function, where individuals would get some fitness for partially solving the task. Instead it was decided to again rely upon an incremental evolutionary approach. The genetic algorithm was seeded with a population consisting of the best individuals from one of the runs of experiment 1 on the simple T-Maze. The results of 10 replications of the experiment are shown in figure 9. In the best run the fitness reached a level of 18 out of 20. During trial 1 the best individual always turned right at the first junction and left at the second. This would take the robot to the upper right corner of the maze. In epochs with reward to the right the robot would soon find and stay in the reward-zone. When the reward was on the left, on the other hand, the robot was not capable of searching for the reward-zone as it did in the simple T-maze case (fig. 4(a)). Instead the robot simply crashed into the upper wall. In the following trials of these epochs, however, the robot would now turn left at the first junction and right at the second, reaching the reward-zone in the upper-left corner. One trial was wasted, but crucial information about the reward position was gathered and used in the following trials, thus resulting in fitness of 18 out of 20. An attempt to further improve this behavior by an additional incremental evolution was now conducted, seeding evolution with a population of individuals from this experiment. This time, however, the attempt was not successful and the fitness level remained at 18 in every replication of the experiment. This results suggests that the maximal problem complexity solvable for the genetic algorithm and neural network used had been reached. This suggestion was confirmed by an unsuccessful attempt to apply reward-switching to the double T-Maze task, as was done in experiment 2 in the simple T-Maze case. No reliable learning behaviors were observed in the evolved controllers in this case.

6   Conclusion

We have shown that evolution of learning-like properties *is* possible without modifications of synapse strengths, but simply by relying on complex internal dynamics of CTRNNs. In experiment 1 the robot had to navigate a simple T-Maze with the reward position fixed during each epoch. Direct evolution of this task was possible and the analysis showed that the employed strategy of an evolved network was to store essential environment information in one of the hidden units. The rest of the neurons would update the activity of this neuron based on current environmental feedback. With a few simulator modifications evolved behaviors were successfully transfered to a real robot. In experiments 2 the reward position would vary within the same epoch forcing evolved robots to keep on adapting their strategy to the current environmental conditions. Successful individuals solving this task were evolved in a two-step incremental evolution. Because of the increased complexity of the maze, direct evolution was not possible in experiment 3. However, by seeding evolution with a population from experiment 1 individuals capable of "learning" were found. It was not possible to perform additional reward switching as in experiment 2 on the simple T-Maze, suggesting that maximal task complexity had been reached.

In this work incremental evolution has proven to be a powerful tool in evolving complex robot behaviors, however evolving CTRNNs as shown here will face evolvability problems if the task complexity is to be further increased. In principle a sufficiently large CTRNN is able to display arbitrarily complex dynamics. However, the problem of how to evolve such networks will have to be addressed in the future. Possible solutions could be to explore new neural mechanisms for information "storage", or to investigate how to preserve the learning capabilities of CTRNNs in networks generally thought of to be easier to evolve such as Hebbian synapse networks or spiking neural networks. Our work will focus on these aspects in the future.

As pointed out by one of the reviewers it can be argued whether the experiments presented in this paper should be classified as learning-like behaviors or simply as internal dynamics investigations. It true that the view of learning presented in this paper is quite different from the traditional computational view of learning, where some update of the control systems always takes place. However neurophysiological experiments have indicated that the way animals and humans perceive, classify, and memorize, for example in the olfactory system, is by transitions between chaotic attractors in dynamical systems formed by large numbers of neurons in the brain [3][7]. These results correspond nicely with the view of memory and learning presented in this paper.

Acknowledgements. The authors wish to thank Tom Ziemke for useful comments and suggestions. This work was supported by the Swiss National Science Foundation, grant nr. 620-58049.

References

1. J. Blynel. Evolving reinforcement learning-like abilities for robots. In *(to appear) Proceedings on the 5th International Conference on Evolvable Systems (ICES'03): From Biology to Hardware*. Springer Verlag, Berlin, 2003.
2. J. Blynel and D. Floreano. Levels of dynamics and adaptive behaviour in evolutionary neural controllers. In Hallam et al. [4], pages 272–281.
3. W. J. Freeman. The physiology of perception. *Scientific American*, 264:78–85, 1991.
4. B. Hallam, D. Floreano, J. Hallam, G Hayes, and J-A Meyer, editors. *From Animals to Animats 7: Proceedings of the Seventh International Conference on Simulation of Adaptive Behavior*. MIT Press-Bradford Books, Cambridge, MA, 2002.
5. N. Jakobi. Evolutionary robotics and the radical envelope-of-noise hypothesis. *Adaptive Behavior*, 6(2):325–368, 1997.
6. O. Miglino, H. H. Lund, and S. Nolfi. Evolving mobile robots in simulated and real environments. *Artificial Life*, 2(4):417–434, 1995.
7. R. Pfeifer and C. Scheier. *Understanding Intelligence*. MIT Press, Cambridge, MA, 1999.
8. J. Tani. Model-based learning for mobile robot navigation from the dynamical systems perspective. *IEEE Transactions on Systems, Man, and Cybernetics - Part B*, 26:421–436, 1996.
9. J. Tani and S. Nolfi. Learning to perceive the world as articulated: An approach for hierarchical learning in sensory-motor systems. *Neural Networks*, 12(7–8):1131–1141, 1999.
10. E. Tuci, I. Harvey, and M. Quinn. Evolving integrated controllers for autonomous learning robots using dynamic neural networks. In Hallam et al. [4], pages 282–291.
11. B. Yamauchi and R. D. Beer. Integrating reactive, sequential, and learning behaviour using dynamical neural networks. In D. Cliff, P. Husbands, J. Meyer, and S. W. Wilson, editors, *From Animals to Animats III: Proceedings of the Third International Conference on Simulation of Adaptive Behavior*, pages 382–391. MIT Press-Bradford Books, Cambridge, MA, 1994.
12. B. Yamauchi and R. D. Beer. Sequential behavior and learning in evolved dynamical neural networks. *Adaptive Behavior*, 2(3):219–246, 1994.
13. T. Ziemke and M. Thieme. Neuromodulation of reactive sensorimotor mappings as a short-term memory mechanism in delayed response tasks. *(to appear) Adaptive Behavior*, 10(3–4), 2003.

Competitive Co-evolution of Predator and Prey Sensory-Motor Systems

Gunnar Búason and Tom Ziemke

Department of Computer Science, University of Skövde
Box 408, 541 28 Skövde, Sweden
{gunnar.buason,tom}@ida.his.se

Abstract. A recent trend in evolutionary robotics research is to maximize self-organization in the design of robotic systems in order to reduce the human designer bias. This article presents simulation experiments that extend Nolfi and Floreano's work on competitive co-evolution of neural robot controllers in a predator-prey scenario and integrate it with ideas from work on the 'co-evolution' of robot morphology and control systems. The aim of the twenty-one experiments summarized here has been to systematically investigate the tradeoffs and interdependencies between morphological parameters and behavioral strategies through a series of predator-prey experiments in which increasingly many aspects are subject to self-organization through competitive co-evolution. The results illustrate that competitive co-evolution has great potential as a method for the automatic design of robotic systems.

1 Introduction

A recent trend in evolutionary robotics and artificial life research is to maximize self-organization in the design of robotic systems in order to reduce the human designer bias. This has been argued to be advantageous from both a scientific and an engineering perspective, since it allows the modeling of natural systems and the design of complex artifacts to be relatively independent of human preconceptions of how a particular behavior is achieved in natural systems or could be achieved in artificial ones (cf. [12, 13]).

Competitive co-evolution (CCE) of two or more robotic species with coupled fitness is of particular interest in this context since it offers a relatively natural approach to the incremental evolution of complex behavior and further reduces the human experimenter's influence on (a) the specification of fitness functions and (b) the design of incremental stages [13]. Furthermore, the trend towards maximization of self-organization has led to experiments with the evolution of robot morphologies, and in some cases the integrated (co-) evolution of robot controllers and morphologies. This maximization of automatic rather than human design addresses what Funes and Pollack called the "chicken and egg" problem of robotics [9, p. 358]: "Learning to control a complex body is dominated by inductive biases specific to its sensors and effectors, while building a body which is controllable is conditioned on the pre-existence of a brain".

S. Cagnoni et al. (Eds.): EvoWorkshops 2003, LNCS 2611, pp. 605–615, 2003.
© Springer-Verlag Berlin Heidelberg 2003

This article summarizes twenty-one simulation experiments, documented in detail elsewhere [1, 2], that incrementally extend Nolfi and Floreano's [6, 8, 12] work on CCE of neural robot controllers in a predator-prey scenario and integrate it with ideas from work on the 'co-evolution' of robot morphology and control systems [5, 11]. The aim of these experiments has been to further systematically investigate the tradeoffs and interdependencies between morphological parameters and behavioral strategies through a series of predator-prey experiments in which increasingly many aspects are subject to self-organization through CCE.

The rest of this paper is structured as follows: The following background section briefly overviews the above previous works. This is followed by a description of our own experiments and the most interesting results, and the final section presents a brief discussion and some conclusions.

2 Background – Previous Work

As mentioned in the previous section, Nolfi and Floreano have in a series of experiments [6, 7, 8, 12] studied the CCE of neural robot controllers in a predator-prey scenario. Some of these experiments showed that the robots' sensory-motor structure, i.e. their morphology as well as their sensory and motor capacities, greatly influenced the evolution of behavioral (and learning) strategies. For example, in experiments where predators and prey were given the possibility to combine evolution and life-time learning [7], it turned out that the predators, which were equipped with long-range cameras, evolved a predisposition to learn behavioral strategies in interaction with the prey they were facing, while the prey, equipped only with short-range infrared sensors that provided little time to observe predators from a distance, evolved general strategies that did not require life-term learning. In more recent experiments Nolfi and Floreano [12] equipped both predator and prey with cameras. This resulted in a well-adapted prey that refined its strategy instead of radically changing it, allowing a more natural 'arms race' to emerge.

Obviously some design decisions in the setting up of such experiments are taken relatively arbitrarily based on what the human experimenter considers appropriate. A more natural evolutionary process that allows the integrated evolution of morphology and control might lead to completely different solutions. Hence, Nolfi and Floreano [14, p. 33] point out that "experiments where both the control system and the morphology of the robot are encoded in the genotype and co-evolved can shed light on the complex interplay between body and brains in natural organisms and may lead to the development of new and more effective artifacts".

The evolution of robot body plans has been addressed by a number of researchers. Lee et al. [10] were among the first to perform experiments that considered co-evolving the structure of a mobile robot, and investigated tradeoffs and dependencies in the robot morphology. Further analysis of the results by Lund et al. [11] showed that the evolved robots cluster in specific regions of the morphological space, e.g. there was a quasi-linear relationship between body size and wheel base for an obstacle avoidance task despite the different sensor ranges. Moreover, the integrated evolution of robot 'brains' and bodies has been addressed by a small, but growing group of researchers (e.g. [5, 15]). Among the first of these were Cliff and Miller [5] who, in simulation, co-evolved 'eye' positions and control networks for pursuing and evading robots with the result that "pursuers usually evolved eyes on the front of their bodies (like cheetahs) while evaders usually evolved eyes pointing sideways or even backwards (like gazelles)."

The aim of our own work presented in the following sections has been to further systematically investigate the tradeoffs and interdependencies between morphological parameters and behavioral strategies through a series of simulated predator-prey experiments in which increasingly many aspects are subject to self-organization through CCE. For this we have adopted the same basic experimental framework as used by Floreano and Nolfi [6, 7, 8, 12] (limiting our experiments to simulated robots though) and incrementally extended it, in a series of twenty-one experiments, taking inspiration mostly from Cliff and Miller's [5] work on the evolution of "eye" positions. However, the focus will not be on evolving the positions of the sensors on the robot alone but instead on investigating the tradeoffs the evolutionary process makes in the robot morphology due to different constraints and dependencies, both implicit and explicit. The latter is in line with the research of Lund et al. [11].

3 Experiments

The following subsections will first give an overview of the experiments performed, followed by details of the robots and controllers used and the parameters that were evolved, and finally the most interesting results are summarized.

Overview

This work did not consider evolution of the architecture and learning rules of the neural network in order to minimize the number of parameters evolved and to allow detailed analysis. Furthermore the experiments focused on evolving a limited number of aspects of robot morphology, i.e. the size of the robot was kept constant, assuming a Khepera-like robot, and all the infrared sensors were used, for the sake of simplicity. The parameters that were evolved in at least some of the experiments were the following (cf. Figure 1):

- the weights of the neural network, i.e. the control system
- the view angle of the camera (0 to 360 degrees)
- the view range of the camera (5 to 500 mm)
- the direction of the camera (0 to 360 degrees)

Moreover, in some experiments further constraints and dependencies were introduced, e.g. by letting the view angle constrain the maximum speed, i.e. the larger the view angle, the lower the maximum speed the robot was allowed to accelerate to. This is in contrast to the experiments of [6, 7, 8, 12], where the predator's maximum speed was always set to half the prey's. Further, each experiment was replicated three times. Table 1 gives a more detailed overview of the individual experiments (see also [1]).

608   G. Búason and T. Ziemke

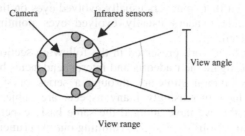

Fig. 1. pera robot equipped with eight short-range infrared sensors and a vision module (a camera [icon] e 'side' of the robot that has six infrared sensors is referred to as the front side. The 'side' that has two infrared sensors is referred to as the rear side. The camera is directed toward the front on the standard robot, but in some of our experiments the direction is evolved.

Table 1. Overview of Experiments

Exp	Purpose	Motivation
1	Replicating the predator-prey experiment in [6] where only the predator had vision.	The framework needs to be verified and for comparisons sake (i.e. so that it is possible to compare the other experiments with these two).
2	Replicating the predator-prey experiment in [12] where both the predator and the prey had vision.	
3	To investigate the impact of allowing the evolutionary process to evolve view angle and view range of the predator's camera.	Demonstrate potential influence of evolving view angle and range of the predator and then later on adding constraints into the behavior of the robot.
4	Investigate the impact of introducing a dependency between the predator's view angle and maximum speed.	
5	Prey can evolve camera view angle and view range. The basic predator from experiment 1 is used.	Similar to the previous experiments, except that in this case the focus is on the prey.
6	Prey can evolve camera angle and range but has speed constraints.	
7	Both predator and prey can evolve camera view angle and view range.	Investigate what happens when the competitive advantage in both species/robots is balanced out by evolving both view angle and view range in both species. In addition, give the robots the same maximum speed, and speed constraints.
8	Both predator and prey can evolve camera view angle and view range. Predator is implemented with speed constraints.	
9	Both predator and prey can evolve camera view angle and view range. Maximum speed of both species is set to the same value. Predator is implemented with speed constraints.	
10	Both can evolve camera view angle and view range, and both have speed constraints. Maximum speed of both species is set to the same value.	
11-20	Repeat all previous experiments investigating the additional possibility of evolving the camera direction.	Would the robots benefit from evolving the direction of the camera?
21	Both can evolve view angle, view range and direction of camera with no speed constraints.	To demonstrate the influence of constraints.

Experimental Setup

For finding and testing the appropriate experimental settings a number of pilot experiments were performed. The simulator that was used in this work is called YAKS [3], which is similar to the one used by [6, 7, 8, 12]. YAKS simulates the popular Khepera robot in a virtual environment defined by the experimenter. The simulation of the sensors is based on pre-recorded measurements of a real Khepera robot's infrared sensors and motor commands at different angles and distances [3].

The experimental framework that was implemented in the YAKS simulator was in many ways similar to the framework used by [6, 7, 8, 12]. Experiments 1 and 2, replications of Floreano and Nolfi's experiments, just served to verify that, despite minor differences, our experimental setup gave sufficiently similar results, which was the case. What differed was that in our work we used real value encoding to represent the genotype instead of direct encoding, and the number of generations was extended from 100 to 250 generations to allow us to observe the morphological parameters over a longer period of time. Beside that, most of the evolutionary parameters were 'inherited' such as the use of *elitism* as a selection method where the 20 best individuals of a population of 100 were chosen to reproduce. In addition, a similar fitness function was used. Maximum fitness was one point while minimum fitness was zero points. The fitness was a simple time-to-contact measurement, giving the selection process finer granularity, where the prey achieved the highest fitness by avoiding the predator for as long as possible while the predator received the highest fitness by capturing the prey as soon as possible. The competition ended if the prey survived for 500 time steps or if the predator made contact with the prey before that.

For each generation the individuals were tested for ten epochs. During each epoch, the current individual was tested against one of the best competitors of the ten previous generations. At generation zero, competitors were randomly chosen within the same generation, whereas in the other nine initial generations they were randomly chosen from the pool of available best individuals of previous generations. This is in line with the work of [6, 7, 8, 12]. In addition, the same environment as in [6, 7, 8, 12] was used (cf. Figure 2).

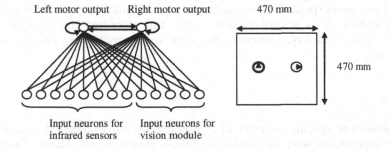

Fig. 2. Left: Neural network control architecture (adapted from [6]). Right: Environment and starting positions. The thicker circle represents the starting position of the predator while the thinner circle represents the starting position of the prey. The triangles indicate the starting orientation of the robots, which is random for each generation.

A simple recurrent neural network architecture was used, similar to the one used in [6, 7, 8, 12] (cf. Figure 2). The experiments involved both robots using the camera in some cases, so depending on whether the robot was using the camera or not, it had eight input neurons for receiving input from the infrared sensors and possibly five additional input neurons for the camera. The neural network had two sigmoid output neurons for each motor of the robot and one bias node that was connected to the output nodes (not shown in Figure 2). The bias weights were also evolved.

The vision module, which was only one-dimensional, was implemented with flexible view range, view angle and camera direction, while the resolution was kept constant (five input neurons covering equal parts of the view angle).

For each experiment, the weights of the neural network were initially randomized and evolved using a Gaussian distribution with a standard deviation of 2.0. When angle, range and camera direction were evolved, the starting values were randomized using a uniform distribution function, and during evolution the values were mutated using a Gaussian distribution with a standard deviation of 5.0. Worth mentioning is that the camera used by Nolfi and Floreano had a view angle of 36 degrees and the camera direction was fixed (pointing 'forward', i.e. in the same direction as most of the infrared sensors). In our experiments the direction of the camera was in the interval of 0 to 360 degrees, where 0 respectively 360 degrees indicated that the camera was pointing in a forward direction on the robot. The view angle could only evolve up to 360 degrees, if the random function generated a value of over 360 degrees then the view angle was set to 360 degrees. The same was valid for the lower bounds of the view angle and also for the lower and upper bounds of the view range. For the camera this was different; if the camera direction evolved over 360 degrees it started again at 0 degrees.

Constraints, such as those used by [6, 7, 8, 12], where the maximum speed of the predator was only half the prey's, were adapted in some of the experiments here where speed was dependent on the view angle. For this, the view angle was divided up into 10 intervals of 36 degrees[1]. The maximum speed of the robot was then reduced by 10% for each additional interval used. That means, if the view angle was between 0 and 36 degrees there were no constraints on the speed, for view angles between 36 and 72 degrees the maximum speed of the robot was limited to 90% of its original maximum, for view angles between 72 and 108 degrees the speed was limited to 80% of its original maximum, etc. This algorithm controlled the output from the neural network, converting the value to the appropriate one for different experiments.

Results

As mentioned before, the results of all 21 experiments are documented in detail in [1, 2]. The experiments were analyzed using fitness measurements, Master Tournament [6] and collection of CIAO data [4]. Furthermore some statistical calculations and behavioral observations were performed. Here a summary of the most interesting results will be given.

---

[1] Alternatively, a linear relation between view angle and speed could be used.

Evolving task-dependent morphologies

In a number of experiments the robots evolved suitable behaviors together with a suitable morphology in order to counter the opponents' strategies. An example of this is experiment 9 where the predator evolved a relatively narrow view angle with a long view range while the prey evolved a wider view angle with a short view range (cf. Figure 3). Experiments 3 to 6 had similar results. This can be compared to nature where predators (such as cheetahs) have to some extent good binocular vision, allowing to spot prey at a long distance and to estimate the distance, while prey (such as gazelles) have to some extent good peripheral vision, allowing to spot predators approaching from different directions.

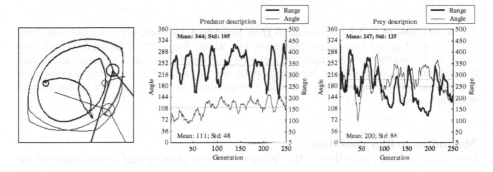

Fig. 3. Experiment 9 – The trajectories on the left are from generation 43 (predator: 57° view angle, 444 mm view range; prey: 136°, 226 mm) after 386 time steps. The predator is marked with a thick black circle and its trajectory with a thick black line. The prey is marked with a thin black circle and its trajectory with a thin black line. Starting positions of both robots are marked with small circles. The view field of the predator is marked with two thick black lines, that of the prey with two thin lines. The angle between the lines represents the current view angle, i.e. 57 degrees for the predator and 136 degrees for the prey, and the length of the lines represent the current view range, i.e. 444 mm for the predator and 226 mm for the prey [1]. The right graphs illustrate the morphological description of view angle (left y-axis, thin line) and view range (right y-axis, thick line) for both species. The values in the upper left corners are the mean and standard deviation for the view range over generations, calculated from the best individual from each generation. The values in the lower left corner are the mean and standard deviation for the view angle over generations, also calculated from the best individual from each generation. The data was smoothed using rolling averages over ten data points [1, 2].

Implementing Constraints

Constraints were introduced into the experiments successively and the effects were analyzed. Particularly interesting was when the camera module of the prey was evolved while the view angle constrained the maximum speed (experiment 10). This resulted in prey that preferred speed to vision (cf. Figure 4); i.e. they preferred to be able to move fast in the environment, relying on short-range infrared sensors instead of the camera to avoid the predators.

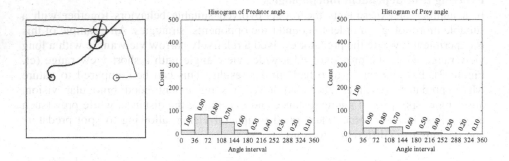

Fig. 4. Experiment 10 – Left: trajectories from generation 134 (predator: 34°, 432 mm; prey: 11°, 331 mm) after 54 time steps are shown [1]. Right: graphs show histograms over view angle, i.e. the number of individuals that preferred a certain view angle. The values above each bin indicate the corresponding maximum speed, e.g. if a robot has evolved a view angle between 36 and 72 degrees then its speed is constrained to be at most 90% of the original maximum speed [1].

Maximizing self-organization

Experiments 11-20 investigated the influence of maximizing self-organization of the robots, by including in the genotype the direction of the camera and then replicating all previously performed experiments. Interestingly, it could be noticed that despite the variety of the experiments, the direction of the camera in most cases evolved towards the rear side of both robots, i.e. the side with fewer infrared sensors (cf. Figure 5). The predator evolved on average over all replications of experiments 11 to 20 a camera direction of 175° with a standard deviation of 64°, and the prey similarly evolved a camera facing in the direction of 178° with a standard deviation of 101°. Although the standard deviation of the prey's camera direction was rather high, it is still possible to conclude that in most cases the prey preferred to have at least parts of the camera facing backwards. Using these morphologies the predator moved in a backward direction while the prey moved in a forward direction (cf. Figure 5). The reason for this behavior is finding the balance between obstacle avoidance and task-dependent behaviors of the robots (i.e. to pursue or evade). Prey used IR sensors to avoid walls while using the camera (when equipped with one) to observe and avoid predators. Predators used the camera to observe and follow prey but did not require the use of IR sensors to capture it. The latter is in line with the results in [6, 8] where the predator performed turns before hitting the prey in order to have a higher probability of hitting it with a side that has no IR sensors.

4   Summary and Conclusion

In our work we have systematically investigated tradeoffs and interdependencies between morphological parameters and behavioral strategies of predator-prey robots through a series of simulation experiments. The results of these experiments were that by using CCE the evolutionary process incrementally built a suitable morphology for the predator and prey respectively, i.e. it found a suitable configuration/combination of the vision module and matching behavioral strategies.

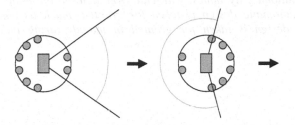

Fig. 5. Schematic illustration of the robots' 'preferred' morphologies and moving directions (indicated by the arrows). The predator (left) has the camera facing towards the 'rear side' and it moves in the same direction, which allows it to keep track of the prey while chasing it. The prey also has the camera facing towards the rear, but it moves in the opposite direction, which allows it to observe the predator while running away from it [1].

More specifically, both robots evolved different hardware configurations depending on the robot's specific task, despite the same basic hardware structure. For example, predator robots typically evolved a focused view angle and a long view range, while the prey evolved a relatively wide view angle and a short view range. This, however, depended on the different constraints implemented. In addition, it was observed that constraints strongly influenced the behaviors of the robots. For example, the prey preferred speed to vision in experiments where the view angle constrained the speed. And finally, the results indicate that implicit constraints, which are results of human designers' biases, can greatly affect the behaviors evolved. For example, Floreano and Nolfi in [6] designed two implicit constraints in their experiments when (a) they assumed the forward direction of the robot and (b) assumed that the robot would benefit from having the camera facing in the same direction. However, the results of our experiments indicate that when the direction of the camera is included in the genotype, both robots prefer to have the camera facing in a backward direction.

Although we have here only been able to overview and exemplify some of the results of our experiments that we believe we have demonstrated the impact of allowing the evolutionary process to adjust and make tradeoffs in the robot morphology in artificial-life simulations. We are, of course, aware of the limitations of simulations and the fact that most of our experiments could not directly be transferred to real robots. Nevertheless, we consider this work as a further step towards removing the human designer from the loop, by suggesting a mixture of CCE and 'co-evolution' of brain and body. This work only reflects on certain parts of evolving robot morphology and the integration of tradeoffs there, but future work would be to extend this work to cover co-evolution of brain and body plus CCE. This is in line with the future expectations of Pollack et al. [15] who considered that our understanding of the dynamics of co-evolutionary learning in the self-organization of complex systems will increase. It is appropriate to end this discussion with a citation from Pollack et al. discussing the future of artificial life [15, p. 221]:

Perhaps the small demonstrations of automatic design, with continued development, and increase in computer speed and simulation fidelity, coupled to increase in basic theory of coevolutionary dynamics, will lead, over time, to the point where fully automatic design is taken for granted, much as computer aided design is taken for granted in manufacturing industries today.

References

1. Buason, G. (2002a). *Competitive co-evolution of sensory-motor systems*. Masters Dissertation HS-IDA-MD-02-004. Department of Computer Science, University of Skövde, Sweden.
2. Buason, G. (2002b). Competitive co-evolution of sensory-motor systems – Appendix. Technical Report HS-IDA-TR-02-004. Department of Computer Science, University of Skövde, Sweden.
3. Carlsson, J. & Ziemke, T. (2001). YAKS – Yet Another Khepera Simulator. In: U. Rückert, J. Sitte & M. Witkowski (eds.), *Autonomous minirobots for research and entertainment – Proceedings of the fifth international Heinz Nixdorf Symposium* (pp. 235–241). Paderborn, Germany: HNI-Verlagsschriftenreihe.
4. Cliff, D. & Miller, G. F. (1995). Tracking the Red Queen: Measurements of adaptive progress in co-evolutionary simulations. In: F. Moran, A. Moreano, J. J. Merelo, & P. Chacon, (eds.), *Advances in Artificial Life: Proceedings of the third European conference on Artificial Life*. Berlin: Springer-Verlag.
5. Cliff, D. & Miller, G. F. (1996). Co-evolution of pursuit and evasion II: Simulation methods and results. In: P. Maes, M. Mataric, J.-A. Meyer, J. Pollack & , S. W. Wilson (eds.), *From animals to animats IV: Proceedings of the fourth international conference on simulation of adaptive behavior (SAB96)* (pp. 506–515). Cambridge, MA: MIT Press.
6. Floreano, D. & Nolfi, S. (1997a). God save the Red Queen! Competition in co-evolutionary robotics. In: J. R. Koza, D. Kalyanmoy, M. Dorigo, D. B. Fogel, M. Garzon, H. Iba, & R. L. Riolo (eds.), *Genetic programming 1997: Proceedings of the second annual conference*. San Francisco, CA: Morgan Kaufmann.
7. Floreano, D. & Nolfi, S. (1997b). Adaptive behavior in competing co-evolving species. In P. Husbands, & I. Harvey (eds.), *Proceedings of the fourth European Conference on Artificial Life*. Cambridge, MA: MIT Press.
8. Floreano, D., Nolfi, S. & Mondada, F. (1998). Competitive co-evolutionary robotics: From theory to practice. In: R. Pfeifer, B. Blumberg, J-A. Meyer, & S. W. Wilson (eds.), *From animals to animats V: Proceedings of the fifth international conference on simulation of adaptive behavior*. Cambridge, MA: MIT Press.
9. Funes, P. & Pollack, J. (1997). Computer evolution of buildable objects. In: P. Husbands & I. Harvey (eds.), *Proceedings of the fourth European conference on artificial life* (pp. 358–367). Cambridge, MA: MIT Press.
10. Lee, W-P, Hallam, J. & Lund, H.H. (1996). A hybrid GP/GA Approach for co-evolving controllers and robot bodies to achieve fitness-specified tasks. In: *Proceedings of IEEE third international conference on evolutionary computation* (pp. 384–389). New York: IEEE Press.
11. Lund, H., Hallam, J. & Lee, W. (1997). Evolving robot morphology. In: *Proceedings of IEEE fourth international conference on evolutionary computation* (pp. 197–202). New York: IEEE Press.
12. Nolfi, S. & Floreano, D. (1998). Co-evolving predator and prey robots: Do 'arms races' arise in artificial evolution? *Artificial Life, 4*, 311–335.

13. Nolfi, S. & Floreano, D. (2000). *Evolutionary robotics: The biology, intelligence, and technology of self-organizing machines*. Cambridge, MA: MIT Press.
14. Nolfi, S. & Floreano, D. (2002). Synthesis of autonomous robots through artificial evolution. *Trends in Cognitive Sciences, 6*, 31–37.
15. Pollack, J. B., Lipson, H., Hornby, G. & Funes, P. (2001). Three generations of automatically designed robots. *Artificial Life, 7*, 215–223.

Evolving Spiking Neuron Controllers for Phototaxis and Phonotaxis

Robert I. Damper and Richard L.B. French

Image, Speech and Intelligent Systems (ISIS) Research Group
Department of Electronics and Computer Science
University of Southampton
Southampton SO17 1BJ, UK.
{rid,rlbf}@ecs.soton.ac.uk

Abstract. Our long-term goal is to evolve neural controllers which reproduce in behaving robots the kind of phonotaxis behaviour seen in real animals, such as crickets. We have previously studied the evolution of neural circuitry which, when implanted in a Braitenberg type 2b vehicle, promoted phototaxis behaviour in the form of movement towards flashing lights of different frequencies. (It was simpler to study light-driven than acoustic-driven behaviour.) Since this is not truly sequential behaviour, we now describe new work to discriminate between particular mark-space ratio patterns of the same basic (flash or on-off) frequency. The next step will be to integrate the two behaviours so that robot taxis is driven by a signal with temporal structure closer to that of the cricket 'song'.

1   Introduction

In previous work (Damper, French, and Scutt 2000; French and Damper 2002), we have used an autonomous robot to study the links between the 'neurophysiology' of a robot and its external behaviour in the belief that this can inform us about the links between neurophysiology and behaviour in real animals. This dictates that we use spiking neurons in the robot controller. Since real nervous systems are evolved rather than designed, we study the evolution of spiking neuron circuitry for our robot. Evolving spiking neural circuitry remains something of an unusual activity; the only other work we know of is that of Floreano and Mattiussi (2001). Our long-term goal is to evolve neural controllers which reproduce in behaving robots the kind of phonotaxis behaviour seen in real animals, such as crickets. Female crickets move towards the 'song' of a male conspecific or an artificially-produced 'song' provided it shares a similar temporal structure. This behaviour has been extensively modelled in robot studies (e.g., Lund, Webb, and Hallam 1997; Kortmann 1998; Webb and Scutt 2000; Reeve and Webb 2002). Since our robot is not currently equipped with acoustic sensors, we have switched modalities to study the evolution of phototaxis behaviour.

We have previously studied the evolution of neural circuitry which, when implanted in a Braitenberg type 2b vehicle, promoted phototaxis behaviour in the form of movement towards flashing lights of different frequencies (French and

S. Cagnoni et al. (Eds.): EvoWorkshops 2003, LNCS 2611, pp. 616–625, 2003.

Damper 2002). However, these simple signals are quite unlike the more complex cricket song. Hence, in the current work, we study a task specifically designed to promote sequence-discrimination behaviour. The approach is very similar to French and Damper (2002), and the reader is referred to this earlier publication for details of the genetic programming (data structures and algorithms) used to evolve the controllers. Specifically, in this new work, we require the robot to move towards an on-off pattern of flashing light (of a fixed flash frequency) with a given mark-space ratio while ignoring an on-off pattern of another mark-space ratio.

2   Evolving a Mark-Space Discriminator

In our recent work (French and Damper 2002), circuits of model spiking neurons were evolved to detect a sensory (light) signal with a particular flash frequency. A simple extension to this work is to evolve a neural circuit to process a signal that varies in duration—it has a variable mark-space ratio.

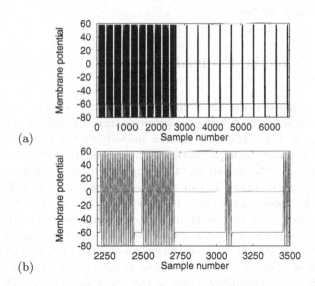

(a)

(b)

Fig. 1. (a) Neuron activity showing two mark-space burst-firing patterns and (b) close-up of the spike train showing individual spikes.

Figure 1 shows the activity of a neuron that alternates between two mark-space patterns. This is taken to represent the output of a sensory neuron that is being depolarised by a stimulus from an environment and forms the input to our evolving neural circuit.

The fitness function for this evolutionary task simply increments a counter by a small amount every time a particular input firing activity coincides with firing of a designated output neuron. The fitness counter is similarly decremented by small amount for an incorrect output response. Because the input neuron's burst trains differ in length (giving rise to two different mark-space patterns), the amount by which the counter is incremented or decremented is weighted to reflect the length of the spike train. For example, when evolving a discriminator for a long spike train the counter is decremented for an incorrect response by an amount greater than that used for an increment. If this safeguard is not used then the discriminator will not filter out the short spike train properly.

A population of 1000 individuals was used running for 500 generations across a MOSIX cluster of nine Athlon-based disk-less PCs. The evolutionary process was repeated 10 times for the short mark-space discrimination task and 10 times for the long mark-space discrimination task. All runs generated circuits that could distinguish between the mark-space patterns.

Figure 2 shows a typical circuit evolved with a fitness function configured to select the long mark-space component of Fig. 1. In the same way, Figure 3 was evolved with a fitness function configured to select the short mark-space component of Fig. 1. Figures 4 and 5 show how the best fitness obtained so far increases during evolution.

3 Analysis of the Evolved Circuits

We will first consider the short mark-space circuit of Fig. 3 because, after so doing, we will be in a better position to understand the workings of the circuit in Fig. 2. The significant element underlying the circuit of Fig. 3 is the habituating excitatory synapse between input and interneuron cells. During the long mark-space firing pattern of the input neuron, the habituating synapse weakens to a point where it can no longer depolarise and fire the output neuron via the interneuron. This is shown in Fig. 3 (b). Critically though, during presentation of the short mark-space firing pattern, the habituating synapse recovers (Rec. = 0.009, a slow recovery rate) to a point where it can depolarise and fire the interneuron. Interestingly, the interneuron has evolved a low threshold of −58 mV, which is close to resting potential. This increases the chances of the interneuron firing during the short spike train (when the degree of temporal summation is low.)

Turning to the long mark-space circuit of Fig. 2, this circuit has a simpler operation. The low value of the synaptic weight between the input and interneuron cells (with respect to that in Fig. 3) means that more spikes from the input neuron are needed to depolarise the interneuron beyond threshold than in the circuit of Fig. 3. Hence, the operation of this circuit depends upon a relatively large degree of temporal summation. Additionally and in contrast with the previous circuit, the interneuron has evolved a high threshold of −43 mV. This requires greater depolarisation of the membrane for the neuron to fire making it less likely to fire during the short spike train.

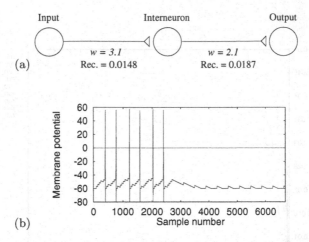

Fig. 2. (a) Nervous system evolved to select the long mark-space input signal (Fig. 1) with synaptic weight and recovery parameters shown and (b) its output. Redundant components have been removed for clarity.

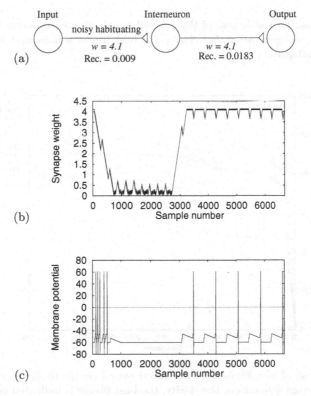

Fig. 3. (a) Nervous system evolved to select the short mark-space input signal (Fig. 1), (b) habituating synapse weight and (c) the circuit's output. Redundant components have been removed for clarity.

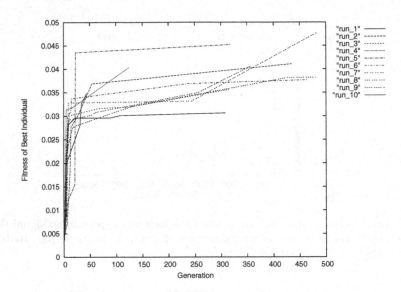

Fig. 4. Increase of best fitness of the evolved neural circuit to detect the long mark-space ratio versus generation. For clarity, the best fitness is indicated only up to point at which it stabilises.

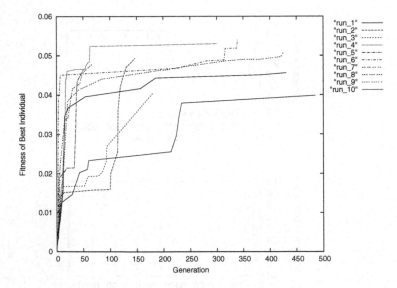

Fig. 5. Increase of best fitness of the evolved neural circuit to detect the short mark-space ratio versus generation. For clarity, the best fitness is indicated only up to point at which it stabilises.

4   Phototaxis in a Braitenberg Vehicle

Now that we have evolved neural circuits 'in isolation' that can discriminate between different mark-space patterns, they were incorporated in a Braitenberg type 2b vehicle (Braitenberg 1984, p. 7). This should result in a vehicle that can seek out the source of a stimulus, e.g., a light source flashing with a particular mark-space ratio, so exhibiting phototaxis.

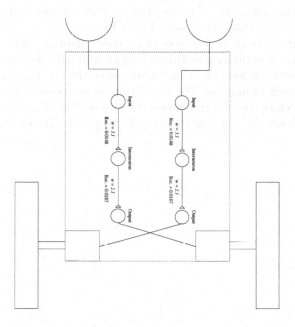

Fig. 6. A Braitenberg type 2b vehicle architecture with spiking neurons that seeks long mark-space signals.

An neural controller describing a type 2b vehicle architecture was created using two of the networks in Fig. 3 to connect contralateral *proximal* sensory-motor pairs (i.e., the sensors respond to changes in light level) as shown in Figure 6. Proximal light sensory neurons were selected to reduce the risk of spurious sensory input. This is because ordinary light-sensor neuron firing is proportional to light intensity.

Minor changes to the evolved nervous system circuitry were made to match the circuit's response to 'live' input from real light sensors. As indicated above, sensory and motor neurons were added in place of input and output neurons. Experiments were carried out to test the expected phototaxis behaviour.

The experimental setup was as follows. A Khepera robot was placed in a test environment facing the mid-point between two 2.5 V filament lamps, approximately 10 cm away. The lamps were positioned 15 cm apart. A driver circuit controlled by software via a PC's parallel printer port switched the lamps (pulsed at 50 Hz via bits D0 and D1) with on/off-times of 1 s/1.02 s and 2 s/0.02 s giving the short and long mark-space patterns, respectively. Software control also allowed toggling of the patterns between the two lamps, i.e., the long mark-space pattern could be presented on the left lamp with the short on the right, or vice versa. This arrangement gave a usable optical signal for the proximal sensory neurons of the Braitenberg vehicle. Although these timings proved adequate for the real-world experiment, however, there are undoubtedly better alternatives that more closely match the simulated sensory-input signals presented during evolution. The nervous system models and lamp controller software ran concurrently on a 80486 Linux PC.

During initial tests it became clear that ambient light influenced the robot's phototaxis performance: Bright background light disrupted the ability to discriminate between long and short mark-space signals. Hence, final runs took place in low levels of lighting. The inability to cope robustly with ambient light is probably because the discrimination circuits of Figs. 2 and 3 were evolved in isolation, without the input noise generated by the Khepera simulator sensor models.

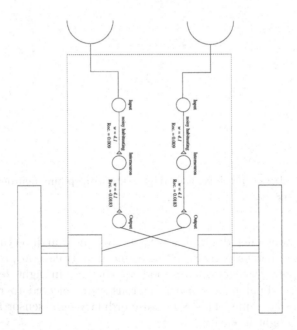

Fig. 7. A Braitenberg type 2b vehicle architecture with spiking neurons that seeks short mark-space signals.

The experiments started with a Braitenberg type 2b vehicle (shown in Figure 7) that favoured the short mark-space signal. Twenty trials were carried out to observe the robot's behaviour with both toggle-settings of the lamp control software. For the first 10 trials the short mark-space signal was on the robot's left; for the second group of 10 trials, the short mark-space signal was moved to the robot's right. This was then repeated with a vehicle that favoured the long mark-space signal (shown in Figure 6).

Fig. 8. Khepera in the test environment used for phototaxis experiments.

Figure 8 shows the vehicle (Khepera) in the test environment. It was noticed that the response of the short mark-space circuit was not as vigorous as its long mark-space counterpart, taking approximately 4 minutes to reach the target lamp. This was entirely due to the slow recovery of the habituating synapses. In contrast, the long mark-space vehicles took approximately 2 minutes to reach the target lamp.

In all trials, the vehicles performed their tasks well. During the short mark-space ratio trials, only three vehicles lost view of the target lamp and halted. This is likely to be an indication that the 'on' time for this target signal is too long. For the long mark-space ratio trials, the vehicles performed without error.

5   Towards the Evolution of Phonotaxis

In cricket phonotaxis, a female insect seeks the calling song of male conspecifics in a noisy environment. This has been extensively studied by Webb and colleagues (Lund, Webb, and Hallam 1997; Webb and Scutt 2000; Reeve and Webb 2002). However, throughout their work, the circuitry underlying the behaviour of their robots has been manually designed. This prompts the question: "Can we evolve neural circuits that allow a robot to perform phonotaxis behaviour?" This is interesting because a neural controller generated through evolutionary computation may operate using principles that had previously not been considered for manual designs.

The work presented here on phototaxis is intended as a stage in the evolution of circuitry to drive cricket phonotaxis. We note that Kortmann (1998) has conducted preliminary work on the evolution of phonotaxis in a robot cricket, but his work merely evolves parameters controlling high-level behaviour (*delay* and *turning time*) in a circuit of predetermined structure rather than, as here, evolving the circuit structure itself. The analysis in Section 3 shows that the operating principles of the evolved spiking-neuron circuits is quite different from that of more conventional 'neural networks' using artificial neurons of the type made popular by (McCulloch and Pitts 1943), based on simple activation functions. By their ability to encode time information in terms of time of firing, and richer endowment of parameters, spiking neuron circuits ought to display enhanced ability to learn and process complex temporal patterns (Bugmann 1997; Koch 1998) such as the cricket song.

6   Conclusions

We have described work to evolve spiking neural circuits that can discriminate between two mark-space patterns. Light-seeking Braitenberg vehicles were constructed that used these circuits as the basis of phototaxis. Although the Braitenberg vehicles were unable to deal with strong ambient light, this is probably because their circuitry was evolved 'in isolation' (for computational convenience) and not onboard a vehicle situated in a real or simulated environment. We intend to extend this work to simulate phonotaxis in the cricket, where the temporal patterning of the relevant signals is considerably more complex. The next stage of this work will focus upon the coupling of evolved frequency discrimination circuitry (French and Damper 2002) and the evolved mark-space discriminators described in this paper.

The evolved controllers in this paper rely on properties of biological neurons which are not modelled in conventional artificial or formal 'neurons' (e.g., recovery rates and/or time of arrival differences between spikes). Hence, as part of future work, we intend to evolve non-spiking neural solutions and to compare them with their spiking counterparts. It also remains to assess the sensitivity of our evolved circuits to small changes in flash frequency and mark-space ratio in their inputs.

Although the circuits evolved in the work so far, and the tasks which they allow the robot to perform, are relatively simple, we envisage that these simple circuits will exploit evolutionary and computational principles which can be combined hierarchically to produce more complex circuits and behaviours. This is much in the flavour of the concept of "a cell biological alphabet" for learning and learned behaviour espoused by Hawkins and Kandel (1984). If so, this would offer at least the prospect of moving from simple small-scale behaviours in insect-like animats to more interesting large-scale behaviours.

References

[Braitenberg (1984)] Braitenberg, V. (1984). *Vehicles: Experiments in Synthetic Psychology*. Cambridge, MA: MIT Press.

[Bugmann (1997)] Bugmann, G. (1997). Biologically plausible neural computation. *BioSystems* 40(1–2), 11–19.

[Damper, French, and Scutt Damper (2000)], R. I., R. L. B. French, and T. W. Scutt (2000). ARBIB: An autonomous robot based on inspirations from biology. *Robotics and Autonomous Systems* 31(4), 247–274.

[Floreano and Mattiussi (2001)] Floreano, D. and C. Mattiussi (2001). Evolution of spiking neural controllers for autonomous vision-based robots. In T. Gomi (Ed.), *Evolutionary Robotics IV*, pp. 38–61. Berlin, Germany: Springer-Verlag.

[French and Damper (2002)] French, R. L. B. and R. I. Damper (2002). Evolution of a circuit of spiking neurons for phototaxis in a Braitenberg vehicle. In *Proceedings of Simulation of Adaptive Behavior 2002 – From Animals to Animats 7*, Edinburgh, UK, pp. 335–344.

[Hawkins and Kandel (1984)] Hawkins, R. D. and E. R. Kandel (1984). Is there a cell biological alphabet for simple forms of learning? *Psychological Review* 91(3), 375–391.

[Koch (1998)] Koch, C. (1998). *Biophysics of Computation: Information Processing is Single Neurons*. Cary, NC: Oxford University Press USA.

[Kortmann (1998)] Kortmann, L. J. (1998). Evolving phonotoaxis in a robot cricket – an investigation in bio-robotics. Master's thesis, University of Groningen, The Netherlands and University of Edinburgh, Scotland.

[Lund, Webb, and Hallam (1997)] Lund, H. H., B. Webb, and J. Hallam (1997). A robot attracted to the cricket species *Gryllus bimaculatus*. In P. Husbands and I. Harvey (Eds.), *Proceedings of Fourth European Conference on Artificial Life, ECAL'97, Brighton, UK*, pp. 246–255. Cambridge, MA: MIT Press/Bradford Books.

[McCulloch and Pitts (1943)] McCulloch, W. S. and W. Pitts (1943). A logical calculus of the ideas immanent in nervous activity. *Bulletin of Mathematical Biophysics* 5, 115–133.

[Reeve and Webb (2002)] Reeve, R. and B. Webb (2002). New neural circuits for robot phonotaxis. In *Proceedings of EPSRC/BBSRC International Workshop on Biologically-Inspired Robotics: The Legacy of W. Grey Walter*, Bristol, UK, pp. 225–232.

[Webb and Scutt (2000)] Webb, B. and T. Scutt (2000). A simple latency-dependent spiking-neuron model of cricket phonotaxis. *Biological Cybernetics* 82(3), 247–269.

Evolving Neural Networks for the Control of a Lenticular Blimp

Stéphane Doncieux and Jean-Arcady Meyer

AnimatLab - LIP6, France
http://animatlab.lip6.fr
{Stephane.Doncieux,Jean-Arcady.Meyer}@lip6.fr

Abstract. We used evolution to shape a neural controller for keeping a blimp at a given altitude, and as horizontal as possible, despite disturbing winds. The blimp has a lenticular shape whose aerodynamic properties make it quite different from a classical cigar-shaped airship. Evolution has exploited these features to generate a neural network that proved to be more efficient than a hand-designed PID-based controller that independently controlled the blimp's three degrees of freedom.

Keywords: neural networks, evolution, lenticular blimp

1 Introduction

Over about the past ten years, attempts at evolving neural controllers for robots have proliferated (see [4] for a review), and this approach is currently the most popular in evolutionary robotics ([6,5]). However, it mostly involves crawling, rolling or walking robots, i.e., robots that move on the ground, and much more rarely swimming or flying robots, probably because such robots are still uncommon in academic laboratories. This situation may well evolve quickly, at least as far as aerial robots are concerned, if only because of the growing military and civilian needs for machines as small and as energetically economical as possible - a set of qualities that perfectly fits academic constraints. Moreover, the control of robots moving in a 3D-environment, whose complex dynamics are likely to be affected by wind or other disturbances, and whose sensory-motor equipment may well be limited by the size and energy constraints just alluded to, raises new and interesting challenges that will certainly trigger a number of future research efforts.

Be that as it may, previous attempts at evolving neural controllers for flying robots have been made by Doncieux [1,2] and by Zufferey et al. [7]. The former work produced neural networks able to combat the effects of wind and to maintain either a constant flying speed and direction in a simulated blimp, or a constant position with respect to the ground in a simulated helicopter. The latter one produced neural controllers that moved a real blimp around a room and used visual information to detect collisions with walls.

In the experiments reported here, we used an evolutionary algorithm to design a neural network that controls roll, pitch and altitude together in a simulated

S. Cagnoni et al. (Eds.): EvoWorkshops 2003, LNCS 2611, pp. 626–637, 2003.

Fig. 1. A 10-meters wide lenticular blimp.

lenticular blimp. Although the ultimate goal of this research is to design an entirely automatic pilot, these experiments mostly contribute to an intermediate objective, i.e., to design a system that will help a human to pilot a real blimp 10-meters wide (figure 1). Because this engine has up to ten effectors, direct control of each motor would be too difficult for a human, and the evolved neural controller will help him in this task. The pilot will just need to set the desired altitude and to control the horizontal position of the aircraft, thus reducing the number of commands from ten to only three.

The article first describes the simulation model and the evolutionary approach that were used. Then, it describes the results that were obtained and provides details about the inner workings of an efficient controller. Directions for future work are also indicated.

2   Simulation Model

A lenticular shape affords several advantages over the traditional cigar-shaped configuration that characterizes most blimps. In particular it renders the aircraft less prone to wind perturbations, thus allowing it, for instance, to be parked outside, directly tied to the ground, whereas cigar-shaped blimps, that need to be linked to the ground by some cable, cannot withstand high winds outside a hangar. However, the dynamic behavior of a lenticular blimp is much more complicated than that of a cigar-shaped blimp, thus providing a richer set of interactions between the robot and its environment, on the one hand, but en-

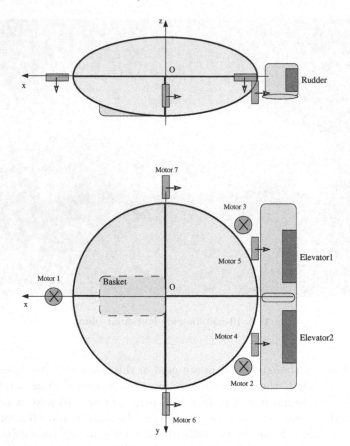

Fig. 2. Side and above views of the blimp's model. In the experiments reported here, only motors 1 to 3 were used.

tailing greater efforts to devise realistic simulation models of such behavior, on the other.

The actual blimp that this research is centered around is equipped with two inclinometers - that are sensitive respectively to roll and pitch -, with an altimeter, a GPS, an anemometer, and a video camera. The blimp is also equipped with seven motors and three control surfaces (figure 2). Motors 1 to 3 are used to control pitch, roll and altitude. Motors 4 to 7 are used for propulsion, but motors 6 and 7 may also control the yaw of the aircraft if their thrusts are set antagonistically. Control surfaces serve the same purpose as motors, although they waste less energy, but they are not usable in all circumstances. In the work to be described here, we did not use them and concentrated on the control of altitude, pitch and roll using only the inclinometers and the altimeter, on the one hand, and motors 1 to 3, on the other.

The general equation that rules the dynamics of the model is:

$$M_t \Gamma = M_t g + P + A_{OXYZ} + F_{OXYZ}$$

A is the aerodynamic force, F is the resultant of command forces, and P is Archimede's thrust. To simplify, we suppose that the aircraft is balanced, i.e., that the Archimede's thrust exactly compensates gravity. The resultant of aerodynamical forces acting on the blimp is computed according to equation:

$$A_{Gxyz} = \begin{bmatrix} T_0 \cos i . \cos j + P_0 . \sin i . \cos j \\ T_0 . \cos i . \sin j + P_0 . \sin i . \sin j \\ -T_0 . \sin i + P_0 . \cos i \end{bmatrix} \tag{1}$$

where T_0 is the drag and P_0 the lift, computed as follows:

$$T_0 = -\frac{1}{2} . \rho . S . V_{rel}^2 . C_x \tag{2}$$

$$P_0 = \frac{1}{2} . \rho . S . V_{rel}^2 . C_{z_i} . \sin(2.i) \tag{3}$$

i is the incidence and j the sideslip of the blimp, computed as follows:

$$i = \arcsin\left(-\frac{Vr_z}{V_{rel}}\right) \tag{4}$$

$$j = \arcsin\left(\frac{1}{\cos i} \frac{Vr_y}{V_{rel}}\right) \tag{5}$$

V_{rel} is the speed of the blimp relative to the surrounding air. Vr_x, Vr_y and Vr_z are its coordinates in the absolute reference. $\rho = 1.2255 kg/m^3$, S is the reference surface: $S = V^{\frac{2}{3}}$, V being the volume of the blimp.

Rotation speed and angles are computed as follows:

$$\psi' = \frac{(q . \sin \varphi + r . \cos \varphi)}{\cos \theta} \tag{6}$$

$$\theta' = q . \cos \varphi - r . \sin \varphi \tag{7}$$

$$\varphi' = p + \psi' . \sin \theta \tag{8}$$

$$p' = \frac{1}{I_{x1}} . (L - (I_z - I_y) . r . q) \tag{9}$$

$$q' = \frac{1}{I_{y1}} . (M - (I_x - I_z) . p . r) \tag{10}$$

$$r' = \frac{1}{I_z} . (N - (I_y - I_x) . p . r) \tag{11}$$

$\Omega = \begin{bmatrix} p \\ q \\ r \end{bmatrix}$ is the rotation vector around the center of gravity. θ is the pitch, φ the roll and ψ the yaw.

$$I = \begin{bmatrix} Ix & 0 & 0 \\ 0 & Iy & 0 \\ 0 & 0 & Iz \end{bmatrix}$$ is the matrix of inertia moments. It takes into account the respective contributions of the motors, the basket and the envelope.

The complete model is too complex to be described here in detail. Suffice it to say that the corresponding code is more than 3000-instructions long, that the mass of the air and the ground effect are taken into account, and that the wind is modelled via a Drydden spectrum. Furthermore, Mt is set to $115 kg$ and V to $141 m^3$, and the equations are integrated using a fourth order Runge-Kutta method.

Numerous simulations have already been performed that validated the qualitative behavior of the airship. Preliminary experiments with the real blimp, which aimed at assessing the correspondence between its behavior and that of the simulated model, have also been carried out and yield encouraging results.

3   Evolutionary Procedure

The procedure that serves to evolve neural networks able to control the blimp calls upon an indirect encoding scheme that favors the discovery of symmetries and the reuse of useful modules. Only a simplified version of this general scheme, which will be described elsewhere, has been used here.

3.1   The Chromosome

A chromosome encodes a list of modules, a list of links, and a list of template-weight values. In this simplified version, a module is composed of just a single connection between an input neuron and an output neuron, and links between modules fuse the output neuron of a module with the input neuron of another module. Other links serve to connect the input neuron of a module with a sensory neuron whose activity level equals the error currently detected by the blimp's corresponding sensor. Likewise, other links serve to connect the output neuron of a module with a motor neuron whose activity level, which varies between +1 and -1, modulates the maximum thrust that the blimp's corresponding motor exerts. By convention, these two categories of links are given connexion weights of +1. Figure 3 illustrates how a chromosome is decoded into a neural network that may exhibit symmetries and modular redundancies.

During the course of evolution, several mutation operators make it possible to create or suppress some template-weights, links or modules. A single crossover operator is used to exchange modules between chromosomes.

3.2   Experimental Set-up

The task to be performed was both to maintain the blimp at a target altitude that changed over time, and to keep it as horizontal as possible, despite wind

List of weight values

   1: w=−3.1   4: w=−6.0

   2: w=4.2   5: w=10.0

   3: w=1.6

List of modules

m1	m2	m3	m4	m5	m6	m7
3	2	1	1	5	2	2

List of links between modules,
inputs and outputs

   m1->m3 m1->m2
   m2->o0 m6->o2
   m3->o1 m7->o1
   i1->m4 m4->m6
   m4->m7 i0->m1

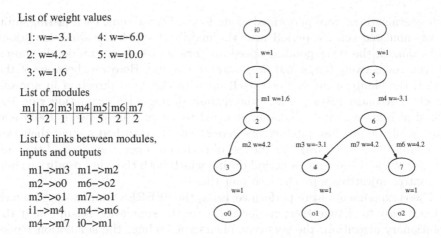

Fig. 3. The chromosome on the left of the figure codes for the neural network on the right, in which i_n are sensory neurons, and o_m are motor neurons. The chromosome includes three lists. The list of modules and the list of template-weights specify the synaptic weight associated with the inner connection of each module. The list of links specifies how a given module is connected to a sensory neuron, a motor neuron or to another module. Note that some modules or some weights may be defined in the corresponding lists, but without being actually used in the final network.

disturbance. Therefore, the fitness function we used takes into account the error on the three degrees of freedom we wanted to control, together with the energy consumption, in order to encourage low cost solutions. Its expression is:

$$f(t) = \frac{\sum_T (1 - \frac{|\delta\theta(t)|}{\delta\theta_{max}}) + \sum_T (1 - \frac{|\delta\varphi(t)|}{\delta\varphi_{max}}) + \sum_T (1 - \frac{|\delta z(t)|}{\delta z_{max}}) + \sum_T (1 - \frac{E(t)}{E_{max}})}{4 \times T}$$

in which $\delta\theta(t)$ and $\delta\varphi(t)$ are the pitch and roll errors with respect to 0, the desired values for these variables, and $\delta z(t)$ is the difference between the real altitude and the current target value. $E(t)$ is the total energy consumed during time step t. E_{max} is the maximum energy motors can consume during one time-step.

The neural networks that were generated could include three sensory neurons whose activity levels were respectively equal to $\delta\theta(t)$, $\delta\varphi(t)$ and $\delta z(t)$, and three motor neurons whose activity levels determined the command forces. In other words, the neural controllers could use the information provided by the blimp's inclinometers and altimeter to set the thrust that each of the three vertical motors should exert.

Using a classical GA algorithm, we ran experiments lasting 1000 generations with a population of 100 individuals to generate efficient controllers. The maximum number of modules that could be included in a given controller was set to 20. The mutation and crossover rates were empirically chosen and did not pretend to optimality. Template-weights were coded on 8 bits and varied between -10 and 10. Each neuron's transfert function was a simple *tanh* function.

Fitness evaluations were performed along four different runs, each lasting 4min 10s of simulated time, a period near the middle of which the wind conditions were changed, the corresponding speed and orientation being randomly chosen between respectively $0m/s$ and $6m/s$ and 0 and 360. However, because of the rudder, the blimp passively lined itself up with the wind direction in approximately 2 seconds. Likewise, at some instant during each evaluation run, the desired altitude was also randomly changed to some value between $100m$ and $200m$, while the roll and pitch errors were abruptly perturbed around their target 0 values. In other words, the fitness of each controller was assessed through 1000 sec of simulated time, a period during which both the wind conditions and the control objectives were changed four times.

These experiments were performed using the SFERES framework [3] which makes it easy to change every major option in the simulation set-up, be it the evolutionary algorithm, the genotype-phenotype coding, the simulation model or the fitness evaluation procedure.

4 Results

4.1 Observed Behavior

Figure 4 shows the behavior generated by the best controller evolved in this manner. It thus appears that, following each imposed disturbance, the altitude (z) and roll (φ) are effectively kept at desired values, whereas the pitch (θ) temporarily deviates from the desired null value, although it eventually does return to it. This behavior exploits the dynamic properties of the blimp, whose lenticular shape makes it behave like a wing in the wind. Thus, when the blimp has to go up or down, the pitch angle must be negative in the former case and positive in the latter[1]. This makes it that, in order to improve fitness, the horizontal trim of the blimp must not be maintained during the whole experiment. Similar behavior has been observed with every controller getting a high fitness rating.

4.2 Comparison with a Hand-Designed Controller

To evaluate the efficiency of this strategy, we hand-designed a reference controller whose behavior was compared to that of the evolved network. This reference controller called upon three PID modules that separately managed each degree of freedom. The input of each such PID module was the error of the degree of freedom it was supposed to keep, while its output was sent to an interface that set the three motor thrusts. Thus, the altitude error generated identical thrusts at the level of each motor, the roll error generated thrusts of opposite signs for motors 2 and 3, and the pitch error generated thrusts of opposite signs for motor 1, on the one hand, and for motors 2 and 3, on the other hand. The three inner parameters of each PID module were also optimized with an evolutionary algorithm.

---

[1] it is traditional in aeronautics to count as positive the angle of an aircraft steering towards the ground.

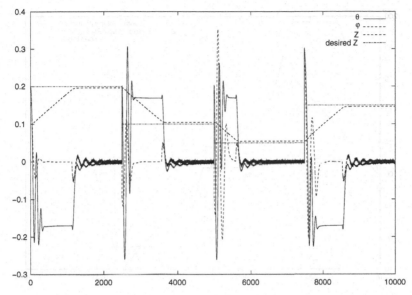

Fig. 4. Typical behavior generated by the best selected neural controller. θ (pitch angle) and φ (roll angle) must be kept at 0 (corresponding to a horizontal trim), while the altitude must be adjusted to a target value. Wind conditions, altitude target, and pitch and roll values are randomly changed every 2500 time-steps. It should be noted that, when the blimp has to go up or down in order to reach the new target value, the pitch is not kept at zero, but around a non-null value (approximately 0.17 radians, i.e. 10 degrees) whose sign depends upon the vertical direction the blimp has to follow. A time-step corresponds to 25ms, angles are in radians and altitude in meters (altitudes are divided by 1000).

A thorough comparison of the behaviors respectively generated by a hand-designed controller and an evolved neural network reveals that the former, which does not call upon any dynamic specificity of the lenticular shape, always maintains the horizontal trim of the blimp, a behavior fundamentally different from that of the evolved neural controller. However, the neural controller is about three times faster than the hand-designed one as far as the control of altitude is concerned (figure 5) because it is able to exploit the dynamic couplings between the blimp's degree of freedom, as explained in the next subsection.

4.3   Experimental Study of the Neurocontroller

Figure 6 shows the evolved neurocontroller that generated the experimental results presented above. It consists of three sensory neurons, three motor neurons and eight hidden neurons, encapsulated into nine modules. Hidden neurons 2 and 3 exhibit a recurrent connection.

As a consequence of this structure, altitude control calls upon the three motors, as might have been expected. Pitch and roll control exploits only one

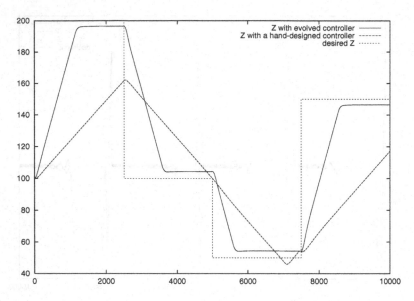

Fig. 5. Comparison of the behaviors generated by an evolved neural network and by a hand-designed system with respect to the control of altitude. The evolved controller is three times faster than the hand-designed one. A time step corresponds to 25ms, altitudes are given in meters.

motor, a solution that would probably not be chosen by a human designer, but which results in nearly the same effect on the blimp control.

In particular, it appears that, when there is no error in altitude, neuron 2 oscillates at a constant frequency, with a constant amplitude. At the level of neuron 1, these oscillations are modulated in amplitude by the activity of neuron 7 (figure 7). On average, these oscillations force the blimp to revert to a null pitch.

When there is an error in altitude, the activity value of neuron 2 always remains saturated, either at +1 - when the blimp is above the target altitude - or at -1 - when the blimp is below the desired altitude. This constant value acts as a bias on neuron 1, thus shifting the equilibirum point of pitch control from zero to a non null value whose sign depends on the sign of the error on altitude. Motor 1 accordingly inclines the blimp in order to generate the force that will reduce the altitude error, as previously explained (figure 7). Similar dynamic couplings explain how roll error is controlled.

Concerning the oscillations, it can be shown that, when a pitch or roll error is detected, because of the influence of the recurrent connection between neurons 2 and 3, a periodic activity is generated at the level of neuron 0, and that the corresponding period (200 ms) is precisely the one that best exploits the time constants of the blimp's dynamic behavior. In other words, evolution discovered a solution that capitalizes on the complex sensory-motor coupling that a lentic-

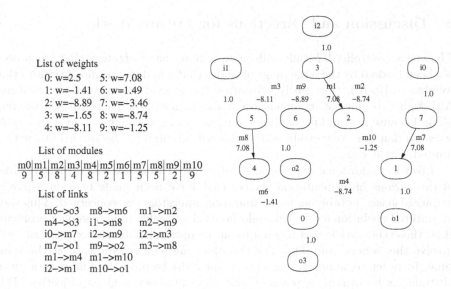

List of weights

0: w=2.5   5: w=7.08
1: w=-1.41   6: w=1.49
2: w=-8.89   7: w=-3.46
3: w=-1.65   8: w=-8.74
4: w=-8.11   9: w=-1.25

List of modules

m0	m1	m2	m3	m4	m5	m6	m7	m8	m9	m10
9	5	8	4	8	2	1	5	5	2	9

List of links

m6->o3   m8->m6   m1->m2
m4->o3   i1->m8   m2->m9
i0->m7   i2->m9   i2->m3
m7->o1   m9->o2   m3->m8
m1->m4   m1->m10
i2->m1   m10->o1

Fig. 6. Structure of the best neural network obtained (right) together with the corresponding chromosome (left). i_0, i_1 and i_2 are sensory neurons that measure pitch, roll and altitude errors. o_1, o_2, and o_3 are motor neurons that are used to set the blimp's horizontal trim and altitude.

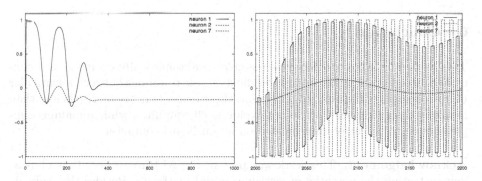

Fig. 7. Outputs of neurons 1, 2 and 7. Left: when the current altitude does not match the target value, the activity level of neuron 2 is saturated (here, it equals -1), thus acting as a bias on neuron 1. The activity level of this neuron, in turn, determines how much, and in which direction, the blimp will be inclined to help getting back to the target altitude. Right: when the altitude is correct, neuron 2 oscillates between its saturation values. Its activity, as well as that of neuron 1, is modulated by the output variations of neuron 7 to keep the blimp horizontal.

ular shape affords, and it ultimately converged towards a periodic signal that minimizes pitch and roll oscillations at the same time.

5   Discussion and Directions for Future Work

The above controller basically calls upon a 'bang-bang' strategy that has probably been favored by the large range of values that could be assigned to connection weights (+10,-10), thus easily saturating the activation levels of the neurons. Although it is efficient in simulation, it should not be implemented on the real blimp because it would probably be too demanding at the motor level and liable to cause damage. New evolutionary runs will accordingly be carried out with a limited weight range.

Likewise, additional experiments are required to assess the respective roles of the various implementation choices that have been made here on a purely empirical basis, notably as far as mutation operators are concerned. This way, an optimal evolutionary set-up could be used to tackle more complex problems than those this article was focused upon. In particular, future experiments will involve the camera, and all available motors and control surfaces, at the same time, in order to minimize the energy spent in keeping the blimp at a given altitude, as horizontal as possible, and directly above a given objective. This will entail using a higher number of sensors and actuators than currently done in evolutionary robotics, in order to control up to five degree of freedom, and will in particular help in assessing the scalability of our approach.

Naturally, our ultimate goal is to implement the resulting controllers on the real blimp and to demonstrate their effectiveness.

6   Conclusion

This work demonstrates that it is possible to automatically evolve neural controllers able to exploit the highly specific dynamic couplings between a lenticular blimp's degrees of freedom. In particular, these controllers are more efficient than humanly designed ones that do not exploit such couplings, while remaining simple enough to be easily implemented on an on-board computer.

Acknowledgement. This work was part of the ALPhA project that aims at demonstrating the potential of lenticular aerial platforms. Besides the Animat-Lab, it brings together Pierre Balaskovic - who designed the blimp blueprints and leads the project -, Sam Tardieu, Gérard Mouret and Bertrand Dupouy from the École Nationale Supérieure des Télécommunications (ENST) - who designed the embedded electronics and the interface software - and AirStar Society - where the aircraft was built.

References

1. S. Doncieux. Évolution d'architectures de contrôle pour robots volants. In Drogoul and Meyer, editors, *Intelligence Artificielle Située*, pages 109–127, Paris, october 1999. Hermes.

2. S. Doncieux. Évolution d'architectures de contrôle pour robots volants. Master's thesis, LIP6 - ENSEA, 1999.
3. S. Landau, S. Doncieux, A. Drogoul, and J.-A. Meyer. Sferes: un framework pour la conception de systèmes multi-agents adaptatifs. *Technique et Science Informatique*, 21(4), 2002.
4. J.-A. Meyer. Evolutionary approaches to neural control in mobile robots. In *Proceedings of the IEEE International Conference on Systems, Man and Cybernetics*, San Diego, 1998.
5. J.A. Meyer, S. Doncieux, D. Filliat, and A. Guillot. Evolutionary approaches to neural control of rolling, walking, swimming and flying animats or robots. In Duro, Santos, and Graña, editors, *Biologically Inspired Robot Behavior Engineering*. Springer Verlag, 2002.
6. S. Nolfi and D. Floreano. *Evolutionary Robotics: Biology, Intelligence and Technology of Self-organizing Machines*. The MIT Press, 2001.
7. J.C. Zufferey, D. Floreano, M. van Leeuwen, and T. Merenda. Evolving vision-based flying robots. In Bülthoff, Lee, Poggio, and Wallraven, editors, *Proceedings of the 2nd International Workshop on Biologically Motivated Computer Vision LNCS*, 2002.

Evolving Symbolic Controllers

Nicolas Godzik[1], Marc Schoenauer[1], and Michèle Sebag[2]

[1] Projet Fractales, INRIA Rocquencourt, France
[2] LRI, Université Paris-Sud, France

Abstract. The idea of *symbolic controllers* tries to bridge the gap between the top-down manual design of the controller architecture, as advocated in Brooks' subsumption architecture, and the bottom-up designer-free approach that is now standard within the Evolutionary Robotics community. The designer provides a set of elementary behavior, and evolution is given the goal of assembling them to solve complex tasks. Two experiments are presented, demonstrating the efficiency and showing the recursiveness of this approach. In particular, the sensitivity with respect to the proposed elementary behaviors, and the robustness w.r.t. generalization of the resulting controllers are studied in detail.

1 Introduction

There are two main trends in autonomous robotics. There are two main trends in autonomous robotics. The first one, advocated by R. Brooks [2], is a human-specified deterministic approach: the tasks of the robot are manually decomposed into a hierarchy of independent sub-tasks, resulting in the the so-called *subsumption architecture*.

On the other hand, evolutionary robotics (see e.g. [13]), is generally viewed as a pure black-box approach: some controllers, mapping the sensors to the actuators, are optimized using the Darwinian paradigm of Evolutionary Computation; the programmer only designs the fitness function.

However, the scaling issue remains critical for both approaches, though for different reasons. The efficiency of the human-designed approach is limited by the human factor: it is very difficult to decompose complex tasks into the subsumption architecture. On the other hand, the answer of evolutionary robotics to the complexity challenge is very often to come up with an ad hoc (sequence of) specific fitness function(s). The difficulty is transferred from the internal architecture design to some external action on the environment. Moreover, the black-box approach makes it extremely difficult to understand the results, be they successes or failures, hence forbidding any capitalization of past expertise for further re-use. This issue is discussed in more details in section 2.

The approach proposed in this work tries to find some compromise between the two extremes mentioned above. It is based on the following remarks: First, scaling is one of the most critical issues in autonomous robotics. Hence, the *same mechanism* should be used all along the complexity path, leading from primitive tasks to simple tasks, and from simple tasks to more complex behaviors.

S. Cagnoni et al. (Eds.): EvoWorkshops 2003, LNCS 2611, pp. 638–650, 2003.

One reason for the lack of intelligibility is that the language of the controllers consists in low-level orders to the robot actuators (e.g. speeds of the right and left motors). Using instead hand-coded basic behaviors (e.g. forward, turn left or turn right) as proposed in section 3 should allow one to better understand the relationship between the controller outputs and the resulting behavior. Moreover, the same approach will allow to recursively build higher level behaviors from those evolved simple behaviors, thus solving more and more complex problems.

Reported experiments tackle both aspects of the above approach. After describing the experimental setup in section 4, simple behaviors are evolved based on basic primitives (section 5). Then a more complex task is solved using the results of the first step (section 6). Those results are validated by comparison to the pure black-box evolution, and important issues like the sensitivity w.r.t. the available behaviors, and the robustness w.r.t generalization are discussed. The paper ends by revisiting the discussion in the light of those results.

2   State of the Art

The trends in autonomous robotics can be discussed in the light of the "innate vs acquired" cognitive aspects – though at the time scale of evolution. From that point of view, Brooks'subsumption architecture is extreme on the "innate" side: The robots are given all necessary skills by their designer, from basic behaviors to the way to combine them. Complex behaviors so build on some "instinctive" predefined simple behaviors. Possible choices lie in a very constrained space, guaranteeing good performances for very specific tasks, but that does not scale up very well: Brooks'initial goal was to reach the intelligence of insects [3].

Along similar lines are several computational models of action-selection (e.g. Spreading Activation Networks [10], reinforcement learning [8,7], ...). Such approaches have two main weaknesses. The first one is that such architecture is biologically questionable – but do we really care here? The second weakness is concerned with the autonomy issue. Indeed, replacing low level reflexes by decisions about high level behaviors (use this or that behavior now) might be beneficial. However, how to program the *interactions* of such reflexes in an open world amounts to solve the *exploration vs exploitation* dilemma – and both Game Theory and Evolutionary Computation have underlined the difficulty of answering this question.

At the other extreme of the innate/acquired spectrum is the evolutionary robotics credo: any a priori bias from the designer can be harmful. Such position is also defended by Bentley [1] in the domain of optimal design, where it has been reinforced by some very unexpected excellent solutions that arose from evolutionary design processes. In the Evolutionary Robotics area, this idea has been sustained by the recent revisit by Tuci, Harvey and Quinn [16] of an experiment initially proposed by Yamauchi and Beer [18]: depending on some random variable, the robot should behave differently (i.e. go toward the light, or away from it). The robot must hence learn from the first epoch the state of that random variable, and act accordingly in the following epochs.

The controller architecture is designed manually in the original experience, whereas evolution has complete freedom in its recent remake [16]. Moreover, Tuci et al. use no explicit reinforcement. Nevertheless, the results obtained by this recent approach are much better than the original results - and the authors claim that the reason for that lies in their complete black-box approach.

However, whereas the designers did decide to use a specifically designed modular architecture in the first experiment, the second experience required a careful design of the fitness function (for instance, though the reward lies under the light only half of the time, going toward the light has to be rewarded more than fleeing away to break the symmetry). So be it at the "innate" or "acquired" level, human intervention is required, and must act at some very high level of subtlety.

Going beyond this virtual "innate/acquired" debate, an intermediate issue would be to be able to evolve complex controllers that could benefit from human knowledge but that would not require high level of intervention with respect to the complexity of the target task.

Such an approach is what is proposed here: the designer is supposed to help the evolution of complex controllers by simply seeding the process with some simple behaviors – hand-coded or evolved – letting evolution arrange those building blocks together. An important side-effect is that the designer will hopefully be able to better *understand* the results of an experiment, because of the greater intelligibility of the controllers. It then becomes easier to manually optimize the experimental protocol, e.g. to gradually refine the fitness in order to solve some very complex problems.

3   Symbolic Controllers

3.1   Rationale

The proposed approach pertains to Evolutionary Robotics [13]. Its originality lies in the representation space of the controllers, i.e. the search space of the evolutionary process. One of the main goals is that the results will be intelligible enough to allow an easy interpretation of the results, thus easing the whole design process.

A frequent approach in Evolutionary Robotics is to use Neural Networks as controllers (feedforward or recurrent, discrete or continuous). The inputs of the controllers are the sensors (Infra-red, camera, ...), plus eventually some reward "sensor", either direct [18] or indirect [16].

The outputs of the controllers are the actuators of the robot (e.g. speeds of the left and right motors for a Khepera robot). The resulting controller is hence comparable to a program in machine language, thus difficult to interpret. To overcome this difficulty, we propose to use higher level outputs, namely involving four possible actions: *Forward, Right, Left* and *Backward*. In order to allow some flexibility, each one of these symbolic actions should be tunable by some continuous parameter (e.g. speed of forward displacement, or turning angle for left and right actions).

The proposed *symbolic controller* has eight outputs with values in $[0,1]$: the first four outputs are used to specify which action will be executed, namely action i, with $i = Argmax(output(j), j = 1..4)$. Output $i+4$ then gives the associated parameter. From the given action and the associated parameter, the values of the commands for the actuators are computed by some simple hard-coded program.

3.2 Discussion

Using some high level representation language for the controller impacts on both the size of the search space, and the possible modularity of the controller.

At first sight, it seems that the size of the search space is increased, as a symbolic controller has more outputs than a classical controller. However, at the end of the day, the high level actions are folded into the two motor commands. On the other hand, using a symbolic controller can be viewed as adding some constraints on the search space, hence reducing the size of the part of the search space actually explored. The argument here is similar to the one used in statistical learning [17], where rewriting the learning problem into a very high dimensional space actually makes it simpler. Moreover, the fitness landscape of the space that is searched by a symbolic controller has many neutral plateaus, as only the highest value of the first outputs is used – and neutrality can be beneficial to escape local optima [6].

On the other hand, the high level primitives of symbolic controllers make room for modularity. And according to Dawkins [4], the probability to build a working complex system by a randomized process increases with the degree of modularity.It should be noted that this principle is already used in Evolutionary Robotics, for instance to control the robot gripper : the outputs of the controllers used in [13] are high-level actions (e.g. *GRAB, RAISE GRIPPER, ...*), and not the commands of the gripper motors.

Finally, there are some similarities between the symbolic controller approach and reinforcement learning. Standard reinforcement learning [15,8] aims at finding an optimal policy. This requires an intensive exploration of the search space. In contrast, evolutionary approaches sacrifices optimality toward satisfactory timely satisfying solutions. More recent developments [11], closer to our approach, handle continuous state/action spaces, but rely on the specification of some relevant initial policy involving manually designed "reflexes".

4   Experimental Setup

Initial experiments have been performed using the Khepera simulator EOBot, that was developed by the first author from the EvoRobot software provided by S. Nolfi and D. Floreano [13]. EvoRobot was ported on Linux platform using OpenGL graphical library, and interfaced with the EO library [9]. It is hence now possible to use all features of EO in the context of Evolutionary Robotics, e.g. other selection and replacement procedures, multi-objective optimization, and even other paradigms like Evolution Strategies and Genetic Programming. However, all experiments presented here use as controllers Neural Networks (NNs) with fixed topology. The genotype is hence the vector of the (real-valued) weights of the NN. Those weights evolve in $[-1, 1]$ (unless otherwise mentioned), using a $(30, 150)$-Evolution Strategy with intermediate crossover and self-adaptive Gaussian mutation [14]: Each one of 30 parents gives birth to 5 offspring, and the best 30 of the 150 offspring become the parents for next generation. All experiments run for 250 generations, requiring about 1h to 3h depending on the experiment.

All results shown in the following are statistics based on at least 10 independent runs. One fitness evaluation is made of 10 *epochs*, and each epoch lasts from 150 to 1000 time steps (depending on the experiment), starting from a randomly chosen initial position.

5 Learning a Simple Behavior: Obstacle Avoidance

5.1 Description

The symbolic controller (SC) with 8 outputs, described in section 3.1, is compared to the classical controller (CC) the outputs of which are the speeds of both motors. Both controllers have 8 inputs, namely the IR sensors of the Khepera robot in active mode (i.e. detecting the obstacles).

Only Multi-Layer Perceptrons (MLP) were considered for this simple task. Some preliminary experiments with only two layers (i.e. without hidden neurons) demonstrated better results for the Symbolic Controllers than for the Classical Controller. Increasing the number of hidden neurons increased the performance of both types of controllers. Finally, to make the comparison "fair", the following architectures were used: 14 hidden neurons for the SC, and 20 hidden neurons for the CC, resulting in roughly the same number of weights (216 vs 221).

The fitness function[12], is defined as $\sum_{epoch} \sum_t |V(t)|(1 - \sqrt{\delta V(t)})$, where $V(t)$ is the average speed of the robot at time t, and $\delta V(t)$ the absolute value of the difference between the speeds of the left and right motors. The difference with the original fitness function is the lack of IR sensor values in the fitness: the obstacle avoidance behavior is here implicit, as an epoch immediately ends whenever the robot hits an obstacle. The arena is similar to that in [12].

5.2 Results

The first results for the SC were surprising: half of the runs, even without hidden neurons, find a loophole in the fitness function: due to the absence of inertia in the simulator, an optimal behavior is obtained by a rapid succession of *FORWARD - BACKWARD* movements at maximum speed - obviously avoiding all obstacles!

A degraded SC that has no *BACKWARD* action cannot take advantage of this bug. Interestingly, classical controllers only discover this trick when provided with more than 20 hidden neurons **and** if the weights are searched in a larger interval (e.g. $[-10, 10]$).

Nevertheless, in order to definitely avoid this loophole, the fitness is modified in such a way that it increases only when the robot moves forward (sum of both motor speeds positive)[1].

This modification does not alter the ranking of the controllers: the Symbolic Controller still outperforms the Classical Controller. This advantage somehow vanishes when more hidden neurons are added (see Table 1), but the results of the SC exhibit a much smaller variance.

Table 1. Averages and standard deviations for 10 independent runs for the obstacle avoidance experiment. * this experiment was performed in a more constrained environment.

Architecture	CC	CS
8-2 / 8-6	861 ± 105	1030 ± 43
8-8-2 / 8-8-6	1042 ± 100	1094 ± 55
8-20-2 / 8-14-6	1132 ± 55	1197 ± 16
8-20-2 / 8-14-6*	1220 ± 41	1315 ± 6

6 Evolution of a Complex Behavior

6.1 Description

The target behavior is derived the homing experiment first proposed in [5], combining exploration of the environment with energy management. The robot is equipped with an accumulator. The robot completely consumes the accumulator energy in 285 times steps. A specific recharge area is signaled by a light in the arena. There are no obstacles in the arena, and the position of the recharge area is randomly assigned at each epoch.

The fitness is increased proportionally to the forward speed of the robot (as described in section 5.2), but only when the robot is not in the recharge area.

In the original experiment [5], the accumulator was instantly recharged when the robot entered the recharge area. We suppose here that the recharge is proportional to the time spent in the recharge area (a full recharge takes 100 times steps). Moreover, the recharge area is not directly "visible" for the robot, whereas it was signaled by a black ground that the robot could detect with a sensor in [5]. These modifications increase the complexity of the task.

---

[1] Further work will introduce inertia in the simulator, thus avoiding this trap – and possibly many others.

6.2 Supervisor Architectures

The *supervisor* architecture is a hierarchical controller that decide at each time step which one of the basic behaviors it supervises will be executed. Its number of outputs is the number of available basic behaviors, namely:

- *Obstacle avoidance.* This behavior is evolved as described in section 5.2;
- *Light following.* The fitness used to evolve this behavior is the number of times it reaches the light during 10 epoch (no energy involved);
- *Stop.* This behavior is evolved to minimize the speed of the center of mass of the robot. Note that a very small number of generations is needed to get the perfect behavior, but that all evolved *Stop* behaviors in fact rotate rapidly with inverse speeds on both motors.
- *Area sweeping.* The arena is divided in small squares, and the fitness is the number of squares that were visited by the robot during 10 epoch[2].

Two types of supervisors have been tested: the Classical Supervisor (CS), using Classical Controllers as basic behaviors, and the Symbolic Supervisor (SS), that uses symbolic controllers (see section 3) as basic behaviors. The NN implementing the supervision are are Elman networks[3] with 5 hidden neurons.

Baseline experiments were also performed using *direct controllers* with the same Elman architecture - either Classical, or Symbolic (see section 3).

All supervisors and direct controllers have 17 inputs: the 8 IR sensors in active mode for obstacle detection, the 8 IR sensors in passive mode for ambient light detection, and the accumulator level.

The number of outputs is 2 for the classical controller, the speeds of the motors, 6 for the symbolic controller using the three hard-coded behaviors *FORWARD, LEFT, RIGHT* (see section 3), and 4 for both supervisors (Classical and Symbolic) that use the evolved basic behaviors *obstacle avoidance, light following, area sweeping* and *stop*. The *obstacle avoidance* behaviors that are used are the best results obtained in the experiments of section 5. Similar experiments (i.e. with the same architectures) were run for the *area sweeper* and the best results of 10 runs were chosen. For the simpler *light following* and *stop*, 2-layers networks were sufficient to get a perfect fitness.

6.3 Results

The statistics over 10 independent runs can be seen on Figure 1. Three criteria can be used to compare the performances of the 4 architectures: the best overall performance, the variance of the results, and how rapidly good performances are

[2] This performance is external, i.e. it could not be computed autonomously by the robot. But the resulting controller only uses internal inputs.

[3] Elman recurrent neural networks are 3 layers MLP in which all neurons of the hidden layer are totally connected with one another.

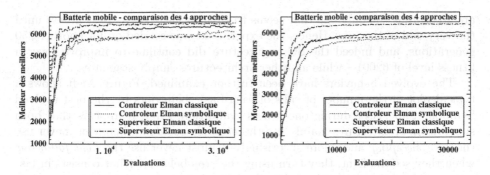

Fig. 1. Maximum (left) and average (right) of best fitness in the population for the energy experiments.

obtained. The sensitivity and generalization abilities of the resulting controllers are important criteria that require additional experiments (sections 6.4, 6.5).

The best overall performance are obtained by the SS (Symbolic Supervisor) architecture. Moreover, it exhibits a very low variance (average best fitness is 6442 ± 28). Note that overpassing a fitness of 6400 means that the resulting behavior could go on for ever, almost optimally storing fitness between the recharge phases).

Next best architecture is the CC (Classical Controller). But whereas its best overall fitness is only slightly less that that of the SS, the variance is 10 times larger (average best fitness is 6044 ± 316, with best at 6354). The difference is statistically significant with 95% confidence using the Student T-test.

The SC (Symbolic Controller) and CS (Classical Supervisors) come last, with respective average best fitness of 5902 ± 122 and 5845 ± 27.

Some additional comments can be made about those results. First, both supervisors architectures exhibit a very good best fitness (≈ 3200) in the initial population: such fitness is in fact obtained when the supervisors only use the obstacle avoidance – they score maximum fitness until their accumulator level goes to 0. Of course, the direct controller architectures require some time to reach the same state (more than 2000 evaluations).

Second, the variance for both supervisor architectures is very low. Moreover, it seems that this low variance is not only true at the performance level, but also at the behavior level: whereas all symbolic supervisors do explore the environment until their energy level becomes dangerously low, and then head toward the light and **stay** in the recharge area until their energy level is maximal again, most (but not all) of the direct controller architectures seem to simply stay close to the recharge area, entering it randomly.

One last critical issue is the low performance of the Symbolic Controller. A possible explanation is the existence of the neutrality plateaus discussed in section 3.2: though those plateaus help escaping local minima, they also slow down the learning process. Also it appears clearly on Figure 1-left that the SC architecture is the only one that seems to steadily increase its best fitness until the very end of the runs. Hence, the experiment was carried on for another 250 generations, and indeed the SC architecture did continue to improve (over a fitness level of 6200) – while all other architectures simply stagnates.

The evolved behaviors have been further examined. Figure 2-left shows a typical plot of the number of calls of each basic behaviors by the best evolved Symbolic Supervisor during one fitness evaluation. First, it appears that both supervisors architectures mainly use the *obstacle avoidance* behavior, never use the *area sweeping*, and, more surprisingly, almost never use the *light following*: when they see the light, they turn using the *stop* behavior (that consists in fast rotation), and then go to the light using the *obstacle avoidance*. However, once on the recharge area, they use the *stop* until the energy level is back over 90%.

Investigating deeper, it appears that the speeds of the *light following* and *area sweeper* are lower than that of the *obstacle avoidance* – and speed is crucial in this experiment. Further experiments will have to modify the speeds of all behaviors to see if it makes any difference. However, this also demonstrates that the supervisor can discard some behavior that proves under-optimal or useless.

6.4   Sensitivity Analysis

One critical issue of the proposed approach is how to ensure that the "right" behavior will be available to the supervisor. A possible solution is to propose a large choice – but will the supervisor be able to retain only the useful ones? In order to assess this, the same energy experiment was repeated but many useless, or even harmful, behaviors were added to the 4 basic behaviors.

First, 4 behaviors were added to the existing ones: *random, light avoiding, crash* (goes straight into the nearest wall!) and *stick to the walls* (tries to stay close to a wall). The first generations demonstrated lower performances and higher variance than in the initial experiment, as all behaviors were used with equal probability. However, the plot of the best fitness (not shown) soon catches up with the plots of Figure 1, and after 150 generations, the results are hardly distinguishable. Looking at the frequency of use of each behavior, it clearly appears that the same useful behaviors are used (see Figure 2-left, and section 6.3 for a discussion). Moreover, the useless behaviors are scarcely used as evolution goes along, as can be seen on Figure 2-right (beware of the different scale).

These good stability results have been confirmed by adding 20 useless behaviors (5 times the same useless 4). The results are very similar - though a little worse, of course, as the useless behaviors are called altogether a little more often.

6.5   Generalization

Several other experiments were performed in order to test the generalization abilities of the resulting controllers. The 10×4 best controllers obtained in the experiments above were tested in new experimental environments.

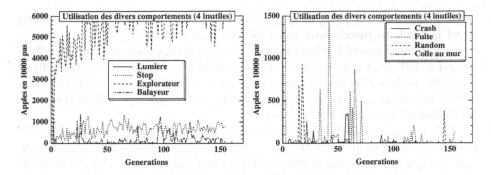

Fig. 2. Number of calls out of 10000 time steps of the **useful** (left) and **useless** (right) behaviors. The plots for the useful behaviors are roughly the same whether or not some useless behaviors are available.

First, some obstacles were added in the environment (the arena was free of obstacle in the experiments described above). But no controller was able to go to the recharge area whenever an obstacle was in the way. However, when the evolution was performed with the obstacles, the overall results are about the same (with slightly worse overall performance, as predicted).

More interesting are the results obtained when the robots are put in an arena that is three times larger than the one used during evolution. The best generalization results are obtained by the Classical Controller architecture, with only a slight decrease of fitness(a few percents). Moreover, 100 additional generations of evolution in the new environment gives back the same level of fitness.

The Symbolic Controllers come next (in terms of generalization capability!): they initially lose around 10% of their fitness, then reach the same level again in 150 generations, and even overpass that level (see the discussion in section 6.3).

Surprisingly, both supervisor architectures fail to reach the same level of performance in the new environment even after a new cycle of evolution. The Symbolic Supervisor lose about 12.5% of their fitness, and only recover half of it, while the Classical Supervisors lose more than 20% and never recover.

These results can be at least partly explained by the behaviors that are obtained by the first experiments: whereas all direct controller architectures mainly stay around the recharge area, and thus are not heavily disturbed by the change of size of the arena, the Supervisor architectures use their exploration behavior and fail to turn back on time. The only surprising result is that they also fail to reach the same level of fitness even after some more generations[4].

This difference in the resulting behaviors also explains the results obtained in the last generalization experiment that will be presented here: the recharge of energy was made 2 times slower (or the energy consumption was made twice faster – both experiments give exactly the same results). Here, the results of the Symbolic Supervisors are clearly much better than those of the other architec-

---

[4] However, when restarting the evolution from scratch in the large arena, the SSs easily reach the 6400 fitness level, outperforming again all other architectures

tures: in all cases, the robot simply stays in the recharge area **until the energy level is back to maximum**, using the *stop* behavior.

Surprisingly, most Classical Supervisors, though they also can use their *STOP* behavior, fail to actually reach the recharge area. On the other hand, both direct controller architecture never stop on the recharge area. However, while the Symbolic Controllers manage to survive more than one epoch for half of the trials, all Classical Controllers fail to do so.

This last generalization experiment shows a clear advantage to the Symbolic Controller architecture: if is the only one that actually learned to recharge the accumulator to its maximum before leaving the recharge area. But the ultimate test for controllers evolved using a simulator is of course to be applied on the real robot. This is on-going work, and the first experiments, applied to the obstacle avoidance behaviors, have confirmed the good performance of the symbolic controllers in any environment.

7   Discussion and Perspectives

The main contribution of this work is to propose some compromise between the pure black box approach where evolution is supposed to evolve everything from scratch, and the "transparent box" approach, where the programmer must decompose the task manually.

The proposed approach is based on a toolbox, or library, of behaviors ranging from elementary hand-coded behaviors to evolved behaviors of low to medium complexity. The combination and proper use of those tools is left to evolution. The new space of controllers that is explored is more powerful that the one that is classically explored in Evolutionary Robotics. For instance, it was able to easily find some loophole in the (very simple) obstacle behavior fitness; moreover, it actually discovered the right way to recharge its accumulator in the more complex homing experiment.

Adding new basic behaviors to that library allows one to gradually increase the complexity of the available controllers without having to cleverly insert those new possibilities in the available controllers: evolution will take care of that, and the sensitivity analysis demonstrated that useless behaviors will be filtered out at almost no cost (section 6.4). For instance, there might be some cases where a random behavior can be beneficial – and it didn't harm the energy experiment. More generally, this idea of a library allows one to store experience from past experiments: any controller (evolved or hand-coded) can be added to the toolbox, and eventually used later on - knowing that useless tools will simply not be used.

Finally, using such a library increases the intelligibility of the resulting controllers, and should impact the way we evolutionary design controllers, i.e. fitness functions. One can add some constraints on the distribution over the use of the different available controllers, (e.g. use the light following action $\varepsilon\%$ of the time); by contrast, traditional evolutionary approach had to set up sophisticated *ad hoc* experimental protocol to reach the same result (as in [16]). Further work will have to investigate in that direction.

But first, more experiments are needed to validate the proposed approach (e.g. experiments requiring some sort of memory, as in [18,16]). The impact of redundancy will also be investigated: in many Machine Learning tasks, adding redundancy improves the quality and/or the robustness of the result. Several controllers that have been evolved for the same task, but exhibit different behaviors, can be put in the toolbox. It can also be useful to allow the overall controller to use all levels of behaviors simultaneously instead of the layered architecture proposed so far. This should allow to discover on the fly specific behaviors whenever the designer fails to include them in the library.

Alternatives for the overall architecture will also be looked for. One crucial issue in autonomous robotics is the adaptivity of the controller. Several architectures have been proposed in that direction (see [13] and references herein) and will be tried, like for instance the idea of auto-teaching networks.

Finally, in the longer run, the library approach helps to keep tracks of the behavior of the robot at a level of generality that can be later exploited by some data mining technique. Gathering the Frequent Item Sets in the best evolved controllers can help deriving some brand new macro-actions. The issue will then be to check how useful such macro-actions can be if added to the library.

References

1. P. J. Bentley, editor. *Evolutionary Design by Computers*. Morgan Kaufman Publishers Inc., 1999.
2. R. A. Brooks. A robust layered control system for a mobile robot. *Journal of Robotics and Automation*, 2(1):14–23, 1986.
3. R. A. Brooks. How to build complete creatures rather than isolated cognitive simulators. In Kurt VanLehn, editor, *Architectures for Intelligence: The 22nd Carnegie Mellon Symp. on Cognition*, pages 225–239. Lawrence Erlbaum Associates, 1991.
4. R. Dawkins. *The blind watchmaker*. Norton, W. W. and Company, 1988.
5. D. Floreano and F. Mondada. Evolution of homing navigation in a real mobile robot. *IEEE Transactions on Systems, Man, and Cybernetics*, 26:396–407, 1994.
6. I. Harvey. *The Artificial Evolution of Adaptive Behaviour*. PhD thesis, University of Sussex, 1993.
7. M. Humphrys. *Action Selection Methods Using Reinforcement Learning*. PhD thesis, University of Cambridge, 1997.
8. Leslie Pack Kaelbling, Michael L. Littman, and Andrew P. Moore. Reinforcement learning: A survey. *Journal of Artificial Intelligence Research*, 4:237–285, 1996.
9. M. Keijzer, J. J. Merelo, G. Romero, and M. Schoenauer. Evolving objects: a general purpose evolutionary computation library. In P. Collet et al., editors, *Artificial Evolution'01*, pages 229–241. Springer Verlag, LNCS 2310, 2002.
10. P. Maes. The dynamics of action selection. In *Proceedings of the 11th International Joint Conference on Artificial Intelligence*, 1989.
11. J.R. Millan, D. Posenato, and E. Dedieu. Continuous-action Q-learning. *Machine Learning Journal*, 49(23):247–265, 2002.
12. S. Nolfi and D. Floreano. How co-evolution can enhance the adaptive power of artificial evolution: implications for evolutionary robotics. In P. Husbands and J.A. Meyer, editors, *Proceedings of EvoRobot98*, pages 22–38. Springer Verlag, 1998.
13. S. Nolfi and D. Floreano. *Evolutionary Robotics*. MIT Press, 2000.

14. H.-P. Schwefel. *Numerical Optimization of Computer Models*. John Wiley & Sons, New-York, 1981. 1995 – 2nd edition.
15. R.S. Sutton and A. G. Barto. *Reinforcement learning*. MIT Press, 1998.
16. E. Tuci, I. Harvey, and M. Quinn. Evolving integrated controllers for autonomous learning robots using dynamic neural networks. In B. Hallam et al., editor, *Proc. of SAB'02*. MIT Press, 2002.
17. V. N. Vapnik. *Statistical Learning Theory*. Wiley, 1998.
18. B.M. Yamauchi and R.D. Beer. Integrating reactive, sequential, and learning behavior using dynamical neural network. In D. Cliff et al., editor, *Proc. SAB'94*. MIT Press, 1994.

Evolving Motion of Robots with Muscles

Siavash Haroun Mahdavi and Peter J. Bentley

Department of Computer Science
University College London
Gower St. London WC1E 6BT, UK
{mahdavi, p.bentley}@cs.ucl.ac.uk

Abstract. The objective of this work is to investigate how effective smart materials are for generating the motion of a robot. Because of the unique method of locomotion, an evolutionary algorithm is used to evolve the best combination of smart wire activations to move most efficiently. For this purpose, a robot snake was built that uses Nitinol wire as muscles in order to move. The most successful method of locomotion that was evolved, closely resembled the undulating motion of the cobra snake. During experimentation, one of the four Nitinol wires snapped, and the algorithm then enabled adaptive behaviour by the robot by evolving another sequence of muscle activations that more closely resembled the undulations exhibited by the earthworm.

1   Introduction

We have grown used to software that crashes several times a day, faulty hardware that results in data loss and Sojourner Mars Rovers that get stuck on rocks. The problems are rare because our technology is very carefully designed. But the problems are significant and often devastating because our technology is not adaptive – it does not have graceful degradation. When something goes wrong, a terminal failure is common. In contrast, natural systems are able to work in many different environments, under very different conditions. Living designs are able to continue working even when damaged. The research described here is inspired by the adaptability of evolved solutions in nature.

The aims of this project are to create a self-adapting snake (SAS). The SAS uses muscles in order to move, made from a smart material call Nitinol. The sequence of muscle activations was controlled by a finite state machine, evolved using a genetic algorithm to maximise the distance travelled by each individual. Technology like this may be used to make devices that need to change their morphology to suit their environments, like a chameleon changing colour to avoid detection. Applications for this type of self-adapting form are obviously vast and so this project seeks to investigate whether a simple 'snake' made out of foam and wood could learn to morph its body in such a way as to move efficiently.

S. Cagnoni et al. (Eds.): EvoWorkshops 2003, LNCS 2611, pp. 651–660, 2003.

2   Background

2.1   Evolutionary Robotics

The field of adaptive robotics is not a new one. Many robots that use standard methods of movement like motors and pistons have been evolved from the design stage and all the way through to the control stage [1]. For a thorough survey, readers are advised to read [5] and [10]. Further examples include Jakobi, who endeavoured to build a simulation within which control systems for real robots could evolve [7]. Hornby evolved L-Systems to generate complicated multi-component virtual creatures [4]. Husbands used evolutionary robotics techniques to develop control networks and visual morphologies to enable a robot to achieve target discrimination tasks under very noisy lighting conditions [6]. Mihalachi and Munerato created a robot snake able to move like an earthworm [11][12].

The use of smart materials in robotics has already been investigated. For example, Kárník looked at the possible applications of walking robots, which use artificial muscles with smart materials as a drive [8]. Mills has written a book aimed at beginners which teaches them how to make simple eight-legged robots using smart materials as actuators [13]. However, no one has used smart materials and evolutionary algorithms together in a robot before. This work uses nitinol wires as muscles within a robot snake, with the muscle activation sequences evolved using genetic algorithms and finite state machines.

2.2   Smart Material

Nitinol, an alloy made of Nickel and Titanium, was developed by the Naval Ordinance Laboratory. When current runs through it, thus heating it to its activation temperature, it changes shape to the shape that it has been 'trained' to remember. The wires used in this project simply reduce in length, (conserving their volume and thus getting thicker), by about 5-8 %.

Shape Memory Alloys, when cooled from the stronger, high temperature form (Austenite), undergo a phase transformation in their crystal structure to the weaker, low temperature form (Martensite). This phase transformation allows these alloys to be super elastic and have shape memory [3].

The phase transformation occurs over a narrow range of temperatures, although the beginning and end of the transformation actually spread over a much larger range of temperatures. Hysteresis occurs, as the temperature curves do not overlap during heating and cooling, see Fig. 1 [3]. With thick wires, this could bring about problems for the SAS as the NiTi wires would take some time before returning to their original lengths, however, due to the very small diameter of the NiTi wires used (~0.15mm), the hysteresis was almost negligible as the cooling to below the Martensite temperature, (Mf), was almost instantaneous [2].

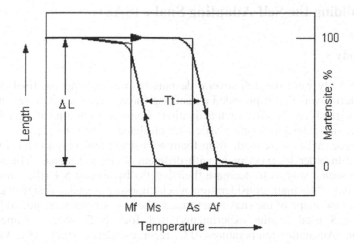

Fig. 1. Graph showing change in length during heating and cooling. The hysteresis is represented by Tt.

2.3   Snake Motion in Nature

This section reviews four different methods of undulation by which real snakes move [12]. An analysis of the most efficient muscle activation sequences is located in the analysis section at the end of the paper.

1. The grass snake undulates in horizontal plane. A horizontal S shape is created in its tail, this then moves along the entire length of the snake's body until the S finally reaches the head. The disadvantage with this motion is that there is a lot of rubbing along the length of the S, which consumes a lot of energy. The advantage with this form of undulation is that the snake is very stable whilst in motion.

2. The adder undulates in a vertical plane. The motion is very similar to that of the grass snake, except that the horizontal S is now replaced with a vertical U which moves from tail to head. The advantages with this motion are medium energy consumption and minimal rubbing. The disadvantage with this motion is with the instability of the snake, which is due to the vertical undulations.

3. The cobra undulates and extends in a horizontal plane. The S starts in the middle of the snake's body, it then repeats in both directions until it covers the entire body. The reverse is then performed until the cobra is once again straight. The advantages are that there is no problem with stability and that the speed is raised. The disadvantages are the high-energy consumption and a lot of rubbing.

4. The earthworm extends in a horizontal plane. This extension is parallel to the body and moves along its length. Unlike the grass snake and adder, the extensions travel from its head to its tail. The advantages are that there is no problem of stability, reduced energy consumption and rubbing is negligible. The disadvantages are that module elongation is necessary.

3   Building the Self-Adapting Snake (SAS)

3.1   Body

The SAS body went through several designs before reaching the final design. The main criterion was that it provided a restoring force great enough to restore the wires to their original lengths after each activation. Foam was chosen for the job for all but one of the snake designs where plastic wire insulators were used [2].

Prototype snakes were made from foam with string replacing the NiTi wires. This was done in order to experiment with different skeletal designs. The strings were pulled in such a way as to decrease their lengths by around 5% (thus mimicking the NiTi wires). The main consideration when choosing a snake design was to see by how much the shape of the snakes deformed when the strings were pulled [2].

The SAS used in the experiments used four NiTi wires (diameter=1.5mm, activation (Austenite) temperature=70°C, recommended current 200mA), that were connected all together at one side with a nut and bolt, which in turn was connected to a piece of normal wire. This wire ran through the middle of the snake and supplied the power, much like a spinal chord carries nerves impulses to muscles through the body, see Fig. 2. The total weight of the SAS was approximately 50g.

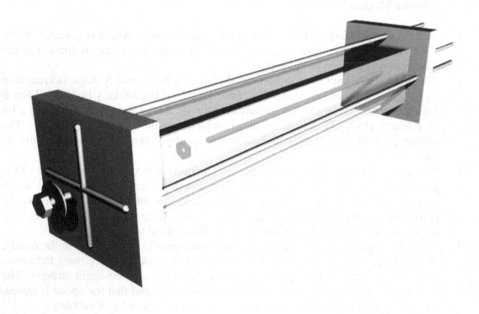

Fig. 2. Four wires were arranged at the top, bottom, left and right of the SAS.

3.2   Brain

A finite state machine was used to determine the wire activation sequences; these were then evolved using a standard single-point crossover genetic algorithm with roulette wheel selection.

Each member of the population was made up of a sting of 0s and 1s. Each string consisted of two segments, 'sequence' and 'next time', for a simple example see Fig. 3. The 'sequence' was the part that was sent to the SAS, and determined which wires were to be activated at that particular time. The 'next time' was the part that told the program to which time slot in the current row it should then jump. The recommended technique for activating the NiTi wires is to pulse them. This is because prolonged heating causes 'hotspots' to occur and these damage the wire and reduce its life span. Therefore, each sequence is in fact pulsed 7 times before moving on to the next sequence [2].

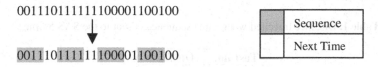

Fig. 3. Example string using 24 instead of 640 bits. Binary string is split into 'sequence' and 'next time' segments.

For the SAS, the 'sequence' length was four bits, and the 'next time' length was six bits, therefore the total length of each string was 2^6 x (4 + 6) = 640 bits.
The program was allowed to send 20 sequences to the SAS before stopping, lasting a total of approximately 40 seconds. Finite state machines that acted as repeating patterns were easily created if the jump points pointed to each other in a loop of some sort. Indeed, very rarely was there a string that didn't loop at all.

The computer interfacing hardware, though quite complex, had two simple tasks to perform. The first was to supply enough power to each NiTi wire (~200mA). This was achieved with the use of some Darlington amplifiers. The second task to perform was the ability to activate muscles in parallel. For this, the microcontroller used was the Motorola MC68H(R)C908JK1. It had 20 pins and with the structure of the circuit board was capable of activating up to 12 pins in parallel (PTBx and PTDx excluding PTB0 and PTB3) [9].

4   Experimental Setup

The SAS was connected to the microcontroller circuit board using a long lead (~600mm) so that it did not have to pull much weight when trying to move. The SAS was placed on grid paper and a starting point was marked out. As the program started, an initial population of randomly generated finite state machines was created. Each individual finite state machine would then be sent to the SAS for evaluation. The fitness of any member was determined solely on how far the snake travelled (in mm) in the forward direction [2].

The final configuration consisted of the SAS being placed on top of a platform about 500mm above the level of the circuit board and power supply. The wires were then draped over the edge of the platform. As the SAS moved forward, the wires were pushed down over the edge and so out of the way of the snake. There was concern over whether the weight of the wires was assisting in pulling the SAS along but much experimentation with the SAS ensured that the wires were pushed along and did not pull. The important factor was that the resistance of the wires was more or less constant no matter how far the SAS moved [2].

The distance travelled by the robot snake could be measured to the nearest millimetre. However, experimentation needed to be done to observe the true accuracy by which the distance travelled should be stated. To test this the SAS was sent the same sequence five times and the distances travelled were noted. The results show that the distance travelled was always within the nearest millimetre, and so the distances travelled by the SAS could be stated to the nearest millimetre without being over accurate, see table 1.

Table 1. Distance travelled when same sequence is sent to the SAS 5 times.

Test no.	Dist/mm
1	18
2	18
3	17
4	18
5	18

Following this, a series of experiments were performed to investigate the evolution of movement for the SAS. The SAS's motion was evolved for 25 generations and the corresponding fitnesses were stored.

5   Results

Because of the length of time required to perform experiments (approx 20 minutes per generation), two runs were executed. The results for the first run are given in Fig 4. As can be seen in the graph below, the distance travelled did not improved at all. In fact there seems to be a downward trend in fitness.

As each member of the population represents a very complicated FSM, finding suitable cut off points that would improve the fitness of the member without completely changing the pattern of sequence activations is very difficult. Such members, once mutated, induce motion that is nothing like the motion of the original parent, and so the fitness maybe considerably lower. Therefore, an individual that travels a particularly long distance has no guarantee of creating offspring that travel anywhere as far. This is likely to be the main cause of the decrease in overall fitness of the population. To overcome this problem, an elitist genetic algorithm was used for the second run.

Fig. 4. Maximum fitness and average fitness plotted at each generation.

This GA works simply by taking the n (n = 2 for the experiment) best members of the population and in addition to using them to generate members of the next generation, they are also placed directly into the next generation. In doing such, they naturally ensure that the maximum distance travelled during the coming generations never decreases, and they are also given another chance influence the next population if they still have the highest fitness. With this change, the maximum fitness never decreased (nearly always increasing), see Fig. 5.

Ideally the second experiment would have been carried out for as many generations as the previous experiment had but it had to be temporarily halted after only 8 generations. This was because one of the NiTi wires snapped!

The NiTi wire snapped at exactly the corner of the hole that it was fed through. It is believed that it snapped due to the sharp corner that it rubbed against and not because of overheating [2]. This would normally spell the end for any other robot, however one of the hypotheses of this research is that such techniques enable self-adaptation through evolution and so it seems only logical that if the SAS was truly self-adapting that it should be able to adapt to the loss of a muscle and learn other ways to move.

To test this, evolution was continued from the point at which the NiTi wire snapped (beginning of 8th generation). Every member of the population probably used the 4th wire to move at some point during its sequence and so the fitness of the whole populations dropped from an average distance of 13.25mm to an average of only 1.95mm. The members that did slightly better than the rest were probably members that had moved quite proficiently and had also used less of wire four. So these sorts of sequences were found more and more abundantly in the following generations. It took around 10 generations for any sequence to be found that moved more than 6mm. Then came a sequence that simply alternated between the top and bottom wire. This caused the SAS to move over 20mm. Since elite selection was being implemented, this sequence was not lost and variations spread throughout the proceeding generations increasing the fitness of the whole population considerably, see Fig. 5.

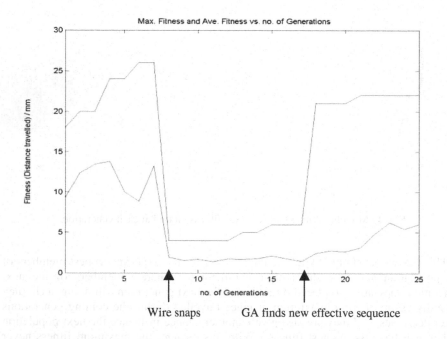

Wire snaps   GA finds new effective sequence

Fig. 5. Maximum fitness and average fitness plotted at each generation.

6   Analysis

During the evolution of the SAS, numerous interesting methods of locomotion were carried out. Though most of them were unsuccessful, this section seeks to analyse the physics behind two of the more efficient evolved sequences.

Though the sequence lengths could vary from a length of one to a maximum of sixty-four, the sequences that travelled the furthest seemed to have short repetitive loops.

The best sequence found before the wire snapped could be compared to a simple but effective version of the cobra undulations, see Fig. 6. The SAS makes S shapes by alternating between activating the top and right wires, with the top and left wires. This S shape is along the whole length of the SAS and so provides the cobra-like undulations.

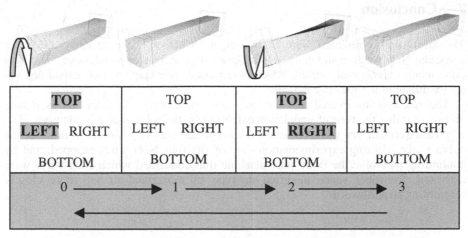

TOP	TOP	TOP	TOP
LEFT RIGHT	LEFT RIGHT	LEFT **RIGHT**	LEFT RIGHT
BOTTOM	BOTTOM	BOTTOM	BOTTOM
0 ────▶ 1 ────▶ 2 ────▶ 3			

Fig. 6. Activation sequence of best sequence found resembles cobra undulations.

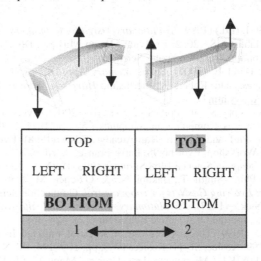

TOP	**TOP**
LEFT RIGHT	LEFT RIGHT
BOTTOM	BOTTOM
1 ◀────▶ 2	

Fig. 7. Activation sequence of best sequence after damage resembles earthworm undulations.

After the NiTi wire snapped, the GA eventually evolved a sequence that simply alternated between the top and bottom wires. Alternating between compression and extension means that the new method of locomotion that had been evolved much resembled the undulations of the earthworm, see Fig. 7. Though this did not propagate the SAS as efficiently as the sequences that had been found previously, this pattern of muscle activations recovered over 85% of the previous mobility when it only had 75% of its muscle intact.

7   Conclusion

The aims of this research were to investigate whether shape memory alloys, in particular NiTi wires, could be used to make a robot snake that could move. This was done using a genetic algorithm, which determined their sequence of activations via the evolution of a finite state machine.

The research discovered that not only can these shape memory alloys induce motion, but that the type of undulation exhibited by the robot snake is adaptive. The type of movement that the robot snake demonstrated was very similar to those of the cobra snake. During experimentation, one of the four NiTi wires snapped, and by continuing to evolve the finite state machine that controlled which NiTi wires were activated, the robot snake modified its type of movement to one more closely resembling an earthworm.

References

1. Bentley, P. J. (Ed.) (1999) *EvolutionaryDesign by Computers*. Morgan Kaufmann Pub.
2. Haroun Mahdavi, S. (2002) Evolving Motion Master's Dissertation, MSc IS, University College London, Dept. Computer Science.
3. Hodgson, Darel E. (2002), *Shape Memory Applications, Inc.*, Ming H. Wu, *Memory Technologies*, and Robert J. Biermann, *Harrison Alloys, Inc.* www.sma.com/Shape Memory Alloys.htm
4. Hornby, Gregory S. and Pollack, Jordan. B. (2001). Evolving L-Systems To Generate Virtual Creatures. Computers and Graphics, 25:6, pp 1041–1048.
5. Husbands, Phil and Meyer, Jean-Arcady (Eds.) (1998) Evolutionary Robotics, First European Workshop, EvoRbot98, Paris, France, April 16–17, 1998, Proceedings. Lecture Notes in Computer Science 1468 Springer 1998, ISBN 3-540-64957-3
6. Husband Phil., Smith T., Jakobi N. & O'Schea M. (1999). *Better living through chemistry: Evolving GasNets for robot control*. Connection Science, (10)3–4:185–210.
7. Jakobi, Nick. (1997), `*Evolutionary robotics and the radical envelope of noise hypothesis*', Journal Of Adaptive Behaviour 6(2), 325–368
8. Kárník, L. (1999) The possible use of artificial muscled in biorobotic mechanism. In ROBTEP'99, Kosice, SF TU Kosice, 1999, pp. 109–114. ISBN 80-7099-453-3.
9. MC68HC908JK1 Microcontroller User's Manual (2002). Motorola Literature Distribution: P.O. Box 5405, Denver, Colorado 80217
10. Meyer Jean-Arcady, Husbands Phil and Harvey Inman (1998) Evolutionary Robotics: A Survey of Applications and Problems. EvoRobots 1998: 1–21
11. Mihalachi, D. and Munerato, F. (1999) Snake-like mobile micro-robot based on 3 DOF parallel mechanism. In Proc. of PKM'99 MILAN November 30, 1999.
12. Mihalachi, D. (2000) *Systeme de Commande Temps Reel Distribue pour un Mini Robot Autonome de type Serpent*. Doctorat de L'Universite de Metz, These Présentée à L'U.F.R. Mathematique, Informatique, Mecanique, Automatique.
13. Mills, Jonathan W. (1999) Stiquito for beginners. ISBN 0-8186-7514-4, 1999.

Behavioural Plasticity in Autonomous Agents: A Comparison between Two Types of Controller

Elio Tuci[1] and Matt Quinn[2]

[1] Institut de Recherches Interdisciplinaires
et de Développements en Intelligence Artificielle (IRIDIA),
Avenue Franklin Roosevelt 50, CP 194/6 - 1050 Bruxelles - Belgium
etuci@ulb.ac.be
[2] Centre for Computational Neuroscience and Robotics (CCNR),
University of Sussex - Falmer - Brighton BN1 9QH, UK
matthewq@cogs.susx.ac.uk

Abstract. Blynel et al. [2] recently compared two types of recurrent neural network, Continuous Time Recurrent Neural Networks (CTRNNs) and Plastic Neural Networks (PNNs), on their ability to control the behaviour of a robot in a simple learning task; they found little difference between the two. However, this may have been due to the simplicity of their task. Our comparison on a slightly more complex task yielded very different results: 70% runs with CTRNNs produced successful learning networks; runs with PNNs failed to produce a single success.

1 Introduction

Recently, a growing amount of research in Evolutionary Robotics has been focusing on the evolution of controllers with the ability to modify the behaviour of autonomous robots by learning [9,8,3,6,7]. This kind of research has been motivated either by an interest in modelling biological phenomena (e.g. the interaction between learning and evolution) or by a more engineering-oriented interest in the design of autonomous agents capable of adapting to unpredictable or contingent aspects of an environment. Several solutions have been already proposed for the implementation of plastic controllers, although there has been little investigation of their comparative capabilities.

Recently, Blynel et al. [2] compared two types of network, Continuous Time Recurrent Neural Networks (CTRNNs: described in [1] and also briefly in section 3 below) and Plastic Neural Networks (PNNs: described in [5,4] and also briefly in section 3 below). The former has fixed synaptic weights and 'leaky integrator' neurons, the latter employs Hebbian-style weight change rules, but neurons exhibit no temporal dynamics. The networks were compared on their ability to control the behaviour of a Khepera mini-robot on a simple learning task. A robot was initially placed in the centre of an arena and required to find a reward area which could be located at either end of that arena. The reward area was a grey stripe at one end of the white arena floor, which the robot could detect by a sensor on its belly. The robot always faced the same end of the arena

S. Cagnoni et al. (Eds.): EvoWorkshops 2003, LNCS 2611, pp. 661–672, 2003.

— it was constrained to only move either forwards or backwards, as if on rails, and could not turn. The position of the reward area remained unchanged for several trials. Within this time, the robot should come to learn the location of the reward area with respect to its starting location, and should thus move towards the reward area without having to explore the opposite end of the arena. After some time, the position of the reward area was switched to the opposite end of the arena, and the robot should consequently switch its behavioural response according to the characteristics of the new environment. Despite the differences in controller architecture, Blynel et al. [2] found the evolved CTRNNs and PNNs to display similar performance, from which they concluded that both architectures are, "*capable of displaying learning behavior and can solve complex tasks which require sequential behavior*" (see [2] p. 279).

Our concern is that the lack of a significant difference in the performance of the two architectures may simply be the result of a *ceiling effect*. That is, potential differences between the two architectures may have been masked by the simplicity of the task with which they were presented. Two main factors contribute to the simplicity of this task. Firstly, the robot is constrained to only move in one of two directions. This reduces the need for sensory-motor coordination to a minimum. The navigational aspect of the task is reduced to a binary choice of direction of movement, moreover the robot will only be exposed a subset of its possible sensory input, thus minimising the possibility of perceptual ambiguity. Clearly, the task would become more complex if the robot were free to roam the arena. Secondly, the learning component of the task is simplified because the robot can use environmental cues to establish the appropriateness of its *current* behaviour. Learning is harder when cues relate to the appropriateness of *past* behaviour, e.g., when there is a delay between a behaviour and its reinforcement. Blynel et al. [2] suggest that both network architectures would be capable of a more complex task but, given the simplicity of their task, it is not obvious that their conclusion is supported by their result. In this paper, we introduce a slightly more complex version of their task, in which the two simplifying factors, mentioned above, are removed. Our results indicate that, contrary to Blynel et al.'s suggestion, there may in fact be significant differences between CTRNNs and PNNs with respect to their ability to solve "complex tasks which require sequential behaviour".

2   Description of the Task

As with the model described in [2], we require a robot to learn which of two types of behaviour is appropriate to its current environment. The robot is tested over a set of trials, with the environment-type remaining constant within a set of trials, but varying between them. The robot — a simulated Khepera — is equipped with 2 motor-driven wheels, 8 infrared sensors, distributed around its body, and a "floor sensor" (see figure 1). This sensor can be conceived of as an ambient light sensor, positioned facing downward on the underside of the robot; this sensor detects the level of grey of the floor. It produces an output which is inversely proportional to the intensity of light, and scaled between 0 — when the robot is positioned over white floor — and 1 — when it is over black floor. For a more detailed description of the implementation of the simulator we refer the reader to [7].

The robot is equipped with 8 infra red sensors (Ir$_0$ to Ir$_7$) and a floor sensor indicated by the central grey circle (F).

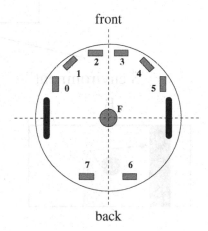

Fig. 1. Plan of a Khepera mini-robot showing sensors and wheels.

At the beginning of each trial, the robot is initially placed, facing forwards, at the left-hand end of a walled arena (see fig. 2A). The robot is free to move in any direction within the arena. In the centre of the arena there is a gradient-grey stripe — the central stripe. If the robot reaches this central stripe, a black stripe — the goal area — will be placed at one end of the arena. Depending on the environment type, the goal will either be at the *same* end as the robot started from (S-env, see fig. 2B), or at the *opposite* end (O-env, see fig. 2C). In a successful trial, the robot should navigate to the central stripe, and then proceed directly to and then stop on the goal. To be able to do this consistently, the robot must first learn whether the environment is one in which it should turn around on the central stripe, or one in which it proceed directly over the central stripe. A robot that does not learn which behaviour is appropriate to its current environment, can do no better than choosing a direction at random.

Notice that the cental stripe contains a gradient: the grey level gradually increases, starting from white on the left end of the stripe to a certain level of grey on the right end. This is intended as a navigational aid. Since the robot may approach the central area from a variety of different angles, this gradient can help the robot determine its orientation and direction of travel (i.e. towards or away from its starting end). The robot can navigate up the gradient to ensure it traverses the central area, or down the gradient to ensure that it returns in the direction it came from. Note, however, that the gradient contains no information about which direction the robot should follow in order to find the goal.

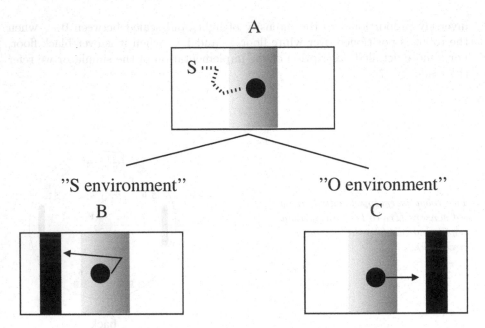

Fig. 2. Depiction of the task. The small circle represents the robot. The black stripe in the arena represents the goal area. The central area is indicated by the stripe in shades of grey — central stripe. Picture A at the top represents the arena at the beginning of each single trial, without the goal. S indicates a possible starting position. The curved dotted lines represent a possible route to the central stripe. The arrows, in picture B and C represent the directions towards which a successful robot should move.

During evolution, each robot undergoes two sets of trials in the S-env. and two sets in the O-env. After each set of trials, the robot's control system is reset, so that there is no internal state or "memory" remaining from the previous environment (see section 3 for details of the networks). Each individual trial within these test sessions is subject to time constraints. Each trail can last no longer than 36 seconds. Each trial can be terminated earlier either because the time limits are exceeded or because the robot crashes into the arena wall. The robot undergoes 15 trials for each test session. During this time, the end at which the reward area is located is kept fixed. At the beginning of each trial, the robot is positioned on the left part of an rectangular arena close to the short side.

Whilst performing the task, the robot's behaviour will be evaluated according to an evaluation function that rewards or penalises certain actions. The evaluation function has been designed in consideration of the selection pressures that might facilitate the evolution of the learning behaviour. The robot increases its fitness by moving from its initial position to the central section of the arena, thereafter by progressing toward the goal. It receives an additional score if it finds the goal and stays over it. The evaluation function penalises the robot for colliding with the arena walls, for failing to reach the central section of the arena, and for moving into the non-goal end of the arena after it has reached the centre. The total fitness score (ϕ) attributed to each individual was simply the average score that it achieved for each trial t of each test session s, after the first trial of each session had been excluded. Performance in the first trial was ignored on the grounds that a robot cannot "know" what kind of environment it is situated within (i.e., O-env. or S-env.), and thus its choice of direction can only be random.

$$\phi = \frac{1}{56} \sum_{s=1}^{4} \sum_{t=2}^{15} F_{st} \quad \text{where} \quad F_{st} = abc \left(3.0 \frac{(d_f - d_n)}{d_f} + \frac{p}{p_{max}} \right) \quad (1)$$

d_f represents the furthest distance that the robot reaches from the goal after having reached the central area; d_n represents the nearest distance that the robot reaches from the goal after having reached the central area. p represents the longest unbroken period of time spent on the goal, and p_{max}, the target length of time for the robot to remain on the goal (i.e., 10 seconds). The value of a, b, c defaults to 1, however a is set to 0 if the robot fails to reach the centre of the arena; b is set to $\frac{1}{3}$ if the robot leaves the central area in the wrong direction (i.e., moves away from the goal); c is set to $\frac{1}{5}$ if the robot crashes into the arena wall. These values were used in initial experiments and proved to work; however there is no reason to believe that other choices may not also work as well or even better.

From the description given above, it should be clear that this task is more complex than that designed by Blynel et al. [2]. Firstly, the task clearly requires more sensorimotor coordination: The robot is not constrained to bi-directional movement, it is exposed to a wider range of sensory input and this input is more ambiguous. For example, in the original task, if a robot encounters a wall while it is on white floor, it is at the wrong end of the arena. Here, however, the robot can encounter any of the four walls whilst being on white floor (note the gap between the black stripe and the wall). Secondly, the robot receives no environmental feedback on its current behaviour, only on its previous behaviour. The robot cannot establish whether its response to the central stripe is correct until it has reached one of the two ends of the arena. Nevertheless, although more complex than Blynel et al.'s task (see [2]), in many respects, the task is still a simple learning task. The robot must distinguish two types of environment, and establish which of two types of behaviour is appropriate to that environment.

3   Two Types of Neural Network: CTRNNs and PNNs

The robots are controlled in one set of experiments by CTRNNs, and in a different set of experiments by discrete time recurrent Platsic Neural Network (PNNs). The implementation of both types of networks matches as much as possible the description which has been given in [2].

The continuous time recurrent neural network robot controller is a fully connected, 10 neuron CTRNN. Each neuron, indexed by i, is governed by the following state equation:

$$\frac{dy_i}{dt} = \frac{1}{\tau_i}\left(-y_i + \sum_{j=1}^{10} \omega_{ji}\sigma_{(y_j+\beta_j)} + gI_i\right) \qquad \text{where,} \qquad \sigma_x = \frac{1}{1+exp\,(-x)} \quad (2)$$

y_i is the neuron activation value, τ_i the decay constant, β_j the bias term, $\sigma(y_j + \beta_j)$ the firing rate, ω_{ji} is the strength of synaptic connections from the jth neuron to the ith neuron, I_i the intensity of the sensory perturbation on sensory neuron i. The neurons either receive direct sensor input or are used to set motor output. There are no interneurons. All but two of the neurons receive direct input from the robot sensors (for the remaining two, $gI_i = 0$). Each input neuron receives a real value (in the range $[0.0 : 1.0]$), which is a simple linear scaling of the reading taken from its associated sensor. The two remaining neurons are used to control the motors. Their cell potential y_i, is mapped onto the range $[0.0 : 1.0]$ by a sigmoid function, and then linearly scaled into $[-10.0 : 10.0]$, set the robot's wheel speeds. The strengths of the synaptic connections, the decay constants, bias terms and gain factors were all genetically encoded parameters. Cell potentials were initialised to 0 each time a network was initialised or reset. State equations were integrated using the forward Euler method with an integration step-size of 0.2 (i.e., the simulation was updated the equivalent of 5 times a second).

The discrete time Plastic Neural Network robot controller is a fully connected 12 neurons network. Each neuron, indexed by i, has an activation value y_i which is updated at every activation cycle, according to the following equation:

$$y_i(n+1) = \sigma\left(\sum_{j=1}^{12} \omega_{ij}y_i(n)\right) + I_i, \qquad \text{where,} \qquad \sigma(x) = \frac{1}{1+exp\,(-x)} \quad (3)$$

ω_{ji} is the strength of synaptic connections from the jth neuron to the ith neuron, I_i the intensity of the sensory perturbation on sensory neuron i. Eight neurons receive input from their associated robot's sensor. Each input neuron receives a real value (in the range $[0.0 : 1.0]$), which is a simple linear scaling of the reading taken from its associated sensor. Two neurons are used to set motor output. One is an interneuron which does not receive any input from the robot's sensor, and one neurons works as a bias neuron. The bias neuron, as defined in [2], is a neuron which receives at any update of the network a fixed input I_i set to 1.

Each synaptic connections ω_{ij} is randomly initialised within the range $[0.0 : 0.1]$ any time the network is initialised or reset, and is updated every sensory-motor cycle according to the following rule:

$$\omega_{ij}(n+1) = \omega_{ij}(n) + \eta\Delta\omega_{ij}(n) \qquad (4)$$

where η is the learning rate ranging within the interval $[0.0 : 1.0]$ and $\Delta\omega_{ij}$ is determined by one of the following four Hebb rules:

Plain Hebb $\Delta\omega_{ij} = (1 - \omega_{ij})x_i y_j$

Post-synaptic $\Delta\omega_{ij} = y\omega_{ij}(-1 + x_i) + (1 - \omega_{ij})x_i y_j$

Pre-synaptic $\Delta\omega_{ij} = x\omega_{ij}(-1 + y_j) + (1 - \omega_{ij})x_i y_j$

Covariance $\Delta\omega_{ij} = \begin{cases} (1 - \omega_{ij})\mathcal{F}(x_i, y_j) & \text{if } \mathcal{F}(x_i, y_j) < 0 \\ \omega_{ij}\mathcal{F}(x_i, y_j) & \text{otherwise} \end{cases}$

where $\mathcal{F}(x_i, y_j) = tanh(4(1 - |x_i - y_j|) - 2)$

x_i refers to the pre-synaptic neuron, and y_j refers to the post-synaptic neuron. The association between each synaptic connection and the corresponding Hebb rule which determines how the strength of the synaptic connection changes, is genetically specified. The sign of each synaptic connection is also genetically specified and can not change during the life-time of the robot. The self-limiting component $(1 - \omega_{ij})$ is maintaining synaptic strengths within the range $[0.0 : 1.0]$.

4   Genetic Algorithm and Encoding Schemes

The same kind of generational genetic algorithm (GA) was used for both sets of runs. The population contained 100 genotypes. Generations following the first one, were produced by a combination of selection with elitism, recombination and mutation. For each new generation, the two highest scoring individuals ("the elite") from the previous generation were retained unchanged. The remainder of the new population was generated by fitness-proportional selection from the 70 best individuals of the old population.

In the experiment with CTRNNs, each genotype was a vector comprising 121 real values (100 connections, 10 decay constants, 10 bias terms, and a gain factor). Initially, a random population of vectors was generated by initialising each component of each genotype to values chosen at uniform random from the range $[0:1]$. New genotypes, except "the elite", were produced by applying recombination with a probability of 0.3 and mutation operator. Mutation entails that a random Gaussian offset is applied to each real-valued vector component encoded in the genotype, with a probability of 0.1. The mean of the Gaussian was 0, and its s.d was 0.1. During evolution, all vector component values were constrained to remain within the range $[0:1]$. Genotype parameters were linearly mapped to produce CTRNN parameters with the following ranges:

- biases $\beta_j \in [-2, 2]$ - weights $\omega_{ji} \in [-4, 4]$ - gain factor $g \in [1, 7]$

Decay constants were firstly linearly mapped onto the range $\tau_i \in [-0.7, 1.3]$ and then exponentially mapped into $\tau_i \in [10^{-0.7}, 10^{1.3}]$

In the experiment with PNNs, each genotype was a vector comprising 60 binary values, 5 bits for each of the 12 neurons. Each neuron has 3 encoded parameter: the sign bit determines the sign of all the outgoing synapses and the remaining four bits determine the properties of all incoming synapses to this neuron—2 bits for one of four Hebb rules and 2 bits for one of four learning rates chosen from the following values: 0.0, 0.3, 0.6 1.0. Consequently, all incoming synapses to a given node have the same properties. Initially, a random population of vectors was generated by randomly initialising each binary component of each genotype. Single-point crossover with 0.04 probability and bit-switch mutation with a 0.02 probability per bit are used. The synaptic strengths are initialised to small random values in the interval $[0.0 : 0.1]$.

5   Results

We conducted 20 evolutionary runs for each type of controller. Each run lasted 5000 generations and used a different random seed. We examined the best (i.e., highest scoring) individual of the final generation of each of these simulation runs in order to establish whether it had evolved an appropriate learning strategy. In order to test the learning ability of the evolved robots, each of these final generation controllers was subjected to 500 test sessions in each of the two types of environment (i.e., O-env. and S-env.). As before, each test session comprised 15 consecutive trials, and controllers were reset before the start of each new session. During these evaluations we recorded the number of times the controller successfully navigated to the target (i.e., found the target directly, without first going to the wrong end of the arena). Note that a controller which does not learn, should be expected to average a 50% success rate. Controllers which consistently produce success rates higher than this can only do so by learning.

The results for the best controller of each of the twenty runs with CTRNNs, can be seen in figure 3. It is clear that in 14 of the runs (i.e., run n. 1, 3, 4, 6, 7, 8, 10, 11, 12, 13, 17, 18, 19 and 20), robots quickly come to successfully use the information provided by the colour gradient of the central stripe to navigate toward the goal, irrespective of whether they are in an O-env. or in a S-env. All these controllers, except run n. 20, employ a default strategy of moving toward the direction in which the level of grey of the central stripe increases, and hence are always successful in the O-env. (see figure 3 dashed lines). Consequently they are initially very unsuccessful in the S-env. (see figure 4 continuous lines). Nevertheless, as can be seen from figure 3, when these controllers are placed in the S-env. they are capable of adapting their behaviour over the course of very few trials and subsequently come to behave very differently with respect to the central stripe. Each of these controllers, when placed in the S-env., initially navigates following the direction of movement indicated by the increasing level of grey, fails to encounter the target and subsequently reaches the wrong end of the arena. At this point each of them turns around and proceeds back toward the correct end of the arena where they ultimately encounter the target (i.e., the black stripe). The fact that in subsequent trials each of these robots moves in

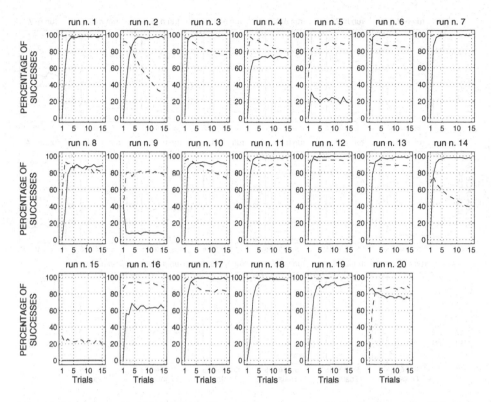

Fig. 3. Experiment I—CTRNNs: each single graph refers to the performance of the best control system of the final generation of a single run, evaluated over 500 test sessions in both types of environment. Dashed lines indicate the percentage of successes per trial in O-env. Continuous lines indicate the percentage of successes per trial in S-env.

the direction indicated by the decreasing level of grey of the central stripe rather than in the opposite direction, demonstrates that these robots have learnt from some aspect, or aspects, of their earlier experience of the current environment. More prosaically, they learn from their mistakes.

The results for the best evolved controller of each of the twenty runs with PNNs, can be seen in figure 4. None of these evolved controllers achieve a success rate significantly above random in both types of environment. The majority of the evolved controllers (i.e., run n. 2, 3, 4, 5, 6, 8, 9, 10, 11, 13, 14, 15, 17), although quite successful in O-env., proved to be unable to adjust their strategies to the characteristics of the S-env. Runs n. 1, 7, 12, 18, 19, do not achieve a success rate significantly above random in either of the environment.

6   Conclusion

Blynel et al. [2] recently compared two different types of control architecture, CTRNNs and PNNs, on a simple learning task. Despite significant architectural

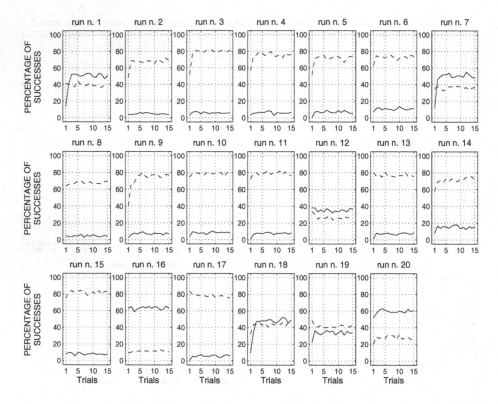

Fig. 4. Experiment II—PNNs: each single graph refers to the performance of the best control system of the final generation of a single run, evaluated over 500 test sessions in both types of environment. Dashed lines indicate the percentage of successes per trial in O-env. Continuous lines indicate the percentage of successes per trial in S-env.

differences between these two types of controller, they found no significant differences in the performance of the evolved controllers. The work described in this paper was designed to investigate the possibility that the simplicity of their task may have masked potential differences between CTRNNs and PNNs. To this end we designed a similar task. Like the task described in [2], this task essentially involved two possible behaviours and two types of environment; the robots were required to learn which of the two types of behaviour was appropriate to the environment in which they found themselves. Our task, however, was more complex. Firstly, it required a greater degree of sensorimotor coordination — in particular the robot was not constrained to bi-directional movement. Secondly, environmental cues which enabled a robot to determine the appropriateness of its behaviour were not available at the time the behaviour was performed. Comparing the two types of network on this task, we were able to evolve successful learning robots in 14 of the 20 CTRNN runs, but in none in the 20 PNN runs. This result is very different from that of [2], and suggests that their result was due to the simplicity of their task.

The results of our comparison suggest that there may actually be significant differences in the ability of CTRNNs and PNNs to perform tasks requiring learning and sequential behaviour. Clearly, it is not advisable to generalise from a single result. Further work will be necessary to establish whether our result holds over a wider range of tasks. (However, we achieved an extremely similar result when we compared the two controllers types on a more complex task variation — the task described in [7]). If CTRNNs prove to be generally better at this type of task, it will clearly be useful to gain some understanding of exactly why this should be the case. This will require establishing which aspects of such tasks CTRNNs are well suited to, and which prove problematic for PNNs. To this end, we are currently performing further experiments and analysis in an attempt to estimate how different elements of our task have impacted on the relative performance of the two types of network.

Acknowledgments. The authors like to thank Jesper Blynel for his help in the design of PNNs, the two anonimous reviewers for their suggestions and comments. We also thank all the members of the Centre for Computational Neuroscience and Robotics (http://www.cogs.sussex.ac.uk/ccnr/) and memebers of the Institut de Recherches Interdisciplinaires et de Développements en Intelligence Artificielle (http://iridia.ulb.ac.be/) for constructive discussion and the Sussex High Performance Computing Initiative (http://www.hpc.sussex.ac.uk/) for computing support. Elio Tuci is supported through a Training Site fellowship funded by the Improving Human Potential (IHP) programme of the Commission of the European Community (CEC), grant HPRN-CT-2000-00032. The information provided in this paper is the sole responsibility of the authors and does not reflect the Community's opinion. The Community is not responsible for any use that might be made of data appearing in this publication.

References

1. R.D. Beer. Dynamical approaches to cognitive science. *Trends in Cognitive Sciences*, 4(3):91–99, 2000.
2. J. Blynel and D. Floreano. Levels of Dynamics and Adaptive Behaviour in Evolutionary Neural Controllers. In B. Hallam, D. Floreano, J. Hallam, G. Hayes, and J-A. Meyer, editors, *FROM ANIMALS TO ANIMATS VII: Proceedings of the* 7th *international conference on the simulation of adaptive behavior (SAB'02)*, pages 272–281, Cambridge, MA, 4–9 August 2002. MIT Press.
3. D. Floreano and J. Urzelai. Evolution and learning in autonomous robotic agents. In D. Mange and M. Tomassini, editor, *Bio-Inspired Computing Systems*, chapter 1. PPUR, Lausanne, 1998.
4. D. Floreano and J. Urzelai. Evolution of plastic control networks. *Autonomous Robots*, 11(3):311–317, 2001.
5. D. Floreano and J. Urzelai. Neural Morphogenesis, Synaptic Plasticity, and Evolution. *Theory in Biosciences*, 120(3–4):223–238, 2001.
6. S. Nolfi and D. Floreano. Learning and evolution. *Autunomous Robots*, 7(1):89–113, 1999.

7. E. Tuci, I. Harvey, and M. Quinn. Evolving integrated controllers for autonomous learning robots using dynamic neural networks. In B. Hallam, D. Floreano, J. Hallam, G. Hayes, and J-A. Meyer, editors, *FROM ANIMALS TO ANIMATS VII: Proceedings of the 7th international conference on the simulation of adaptive behavior (SAB'02)*, Cambridge, MA, 4–9 August 2002. MIT press.
8. B. M. Yamauchi and R. D. Beer. Integrating Reactive, Sequential, and Learning Behavior Using Dynamical Neural Networks. In D. Cliff, P. Husbands, J.-A. Meyer, and S. W. Wilson, editors, *From Animals to Animats III: Proceedings of the 3rd International Conference on Simulation of Adaptive Behavior*. MIT Press, 1994.
9. B. M. Yamauchi and R. D. Beer. Sequential behavior and learning in evolved dynamical neural networks. *Adaptive Behavior*, 2(3):219–246, 1994.

DNA Based Algorithms for Some Scheduling Problems

Jacek Błażewicz[1,2], Piotr Formanowicz[1,2], and Radosław Urbaniak[1]

[1] Institute of Computing Science, Poznań University of Technology,
Piotrowo 3A, 60-965 Poznań, Poland.
blazewic@put.poznan.pl, {piotr, Radoslaw.Urbaniak}@cs.put.poznan.pl
[2] Institute of Bioorganic Chemistry, Polish Academy of Sciences,
Noskowskiego 12/14, 61-704 Poznań, Poland.

Abstract. DNA computing is an alternative approach to performing computations. In general it is possible to design a series of biochemical experiments involving DNA molecules which is equivalent to making transformations of information. In classical computing devices electronic logical gates are elements which allow for storing and transforming information. Designing of an appropriate sequence or a net of "store" and "transform" operations (in a sense of building a device or writing a program) is equivalent to preparing some computations. In DNA computing the situation is analogous, and the main difference is that instead of electronic gates DNA molecules are used for storing and transforming information. From this follows that the set of basic operations is different in comparison to electronic devices but the results of using them may be similar. In this paper DNA based algorithms for solving some single machine with limited availability scheduling problems are presented. To our best knowledge it is the first attempt to solve scheduling problems by molecular algorithms.

1 Introduction

Richard Feynman was the first who spoke about possibility of making computations at a molecular level. His ideas had to wait more than twenty years to be implemented. They evolved in two directions one of them being *quantum computing* and the other *DNA computing*. The latter one has been established by Leonard Adleman who showed how DNA computing might be used to solve a simple instance of the well-known hard combinatorial problem of finding Hamiltonian path in a directed graph [1]. His seminal paper influenced many researchers who realized that DNA molecules may serve as potential new computing devices. He also founded a new way of thinking about computations since in DNA computing some new paradigms are needed. One of them is a (real) massive parallelism where (almost) all encoding strings are processed at the same time. (To tell the truth it is a kind of simplification because in practice it is obviously impossible to process all strings encoding potential solutions to a given problem. But on the other hand the number of these strings encoded in DNA molecules used in DNA computations is usually so huge that in practice it may be considered as

S. Cagnoni et al. (Eds.): EvoWorkshops 2003, LNCS 2611, pp. 673–683, 2003.

processing of all of them.) Shortly after the publication of Adleman's paper for many classical hard combinatorial problems DNA based algorithms have been proposed (cf. [6,8]).

Nowadays the research effort in the area of DNA computing concentrates on three main issues: designing algorithms for some known combinatorial problems, designing new basic operations of "DNA computers" (they are some biochemical procedures whose results may be interpreted as results of some computations) and developing new ways of encoding information in DNA molecules (cf. [7,5]).

In this paper we propose DNA based algorithms for two problems of *scheduling theory* which is a part of *operational research*. Generally speaking, scheduling theory deals with problems of allocation of limited amount of resources (usually discrete, but not necessarily) to a set of tasks in such a way that the set is processed optimally from a view point of some criterion or criteria (good overviews of scheduling theory are given in [2,10]). Usually the resources are machines or processors and the tasks may be some operations performed in a factory or computer systems. One of the most important and natural criteria is the schedule length, i.e. the length of the period necessary for performing all tasks from the considered set.

The problems considered in this paper deal with a single machine. Given the machine some set of non-preemptable tasks has to be processed in such a way that the time span between the beginning of processing the first task and the completion of the last one is minimal (task is non-preemptable if its processing when started cannot be interrupted until it is completed). These types of problems are the classical ones in scheduling theory, but we solved an important extension of the classical variant, i.e. problems where the machine is not continuously available for processing tasks. Problems with such a *limited machine availability* have recently became a subject of an intensive research in scheduling theory, since they are good approximates of some phenomena occurring in manufacturing and computer environments. They are usually computationally hard, especially in the case of non-preemptable tasks, hence there is a need for developing some heuristic polynomial-time approaches or efficient in an average case enumeration methods. Another possibility to attack such problems are some non-classical approaches, like DNA computing.

To our best knowledge the algorithms presented in this paper are the first attempts to solving scheduling problems by DNA based algorithms.

The structure of the paper is as follows. In Section 2 we describe the main idea of DNA computing and basic operations used in DNA based algorithms. In Section 3 the scheduling problems are formulated, while in Section 4 the algorithms solving them are described. The paper ends with conclusions in Section 5.

2   Basics of DNA Computing

DNA based algorithms are composed of some basic operations, analogously like their classical counterparts. In this section we briefly describe the general idea of DNA computing and some of these operations.

The fundamental phenomenon for DNA computing is the structure of the DNA molecule [12]. This molecule is composed of two strands, each of them being a sequence of *nucleotides*. There are four types of nucleotides denoted by A, C, G, T. So, from computer science point of view such a strand is simply a word over an alphabet $\Sigma_{DNA} = \{$A, C, G, T$\}$. One of the most important properties of nucleotides is their ability to join by hydrogen bonds. This property allows for the creation of the double stranded structure of the molecules. To be more precise, in one DNA strand nucleotides are joined by strong phosphodiester bonds. But additionally, every nucleotide A may be joined with nucleotide T by two hydrogen bonds and C may be joined with G by three hydrogen bonds (this rule is called *Watson-Crick complementarity* and we say that A-T and C-G are pairs of *complementary* nucleotides). These bonds are weaker than the phosphodiester ones and they allow for forming the double stranded DNA molecules. For example, if we have a single stranded molecule 5'-CTTAGTCA-3' it is possible that another molecule 3'-GAATCAGT-5' will *hybridize* to it, i.e. it will form a duplex:

5'-CTTAGTCA-3'
3'-GAATCAGT-5'

(Note, that in the above example we denote the ends of the strands by 5' and 3'. In fact, DNA strands have orientation and by convention the left end of the molecule is denoted by 5' and the right one by 3'. In the example 3'-GAATCAGT-5' is a sequence complementary to 5'-CTTAGTCA-3' - such a sequence has an opposite direction and at each position of the duplex nucleotides are complementary to each other.)

In all DNA based algorithms Watson-Crick complementarity allows for creating solutions encoded in double stranded DNA molecules composed of single stranded ones used for encoding a problem instance.

The idea of DNA computing is in some sense similar to the non-deterministic Turing machine. Indeed, in many DNA based algorithms at the beginning there are created DNA sequences encoding all feasible (not necessarily optimal) solutions to a given problem and a lot of other sequences which do not encode feasible solutions. In the next steps the algorithm eliminates from this input set of sequences those which are not solutions to the considered problem and also those ones which do not encode optimal solutions. This elimination is performed successively and the power of DNA computers lies in their massive parallelism - (almost) all encoding sequences are processed simultaneously. At the end of the computations in the set of DNA molecules there remain only those which encode the optimal solution.

The elimination of "bad" sequences is usually done in such a way that the algorithm checks successively if the sequences posses some properties necessary for every string which potentially could encode the problem solution. If they do not have these properties, they are eliminated from the set.

Hence, the basic operations of DNA algorithms are usually designed for selecting sequences which satisfy some particular conditions. On the other hand, there may be different sets of such basic operations. In fact, any biochemical procedure which may be interpreted as a transformation (or storing) information encoded in DNA molecules may be treated as a basic operation of DNA based algorithms.

In this paper we adopt a set of operations described in [9]. This set includes the following operations (here a *test tube* is a mixture of DNA molecules):

MERGE - given two test tubes N_1 and N_2 create a new tube $N_1 + N_2$ containing all strands from N_1 and N_2.
AMPLIFY - given tube N create a copy of them.
DETECT - given tube N return *true* if N contains at least one DNA strand, otherwise return *false*.
SEPARATE - given tube N and word w over alphabet {A, C, G, T} create two tubes $+(N, w)$ and $-(N, w)$, where $+(N, w)$ consists of all strands from N containing w as a substring and $-(N, w)$ consists of the remaining strands.
LENGTH-SEPARATE - given tube N and positive integer n create tube $(N, \leq n)$ containing all strands from N which are of length n or less.
POSITION-SEPARATE - given tube N and word w over alphabet {A, C, G, T} create tube $B(N, w)$ containing all strands from N which have w as a prefix and tube $E(N, w)$ containing all strands from N which have w as a suffix.

The crucial issue in DNA computing is a set of DNA molecules which encode the problem instance. The nucleotide sequences of these molecules usually have to be designed in such a way that when poured into one test tube in certain physical conditions they will create double stranded DNA molecules encoding potential solutions to the problem. For example, let us consider a hypothetical combinatorial problem where certain permutation of some elements is looked for. For the sake of simplicity let us assume that the problem instance consists of only four elements $\alpha, \beta, \gamma, \delta$ and the optimal permutation is $\beta\gamma\alpha\delta$. One of the possible way of encoding the instance in DNA molecules could be as follows. To the elements of the instance there are assigned the following oligonucleotides:

α – 5'-AACTGCTA-3', β – 5'-TGGGTACC-3', γ – 5'-GTCGTCGT-3',
δ – 5'-GAGGGACC-3'.

This set of oligonucleotides does not suffice for the above mentioned process of forming DNA duplexes. It is necessary to add some oligonucleotides which would be able to hybridize to right half of some of the oligonucleotides shown above and to left half of some other. Such oligonucleotides will bind consecutive oligonucleotides in the double stranded DNA molecule encoding potential solution. In our example one could design such "binding" oligonucleotides for every pair of elements. They should be as follows:

$\alpha - \beta$ – 3'-CGATACCC-5', $\alpha - \gamma$ – 3'-CGATCAGC-5', $\alpha - \delta$ – 3'-CGATCTCC-5',
$\beta - \alpha$ – 3'-ATGGTTGA-5', $\beta - \gamma$ – 3'-ATGGCAGC-5', $\beta - \delta$ – 3'-ATGGCTCC-5',
$\gamma - \alpha$ – 3'-AGCATTGA-5', $\gamma - \beta$ – 3'-AGCAACCC-5', $\gamma - \delta$ – 3'-AGCACTCC-5',
$\delta - \alpha$ – 3'-CTGGTTGA-5', $\delta - \beta$ – 3'-CTGGACCC-5', $\delta - \gamma$ – 3'-CTGGCAGC-5'

When all these oligonucleotides are poured into test tube they will form complexes according to the complementarity rule. The complexes will be of various lengths - they will be composed of various numbers of oligonucleotides. Some of them may be as follows (it the parentheses there are shown sequences of the elements which correspond to the molecules):

5'-AACTGCTATGGGTACCGAGGGACC-3' $(\alpha\beta\delta)$
  3'-CGATACCCATGGCTCC-5'
5'-GTCGTCGTAACTGCTA-3' $(\gamma\alpha)$
  3'-AGCATTGA-5'
5'-GTCGTCGTAACTGCTAGAGGGACC-3' $(\gamma\alpha\delta)$
  3'-AGCATTGACGATCTCC-5'
5'-TGGGTACCGTCGTCGTAACTGCTAGAGGGACC-3' $(\beta\gamma\alpha\delta)$
  3'-ATGGCAGCAGCATTGACGATCTCC-5'
5'-TGGGTACCGTCGTCGTAACTGCTAGTCGTCGTGAGGGACC-3' $(\beta\gamma\alpha\gamma\delta)$
  3'-ATGGCAGCAGCATTGACGATCAGCAGCACTCC-5'

Of course in the real biochemical experiment there would be created much more various DNA duplexes, but only one of these many types encodes the solution to the problem. As was mentioned earlier, DNA based algorithms are usually designed to eliminate from the huge number of molecules in the initial mixture those which do not encode the solution. This is usually done successively by eliminating molecules which do not posses some properties which the solution has to have. In our example some of the initial steps of the algorithm could eliminate the molecules which have length smaller than four oligonucleotides, because they cannot encode any permutation of the elements from the instance. Next, molecules whose length is bigger than sum of length of four oligonucleotides could be removed because they must consist of at least one repetition of some of the elements. Obviously, there must be also a possibility of detection the molecules which encode the solution which maximizes (or minimizes) the value of the criterion function. This issue is usually strongly problem-dependent and will be illustrated in Section 4 where our algorithms are described.

3 Formulations of the Problems

In this section we formulate the scheduling problems which will be solved by the proposed DNA based algorithms.
To denote the problems we used the standard three-field notation (cf. [2,3]).

Problem $1, h_1||C_{max}$
INSTANCE: Set of n tasks $\mathcal{T} = \{T_1, T_2, \ldots, T_n\}$ to be processed, processing time p_j for each $T_j \in \mathcal{T}$, starting time of a period of the machine non-availability s and length of this period h.
ANSWER: A feasible schedule of minimal length.

Problem $1, h_k||C_{max}$

INSTANCE: Set of n tasks $\mathcal{T} = \{T_1, T_2, \ldots, T_n\}$ to be processed, processing time p_j for each $T_j \in \mathcal{T}$, starting times of the periods of machine non-availability $s_1, s_2, \ldots, s_K$ and lengths of these periods $h_1, h_2, \ldots, h_K$.

ANSWER: A feasible schedule of minimal length.

In the above formulations feasible schedule means such a schedule where the machine processes at most one task at any time and no task is processed when the machine is not available.

The two problems differ only in the number of machine non-availability periods. In the former one there is only one such a period and in the latter one the number of such periods is arbitrary. Both of them are hard - the first one is NP-hard in the ordinary sense and the second one is strongly NP-hard [4]. As can be seen in the next section it is relatively easy to propose an algorithm for the problem with one period of non-availability, while many such periods cause the algorithm to be much more complicated.

4   The Algorithms

In this section algorithms for the problems previously formulated are presented (see also [11]). One of the important properties of DNA computing are strong connections between encoding scheme, i.e. the ways in which the instance of the considered problem is encoded in DNA molecules and the algorithm itself. Obviously, such connections exists also in the case of classical algorithms but in DNA computing the encoding scheme influences the algorithm usually much more than in the classical case. Hence, the scheme should be developed very carefully.

In the following subsections first there will be described the encoding schemes and then the algorithms for the problems under consideration.

4.1   The Algorithm for Problem $1, h_1||C_{max}$

The general idea of the algorithm for this problem follows from the obvious observation that in order to solve it one should construct a partial schedule for the time period between moment 0 and the starting time of the non-availability period. In this partial schedule the idle time should be as short as possible and the sequence of tasks scheduled before the non-availability period may be arbitrary (the sequence of the remaining tasks, i.e. those scheduled after the period, obviously also may be arbitrary). It is easy to see that the criterion function, i.e. the schedule length, is minimized when the idle time is minimized. Hence, in order to solve the problem optimally it suffices to choose a subset $\mathcal{T}' \in \mathcal{T}$ such that $\sum_{T_i \in \mathcal{T}'} p_i$ is maximal, but not greater than s.

The algorithm follows the "standard" framework of DNA computing, i.e. at the beginning oligonucleotides encoding all tasks and some auxiliary oligonucleotides are poured to a test tube. As a result of biochemical reactions in the tube there are created longer DNA molecules which are concatenations of oligonucleotides encoding the tasks (in one strand) and the auxiliary oligonucleotides (in the complementary strand). Most of these molecules do not encode any feasible solution to the problem but some of them do. Moreover, there is also a small fraction of molecules which encode the optimal solutions (it follows from the fact that almost all encoding sequences are created, so among them there are also those which encode the optimal solution). The goal of the algorithm is to remove from the test tube all DNA molecules which do not encode those solutions. In order to read the information encoded in the resulting DNA molecule a standard *DNA sequencing* procedure can be applied.

Encoding scheme. The general principle of the encoding scheme is that each task T_i, $i = 1, 2, \ldots, n$ is encoded by a unique DNA sequence (oligonucleotide). (Of course, the oligonucleotides described here should be used in many copies in the biochemical experiment.) We will denote such a sequence by O_i. The length of O_i is equal to dp_i, where d is an even integer constant. Moreover, for building the solution some auxiliary oligonucleotides are necessary which join two consecutive tasks in the sequence encoding the solution. Each task from set T has to appear exactly ones in the solution. In order to avoid a repetition of some tasks in the schedule for each T_i, $i = 1, 2, \ldots, n$ it is necessary to create oligonucleotides joining O_i with O_j for $j = i + 1, i + 2, \ldots, n$. It means that it is necessary to create DNA sequences complementary to the right half of O_i and the left half of O_j. We will denote such a sequence by $U_{i,j}$.

All these oligonucleotides are poured into test tube N where they form a variety of double stranded DNA molecules (some of them will encode the optimal solution of the problem).

Algorithm 1. The steps of the algorithm are as follows:
1. input (N)
2. $N \leftarrow (N, \leq ds)$
3. $N \leftarrow (N, max)$

In step 1. the algorithm "reads" an input, i.e. the test tube N containing DNA strands encoding all potential solutions to the problem.
In step 2. from the solution N there are extracted only those sequences whose lengths do not exceed ds nucleotides. The other ones are removed and the result of this operation is assigned again to tube N.
In step 3. only the longest sequences are kept in tube N.

As one can notice in the algorithm there is used only one type of the basic operations described in Section 2. In step 2. LENGTH-SEPARATE is applied in order to remove those molecules which are to long for encoding the partial schedule which could fit into the time slot before the non-availability period. In step 3. the variant of this operation is used which selects the longest molecule (this molecule is selected from the set containing those ones which have lengths not exceeding ds and thus corresponding to partial schedules which fit into the time slot before the non-availability period). This molecule encodes the optimal subset of tasks which should be scheduled before the non-availability interval. The sequence of these tasks does not affect the value of the criterion function. The remaining tasks can be scheduled arbitrarily after the non-availability period. The schedule length is equal to $s + h + \sum_{T_i \in \mathcal{T} - \mathcal{T}'} p_i$.

4.2   The Algorithm for Problem $1, h_k || C_{max}$

Encoding scheme. 1) for each task T_i $i = 1, 2, \ldots, n$ there are created oligonucleotides of length dp_j. For encoding any pair of tasks T_i and T_j for which $p_i = p_j$ different oligonucleotides encoding them should be used.
2) there is created oligonucleotide I of length equal to d nucleotides corresponding to a unit idle time (the sequence of I has to be different from all O_i, $i = 1, 2, \ldots, n$.
3) for each task T_i $i = 1, 2, \ldots, n$ it is necessary to create oligonucleotides joining O_i with O_j for $j = 1, 2, \ldots, n$ and $i \neq j$. It means that it is necessary to create DNA sequences complementary to the right half of O_i and the left half of O_j. We will denote such a sequence by $U_{i,j}$.
4) for each task T_i, $i = 1, 2, \ldots, n$ oligonucleotides joining O_i with I and I with O_i are created. These oligonucleotides are denoted by $U_{i,I}$ and $U_{I,i}$, respectively. ($U_{i,I}$ is complementary to the right half of O_i and the left half of I and $U_{I,i}$ is complementary to the right half of I and the left half of O_i).
5) there is generated an oligonucleotide joining I with I, denoted by U_I. This oligonucleotide is complementary to the right and the left half of I.
6) for each task T_i, $i = 1, 2, \ldots, n$ oligonucleotide U_i complementary to the left half of O_i is created.
7) oligonucleotide U_{I-} complementary to the left half of I is created.

Algorithm 2. According to the described above encoding scheme the lengths of the non-availability intervals are not represented by any oligonucleotides. Indeed, it is assumed that all of these lengths are equal to zero. This assumption simplifies the encoding scheme and the algorithm. On the other hand, the lengths may be easily added to the resulting schedule, so the final schedule will be the real solution to the problem.

The reduction of all non-availability periods lengths to zero causes the change of the periods' starting times (except the first one). The new ones may be computed according to the following formulae:

$$q_1 = s_1 \text{ and } q_i = s_i - \sum_{k=1}^{i-1} h_k, \quad i = 2, 3, \ldots, K$$

The algorithm also uses the lengths of time interval between any two consecutive non-availability periods. They may be determined using formulae:

$$v_1 = s_1 \text{ and } v_i = s_i - (s_{i-1} + h_{i-1}), \quad i = 2, 3, \ldots, K$$

Moreover, the algorithm uses four test tubes:
N - it consists of O_i for $i = 1, 2, \ldots, n$ and I,
N_0 - it consists of O_i for $i = 1, 2, \ldots, n$,

N_1 - it consists of U_i for $i = 1, 2, \ldots, n$ and U_{I-},
N_2 - it consists of $U_{i,j}$ for $i = 1, 2, \ldots, n$, $j = 1, 2, \ldots, n$, $i \neq j$, $U_{i,I}$, $U_{I,i}$ for
$i = 1, 2, \ldots, n$ and U_I.

The steps of the algorithm are as follows:

1. input(N, N_0, N_1, N_2)
2. for $i = 1$ to K do begin
3. $N_{temp} \leftarrow$ amplify(N)
4. $M_i \leftarrow (N_{temp}, = dv_i)$
5. end
6. $M_1 \leftarrow$ merge(M_1, N_1)
7. $M = \emptyset$
8. for $i = 1$ to K do begin
9. $M \leftarrow$ merge(M, M_i, N_2)
10. $M \leftarrow (M, = dq_i)$
11. end
12. $N \leftarrow$ merge(M, N_0)
13. for $i = 1$ to n do begin
14. $N \leftarrow +(N, O_i)$
15. end
16. $N \leftarrow (N, min)$

In line 1. of the algorithm the input data is read, i.e. tubes N, N_0, N_1 and N_2. The loop in lines 2. – 5. is performed K times (K is the number of non-availability intervals). In this loop tubes $M_1, M_2, \ldots, M_K$ are created. These tubes contain partial schedules corresponding to the time slots between any pair of two consecutive non-availability intervals (except the first one, which contains the partial schedules for the time slot before the first non-availability interval). These partial schedules will be merged in the next steps of the algorithm. In line 6. the terminators, i.e., oligonucleotides U_i for $i = 1, 2, \ldots, n$ and U_{I-} are joined to the ends of the partial schedules for the first time slot (between the starting point of the schedule and the first non-availability period). The terminators guarantee that in the following steps of the algorithm potential partial schedules corresponding to the first time slot will be at the beginning of every schedule generated by the algorithm (the terminators block left ends of these partial schedules and no DNA molecule can hybridize there). If the terminators were not joined at this stage of the algorithm it could happen that partial schedules corresponding to the second time slot would joined to the left ends those ones which correspond to the first slot. In this way unfeasible schedule would be created (because such a schedule would not correspond to the time slots between the non-availability periods). In line 7. there is prepared a new test tube M, which at the beginning consists of no DNA strands. The loop in lines 8. – 11. is performed K times. In each iteration i of this loop a partial solution is extended by a partial schedule which fits to the ith time slot. The solution which is built in the loop is kept in tube M. At iteration i all DNA strands whose lengths are not equal to the sum of lengths of i first time slots are removed from M. In line 12. the partial solution is extended by the partial schedule corresponding to the last time slot, i.e. time interval which begins at the end of the last non-availability interval and the solutions are kept in tube N. In the loop in lines 13. – 15. all strands which do not contain at least one occurrence of every task are removed from tube N. If some task appears more than once every "additional" copies of the task should be interpreted as idle times. In step 16. DNA strands with minimal length are extracted and the others are removed from N. This test tube after the step contains optimal solutions to the problem. Indeed, at the end of the algorithm tube N contains only those strands which code each task at least once. Moreover, the way of creation the partial schedules in lines 2. – 5. and joining them together in lines 8. – 11. ensures that in the final schedule there are taken into account the moments when non-availability periods begin, i.e. this moments corresponds to the ending point of some task or idle time in the schedule. Obviously, in order to obtain the real length of the schedule it is necessary to add the sum of all non-availability intervals to the length of the molecules from N divided by d. The number of steps in this algorithm is proportional to the number of non-availability intervals (the loops in lines 2. – 5. and 8. – 11.) and the number of tasks (the loop in lines 13. – 15.).

5   Conclusions

In this paper new DNA based algorithms for scheduling problems have been presented. To our best knowledge it is the first attempt to construct this kind of algorithms for scheduling problems. The problems solved by the presented methods belongs to the relatively new branch of scheduling theory where only limited availability of machines is assumed. This kind of problems better approximate the manufacturing or computer systems reality than their classical counterparts. On the other hand, they are usually computationally hard. From this intractability it follows a need for developing efficient heuristic methods or efficient in an average case enumeration algorithms. The third possibility is the development of some non-standard approaches. DNA computing is one of them and the real massive parallelism (billions of simple "processors") allows, at least in principle, for solving computationally hard problems in a polynomial number of steps. Obviously, the size of problem instances for which this approach could be applied is limited by the amount DNA molecules needed for encoding all potential solutions to the problem under consideration. The presented algorithms are probably the first attempts to apply DNA computing to scheduling problems. In principle, many other scheduling problems might be solved using this approach, although a simple adaptation of the algorithms presented in the paper to the other problems is hardly possible. It follows from the nature of the general DNA computing framework (at least from the current one) where the encoding scheme is strongly dependent on the problem (e.g. it is not obvious how to eventually change the algorithms in order to solve similar problems with L_{max} or $\sum C_j$ criteria - that is the subject of our future research). It is also worth mentioning that the encoding scheme might affect the sizes of the instances which may be effectively solved by this technique. Nevertheless, DNA computing seems to be very promising and it is the subject of an intensive research in the area of new models of computations.

References

1. Adleman, L.: Molecular computations of solutions to combinatorial problems. Science **266** (1994) 1021–1024
2. Błażewicz, J., Ecker, K.H., Pesch, E., Schmidt, G., Węglarz, J.: Scheduling Computer and Manufacturing Processes. Springer Verlag, Berlin (1996)
3. Formanowicz, P.: Selected deterministic scheduling problems with limited machine availability. Pro Dialog **13** (2001) 91–105
4. Lee, C.-Y.: Machine scheduling with an availability constraint. Journal of Global Optimization **9** (1996) 395–416
5. Landweber, L.F., Baum E.B. (eds.): DNA Based Computers II. American Mathematical Society, (1999)
6. Lipton, R.J.: DNA solution of hard computational problems. Science **268** (1995) 542–545
7. Lipton, R.J., Baum, E.B. (eds.): DNA Based Computers. American Mathematical Society, (1996)
8. Ouyang, Q., Kaplan, P.D., Liu, S., Libchaber, A.: DNA solution of the maximal clique problem. Science **278** (1997) 446–449
9. Păun, G., Rozenberg, G., Salomaa, A.: DNA Computing. New Computing Paradigms. Springer-Verlag, Berlin (1998)
10. Pinedo, M.:. Scheduling. Theory, Algorithms, and Systems. Prentice Hall, Englewood Cliffs, New Jersey (1995)
11. Urbaniak, R.: New Models of Computation: DNA Computers and Quentum Computers (in Polish). Master thesis, Poznań University of Technology, Poznań (2002)
12. Watson, J.D., Crick, F.H.C.: A structure of deoxiribose nucleic acid. Nature **173** (1953) 737–738

Learning Action Strategies for Planning Domains Using Genetic Programming

John Levine and David Humphreys

Centre for Intelligent Systems and their Applications,
School of Informatics, University of Edinburgh,
80 South Bridge, Edinburgh, EH1 1HN
johnl@inf.ed.ac.uk

Abstract. There are many different approaches to solving planning problems, one of which is the use of domain specific control knowledge to help guide a domain independent search algorithm. This paper presents L2Plan which represents this control knowledge as an ordered set of control rules, called a *policy*, and learns using genetic programming. The genetic program's crossover and mutation operators are augmented by a simple local search. L2Plan was tested on both the blocks world and briefcase domains. In both domains, L2Plan was able to produce policies that solved all the test problems and which outperformed the hand-coded policies written by the authors.

1 Introduction

This paper presents L2Plan (learn to plan) as a genetic programming based method for acquiring control knowledge. L2Plan produces strategies similar to those produced by [6] using a GP similar to that used by [1]. L2Plan is a complete system that generates its own training examples, testing problems and contains its own simple planner. L2Plan uses a mutation-based local search to augment the genetic program's crossover and mutation operators.

L2Plan was tested on two domains, the blocks world domain and the briefcase domain. In both domains, L2Plan was able to produce control knowledge that allowed the planner to find solutions to all of the test problems.

2 Previous Work

There have been many methods [1,3,4,8] used for learning control knowledge. The two most relevant works are Khardon's L2Act system [5,6], and Aler et al's EvoCK system [1,2].

Khardon's system, L2Act [5,6], represented the control knowledge as "an ordered list of existentially quantified rules" [6], known as a Production Rule System (PRS). The learning algorithm used was a variation of Rivest's [9] learning algorithm.

The PRS strategies produced by L2Act were able to demonstrate some ability to generalize as they could solve some problems of greater complexity than the examples used to generated them. L2Act was able to solve:

S. Cagnoni et al. (Eds.): EvoWorkshops 2003, LNCS 2611, pp. 684–695, 2003.

"roughly 80 percent of the problems of the same size [as the training examples] (8 blocks), and 60 percent of the larger problems (with 20 blocks)." [5]

Since the PRS-based planner used always adopted the first action suggested, the solutions were found very efficiently. The strategies, however, failed to find solutions to some problems and the solutions that were found were often sub-optimal.

L2Act uses the simplest planning algorithm possible: given a set of production rules, apply first rule that fires to the current state and continue until the goal is reached or no further progress is possible. In contrast, Aler et al's EvoCK [1,2] uses genetic programming (GP) [7] to evolve the heuristics of the Prodigy4.0 [11] planner. Prodigy4.0 is a sophisticated domain independent search-based planner. One of its features is that it allows the user to supply domain specific control knowledge to be used to guide its decision making process. It is this control knowledge that EvoCK generates.

EvoCK uses heuristics generated by HAMLET [3] for Prodigy4.0 and evolves them to produce better heuristics. These heuristics are converted by EvoCK into control rules which are then used to generate the initial population for EvoCK's GP. The candidates are sets of different control rules. The EvoCK GP then uses various mutation and crossover operators to evolve the candidates. The candidates are evaluated using Prodigy4.0 to solve example problems. The fitness function takes into account how much improvement they achieve over Prodigy4.0 with no control knowledge, how many problems they solve and the lengths of the plans.

EvoCK was tested on two domains, the blocks world domain and the logistics domain. For both domains populations of 2 and 300 candidates were used and the best results are shown in Table 1. These results refer to the number of test problems solved, rather than the number of problems solved optimally. Overall, EvoCK outperformed both Prodigy4.0 on its own, and HAMLET.

Table 1. HAMLET-EvoCK Results in the Blocks World Domain

Problems Solved			
10 blocks, 5 goals	20 blocks, 10 goals	20 blocks, 20 goals	50 blocks, 50 goals
95%	85%	73%	38%

L2Plan evolves a domain specific planner similar to [7,10], but does so by evolving the domain specific control knowledge (hereafter called a policy) rather than the planner itself. The policy is represented similarly to Khardon [5,6] while the learning algorithm is a GP similar to that used by EvoCK [1,2]. The policy can either be interpreted as a set of production rules, as in L2Act, or as a set control rules to guide a breadth-first search algorithm. L2Plan can take advantage of background theory, or support predicates, if this is available.

3   Policy Restricted Planning

Planning can be viewed as a tree search, where the tree's nodes are states and the branches connecting them are actions with specific bindings, as shown in Figure 1.

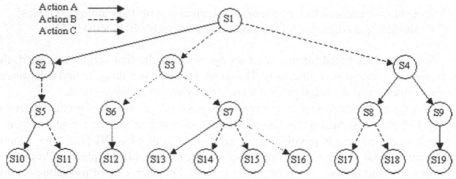

Fig. 1. Planning as a Tree Search

Planning in this manner involves searching the tree for a state that achieves the desired goals and the path of actions from the initial state to that state is the plan. Finding optimal plans is more complicated, as all paths must be searched to ensure that no shorter paths lead to a state that achieves the desired goals. The simplest way to perform optimal searching is to breadth-first search of all states, as shown by the state number in Figure 1. This method of searching, however, requires the planner to look at many, many states. The number of states increases exponentially with the complexity of the problems, which makes this search method infeasible for large problems.

Policy restricted planning involves using a policy to limit the search by restricting which branches are searched. The light grey area in Figure 2 shows an example of this restriction. With policy restriction the number of states to be examined using breadth-first searching can be reduced significantly. In the ideal case, this should be done without affecting the planner's ability to find an optimal plan.

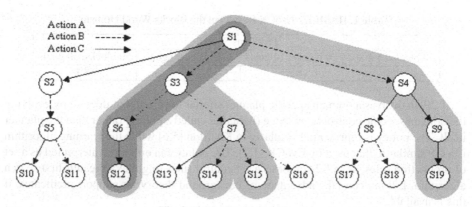

Fig. 2. Policy Restricted Planning

The extreme case of policy restricted planning is where the policy is trusted to generate a good action at its first attempt, and no backtracking is performed. This is the same as treating the policy as a set of production rules, as in L2Act, and is shown in dark grey in Figure 2. We use the term *first-action* planning to refer to this. Because finding optimal plans is NP-hard, optimality is not guaranteed; a good policy for first-action planning should solve all problems and keep the length of the plans as near to the optimum as possible.

4   L2Plan

L2Plan is a system that takes a domain as an input and generates control knowledge, in the form of a policy, specific for that domain as an output. The domains are specified in the untyped STRIPS version of the Planning Domain Description Language (PDDL) and consist of the base predicates and planning operators. An example domain is shown in Figure 3.

```
(define (domain blocksworld)
(:predicates (clear ?x)
             (on-table ?x)
             (on ?x ?y))
(:action move-block-to-block
  :parameters (?bm ?bf ?bt)
  :precondition (and (clear ?bm) (clear ?bt) (on ?bm ?bf))
  :effect (and (not (clear ?bt)) (not (on ?bm ?bf))
               (on ?bm ?bt) (clear ?bf)))
(:action move-block-to-table
  :parameters (?bm ?bf)
  :precondition (and (clear ?bm) (on ?bm ?bf))
  :effect (and (not (on ?bm ?bf))
               (on-table ?bm) (clear ?bf)))
(:action move-table-to-block
  :parameters (?bm ?bt)
  :precondition (and (clear ?bm) (clear ?bt) (on-table ?bm))
  :effect (and (not (clear ?bt)) (not (on-table ?bm))
               (on ?bm ?bt))))
```

Fig. 3. PDDL Definition of the Blocks World Domain

The policies generated by L2Plan are then used by L2Plan's two policy restricted planners to solve planning problems. The policies are specified as ordered sets of control rules, similar to the PRS used by Khardon. A control rule consists of a condition, a goal condition and an action in the form:

if condition *and* goal condition *then* perform action

A policy is used to determine which action to perform in a given situation. For a given situation, the first rule in the list that can be used is used. In order to use a rule, both its condition, **and** its action's precondition, must be valid in the current state and its goal condition valid in the goal. The condition and goal condition are allowed to refer to all the variables present in the action, together with up to n non-action variables, where n is set as one of the parameters of the GP.

An example hand-coded policy for the blocks world is shown in Figure 4. As well as the base predicates provided by the PDDL domain definition, the support predicate wp is used to denote well-placed blocks: a block is well-placed if it is on the table in both the current state and the goal state, or if it is on the correct block and all blocks below it are also well-placed. In the policy, the first three rules add well-placed blocks and the final rule places any non-well-placed block onto the table. Under policy restricted breadth-first planning, this policy solves all problems optimally, with the search increasing in size with the size of the problem. Under policy restricted first-action planning, it solves all problems, with the number of non-optimal solutions generated increasing with the size of the problem.

```
(define (policy blocks1)
(:rule make_well_placed_block_1
  :condition (and (on ?bm ?bf) (wp ?bt))
  :goalCondition (and (on ?bm ?bt))
  :action move-block-to-block ?bm ?bf ?bt)
(:rule make_well_placed_block_2
  :condition (and (wp ?bt))
  :goalCondition (and (on ?bm ?bt))
  :action move-table-to-block ?bm ?bt)
(:rule make_well_placed_block_3
  :condition (and (on ?bm ?bf))
  :goalCondition (and (on-table ?bm))
  :action move-block-to-table ?bm ?bf)
(:rule move_non_wp_block_to_table
  :condition (and (on ?bm ?bf) (not (wp ?bm)))
  :goalCondition (and )
  :action move-block-to-table ?bm ?bf))
```

Fig. 4. Example Hand-Coded Policy for the Blocks World

In order to generate a policy for a domain, L2Plan performs three major functions: the generation of problems and examples, the evolution of the policy and the evaluation of that policy.

4.1   Generation of Problems and Examples

Problems are generated using domain specific problem generators, one for the blocks world domain, and another for the briefcase domain. The problems consist of an initial state, a set of goals and the optimal plan length to solve the problem. The initial state is a list of facts describing the state completely and the goal is a conjunction of facts that describes the goal. An example problem for the briefcase domain is shown in Figure 5.

Examples are extensions of problems, with the addition of a list of all of the possible actions that are valid from the initial state and a corresponding cost for taking those actions. They represent single action decisions. For a given situation an example consists of a state, a goal state and a list of possible actions with associated costs. For each action, the shortest plan starting with that action is found. The action (or actions) with the shortest path is obviously the optimal action (or actions) to take. All other actions are given a cost indicating how many steps more their paths have in them than the optimal path. These actions fall into three categories:

```
(define (problem bc_12)
(:domain briefcase)
(:length 7)
(:objects bc_1 obj_1 obj_2 loc_1 loc_2 loc_3 loc_4 loc_5)
(:init
   (at bc_1 loc_2) (at obj_1 loc_3) (at obj_2 loc_5)
   (briefcase bc_1) (object obj_1) (object obj_2)
   (location loc_1) (location loc_2) (location loc_3)
   (location loc_4) (location loc_5))
(:goal (and (at obj_2 loc_3) (at obj_1 loc_5)))))
```

Fig. 5. Example Problem from the Briefcase Domain

Optimal Actions: There will always be at least one optimal action, but more than one may exist. These actions are given a cost of 0.

Neutral Actions: These actions have plans that are only 1 step longer than the optimal plan. They have a cost of 1 since they result in the plan being one step longer, but these actions don't need to be "undone" to reach the solution.

Negative Actions: These actions have plans that are more than 1 step longer than the optimal plan. They have a cost of 2 or more. In the blocks world and briefcase domains all actions are reversible so negative actions will always have a cost of exactly 2, but in some domains where actions are not reversible (e.g. driving and running out of fuel or shooting a missile) this cost may be higher.

Examples are generated from problems, with each problem providing an example for each step along its optimal path. As shown in Figure 6, this was done by starting with the initial state and determining all of the possible actions that could be taken, as determined by the preconditions of the actions in the domain, not by any particular policy. Each action was taken and the cost to solve each resulting situation was determined. The optimal action was taken, and the process was repeated until the goal was reached. In the cases where there were more than one optimal actions in a situation, the first one found was used.

4.2   Evolution of Policies

The evolution of the policies is performed using genetic programming (GP). An initial population of policies is generated randomly. Each policy generated contains a random number of independently generated rules. That is, there is no effort made to ensure each policy fits a predetermined pattern (e.g. having a rule using each action in the domain). Also, there are no guarantees on the sanity of the generated rules (i.e. that they aren't self-contradictory). The GP then uses these policies to evolve a policy with a fitness of 1.0, training against a set of generated examples.

The fitness of a policy is determined by evaluating the policy against the set of training examples. The policy is evaluated against each of the examples in turn and averaged to give an overall evaluation. An evaluation involves using the policy to determine what action should be taken in the situation of the example. The cost of that action is then retrieved from the example. In the case where a policy is non-deterministic (i.e. it recommends more than one action for a given example) the policy evaluator selects the first

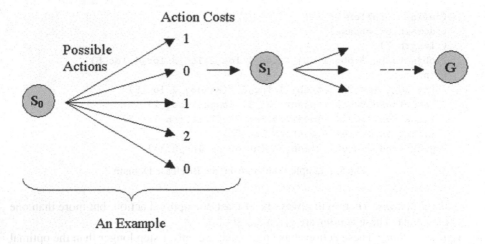

An Example

Fig. 6. Generation of Examples from Problems

action in the list of actions, and uses its action's cost. The lists of actions are sorted to ensure consistency.

Once the action's cost has been determined for each of the examples in the training set, the fitness of the policy is calculated as:

$$F(p_i) = \frac{1}{1 + \left(\sum_{j=1}^{n} C(p_i, e_j)\right)/n}$$

where $F(p_i)$ is the fitness of policy p_i, $C(p_i, e_j)$ is the cost of the action returned by applying policy p_i to example e_j and n is the number of examples in the training set.

L2Plan's GP uses three crossover and four mutation operators to perform the evolution. Selection was performed using tournament selection with a size of 2. The result of the crossover or mutation was always a single policy: the fittest of all of the newly created policies and ones selected for crossover or mutation.

Single Point Rule Level Crossover: a single crossover point is selected in each of the two selected policies' rule sets. Possible crossover points are before of any of the rules, thus resulting in the same number of possible crossover points in each policy as there are rules. This also prevents a crossover in which one of the policies ended up with no rules. Single point crossover is then performed in the usual manner [7].

Single Rule Swap Crossover: a rule is selected from each policy and swapped. The replacing rule is placed in the same location as the one begin removed.

Similar Action Rule Crossover: single rules with the same action are selected from each policy. From these two rules, two more are created by using the condition from one and the goal condition from the other. These original rules in each policy are then replaced by both of the new rules, resulting in four new policies.

Rule Addition Mutation: a new rule is generated and inserted at a random position in the policy.

Rule Deletion Mutation: a rule is selected at random and removed from the policy. If the policy contains only one rule, this mutation is not performed.

Rule Swap Mutation: two rules are selected at random and their locations in the rule set are swapped. If the policy has only one rule, this mutation is not performed.

Rule Condition Mutation: a rule is selected at random and its condition and/or goal condition are mutated. Each of the conditions is subjected, with equal probability, to one of four different mutations:

- Add a predicate to the conditional conjunction
- Remove a predicate from the conditional conjunction
- Replace the condition with a new, randomly generated one
- Do nothing

L2Plan also uses a local search to augment the GP. It is run on each policy prior to it being added to the population. L2Plan performs local search by using random mutations, using the Rule Condition Mutation, to look "around" the candidate in the search space. Since it is infeasible to look at all possible permutations due to the number of these permutations, a few are selected randomly.

The amount of searching performed is determined by a branching factor and a maximum depth, as shown in Figure 7. In this example, the initial candidate is mutated 4 times to produce mutations 1a, 1b, 1c and 1d. Each of these mutation is evaluated and since 1d is the fittest, this replaces the original candidate. The search continues until no improvement is found or the maximum depth limit is reached.

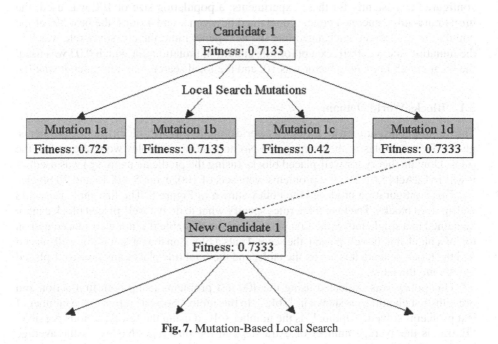

Fig. 7. Mutation-Based Local Search

As soon as a policy is found that selects the optimal action for each of the examples, thus giving the policy a fitness of 1.0, evolution stops.

4.3   Evaluation of Control Knowledge

The optimal policy found during evolution is tested against problems generated by L2Plan. The policy is tested on each problem using two forms of policy restricted planning: breadth-first and first-action. Breadth-first planning examines all of the possible plans allowed by the policy at each length until a solution is found. First-action planning only explores the first possible action at each state, as shown in Figure 2.

Testing can produce one of three results: failure, solved or optimal. A problem is considered solved by a policy if the planner produces a plan that achieves the goal. To be considered optimal, the plan must be the same length as the optimal plan. Failures occur when the policy recommends no action for a given state/goal combination, or when the action recommended results in a state which has already been visited, thus resulting in a non-terminating loop.

There are three metrics tracked by the policy tester. First is the number of problems that are either solved or optimal. The second is the average number of extra steps taken by all solved and optimal solutions. This second metric provides a measure of the quality of solutions. The third metric tracked was the number of states examined during the planning. This is used as a measure of the cost of finding the solutions.

5   Experiments and Results

Experiments were performed on the blocks world and the briefcase domain. The GP was configured consistently for these experiments: a population size of 100 was used; the initial randomly generated policies contained between 1 and 4 rules; the best 5% of the population was copied unchanged into the next generation; the crossover rate was 0.9; the mutation rate was 0.01, except for rule condition mutation, for which 0.03 was used; the local search branching factor was 10; and the local search maximum depth was 10.

5.1   Blocks World Domain

The training configuration for the blocks world domain was 30 5-block training problems, giving 135 examples in the training set. No non-action variables were allowed in the rules. Domain theory for well-placed blocks (using the predicate name wp) was used, as it was in L2Act [5,6]. The test problems were sets of 100, using 5, 10, 15 and 20 blocks.

This configuration produced the policy shown in Figure 8. The first three rules add well-placed blocks. The final three rules specify what to do if a well-placed block cannot be gained in a single move: the first puts a block on the table if it not well-placed and on top of a block that is well-placed, the second puts a block on the table it is not well-placed and the block beneath it is not on the table, and the final rule places any non-well-placed block onto the table.

This policy was evaluated using the 400 test problems using both first-action and breadth-first planning, as shown in Table 2. In this table, "Solved" refers to the number of test problems solved, "Optimal" is the number solved using the fewest actions possible, "Extra" is the average number of extra steps in the plan and "Nodes" is the average number of nodes visited in the search tree.

For comparison, results are given for the 20-block test problems using the hand-coded policy shown in Figure 4. The policy found by L2Plan outperforms this policy under

```
(define (policy Policy_19683)
(:rule GenRule_8949
  :condition (and (on ?bm ?bf))
  :goalCondition (and (on-table ?bm))
  :action move-block-to-table ?bm ?bf)
(:rule GenRule_7388
  :condition (and (wp ?bt) (not (wp ?bm)))
  :goalCondition (and (on ?bm ?bt))
  :action move-table-to-block ?bm ?bt)
(:rule GenRule_12975
  :condition (and (wp ?bt))
  :goalCondition (and (on ?bm ?bt))
  :action move-block-to-block ?bm ?bf ?bt)
(:rule GenRule_17355
  :condition (and (wp ?bf) (not (wp ?bm)))
  :goalCondition (and )
  :action move-block-to-table ?bm ?bf)
(:rule GenRule_8980
  :condition (and (not (on-table ?bf)) (not (wp ?bm)))
  :goalCondition (and )
  :action move-block-to-table ?bm ?bf)
(:rule GenRule_15502
  :condition (and (clear ?bm) (not (wp ?bm)))
  :goalCondition (and )
  :action move-block-to-table ?bm ?bf))
```

Fig. 8. L2Plan Policy for the Blocks World

first-action planning. Under breadth-first planning, the hand-coded policy is superior in terms of the number of optimal solutions generated, but the learnt policy is near-optimal and manages to halve the amount of search used.

A configuration with one non-action variable was also tried: this produced very similar results, except that the policy produced was slightly less readable, due to presence of the non-action variables.

5.2 Briefcase Domain

In the briefcase domain, the training configuration was 30 2-object, 5-city training problems, giving 167 examples in the training set. One non-action variable was allowed in the rules. No domain theory was used. The test problems were sets of 100, with 2 or 4 objects and 5 or 10 cities.

This configuration produced the policy shown in Figure 9. The first rule takes an object out of the briefcase if it has reached it goal location. It should be noted that the preconditions of the action also have to be true for the rule to fire, which means that in order for the first rule to fire, the object has to be in the briefcase. The second rule puts an object in the briefcase if it is not at its goal location. The fourth rule takes the briefcase to an object that needs to be moved and the final rule moves the briefcase to an object's goal location. The third rule is a refinement of the fourth rule: it takes the briefcase to an object that needs to be moved if the place the briefcase is coming from is

Table 2. Blocks World Test Problem Results

	first-action planning				breadth-first planning			
	Solved	Optimal	Extra	Nodes	Solved	Optimal	Extra	Nodes
5 blocks	100	100	0.00	5.76	100	100	0.00	7.14
10 blocks	100	86	0.15	12.51	100	94	0.07	17.60
15 blocks	100	65	0.57	21.02	100	88	0.15	41.12
20 blocks	100	46	0.91	29.63	100	84	0.21	99.02
hand-coded	100	34	1.26	29.98	100	100	0.00	197.42

```
(define (policy Policy_62474)
(:rule GenRule_7074
  :condition (and (not (at ?obj ?loc)))
  :goalCondition (and (at ?obj ?loc))
  :action takeout ?obj ?bc ?loc)
(:rule GenRule_33919
  :condition (and (at ?bc ?loc))
  :goalCondition (and (not (at ?obj ?loc)))
  :action putin ?obj ?bc ?loc)
(:rule GenRule_26811
  :condition (and (object ?x) (at ?x ?to))
  :goalCondition (and (not (at ?x ?to)) (not (at ?x ?from)))
  :action movebriefcase ?bc ?from ?to)
(:rule GenRule_44204
  :condition (and (object ?x) (at ?x ?to))
  :goalCondition (and (not (at ?x ?to)))
  :action movebriefcase ?bc ?from ?to)
(:rule GenRule_52350
  :condition (and (object ?x) (in-briefcase ?x ?bc))
  :goalCondition (and (at ?x ?to))
  :action movebriefcase ?bc ?from ?to))
```

Fig. 9. L2Plan Policy for the Briefcase Domain

not the object's goal location. Our hand-coded policy for this domain omitted the third rule, but was otherwise identical to the policy found by L2Plan.

This policy was evaluated using the 400 test problems using both first-action and breadth-first planning, as shown in Table 3.

For comparison, results are given for the test problems with 4 objects and 10 cities using the hand-coded policy referred to above. The extra rule in the policy found by L2Plan enables it to outperform the hand-coded policy under both first-action and breadth-first planning.

6   Conclusions and Future Work

L2Plan has successfully shown that polices, of the nature shown in this paper, can be learned using genetic programming. We are now working to apply L2Plan to more

Table 3. Briefcase Domain Test Problem Results

	first-action planning				breadth-first planning			
	Solved	Optimal	Extra	Nodes	Solved	Optimal	Extra	Nodes
2 objects, 5 cities	100	95	0.05	6.05	100	100	0.00	9.37
2 objects, 10 cities	100	96	0.04	6.87	100	100	0.00	11.97
4 objects, 5 cities	100	80	0.20	10.38	100	100	0.00	27.98
4 objects, 10 cities	100	76	0.25	12.91	100	100	0.00	62.04
hand-coded	100	74	0.28	12.94	100	100	0.00	68.76

planning domains and to refine the learning method used. In doing the former, we will modify the system to support PDDL with typing: this will restrict the space of possible rules and should make the learning task easier.

References

1. Aler, R., Borrajo, D. and Isasi, P. (1998). Genetic programming of control knowledge for planning. In Proceedings of AIPS-98, Pittsburgh, PA.
2. Aler, R., Borrajo, D. and Isasi, P. (2001). Learning to Solve Problems Efficiently by Means of Genetic Programming. Evolutionary Computation 9(4), 387–420.
3. Borrajo, D., and Veloso, M. (1997) Lazy incremental learning of control knowledge for efficiently obtaining quality plans. AI Review 11(1-5), 371–405.
4. Katukam, S. and Kambhampati, S. (1994). Learning explanation-based search control rules for partial order planning. In Proceedings of the Twelfth National Conference on Artificial Intelligence, 582–587, Seattle, WA. AAAI Press.
5. Khardon, R. (1996). Learning to take actions. In Proc. National Conference on Artificial Intelligence (AAAI-96), 787–792. AAAI Press.
6. Khardon, R. (1999). Learning action strategies for planning domains. Artificial Intelligence 113(1-2), 125–148.
7. Koza, J.R. (1992). Genetic Programming: On the Programming of Computers by Means of Natural Selection. The MIT Press.
8. Leckie, C. and Zukerman, I. (1991). Learning search control rules for planning: An inductive approach. In Proceedings of Machine Learning Workshop, 422–426.
9. Rivest, R. L. (1987). Learning decision lists. Machine Learning 2(3), 229–246.
10. Spector, L. (1994). Genetic programming and AI planning systems. In Proceedings of Twelfth National Conference on Artificial Intelligence, Seattle, Washington, USA, 1994. AAAI Press/MIT Press.
11. Veloso, M., Carbonell, J., Perez, M., Borrajo, D., Fink, E., and Blythe, J. (1995). Integrating planning and learning: The PRODIGY architecture. Journal of Experimental and Theoretical Artificial Intelligence 7, 81–120.

Routing Using Evolutionary Agents and Proactive Transactions

Neil B. Urquhart, Peter Ross, Ben Paechter, and Ken Chisholm

School of Computing, Napier University, Edinburgh
n.urquart@napier.ac.uk

Abstract. The authors have previously introduced the concept of building a delivery network using an agent-based system. The delivery networks are built in response to a real-world problem that involves delivering post to a large number of households within an urban area. The initial agent based system worked to primarily resolve hard constraint violations. To further improve the solution obtained by the agents, we propose to allow agents to negotiate exchanges of work. We demonstrate the solution obtained may be further improved by allowing such negotiated transactions.

1 Introduction

Evolutionary Algorithms (EAs) have been utilised with success on scheduling and routing problems [1,7,4,3,2]. A problem that has been previously investigated by the authors is that of postal routing. The authors have developed a representation that allows urban delivery routes to be incorporated within an EA [10,9,8]. This technique is known as street based routing (SBR). Houses requiring deliveries are grouped into street sections, each street section being a length of road that runs between two junctions. The genotype used within the EA is a permutation of street sections. The phenotype route is constructed by applying simple routing heuristics to each street section to determine the optimum route to visit each of the households within the street section. This representation has a much smaller search space than the equivalent TSP style solution that considers every household individually.

To allow the construction of delivery networks, an agent-based implementation has been developed. The use of Agents allows the problem to be broken down into sub problems, each of which may be solved by a particular agent. An advantage of using a collection of agents is the ability of Agent based architectures to support collaboration, allowing the Agents to exchange work as required for them each to solve their part of the problem.

2 The Agent Architecture

The task of building up a network of routes neatly subdivides into a number of smaller problems each of which is self-contained and has conflicting objectives.

S. Cagnoni et al. (Eds.): EvoWorkshops 2003, LNCS 2611, pp. 696–705, 2003.

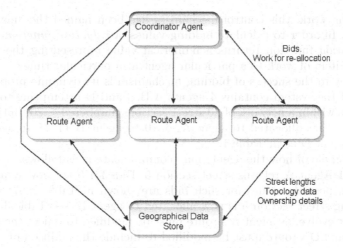

Fig. 1. The agent architecture used.

The architecture is designed to allow each individual agent to optimise its route and only interact with the rest of the system as required. Using a single evolutionary algorithm to build the network of routes concurrently suffers from the problem of finding a suitable representation. For instance, if a single chromosome encodes all of the routes, but only two of these routes require modifications (the rest being within the constraints) it becomes difficult to ensure that only the specific routes within the chromosome are targeted by mutation and crossover. The agent architecture allows each element of the problem to modified or left alone as required. A diagrammatic representation of the architecture that allows this may be seen in Figure 1.

Each Agent incorporates an EA that is used to construct a route based on the street sections allocated to that agent. This EA uses the SBR representation combined with tournament selection and replacement scheme, 2 point crossover (subject to a repair to eliminate duplicated genes) and a simple mutation operator that randomly selects a gene and moves it to a randomly selected point in the chromosome.

The concept of agents bidding for surplus work is used as the basis for the control mechanism. The system is initialised using the output of a grouping-EA, which allocates each agent an initial group of adjacent street sections. The initial number of agents a may be calculated by $a = \frac{TOTDELS}{MAXDELSperROUND}$. Each agent may now construct the delivery round based on this initial allocation and subsequently continue to evolve the delivery round based on any street sections subsequently allocated to it. The co-ordinator requests that the agent with the greatest violation of hard constraints surrenders a single street section back to the co-ordinator for re-allocation to another agent. The coordinator then re-allocates the work by means of a "sealed-bid auction", the agent submitting the lowest bid has the section allocated to it. Because it is based around the agents bidding for, and trading work, this control mechanism has been named the "marketplace" algorithm. In order to calculate bidding values, a *valuation function* is utilised. The valuation function returns a numerical value representing the value of a particular item of work to a particular agent at a particular time.

The key to the success of bidding mechanism is its dynamic properties. For example if the system contains 4 agents A,B,C and D attempting to construct four routes within a network of 20 street sections numbered 1..20. Initially street sections 1..5 are allocated to agent A, 6..10 to agent B 11..15 to agent C and 16..20 to agent D as in Table 1.

An example of how the agents may communicate is as follows:

Step 1 Agent A returns street section 5. Bids for 5 are now requested and submitted, possible values for such bids are shown in Table 2. As a result of the bidding street section 5 is now allocated to agent D, see Table 3. Agents A and D now evolve, as agent A's route must be modified to reflect the removal of section 5 and D's route must be modified to include the addition of section 5.

Step 2 Agent B returns street section 7. Bids are requested and submitted for 7 as in Table 4. Street section 7 is now allocated to agent C, as agent C submitted the lowest bid, see Table 5. Agents C and B now evolve, because they have both been affected by the transaction of section 7.

Step 3 Agent D returns street section 5. Bids for 5 are submitted as in Table 6 Street section 5 is now allocated to agent B. Because the transfer of section 5 from agent D to agent B will require them both to produce updated routes, they both now evolve.

Note that in the above (fictional) example, when street section 5 is auctioned a second time in step 3 the bids submitted are different than those submitted in step 1. This is because the exchange of work has altered the agent's routes and therefore value placed on items of work by the agent's changes. This dynamic market place means that the value that an Agent places on a specific section of work changes as the further auctions take place.

Table 1. The initial allocation of street sections (1-20) amongst agents A-D

Agent	street sections				
A	1	2	3	4	5
B	6	7	8	9	10
C	11	12	13	14	15
D	16	17	18	19	20

The bidding logic used within the marketplace algorithm is shown in Figure 2.

The bidding logic operates using one of two strategies, a relaxed strategy that submits a bid even where the agent may break a hard constraint by accepting the work and a cautious strategy at other times. The cautious strategy declines to participate in the auction if the agent estimates that the addition of the street section will violate the maximum length constraint. All agents initially

Table 2. Step 1, bids for street 5

Agent	Bid
A	25.8
B	16.2
C	14.3
D	12.5

Table 3. Work allocation after step 1

Agent	street sections					
A	1	2	3	4		
B	6	7	8	9	10	
C	11	12	13	14	15	
D	16	17	5	18	19	20

Table 4. Step 2, bids for street 7

Agent	Bid
A	8.4
B	26.2
C	4.8
D	16.9

Table 5. Work allocation after step 2

Agent	street sections					
A	1	2	3	4		
B	6		8	9	10	
C	11	12	13	14	15	7
D	16	17	5	18	19	20

Table 6. Step 3, bids for street 5

Agent	Bid
A	14.2
B	12.2
C	22.4
D	25.0

```
{leastDist(st)   returns the distance to the closest street }
{                to st within the current agent }
{dels(st)        returns the number of deliveries to st }
{oldR(st)        returns the id of the previous agent that }
{                st was allocated to }
{rdLength(st)    returns the length of street (st) }
{st              street to bid for }
{MAXDELS         the maximum deliveries constraint}
{MAXLEN          maximum length constraint}
{cLen            the length of the agents current round }
{cDels           the no of deliveries in the agents current round }
{avgD(st)        returns the average distance between each street }
{                section allocated to the agent and st }
{me              the id of the current agent }

theBid = avgD(st)
newDels = (cDels + dels(st))
if (newDels > MAXDELS) then
  theBid = theBid + (newDels*2)
if (oldR(st) = me) then
  theBid = DECLINE

if (!RELAXED) then
  if (cLen > MAXLEN) then
    theBid = DECLINE
  least = (leastDist(st) + rdLength(st))*2
  if ((cLen + least) > MAXLEN) then
    thebid = DECLINE
```

Fig. 2. The bidding logic used within the initial marketplace algorithm

commence bidding using the relaxed strategy. When a situation arises where no agent is bidding for a street, a new agent is allowed into the marketplace. The initial allocation of streets to the new agent is made up from streets from the agent with the longest round (each street section of has a probability of 0.5 of being transferred to the new round).

The system reaches equilibrium with the same street section being exchanged between the same agents. Once this stage has been reached the bidding strategy of agents may then be set back to the cautious strategy. This allows the rejection of surplus work to create one or more new agents, and reduce each agent's workload to an acceptable level. Allowing each agent to keep a list of those street sections last surrendered by the agent facilitates this change in strategy. This list is of length 3 and initially empty, but when it contains 3 identical items, the agent realising that it no longer engaged in productive transactions and changes to a cautious strategy. Using the cautious strategy no work should be accepted until some work has been surrendered allowing the agents hard constraints to be fulfilled. The trading list is similar in concept to the tabu list used in [5,6], in that they both keep a form of record concerning the algorithms recent actions. In a tabu list this information is used to prevent the algorithm re-examining solutions too frequently. In this context the trading list is used to indicate when the system has found an interim relaxed solution.

3 Postactive versus Proactive Transactions

The initial version of the market place algorithm ([10]) only utilised the auction-based transactions, which are a direct reaction to the violation of a hard constraint. The coordinator forces the agents to surrender if they are breaking any of the hard constraints. Such transactions are a reaction to hard constraint violations. The final solution produced by this scheme is therefore a feasible solution, but not necessarily a particularly optimal solution. Some benefit may be gained by allowing agents to negotiate transactions directly with other agents, after a feasible solution has been found. Such transactions may be termed "proactive" transactions.

Such proactive transactions take place in the steps remaining between a feasible solution being found and maximum number of steps being utilised. In more detail the four steps are:

1. agent A examines (via the Geographical Data Agent) the street sections in the current problem, and calculates the bid that they (agent A) would submit for each piece of work in turn. The street section that would have attracted the lowest bid (that is not already owned by agent A) is selected and may be termed ST1. The ownership of ST1 is established and the owner (agent B) is sent a message requesting ST1.
2. agent B having a received a request for ST1, then examines the street sections owned by Agent A and selects the section that would attract the lowest bid from agent B if it were auctioned, this section is now referred to as ST2. If the agent B's bidding value for ST1 is greater than that for ST2, B responds to A agreeing to transact ST2 for ST1. If however the bidding value for ST1 is less than ST2, the response is negative and the transaction does not go ahead.
3. Assuming agent B has responded positively, agent A calculates bid values for ST1 and for ST2 if the value for ST2 is less than that for ST1, it commences the transaction by transferring ST1, else the transaction is halted.
4. Upon receipt of ST1, Agent B transfers ST2. There is no decision to be made at this step as agent B gave a commitment to transacting ST1 and ST2 at step 2.

This scheme makes use of the bidding mechanism to assess relative the value (or usefulness) of items of work to particular agents. Results achieved by implementing this scheme may be seen in Table 8 (the parameters used are given in Table 10). Note that this is essentially a win-win scheme, in that the transaction must benefit **both agents**. In a typical run only 5% of transactions are aborted.

4   Results

The system was tested on data set based on housing scheme in the South East of Edinburgh. It is based on the graph structure shown in Figure 3. It represents the Moredun and Craigour housing estates located in south Edinburgh. Unlike previous data sets, this set 6 incorporates tower blocks that, although only appearing as one delivery point within the data set actually containing 100 households within any route they are incorporated into. A smaller set of flats containing 9 households is also incorporated within the dataset.

The effect of the extra transactions may be seen in Table 8, these results are an average of 10 runs. The average number of delivery routes present in the final solution decreases by 2 for the 2 km problem and 1 for the 2.5km problem.

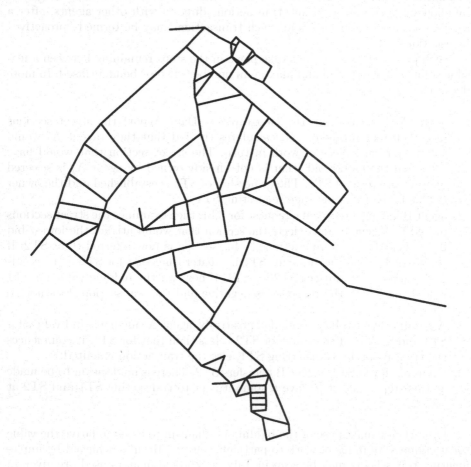

Fig. 3. The street graph for the ScoutPost data set

Table 7. Statistics for the ScoutPost dataset

Total Delivery points (dp)=	1770
Total street sections =	165
Ratio dp/street sections =	7.04
Total length of streets (m) =	10856.9
Average Length/width ratio =	8.2
Total Streets =	38
Average streets/junction	2.8
Average sections/street	4.4

Although the average route length increases, it does not exceed the maximum route length allowed.

Table 8. The effect of allowing additional transactions. * 4.4% of routes produced exceeded the maximum length constraint of 2k.

	max round length	initial version	extra trans
Average round Dist	2K	1553.6	1617.7
	2.5K	1955.8	1911.8
	3K	2149.7	2097.9
Maximum round Dist	2K	1999.8	2674.9 *
	2.5K	2495.8	2498.4
	3K	2998.2	2995.3
Average No of routes	2K	12.4	10.3
	2.5K	10.1	9.3
	3K	9.1	9.1

Table 9 shows the results of running a t-test on the average length of the delivery routes produced with the maximum length constraint set to 2, 2.5 and 3km. The t-test results may be interpreted in the following manner; as the problem becomes less constrained (i.e. the maximum length constraint is increased) so the difference made by the extra transactions increases. The use of extra transactions is of questionable value for the 2km problem, but becomes more significant for the 2.5 and 3km problems. When assessing the overall effectiveness of extra transactions the ability to reduce in the quantity of delivery rounds in the final solution is also significant (see Table 8).

5 Conclusions

The combination of evolutionary algorithms and agents has been proven by the authors to be an effective method of solving routing problems [10]. The use of pro-active transactions goes some way towards negating the criticism that the

Table 9. A comparison of the average round lengths from each of the 20 runs using a t-test to establish the degree of difference between the runs with and without extra transactions.

2	2.5	3
0.62	0.89	0.9

Table 10. The parameters used for additional transactions. *The maximum length value was adjusted as per the details given in the text

SBR parameters	
Population Size	50
Maximum round length	2000/2500/3000 *
New individuals/generation	25
Tournament size for recombination	2
Heuristic application Interval	20
Tournament size for re-inclusion in pop	2
Mutation rate	0.2
Building block mutation rate	0.5
Recombination rate	0.5
No of evaluations per step	2750

original marketplace algorithm only finds a legal solution, and does not continue to optimise the solution further.

The routing application used to test the marketplace algorithm is based on a real-world problem that has been explained elsewhere [10]. It is the authors' opinion that the potential of the marketplace algorithm has not yet been fully exploited. There exist a number of other timetabling and scheduling problems that may be sub-divided and tackled using agents in the manner described. What is important is the valuation function used by the agents to value work with regards to placing bids, selecting work for disposal and negotiating pro-active transactions.

References

1. C. Bierwirth, D. C. Mattfeld, and H. Kopfer. On permutation representations for scheduling problems. In H.-M. Voigt, W. Ebeling, I. Rechenberg, and H.-P. Schwefel, editors, *Parallel Problem Solving from Nature – PPSN IV*, pages 310–318, Berlin, 1996. Springer.
2. T. Bousonville. Local search and evolutionary computation for arc routing in garbage collection. In L. Spector, editor, *Proceedings of the Genetic and Evolutionary Computation Conference 2001*. Morgan Kaufman Publishers, 2001.

3. Dan Bonachea, Eugene Ingerman, Joshua Levy and S. McPeak. An improved adaptive multi-start approach to finding near-optimal solutions to the euclidean TSP. In D. Whitley, D. Goldberg, E. Cantu-Paz, L. Spector, I. Parmee, and H.-G. Beyer, editors, *Proceedings of the Genetic and Evolutionary Computation Conference (GECCO-2000)*, pages 143–150, Las Vegas, Nevada, USA, 10–12 2000. Morgan Kaufmann.
4. B. Freisleben and P. Merz. New genetic local search operators for the travelling salesman problem. In H.-M. Voigt, W. Ebeling, I. Rechenberg, and H.-P. Schwefel, editors, *Proceedings of the Fourth Conference on Parallel Problem Solving from Nature(PPSN IV)*, volume 1141, pages 890–899, Berlin, 1996. Springer.
5. F. Glover. Tabu search - part i. *ORSA Journal of Computing*, 1(3):190–206, 1989.
6. F. Glover. Tabu search - part ii. *ORSA Journal of Computing*, 2:4–32, 1990.
7. S. R. Thangiah, J.-Y. Potvin, and T. Sun. Heuristic approaches to vehicle routing with backhauls and time windows. *International Journal of Computers and Operations Research*, 23(11):1043–1057, 1996.
8. N. Urquhart, B. Paechter, and K. Chisholm. Street based routing using an evolutionary algorithm. In E. J. W. Boers, editor, *Real World Applications of Evolutionary Computing*, pages 495–504. Springer-Verlag, 2001.
9. N. Urquhart, P. Ross, B. Paechter, and K. Chisholm. Improving street based routing using building block mutations. In *Applications of Evolutionary Computing, proceedings of EvoWorkshops 2002*. Springer-Verlag, 2002.
10. N. Urquhart, P. Ross, B. Paechter, and K. Chisholm. Solving a real world routing problem using multiple evolutionary algorithms. In *Proceedings of Parallel Problem Solving from Nature (PPSN) VII*, pages 871–882. Springer-Verlag, 2002.

Author Index